GENETICS AND BIOGENESIS OF CHLOROPLASTS AND MITOCHONDRIA

GENETICS AND BIOGENESIS OF CHLOROPLASTS AND MITOCHONDRIA

Interdisciplinary Conference on The Genetics and Biogenesis of Chloroplasts and Mitochondria, Munich, Germany, August 2–7, 1976, held under the auspices of the Bayerische Akademie der Wissenschaften, Mathematisch-Naturwissenschaftliche Klasse.

Editors
Th. Bücher
W. Neupert
W. Sebald
S. Werner

1976

NORTH-HOLLAND PUBLISHING COMPANY
AMSTERDAM · NEW YORK · OXFORD

North-Holland ISBN: 0 7204 0604 8

Publishers:
North-Holland Publishing Company-Amsterdam
North-Holland Publishing Company, Ltd.-Oxford

Sole distributors for the U.S.A. and Canada:
Elsevier/North-Holland Inc.
52 Vanderbilt Avenue
New York, N.Y. 10017

Library of Congress Cataloging in Publication Data

Interdisciplinary Conference on the Genetics and Biogenesis of Chloroplasts and Mitrochondria, Munich, 1976.
Genetics and biogenesis of chloroplasts and mitochondria.

Includes indexes.
1. Plant genetics--Congresses. 2. Chloroplasts --Congresses. 3. Mitochondria--Congresses. I. Bücher, Theodor. II. Akademie der Wissenschaften, Munich. Mathematisch-Naturwissenschaftliche Klasse. III. Title.
QH433.I57 1976 581.8'733 76-45459
ISBN 0-7204-0604-8

PRINTED IN GREAT BRITAIN

Preface

Six and a half years ago the Australian Academy of Science organized a first interdisciplinary Symposium on "Autonomy and Biogenesis of Mitochondria and Chloroplasts" (1). This meeting conceived by the late Dr. Falk together with Drs. Boardman, Linnane and Smillie served as a model for a similar conference promoted by the Bavarian Academy of Sciences. The Canberra meeting took place at a time when this area of organelle research was entering an exponential phase. The research being now truly into log-phase it was appropriate to reassemble the veterans and to introduce the newcomers.

A little over two months prior to this Munich conference, the "traditional" Bari conference was held on "The Genetic Function of Mitochondrial DNA" (2). Also a CNRS colloquium and a Nato Advanced Study Institute on "Nucleic Acid and Protein Synthesis in Plants" was held at Strasbourg in July this year (3). Even so, a remarkable number of titles were submitted for presentation at Munich.

A novel format for the conference was devised in order to accomodate the overwhelming amount of submitted presentations. The number of talks was reduced drastically to the minimum afforded by the interdisciplinary character of the meeting. Each active participant produced at least one poster. The posters were displayed for the full meeting and served as a forum for personal discourse and exchange of information. The major portion of the conference time was devoted to plenary discussion on the poster presentations. The organizers regret that the discussions could not also be printed. In order to expedite publication of the proceedings the participants manuscripts are reproduced as submitted.

---

(1) Autonomy and Biogenesis of Mitochondria and Chloroplasts eds. N.K. Boardman, Anthony W. Linnane, Robert M. Smillie North-Holland Publishing Co., Amsterdam-London (1971)

(2) The Genetic Function of Mitochondrial DNA, eds. C. Saccone, A.M. Kroon, Elsevier/North-Holland Biomedical Press, Amsterdam, The Netherlands (1976)

(3) To be published by Plenum Press, New York, London and by the CNRS

The conference could not have taken place without generous assistance from many quarters. Firstly from the Advisory Board, who discussed and decided all essential matters on a joint basis since an inaugural meeting held at Gif-sur-Yvette in July, 1975 under the chairmanship of Piotr Slonimski. The Advisory Board were joined and assisted in their efforts by Drs. Saccone and Kroon, experienced organizers of the Bari conferences.

On the financial side the conference was sponsored mainly by the "Stiftung Volkswagenwerk". The National Institute of Health awarded travel support to participants from the United States with grant GM 23355 procured by Dr. Gillham. Also friends of the Bavarian Academy subscribed in very generous measure.

Last, but not least, our sincere thanks are due to the numerous devoted ladies and gentlemen of the Academy, the University and the Sonderforschungsbereich 51, who gave of their time and energy during the meeting, and for many weeks before, to ensure the smooth running of the conference.

München,
August 18, 1976

Theodor Bücher

Contents

MITOCHONDRIAL DNA: GENES AND MOLECULES

GENERAL ASPECTS OF MITOCHONDRIAL BIOGENESIS

ROLE OF CHLOROPLAST AND NUCLEAR GENES IN THE PRODUCTION OF CHLOROPLAST PROTEINS

Genetics and Biogenesis of Chloroplasts and Mitochondria, Th. Bücher et al. eds.

ROLE OF CHLOROPLAST AND NUCLEAR DNA GENES DURING EVOLUTION OF FRACTION I PROTEIN

Kevin Chen, Sarjit Johal and S. G. Wildman
Department of Biology
Molecular Biology Institute
University of California at Los Angeles

Abstract. Crystalline Fraction I protein consists of 8 large and 8 small subunits. The large subunits coded by chloroplast DNA resolve into a cluster of 3 polypeptides by electrofocusing. The small subunits coded by nuclear DNA resolve into 1 to 4 polypeptides depending on the species of enzyme. The protein was electrofocused from 63 of the 65 plant species constituting the genus *Nicotiana*. Only 4 different clusters of large subunit polypeptides were found and the view advanced that no mutations have occurred in about 100 million years. Thirteen different small subunit polypeptides were found whose rate of mutation is close to that of cytochrome C. Computerized numerical taxonomic analysis indicates 6 branches to have developed on the Fraction I protein evolutionary tree. Proteins from two species of *Nicotiana* at the tip of South America are likely relics of the proteins existing before continental drift separated Australia from Antarctica and the Western Hemisphere. *N. tabacum* protein contains two small subunit polypeptides, one having been donated by the paternal parent, the other by the maternal parent when the new plant species was created by interspecific hybridization. Creation of new species of plants by interspecific hybridization may also produce isozymes as illustrated by ferredoxin.

INTRODUCTION

Macromolecular Structure and Function of Fraction I Protein. Fraction I protein (F-I-p) is an enzyme possessing ribulose diphosphate carboxylase and oxygenase activities[1] whose probable structure is shown in Fig. 1. The protein is composed of 8 large subunits (LS) and 8 small subunits (SS). Each monomeric LS has a mol. wt. of 56,000 daltons; each monomeric SS 12,500. The LS and SS dissociate and resolve during SDS gel electrophoresis. They also dissociate in 8M urea and isoelectric focusing shows the LS to be composed of three polypeptides whose isoelectric points are approximately 0.1 of a pH unit apart[3]. F-I-p from representatives of all Phyla of the Plant Kingdom have been shown to have LS composed of three polypeptides[4]. However, the SS is found to resolve into from 1 to 4 polypeptides depending upon the species of plant from which the F-I-p was obtained. Chloroplast DNA (c-DNA) contains the genetic information for the polypeptides of the LS of F-I-p; nuclear DNA (n-DNA)

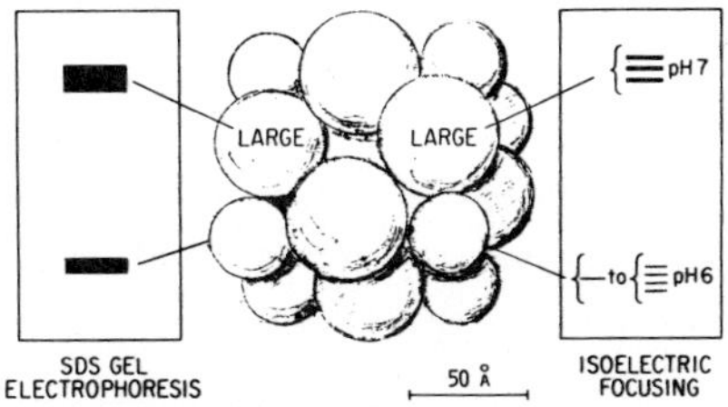

Fig. 1. A possible macromolecular structure of tobacco F-I-p and separation of LS and SS by two methods. (Structure adapted from Baker et al.[2])

contains the code for the SS polypeptide(s)[5].

Evolution of Fraction I Protein in the Genus Nicotiana. T. H. Goodspeed and N. Burbidge[6] have shown 65 species to comprise the genus *Nicotiana*. 20 species are located in Australia and 45 species are confined to the Western Hemisphere but broadly distributed from Canada to Argentina. It is Goodspeed's contention (Fig. 2) that the restriction of the Nicotianas to these regions resulted from continental drift having prevented further spread of evolving species into regions of the earth which became Europe, Asia and Africa. However, before isolation of continents, the ancestors of today's Nicotianas had traversed the Antarctic land bridge and entered Australia prior to its separation as a continent. The continents commenced to drift apart about 75-100 million years ago. Consequently, the F-I-p in Australian species have been separated in their evolution for 10^8 years from F-I-p species undergoing evolution in the Western Hemisphere.

The mode of origin of the *Nicotiana* species typifies the origin of about 40% of all plants now occupying the face of the globe[7]. New species arise by interspecific hybridization. Because of chromosome pairing anomalies most interspecific hybrids cannot complete alternation of generations and are eliminated as potential new species. If somatic doubling of the interspecific hybrid's chromosomes occurs proper meiosis may thereby create a new species of plant capable of alternation of generations. *N. tabacum*, the tobacco of commerce, arose by interspecific hybridization. Analysis of the *N. tabacum* F-I-p composition and the protein composition of the two species which hybridized to produce *N. tabacum* has been most instructive in regard to elucidating the mechanism of F-I-p evolution.

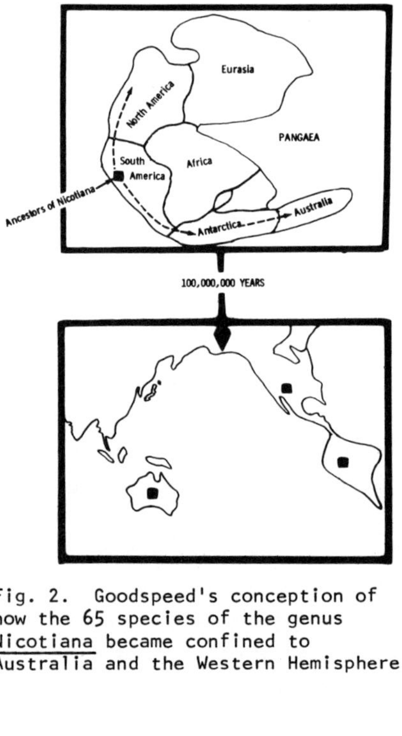

Fig. 2. Goodspeed's conception of how the 65 species of the genus *Nicotiana* became confined to Australia and the Western Hemisphere.

Total Number of Large and Small Subunit Polypeptides. Crystalline F-I-p have been obtained from 63 of the 65 species of *Nicotiana*. The results of electrofocusing 15 species of protein in 8M urea are shown in Fig. 3. These 15 species collectively display all but 2 of the 13 different polypeptides encountered among the 63 species of protein.

The LS of each species of F-I-p resolves into a cluster of three polypeptides whose isoelectric points are about 0.1 pH unit separated from each other. All 20 proteins from Australian species resolve into the same cluster of LS polypeptides with respect to their isoelectric points. The Australian cluster of polypeptide isoelectric points is different, however, from the clusters obtained by resolving the LS of proteins obtained from species of Nicotiana confined to the Western Hemisphere.

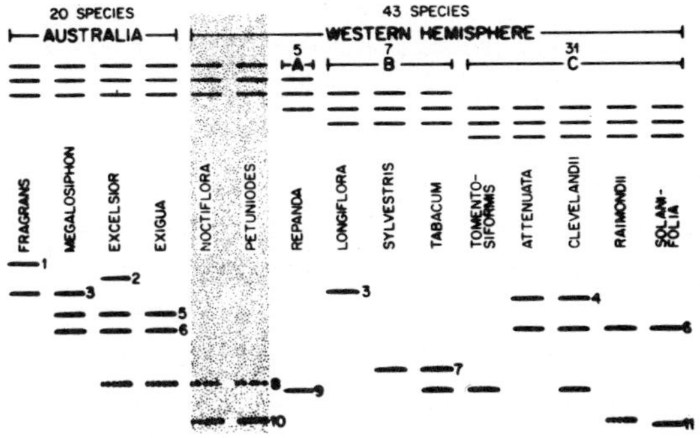

Fig. 3. Diagrammatic representation of the number and position of the LS and SS polypeptides of F-I-p contained in 63 species of Nicotiana. Peptides resolved by electrofocusing in 8M urea. SS peptide 7' found in N. glauca and 5' found in N. glutinosa are not shown. Peptides 1 and 2 found only in Australian species; 4, 7 and 9 found only in Western Hemisphere species. Peptides 8 and 10 dotted to emphasize probability of N. noctiflora and N. petunioides containing proteins closely similar in composition to those existing before continental drift occurred to separate the Nicotianas.

The genetic basis of the origin of the cluster of three polypeptides of the same molecular weights but different isoelectric points comprising the LS is not known. It is uncertain as to whether each polypeptide of different isoelectric point is coded by a gene in c-DNA or if one gene codes for the cluster of three polypeptides, two of the three polypeptides arising as post-translational events.

In the 63 species of proteins, there are only 4 different clusters of LS polypeptides as shown in Fig. 3. One cluster consists of the LS of proteins from the 20 Australian Nicotiana species plus two from the Western Hemisphere; 5 proteins comprise the Group A cluster; 7 proteins, the Group B cluster; and 31 proteins belong to the Group C cluster. If the cluster of three polypeptides proves to be the genetic unit, then $2^x=4$ is an indication that only two mutations in the c-DNA coding information for the LS of F-I-p have survived during the entire life span of the genus Nicotiana. However, the Australian F-I-p have no differences in LS. This suggests that no mutations in the c-DNA have survived in the 10^8 years since the ancestors of the Australian Nicotianas lost contact with the ancestors of Western Hemisphere Nicotianas. The primary structure of a LS polypeptide contains around 375 amino acids. Therefore, the rate of mutation of c-DNA coding for isoelectric points of F-I-p is less than 0.27 mutations/100 amino acids/10^8 years. This value may be compared with that of Histone IV (0.167 mutations/100 amino acids/10^8 years)[8] which holds the record for being the least mutated biological material.

In the 63 species of Nicotiana F-I-p 13 polypeptides of different isoelectric points comprising the SS have made their appearance. The relative isoelectric points of 11 of the 13 polypeptides are shown in Fig. 3. No species of Australian F-I-p contains less than 2 polypeptides composing the SS coded by n-DNA, and most of the species have three or four polypeptides. The situation is different for Western Hemisphere species of protein. Only one out of the 45 proteins has three polypeptides; most have two, but several species have only a single polypeptide. The SS polypeptide has about 85 amino acids. For a total of 13 polypeptides, $2^x=13$ would require only 3.7 mutations during the entire span of the genus Nicotiana. Six of the peptides are found in the Australian species so the mutation rate of the n-DNA coding for the SS can be estimated as 3.04 mutations/100 amino acids/10^8 years which may be compared to the rate for cytochrome C (5 mutations/100 amino acids/10^8 years)[8].

Evolution of Fraction I Protein in the Genus Nicotiana as Judged by Numerical Analysis. Figure 4 is a diagram generated by a computer showing the relative degree of similarity of one F-I-p to any of the other 63 proteins contained in the genus Nicotiana. To construct the diagram, the isoelectric points of the 6 different polypeptides of the LS together with the 13 SS polypeptides were arranged in an order beginning with the polypeptide of greatest neutrality and descending to the polypeptide of greatest acidity. The 6 LS and 13 SS polypeptides comprise the total number encountered among the 63 species of F-I-p, 17 of the 19 polypeptides having been shown in Fig. 3. One was used to designate the presence of a polypeptide of a particular isoelectric point exhibited

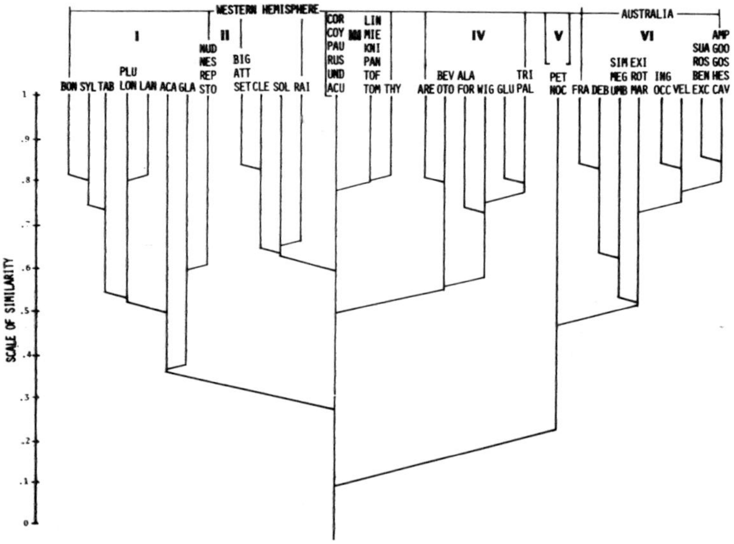

Fig. 4. Computer generated evolutionary tree for 63 species of F-I-p contained in the genus Nicotiana. Plant species indicated by 3 letter abbreviations.

by a F-I-p; its absence by zero. These symbols were fed into a numerical taxonomic computer program[9] which computed similarity coefficients in a range from 0 to 1, and generated the phenogram[10] in Fig. 4. The computer compared those proteins with polypeptides which were most alike. The highest degree of similarity would be species having coefficients between 0.9 and 1. Thus, the F-I-p designated by the symbols LIN, MIE, KNI, PAN, TOF, TOM in Group III become stacked on top of each other because they are identical in regard to isoelectric points of their 3 LS and 1 SS polypeptides. The proteins in the same Group III marked COR, COY, PAU, RUS, UND, ACU are stacked because they are also identical with each other in their 3 LS and 2 SS polypeptides. The analysis distinguishes the latter from the former cluster because of the presence of this additional SS polypeptide. However, the computer sees these two clusters as also being very closely related to each other because all 12 proteins become indistinguishable when the computer examines similarities at the 0.75 level. To bring these 12 proteins into a condition of similarity with any of the 50 other proteins requires the scale of similarity to be reduced to 0.58. Then the 6 proteins to the immediate left of the 13 in Group III become indistinguishable from the Group III proteins. Another way of visualizing the meaning of the analysis is to say that the F-I-p of BON and SYL (Group I) are different at a similarity coefficient of 0.8 but more alike to each other than they are to the proteins of TAB, PLU, LON, etc.

The numerical analysis gives evidence of the emergence of six major branches in the F-I-p evolutionary tree during the preceding 10^8 years. One branch terminates with just two species, PET and NOC (Group V) which appear to be the "bridge" between the Australian and Western Hemisphere proteins. The analysis further indicates the Australian F-I-p (Group VI) to be closely similar to PET and NOC in Group V, but very dissimilar to the other proteins. The Australian proteins are most unlike the 13 proteins in Group III.

The Group VI proteins contain from 2 to 4 SS polypeptides in contrast to those in Group III, 6 of which contain only a single polypeptide and 7 contain two. It appears therefore that the evolution of the Australian proteins has progressed at a faster pace than the proteins placed by the computer in Group III.

Mechanism of Evolution of Fraction I Protein. The data in Fig. 3 also illustrate the manner of evolution of F-I-p as exemplified by the polypeptides of N. tabacum protein. The tobacco plant is an allotetraploid which arose as N. tabacum after N. sylvestris served as the female parent during interspecific hybridization with N. tomentosiformis[11]. The F-I-p in N. tabacum acquired its LS polypeptides belonging to cluster B (Fig. 3) from N. sylvestris since this genetic information is transferred only by the maternal line[12]. However, both parents made an equal contribution to the composition of the SS polypeptides.

N. tabacum F-I-p acquired polypeptide 9 (Fig. 3) from genetic information transferred by the pollen of N. tomentosiformis when it fertilized the egg cell of N. sylvestris. Coding information for polypeptide 7 was already present in the female parent.

Kawashima et al.[13] have analyzed the chymotryptic peptides of the LS and SS of F-I-p contained in N. tabacum and the two parents of this species. Out of 38 peptides separated after chymotrypsinolysis of the LS the proteins of N. tabacum and N. sylvestris were found to be identical in composition, but N. tomentosiformis lacked one of the peptides found in N. tabacum. Of 25 peptides obtained for the SS of N. tabacum protein, N. sylvestris SS contained 2 peptides not found in N. tomentosiformis whereas the latter contained 2 not found in the former species. However, complementation of genetic information contained in n-DNA of both parents supplied the code necessary to produce the full complement of chymotryptic peptides of the SS of N. tabacum protein. The same study established that F-I-p in N. tabacum had not undergone further evolution by additional interspecific hybridizations between the newly-arisen N. tabacum and either of its two parents.

Formation of Isozymes in New Species of Plants Arising from Interspecific Hybridization. It is still a question as to whether the protein in N. tabacum is a hybrid in the sense of whether it consists of 8 LS of the N. sylvestris type combined with 8 SS, four of whose polypeptides are of the N. sylvestris type and the other four are of the N. tomentosiformis type. Or, whether N. tabacum contains two kinds of protein, both with N. sylvestris type LS, but one having 8 SS of N. sylvestris type polypeptides and the other macromolecule having 8 of the N. tomentosiformis type to produce two species of ribulose diphosphate carboxylases, or isozymes. Reasons for favoring the hybrid protein idea have been advanced[14]. By analogy, more compelling evidence for hybrid proteins comes from the results of Steer[15] on inheritance and evolution of Avena F-I-p. Protein from one species of Avena may have a electrophoretic mobility distinctly different from the protein obtained from another species of Avena. When the two species of Avena are hybridized, a protein of only one mobility is found rather than a mixture of two proteins as would have been the case if isozymes of F-I-p had been formed.

The F-I-p SS of N. tabacum has been partially sequenced[16]. However, the two polypeptides were not separated before sequencing. At the 7th residue an ambiguity appeared wherein both tyrosine and isoleucine appeared to occupy this position. However, the coding triplet for tyrosine, (UAU, UAC) is different by two nucleotides from the triplet for isoleucine (AUU, AUC), showing that the ambiguity could not have arisen by a point mutation. More recent analysis of the SS of N. sylvestris protein[17] shows tyrosine at residue 7 so that it seems likely that when the sequence of the N. tomentosiformis SS is available, isoleucine will be found at residue 7. Whereas the probability of two successive mutations producing the tyrosine/isoleucine ambiguity is remote, production of the ambiguity is the natural consequence of producing two polypeptides during interspecific hybridization of the two parents to produce *N. tabacum* as a new species.

In the case of proteins coded entirely by n-DNA, the production of two or more kinds of proteins, or isozymes, is always a significant possibility during speciation following interspecific hybridization. Ferredoxin is a good example of the situation of a protein coded by nuclear DNA[18]. Ferredoxin isolated from *N. glauca* contains 2 methionine residues whereas no methionine is contained in protein from *N. glutinosa*. The ferredoxin in the interspecific F_1 hybrid, *N. glutinosa* ♀ x *N. glauca* ♂, contains 1 methionine. Furthermore, the ferredoxin in the F_1 hybrid resolves into two polypeptides by electrofocusing, showing that genetic information determining the isoelectric point of one of the polypeptides had been conveyed by the pollen of *N. glauca* during creation of the interspecific hybrid.

Since so many present day flowering plants have arisen by interspecific hybridization, it seems highly probable that their evolutionary process has been repeatedly accompanied by formation of isozymes by the mechanism found for ferredoxin. On this account, as more and more plant proteins are subjected to sequencing, discovery of ambiguities may be a signal that more than one polypeptide is present. The ferredoxin of spinach leaves has been completely sequenced and contains two ambiguities in a polypeptide of 97 residues[19]. Whereas both ambiguities could have evolved from point mutations, the SS of spinach F-I-p resolves into two polypeptides by electrofocusing (4) which is a strong indication that the spinach plant evolved by interspecific hybridization. So the possibility exists that the ferredoxin in spinach consists of two polypeptides and the ambiguities exist because the polypeptides were not separated before being sequenced.

Evolution of Australian Fraction I Proteins. The F-I-p of *N. noctiflora* and *N. petunioides* are the only proteins among the 43 Western Hemisphere *Nicotiana* species which have LS polypeptides whose isoelectric points are identical with F-I-p in Australian *Nicotiana* species (Fig. 3). They are therefore the only species of *Nicotiana* in the Western Hemisphere whose c-DNA coding for F-I-p is the same as that in the 20 Australian species. These c-DNA genes have therefore been preserved without further mutation since continental drift began about 10^8 years ago. It is of considerable interest that *N. noctiflora* and *N. petunioides* are native plants in a highly restricted habitat at the extreme tip of South America, as shown in Fig. 5, which also serves as a reminder of the resemblance between the physiognomy of Australia, Antarctica, and South America.

That the F-I-p of *N. noctiflora* and *N. petunioides* are not greatly changed from the proteins contained in those plants which reached Australia via Antarctica is also borne out by their SS polypeptides. They contain SS polypeptide number 8 which is found in some Australian proteins but in no Western Hemisphere proteins. They also contain polypeptide 10 which is absent in the 20 Australian proteins but is found in some Western Hemisphere proteins. It is probable therefore, that the Noctiflora-Petunioides type F-I-p is the relic of the type of protein which was contained in the plants which evolved into the Australian species of Nicotiana. We deduce therefore that the Australian F-I-p are less aged than those of the Western Hemisphere. SS polypeptides 3 and 6 are found in both Australian and Western Hemisphere species. Their existence points to the need to be cautious in speculating further on F-I-p evolution in the absence of substantial sequence data. Having the same isoelectric points could be coincidental and have little to do with the actual arrangement of the amino acids forming the primary structure of the SS polypeptides in question. However, it is clear that sequencing could produce the information required for an in depth understanding of the molecular events concerned with the evolution of Fraction I protein.

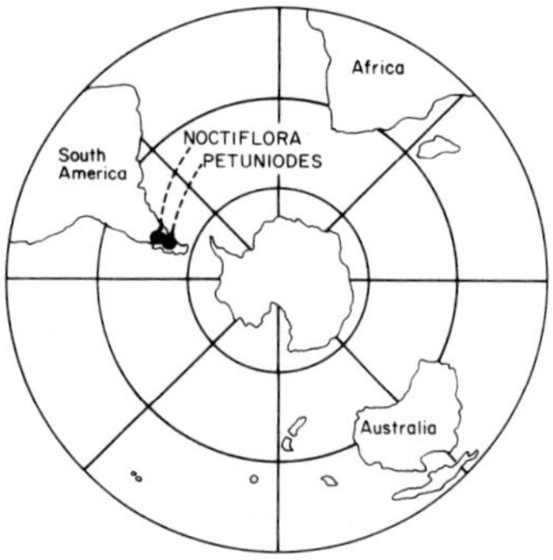

Fig. 5. Native habitat of the Nicotiana species which contain F-I-p thought to be relics of those that accompanied the species which colonized Australia.

ACKNOWLEDGEMENT. Research supported by Contract E (04-3) 34, Project 8, U.S. Energy Research and Development Administration, U.S. National Science Foundation grant 75-07368, and a grant from Monsanto Commercial Products Corporation.

REFERENCES

1. Andrew, T.J., Lorimer, G.H., and Tolbert, N.E. (1973) Biochem. 12: 11-18.
2. Baker, T.S., Eisenberg, D., and Eiserling, F.A. Personal communication.
3. Kung, S.D., Sakano, K., and Wildman, S.G. (1974) Biochim. Biophys. Acta. 365: 138-47.
4. Chen, K., Gray, J., Kung, S.D., and Wildman, S.G. (1976) Plant Science Letters In press.
5. Kung, S.D. (1976), Science, 191: 429-434.
6. Goodspeed, T.H. (1954) The Genus Nicotiana, Chronica Botanica, Waltham, Mass. 636 pp.
   Burbidge, N.T. (1960) Aust. J. Bot., 8: 342-380.
7. Solbrig, O.T. (1966) Evolution and Systematics, The Macmillan Company.

8. Dickerson, R.E. (1972) Sci. American, 226: 58-72.

9. Rohlf, F.J., Kishpaugh, J., and Kirk, D. (1972) Numerical Taxonomy: System of Multivariate Statistical Programs, The State University of New York at Stony Brook.

10. Sneath, P.A., and Sokal, R.P. (1973) Numerical Taxonomy, W.H. Freeman and Company, San Francisco, California.

11. Sheen, S.J. (1972) Evolution, 26: 143-154.
Gray, J.C., Kung, S.D., Wildman, S.G., and Sheen, S.J. (1974) Nature, 252: 226-227.

12. Chan, P.H., and Wildman, S.G. (1972) Biochim. Biophys. Acta. 227: 677-680.
Sakano, K., Kung, S.D., and Wildman, S.G. (1974) Molec. gen. Genet. 130: 91-97.

13. Kawashima, N., Tanabe, Y., and Iwai, S. (1976) Biochim. Biophys. Acta. 427: 70-77.

14. Wildman, S.G., Chen, K., Kung, S.D., Sakano, K., and Kwanyuen, P. (1975) in _Genetics and Biogenesis of Mitochondria and Chloroplasts_, C.W. Birky, P.S. Perlman, T.J. Byers, Eds., Ohio State University Press, Chap. 9.

15. Steer, M.W. (1975) Can. J. Genet. Cytol. 17: 337-344.

16. Gibbons, G.C., Strobaek, S., Haslett, B., and Boulter, D. (1975) Experientia, 31: 1040-1041.
Haslett, B., Yarwood, A., Evans, I.M., and Boulter, D. (1976) Biochim. Biophys. Acta. 420: 122-132.

17. Gibbons, G.C., Personal communication.

18. Kwanyuen, P., and Wildman, S.G. (1975) Biochim. Biophys. Acta. 405: 167-174.

19. Matsubara, H., and Sasaki, R.M. (1968) J. Biol. Chem. 243: 1732.

Genetics and Biogenesis of Chloroplasts and Mitochondria, Th. Bücher et al. eds.

POLYPEPTIDE CHAINS OF THE LARGE AND SMALL SUBUNITS OF FRACTION I PROTEIN

J.C.Gray*, S.D. Kung** and S.G. Wildman
Molecular Biology Institute & Department of Biology,
University of California at Los Angeles,
California 90024, U.S.A.

INTRODUCTION

Fraction I protein (ribulose 1,5-bisphosphate carboxylase) is the major soluble protein in chloroplasts, where it catalyses CO_2 fixation in photosynthesis. In the presence of denaturing agents, such as dodecylsulphate or urea, the protein dissociates into its component polypeptides which may be separated by a variety of techniques. Separation on the basis of size produces two fractions of molecular weights 55,000 and 13,000[1-3], currently known as the large and small subunits. However recent separations by isoelectric focusing have shown a wider range of polypeptides[4]. Fraction I protein from tobacco leaves was separated by isoelectric focusing in 8M urea into five polypeptides, three grouped near pH 7 each with a molecular weight of 55,000 and two near pH 6 with molecular weights of 13,000. The nature of the differences between these subunit polypeptides is the subject of the present investigation.

SUBUNIT POLYPEPTIDES

Isoelectric focusing of S-carboxymethylated Fraction I protein from tobacco leaves gives a pattern of five polypeptide bands as shown in Fig. 1. Preliminary experiments indicated that the five polypeptides were not the products of proteolytic degradation nor the result of interactions of the protein with polyphenols during preparation. Similarly, it was demonstrated that the five polypeptides were not the result of incomplete carboxymethylation of the protein nor of carbamylation of the protein by cyanate present in urea solutions. It therefore appears that the multiple polypeptide composition of the protein revealed by isoelectric focusing is a true indication of the structural complexity of Fraction I protein.

SEPARATION OF SUBUNIT POLYPEPTIDES

Separation of the subunit polypeptides was achieved by isoelectric focusing in polyacrylamide gel slabs[4], The gels, similar to the analytical gels but with a higher loading of protein, were run daily for several weeks until several mg of each polypeptide were separated. The polypeptides were visualised at the end of

---

*Present address: Department of Biochemistry, University of Cambridge, Cambridge CB2 1QW, U.K.
**Permanent address: Department of Biological Sciences, University of Maryland Baltimore County, Catonsville, Maryland 21228, USA.

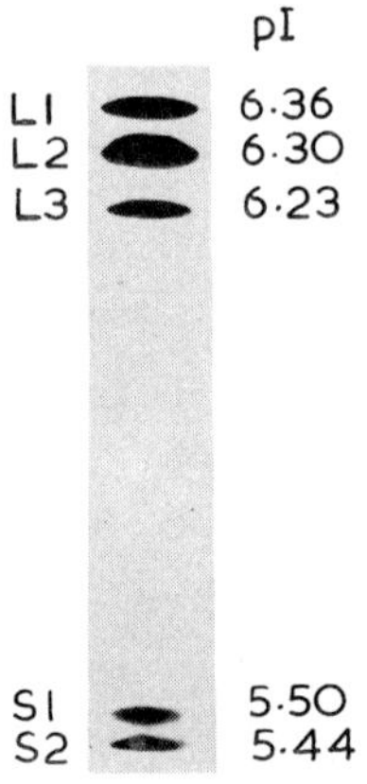

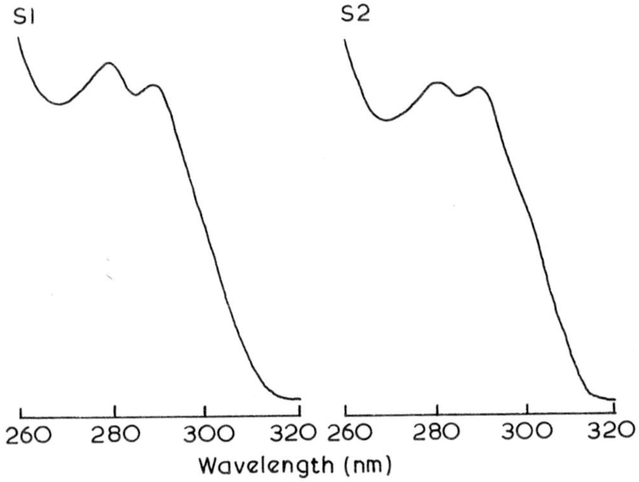

Fig. 1. Separation of subunit polypeptides of S-carboxymethylated Fraction I protein from tobacco leaves, by isoelectric focusing in 5% polyacrylamide gel containing 8M urea and 1% Ampholine pH 5-7.

Fig. 2. UV spectra of small subunit polypeptides S1 and S2 in 0.1M NaOH.

each run by placing the gels in water; the polypeptides precipitated in the gel as white bands as the urea diffused away. The bands were cut out and placed in 6M guanidine HCl to solubilise the protein. The protein was extracted from the gel by fragmentation through a syringe and elution with 6M guanidine HCl. The subunit polypeptides were then exhaustively dialysed against water and freeze-dried. Examination of the polypeptide preparations by isoelectric focusing in 8M urea showed them to be substantially free from contamination by other polypeptides.

SMALL SUBUNIT POLPYEPTIDES

Polypeptides S1 and S2 were shown to be different by their UV spectra in 0.1M NaOH, by amino acid analysis and by fingerprinting of peptides produced by trypsin hydrolysis. The UV spectra of the polypeptides, shown in Fig. 2., indicate that the two polypeptides have different ratios of tyrosine to tryptophan; S1 had a ratio of 1.5:1 whereas S2 had a ratio of 2.5:1. A difference in the content of tyrosine in S1 and S2 was also demonstrated by amino acid analysis. As shown in Table I there is very little difference between S1 and S2 except for the tyrosine content. Two dimensional chromatography/electrophoresis of tryptic peptides of S1 and S2 indicated that each polypeptide contains (at least) one unique peptide; these are marked with arrows in Fig. 3. These results indicate that the small subunit polypeptides S1 and S2 are distinct polypeptides although

TABLE I

AMINO ACID COMPOSITION OF SUBUNIT POLYPEPTIDES

Amino acid	S1	S2	L1	L2	L3
		residues/100 residues			
Asp	8.1	7.2	9.7	9.1	9.6
Thr	4.7	4.2	5.0	5.8	5.4
Ser	6.0	5.8	3.9	3.3	3.3
Pro	5.4	6.3	4.4	5.1	4.0
Glu	17.2	16.1	10.7	10.2	10.1
Gly	9.6	10.5	11.1	10.6	10.7
Ala	6.5	6.1	9.7	9.6	9.8
Val	6.9	6.0	7.6	7.7	7.5
Met	1.1	1.4	1.2	1.1	1.0
Ile	4.7	4.4	4.5	4.4	4.4
Leu	8.9	8.9	10.4	9.1	9.5
Tyr	2.6	6.6	3.2	2.7	2.5
Phe	5.4	4.4	4.3	4.9	4.9
Lys	7.5	7.4	6.0	5.4	5.5
His	0.6	1.3	3.0	3.4	3.3
Arg	4.9	5.0	8.6	8.8	8.6

very similar in many respects. Gibbons _et al_[5] have reported a partial N-terminal amino acid sequence of total small subunit of tobacco showing heterogeneity at position 7, where equal amounts of tyrosine and isoleucine were found in all runs. It is therefore most likely that the sequence heterogeneity is due to the presence of two slightly different subunit polypeptides. Phylogenetic studies on the origin of _N. tabacum_ have shown that the two small subunit polypeptides were inherited from two separate parental species[6] and must therefore be regarded as separate gene products.

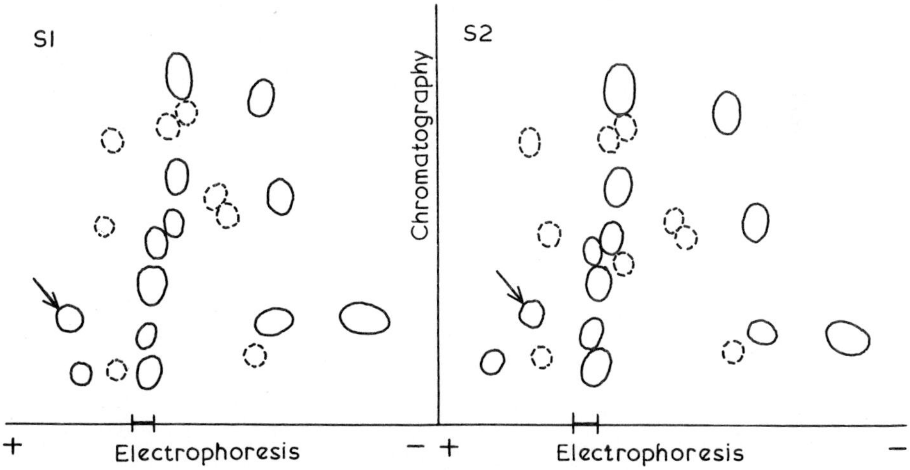

Fig. 3. Fingerprints of tryptic peptides of small subunit polypeptides S1 and S2. Faint spots are shown by broken circles.

LARGE SUBUNIT POLYPEPTIDES

In the present study it was not possible to find any differences between the large subunit polypeptides L1, L2 and L3, except for their separation by isoelectric focusing. The three polypeptides had very similar amino acid compositions, as shown in Table I. Fingerprinting of tryptic or chymotryptic peptides, employing chromatography and electrophoresis at pH 3.6 or pH 6.5, failed to demonstrate reproducible differences between the large subunit polypeptides. This would indicate that the large subunit polypeptides are very similar and a full understanding of the differences might not be forthcoming until a complete amino acid sequence of the large subunit is available. The three large subunit polypeptides might be due to either three separate gene products or to modification of a single gene product. Several pieces of evidence seem to point towards the latter alternative as the more probable. Examination of Fraction I proteins from a wide range of plants has shown that all proteins are composed of three equally-spaced large subunit polypeptides[7]; with three separate gene products it might be expected that independent mutation of the separate genes during the course of plant evolution would give rise to a polypeptide pattern in which the bands were unequally spaced. This was not seen in any of the wide variety of plants examined.

On the basis of a 0.5 pH shift of all three bands on modification of the protein with iodoacetamide instead of iodoacetic acid, it appears that the three polypeptides are one unit charge apart, which is compatible with the modification of a single gene product. However such modification would have to occur very soon after the synthesis of the large subunit, for newly synthesised multiple polypeptides have been observed in an isolated chloroplast system (R.J. Ellis, personal communication) and in a wheat germ system for the translation of mRNA for the large subunit[8]. This might suggest that the differences are due to variation in removal of the initiation sequence at the N-terminal of the polypeptide, as has been suggested for the small subunit[5].

REFERENCES

1. Rutner, A.C. and Lane, M.D. (1967) Biochem. Biophys. Res. Commun. 28, 531-537
2. Kawashima, N. (1969) Plant Cell Physiol. 10, 31-40
3. Rutner, A.C. (1970) Biochem. Biophys. Res. Commun. 39, 923-929
4. Kung, S.D., Sakano, K. and Wildman, S.G. (1974) Biochim. Biophys. Acta 365,138-147
5. Gibbons, G.C., Strobaek, S., Haslett, B. and Boulter, D. (1975) Experientia 31, 1040-1041
6. Gray, J.C., Kung, S.D., Wildman, S.G. and Sheen, S.J. (1974) Nature 252,226-227
7. Chen, K., Kung, S.D., Gray, J.C. and Wildman, S.G. (1976) Plant Sci. Lett., in press
8. Sagher, D., Grosfeld, H. and Edelmann, M. (1976) Proc. Nat. Acad. Sci.U.S.73, 722-726

Genetics and Biogenesis of Chloroplasts and Mitochondria, Th. Bücher et al. eds.

STUDIES ON THE PRIMARY STRUCTURE OF THE SMALL SUBUNIT OF RIBULOSE-1,5-DIPHOSPHATE CARBOXYLASE

Carsten Poulsen and Steffen Strøbæk
Department of Physiology, Carlsberg Laboratory
Copenhagen, Denmark

and

Barry G. Haslett
Department of Botany, University of Durham
Durham, England

INTRODUCTION

The CO_2 fixing enzyme, RudP carboxylase+ (E.C. 4.1.1.39) of the chloroplasts has a molecular weight of about 550,000 and consists of large subunits with a molecular weight of 55,000 and small subunits with species specific differences in molecular weight ranging between 12,000 and 15,000[1,2,3]. The large subunit appears to carry the catalytic site[4] and probably is coded for by chloroplast DNA[5]. The small subunit is coded for by a gene in the nucleus[6] and considered to have regulatory functions[7]. For future studies on the function and biosynthesis of this enzyme, it is desirable to know the amino acid sequences of the two subunits. Knowledge of the primary structure will also aid in finding suitable mutants for inquiries into the genetic regulation of this enzyme. In this paper we report on the progress in the determination of the primary structure of the small subunit of RudP carboxylase of barley and compare the partial sequences obtained with those known from other plants.

MATERIALS AND METHODS

Preparation of RudP carboxylase and separation of subunits: The enzyme was prepared from 7-8 days old light grown barley seedlings (*Hordeum vulgare* L. cv. Svalöf's Bonus) according to Strøbæk and Gibbons[3] with the following modifications. In order to purify the protein from 200 g of leaves a 5 x 50 cm column of Sephadex^R G-25 Medium (Pharmacia) and a 10 x 80 cm column of Sepharose^R 6B was used. The solution of the purified protein was concentrated to 10 ml in an Amicon 402 ultrafiltration cell supplied with a PM30 filter. 10 ml of 0.1 M Tris-HCl pH 8.5, containing 1 % SDS and 0.2 M in 2-mercaptoethanol was added and the subunits were separated by SDS column chromatography as described by Rutner and Lane[1]. Fractions containing the two subunits were pooled and freezedried separately. The dried powders were resuspended in 10 ml of 10 mM HCl, 40 ml of icecold acetone were

+Abbreviations: RudP, ribulose-1,5-diphosphate; SDS, sodium dodecyl sulfate; CNBr, cyanogen bromide.

added and the precipitated protein was collected by centrifugation. The pellets were resuspended in 0.1 M NaOH, dialyzed against 3 x 5 l of 50 mM NH_3 for 24 hours and subsequently freezedried.

Amino acid analysis: Protein was hydrolysed in 6 M redestilled HCl containing 0.1 % phenol for 24, 48, and 72 hours at 110°C in evacuated ampoules and analyzed on a Durrum D-500 automatic amino acid analyzer. For the determination of tyrosine and tryptophan 0.5-1.0 mg of the subunits were dissolved in 3.0 ml of 0.1 M NaOH, ultraviolet spectra recorded in the 250-350 nm region, and the concentration of the two amino acids calculated after Bencze and Schmid[8]. Tyrosine and protein content of the solutions were independently determined by amino acid analysis.

Polyacrylamide gel electrophoresis: Molecular weight and homogeneity of the subunits was analyzed with the system of Chua and Bennoun[9]. Samples of 5-10 µg subunit were run on gels consisting of a 10-15 % gradient of acrylamide with 2.7 % bis-acrylamide. A current of 17.5 mA per gel was applied for 7 hours at room temperature. For molecular weight calibration the following standards were used: bovine serum albumin, catalase, ovalbumin, aldolase, chymotrypsinogen A and cytochrome c from Boehringer's Combithek^{R}, and sperm whale myoglobin from Koch-Light.

Reduction and carboxymethylation: Samples of the small subunit were reduced and carboxymethylated as described by Hirs[10]. After the reaction, the samples were dialyzed against 3 x 5 l of 50 mM NH_3 for 18 hours at 5°C and then freezedried.

Citraconylation: Carboxymethylated small subunit was citraconylated as described by Atassi and Habeeb[11]. The citraconylated, carboxymethylated small subunit was then desalted and freezedried.

Tryptic digestion: 20 mg of citraconylated, carboxymethylated small subunit was dissolved in 2.5 ml of 0.2 M ammoniumbicarbonate pH 8.5. 75 µl of a solution of tosyl-L-phenylalanine chloromethylketone trypsin from Worthington (5 mg/ml 1 mM HCl) was added. After digestion for 1 hour at 20°C the solution was freezedried.

Separation of tryptic peptides: The tryptic digest was redissolved in 2.5 ml of 0.2 M ammoniumbicarbonate pH 8.5 and applied to a 2.5 x 90 cm column of Sephadex G-25 Superfine and eluted with the same buffer at a rate of 45 ml per hour. 3 ml fractions were collected. Appropriate peptide containing fractions were pooled and freezedried. The material of the first peak was subjected to ion exchange chromatography on a 0.9 x 15 cm column of DEAE-cellulose (DE-52 from Whatman) with a 600 ml linear gradient from 0.1 M to 1 M ammoniumacetate pH 8.5. Appropriate fractions were pooled and freezedried.

Cyanogenbromide cleavage: Cleavage of the peptide chain after methionine residues was carried out in 70 % formic acid as described by Gross and Witkop[12]. A

100 fold molar excess of CNBr over methionine was used. After 20 hours of reaction at room temperature, the samples were deluted with water and dried in vacuo over solid NaOH, resuspended in destilled water and freezedried.

Separation of cyanogenbromide fragments: 20 mg of CNBr cleaved, unmodified small subunit were suspended in 4 ml of 0.5 M NaCl, 1 % in 2-mercaptoethanol and incubated under N_2 at 40°C for 1 hour. Addition of 1 ml of 5 M NaOH solubilized all material. The sample was applied to a 2.5 x 90 cm column of Sephadex G-75 Superfine, equilibrated and eluted with 50 mM NH_3, 0.5 M NaCl (20 ml per hour, 3 ml fractions).

Alternatively, 20 mg of CNBr cleaved, carboxymethylated small subunit were citraconylated and applied directly to a 2.5 x 90 cm column of Sephadex G-75 Superfine, equilibrated and eluted with 0.2 M ammoniumbicarbonate pH 8.5.

End group identification: The dansyl chloride procedure, as described by Gray[13], was carried out on the intact subunits or purified peptides. The released amino terminal dansyl derivatives were identified on 5 x 5 cm polyamide plates (Schleicher-Schüll) as described by Hartley[14].

Manual sequencing: Dansyl-Edman degradation as described by Gray[15] was carried out on 100 nmoles quantities. The released anilinothiazolinones were converted to the corresponding free amino acids by hydrolysis in 57 % hydriodic acid for 20-24 hours in evacuated ampoules at 120-130°C. Samples were then dried and subsequently identified and quantified on the amino acid analyzer[16].

Automatic sequencing: Edman degradation of 10 mg small subunit of barley carboxylase was carried out with a Beckman 890C automatic sequencer using the fast protein program (No. 072172C).

Solid-phase Edman degradation of 1 mg of a CNBr fragment was carried out with an Anachem APS2400 automatic sequencer using the standard program supplied with the sequencer. 3-aminopropyl glass (Corning) was used as a solid support[17] to which the peptide was attached by their lysine ε-amino groups using p-phenyl diisothiocyanate[18].

Resultant anilinothiazolinones were converted to the corresponding phenylthiohydantoins and identified[19].

RESULTS

The amino acid compositions of the two subunits of RudP carboxylase in barley have been reported earlier[3]. The ratio of tyrosine to tryptophan was determined to be 1.00 for the large subunit and 1.20 for the small subunit. The amino acid analysis gave a tyrosine recovery of 90-95 %. Extinction coefficients ($E^{1\%}_{280\ nm}$) were accordingly determined to be 18.0 cm^{-1} for the large and 26.9 cm^{-1} for the small subunit. These have been used to determine the amount of large and small subunits in the carboxylase molecule. The ratio of OD_{280} units for subunits separated in single experiments was 1.9-2.0. Thus, 75 % of the weight of the carboxylase is contributed by the large and 25 % by the small subunit. In terms of yield 200 g of leaves gave 100-105 mg of the large and 30-35 mg of the small subunit. The recovery of the acetone precipitation step and the following dialysis was 60-70 % for the two subunits as determined by amino acid analysis.

Gel electrophoresis of the small subunit resulted in a single band corresponding to a protein with a molecular weight between 13,000 and 16,000. Electrophoresis of the large subunit yielded a major band (molecular weight 54-56,000) and four faint bands (molecular weights 50-60,000).

Automatic sequencing of the small subunit resulted in the N-terminal sequence given below. Deviations in sequence for pea, broadbean and tobacco[20,21] are shown:

```
                             5                  10                  15
Barley  :  Met-Gln-Val-Trp-Pro-Ile-Glu-Gly-Ile-Lys-Lys-Phe-Glu-Thr-Leu-
Pea     :                     -Pro-Ile-   -Lys-
Bean    :                     -Pro-Ile-   -Lys-
                                   Tyr
Tobacco :                     -Pro-   -Asn-Lys-        -Tyr-
                                   Ile

                           20                  25
Barley  :  Ser-Tyr-Leu-Pro-Pro-Leu-Ser-Thr-Glu-Ala-
Pea     :     -Trp-                -Thr-Pro-Asp-Gln-
Bean    :                          -Thr-Gln-Asp-Gln-
Tobacco :                     -Asn-       -Gln-Gln-Gln-
```

Gelfiltration of the tryptic digest on Sephadex G-25 Superfine gave rise to four well separated peaks, eluting with K_{av}-values of 0, 0.23, 0.41 and 0.55. End group identification of the first peak gave three different N-terminal residues, whereas the others gave only one. The peptides of the first peak were subjected to ion exchange chromatography. Three peaks eluted at 0.32 M, 0.38 M and 0.55 M ammonium acetate, having end groups Glx, Ala and Tyr, respectively. Thus, cleavage of the peptide chain after arginine residues with trypsin resulted in six peptides. Of these, the two peptides with Glx and Tyr as N-terminals, contributed with 80 % to the total absorbance at 280 nm, indicating their relative high content of aromatic amino acids. These two peptides were subjected to manual Edman degradation with the following results:

I   Glx-Val-(Trp)-Pro-Ile-Glx-

II   Tyr-(Trp)-Thr-Leu-(Trp)-Lys-Leu-Pro-(Met)-Phe-Gly-Ile-Pro-

After cyanogen bromide cleavage of the unmodified subunit only Phe and homoserine (N-terminal Met or Met-Met sequence) were identified as N-terminals. Gel chromatography resolved the CNBr-peptides into three peaks (Fig. 1), all of which yielded Phe in end group analysis. Automatic and manual sequence analysis of the first peak gave one N-terminal sequence:

CNBr-1: Phe-Thr-Ser-Ile-Val-Gly-Gln-Val-Phe-Gly-Phe-(Arg)-Ala-Leu-Leu-Ala-

One single N-terminal sequence was found for the peptides of the second and third peak:

CNBr-2: Phe-Gly-Ile-Pro-Asx-Ala-

A CNBr-peptide from the small subunit of pea has the following N-terminal sequence[21]:

Phe-Gly-Pro-Pro-Asp-Ala-Ser

According to amino acid analysis, the peptide CNBr-2 did not contain homoserine, which places this peptide as the C-terminal.

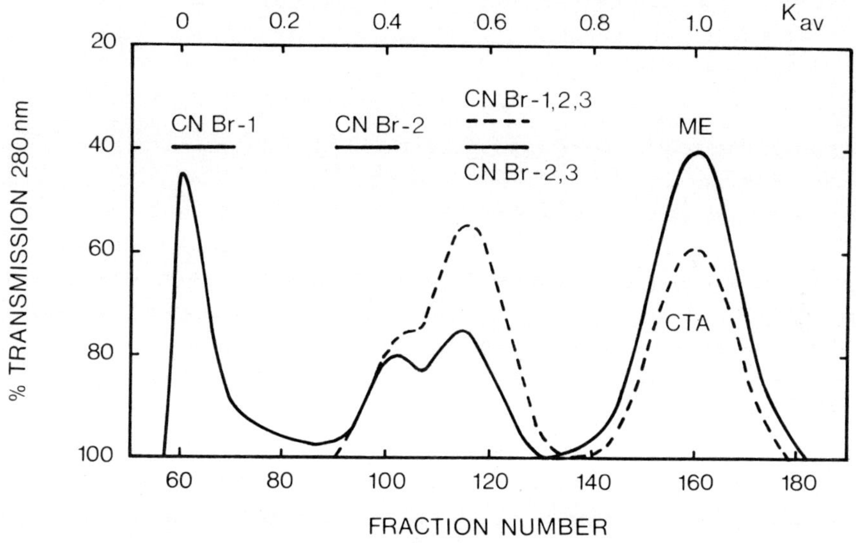

Fig. 1. Chromatography of CNBr-fragments of RudP carboxylase small subunit from barley.
—— : Unmodified small subunit fragments.
- - - : Citraconylated fragments of carboxymethylated small subunit.
Bars show how fractions were pooled for further analysis. Contents of total volume are 2-mercaptoethanol (ME) and homoserine or citraconic acid (CTA) and homoserine.

The chromatography of citraconylated CNBr-fragments from carboxymethylated small subunit shows an elution profile with only two peaks. The net increase in negative charges, accomplished by citraconylation, eliminates the presence of peptides eluting in the front, indicating that these are aggregated without side-group modification. After modification of the subunit and the peptides, these elute with K_{av}-values corresponding to the second and third peak in the elution profile of peptides from the unmodified subunit (Fig. 1).

These results can be interpreted as follows: CNBr-cleavage results in three peptides of roughly equal size. Peptide CNBr-1 aggregates if unmodified and elutes with a K_{av} of 0.55 if modified. The C-terminal peptide CNBr-2 elutes with a K_{av} of 0.42 whether modified or not.

From the known N-terminal sequence of the intact small subunit, a third CNBr-peptide (CNBr-3) having the Gln of position 2 as N-terminus, can be predicted. This peptide has not been localized in the elution profile by its N-terminus, probably due to cyclization of the Gln into a pyrrolidone ring, which is not accessible to dansyl chloride or Edman's reagent[22]. Using unmodified subunit, the amino acid composition determined from eluates with a K_{av} of 0.42 and known to contain only CNBr-2, differs from the eluates with a K_{av} of 0.55, in which no CNBr-1 was found. This indicates that CNBr-3 elutes with a K_{av} of 0.55 and thus could not be purified in these experiments.

DISCUSSION

The molecular weight of the small subunit of RudP carboxylase from barley varies with the system used for its determination[3]. A molecular weight of 14,000 is in good agreement with our data on the mobility of the small subunit by SDS gel-electrophoresis and a good stoichiometric amino acid composition with five Arg's and three Met's. Five arginines are required to yield six tryptic peptides and three methionines are needed to give three CNBr-peptides in addition to a homoserine derived from the N-terminus of the subunit.

The N-terminal amino acid sequence as determined for the tryptic peptide (I) is identical to position 2-7 of the intact small subunit. As observed earlier[20,21] the N-terminal methionine of the small subunit is easily lost during purification procedures. The tryptic peptide (II) overlaps the N-terminus of CNBr-2.

When the N-terminal sequences of the small subunit of the four species comprising barley, pea, bean and tobacco are compared, differences were found in 9 positions out of 25. Amino acid substitutions are concentrated in two regions, namely positions 6-9 and 20-25. The most striking substitutions are those of the prolines in positions 6, 20, and 23. Prolines are commonly placed at the bends of α-helices and β-sheets.

Comparing CNBr-2 from barley with the CNBr-fragment of pea, identical sequences is seen if an Ile in the barley fragment is changed into a Pro. Thus three Pro-Pro sequences are present in the small subunit of pea carboxylase, whereas only one Pro-Pro sequence is found in barley. These differences in the distribution of prolines are likely to cause differences in secondary and tertiary structure of the small subunit. Some of the amino acids which substitute for proline or are placed in the vicinity of proline comprise Tyr, Glu, Gly and Asn. These amino acids are often placed at or near bends of α-helices and β-sheets[23]. Thus, possible bends in the secondary structure at positions 5-8 and 17-20 could be conserved in spite of extensive amino acid substitutions.

ACKNOWLEDGEMENT

We are indebted to professor B. Foltmann and the staff of the Institute of Biochemical Genetics, University of Copenhagen, and dr. G.C. Gibbons, Carlsberg Research Laboratory, for kind advice and help, and to cand.scient. Gunilla Høyer-Hansen for the molecular weight determinations carried out by SDS gelelectrophoresis. This work was supported in part by Grant GM-22051 of the National Institutes of Health, United States Public Health Services to professor D. von Wettstein.

REFERENCES

1. Rutner, A.C. and Lane, M.D. (1967) Biochem. Biophys. Res. Comm. 28, 531-537.
2. Börner, T., Jahn, G. and Hagemann, R. (1976) Biochem. Physiol. Pflanzen 169, 179-181.
3. Strøbæk, S. and Gibbons, G.C. (1976) Carlsberg Res. Commun. 41, in press.
4. Sugiyama, T., Matsumoto, C. and Akazawa, T. (1970) J. Biochem. 68, 821-831.
5. Chan, P.H. and Wildman, S.G. (1972) Biochim. Biophys. Acta 277, 677-680.
6. Kawashima, N. and Wildman, S.G. (1972) Biochim. Biophys. Acta 262, 42-49.
7. Takabe, T. and Akazawa, T. (1975) Plant Cell Physiol. 16, 1049-1060.
8. Bencze, W.L. and Schmid, K. (1957) Anal. Chem. 29, 1193-1196.
9. Chua, N.H. and Bennoun, P. (1975) Proc. Nat. Acad. Sci. U.S.A. 72, 2175-2179.
10. Hirs, C.H.W. (1967) Methods in Enzymology, Vol. XI, 199-203.
11. Atassi, M.Z. and Habeeb, A.F.S.A. (1972) Methods in Enzymology, Vol. XXV, 546-553.
12. Gross, E. and Witkop, B. (1962) J. Biol. Chem. 237, 1856-1860.
13. Gray, W.R. (1967) Methods in Enzymology, Vol. XI, 139-151.
14. Hartley, B.S. (1970) Biochem. J. 119, 805-822.
15. Gray, W.R. (1967) Methods in Enzymology, Vol. XI, 469-475.
16. Smithies, O., Gibson, D., Fanning, E.M., Goodfliesh, R.M., Gilman, J.G. and Ballantyne, D.L. (1971) Biochemistry 10, 4912-4921.
17. Wachter, E., Machleidt, W., Hofner, H. and Otto, J. (1973) FEBS Letters 35, 97-102.
18. Laursen, R.A., Horn, M.J. and Bonner, A.G. (1972) FEBS Letters 21, 67-70.

19. Haslett, B.G. and Boulter, D. (1976) Biochem. J. 153, 33-38.

20. Gibbons, G.C., Strøbæk, S., Haslett, B. and Boulter, D. (1975) Experientia 31, 1040-1041.

21. Haslett, B.G., Yarwood, A., Evans, I.M. and Boulter, D. (1976) Biochim. Biophys. Acta 420, 122-132.

22. Kawasaki, I. and Itano, H.A. (1972) Anal. Biochem. 48, 546-556.

23. Chou, P.Y. and Fasman, G.D. (1974) Biochemistry 13, 222-245.

Genetics and Biogenesis of Chloroplasts and Mitochondria, Th. Bücher et al. eds.

ONTOGENY, INSERTION, AND ACTIVATION OF TWO THYLAKOID PEPTIDES REQUIRED FOR PHOTOSYSTEM II ACTIVITY IN THE NUCLEAR, TEMPERATURE SENSITIVE T4 MUTANT OF CHLAMYDOMONAS REINHARDI

Frank Kretzer, Itzhak Ohad, Pierre Bennoun[+]
Departement of Biological Chemistry, The Hebrew University of Jerusalem,Israel [+] Inst. de Biologie Physico-Chimique, 13, rue Pierre et Marie Curie 75005-Paris-France

INTRODUCTION

The contribution of both the chloroplast genome and protein synthesizing machinery to the development of the photosynthetic apparatus has been investigated by the use of appropriate mutants (1) and specific translation and transcription inhibitors (2). The degree of chloroplast autonomy in the process of photosynthetic membrane biogenesis is rather restricted since many membrane components are coded by nuclear genes (3). In the green alga, Chlamydomonas reinhardi, it has been demonstrated that several peptides which are synthesized by cytoplasmic ribosomes are essential for membrane formation, while peptides synthesized in the chloroplast participate in the formation of photosystem I (PSI) and photosystem II (PSII) reaction centers (4). So far, it appears that the development of the photosynthetic membrane depends on the import of translation products from the cytoplasm. The question arises whether nuclear messages are also transported and subsequently translated in the chloroplast. Indirect evidence for such a phenomena was previously suggested (5).
Information relevant to this problem can be obtained by studying conditional mutants affecting thylakoid membranes. The present study is concerned with the repair process of SystemII activity under permissive conditions in the nuclear thermosensitve mutant of Chlamydomonas T4 (6).

MATERIALS AND METHODS

The C. reinhardi T4 mutant was selected on the basis of its high fluorescence when grown at 37°C and was previously described (6). Cells were grown in continuous cultures (7) either at 25°C (permissive temperature)or for 4-5 generations at 37.5°C (restrictive temperature). Photosynthetic activity of PSII was measured using H_2O or 1,5-diphenylcarbazide (DPC) as electron donors and 2,6-dichlorophenolindophenol (DCIP) as an electron acceptor (8). Activity of PSI was measured polarigraphically using ascorbate-DCIP as electron donors and methyl viologen as an electron acceptor (8). For gel electrophoresis, membranes were isolated by centrifugation on sucrose gradients (9), solubilized in 4% SDS, and run on 7.5% or 10% acrylamide gels (10). For repair of PSII activity, cells grown at 37°C were transferred to fresh medium at a final concentration of 10^7 cells/ml (non-dividing conditions) and incubated at 25°C either in the light (13000 erg x cm^{-2} x hr^{-1}) or dark. For radioactive labeling, ^3H-acetate (45 μCi/μmol; 2 mM) was added after 4 hrs of incubation at 25°C and the incubation continued for 6 hrs. Inhibitors were used at the following concentrations: α-amanitin, 50; actinomycin D (AcD), 33; anisomycin, 80; chloramphenicol (CAP), 200; cycloheximide (CHI), 2; rifampicin (Rif), 350. The concentration of 3-(3,4-dichlorophenyl)-1,1-dimethylurea (DCMU) was 10^{-5} M.

RESULTS AND DISCUSSION

The photosynthetic activity and chlorophyll (chl) content of cells grown at permissive or restrictive temperatures are shown in Table I. While the activity of PSI was constant, both water-splitting and photo-oxidation of DPC were lost when the cells divided at 37°C. The decay of both partial reactions of PSII was found to be exponential, indicating that the activity initially present in cells grown at 25°C was diluted during growth at 37°C and not destroyed. This interpretation was further supported by the fact that PSII activity was stable in cells incubated at 37°C in non-dividing conditions. Cells grown at 37°C completely recovered their PSII activity following incubation for 10 hrs in the light at 25°C with no cell division, chlorophyll synthesis, or change in PSI activity. These data suggested that the recovery of PSII activity at 25°C involved repair of pre-existing membranes.

Previously it was shown that specific thylakoid polypeptides required for System II activity where deficient in T4 grown at 35°, as well as in another PS II deficient mutant F 34 (6). However it was not known if the repair of activity in T4 at 25° requires *de novo* protein Syntesis.

Inactive membranes from cells grown at the restrictive temperature lacked 2 peptides (II_2 and II_4) in the region previously identified as being associated with PSII activity (4) (Table II). However, non-dividing cells kept at the restrictive temperature for 48 hrs had active membranes with both II_2 and II_4

TABLE I

PHOTOSYNTHETIC CHARACTERISTICS OF C. REINHARDI (T4) MUTANT

measured parameter \ cell source	25°C Continuous Culture	37°C 5 generations	37°C non-dividing conditions 48 hrs	37°C grown transferred to 10 hr, 25°C
PSII ACTIVITY				
H_2O-DCIP	100	0	100	100
DPC-DCIP	50	0	50	50
PSI ACTIVITY	260	260	260	260
μg chl/10^7 cells	17	22	22	22

Activities are given in μmoles x mg chl^{-1} x hr^{-1} of substrate reduced. All incubations were carried out in the light.

present. Recovery of the partial reactions of PSII was blocked by CAP but not by CHI or anisomycin, indicating that de novo protein synthesis on 70s ribosomes was required. Peptides II_2 and II_4 were absent in cells treated with CAP. Inhibition of transcription by AcD had identical results. However, membranes from cells repaired in the presence of Rif could oxidize DPC (associated with PSII reaction center) and had II_4 present, while water-splitting activity and peptide II_2 were absent. Both II_2 and II_4 and the partial reactions of PSII were absent in membranes from cells incubated with α-amanitin, which inhibits transcription of nuclear DNA. These results (Table II) and the fact that the mutant is nuclear suggested that the presence of II_2 and II_4 in the membrane and the repair of associated activities necessitated chloroplast and nuclear transcription respectively, but only chloroplast translation. A plausible working hypothesis could be that the II_4 peptide was a product of the T4 mutated gene and controls in some way the synthesis or integration into the membrane of the II_2 peptide (chloroplastic transscribed and translated).

When cells were radioactively labeled during repair at 25°C in the presence or absence of Rif and CAP, and the radioactive pattern of membrane polypeptides analyzed (Figure I), it was found that both II_2 and II_4 were labeled in the control cells showing that they were synthesized de novo. Only II_4 was labeled in the presence of Rif (PSII reaction center was active), and no peptides were labeled in the presence of CAP (no repair of activities).

Translation by 70s ribosomes was required throughout the repair process, while the transcription phase which was sensitive to both AcD and Rif was restricted to the initial 4-5 hrs which corresponded to a lag in the repair of

TABLE II

PHOTOSYNTHETIC ACTIVITIES AND ELECTROPHORETIC PATTERN OF MEMBRANES FROM CELLS GROWN AT 37°C AND INCUBATED AT 25°C WITH OR WITHOUT ADDITION OF TRANSCRIPTION AND TRANSLATION INHIBITORS

	Treatment	Peptides in the "II" region	PSII Activity µmoles x mg chl^{-1} x hr^{-1}	
		1 3 5	H_2O-DCIP	DPC-DCIP
37°C incubation	5 generations		0	0
	48 hr non-dividing		100	50
Repair in the presence of inhibitors (25°C)	CHI or anisomycin		70*	35*
	CAP		0	0
	α-amanitin		0	0
	AcD		0	0
	Rif		0	50
	None		100	50

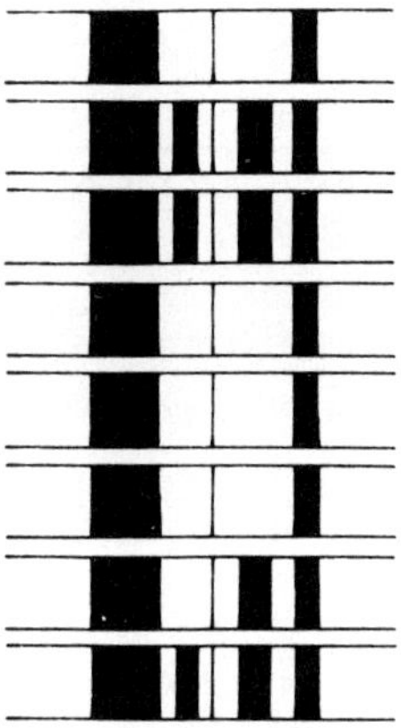

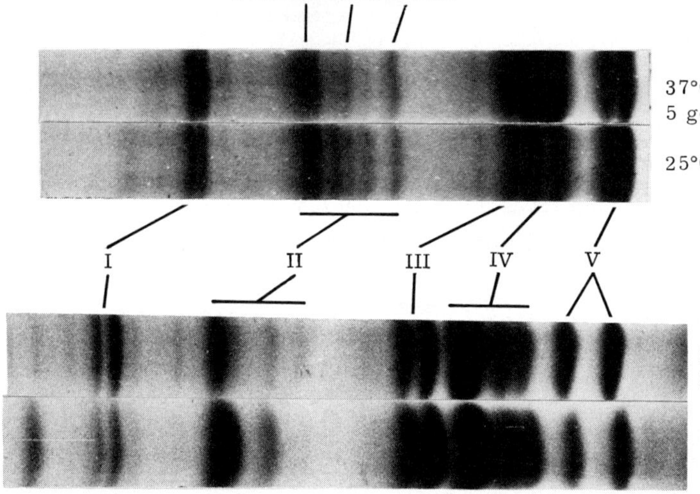

The schematic electrophoretic pattern of the "II" region on 7.5% gels depicted in the table is based on actual electrophoretograms similar to those shown. No detectable changes were observed in the pattern of regions I, III, IV or V (4) when these regions were properly resolved on 10% gels. All incubations were carried out in the light.

* These activities increased to the control level after an additional incubation of 5 hr.

activity (Figure II). Basically, the labeling of II_2 and II_4 followed the same time course as that of the activity repair. These data indicated that formation of membrane bound, stable polyribosomes which synthesize peptides required for the repair process might take place during the lag phase. This was further supported by the fact that cells grown at 37°C displayed extensive stacking of thylakoids which was reduced during growth at 25°C, thus exposing large areas

FIGURE I

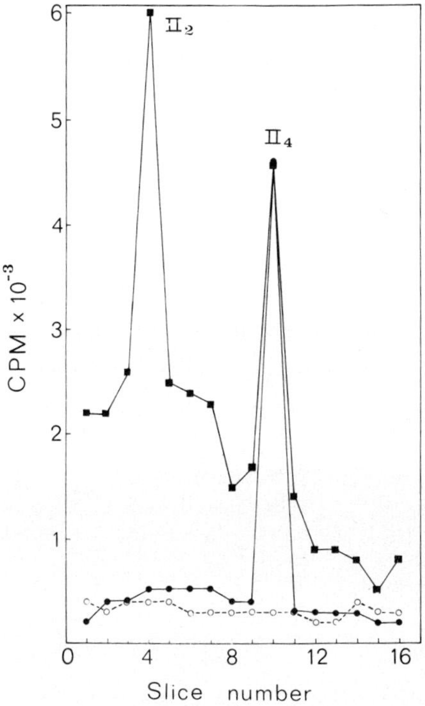

FIGURE II

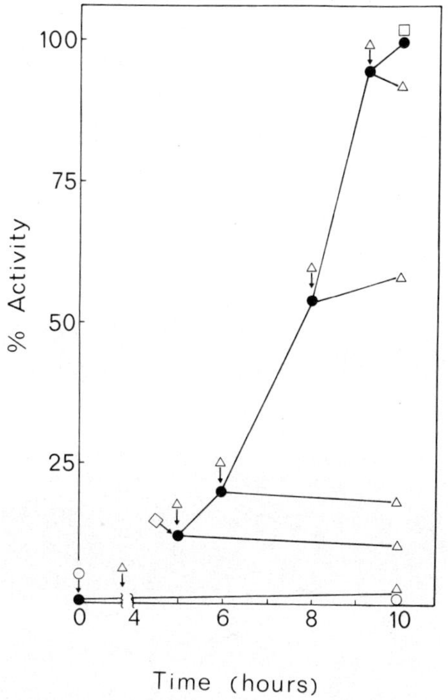

FIGURE I- LABELING OF MEMBRANE PEPTIDES DURING 25°C INCUBATION IN THE PRESENCE OF CAP AND RIF
Cells grown at 37°C were incubated in the light for 10 hrs and labeled with ^3H-acetate. Only the "II" region of the gel was analyzed for radioactivity. The total radioactivity in peaks II_2 and II_4 amounted to more than 50% of the radioactivity loaded on the gel. The activities (μmoles x mg chl^{-1} x hr^{-1}) measured at the end of the incubation were: Control (■), 100; CAP (O), nondetectable; Rif (●), 50 (measured with DPC as electron donor).

FIGURE II - EFFECT OF TRANSCRIPTION AND TRANSLATION INHIBITORS ADDED AT DIFFERENT TIMES OF INCUBATION ON REPAIR OF ACTIVITY
Cells grown at 37°C were incubated at 25°C in the light and the kinetics of PSII repair were followed (●). At the times indicated by the arrows, inhibitors were added. The final activities obtained were shown by the corresponding symbols marked on top of the arrows: (O), AcD or Rif (added at 0 time); (Δ), CAP; (□), AcD, Rif, or CHI (added at 5 hrs).

of thylakoid surfaces to the matrix which contained many ribosomes or polyribosome clusters. The unstacking process was completed during the lag phase and was inhibited by CAP but not by Rif (Figure III).

The synthesis of II_2 and II_4 and their membrane association was not sufficient for the formation of an active, stable PSII complex. When the repair process was carried out at 25°C in the dark, both II_2 and II_4 were synthesized and associated with the membrane but only oxidation of DPC could be detected (Table III). Following a short illumination in absence of additional protein synthesis, water-splitting activity was also re-established. This activation step apparently required a light dependent electron flow through PSII (DCMU inhibited). Based on results of experiments in which membranes which were repaired in the dark were subsequently temperature shocked (38°C) or treated with 0.8 M Tris or low concentrations of deoxycholate (DOC), one could regard the process of insertion and activation of II_2 and II_4 as including the following steps: Peptide II_4 can bind directly to the membrane and establishes electron flow (DPC oxidation) in the absence of II_2. However in this case, the binding is not strong enough to withstand denaturation and dissociation from the membrane by temperature shock, Tris or DOC treatment. The binding of II_2 to the membrane is

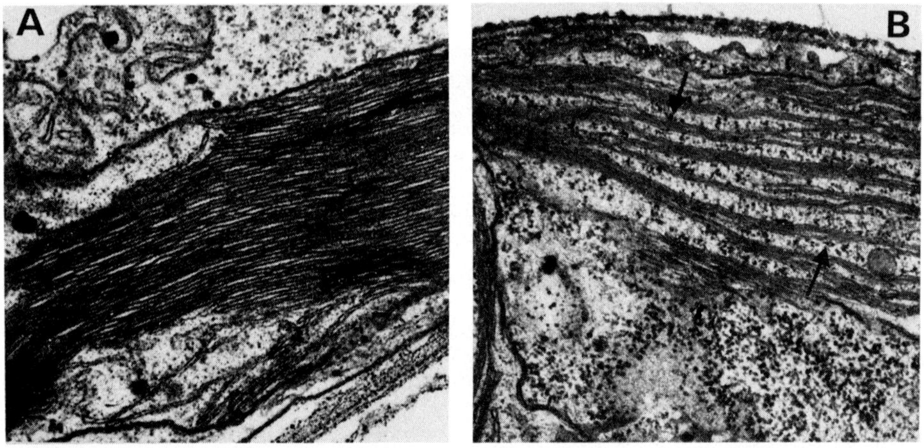

FIGURE III - ELECTRON MICROGRAPHS OF CHLOROPLAST REGIONS OF CELLS GROWN AT 37°C BEFORE (A) AND AFTER REPAIR OF PSII ACTIVITY AT 25°C (B)
The extensive stacking shown in A was not changed when CAP was added at the beginning of incubation. Unstacking in the Rif treated cells was similar to that shown in B. Arrow show ribosome like clusters in close association with thylakoid surfaces.

mediated by the presence of II_4. However, following the light activation step, the proper connections of this peptide to the rest of the electron chain are established (water-splitting activity) and the final organization of the PSII complex obtained. This renders both peptides stable toward temperature shock and treatment with 0.8 M Tris or DOC.

TABLE III - EFFECT OF LIGHT DEPENDENT ELECTRON FLOW ON THE ACTIVATION AND STABILITY OF PEPTIDES II_2 AND II_4 SYNTHESIZED AND BOUND TO THE MEMBRANES DURING REPAIR IN THE DARK

Peptide pattern in the "II" region

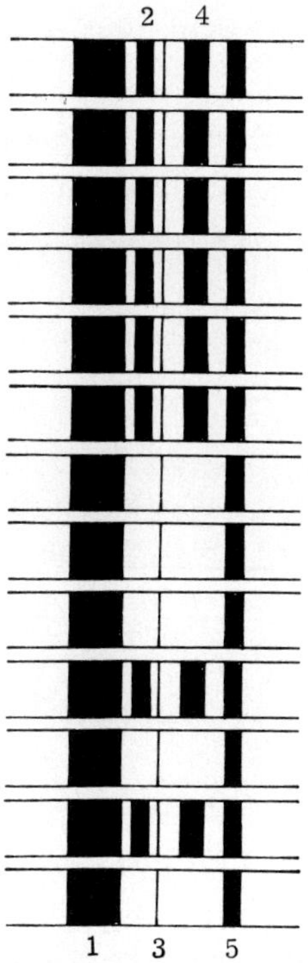

Treatment	PSII Activity μmoles x mg chl^{-1} x hr^{-1} H_2O-DCIP	DPC-DCIP
10 L (10 hours/light)	100	50
10 D (10 hours/dark)	0	50
9D-1L	100	50
9D - 1L (DCMU)	0	50
9D - 1L (CAP + CHI)	92	45
10L - 38°C shock	100	50
10D - 38°C shock	0	0
10L Rif - 38°C shock	0	0
10D Rif - 38°C shock	0	0
10L - 0.8 M Tris	0*	50
10D - 0.8 M Tris	0	0
10L - 125 μg DOC/ml	100	50
10D - 25 μg/DOC/ml	0	0

Cells grown at 37°C were incubated at 25°C either in the light (L) or dark (D). The inhibition of PSII activity by addition of DCMU could be completely reversed by reducing its concentration to less than 10^{-11} M. The temperature shock and Tris or deoxycholate treatment were carried out on isolated membranes (2.5 mg protein/ml). The schematic representation of the electrophoretic pattern is based on 7.5% gels similar to those shown in Table II.

* Tris treatment of active membranes reversibly removes only Mn required for water splitting activity.

The data reported here can tentatively be explained as follows:

Peptide II_4, which is an essential component of the PSII reaction center, is coded by a nuclear gene but translated on chloroplast ribosomes. Its presence regulates the synthesis (chloroplast transcription or translation) or insertion into the membrane of peptide II_2, which is required for the water-splitting activity of the PSII complex. In the T4 mutant, the affected peptide (II_4) assumes an altered configuration at 37°C. This prevents not only its assembly into membrane but also that of peptide II_2, thus resulting in a complete inhibition of active PSII complex formation during growth at 37°C. The synthesis and insertion of the peptides is light independent. However, the activation and stabilization of the PSII complex during the repair of otherwise intact membranes requires light-dependent PSII-mediated electron flow.

The basic biological phenomena exemplified in this work is the high degree of control acquired by the nucleus in Chlamydomonas over the formation of the PSII complex, a chloroplast component common to all eukaryotic photosynthetic organisms. Experiments are in progress in order to check the validity of the scheme proposed above.

This work was supported by grants from the U.S. - Israel Binational Science Foundation (No. 184), and Deutche Forschungsgemeinschaft (No. DR 29/17).

REFERENCES

1. Goodenough, U.W. and Levine, R.P. (1969) Plant Physiol. 44 990.
2. Eytan, G. and Ohad, I. (1970) J. Biol. Chem. 245 4297.
3. Ohad, I. (1975) in Membrane Biogenesis Mitochondria, Chloroplasts, and Bacteria, Tzagoloff (ed.) 279.
4. Bar-Nun, S. and Ohad, I. (1974) in Proc 3rd International Congress on Photosynthesis, Avron (ed.), 1627.
5. Jennings, R.C. and Ohad, I. (1972) Archives Biochem. Biophys. 153(1) 79.
6. Chua, N.H. and Bennoun, P. (1975) Proc. Nat. Acad. Sci. USA 72 2175.
7. Ohad, I., Sikevitz, P. and Palade, G.E. (1967) J. Cell Biol. 35 521.
8. Regitz, G. and Ohad, I. (1976) J. Biol. Chem. 251(1) 247.
9. Eytan, G. and Ohad, I. (1972) J. Biol. Chem. 247 122.
10. Laemmli, U.K. (1970) Nature 227 680.

Genetics and Biogenesis of Chloroplasts and Mitochondria, Th. Bücher et al. eds.
© 1976 Elsevier/North-Holland Biomedical Press, Amsterdam, The Netherlands.

METHODS FOR THE DETECTION AND CHARACTERIZATION OF PHOTOSYNTHETIC MUTANTS IN CHLAMYDOMONAS REINHARDI

Pierre Bennoun and Nam Hai Chua
Institut de Biologie Physico-Chimique
13, rue Pierre et Marie Curie
75005 - Paris - France
The Rockefeller University
New York, N. Y., 10021

INTRODUCTION

The photosynthetic process takes place in thylakoid membranes, a highly integrated system. The final elucidation of the mechanisms involved in this process requires a detailed knowledge of how the structure of the membrane relates to its functions. One approach to solving this problem is through biophysical and biochemical dissection of mutants of *Chlamydomonas reinhardi* which have lesions in the photosynthetic apparatus. Direct correlation of functional and biochemical defects can be obtained by studying acrylamide gel profiles of photosynthetic mutants characterized by fluorescence methods. This paper summarizes data on mutants defective in photosystem I (PSI), photosystem II (PSII), and in a terminal stage of photosynthetic phosphorylation. We describe the methods we developed for screening photosynthetic mutants.

RESULTS

1) *Screening of mutants unable to carry out normal photosynthesis.*

Thylakoid membranes are characterized by their high content of chlorophylls and carotenoids. The absorbance and fluorescence emission characteristics of these pigments are very sensitive to any changes occuring within the membranes. For instance, the fluorescence yield of chlorophyll *in vivo* is related to various parameters including the redox state of several electron carriers (1, 2, 3, 4, 5), the energy state of the membrane (6, 7), the mode of excitation transfer between different species of chlorophylls (8, 9).

Therefore, the measurement of the fluorescence yield permits screening of a wide variety of mutations affecting photosynthesis (10, 11, 12).

Plate 1 shows the fluorescence emitted in the far red part of the spectrum by individual colonies of *C. reinhardi* grown on agar plates and illuminated with blue light. The colonies are the offspring of a cross between the wild type and F14, a Mendelian mutant deficient in PSI activity. Two types of colonies of different

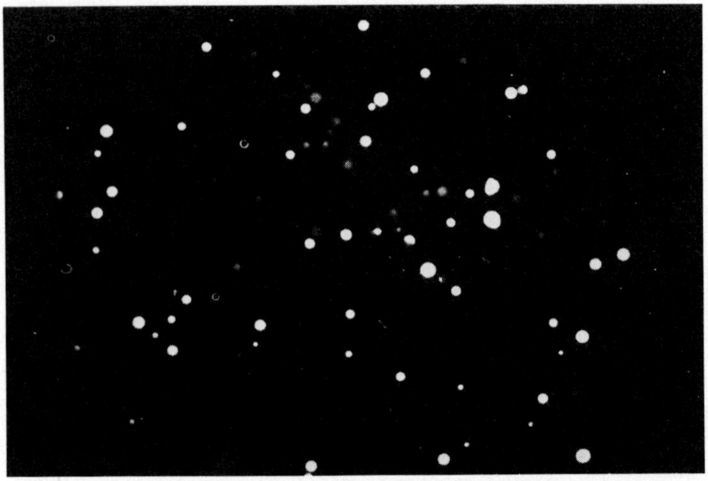

Plate .1. Chlorophyll fluorescence emitted by colonies of _C. reinhardi_ grown on agar plates. The colonies are meiotic progeny from the cross F14 mt+ x mt-. Two types of colonies showing different levels of fluorescence yield are present in equal amounts. The mutant colonies are highly fluorescent since they are blocked in the PSI reaction centers.

fluorescence yields are present in equal amounts : mutant colonies are highly fluorescent compared to those of wild type. With this method it is possible to compare the relative fluorescence quantum yield of the illuminated colonies and to screen rapidly a large number of different colonies for possible defects in photosynthesis.

Using the fluorescence technique, a large number of photosynthetic mutants have been isolated. To characterize more precisely a limited number of mutant colonies an apparatus which records the kinetics of fluorescence emitted by individual colonies was constructed (13). This apparatus permits rapid initial characterization of mutants on the basis of their fluorescence induction kinetics. It is currently used for instance for the tetrad analysis of crosses involving photosynthetic mutants. Fig. 1. shows typical results obtained with wild type, wild type poisoned with dichlorophenyl dimethylurea (DCMU) and mutants blocked in PSI and PSII reaction centers.

In one of the Mendelian mutants F54 which is defective in photophosphorylation, the decay of the transmembrane electric field is slowed (10, 14, 15). This is consistent with the concept that a membrane potential is utilized to drive photophosphorylation as proposed by Mitchell (16). It has been shown, moreover, that delayed light emission is enhanced by a transmembrane electric field (17).

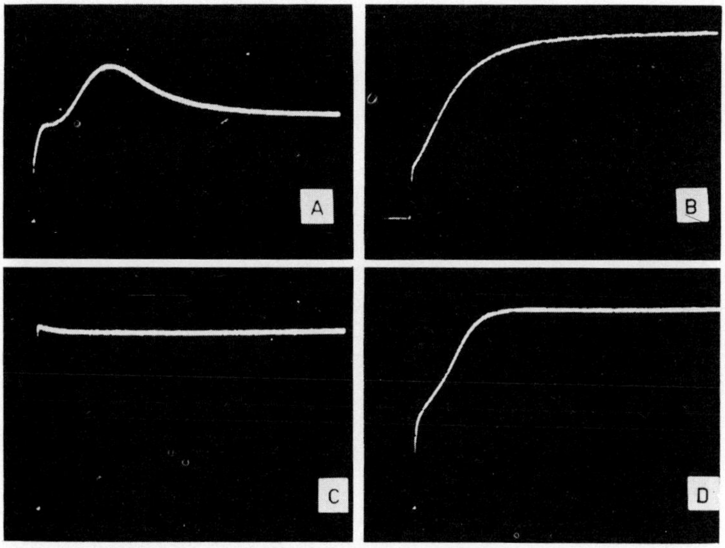

Fig. 1. Fluorescence induction kinetics of single, dark adapted colonies of wild type and mutant strains of *C. reinhardi*. Tracings are taken directly from oscillographic recordings. (a) wild type ; total sweep, 2sec. (b) wild type with a drop of DCMU (1×10^{-5}M). In the presence of DCMU there is a fast increase of fluorescence yield which reflects the photoreduction of the primary electron acceptor of PSII. Total sweep, 100msec. (c) F34. This Mendelian mutant has no PSII activity but normal PSI activity. The absence of active PSII centers leads to the disappearance of variable fluorescence and a high fluorescence yield. Total sweep, 2 sec. (d) F14. This Mendelian mutant is blocked at the level of the PSI reaction center but has normal PSII activity. The fluorescence yield remains at a high level because of the inactivity of PSI. Total sweep, 2 sec.

Therefore, the increase in delayed light emission can be used to screen for mutants which are blocked in photophosphorylation. Results from experiments of this type are shown in Fig. 2. One can expect that another class of mutants defective in photophosphorylation should exist in which the decay of the transmembrane electric field is accelerated.

2) Characteristics of some photosynthetic mutants of *C. reinhardi*.

Thylakoid membranes of C. reinhardi contains at least 33 polypeptides ranging in molecular weight from 70 000 to less than 10 000 (18, 19). In an attempt to correlate these polypeptides with specific electron transport function, experiments were carried out with

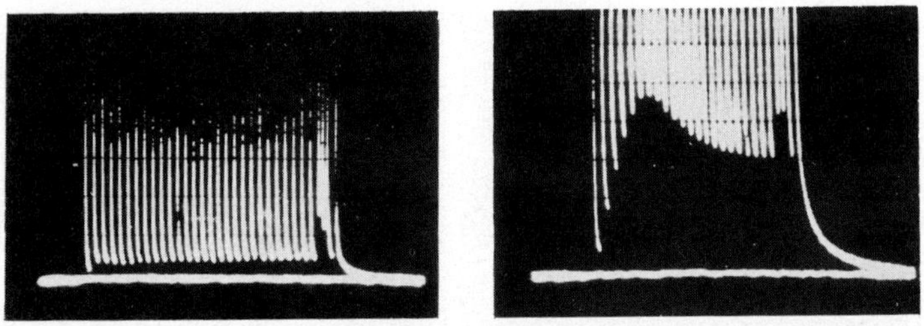

Fig. 2. Delayed luminescence emitted by C. reinhardi wild type (left) and mutant F54 (right) grown on agar plate. The petri dish is illuminated by short (2μsec) saturating flashes fired every 100msec for 20 sec, then kept 10 sec in the dark and illuminated by a new series of flashes 100 msec apart. Luminescence is measured during this second series. The bottom of each vertical trace on the oscilloscope represents a measurement of the luminescence emitted 100 msec after each flash. The comparison of the sequence of luminescence in wild type and F54 shows that for any flash of the sequence, the luminescence yield at 100 msec is increased in the mutant compared to wild type. Furthermore, in the course of the sequence a gradual increase in luminescence yield occurs during the first 6 flashes in F54 whereas the wild type sequence is flat. The increase in luminescence yield occuring at the beginning of the sequence in F54 is explained as follows. The transmembrane electric field generated in the light stimulates the luminescence yield and is normally dissipated between two consecutive flashes to form ATP. Because F54 is blocked at the terminal stage of photosynthetic phosphorylation, the transmembrane electric field cannot be dissipated during this period. Consequently, an additive effect of each flash is observed resulting in the accumulation of an appreciable transmembrane electric field.

some photosynthetic mutants isolated by methods described in the earlier section. Three Mendelian mutants, F14, F34 and F54 (10) and one non Mendelian mutant, C1 (Bennoun and Girard, unpublished) were investigated. F14 and F34 are blocked at the level of PSI and PSII reaction centers, respectively and their fluorescence properties have been described Fig.1. Fluorescence induction kinetics of F54 and C1 are illustrated in Figs. 3 and 4, respectively. C1 is specifically affected at the level of PSI as F14, but it is slightly leaky. Thylakoid membranes were prepared from all these mutants and their polypeptides profiles displayed by SDS gels containing a 7.5-15% acrylamide concentration gradient (Plate 2). The results of parallel biophysical and biochemical studies are summarized in Table I. As has been shown previously (18, 19) polypeptide 2 and po-

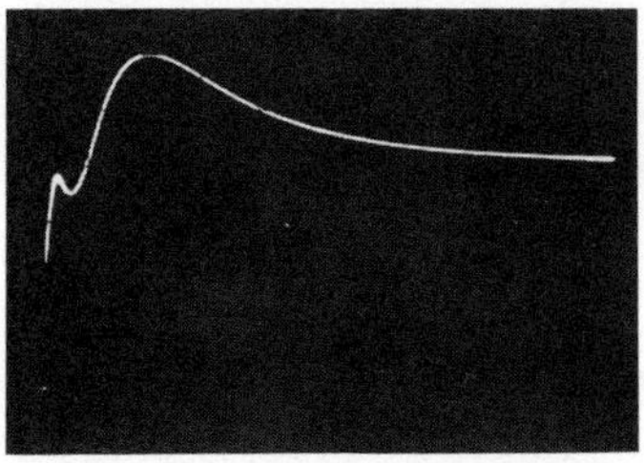

Fig. 3. Fluorescence induction kinetics of a single, dark-adapted colony of F54. The oscillation occuring in the course of the fluorescence rise is typical of this mutant which is defective in photosynthetic phosphorylation. Total sweep, 2 sec. Compare with Fig. 1A.

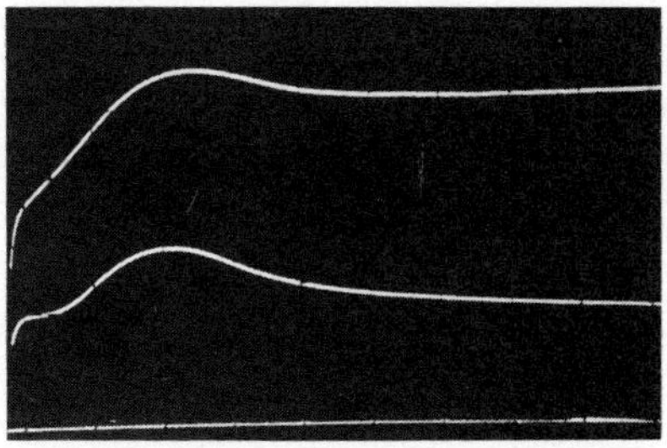

Fig. 4. Fluorescence induction kinetics of single, dark-adapted colonies of wild type (lower) and C1 (upper). The relative amplitude of the decrease phase of fluorescence in C1 is reduced. This is interpreted as a deficiency in PSI activity. Unlike F14 (Fig. 1.) C1 is leaky.

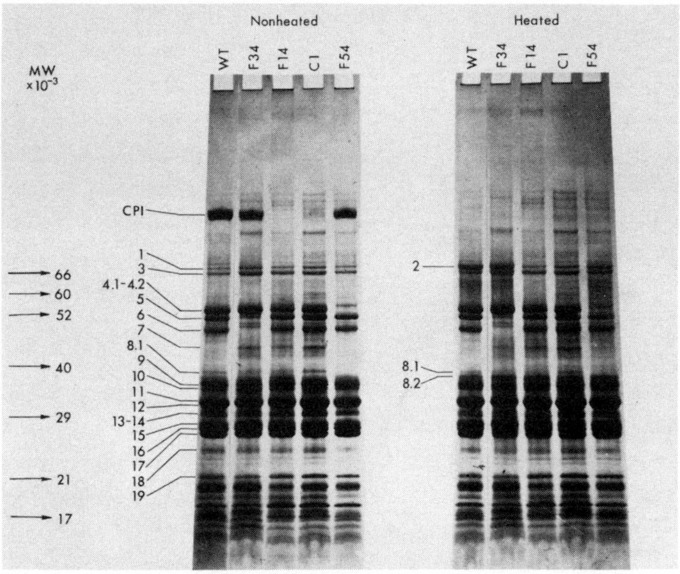

Plate. 2. Thylakoid polypeptides profiles of wild type and mutants of *C. reinhardi*. Conditions of sample preparation have been described previously (18). Each sample was divided into two portions : one portion was incubated at 100° for 1 min. to dissociate chlorophyll protein complexes (heated) whereas the other portion was not treated further (nonheated). The load on each slot is equivalent to 20µg chlorophyll.

TABLE I

Strain	Thylakoid membrane phenotype	Photosynthetic lesion	Heredity
F14	Lacking CP1* and polypeptide 2	No system I activity No detectable P700	Mendelian
C1	Deficient in CP1 and polypeptide 2	Deficient in Photosystem I activity	non Mendelian
F34	Lacking polypeptide 6 and deficient in polypeptides 5 and 12	Absence of active PSII reaction centers	Mendelian
F54	Lacking polypeptides 4.1, 4.2 and 8.1.	Unable to do photophosphorylation Non-latent ATPase	Mendelian

*CP1 = chlorophyll-protein complex 1 (19)

lypeptide 6 are associated with PSI and PSII reaction center activity, respectively. C1, a non-Mendelian mutant, deficient in PSI activity has similar thylakoid membrane phenotype as F14 in that it is also deficient in CP I and its constituent polypeptide (polypeptide 2). Polypeptides 4.1, 4.2 and 8.1 are missing in F54 which is unable to do photosynthetic phosphorylation (15). Since this mutant has an active, but non-latent ATPase the missing polypeptides could be the membrane-sector polypeptides which interact with CF1 (20). Work is in progress to verify this hypothesis.

ACKNOWLEDGEMENTS

This work is supported in part by NIH grant GM 21060 to N.H. Chua and by NATO research grant n° 1135 to the authors and in part by the Centre National de la Recherche Scientifique (Equipe de Recherche N°16).

REFERENCES

1. Duysens, L.N.M. and Sweers, H.E. (1963) Studies on Microalgae and Photosynthetic Bacteria. Japan Soc. of Plant Physiologists University of Tokyo Press, pp. 353-372.
2. Delosme, R. (1967) Biochim. Biophys. Acta. 143, 108-128.

3. Delosme, R. (1972) Proceedings of the IInd International Congress on Photosynthesis Research. Junk Publishers. The Hague. pp. 187-195.
4. Butler, W.L., Visser, J.M.W. and Simons, H.L. (1969) Biochim. Biophys. Acta. 172, 242-251.
5. Joliot, P., Joliot, A., Bouges, B. and Barbieri, G. (1971) Photochem. Photobiol. 14, 287-305.
6. Diner, B. and Joliot, P. (1976) Biochim. Biophys. Acta. 423, 479-498.
7. Papageorgiou, G. (1975) in Bioenergetics of Photosynthesis. Govindjee ed. Academic Press. pp. 345.
8. Murata, N. (1969) Biochim. Biophys. Acta. 172, 242-251.
9. Bonaventura, C. and Myers, J. (1969) Biochim. Biophys. Acta. 189, 366-383.
10. Bennoun, P. and Levine, R.P. (1967) Plant Physiology, 42, 1284-1287.
11. Garnier, J. (1967) C. R. Acad. Sc. Paris, 265, 874-877.
12. Epel, B.L. and Levine, R.P. (1971) Biochim. Biophys. Acta. 226, 154-160.
13. Bennoun, P. (1974) C. R. Acad. Sc. Paris, 279, 1915-1917.
14. Joliot, P. and Delosme, R. (1974) Biochim. Biophys. Acta. 357, 267-284.
15. Sato, V.L., Levine, R.P. and Neumann, J. (1971) Biochim. Biophys. Acta. 253, 437-448.
16. Mitchell, P. (1967) Nature. 144-148.
17. Joliot, P. and Joliot, A. (1974) Proceedings of the IIIrd International Congress on Photosynthesis, Rehovot, Israël. Elsevier Publishing Company, Amsterdam. pp. 25-39.
18. Chua, N.H. and Bennoun, P. (1975) PNAS, 72, 2175-2179.
19. Chua, N.H., Matlin, K. and Bennoun, P. (1975) J. of Cell. Biology. 67, 361-377.
20. Senior, A.G. (1973) Biochim. Biophys. Acta. 301, 249- 277.

Genetics and Biogenesis of Chloroplasts and Mitochondria, Th. Bücher et al. eds.

BIOCHEMICAL STUDIES ON A PLASTID RIBOSOME-DEFICIENT MUTANT OF HORDEUM VULGARE

Thomas Börner, Brigitte Schumann and Rudolf Hagemann
Department of Genetics, Section Biosciences
Martin-Luther-University, Domplatz 1
DDR - 402 H a l l e / S., German Dem. Rep.

SUMMARY

The plastids of light-grown and etiolated white leaves of a barley mutant contain neither RNA and ribosomes nor fraction I protein. Plastid protein synthesis is totally blocked at all stages of plastid development. In spite of this defect numerous plastid membrane proteins are found which apparently are synthesized on cytoplasmic ribosomes. The pattern of electrophoretically separated membrane proteins of mutated plastids is similar to that of etiolated wild-type plastids. The protein bands of pigment-protein complexes I and II and of the two largest subunits of the coupling factor 1 (CF_1) are lacking or strongly reduced.

INTRODUCTION

Chloroplast proteins are synthesized either on cytoplasmic or chloroplast ribosomes. Different methods are used to determine which components of the chloroplasts are of cytoplasmic origin and which components are synthesized inside the organelles. First, but in many cases controversial results were obtained using selective inhibitors of translation. Now it has been clearly demonstrated that isolated chloroplasts are able to incorporate labelled amino acids into discrete polypeptides[1,2,3,4]. An additional approach to this problem is the analysis of mutants with deficient plastid protein synthesis[5,6,7]. The rationale of studies on such mutants is that all plastidal proteins found in the mutant in spite of the deficiency in protein synthesis on 70S ribosomes have to be synthesized outside the plastids, most probably on 80S cytoplasmic ribosomes. So fare there has no been any reason to assume that mitochondrial ribosomes contribute to chloroplast biogenesis as well. On the other hand, plastidal proteins lacking in these mutants may normally be products of the protein synthesis inside the plastids. But it is also possible that their absence is a secondary effect of the deficient protein-synthesizing apparatus, e.g. caused by a regulatory mechanism.

MATERIAL AND METHODS

The mutant line "albostrians" (= Mut. 4205) of Hordeum vulgare L. is described by HAGEMANN and SCHOLZ[8]. White seedlings and plastids of this line represent a gene-induced plastom mutant.

Nucleic acids were isolated from 10 days old barley seedlings as previously described[5] and separated by polyacrylamide gel electrophoresis in the semi-micro scale[7]. 1mg/ml bentonite was added to the isolation buffer.

Soluble proteins were extracted from leaves by homogenization in 0.05M tris-glycine buffer[9]. The protein solutions were concentrated twofold by ultrafiltration and subjected to immunoelectrophoresis. Two-dimensional immunoelectrophoresis was performed in 1.5% agarose gels containing in the second dimension 12% antiserum to fraction I protein (the antiserum was kindly given by Dr. D. Reichenbächer).

Preparation and electrophoresis of plastid membrane proteins were carried out essentially according to HERRMANN and MEISTER[10].

For the isolation of the coupling factor 1 (CF_1) plastid membranes isolated from 5g leaf material were stirred 20min with 10ml 1mM EDTA, pH 8.0, at room temperature[11]. After centrifugation (1hr, 20,000 x g) the supernatants were concentrated tenfold by ultrafiltration. Two-dimensional immunoelectrophoresis of EDTA-extracts was performed in 1.5% agarose gels. The agarose contained 2.5% antiserum to CF_1 of Vicia faba in the second dimension (the antiserum was kindly provided by K.-H. SÜSS).

RESULTS

Ribosomal RNA

Nucleic acids were extracted from green leaves and etiolated wild-type leaves as controls and from white light-grown leaves and etiolated white leaves of the mutant line "albostrians" of Hordeum vulgare. After electrophoretic separation of the nucleic acids of green leaves 4 bands of high molecular-weight rRNA are detectable: the 25S and 18S RNA of cytoplasmic ribosomes and the 23S and 16S RNA of plastid ribosomes (fig.1A). A similar pattern is obtained with nucleic acids of etiolated wild-type leaves. In contrast, plastid rRNA could be isolated neither from white light-grown leaves nor from etiolated white leaves, whereas the 25S and 18S RNA of cytoplasmic ribosomes are found in normal amounts (fig.1B).

These results are in agreement with the lack of plastid ribosomes in the mutant leaves as shown by electron microscopy[12,13].

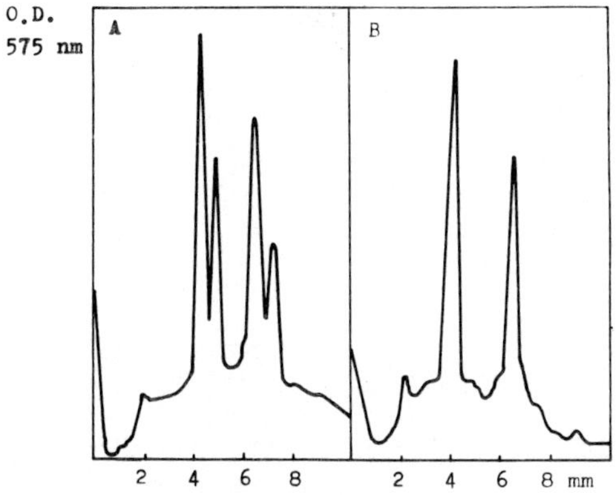

Fig. 1. Separation of high molecular-weight rRNA in 3% polyacrylamide gels at 0.5mA/gel for 1hr. A: RNA of green leaves, B: RNA of white light-grown leaves.

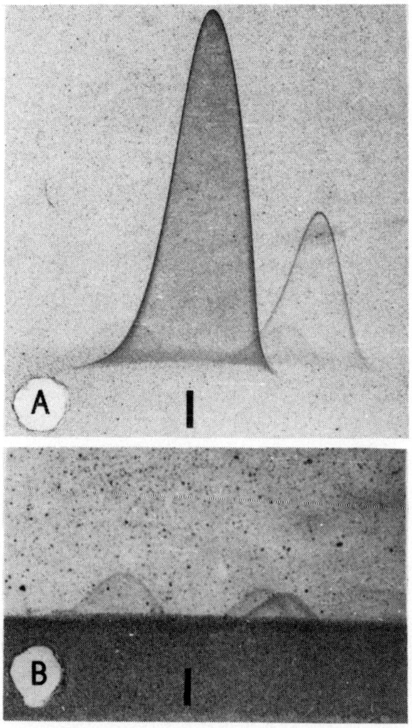

Fig. 2. Two-dimensional immunoelectrophoresis of soluble proteins of etiolated wild-type (A) and etiolated mutant leaves (B). Conditions: 4.0 V/cm, 4°C, 1st dimension 3hr, 2nd dimension 16hr.

The absence of plastid rRNA and plastid ribosomes, respectively, should cause a total deficiency in plastid protein synthesis. If this conclusion is true fraction I protein should be lacking in these mutants as well, since the large subunits of this protein are normally synthesized on plastid ribosomes (cf. 14).

Fraction I protein

We demonstrated by polyacrylamide gel electrophoresis of soluble proteins that fraction I protein is not detectable in a plastid ribosome-deficient Pelargonium mutant ("Mrs.Parker")[9].

In order to detect even smallest quantities of fraction I protein which would indicate a still existing low rate of protein synthesis in mutant plastids we tried to find fraction I protein in wild-type and mutant leaves by immunological tests. The antiserum to fraction

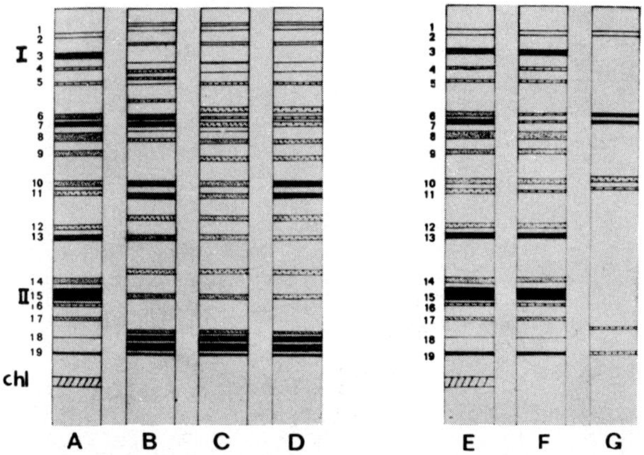

Fig. 3. Separation of plastid membrane proteins in 9% polyacrylamide gels containing 0.1% SDS at 1mA/gel for 30min and at 3mA/gel for 4.5hr. A: green wild-type, B: etiolated wild-type, C: light-grown mutant, D: etiolated mutant, E: green wild-type, F: EDTA-treated membrane proteins of green wild-type, G: coupling factor 1 (CF_1) (I = PPC I, II = PPC II, chl = free chlorophyll)

I protein was produced by immunization of rabbits with roughly purified fraction I protein (cf. 15). This antiserum reacts in two-dimensional immunoelectrophoresis with fraction I protein and three other unknown soluble proteins of green and etiolated wild-type leaves (fig.2A). No fraction I protein is detectable in protein solutions extracted from white light-grown and white etiolated leaves of the Hordeum mutant which indicates that actually no protein synthesis is carried out in the mutated plastids. The three other proteins reacting with the antiserum are present, but one of them strongly reduced in amount (fig.2B).

Membrane proteins and coupling factor 1 (CF_1)

We studied the pattern of SDS-solubilized plastid membrane proteins by electrophoresis in SDS-containing polyacrylamide gels (fig.3). The membrane proteins of normal chloroplasts can be separated into 19 different moving polypeptides: two pigment-protein complexes (PPC I and PPC II) corresponding to photosystem I and as "light harvesting protein" to photosystem II, respectively[16], and a number of pigment-free protein bands of unknown function. In addition, a fast moving band of free chlorophyll can be observed (fig.3A). Membrane proteins of etiolated wild-type plastids produce quite a different pattern (fig.3B). The protein bands of the pigment-protein complexes dominating in the pattern of membrane proteins of green chloroplasts are either lacking (PPC I) or at least strongly reduced (PPC II). Other proteins are reduced (e.g. band 8) or increased (e.g. bands 10, 11) as compared to the control and several new proteins are seen. The membrane proteins of light-grown and etiolated plastids of the Hordeum mutant show a surprisingly similar pattern (fig.3C,D). No remarkable differences are found between the membrane proteins of light-grown und dark-grown mutant plastids, and there is only one striking difference between etiolated wild-type and mutant plastids: Two proteins (bands 6 and 7) of etiolated wild-type plastids appear strongly reduced and are even sometimes lacking in the pattern of the two types of mutant plastids. We found that these two proteins correspond to the two largest subunits of the coupling factor of photophosphorylation (cf.17).

Treatment of the membranes with 1mM EDTA which removes the CF_1[11] results in a reduced amount of protein in bands 6 and 7 (fig.3F). The extracted protein can be precipitated by antiserum to CF_1 (s. below) and shows ATPase activity (BÖRNER unpubl.). The extracted CF_1 was further purified by preparative electrophoresis and after incubation with 0.1% SDS and 1% mercaptoethanol co-electrophoresed with chloroplast membrane proteins (fig.3G). The two largest of the five subunits of the CF_1 migrate in the gels exactly to the position of the two membrane proteins which are reduced by EDTA-treatment. We may conclude from these findings, that the two membrane proteins (6 and 7 in fig.3A and E) are identical with the two largest subunits of the CF_1. We cannot rule out that bands 6 and 7 contain additional polypeptides different from the subunits of the coupling factor.

It was not possible by this method to show unequivocally a similar correlation between the other three subunits and certain membrane proteins. This can be done by separating the membrane proteins in more protein bands by means of two-dimensional electrophoresis (K.-H. SÜSS, pers. commun.).

The absence or quantitative reduction of the two largest subunits of CF_1 made us test EDTA-extracts from membranes of mutant plastids for reaction with antiserum to CF_1. EDTA-extracts of normal green membranes exhibit good immunological reaction with antiserum to the CF_1 of Vicia faba. But no agglutination was obtained between the antiserum and EDTA-extracts of plastid membranes from light-grown and etiolated mutant leaves even if a fivefold amount (related to the extract of green membranes) of the extracts was applied to the agarose gels (fig.4).

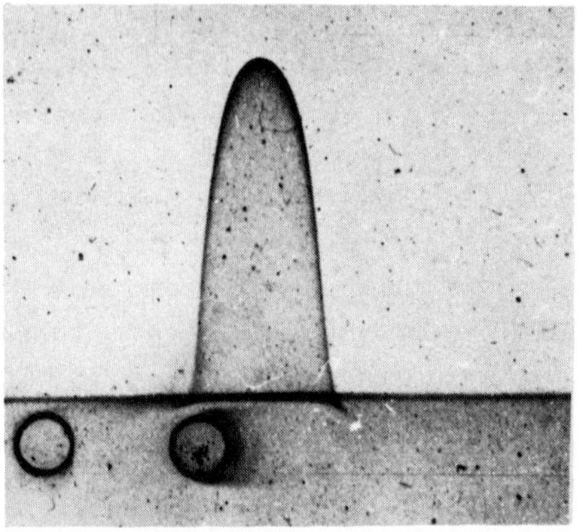

Fig. 4. Two-dimensional immunoelectrophoresis of EDTA-extracts from from green wild-type membranes (left well, 5ul) and mutant membranes (right well, 50ul).
Conditions: 20° C; 1st dimension: 200V, 1.5hr; 2nd dimension: 50V, 16hr. (Two peaks would be expected if also mutant extract contained CF_1).

DISCUSSION

There are three lines of evidence for a total deficiency in protein synthesis in the white plastids of the gene-induced plastom mutant "albostrians" of Hordeum vulgare:

(1) In the white leaves no 16S and 23S RNA of plastid ribosomes are found.

(2) No ribosomes are abserved in the white plastids by electron microscopy whereas in the cytoplasm numerous ribosomes are seen (cf.13).

(3) No fraction I protein is detectable by means of two-dimensional immunoelectrophoresis. In agreement with this, no activity of ribulose-1,5-diphosphate carboxylase was found in extracts of soluble proteins from white leaves (BÖRNER and LIPPMANN unpubl.).

Ribosome deficiency concerns all developmental stages of the plastids beginning with proplastids[13]. The lack of ribosomes in our mutant is not due to photodestruction as assumed in the case of carotene-deficient Helianthus mutants[18] or of herbicide-treated wheat[19,20] since also mutated etioplasts do neither contain rRNA nor fraction I protein.

These ribosome-deficient mutants are useful objects in the study of the function of cytoplasmic protein synthesis in the biogenesis of chloroplasts. It is an advantage of our Hordeum mutant that plastid protein synthesis is not only partially inhibited and that it is blocked at all stages of plastid development. Thus, we can assume that all plastid proteins still found in the mutant must have their origin on cytoplasmic ribosomes, and all structures and functions of the plastids still observed must be independent of the existence of a functioning protein-synthesizing apparatus in the plastids.

Since the cells of white mutant leaves contain numerous plastids we must conclude that the existence and multiplication of plastids do not need proteins synthesized on plastid ribosomes (cf.5). This is also true for the replication of plastid DNA[21]. Like normal chloroplasts ribosome-deficient plastids are also surrounded by a double-layered envelope[5,22,23,24]. The membranes of the envelope seem to be sufficiantly intact as tested by their capacity to retain certain enzymes inside the organelle[24]. Therefore, at least the major part of the envelope proteins must be synthesized in the cytoplasm.

Also the proteins of internal lamellar structures seen in the electron-microscope as stroma lamellae or as prolamellar bodies in etiolated mutant plastids must be of cytoplasmic origin[12,13]. We could isolate a remarkable number of membrane proteins from mutant plastids. Most probably the patterns of membrane proteins of mutant plastids and of etioplasts include also envelope proteins which are not visible in the case of normal chloroplasts because of an excess of internal lamellar structures. The similarities in the membrane protein patterns of etioplasts and of mutant plastids (light- or dark-grown) suggest that cytoplasmic protein synthesis alone allows the plastids to develope to the stage of etioplasts. Indeed, in etioplasts of the Hordeum mutant numerous stroma lamellae are visible, and even highly-organized prolamellar body-like structures can be formed by ribosome-deficient mutant plastids of the variety "Mrs.Parker" of Pelargonium[12]. The further light-induced development apparently needs the participation of plastid protein synthesis

Only a few membrane proteins are found among the products of protein synthesis performed by isolated chloroplasts[2,3,4]. MENDIOLA-MORGENTHALER et al. demonstrated that isolated chloroplasts incorporate amino acids into the two largest subunits and possibly also into the smallest subunit of CF_1[25]. This is in agreement with our findings that in ribosome-deficient plastids no coupling factor is detectable and its two largest subunits are reduced or lacking in the pattern of membrane proteins. Nevertheless, it should be emphasized that the absence of CF_1 and of the pigment-protein complexes or of any other plastidal component in this mutant is not necessarily an immediate result of the deficient protein synthesis, it may as well be a regulatory effect.

ACKNOWLEDGEMENTS

We thank Dr. Manteuffel and Dipl.-Biol. K.-H. SÜSS (Gatersleben) and Dr. Reichenbächer (Aschersleben) for helping us in the immunological studies and for generous gifts of antiserum.

REFERENCES

1. Blair,G.E. and Ellis,R.J. (1973) Biochim.Biophys.Acta 319, 223-234.
2. Bottomley,W., Spencer,D. and Whitfeld,P.R. (1974) Arch.Biochem. Biophys. 164, 106-117.
3. Eaglesham,A.R.J. and Ellis,R.J. (1974) Biochim.Biophys.Acta 335 396-407.
4. Morgenthaler,J.-J. and Mendiola-Morgenthaler,L. (1976) Arch. Biochem.Biophys. 172, 51-58.
5. Börner,Th., Knoth,R., Herrmann,F. and Hagemann,R. (1972) Theor.Appl.Genetics 42, 3-11.
6. Boynton,J.E., Gillham,N.W. and Chabot,J.F. (1972) J.Cell Sci. 10, 267-305.
7. Börner,Th., Herrmann,F. and Hagemann,R. (1973) FEBS Lett. 37, 117-119.
8. Hagemann,R. and Scholz,F. (1962) Züchter 32, 50-59.
9. Börner,Th., Jahn,G. and Hagemann,R. (1976) Biochem.Physiol. Pflanzen 169, 179-181.
10. Herrmann,F. and Meister,A. (1972) Photosynthetica 6, 177-182.
11. Horak,A. and Hill,R.D. (1971) Can.J.Biochem. 49, 207-209.
12. Knoth,R. Dr.rer.nat. thesis (1974) University of Halle/S.
13. Knoth,R. and Hagemann,R. (1976) Biol. Zbl. in press.
14. Ellis,R.J. (1973) Comment.Plant.Sci. 4, 28-38.
15. Richter,J., Reichenbächer, D. and Spohn,H.-J. (1975) Biochem. Physiol.Pflanzen 167, 285-290.
16. Thornber,J.P. and Highkin,H.R. (1974) Eur.J.Biochem. 41, 109-116.
17. Racker,E., Hauska,G.A., Lien,S., Berzborn,R.J. and Nelson,N. (1972) Proc. IInd Internat.Congr.Photosynth.Res. (Forti,G., Avron,M. and Melandri,A. eds.) pp. 1097-1113, Junk Publ., The Hague.
18. Walles,B. (1972) Protoplasma (Wien) 75, 215-227.
19. Bartels,P.G., Matsuda,K., Siegel,A. and Weier,T.E. (1967) Plant Physiol. 42, 736-741.
20. Bartels,P.G. and Hyde,A. (1970) Plant.Physiol. 46, 825-830.
21. Knoth,R., Herrmann,F.H., Böttger,M. and Börner,Th. (1974) Biochem.Physiol.Pflanzen 166, 129-148.
22. Dyer,T.A., Miller,R.H. and Greenwood,A.D. (1972) J.Exp.Bot. 22, 125-136.
23. Sprey,B. (1972) Z.Pflanzenphysiol. 67, 223-243.
24. Feierabend,J. and Schrader-Reichardt,U. (1976) Planta (Berl.) 129, 133-145.
25. Mendiola-Morgenthaler,L.R., Morgenthaler,J.-J. and Price,C.A. (1976) FEBS Lett. 62, 96-100.

Genetics and Biogenesis of Chloroplasts and Mitochondria, Th. Bücher et al. eds.

SITES OF SYNTHESIS OF CHLOROPLAST MEMBRANE PROTEINS IN VICIA FABA

Wolfgang Hachtel
Botanisches Institut der Universität,
5300 Bonn, Kirschallee 1 (FRG)

SUMMARY

Isolated chloroplasts of Vicia faba incorporate ^{14}C-amino acids into 5 chloroplast membrane polypeptides as revealed by sodium dodecylsulfate polyacrylamide gel electrophoresis. Labelling of four of these polypeptides including a photosystem I protein is inhibited by chloramphenicol. Presumably they are synthesized on the 70 S ribosomes of the chloroplasts. The incorporation into a chlorophyll-protein complex derived from photosystem II is blocked only by cycloheximide. It is concluded that this protein is synthesized either on 80 S cytoplasmic or 80 S plastid ribosomes.

INTRODUCTION

From in vivo inhibitor experiments and from the analysis of mutants with deffective plastid protein synthesis it was suggested that some of the insoluble polypeptides bound to chloroplast membranes are synthesized in the chloroplast itself, whereas some others are made in the cytoplasm (for a review see 1). Direct evidence that some membrane polypeptides are synthesized in the chloroplasts was provided by investigations concerning protein synthesis in isolated chloroplasts (2-4). In the present study the influence of specific inhibitors of protein synthesis on 80 S and 70 S ribosomes, respectively, cycloheximide and chloramphenicol, upon the in vitro incorporation of ^{14}C-amino acids into membrane polypeptides by isolated chloroplasts of Vicia faba is investigated.

MATERIALS AND METHODS

Isolation of chloroplasts: Leaves (20 g) from two week old plants of Vicia faba were blended in 120 ml isolation buffer (0,35 M sucrose, 20 mM HEPES-NaOH pH 7,3, 2 mM sodium isoascorbate, 1 mM dithiothreitol, 10 mM $MgSO_4 \times 7\ H_2O$, 100 mM KCl, 0,2 % bovine serum albumin) in a Sorvall Omnimixer for 2 x 5 sec at full voltage. The homogenate was filtered, the debris were removed by centrifugation at 250 x g for 2 min and the chloroplasts sedimented by centrifugation at 1 500 x g for 10 min, resuspended in isolation buffer, again centrifuged at 1 500 x g for 10 min and again resuspended in isolation buffer.

Incubation of isolated chloroplasts: Chloroplasts were incubated in a final volume of 3 ml (equal to 150-400 μg chlorophyll) with 6 μCi of a ^{14}C-amino acid mixture. The medium was supplemented with ATP (0,8 mM), P-enolpyruvate (2 mM), pyruvate kinase (40 μg/ml) and GTP (0,04 mM). The suspensions were illuminated with white light at 3×10^5 erg x cm^{-2} x sec^{-1}. After 45 min the suspension was centrifuged at 2 500 x g for 1 min. The supernatant was discarded and the chloroplast sediment resuspended in incubation medium containing in addition 2 % unlabelled casein hydrolysate but no ^{14}C-amino acids and reincubated under the same conditions as before ("chase"). In experiments with inhibitors the medium was supplemented with either cycloheximide (100 μg/ml) or chloramphenicol (500 μg/ml).

Separation of membrane polypeptides: After 2 h of incubation the suspensions were centrifuged and the chloroplast pellets homogenized in destilled water using a Potter-Elvehjem homogenizer. The chloroplast fragments were made free of soluble proteins and solubilized with 2 % sodium dodecylsulfate (SDS); the membrane polypeptides were separated by SDS-polyacrylamide gel electrophoresis, the gels stained with Coomassie blue and scanned at 600 nm with a Gilford gel scanner (5). For the determination of radioactivity, 1 mm-gel slices were digested with 30 % hydrogen peroxide at 50 ^{o}C for 3 h and counted in a Packard Tri-carb liquid scintillation spectrometer.

RESULTS

Before staining, the gels show three green bands. After staining 11 polypeptide bands (A-L) of very different intensities can be resolved (Fig. 1). The green bands correspond to the protein bands A and I and to the unstained peak M. The protein-chlorophyll complexes A and I are derived from photosystem I and photosystem II, respectively, and the peak M consists of free chlorophyll (5).

After the incubation of isolated chloroplasts with ^{14}C-amino acids the gels reveal 5 radioactive peaks (Fig. 1). In chase experiments the labelled acids are replaced by an excess of unlabelled amino acids after 45 min of incubation. Inspite of the chase the amount of radioactivity incorporated after 2 h into the 5 peaks is more than doubled as compared with experiments without chase. During the chase

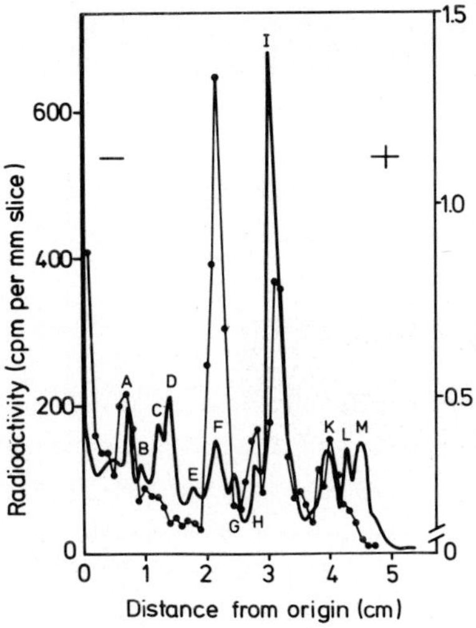

Fig. 1. Electrophoretic analysis of the chloroplast membrane polypeptides of Vicia faba after incubation of isolated chloroplasts with a mixture of ^{14}C-amino acids. Absorbance at 600 nm ——— , ^{14}C-radioactivity ●——● .

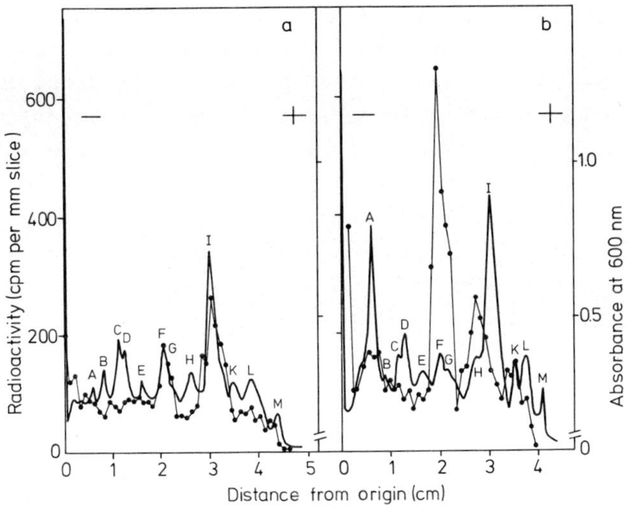

Fig. 2. Electrophoretic analysis of the chloroplast membrane polypeptides of Vicia faba after incubation of isolated chloroplasts with ^{14}C-amino acids and with 500 µg/ml chloramphenicol (a) and 100 µg/ml cycloheximide (b). Absorbance ——, radioactivity ●——●.

time no further incorporation into the total protein occurs. The radioactive peaks coincide with the protein peaks A, F, H, I and K.

In the presence of chloramphenicol the ^{14}C- incorporation into the peaks A (photosystem I protein), F, H and K is drastically reduced, whereas the synthesis of photosystem II protein (peak I) is not inhibited (Fig. 2a). Cycloheximide does not reduce labelling of A, F, H and K but blocks synthesis of photosystem II protein (Fig 2b).

DISCUSSION

From in vivo inhibitor experiments with excised shoots of Vicia faba (5) one can suggest that the chlorophyll associated photosystem I protein but not the photosystem II protein is made on 70 S plastid ribosomes. On the other hand neither of these proteins was labelled in isolated chloroplasts of Pisum sativum (3). The fact that isolated chloroplasts of Vicia faba incorporate radioactive amino acids into photosystem I protein now provides direct support for the inference from the in vivo inhibitor studies. But in addition photosystem II protein is labelled in preparations of isolated Vicia faba chloroplasts. Inhibition of this labelling by cycloheximide but not by chloramphenicol indicates that this protein is synthesized on 80S ribosomes. At the moment I cannot exclude that the chloroplast suspensions are contaminated by 80 S ribosomes of the cytosol due to a preparation artefact. But I wonder why the other proteins which are synthesized on 80 S ribosomes as infered from the in vivo inhibitor experiments are not labelled in isolated chloroplasts. Therefore it is more likely that specific 80 S ribosomes engaged in the synthesis of the photosystem II protein are associated with the chloroplasts in situ.

REFERENCES

1. BÖRNER, T. (1973) Biol. Zbl. 92, 545-561
2. HOOBER, J. K. (1970) J. Biol.Chem. 245, 4327-4334
3. EAGLESHAM, A. R. J. and ELLIS, R. J. (1974) Biochim. Biophys. Acta 335, 396-407
4. BOTTOMLEY, W., SPENCER, D. and WHITFELD, P. R. (1974) Arch. Biochem. Biophys. 164, 106-117
5. MACHOLD, O. and AURICH, O. (1972) Biochim. Biophys. Acta 281, 103-112.

Genetics and Biogenesis of Chloroplasts and Mitochondria, Th. Bücher et al. eds.

Synthesis of chlorophyll-free thylakoids in Chlorella after clindamycin-treatment and in a temperature sensitive mutant of Chlorella.

Gottfried Galling
Institut für Allgemeine Botanik,
Arbeitsbereich Genetik
Hamburg, Germany

SUMMARY.

In Chlorella pyrenoidosa chlorophyll synthesis is prevented by very low concentrations of clindamycin, a lincomycin derivative. Cell growth is not markedly influenced by the antibiotic, and also thylakoid formation is almost normal as shown in the electron microscope. 70 S ribosomes cannot be observed in clindamycin-treated cells. Analysis of membrane proteins using electrophoresis on gels shows that at least 4 bands are missing and 2 new bands appear in the clindamycin-treated Chlorella cells.
From the same algal strain, a temperature-sensitive UV-mutant has been isolated which, at 30°C, is not able to synthesize chlorophyll. At 22°C, chlorophyll is synthesized in the light but not in the dark. The level of 70 S ribosomes of the mutant grown at 30°C is identical to that of the wild type. In chlorophyll-free mutant cells, large thylakoids are found. Also from these membranes, differences in the protein patterns are described.

INTRODUCTION

Many attempts have been undertaken to study the chloroplast development independently of its photosynthetic activity. In genetic experiments, parts of higher plants without functional chloroplasts have been analyzed, and the complete loss of 70 S ribosomes has been observed (see BÖRNER et al, 1973). Also with physiological methods, ribosome-free chloroplasts can be generatedin rye seedlings, which are called "white chloroplasts" (FEIERABEND and SCHRADER-REICHARDT, 1976). In both cases, the inner structure of the plastids is degenerated. No decision is possible whether the loss of thylakoids is caused by degradation or by non-formation of the chloroplasts membranes since light may cause a photodestruction of chlorophyll, or darkness may prevent the synthesis of this pigment.

Many algal strains are able to synthesize chlorophyll in the dark. In our laboratory, the strain 211/8b of Chlorella pyrenoidosa has been extensively studied to show the site of synthesis of chlorophyll and thylakoid membranes. The results obtained inhibiting transcription with rifampicin and translation with spectinomycin or lincomycin indicate that synthesis of chlorophyll and of chloroplast membranes is not affected by these antibiotics (GALLING, et al, 1973). Even, an enlargement of the thylakoid stacks can be observed after application of the drugs. These results are in good agreement with data obtained on Chlamydomonas reinhardi by many other groups (see e.g. HOOBER et al, 1969, OHAD et al, 1971, SURZYCKI et al, 1970).
Here, we describe two ways to obtain Chlorella cells, obviously free of chlorophyll, but with large stacks of thylakoids inside the plastid. The bleaching is observed either after treatment of wild-type Chlorella cells with clindamycin or after mutagenesis with UV-light.

MATERIALS AND METHODS

Experiments have been carried out using the strain 211/8b of Chlorella pyrenoidosa Chick (Göttingen algal collection) and an UV-irradiated mutant.
Liquid culture was performed in mineral medium + 1% of glucose in light thermostates at 22°C or 30°C in constant light or constant darkness, according to LORENZEN and KUHL (1964). Clindamycin, the 7-Cl derivative of lincomycin was used at concentrations of 1 - 5 µg/ml. The drug was a generous gift of The Upjohn Cie, Kalamazoo, Mich., U.S.A.
Cell growth was determined after OD-measurement at 600 nm. Chlorophyll was extracted with methanol according to McKINNEY (1941), and spectra were recorded with a GILFORD double beam photometer 243. Cell homogenization and ribosome isolation were performed as previously described (GALLING et al, 1973). Ribosomes were separated either in sucrose gradients or on agarose gels.
Electron microscopy was performed after fixation of the cells with glutaraldehyde, embedding them with EPON resin and doublestaining with OsO_4 and lead citrate according to PICKETT-HEAPS (1972). Cuttings were made with a LKB ultratome, and microscopy was performed on an ELMISKOP I (SIEMENS) or, in some cases, on a ZEISS EM 10.
For separation of membrane proteins, cells were homogenized with glass beads, and thylakoid membranes were purified in Ficoll-gradients. Freeze-dried powder of membrane proteins were separated on acrylamide gels in SDS-buffer-solutions. Gels were stained with amido black and destained in acetic-acid-methanol. Scanning was performed in a JOYCE-LOEBL Scan 400.

RESULTS AND DISCUSSION

1. Clindamycin-treatment of wilde-type Chlorella cells.

The heterotrophic growth of Chlorella pyrenoidosa is not markedly reduced by the antibacterial antibiotic clindamycin in concentrations between 1 and 5 μg/ml. As shown in fig. 1 a, the shape of growth curves in the presence or absence of 2 μg/ml is rather similar. Clindamycin-treated cultures, however, are free of chlorophyll, as shown by methanol extraction of the cells (fig. 1, b). Ribosomes were obtained from cells harvested at the end of the log. phase. Using sucrose gradients or agarose gel techniques, no 70 S ribosomes were found in clindamycin-treated cells.

To obtain better electron micrographs, starch content of the cells was reduced transferring the cells for 24 hrs into mineral medium prior to fixation with glutaraldehyde. In plate I, a comparison can be made between normal and clindamycin-treated Chlorella cells. It is obvious that both types of cells contain stacks of thylakoids. However, no plastidal ribosomes can be seen inside the clindamycin-treated cells. Clindamycin induces the formation of "pseudograna" whereas in most of the control cells thylakoids are found only in duplicate or triplicate strands.

In plate I, also the protein pattern of purified thylakoid particles is shown after separation on acrylamide gels. For comparison, also membranes of spectinomycin- and lincomycin-treated cells are shown. The patterns show clearly that treatment of Chlorella with antibiotics causes a significant alteration of the protein composition of thylakoids. Concerning the main protein bands, similar changes are observed after treatment of the cells with spectinomycin, lincomycin or clindamycin. The difference between clindamycin and the other drugs was found in the region of smaller protein bands of high molecular weight where at least one more band is missing, and in the region of proteins with low molecular weight where two new protein bands appear.

From our results with rifampicin, it is obvious that the de novo ribosome synthesis is suppressed totally but the formation of thylakoid proteins and chlorophyll takes place (GALLING et al, 1973). This formation may be performed at least partially by membrane bound "old" 70 S ribosomes. Such ribosomes have been described in Chlamydomonas by CHUA and SIEKEVITZ (1974) and they may also be present in other algae as Euglena (LEDOIGT, pers. commun) and Chlorella. Spectinomycin and lincomycin have been shown to be very effective inhibitors of chloroplast protein synthesis in algae (GOODENAUGH, 1971). Since the concentration needed for complete inhibition is rather high we assume that free 70 S ribosomes may be easier affected by these drugs than membrane bound ribosomes.

Fig. 1, a - c. Cell growth, and chlorophyll production of <u>Chlorella</u> after clindamycin-treatment and in the mutant under different temperatures. Growth was measured as OD directly in the cell suspension. Chlorophyll was extracted with methanol, and OD_{665} was measured.

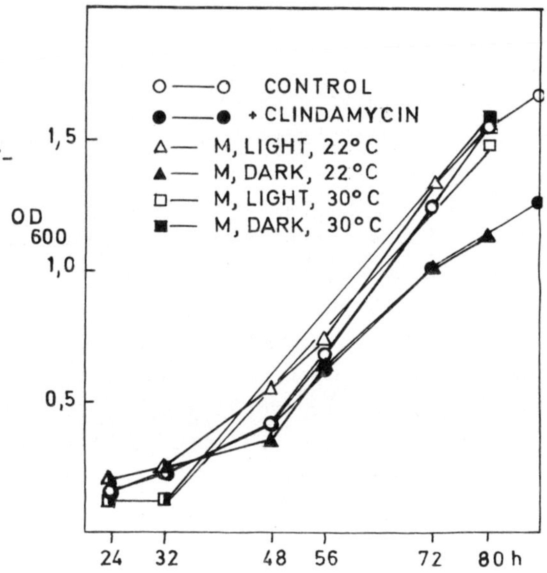

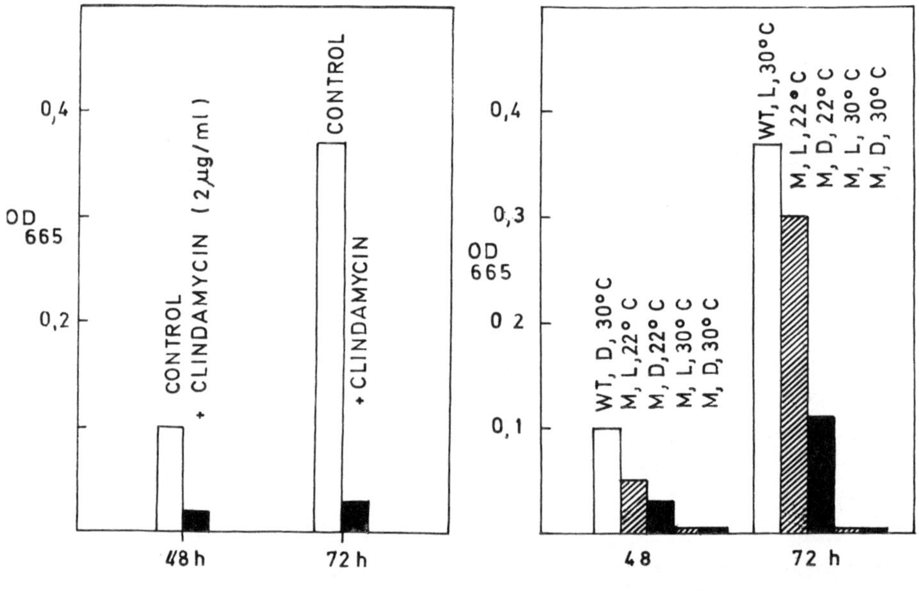

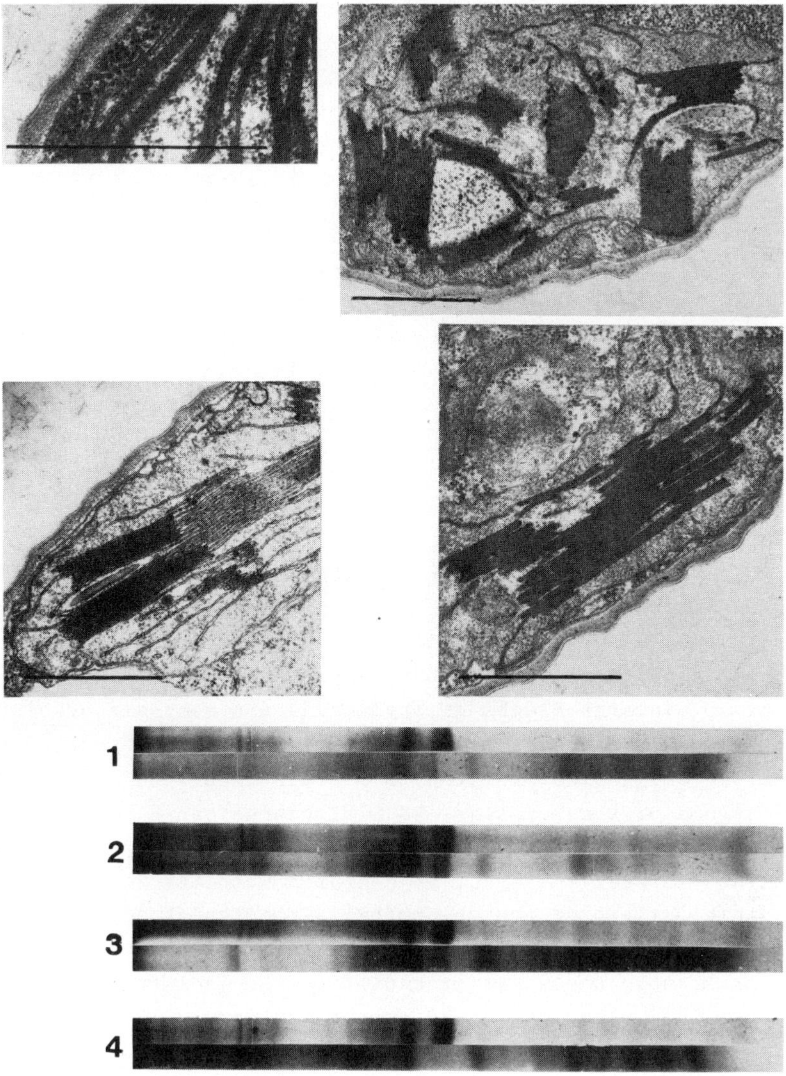

Plate I. Electron micrographs obtained from Chlorella pyrenoidosa after heterotrophic culture in presence or absence of 2 µg/ml of clindamycin. Left side, top: untreated control. The bar represents 1 µ.
Lower row: Electrophoretic patterns of thylakoid proteins obtained from Chlorella pyrenoidosa after treatment with clindamycin (2 µg/ml) (1), lincomycin (50 µg/ml) (2), spectinomycin (20 µg/ml) (3) and of the mutant of Chlorella. For comparision, in each case, the protein pattern of the heterotrophic grown control is given. Electrophoresis from left to right.

Clindamycin, however, is the most effective inhibitor of plastidal protein synthesis in Chlorella. Therefore, it may also inhibit the protein synthesis on membrane bound 70 S ribosomes. Some more data are needed to prove the hypothesis that protein bands missing in the clindamycin pattern are really products of membrane bound 70 S ribosomes.

2. The yellow mutant of Chlorella pyrenoidosa.

The growth rate of the yellow mutant of Chlorella is comparable to that of the wild type under heterotrophic conditions.
At 30°C, the mutant has no chlorophyll production, neither in the light nor in the dark. The spectra show that yellow pigments are synthesized which are thought to be mainly secondary carotenoids.
Chlorophyll production was only observed in cultures grown with glucose in the light at 22°C. In such cells, the chlorophyll level is as high as in wild-type cells, grown under heterotrophic conditions. Ribosome analysis of the mutant grown under different conditions shows that 70 S ribosomes are present. Electron microscopy of cells, grown at 30°C and 22°C shows that thylakoid formation is quite similar in both cases (see plate II). In cells grown at 30°C, large stacks of thylakoids are seen although no chlorophyll is synthesized. "Green" thylakoids from the 22°C cells and "white" thylakoids from the 30°C cells cannot be distinguished. In the stroma of the plastids, 70 S ribosomes can be seen in both cases. Also, the other submicroscopic architecture of the mutant cell is not markedly altered, as compared with the wild type.
The protein pattern of the chlorophyll-free thylakoids shows that one major protein band is missing. Further differences may be in the minor protein bands. At present time the isolation of thylakoids from the mutant cannot be reproduced very well. Thus, the proteins, altered by the mutational step cannot be identified in total.
From all these data, it is obvious that in the mutant the formation of chlorophyll-free thylakoids is not caused by the lack of 70 S ribosomes as in the clindamycin-treated cells. Apparently, one step in the formation of chlorophyll containing parts of the thylakoids is temperature sensitive. Since, at 22°C, the formation of chlorophyll takes place in the light but not in the dark, the sensitivity cannot be located in the chlorophyll biosynthesis pathway. In our opinion, one thylakoid proteins is not synthesized

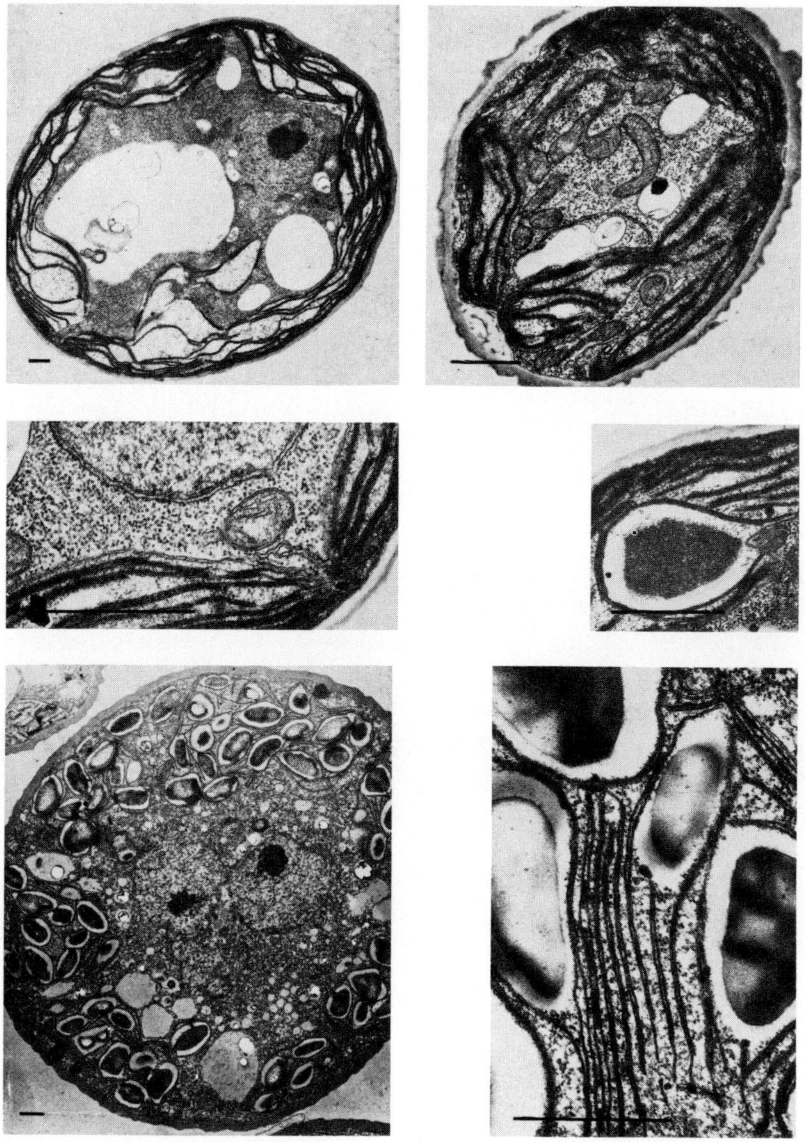

Plate II. Electron micrographs obtained from the mutant of *Chlorella pyrenoidosa* after heterotrophic growth at 30°C and 22°C. Upper row: Chlorophyll-free cells, grown at 30°C. One inset shows that pyrenoid is present in some cells. Lower row: green cells, grown at 22°C in the light. These cells were not starved, and therefore contain some starch. The bar represents 1 μ.

at 30°C, and this causes a feed-back inhibition of the chlorophyll synthesis, and, finally, results in non-chlorophyllous thylakoids at high temperature. At 22°C, this protein in question is synthesized but, at the same time, chlorophyll synthesis needs light. The reason for this is still unknown.

REFERENCES

1. BÖRNER, T., HERRMANN, F., HAGEMANN, R., FEBS Lett. 37, 117 - 119 (1973)
2. CHUA, N.-H., BLOBEL, G., SIEKEVITZ, P., PALADE, G.E., Proc. Nat. Acad. Sci. U.S. 70, 1554 - 1558 (1974)
3. FEIERABEND, J., SCHRADER-REICHHARDT, U., Planta 129, 133 - 145 (1976)
4. GALLING, G., SALZMANN, Ch., SPIESS, E. Planta 114, 269 - 284 (1973)
5. GOODENAUGH, U.W., J. Cell Biol. 50, 35 - 49 (1971)
6. HOOBER, J.K., SIEKEVITZ, P., PALADE, G.E., J. biol. Chem. 244, 2621 - 2631 (1969)
7. LORENZEN, H., KUHL, A., In: Methods in Cell Physiology, (ed. D.M. PRESCOTT) 1, 159, N.Y. Acad. Press, 1964
8. McKINNEY, G., J. biol. Chem. 140, 315 - 322 (1941)
9. OHAD, I., EYTAN, G., JENNINGS, R.C., GOLDBERG, J., BAR-NUN, S., Proc. IInd Int. Congr. Photosynthesis Res., 2563 - 2584, (1972)
10. PICKETT-HEAPS, J.D., Cytobios 5, 59 - 77 (1972)
11. SURZYCKI, S.J., GOODENAUGH, U.W., LEVINE, R.P., ARMSTRONG, J.J., Symp. Soc. exp. Biol. 24, 13 - 27 (1970).

Genetics and Biogenesis of Chloroplasts and Mitochondria, Th. Bücher et al. eds.

GENES AFFECTING ERYTHROMYCIN RESISTANCE AND SENSITIVITY OF CHLAMYDOMONAS REINHARDI CHLOROPLAST RIBOSOMES

Lawrence Bogorad, Jeffrey N. Davidson and Maureen R. Hanson
The Biological Laboratories, Harvard University
Cambridge, Massachusetts 02138 USA

INTRODUCTION

Extranuclear, or cytoplasmic, inheritance of chloroplast characters in plants was first recognized by Bauer[1] and by Correns[2] at the begining of this century and is now a well-documented phenomenon. The discovery of DNA, ribosomes and other biological information processing equipment (e.g. review by Woodcock and Bogorad, ref. 3) in chloroplasts supported the long extant idea that these organelles may be genetically independent structures that depend upon the nuclear-cytoplasmic system only for small metabolites. The same view, of course, has been advanced for mitochondria.

What genes are in the plastid DNA? At the time this work was begun it was known from molecular hybridization experiments that chloroplasts contain genes for the RNAs of their ribosomes and that rDNA genes comprise between 1% and 2% of the chloroplast DNA (e.g. ref. 3 for review). Another 0.75% or so of higher plant chloroplast DNA codes for tRNAs[4]; recently tRNAs for at least 16 different amino acids have been found to hybridize to maize chloroplast DNA[5].

SELECTION AND CHARACTERIZATION OF ERYTHROMYCIN-RESISTANT MUTANTS

A few years ago we undertook to locate genes for proteins of chloroplast ribosomes in the single-celled alga _Chlamydomonas_ _reinhardi_. Growth of wild-type _C. reinhardi_ in liquid culture is completely inhibited by 10-20 μg/ml erythromycin A. This antibiotic, which inhibits protein synthesis of prokaryotes, binds to the large (i.e. 52S) subunit of _Chlamydomonas_ chloroplast ribosomes[6, 7]. The binding constant of these ribosomes for erythromycin A is 3-4 x 10^6 M^{-1}. The antibiotic does not bind to the small chloroplast subunit nor to either the large or small subunits of the cytoplasmic ribosomes of _Chlamydomonas_.

In the first portion of this research program [6,8], we selected mutants of Chlamydomonas resistant to at least 350 μg/ml erythromycin. The mutant strains we worked with had chloroplast ribosomes which were altered in such a way that, unlike ribosomes from wild-type cells, they could not bind erythromycin _in_ _vitro_ in low concentrations of KCl.

With the knowledge gained from _in_ _vitro_ binding studies that the ribosomes of the erythromycin-resistant mutants we had selected were in fact altered, we went on to use erythromycin resistance as a genetic marker. Of nine mutants we selected for further analysis, resistance was transmitted according to Mendel's rules in eight cases and each of these was judged to be the result of a mutation in a nuclear gene. The ninth mutation is transmitted uniparentally, as expected for a chloroplast gene. From the results of series of crosses among the nuclear mutants, this group was found to be composed of two genetically unlinked subgroups designated by us as ery-M1a, b, c, d and ery-M2a, b, c, d[6, 8]. The mutation in the chloroplast gene was designated ery-U1a. The letters "M" and "U" designate Mendelian and uniparental transmission respectively.

A method was developed for the separation of the large and small chloroplast and cytoplasmic ribosomal subunits by sucrose density centrifugation[8] and for the analysis of proteins of these subunits by one- or two-dimensional polyacrylamide gel electrophoresis[9]. The first dimension is urea at pH 5 and the second on a gel slab in the presence of the detergent sodium dodecyl sulfate[9, 10]. An altered protein in the 52S subunit of a strain carrying ery-M2d was detected electrophoretically using a β-alanine-acetate buffer in 7.5% polyacrylamide gel[8]. Evidence for alteration in a different chloroplast ribosomal protein in the strain carrying chloroplast gene mutation ery-U1a was obtained from two-dimensional gel electrophoresis system[10] (Fig. 1), the protein with a possible alteration in strain ery-U1a was identified as LC4. (LC is the designation for proteins of the large subunit of the chloroplast ribosome.) Thus, the mutation to erythromycin resistance in a chloroplast gene could be correlated with the alteration of chloroplast ribosomal protein LC4. The mutation to resistance in the nuclear gene we have designated ery-M2 could be correlated with an alteration in another protein of the 52S subunit.

During the next stage of this work, carried out largely by Jeffrey Davidson[7, 11,12], two characteristics of mutants of the type ery-M1 were studied intensively. First, all four of the ery-M1 mutations we had available were mapped genetically and traced to a point about 12-15 map units to the right of the marker paralyzed flagellae (pf)2 on nuclear linkage group XI. The proteins prepared from 52S subunits of chloroplast ribosomes of mutants carrying ery-M1a, b, c and d were examined by polyacrylamide gel electrophoresis, all four were found to have protein LC6 altered (Fig. 2). Three different types of alterations in charge or size were seen among these four mutants. We have found that strains designated ery-10, -11 and -14 by Surzycki and Gillham[13] also map to the ery-M1 locus. (Each has an altered form of LC6 but each of these forms corresponds to one of the alterations in LC6 already identified in one of the other members of the ery-M1 group.) These data strongly supported the view that the locus ery-M1 on Mendelian linkage group XI is the site of the structural gene for chloroplast ribosomal protein LC6 of *C. reinhardi*.

This view has been bolstered by a study of diploid strains of *C. reinhardi* constructed and analyzed by Maureen Hanson[7]. Using methods previously described[14, 15] diploids were constructed between wild-type cells and various ery-M1 mutants. As shown in Fig. 3 the diploid +/ery-M1b contains both the wild-type and ery-M1b

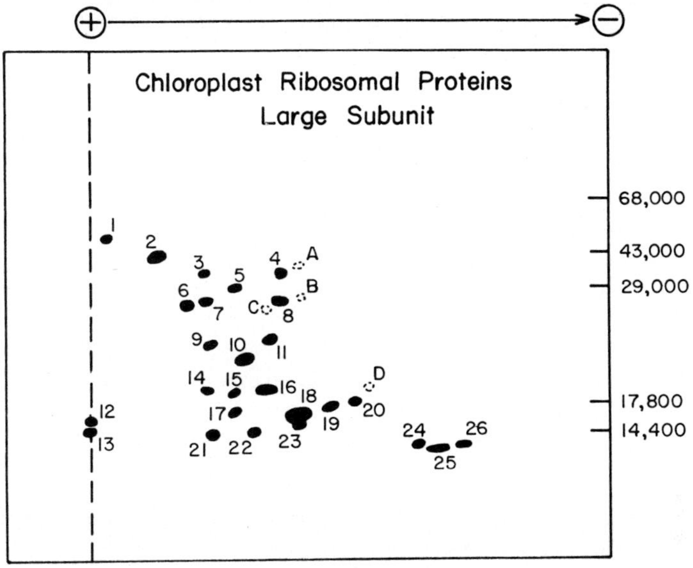

Fig. 1 Schematic diagram of the positions of proteins from the 52S subunit of *C. reinhardi* chloroplast ribosomes after two-dimensional polyacrylamide gel electrophoresis[7]. Electrophoresis from left to right was in 4% polyacrylamide at pH. 5 Electrophoresis from top to bottom was into a 10% polyacrylamide gel slab in the presence of sodium dodecyl sulfate. The positions of molecular weight markers included during the second dimension electrophoresis are shown on the right-hand margin.

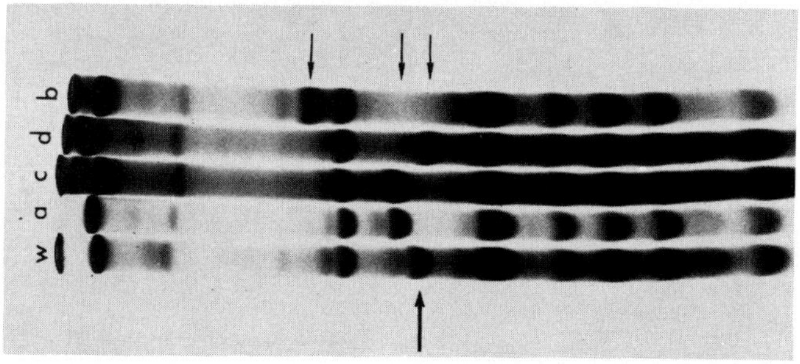

Fig. 2 Portions of polyacrylamide gels showing electrophortic separation of the 52S ribosomal proteins from wild-type (W), *ery*-M1a (a), ery-M1c (c), ery-M1d (d), and ery-M1b (b) strains of C. reinhardi[11]. Arrows show positions of wild-type and mutant forms of chloroplast ribosomal protein LC6.

forms of LC6. In each case studied both the wild-type and mutant forms of LC6 were present in the ribosomal subunits isolated from the diploid although the ratio of wild-type LC6 to the mutant form was generally about 3:2. This may result from the mutant forms having lower affinities than the normal one for the site on the ribosome.

The work carried out with mutations to erythromycin-resistance has shown: a) the location of some genes specifying chloroplast ribosomal proteins; and b) that, at least in C. reinhardi, some chloroplast ribosomal components are specified by nuclear genes and others by chloroplast genes.

REVERTANTS TO ERYTHROMYCIN SENSITIVITY

To extend our knowledge of the genetics of chloroplast ribosomal proteins in Chlamydomonas a new experimental approach to the problem was undertaken by Davidson[12]. Strains carrying the well-characterized mutation ery-M1b were mutagenized and erythromycin-sensitive strains were identified and selected using a replica plating technique. Reversion to erythromycin sensitivity could result from additional mutations in the ery-M1 gene leading to protein LC6 of ery-M1b being changed back to a form which yields an erythromycin-binding ribosome. Alternatively, a change in some other ribosomal protein might eliminate the effect of the ery-M1b form of LC6 with regard to erythromycin binding. The latter more probable case would be the basis for further genetic analysis of chloroplast ribosomal proteins.

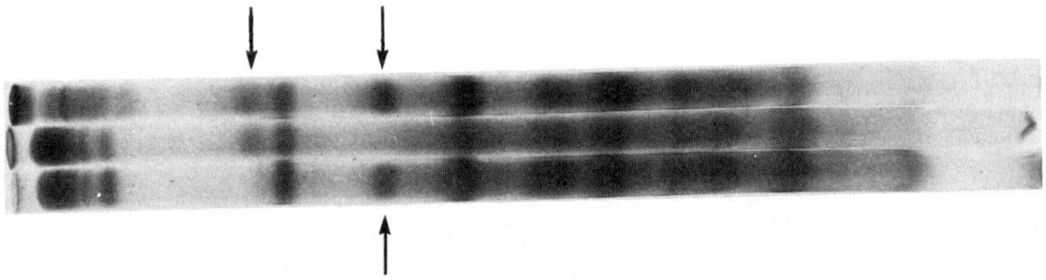

Fig. 3 Gel electrophoretic separation of proteins of 52S ribosomes from wild-type (lower), ery-14 which has a form of LC6 identical to that in ery-M1b[12] (center), and the diploid +/ery-M1b (upper) C. reinhardi[7]. Arrows show positions of LC6. Electrophoresis was in 8 M urea pH 5 in 4% polyacrylamide gels. This is the first dimension of the two dimensional electrophoretic system used in these studies[9,10].

Three of the phenotypic revertants selected which have been studied most thoroughly are designated es-M1, es-M2 and es-M3. All of these, as indicated by "M", are alterations in nuclear genes.

Chloroplast ribosomes from ery-M1b strains have higher binding constants for erythromycin when es-M1 or -M2 is present in the genome. (The combination with es-M3 has not yet been tested.) The binding behavior demonstrates that the mutation and reversion involves a change in the chloroplast ribosomes. However, LC6 of es-M1ery-M1b was found to be identical to the ery-M1b form on two-dimensional polyacrylamide gel electrophoresis. Thus, the return to erythromycin sensitivity can best be accounted for by a change in another ribosomal protein although the latter has not yet been detected.

The es-M1 locus has been traced to linkage group XI and found to lie between pf2 and ery-M1 as shown below:

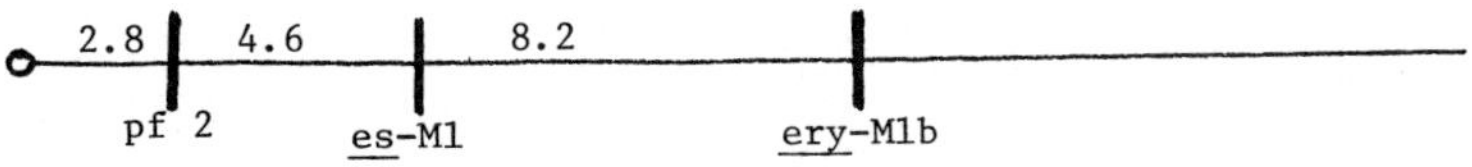

Thus, two genes affecting proteins of 52S subunits of Chlamydomonas chloroplast ribosomes have been mapped to nuclear linkage group XI. Es-M1 and -M2 are unlinked. Es-M3 may be linked or allelic to es-M2 but this relationship has not been resolved yet.

In all, our work has led to unearthing at least 5 genes which affect erythromycin resistance of C. reinhardi chloroplast ribosomes. One of these is uniparentally transmitted and thus is in the chloroplast genome; the others are in the nuclear genome. Within the nuclear genome two genes appear to be on the same linkage group while the others are spread out over the remaining linkage groups.

Conde et al.[16] have isolated another erythromycin-resistant strain of C. reinhardi, ery-U37, which is linked but is not allelic with ery-U1a (Gillham, personal communication).

RIBOSOMAL PROTEIN INTERACTIONS. GENOME-GENOME INTERACTIONS.

The mutations to erythromycin sensitivity are not equivalent. First, es-M1 and -M2 are not linked. Also, a study of resistance of double mutants to a number of antibiotics (Table 1) shows some other differences. For example es-M1ery-M1b is characterized by sensitivity to lincomycin; es-M2ery-M1b by its resistance to high levels of oleandomycin; and es-M3ery-M1b by its complete insensitivity to carbomycin. In all of these cases, just as in the original ery-M mutations to erythromycin resistance, an alteration in one nuclear gene affects the functioning of another.

Table 1 also shows that the combination ery-M1aery-U1a yields a strain which is sensitive to lincomycin although both ery-M1a and ery-U1a are resistant to varying degrees. The interaction between effects of a nuclear and a chloroplast gene is striking. Similarly, the nuclear suppressor es-M2 increases the sensitivity of ery-U37 to erythromycin.

These latter interactions between the products of genes from two genomes on ribosome function have interesting and important implications for considering some stages in the evolution of organelle ribosomes[17].

TABLE 1

Resistance of Single and Double Mutants

Strain	Erythro-mycin	Oleando-mycin	Carbo-mycin	Linco-mycin	Cleocin
wild-type	S	S	S	S	S
es-M1ery-M1b	r	r	r	S	S
es-M2ery-M1b	r	R	r	R	S
es-M3ery-M1b	r	r	R̲	r	S
ery-M1b	R	R	R	R	S
ery-M1a	R	R	R	R	S
ery-M1aery-U1a	R	R	-	S	S
ery-U1a	R̲	R	R	r	S
ery-M2dery-U1a	R	R	-	R̲	R
ery-M2d	R	R	R	R̲	r

Growth was scored on agar plate cultures.

Symbols: r, low level resistance
R, high level resistance
S, sensitive
R̲, growth not affected by antibiotic

ACKNOWLEDGEMENTS

The work described here was supported in part by research grant GM 20470 from the National Institute of General Medical Sciences and in part by the Maria Moors Cabot Foundation of Harvard University.

REFERENCES

1. Bauer, E. (1909) Z. Vererbungshehre 1, 330.
2. Correns, C. (1909) Z. Vererbungshehre 2, 331.
3. Woodcock, C.L.F. and Bogorad, L. (1971) In, Structure and Function of Chloroplasts. M. Gibbs, ed. pp. 89-128. Springer-Verlag, New York.
4. Tewari, K.K. and Wildman, S.G. (1970) Symp. Soc. Exptl. Biol 24, 147.
5. Haff, L.A. and Bogorad, L. (1976) Biochemistry. In press.
6. Mets, L.J. and Bogorad, L. (1971) Science 174, 707.
7. Hanson, M.R. (1976) The genetics and Biochemistry of Chloroplast Ribosome Mutants of Chlamydomonas reinhardi. Ph.D. Thesis. Department of Biology, Harvard University, Cambridge.
8. Mets, L.J. and Bogorad, L. (1972) Proc. Nat. Acad. Sci. 69, 3779.

9. Mets, L.J. and Bogorad, L. (1974) Anal. Biochem. 57,200.

10. Hanson, M.R., Davidson, J.N., Mets, L.J. and Bogorad, L. (1974) Molec. Gen. Genet. 132, 105.

11. Davidson, J.N., Hanson, M.R. and Bogorad, L. (1974) Molec. Gen. Genet. 132, 119.

12. Davidson, J.N. (1976) Genes Affecting Erythromycin Resistance in ery-M1 Mutants of Chlamydomonas reinhardi. pH.D. Thesis. Department of Biology, Harvard University, Cambridge.

13. Surzycki, S.J. and Gillham, N.W. (1971) Proc. Nat. Acad. Sci. 68, 1301.

14. Ebersold, W.T. (1967) Science 175, 447.

15. Gillham, N.W. (1963) Nature 200, 294.

16. Conde, M.F., Boynton, J.E., Gillham, N.W., Harris, E.H., Tingle, C.L. and Wang, W.L. (1975) Mol. Gen. Genet. 140, 183.

17. Bogorad, L. (1975) Science 188, 891.

Genetics and Biogenesis of Chloroplasts and Mitochondria, Th. Bücher et al. eds.
© 1976 Elsevier/North-Holland Biomedical Press, Amsterdam, The Netherlands.

GENETIC CONTROL OF CHLOROPLAST RIBOSOME BIOGENESIS IN CHLAMYDOMONAS

Nicholas W. Gillham, John E. Boynton, Elizabeth H. Harris, Sue B. Fox, and Paul L. Bolen, Duke University, Durham, NC 27706

Chloroplast ribosome biogenesis in *Chlamydomonas reinhardtii* requires the participation of both nuclear and chloroplast genes[1]. Over the past few years some 20 genes in two major phenotypic groups have been identified through use of appropriate mutants (Table 1). In this paper we briefly review the effects of these mutations on chloroplast ribosome structure and function in *Chlamydomonas*.

TABLE 1. Genes involved in chloroplast ribosome formation in *Chlamydomonas*. Data summarized from references 1-6.

TYPE OF ALTERATION	MUTANT PHENOTYPE	CHLOROPLAST GENES	NUCLEAR GENES
Antibiotic resistance	erythromycin resistant	2	2
	carbomycin resistant cleocin resistant	2?	0
	streptomycin resistant	4	1
	spectinomycin resistant neamine resistant	1?	0
Assembly defective	both subunits affected	0	3
	small subunit affected	0	5?

Since chloroplasts are very difficult to isolate from *Chlamydomonas*, chloroplast and cytoplasmic ribosomes are generally separated from whole cell supernatants on sucrose gradients[7]. We will follow the classification of Bourque *et al.*[8], who identified 5 principal generic classes of ribosomes with sedimentation velocities of 83, 70, 66, 54, and 41S as determined by linear extrapolation against the 83S cytoplasmic monomer peak. The 70S ribosomes are derived from the chloroplast and contain large (54S) and small (41S) subunits. In certain mutants, 66S ribosomes virtually replace the 70S chloroplast ribosomes[9]. The large ribosomal subunit contains 5S and 23S rRNA, while the small subunit contains 16S rRNA; these ribosomal RNAs are thought to be transcribed from chloroplast DNA[10,11]. Electrophoretic and chromatographic analyses indicate that the large subunit comprises 26-34 proteins, whereas the small subunit contains 19-25[12-14]. Using one-dimensional polyacrylamide gradient gel electrophoresis in SDS, we have resolved about 20 proteins of the small subunit, of which 17 are stable to dialysis against 850 mM KCl and include 6 or 7 with molecular weights > 50,000. Similar large proteins were reported by Brügger and Boschetti[13] but not by Hanson *et al.*[12]. Isoelectric focusing followed by gradient gel electrophoresis in the second dimension resolves an additional 3 proteins of molecular weight 12,000-19,000. Of the 23 small subunit proteins thus resolved by our two-dimensional system, 14 are extremely basic. In the large subunit we can distinguish about 21 proteins, most of which appear to be salt stable, with the highest molecular weight being about 37,000. Isoelectric focusing reveals that most of the large subunit proteins are also very basic.

Seven or possibly eight nuclear genes involved in chloroplast ribosome biogenesis have now been identified through isolation of assembly-defective mutants[1,6], all designated cr- with the exception of ac-20, the first mutant described as belonging to this class[15]. These mutants grow normally when acetate is supplied as a carbon source, but grow slowly under photosynthetic conditions. This poor photosynthetic growth results from an impairment of chloroplast protein synthesis which produces a specific syndrome of photosynthetic defects, including inability to synthesize the first enzyme of CO_2 fixation, ribulose diphosphate carboxylase (RuDPCase), and to carry out the Hill reaction, a measure of photosystem II activity. The large subunit of RuDPCase is known to be translated on chloroplast ribosomes[1,16], and the photosystem II deficiency can be traced, in part, to the deficiency of a specific thylakoid membrane polypeptide (polypeptide 6), which is synthesized in the chloroplast[17,18]. Assembly defective mutants make normal amounts of chlorophyll and membranes[6], but their membranes are structurally defective and improperly assembled into grana.

With the possible exception of cr-7 and ac-20, all of the assembly defective mutants are nonallelic by recombination analysis and all including ac-20 and cr-7 appear to complement in diploids. One mutant, cr-1, is semidominant, as shown by the ribosomal phenotype of diploids with wild type. The mutants ac-20, cr-4, and cr-6 cannot synthesize normal amounts of either ribosomal subunit (Fig. 1) and are probably blocked in ribosomal assembly at a very early point (e.g. at the level of rRNA precursors). cr-1, -2, -3, -5, and -7, which accumulate what appear to be 54S subunits and are deficient to varying degrees in small subunits, may be blocked specifically at various points in small subunit assembly (Fig. 1). So far, no mutants have been found which are blocked only in large subunit formation.

Mutations in three other nuclear genes confer antibiotic resistance on chloroplast ribosomes. The demonstration by Davidson et al.[2] that 4 erythromycin resistance alleles at the ery-M1 locus alter a single protein (LC6) of the large chloroplast ribosomal subunit, is the most rigorous evidence to date that any gene, chloroplast or nuclear, is a structural gene for a chloroplast ribosomal protein. Three of the ery-M1 mutant alleles alter the charge of the protein, while in the fourth some 30% of the protein is deleted. The ery-M2 locus may specify a different protein of the large subunit[2]. Chloroplast ribosomes from mutants at the ery-M1 and ery-M2 loci, unlike those of wild type, do not bind radioactive erythromycin[19,20]. We have demonstrated that mutants at a third nuclear gene locus, sr-1, confer streptomycin resistance on chloroplast ribosomes (Table 2).

Chloroplast gene mutations confer resistance directly on chloroplast ribosomes to the antibiotics streptomycin, spectinomycin, neamine, erythromycin, carbomycin, and cleocin[3,5,14,21]. By mapping and allele testing, we have identified 4 chloro-

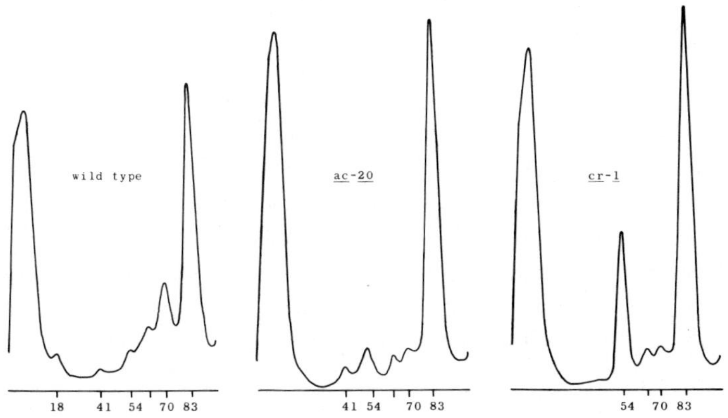

FIGURE 1. A_{254} profiles of whole cell ribosomes displayed on sucrose gradients. Ribosomal peaks are identified by sedimentation velocity[8]. The ribosomal profile of ac-20 resembles that of cr-4 and cr-6; cr-1 is similar to cr-2, -3, -5 and -7. From Harris *et al*. 1974[6].

plast loci for streptomycin resistance, two for erythromycin resistance, and one for spectinomycin resistance[22]. A neamine resistant mutant is very tightly linked to the spectinomycin locus, and may in fact be part of it.[25]

Mutations at each locus have been tested for antibiotic resistance *in vitro* using methods adapted for *Chlamydomonas* by Schlanger *et al*.[5,23] and modified by Conde *et al*.[3]. Chloroplast ribosomes are isolated by sucrose gradient sedimentation and resistance is assayed using synthetic polynucleotides and radioactive amino acids. A supernatant fraction from wild type cells is used with mutant as well as wild type ribosomes, so antibiotic resistance of the ribosomes themselves is the only thing being measured. For spectinomycin or neamine resistance, we assay inhibition of poly-U directed phenylalanine incorporation. To assay erythromycin resistance, we substitute poly-C and proline, since wild type chloroplast ribosomes appear resistant to erythromycin in the poly-U phenylalanine assay[3]. Streptomycin resistance is assayed by the misreading reaction, first described by Davies *et al*.[24] for *E. coli* ribosomes, which employs poly-U, cold phenylalanine, and radioactive isoleucine. Streptomycin dramatically stimulates isoleucine incorporation in response to a poly-U template by wild type chloroplast ribosomes, and each streptomycin resistant mutant has a characteristic lowered level of isoleucine incorporation which we call the misreading index.

While all of the mutants tested in these assays can be distinguished from wild type, some are more antibiotic resistant than others (Fig. 2). Chloroplast ribosomes from spr-u-1-6-2 are completely resistant to spectinomycin in this assay, while those from spr-u-1-27-3 are only slightly resistant (Fig. 2). In a third allelic mutant, spr-u-sp-23, spectinomycin resistance has been localized to the small subunit by Schlanger and Sager[23]. The neamine resistant mutant nr-u-2-1 is also highly resistant in the poly-U phenylalanine assay, and this resistance is also a property of the small subunit[23]. nr-u-2-1 is very tightly linked to the spectinomycin resistance alleles, and we have postulated that these mutations fall within a single gene, coding for one ribosomal protein[25]. Since assays in vitro show very little cross resistance of the chloroplast ribosomes from these two classes of mutants, we further suppose that mutations in one region of the gene confer spectinomycin resistance, while those in a second region lead to neamine resistance[25]. This hypothesis is supported by the recent finding that mutations for neamine resistance in E. coli may affect S5, the ribosomal protein implicated in spectinomycin resistance[26].

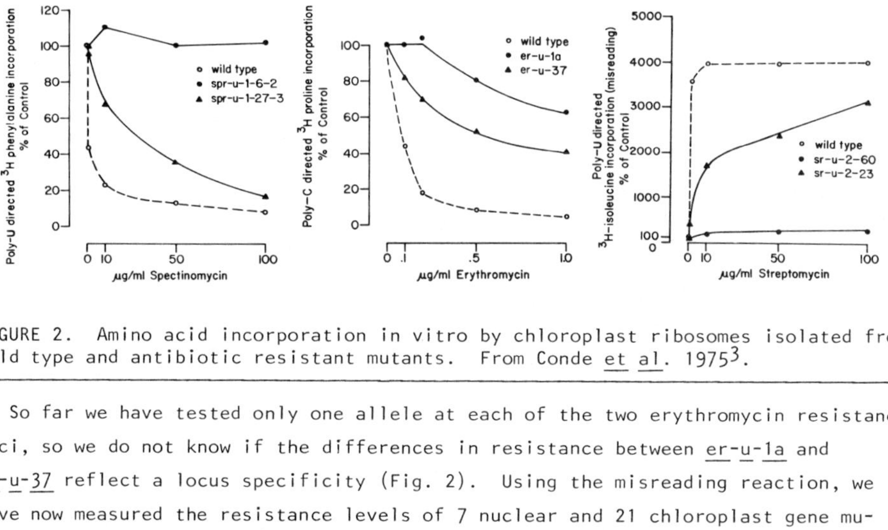

FIGURE 2. Amino acid incorporation in vitro by chloroplast ribosomes isolated from wild type and antibiotic resistant mutants. From Conde et al. 1975[3].

So far we have tested only one allele at each of the two erythromycin resistance loci, so we do not know if the differences in resistance between er-u-1a and er-u-37 reflect a locus specificity (Fig. 2). Using the misreading reaction, we have now measured the resistance levels of 7 nuclear and 21 chloroplast gene mutants resistant to streptomycin[27]. Results from sr-u-2-60 and sr-u-2-23 are shown in Fig. 2, and in Table 2, data are summarized for the mutants we have so far mapped. Mutants at the sr-u-sm-3a locus are resistant in vitro to relatively low levels of streptomycin and have relatively high, but discretely different, misreading indices, as do the nuclear sr-1 mutants (Table 2). The two alleles at the sr-u-2-60 locus exhibit high levels of resistance in vitro, and are also assembly defective, forming subnormal amounts of 70S ribosome monomers and accumulating 54S subunits. Thus they appear to be partially defective in small subunit formation like certain of the nuclear cr- mutants. Interestingly, the streptomycin dependent mutant sd-u-3-18 has a similar assembly defect[9], and we plan to test whether this and other dependent mutants might not be allelic with sr-u-2-60. In E. coli, streptomycin resistance and dependence mutations map at a single locus and affect the same ribosomal protein[28]. Two other streptomycin resistance loci in the chloroplast genome are so far identified only by single mutations isolated by Sager[29]. sr-u-sm3 is resistant to low levels of streptomycin in vivo, whereas sr-u-sm2 is resistant to high levels of streptomycin in vivo and has chloroplast ribosomes with the highest level of resistance yet encountered in the misreading reaction (Table 2). Resistance in this mutant is a property of the small subunit[23] and is correlated with the alteration of a specific small subunit protein[14].

TABLE 2. Streptomycin resistance in vitro and chloroplast ribosome profile of streptomycin resistant mutants. The misreading index = % of wild type incorporation of isoleucine in response to a poly-U template, averaged from assays at 1, 10, and 100 μg/ml streptomycin. Ribosome profile data are % of total ribosomes (A_{254} after sucrose gradient sedimentation) in large subunit (54S) and two chloroplast ribosome classes (66S and 70S).

LOCUS	ALLELES	MISREADING INDEX	CHLOROPLAST RIBOSOME PROFILE 54S	66S	70S
sr-u-sm-3a	sr-u-sm-3a	26	3	11	22
	sr-u-2-23	58	2	6	26
	sr-u-2-NG-1	77	2	6	26
	sr-u-sm5	--	--	--	--
sr-u-2-60	sr-u-2-60	18	12	6	16
	sr-u-2-281	9	18	9	12
sr-u-sm3	sr-u-sm3	--	--	--	--
sr-u-sm2	sr-u-sm2	3	3	5	19
sr-1	sr-1-H-5	40	5	7	20
	sr-1-18-3	79	4	6	18
	sr-1-3-3-9	51	3	6	21
wild type	-----------	100	2	6	24

We have also measured chloroplast ribosomal resistance in vivo by two methods[3,21]. The first, devised by Chua et al.[30] to study wild type membrane-bound chloroplast ribosomes which normally run-off during membrane isolation, measures the ability of antibiotics to "freeze" chloroplast polysomes on thylakoid membranes. When ribosome translocation is blocked by antibiotics, chloroplast ribosomes remain bound to the thylakoid membranes, presumably by nascent chains, and can be released by treatment with puromycin and high concentrations of salt. In a mutant with ribosomes resistant to an antibiotic in vivo, run-off should not be blocked and resistant ribosomes should not be found attached to the thylakoid membranes. The two spectinomycin resistant mutants spr-u-1-6-2 and spr-u-1-27-3, which differ widely in their resistance levels both in vitro (Fig. 2) and in vivo (Fig. 3), also differ greatly in this binding assay[3].

In the second assay, increase in RuDPCase activity in a short-term culture is assayed in the presence of varying antibiotic concentrations. This increase results from active synthesis of the enzyme[21]. Thus we would expect antibiotic inhibition in wild type cells but not in a mutant with chloroplast ribosomes which show high levels of antibiotic resistance _in vitro_, and in fact we find that this assay _in vivo_ (Fig. 3) closely reflects sensitivity levels measured _in vitro_ (Fig. 2). Thus all our studies indicate that resistance _in vivo_ as well as _in vitro_ is a ribosomal character and is not due to secondary effects.

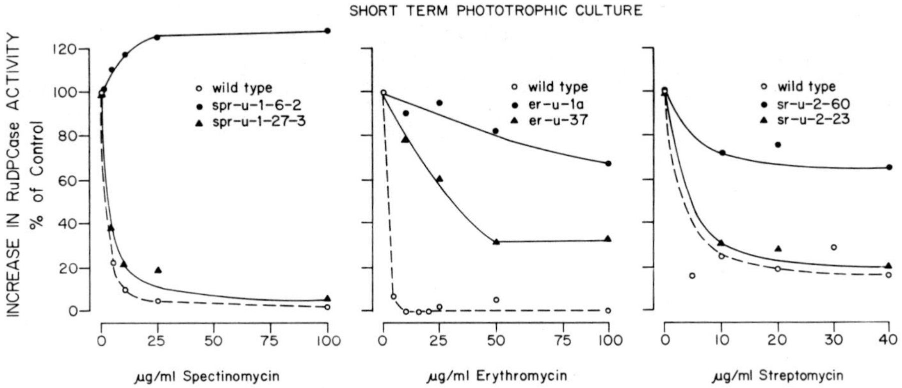

FIGURE 3. Resistance of chloroplast gene mutants to antibiotic inhibition of RuDPCase synthesis in short term phototrophic culture. From Conde _et al_. 1975[3].

In conclusion, a significant number of chloroplast and nuclear genes involved in chloroplast ribosome formation have now been identified in _Chlamydomonas_, and their functions are beginning to be understood. Chloroplast rRNA is thought to be a chloroplast gene product[10,11]. Although some of the chloroplast genes identified by antibiotic resistance mutations could conceivably directly affect rRNA or its processing, their phenotypes are strikingly similar to bacterial mutations known to alter ribosomal proteins. In certain cases there is evidence that ribosomal proteins are specifically affected by chloroplast gene mutations[14,21,31]. Assuming, therefore, that chloroplast ribosomal proteins may be products of both nuclear and chloroplast genes, there would appear to be no correlation between the location of the gene and either the ribosomal subunit affected, or the type of antibiotic resistance conferred. We are presently examining in detail proteins of the small subunit of the chloroplast ribosomes from the several mutant alleles known at the one chloroplast locus for spectinomycin resistance, and at 2 of the 4 chloroplast loci for streptomycin resistance, to determine whether we can detect systematic genetic changes of the type already demonstrated in the case of mutants at the ery-M1 locus. We feel that this sort of evidence is essential to establish unequivocally that a particular gene codes for a given ribosomal protein. To account for the ability of mutations at any one of 4 gene loci (2 nuclear and 2 chloroplast) to confer erythromycin resistance on chloroplast ribosomes, and mutations at any one of 5 gene loci (1 nuclear and 4 chloroplast) to confer streptomycin resistance, we postulate that there are ribosomal "neighborhoods" on the large and small subunits, respectively, where these two drugs bind and block protein synthesis. Assuming that these ribosomal neighborhoods are composed of a number of proteins, each of which is coded by a separate gene, an appropriate alteration of any one of these several ribosomal proteins by gene mutation, must be sufficient to change the conformation of the binding site and thus confer antibiotic resistance on the chloroplast ribosome. This sort of analysis may allow one to begin to visualize the arrangement of individual ribosomal proteins within the three-dimensional structure of the ribosomal subunits themselves.

REFERENCES

1. Harris, E.H., J.E. Boynton & N.W. Gillham, 1976, in The Genetics of Algae, R.A. Lewin, ed., Botanical Monographs Vol. 12, Blackwell Scientific, Oxford, pp. 119-144.
2. Davidson, J.N., M.R.Hanson & L. Bogorad, 1974, Molec. gen. Genet. 132:119-129.
3. Conde, M.F., J.E. Boynton, N.W. Gillham, E.H. Harris, C.L. Tingle & W.L.Wang, 1975, Molec. gen. Genet. 140: 183-220.
4. Sager, R. & Z. Ramanis, 1976, Genetics 83: 303-321.
5. Schlanger, G., R. Sager & Z. Ramanis, 1972, Proc. Nat. Acad. Sci. USA 69: 3551-3555.
6. Harris, E.H., J.E. Boynton & N.W. Gillham, 1974, J. Cell Biol. 63: 160-179.
7. Hoober, J.K. & G. Blobel, 1969, J. Mol. Biol. 41: 121-138.
8. Bourque, D.P., J.E. Boynton & N.W. Gillham, 1971, J. Cell Sci. 8: 153-183.
9. Gillham, N.W., J.E. Boynton & B. Burkholder, 1970, Proc. Nat. Acad. Sci. USA 67: 1026-1033.
10. Lambowitz, A.M., C.R. Merril, E.A. Wurtz, J.E. Boynton & N.W. Gillham, 1976, abstract submitted to J. Cell Biol.
11. Surzycki, S.J. & J.D. Rochaix, 1971, J. Mol. Biol. 62: 89-109.
12. Hanson, M.R., J.N. Davidson, L.J. Mets & L. Bogorad, 1974, Molec. gen. Genet. 132: 105-118.
13. Brügger, M. & A. Boschetti, 1975, Eur. J. Biochem. 58: 603-610.
14. Ohta, N., M. Inouye & R. Sager, 1975, J. Biol. Chem. 250: 3655-3659.
15. Goodenough, U.W. & R.P. Levine, 1970, J. Cell Biol. 44: 547-562.
16. Gooding, L.R., H. Roy & A.T. Jagendorf, 1973, Arch. Biochem. Biophys. 159: 324-335.
17. Chua, N.-H. & N.W. Gillham, 1976, in preparation.
18. Chua, N.-H. & P. Bennoun, 1975, Proc. Nat. Acad. Sci. USA 72: 2175-2179.
19. Mets, L.J. & L. Bogorad, 1971, Science 174: 707-709.

20. Bogorad, L., J.N. Davidson, M.R. Hanson & L.J. Mets, 1975, in Molecular Biology of Nucleocytoplasmic Relationships, Elsevier Scientific, Amsterdam.

21. Boynton, J.E., W. Burton, N.W. Gillham & E.H. Harris, 1973, Proc. Nat. Acad. Sci. USA 70: 3463-3467.

22. Boynton, J.E., N.W. Gillham, E.H. Harris, C.L. Tingle, K. Van Winkle-Swift & G.M.W. Adams, this volume.

23. Schlanger, G. & R. Sager, 1974, Proc. Nat. Acad. Sci. USA 71: 1715-1719.

24. Davies, J., W. Gilbert & L. Gorini, 1964, Proc. Nat. Acad. Sci. USA 51: 883-890.

25. Boynton, J.E., E.H. Harris, C.L. Tingle, S.B. Fox & N.W. Gillham, 1975, J. Cell Biol. 67: 41a.

26. DeWilde, M., T. Cabezón, R. Villarroel, A. Herzog & A. Bollen, 1975, Molec. gen. Genet. 142: 19-33.

27. Boynton, J.E., E.H. Harris, C.L. Tingle, S.B. Fox & N.W. Gillham, 1975, Genetics 80: s15-s16.

28. Breckenridge, L. & L. Gorini, 1970, Genetics 65: 9-25.

29. Sager, R., 1972, Cytoplasmic Genes and Organelles, Academic Press, New York.

30. Chua, N.-H., G. Blobel, P. Siekevitz & G.E. Palade, 1973, Proc. Nat. Acad. Sci. USA 70: 1554-1558.

31. Mets, L. & L. Bogorad, 1972, Proc. Nat. Acad. Sci. USA 69: 3779-3783.

ACKNOWLEDGEMENTS

We thank Connie Tingle for her able assistance. This work was supported by grants NSF-GB22769, NSF-GB43559, and NIH-GM19427, and by NIH RCDA awards GM70437 to N.W.G. and GM70453 to J.E.B.

CONTROL OF FORMATION AND ASSEMBLY OF CHLOROPLAST CONSTITUENTS

Genetics and Biogenesis of Chloroplasts and Mitochondria, Th. Bücher et al. eds.

CELLULAR ORIGINS OF PLASTID MEMBRANE POLYPEPTIDES IN EUGLENA*

Scott Bingham** and Jerome A. Schiff
Institute for Photobiology of Cells and Organelles
Brandeis University, Waltham, MA 02154 U. S. A.

INTRODUCTION

When grown continuously in the light, *Euglena gracilis* var. *bacillaris* contains about ten chloroplasts which divide in a manner that keeps the plastid number relatively constant from one generation to the next[1,2,3]. When grown in the dark, the plastids become reduced to proplastids which also remain fairly constant in number as the cells divide. If dark-grown cells are placed on a resting medium lacking nutrients essential for growth, cell division ceases and, on exposure to light, the proplastids of these non-dividing cells develop into chloroplasts within 72-96 hrs. providing an excellent system in which to study the details of chloroplast development[1,2,3].

On exposure of resting dark-grown cells to light, chlorophyll and most other parameters increase slowly for the first twelve hrs. after which all parameters increase linearly until completion of the plastid at 72-96 hrs. The formation of new plastid membranes begins 12 hrs. after light induction and proceeds during the linear increase in chlorophyll[1,2,3]. Although many new proteins are made during this period, photosynthesis is not necessary and small molecules and energy are supplied by the rest of the cell including the mitochondria which are in close association with the developing plastid. Plastid proteins arise in two patterns during development: 1) those that appear without a lag are synthesized on cytoplasmic ribosomes from information supplied by the nucleus (or, possibly, the mitochondrion) since their synthesis is inhibited by cycloheximide and they are present in W_3BUL, a mutant lacking detectable plastid DNA, at the same levels as in dark-grown wild-type cells and 2) those that appear after a 12 hr. lag and whose synthesis is streptomycin or chloramphenicol-sensitive and the levels of which are below detection in mutant W_3BUL lacking detectable plastid DNA, indicating that they are synthesized on plastid ribosomes and may be coded in plastid DNA[1,2,3].

The problem, then, in using this system to locate the source of information for plastid thylakoid polypeptides and the site of their synthesis, is to find a marker that will permit the localization of these membranes in cells of light-grown wild-type cells (containing chloroplasts) (WTL), dark-grown wild-type cells (containing proplastids) (WTD) and mutant W_3BUL containing nuclear (or mitochondrial) coded plastid remnants. Chloroplasts can be readily isolated from *Euglena* but the isolation of proplastids or their remnants has not been achieved in quantity. We have found, however, that the sulfolipid of the plastid membranes is synthesized in WTL, WTD and W_3, indicating that the enzymes forming it are coded outside the plastid DNA (D. Safer, unpublished; see also 4). Since this lipid can be readily labeled with $^{35}SO_4^{2-}$, it provides an excellent marker for locating plastid-related membranes in gradients prepared from whole cells of WTL, WTD and W_3. In this paper we will describe these methods and use them to determine the plastid polypeptides present in all three types of cells leading to inferences about where they arise in the Euglena cell.

MATERIALS AND METHODS

Euglena gracilis Klebs var. bacillaris Pringsheim was maintained in light or darkness on Hutner's pH 3.5 medium[5] as previously described[6]. Non-dividing cells used for chloroplast development were obtained as described previously[7] using pH 5 resting medium. Three times washed chloroplasts were obtained as previously described[12]. These plastids were suspended in either buffer IV or buffer IV minus Mg^{2+}; the same results as far as plastid membrane polypeptides are concerned are obtained with either buffer. After passage through the French pressure cell[12] and centrifugation, the pellet was washed in either 0.15M EDTA pH 7.0, or in 50mM Tris-HCl pH 7.6 at 4° and the pellet was used for solubilization of polypeptides. Alternatively, after passage through the French pressure cell, deoxycholate was added to a concentration of 0.1% (w/v), the suspension was shaken for 15 min. at 2°C and the pellet after centrifugation[12] was used for solubilization of membrane polypeptides.

All steps in membrane isolation from whole cells were performed at 4°. One liter of growing or resting cells (1 to 1.5 x 10^6 cells/ml) were harvested and washed in 50 mM Tris-HCl buffer pH 7.6. They were then resuspended in 5 ml of this buffer and passed through a French pressure cell at 4000 PSI. In order to break organelles and provide reasonably uniform sizes of membrane fragments, these homogenates were sonicated in a Branson Sonifier for 5 sec. at 4 amps. This treatment completely disrupted chloroplasts and mitochondria as determined by the release of ribulose diphosphate carboxylase[8] and malate dehydrogenase[9] activity respectively. Five ml of sonicate were layered on a step gradient composed of 35, 43, 50 and 60% (w/v) sucrose in the above buffer and were centrifuged for 3 hr. at 21,000 rpm in a Beckman SW-27 swinging bucket rotor in an L5-50 preparative ultracentrifuge. One ml fractions were collected, and to isolate membranes, appropriate fractions were diluted with 0.15M EDTA pH 7.0, centrifuged at 100,000g for 30 min. and washed by centrifugation with 50mM Tris-HCl pH 7.6 to remove the EDTA.

To label the cells, Hutner's pH 3.5 medium with 0.1 the normal level of sulfate (achieved by replacing $MgSO_4$ with $Mg(NO_3)_2$ as necessary) was fortified with 1.0 mCi of carrier-free $^{35}SO_4^{2-}$ per liter of medium. After growth, gradient fractions were prepared as above, one ml fractions were collected, and lipids for lipid-soluble radioactivity determinations were extracted as described by Bligh and Dyer[10]. To each 1.0 ml fraction from the gradient, 2.50 ml of methanol and 1.25 ml chloroform were added. 1.25 ml of water was then added and after shaking, another 1.25 ml of chloroform was added. After brief centrifugation in a table centrifuge, the upper aqueous phase was removed and 1.0 ml of the lower organic phase was placed in a scintillation vial together with a drop of 1% benzoyl peroxide in chloroform to decolorize the pigments and the solvents were evaporated off. Ten ml of scintillation mixture (5g/1 PPO in Toluene) were then added and the samples were counted in a Beckman liquid scintillation counter.

The pelleted EDTA and buffer-washed membrane fractions (see above) were solubilized by heating at 100° for 2 min. in 0.05M Tris-HCl pH 8.3, 8M with respect to urea, 2% (w/v) with respect to sodium dodecyl sulfate and 5% (v/v) with respect to mercaptoethanol. Glycerol to a final concentration of 5% (v/v) was added to the cool solution to facilitate application to the gels. Gel electrophoresis on polyacrylamide employed the discontinuous buffer system of Laemmli[13] using a slab 0.75mm thick with a 20 cm linear 7.5-15% acrylamide gradient accompanied by a 5-17.5% sucrose gradient[14]. The samples were subjected to electrophoresis at 50 v through the stacking gel for one hr. and 300v through the running gel for six hrs. The gels were stained with 0.25% coomassie blue in 50% methanol-7% acetic acid and were destained in 10% acetic acid.

The following enzymes were assayed by diluting ca. 0.1 ml of each 1.0 ml gradient fraction with the appropriate reaction mixture and buffer given in the original methods: succinate dehydrogenase[15,16], NADH-ferricyanide reductase[16], antimycin A-insensitive NADPH-cytochrome c reductase[16], NADH cytochrome c reductase[16], and dicyclocarbohexyldimide-insensitive Mg^{2+}-dependent ATPase[17]. Inorganic phosphate was determined by the method of Chen et al.[18] and protein by the method of Lowry et al.[19]. Chlorophyll was determined as described previously[7].

RESULTS AND DISCUSSION

The Euglena system has been especially useful in determining where plastid constituents are coded and synthesized because dark-grown cells contain proplastids which can be induced by light to form chloroplasts in non-dividing cells where complications due to cell divisions are not present[1,2,3]. In addition, mutants such as W_3BUL are available which lack detectable plastid DNA and plastid pigments and probably contain only plastid constituents coded outside of plastid DNA and synthesized on non-chloroplast ribosomes[1,2,3]. Reasonably good chloroplasts can be prepared from Euglena, especially if only thylakoid membranes are desired, but methods for obtaining large quantities of clean proplastids from dark-grown cells and nuclear coded plastid membrane remnants from W_3 are not available. (A method for obtaining small amounts of proplastids from dark-grown cells has been reported [20]). The problem, then, in determining where plastid membrane constituents are coded and synthesized is to locate and isolate these membranes when a fully-formed intact plastid is not present in the cells. Since we have found (consistent with earlier work of others) that the sulfolipid of Euglena is synthesized in light

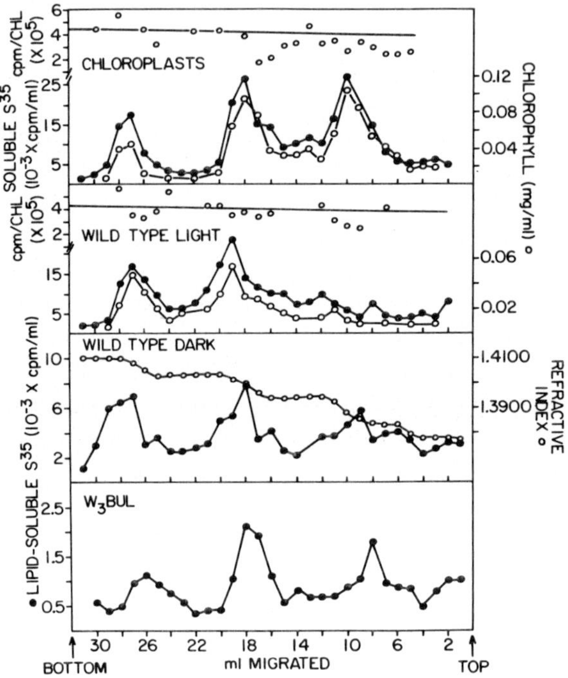

Fig. 1. Membranes from purified chloroplasts and whole cells of light-grown wild-type, dark-grown wild-type and light-grown W_3BUL separated on sucrose step gradients. Lipid-soluble ^{35}S estimates sulfolipid; chlorophyll content and refractive index are also shown. A lower, middle and upper band is found in each case; the middle band contains fairly clean plastid thylakoids. The plots of cpm/chl show the constancy of the ratio of lipid-soluble ^{35}S to chlorophyll across the gradients.

grown wild-type cells containing chloroplasts (WTL), dark-grown wild-type cells (containing proplastids) (WTD) and W_3 cells (lacking detectable plastid DNA), this membrane constituent must be synthesized by enzymes coded outside of plastid DNA, probably in nuclear DNA and provides a useful marker for plastid thylakoid membranes from all three types of cells. Therefore, the three types of cells were grown on $^{35}SO_4^{2-}$ to label the sulfolipid which they all contain. The total membrane fractions of these cells and of isolated chloroplasts were isolated after mild sonication which was found to improve the uniformity of the preparations, probably by reducing fragments to a more or less uniform size distribution.

After centrifugation of these membrane preparations through sucrose step gradients, three peaks of lipid-soluble ^{35}S were observed in membranes prepared from purified chloroplasts, WTL, WTD and cells of W_3BUL (light or dark-grown) (Fig. 1). The fact that pure chloroplasts also give three peaks coupled with the observation (Fig. 1) that the ratio of lipid-soluble ^{35}S to chlorophyll is relatively constant across the gradients from chloroplasts and WTL, suggests that all three fractions contain plastid membranes and that ^{35}S and chlorophyll move together through these gradients establishing the usefulness of lipid-soluble ^{35}S as a plastid thylakoid membrane marker.

Experiments aimed at determining which fractions contained pure plastid membranes yielded the data of Table I which shows that markers[15,16,17] of mitochondrial membranes in other systems such as succinic dehydrogenase, NADH ferricyanide reductase, and NADH cytochrome c reductase, microsomes (antimycin a-insensitive NADPH-cytochrome c reductase and NADH ferricyanide and cytochrome c reductases), and enzymes usually found in the plasma membrane and plastid envelope (Mg^{2+} dependent ATPase) are high in the upper fraction, appreciable in the lower fraction, but show minimum values in the middle fraction. The lowermost fraction probably contains clumped organelles and membranes while the upper fractions contain endoplasmic reticulum, mitochondrial membranes and, perhaps, plastid envelopes and microbody membranes[21]. The mitochondrial membrane fraction bands at the interface between 35% and 43% sucrose (density about 1.17 g/cm^3) while the microsomes band above this at a density of about 1.13 g/cm^3. The purest plastid thylakoid membranes are found at the interface of 43% and 50% sucrose (density about 1.21).

TABLE I

Enzyme Activities in Upper, Middle (Plastid Thylakoid) and Lower Gradient Fractions and in Purified Chloroplasts

		Succinate Dehydrogenase Cytc	Succinate Dehydrogenase Ferri-cyanide	NADH Cytc Reductase	NADH Ferricyanide Reductase	NADPH Cytc Reductase	Mg^{2+} dependent ATPase
		u moles cytochrome c or Ferricyanide reduced ml^{-1} min^{-1}					u moles Pi ml^{-1} hr^{-1}
WTL	U	6.6	-	60.0	660	8.4	18
	M	1.3	-	1.1	2	0.84	1.9
	L	5.6	-	12.0	51	2.1	0
WTD	U	8.4	4.1	62.0	295	11.4	6.7
	M	0.47	0	0.62	29	1.6	0.87
	L	8.4	0	4.0	53	2.2	0.35
W_3	U	7.0	4.4	20.0	233	14.2	0.29
	M	0.14	0	1.0	29	0.7	0.16
	L	2.9	0	4.5	46	1.9	0.13
Chloroplasts		0	0	0	0	0	-

In the experiments to be described this middle fraction was used as a source of thylakoid membranes from cells of WTL, WTD and W_3. These thylakoid membranes were solubilized and polypeptides were prepared by treatment with SDS and urea. Plates 1 and 2 show acrylamide gel patterns obtained in various experiments and the relevant distinctions among the various types of cells is summarized in Fig. 2. More than 40 bands can be resolved in patterns from thylakoid membranes obtained from chloroplasts and WTL. The patterns from chloroplasts and WTL are very similar to or identical with each other. There are a few bands in preparations from WTL chiefly in the molecular weight range beyond 60,000 daltons which are variable from one preparation to the next and may represent contaminants since they appear in all preparations except those from purified chloroplasts. Plate 1 also shows that the patterns from W_3 and WTD are very similar to or identical with each other indicating that all of the plastid thylakoid polypeptides in proplastids are coded

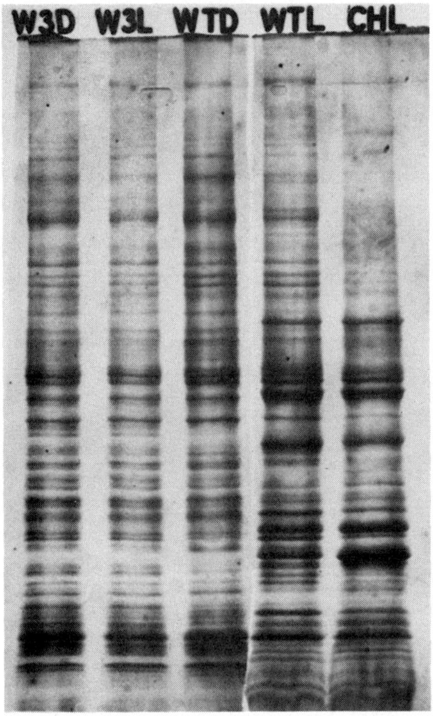

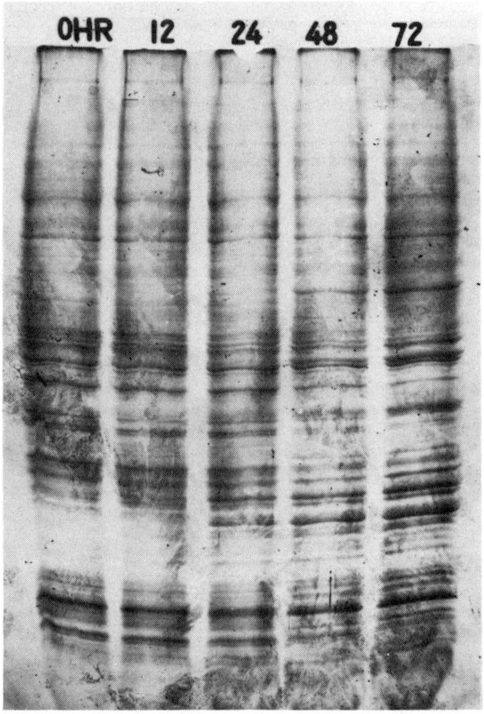

Plate 1. Polypeptide patterns from plastid thylakoid (middle) fractions of dark-grown W_3, light-grown W_3, dark-grown wild-type, light-grown wild type and from thylakoid membranes of purified chloroplasts separated on SDS acrylamide gels.
Plate 2. Polypeptide patterns from plastid thylakoid (middle) fractions prepared from equal numbers of dark-grown resting cells exposed to light for various times, separated on SDS acrylamide gels.

outside of plastid DNA, probably in nuclear DNA. W_3 cells grown in light or darkness give identical patterns indicating that light does not induce the formation of plastid thylakoid polypeptides in this mutant.

The difference between WTD and WTL patterns include certain bands which are present in WTL but not WTD and certain bands which are present in WTD but not in WTL. These bands can be seen to appear or disappear, respectively, when dark-grown resting cells are exposed to light to induce them to form chloroplasts from the proplastids (Fig. 2 and Plate 2). In addition, certain bands already present in WTD and W_3 increase during greening.

To summarize (Fig. 2) there are more than 40 bands detectable in plastid thylakoid membranes from chloroplasts, WTL, WTD and W_3. Of these, a few may be contaminants since they are variable in WTL, WTD and W_3 preparations and low or absent in chloroplast preparations. Of the remaining bands, the greater number of them appear to be coded outside plastid DNA since they are present in membranes from W_3

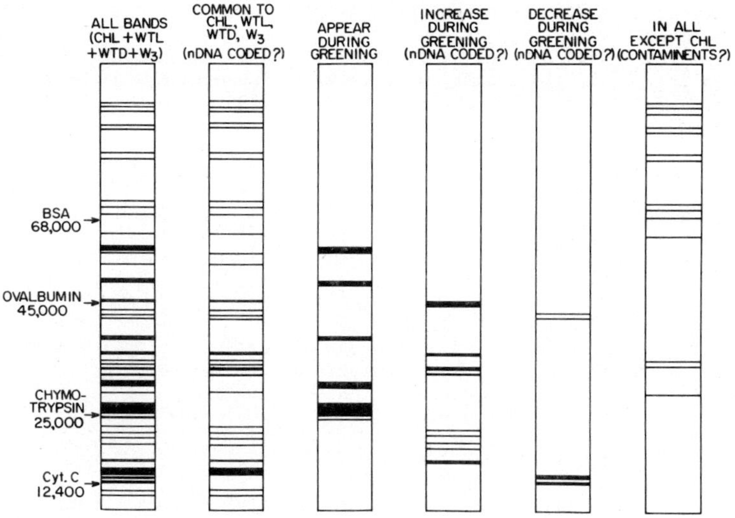

Fig. 2. Summary of differences among plastid thylakoid polypeptide patterns from the various preparations shown in Plates 1 and 2.

which lacks detectable plastid DNA. There are at least five components which appear during greening and are found only in WTL but not in WTD or W_3. These may be coded in plastid DNA, their structural genes being deleted along with the DNA. This is not unambiguous, however, since plastid DNA gene products might regulate the transcription of structural genes for these proteins in nuclear DNA and when the plastid DNA is deleted, these nuclear genes would not be expressed. At least one of these bands is very prominent at a m.w. of about 27,000 daltons and appears to be very similar to a component from Chlamydomonas observed to increase during greening and said to be coded in nuclear DNA[22,23]. Those bands which are already present in WTD and W_3 but which increase during greening are coded outside plastid DNA, probably in nuclear DNA, but are derepressed by light during plastid development, like several soluble chloroplast enzymes already studied[1,2,3] which are synthesized on cytoplasmic ribosomes. The peptides which decrease or disappear during greening and are therefore present in WTD but absent in WTL may have functions peculiar to proplastid function and to the greening process but are not necessary for photosynthesis and function of the mature plastid. They, too, are coded outside plastid DNA probably in nuclear DNA.

*Work supported by NIH GM-14595. **Supported by NIH GM-1586.

REFERENCES

1. Schiff, J. A. (1974) in M. Avron, Ed., "Proc. Third Int. Congress on Photosynthesis", Elsevier, Amsterdam, pp. 1691-1717.
2. Schiff, J. A. (1975) in "Les Cycles Cellulaires et Leur Blocage Chez Plusieurs Protistes", CNRS, Gif-sur-Yvette, France, pp. 79-93.

3. Schiff, J. A. (1973) in M. Abercrombie and J. Brachet, Eds., "Advances in Morphogenesis", Vol. 10, Academic Press, N. Y., p. 265.
4. Abraham, A. and Bachhawat, B. K. (1963) Biochim. Biophys. Acta 70:104-106.
5. Greenblatt, C. L. and Schiff, J. A. (1959) J. Protozool. 6:23-28.
6. Lyman, H., Epstein, H. T. and Schiff, J. A. (1961) Biochim. Biophys. Acta 50:301-309.
7. Stern, A. I., Schiff, J. A. and Epstein, H. T. (1964) Pl. Physiol.39:220-226.
8. Bovarnick, J. G., Chang, S., Schiff, J. A. and Schwartzbach, S. D. (1974) J. Gen. Microbiol. 83:51-62.
9. Ochoa, S. (1955) in S. P. Colowick and N. O. Kaplan, Eds., "Methods in Enzymology" Vol. 1, Academic Press, N. Y. p. 735.
10. Bligh, E. G. and Dyer, W. J. (1959) Can. J. Biochem. Physiol. 37:911-917.
11. Schwartzbach, S. D., Freyssinet, G. and Schiff, J. A. (1974) Pl. Physiol. 53:533-542.
12. Laemmli, V. (1970) Nature 227:680-685.
13. Chua, N. and Bennoun, P. (1975) Proc. Nat. Acad. Sci. (USA) 72:2175-2179.
14. Bonner, W. D. (1955) in S. P. Colowick and N. O. Kaplan, Eds., "Methods in Enzymology", Vol. 1, Academic Press, N. Y., p. 722.
15. Douce, R., Holtz, R. B. and Benson, A. A. (1973) Biochim. Biophys. Acta 292: 105-116.
16. Douce, R., Holtz, R. B. and Benson, A. A. (1973) J. biol. Chem. 218(20): 7215-7222.
17. Chen, P. S., Toribera, T. Y. and Warner, H. (1956) Analyt. Chem. 28:7756-1758
18. Lowry, O. H., Risebrough, J. J., Fair, A. L. and Randall, R. J. (1951) J. biol. Chem. 193:265-275.
19. Ophir, I., and Ben-Shaul, Y. (1973) Pl. Physiol. 51:1109-1116.
20. Donaldson, R. P., Tolbert, N. E. and Schnarrengerger, C. (1972) Archs. Biochem. Biophys. 152:199-215.
21. Hoober, J. K. (1970) J. biol. Chem. 245:4327-4334.
22. Eytan, G. and Ohad, I. (1970) J. biol. Chem. 245:4297-4307.

Genetics and Biogenesis of Chloroplasts and Mitochondria, Th. Bücher et al. eds.

SYNTHESIS OF THE MAJOR THYLAKOID POLYPEPTIDES DURING GREENING OF CHLAMYDOMONAS REINHARDTII Y-1

J. Kenneth Hoober
Department of Biochemistry
Temple University Medical School
Philadelphia, Penna. U.S. A. 19140

The formation of thylakoid membranes during greening of etiolated cells of *Chlamydomonas reinhardtii* y-1 involves the synthesis of polypeptides both on chloroplast and on cytoplasmic ribosomes[1-3]. Of particular interest are the major polypeptides of the membranes, which are synthesized on cytoplasmic ribosomes[1,3]. However, the production of these polypeptides seems to be controlled by the photoreduction of protochlorophyll(ide)[3,4], a reaction that occurs in the chloroplast[5,6]. The mechanisms of this control and of how these polypeptides become integrated into the membrane structure are not known. The greening of *C. reinhardtii* y-1 provides a favorable system with which to examine these processes. This article describes some aspects of this system and offers a suggestion on how polypeptides made in the cytoplasm become integral components of thylakoid membranes.

Thylakoid Polypeptides in Chlamydomonas

These polypeptides have been studied by polyacrylamide gel electrophoresis in the presence of sodium dodecyl sulfate[1,3,7,8]. It was reported previously[1,3] that *Chlamydomonas* thylakoid membranes contain two predominant polypeptide fractions, apparently homogeneous in size, with masses of about 28,000 and 24,000 daltons (designated *b* and *c*, respectively). Electrophoretic patterns of polypeptides in urea-washed membranes, representative of these results, are shown in Fig. 1. An unusual feature of fraction *c* is its sensitivity to the conditions of electrophoresis. If the pH of the separating gel was 9 or above, three distinctly resolved, roughly equally staining fractions were found in the place of *c* (see also ref. 8), rather than one major fraction as shown in Fig. 1. The presence of urea in the gel and of thioglycolate in the reservoir buffer also favored a single major fraction in this position. The nature and cause of this

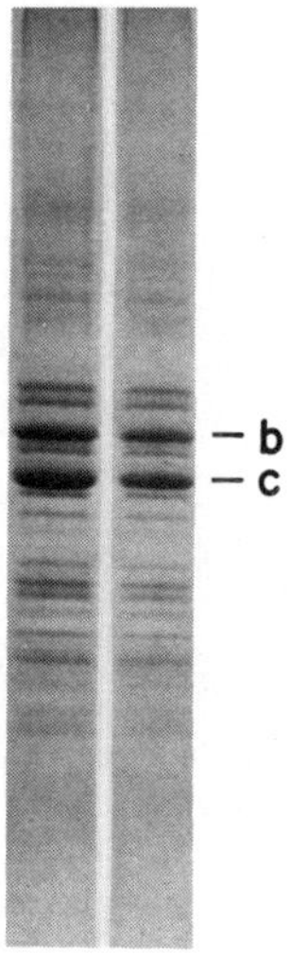

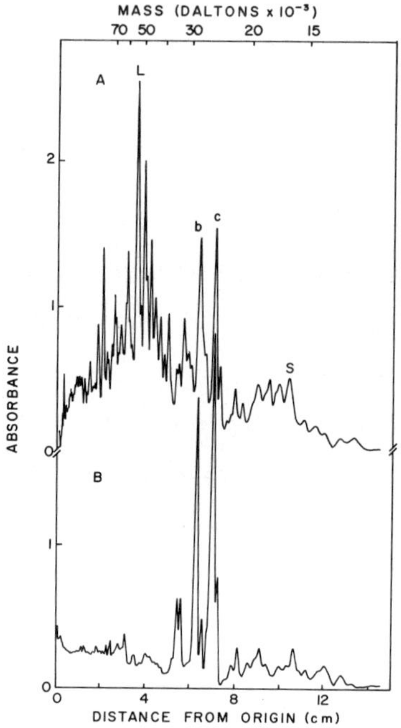

Fig. 1. (left) Pattern of protein stain after electrophoresis of polypeptides in purified thylakoid membranes from C. reinhardtii y-1. Membranes were prepared as before[3] and extracted with 90% acetone. Polypeptides were dissolved in Tris buffer (pH 8.4) containing 1.5% sodium dodecyl sulfate, treated with mercaptoethanol[3] and subjected to electrophoresis through a 10 to 20% gradient polyacrylamide slab gel according to Neville[9]. The separating gel (pH 8.7) contained 1 M urea. The reservoir buffer contained 5 mM thioglycolate.

Fig. 2. (right) Patterns of absorbance after electrophoresis of (A) total protein of light-grown C. reinhardtii y-1 cells and of (B) polypeptides in purified thylakoid membranes. Both samples were extracted with 90% acetone. The polypeptides were subjected to electrophoresis on 11% polyacrylamide gels as described before[3].

variability are not known. Polypeptide b was not affected by changes in the conditions of electrophoresis.

In membranes washed with 8 M urea, polypeptides b and c account for about 20 and 25%, respectively, of the total stain bound to protein after electrophoresis. Kan and Thornber[10] found that these polypeptides are subunits of the apoprotein of the light-harvesting, chlorophyll (chl) a/b-protein complex in C. reinhardtii. Nearly all the chl b in the membrane was recovered in this complex, which contained 3 molecules each of chl a and chl b per 1 polypeptide. As calculated from the ratio of protein to chl in whole membranes[11], about 4 moles of polypeptide b and 6 moles of polypeptide c are present per 30 moles of chl b. Therefore, to account for the observed ratio in the chl a/b-protein complex in Chlamydomonas, both polypeptides must participate. The reason for the apparently uniform subunit size of the complex[10], if polypeptides of different size are involved, must await further work.

Synthesis of Polypeptides b and c

Fig. 2A illustrates that polypeptides b and c can be detected after electrophoresis of total protein of green cells. The positions of these polypeptides in the pattern for total protein were determined by electrophoresis of membrane protein on companion gels (Fig. 2B). Peaks corresponding to polypeptides b and c were absent in patterns for total protein from etiolated cells, which correlates with the observation that etiolated cells have minimal amounts of intrachloroplast membranes[12].

Synthesis of polypeptides b and c can be measured readily by the incorporation of labeled amino acids. Fig. 3 shows patterns of radioactivity after electrophoresis of total protein from cells labeled with 3 different amino acids. The incorporation of [^{3}H]leucine and in particular of [^{3}H]phenylalanine resulted in prominent labeling of polypeptides b and c. In contrast, the incorporation of [^{3}H]arginine, the amino acid generally used for the assay of protein synthesis

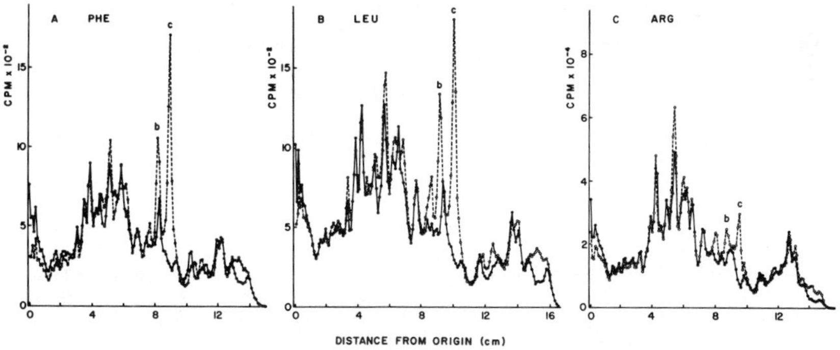

Fig. 3. Patterns of radioactivity after electrophoresis of total cellular protein of C. reinhardtii y-1 cells labeled with (A) L-[^{3}H]phenylalanine (16 Ci/mmole, 25 μCi/ml), (B) L-[^{3}H]leucine (55 Ci/mmole, 25 μCi/ml) or (C) L-[^{3}H]arginine (20 Ci/mmole, 5 μCi/ml). Conditions of incubation were as described before[3]. •—•, incorporation of amino acid in the dark; o---o, incorporation of amino acid in the light.

in Chlamydomonas, resulted in relatively minor peaks for these two polypeptides. To achieve adequate labeling of cellular protein with [^{3}H]leucine and [^{3}H]phenylalanine, for which apparently no facilitated mechanisms of transport are present on the plasma membrane[3], a relatively high concentration of amino acid in the medium was necessary. The incorporation also was markedly favored by increasing the temperature to 38°C.

An enrichment of phenylalanine and leucine in the membrane polypeptides is apparently the reason for the prominent labeling of polypeptides b and c. The relative amounts of these amino acids in purified polypeptides are shown in Table I. Phenylalanine is the amino acid most enriched in the polypeptides, whereas among all the amino acids, the ratio for arginine has the lowest value (complete data unpublished).

Table I. Ratios of Concentrations (residues/100 residues) of Amino Acids in Purified Polypeptides versus Total Protein.

Polypeptide	Phenylalanine	Leucine	Arginine
b	1.7	1.1	0.54
c	1.9	1.2	0.56

This assay for the synthesis of polypeptides b and c, in which total cellular protein is subjected to electrophoresis, is not dependent upon formation of membranes and allows measuring the full extent of synthesis of these polypeptides. Also, the presence of other, non-membrane polypeptides on the gels provides an internal control for monitoring the effects of various conditions on general protein synthesis. These advantages are particularly important in determining whether synthesis of the membrane polypeptides occurs in the dark. As shown in Fig. 3, no synthesis of polypeptides b and c could be detected in cells unexposed to light. Upon exposure to light, the wavelengths most effective in promoting synthesis of these polypeptides are the same as those effective in the photoreduction of protochlorophyll(ide)[3].

Conditions have been found that permit a rapid course of greening of C. reinhardtii y-1 upon exposure to light[3]. When etiolated cells are incubated for 1-2 hr in the dark at 37°-40°C, and then exposed to light at the same temperature, synthesis both of chl and of polypeptides b and c begins immediately and exhibits linear kinetics during a subsequent hour of illumination. The initial rate of greening at 38°C is greater than the maximal rate achieved at 25°C.

The absence of a lag in the synthesis of polypeptides b and c upon exposure to light suggested that mRNA for these polypeptides was transcribed during incubation in the dark at the elevated temperatures. This suggestion was further supported by the effects of actinomycin D, which did not inhibit the initial rate of synthesis of polypeptides b and c if added 10 min before light was given, but did inhibit strongly this synthesis if added at the beginning of the 2-hr preincubation period in the dark at 38°C[3]. Apparently a mechanism, which is influenced by the absorption of light by protochlorophyll(ide), operates at a post-transcriptional level to control production of these polypeptides.

Inhibitors of chloroplast 70S ribosomes have no effect on the synthesis of polypeptides b and c[1,3]. However, cycloheximide, an inhibitor of cytoplasmic 82S ribosomes, completely blocks synthesis of these polypeptides. The conclusion that these polypeptides are made on cytoplasmic ribosomes is supported by similar work on higher plants [13-15] and by the observations that isolated chloroplasts do not make the analogous polypeptides *in vitro*[16-18].

The available evidence indicates that these major thylakoid polypeptides of cytoplasmic origin are not needed for photosynthetic activity but may be required for the production of chl b. In several mutant strains of higher plants and algae[7,19,20], a deficiency of chl b is accompanied by a corresponding decrease in the amount of the major polypeptides. Lighting regimes that allow chloroplast development but no production of chl b[21,22] are unfavorable also to the synthesis of these polypeptides[3,23]. And the classes of algae that have chloroplasts surrounded by a chloroplast endoplasmic reticulum[24], a structure that should severely restrict polypeptides made by the nuclear-cytoplasmic system from entering the plastid, by nature lack chl b[25].

Incorporation of Polypeptides b and c into Thylakoid Membranes - a Hypothesis

Exposure of the inner faces of thylakoid membranes by freeze-fracture reveals a characteristic but asymmetric array of intramembrane particles[26-28]. The inner or E[29] half of the membrane exhibits a fracture face studded with large particles, whereas smaller particles are associated with the exposed face of the outer or P half of the membrane. After fractionation of the membrane, the large particles and the major polypeptides are recovered in fractions enriched in chl b and in photosystem 2 activity[30,31]. Since the large particles have been found only in membranes containing chl b and the associated apoprotein, Anderson[32] has concluded that these particles contain the chl a/b-protein complex. Polypeptides b and c, as the apoprotein subunits of this complex in *Chlamydomonas*[10], are major components of the membrane and so are appropriate candidates for the factors responsible for formation of the large particles. If the large particles are indeed markers of this complex, then the polypeptides are associated primarily with the E half of the membrane.

The task now is to determine how polypeptides synthesized on cytoplasmic ribosomes become positioned in the E or inner half of thylakoid membranes. Observations made during electron microscopic studies of developing chloroplasts[33-35] seemed to indicate that the inner membrane of the chloroplast envelope invaginated and formed vesicles, which then fused with growing thylakoids. These findings suggested that the envelope inner membrane may be a site of association of thylakoid membrane components. In support of this suggestion, the envelope inner membrane is apparently the site of synthesis of galactolipids[36], the major lipid components of thylakoid membranes[37,38].

It is conceivable that the major thylakoid polypeptides, of cytoplasmic origin, cross the outer membrane of the chloroplast envelope during synthesis by the process extensively documented for polypeptides synthesized on membrane-bound ribosomes[39]. Interaction of the polypeptides with components of the inner membrane then may lead to the production of chl b and generation of the chl a/b-protein complex at the E surface of this membrane, which faces the cytoplasm. Subsequently, this complex would become positioned with the E half

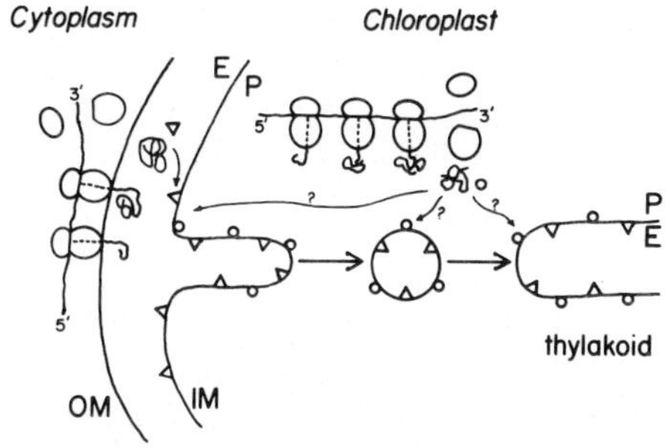

Fig. 4. Scheme depicting the association of the apoproteins of the chl-protein complexes with thylakoid membranes in green algae and higher plants, as proposed in the text. The chl a/b complex apoproteins (Δ) are released, after synthesis on cytoplasmic ribosomes, into the space between the chloroplast outer (OM) and inner (IM) envelope membranes, and then associate with chl in the E half of the inner membrane. The production of vesicles, which fuse to form thylakoids, leads to the positioning of the chl a/b-protein complex within the E half of the thylakoid membrane. After synthesis on chloroplast ribosomes, the apoprotein of the chl a complex (o) associates with chl in the P half of the membrane. The positions of the symbols (Δ and o) are not intended to imply that the complexes project from the membrane surfaces. Rather, they only indicate the half of the membrane with which these complexes are primarily associated, as determined by freeze-fracture studies.

of thylakoid membranes as a consequence of the fusion of vesicles derived from the inner envelope membrane (Fig. 4).

The smaller particles, 80-100 Å in diameter, on the P fracture face of thylakoid membranes apparently contain the chl a-protein complex associated with photosystem 1[30]. Results of several studies suggest that the apoprotein of this complex in higher plants is synthesized on chloroplast ribosomes[13-15]. It seems reasonable that the association of this protein with the P half of the membrane is determined by its approach to that surface after synthesis (Fig. 4).

Since chl synthesis is necessary for association of the apoproteins of the chl a/b- and the chl a-protein complexes with the membrane, chl a may be the receptor in the membrane for both proteins. The asymmetrical arrangement of the complexes within the membrane, therefore, may be a consequence of the direction from which the polypeptides approach the membrane. In this manner, transport of polypeptides across one or more membranes is not required before they become properly oriented, integral components of the thylakoid membrane structure.

Supported by National Science Foundation Grant No. BMS-71-01550.

References

1. Hoober, J.K. (1970) J. Biol. Chem. 245: 4327-4334.
2. Eytan, G., and I. Ohad (1970) J. Biol. Chem. 245: 4297-4307.
3. Hoober, J.K., and W.J. Stegeman (1976) J. Cell Biol. in press.
4. Ohad, I., and G. Drews (1974) Proc. Third Internat. Cong. Photosyn. (M. Avron, editor), Elsevier, Amsterdam. 1907-1912.
5. Rebeiz, C.A., J.R. Mattheis, B.B. Smith, C.C. Rebeiz, and D.F. Dayton (1975) Arch. Biochem. Biophys. 171: 549-567.
6. Griffiths, T. (1975) Biochem. J. 152: 623-635.
7. Levine, R.P., and H.A. Duram (1973) Biochim. Biophys. Acta 325: 565-572.
8. Chua,N.-H., and P. Bennoun (1975) Proc. Nat. Acad. Sci. U.S.A. 72:2175-2179.
9. Neville, D.M., Jr. (1971) J. Biol. Chem. 246: 6328-6334.
10. Kan, K.-S., and J.P. Thornber (1976) Plant Physiol. 57: 47-52.
11. Hoober, J.K. and W.J. Stegeman (1975) In Genetics and Biogenesis of Chloroplasts and Mitochondria (C.W. Birky, Jr., T.J. Byers and P.S. Perlman, editors), The Ohio State Univ. Press, Columbus, Ohio. 225-251.
12. Ohad, I., P. Siekevitz and G.E. Palade (1967) J. Cell Biol. 35: 553-584.
13. Machold, O., and O. Aurich (1972) Biochim. Biophys. Acta 281: 103-112.
14. Ellis, R.J. (1975) Phytochemistry 14: 89-93.
15. Cashmore, A.R. (1976) J. Biol. Chem. 251: 2848-2853.
16. Eaglesham, A.R.J., and R.J. Ellis (1974) Biochim. Biophys. Acta 335: 396-407.
17. Bottomley, W., D. Spencer and P.R. Whitfeld (1974) Arch. Biochem. Biophys. 164: 106-117.
18. Morgenthaler, J.-J. and L. Mendiola-Morgenthaler (1976) Arch. Biochem. Biophys. 172: 51-58.
19. Thornber, J.P., and H.R. Highkin (1974) Eur. J. Biochem. 41: 109-116.
20. Anderson, J.M., and R.P. Levine (1974) Biochim. Biophys. Acta 333: 378-387.
21. Argyroudi-Akoyunoglou, J.H., Z. Feleki and G. Akoyunoglou (1971) Biochem. Biophys. Res. Commun. 45: 606-614.
22. DeGreef, J., W.L. Butler and T.F. Roth (1971) Plant Physiol. 47: 457-464.
23. Hiller, R.G., D. Pilger and S. Genge (1973) Plant Sci. Lett. 1: 81-88.
24. Gibbs, S. (1970) Ann. New York Acad. Sci. 175: 454-473.
25. Strain, H.H., B.T. Cope and W.A. Svec. (1971) Meth. Enzymol. 23: 452-476.
26. Branton, D., and R.B. Park (1967) J. Ultrastruct. Res. 19: 228-303
27. Goodenough, U.W., and L.A. Staehelin (1971) J. Cell Biol. 48: 594-619.
28. Staehelin, L.A. (1975) Biochim. Biophys. Acta 408: 1-11.
29. Branton, D., S. Bullivant, N.B. Gilula, M.J. Karnovsky, H. Moor, K. Mühlethaler, D.H. Northcote, L. Packer, B. Satir, P. Satir, V. Speth, L.A. Staehelin, R.L. Steere and R. Weinstein (1975) Science 190: 54-56.
30. Arntzen, C.J., R.A. Dilley and F.L. Crane (1969) J. Cell Biol. 43: 16-31.
31. Levine, R.P., W.G. Burton and H.A. Duram (1972) Nature New Biol. 237: 176-177.
32. Anderson, J.M. (1975) Biochim. Biophys. Acta 416: 191-235.
33. von Wettstein, D. (1959) Brookhaven Symp. Biol. 11: 138-157.
34. Mühlethaler, K., and A. Frey-Wyssling (1959) J. Biochem. Biophys. Cytol. 6: 507-512.
35. Menke, W. (1962) Ann. Rev. Plant Physiol. 13: 27-44.
36. Douce, R. (1974) Science 183: 852-853.
37. Goldberg, I., and I. Ohad (1970) J. Cell Biol. 44: 563-571.
38. Leese, B.M., and R.M. Leech (1976) Plant Physiol. 57: 789-794.
39. Blobel, G., and B. Dobberstein (1975) J. Cell Biol. 67: 835-851.

Genetics and Biogenesis of Chloroplasts and Mitochondria, Th. Bücher et al. eds.

RELATIONSHIP BETWEEN CHLOROPLASTIC METABOLISM AND CYTOPLASMIC TRANSLATION

Gérard Ledoigt et Marcelle Lefort-Tran

Cytophysiologie de la Photosynthèse
C.N.R.S. 91190 Gif sur Yvette
France

INTRODUCTION

RNA and protein synthesis in different cell compartments, their type of regulation and their sequential appearance are analyzed during organelle biogenesis, in particular that of chloroplasts.

METHODOLOGY

Euglena gracilis, strain Z are grown in the dark on lactate medium (1) and then transferred on a resting medium according to Stern et al. (2).

Protein synthesis inhibitors acting on 68 S plastid ribosomes : streptomycin (Specia), rifampin (Calbiochem), spectinomycin and clindamycin (Upjohn) are added at the same time as tritiated uracil (U^3 H-5, 24 Ci/mM) during the greening process. The zero time is defined as the time of light exposure (1 500 lux).

The RNAs are extracted from cells according Heizmann (3) and analyzed by polyacrylamide gel electrophoresis as described by Loening (4). Chlorophylls are determined by the Mac Kinney method (5).

RESULTS

1 - Synthesis : Labelling of control cell RNA during greening (t 0, t 6, t 24) and analysis of specific radioactivity, at t 72 hours, of chloroplastic (16 S) and cytoplasmic (20 S) ribosomal RNA are given in table 1.

It is shown that the cytoribosomal RNA specific radioactivity is 8 to 9 times lower with 2 μCi/ml than with 5 μCi/ml whereas for the total RNA, a value 13 to 14 times, lower is obtained when 2 μCi are given. From these results ribosomal RNA types seems to be preferentially synthesized.

It appears that a preferential synthesis of the cytoribosomal RNAs occurs during the first 6 hours of greening, while chloroplastic ribosomal RNAs are synthesized later, between 6 and 24 h.

A competition for labelling appears between these two types of r RNA since the specific radioactivity is higher for chloroplastic r RNA at t 6 than at t 0, whereas the opposite is noted for the cytoplasmic r RNA in 5 μci labelling conditions. An increase in the cytoplasmic r RNA (20 S) specific radioactivity is observed in the presence of sm compared to the control, namely the r RNA radioactivity between 6 and 72 h increases by 2.5 while this value doubles between 24 and 72 h.

During these periods, the 16 S RNA radioactivity is essentially constant, suggesting that sm blocks preferentially the incorporation in chloroplastic r RNA, involving a competition between cytoplasmic and chloroplastic RNA syntheses.

Since in the absence or presence of spectinomycin (table 1) the specific radioactivity is essentially stable, one can infer that the lower concentration of radioactivity (2 µCi/ml) is not suffisant to supply the ribosomal RNA pool.

TABLE 1

Specific radioactivity of Euglena RNA (c.p.m./1 OD 260 nm)

Samples		t 0 → t 72		t 6 → t 72		t 24 → t 72	
c.p.m. x 10^3		2 µCi	5 µCi	2 µCi	5 µCi	2 µCi	5 µCi
Total RNA		13.5	180	9.0	118	5.7	82
Control	16 S	1.44	3.5	0.92	7.8	0.23	1.7
	20 S	1.07	9	0.85	6.5	0.16	1.3
sm	16 S		5.5		5.5		1.85
	20 S		11.6		14.8		2.6
spc	16 S	0.25		0.78		0.59	
	20 S	0.4		0.80		0.21	

- Tritiated uracil (U^3 H-5, 24 Ci/mM) was added to cell cultures at two concentrations, 2 µCi/ml or 5 µCi/ml, at various greening times 0, 6 or 24 h. RNAs were separated and measured from 72 h light-grown cells. Data are given from the most stable ribosomal RNA which are 16 S for chloroplast and 20 S for cytoplasm. Antibiotics, streptomycin (sm) or spectinomycin (spc) are added at the same time as the precursor.

2 - Turn-over : The cytoplasmic r RNA labelling ratio 25 S / 20 S increase by a factor of 3 to 4 between t 0 to t 24 (table 2) while this ratio measured by optical density at 260 nm is always close to 1. Thus, one can assume a preferential labelling of RNA 25 S corresponding to a turn-over. At the same time a degradation of cytoribosomal RNA is noted which begins around of the sixth hour of

greening and becomes important at the 24th hour.

The antibiotics (streptomycin and spectinomycin) (table 2) decrease the ratio of 25 S / 20 S when added with tritiated uracil. In resting medium the partial inhibition by chloroplastic synthesis inhibitors on turn-over 25 S suggests a regulation system between chloroplastic and cytoplasmic metabolisms.

TABLE 2

Streptomycin (sm) and spectinomycin (spc) actions during the greening process on the radioactivity ratio of cytoplasmic ribosomal RNA $\frac{25\ S}{20\ S}$.

Samples / $\frac{25\ S}{20\ S}$ in R.A.	t 0 → t 72	t 6 → t 72	t 24 → t 72
Control	1	2	3.9
sm	1.2	1.4	2.5
Control	0.8	1.5	2.3
spc	1.1	1.3	1.8

<u>3 - Compartment relations</u> : During greening (0 - 72 h), the action of spectinomycin, clindamycin and streptomycin at different concentrations on the 16 S/20 S ratio which is a measure of chloroplast RNA synthesis, on the chlorophyll content and on the plastid structures is shown in table 3. At lower concentrations (4.10^{-5}) the chloroplast RNA synthesis is slightly reduced and chlorophyll synthesis is increases. The plastid structures are nearly the same as the control (in the case of spectinomycin) or only disturbed in the organization of thylakoïds (in the case of clindamycin). At higher concentrations (4 and 9.10^{-4}M), chlorophyll synthesis could be completely inhibited except for the phototransformed part, while chloroplastic RNA synthesis is only partially inhibited. Streptomycin ($1.4.10^{-3}$M) added at the greening inhibits completely all types of chloroplast synthesis.

In the presence of sm and / or high concentration of clindamycin, organelle structures are of the proplastid type but only with clindamycin 50 % of r RNA synthesis occurs.

On the other hand high concentration of spectinomycin and intermediate concentration of clindamycin allow subnormal chloroplast development with discrete modifications concerning the absence of stacking of thylakoïds (pseudograna).

These results suggest that lamellar proteins, chlorophylls and chlororibosomal RNA syntheses can be uncoupled.

TABLE 3

Antibiotics action (spectinomycin : spc ; clindamycin : clind ; streptomycin : sm) on the chloroplast r RNA synthesis ($\frac{16\ S}{20\ S}$ ratio in OD 260 nm), chlorophyll synthesis and plastid structures, during the greening of *Euglena*, grown on resting medium. The values are given in % of a control sample after 72 h light.

Antibiotics	Low concentrations			High concentrations		
	spc (4.10^{-5}M)	clind (4.10^{-5}M)	spc (9.10^{-4}M)	clind (4.10^{-4}M)	(10^{-3}M)	sm ($1.4.10^{-3}$M)
$\frac{16\ S}{20\ S}$	90 %	90 %	40 %	70 %	50 %	∿ 0
Chlorophyll	120 %	110 %	10 %	12 %	∿ 0	5 %
Plastid structures	normal	abnormal lamellar organi-zation	undiffe-renciated plastids	plastids with some indepen-dant thylakoïds	proplasts	proplasts

DISCUSSION AND CONCLUSION

The experimental conditions and parameters show, during *Euglena* greening on resting medium, a sequential appearance of r RNA syntheses, cytoribosomal degradation and turn-over of 25 S r RNA.

By using certain chloroplast inhibitors (rifampin, spectinomycin, streptomycin) it is suggested that this turn-over is regulated by the plastid genome.

The differential sensibility to the inhibitors of various chloroplast macromolecular components synthesis involves some coupling factors between chloroplastic and cytoplasmic syntheses for the plastid differentiation which could be the translated on different ribosome types.

In the chloroplasts, this functional segregation can be correlated with the localization of ribosomal particles in the stroma and on the thylakoïds.

REFERENCES

1. Calvayrac, R. & Douce, R. (1970) FEBS Letters, 7, 259.
2. Stern et al., (1964) Plant Physiol., 39, 220.
3. Heizmann, P. (1970) Biochim. Biophys. Acta, 224, 144.
4. Loening, V.E. (1967) Biochem. J., 102, 251.
5. Mac Kinney (1942) J. Biol. Chem., 140, 315.

- This work was supported by C.N.R.S. and a grant from C.E.A.

Genetics and Biogenesis of Chloroplasts and Mitochondria, Th. Bücher et al. eds.

TEMPERATURE-SENSITIVITY OF CHLOROPLAST RIBOSOME FORMATION IN HIGHER PLANTS

Jürgen Feierabend
Ruhr-Universität Bochum, Arbeitsgruppe Pflanzliche Zellphysiologie, Postfach 2148, 4630 Bochum, Germany

Systems which have been mainly used for studying the role of cytoplasmic and plastidic protein synthesis during the biogenesis of chloroplasts are the use of specific antibiotics, in vitro protein synthesis of isolated chloroplasts with subsequent identification of the translation products, and 70S ribosome-deficient mutants.[1] With regard to investigations on the mutual regulation and long-term cooperation of the two sites of protein synthesis antibiotics have, especially in higher plants, the disadvantage of unspecific and toxic side effects, and the capacity of isolated chloroplasts may be rather restricted because of the lack of products or signals needed from the cytoplasm. In higher plants we have observed and characterized a temperature-sensitivity of chloroplast ribosome formation which offers a further convenient and obviously rather specific means for studying the role of the cytoplasm in the biosynthesis and transfer of chloroplast proteins.

In young leaves of rye seedlings the formation of chloroplast 70S ribosomes was specifically prevented when the plants were grown at an elevated temperature of 32° instead of 22° C which was used for normal development.[2] Accordingly, also the 23S and 16S rRNAs of the chloroplasts were no longer formed at the higher temperature (Fig. 1).

We have shown that in rye leaves the high-temperature effect is highly specific, since no major defects were observed in other organelles, or in other biosynthetic and metabolic activities, except for those of the chloroplasts. At appropriate stages rye leaves were of almost normal size and reached as high contents of total amino nitrogen compounds at 32° as at 22°. Cytoplasmic constituents, such as 80S ribosomes, activities of enzymes predominantly localized in the cytoplasm, or the contents of sterols reached amounts at 32° which were comparable to or even higher than at 22° [2,3]. Mitochondria were ultrastructurally unchanged at the higher temperature and mitochondrial marker enzymes reached equal

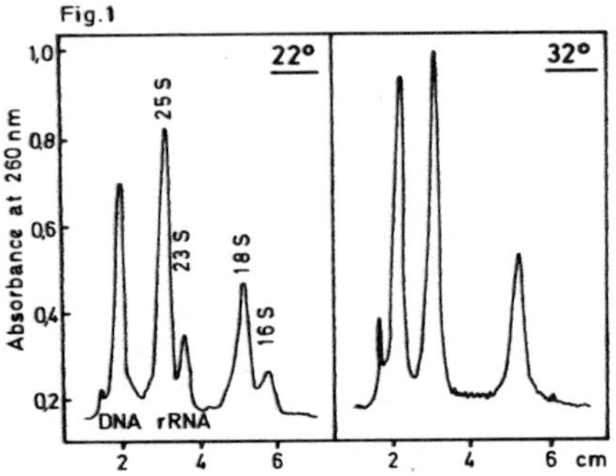

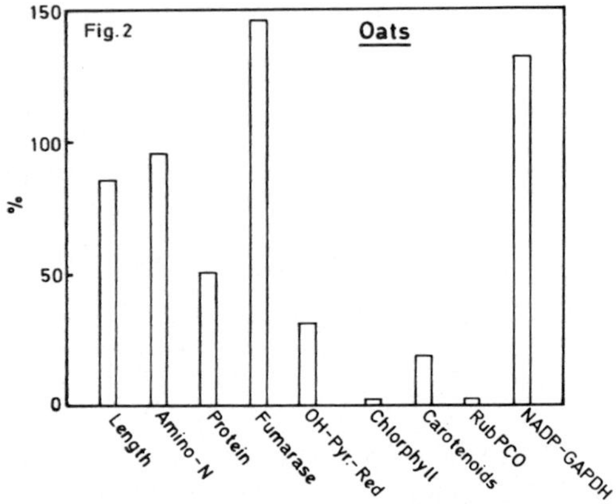

Fig.1. Electrophoretic separation of nucleic acids from green 22°-grown leaves and chlorotic parts of 32°-light-grown leaves of 6-day-old rye seedlings on polyacrylamide gels.

Fig.2. Growth parameters (leaf length, total amino-nitrogen and total protein contents), pigments and activities of several marker enzymes in the first leaves of 9-day-old oat seedlings grown at 32° in light, as % of corresponding measurements in control plants grown at 22°. OH-Pyruv.-Red.: OH-pyruvate reductase: RubPCO: ribulose-1,5-bisphosphate carboxylase; NADP-GAPDH: NADP-glyceraldehyde-3-phosphate dehydrogenase.

(cytochrome c oxidase) or even higher (fumarase) activities at 32° than at 22°. Similiar respiration rates were measured in 32°- and 22°-grown leaf tissues. As a major modification, only the 32°-grown tissue possessed the cyanide-resistant pathway of respiration in addition to the cyanide-sensitive respiration.[2] Even leaf peroxisomes which are thought to be metabolically dependent on functional chloroplasts were formed in the cells with the defective ribosome-less plastids. Peroxisomes were frequently seen on electron micrographs. Their marker enzymes reached high activities in the 32°-grown rye leaves, and a typical peroxisome fraction was isolated on sucrose gradients.[2]

As may be expected, major but specific defects occurred in the biosynthesis of the chloroplasts when their ribosome formation was prevented at the higher temperature. The rye leaves became chlorotic, since chlorophyll accumulation was prevented at 32°. The chlorotic leaf cells still contained rudimentary plastid structures which must have been synthesized on cytoplasmic ribosomes. They were much smaller than normal chloroplasts and had a proplastid-like appearance. Such plastids were surrounded by an envelope of typical double membrane structure and contained some vesicles or tubule-like structures and groups of osmiophilic globules. Grana and major thylakoid structures, however, were absent. Though the major part of carotenoids which is normally associated with the thylakoid membranes of the chloroplasts was missing, such plastids still contained appreciable amounts of carotenoids as their main pigments.[3]

It is now a well documented notion that the large subunit of ribulose-1,5-bisphosphate carboxylase, or fraction-I protein, is synthesized on chloroplast ribosomes.[1] Accordingly, this enzyme was not formed in the 70S ribosome-deficient leaf tissue at 32°. Consequently, no significant photosynthetic CO_2-fixation was observed in the chlorotic leaf tissue. However, at least a dozen other soluble chloroplast enzymes, or chloroplast-specific isoenzymes which have been so far investigated, reached 80-100 % of the activities present in normal 22°- grown leaves also in the 70S ribosome-deficient leaves at 32° and must be of cytoplasmic origin.

A similar high-temperature-sensitivity as in rye leaves has now been found to be quite common among higher plants. At growth temperatures between 32° and 34°, also in oats, wheat, barley and peas the formation of chloroplast ribosomes, as estimated from their rRNAs, was specifically prevented and chlorotic leaves were formed. Growth of the cereal leaves (length, expansion, amino-nitrogen content) was fairly unaffected by the higher temperature, while that of the pea leaves suffered from a general retardation. Typically, as shown in Fig. 2 for oats and similar to the behavior of rye, chlorophyll and ribulose-1,5-bisphosphate carboxylase activity were virtually absent in 32°-grown leaves. The carotenoid contents were still 20-40 % of the amounts found in normal green leaves. The activity of NADP-glyceraldehyde-3-phosphate dehydrogenase, an example for an enzyme of cytoplasmic origin, was always similar or even higher at the elevated temperature, as compared to 22°. Thus, a high-temperature-induced deficiency of 70S ribosomes may be used in many plant species as a specific and readily available tool for studying several questions of chloroplast biogenesis.

Major aspects which we have so far followed by the use of this method in rye leaves are:

1. Evaluation of the site of synthesis of chloroplast proteins.[2] In contrast to the use of antibiotics this procedure is non-toxic and does not interfere with mitochondrial development.

2. Transfer of chloroplast proteins. By appropriate gradient centrifugations ribosome-deficient plastids were isolated from 32°-

grown chlorotic rye leaves. They contained high yields of several soluble chloroplast enzymes. From this we conclude that the cytoplasm must contribute the mechanism for the discrimination and translocation of proteins formed at 80S ribosomes into the plastids[2].

3. Coordination of cytoplasmic and plastidic protein synthesis. With specific antibodies, prepared against ribulose-1,5-bisphosphate carboxylase (fraction-I) and its small subunit, we obtained evidence that the protein of the small subunit of this enzyme was formed in the absence of the large subunit in 32°-grown rye leaves (Fig. 3).

Fig.3. Sodium dodecylsulfate-polyacrylamide electrophoresis of the protein eluted from a column with immobilized antibody against ribulosebisphosphate carboxylase (according to Gray and Wildman[4]). The protein had been absorbed from crude extracts of either 22°-, or 32°-light-grown rye leaves. SDS-dissociated purified fraction-I protein was run as a reference. Location of large (LS) and small suunit (SS).

Acknowledgements

I thank Dr. G. Wildner and Dr. R. Berzborn for the preparation of the antisera. Parts of the unpublished results included in this article were obtained under the cooperation of M. Mikus and H.-A. Schäfers. Support by the Deutsche Forschungsgemeinschaft is appreciated.

References

1. Ellis, R.J., Blair, G.G., Hartley, M.R. (1973) Biochem. Soc. Symp. 38, 137-162.
2. Feierabend, J., Schrader-Reichhardt, U. (1976) Planta (Berl.) 129, 133-145.
3. Rademacher, E., Feierabend, J. (1976) Planta (Berl.) 129,147-153
4. Gray, J.C., Wildman, S.G. (1976) Plant Sci. Lett. 91-96.

Genetics and Biogenesis of Chloroplasts and Mitochondria, Th. Bücher et al. eds.

TEMPERATURE CONTROL OF CHLOROPLAST DEVELOPMENT

Robert M. Smillie
Plant Physiology Unit, CSIRO Division of Food Research, and School of Biological Sciences, Macquarie University, North Ryde 2113, Sydney, Australia.

SUMMARY

Environmental regulation by temperature of chloroplast membrane assembly has been studied in barley. It is shown how chloroplast membrane assembly can be utilized as a model system to indicate how plant growth and development may be regulated through temperature-induced changes in the properties of membrane systems. A similar study made with cucumber, a plant sensitive to chilling temperatures (0° to around $12^{o}C$), has led to a hypothesis to account for the contrasting effects of chilling temperatures on developmental processes in chilling-sensitive and chilling-resistant plants.

INTRODUCTION

The assembly of chloroplast and mitochondrial membranes requires the interaction of nuclear, cytoplasmic and organelle-localized systems. In spite of the complexities involved in this process our knowledge of it has advanced to the point where it is possible to consider how it may be regulated by physiological and environmental factors. It is already evident that in some plants the joint expression of nuclear and chloroplast genes to produce chloroplast lamellar systems is subject to control by at least one environmental factor, namely light, since the duration and intensity of light can alter not only the rate of formation of chloroplast lamellar membranes but also their final ultrastructure and composition and, as a consequence of this, photochemical activity[1,2]. Studies with a mutant barley, *xantha-b*^{12}, suggest that temperature too can exert an influence on the final composition of chloroplast membranes and their photochemical capability[1]. However, there is little information available about control of chloroplast development by temperature or the mechanisms involved. In this paper studies on chloroplast development in barley at different temperatures are described and comparisons are made with chloroplast development in a chilling-sensitive plant, cucumber, whose growth compared with barley is severely restricted by low temperatures.

MATERIALS AND METHODS

Plant material: Barley (Hordeum vulgare L. cv. Abyssinian) and cucumber (Cucumis sativus L. var. Palomar) were used.

Chlorophyll production at different temperatures: Seeds of barley or cucumber were germinated for 5 days at 23° to 25°C under far-red light (>700 nm) to promote etioplast development. The upper 4-cm of the primary leaf of barley seedlings or pairs of cucumber cotyledons were transferred to compartments in an aluminium block along which a temperature gradient was maintained by heating one end and cooling the other. The plant tissues were illuminated with white light. Chlorophyll production was measured spectrophotometrically after extracting the tissues in 80% acetone. The design of the block allowed for up to ten temperatures and in order to adequately cover the physiological temperature range for chloroplast development (i.e. from 0° to about 40°C), four separate experiments incorporating different temperature ranges were necessary.

Growth experiments: The temperature gradient block was also utilized for measurements of the elongation of the primary leaf of barley and cucumber hypocotyl. Seedlings were raised in the dark at 22°C until the primary leaf (barley) or hypocotyl (cucumber) was approximately 1 cm long and were then transferred to the block. Growth in the dark at different temperatures was then measured.

The isolation of barley chloroplasts and measurements of Hill activity were carried out as described by Nolan and Smillie[3].

RESULTS AND DISCUSSION

The effect of temperature on chloroplast development: Fig. 1 shows chloroplast development at different temperatures in barley. The logarithm of the chlorophyll produced is plotted against temperature. The resulting graph is very similar to an Arrhenius plot (i.e. as a function of decreasing 1/T where T is the absolute temperature). The rate of chlorophyll increase with increasing temperature was not constant and changes occurred at 10°, 19°, 28° and 36°C. The plot was almost linear below 10°C and again above 10°C, but the slopes differed. This change in temperature coefficient has been observed repeatedly starting with either completely etiolated or partially greened leaves. A second pronounced change was found around 19°C. Above this temperature the slope decreased again but the plot remained almost linear with increasing temperature. A third change at 28°C was slight but observed consistently. Above 36°C chlorophyll production rapidly declined.

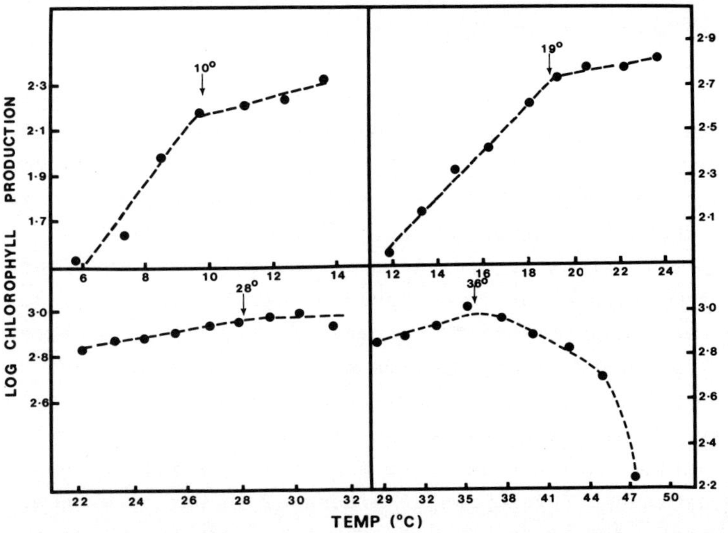

Fig. 1. The effect of temperature on chloroplast development (ng chlorophyll h^{-1}.primary leaf $section^{-1}$) in barley.

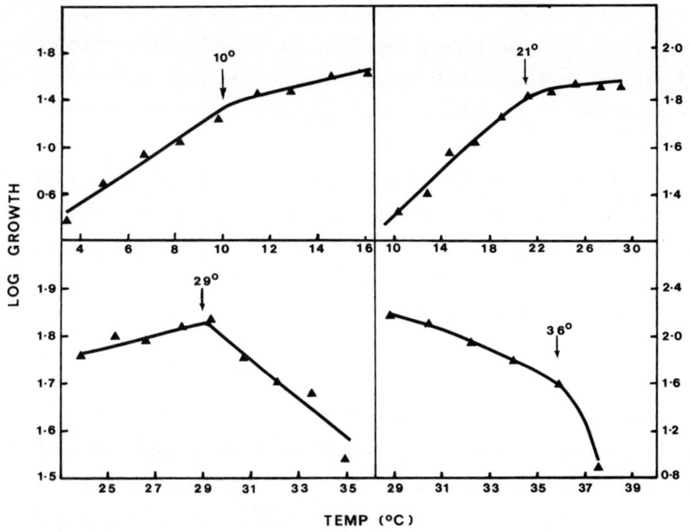

Fig. 2. The effect of temperature on elongation (mm x $10^{-2}.h^{-1}$) of the primary leaf of barley.

Correlation with leaf growth: In order to determine if the temperature effects at 10°, 19°, 28° and 36°C were unique to chloroplasts in barley leaves, elongation of the etiolated primary leaf at different temperatures was measured. Fig. 2 shows that the temperature coefficient for leaf growth changed at about these same temperatures. The plot of the logarithm of elongation rate against temperature was almost linear below 10°C and also between 10° and 21°C, 21°-29° and 29°-36°C. There may have been a very gradual change in temperature coefficient within these temperature ranges, but major changes occurred at the indicated temperatures.

On a practical note, the close correspondence between effects of specific temperatures on chlorophyll production and leaf elongation suggests that measurement of the rate of chlorophyll synthesis at different temperatures may provide a relatively simple and rapid method for predicting temperature-growth responses of young plants.

Temperature-induced changes in isolated barley chloroplasts: The observation that the same temperatures induce like changes in two non-interdependent developmental processes in barley suggest that these temperature effects operate through a mechanism common to both processes. Temperature-induced changes in membrane properties, attributed to phase changes in the membrane lipids, have been described for mitochondrial and chloroplast membranes of chilling-sensitive plants[4,5]. Although barley is a chilling-resistant plant, it is nevertheless conceivable that a temperature-induced change, such as in the packing or structural orientation of plastid membrane lipids or proteins, could be the primary effector mechanism responsible for the observed effects of temperature on chlorophyll synthesis. As in the case of mitochondria from chilling-sensitive plants[6], such changes might be monitored in isolated barley chloroplasts by temperature-induced changes in the Arrhenius activation energy (Ea) for membrane-catalyzed electron transport reactions. Dr. W.G. Nolan and I have measured Hill activity of isolated barley chloroplasts at different temperatures and found changes in Ea at 9°, 20° and 29°C (Fig. 3) i.e. at around the same temperatures causing changes in chloroplast development and leaf growth described above. Above 36°C irreversible thermal inactivation occurred. Assays were carried out under conditions yielding high rates of electron flow otherwise, as shown in the right-hand side of Fig. 3, the changes at 9°, 20° and 29°C may be missed. High flow rates were achieved by using a photophosphorylation uncoupler, by prior exposure of the barley chloroplasts to a low

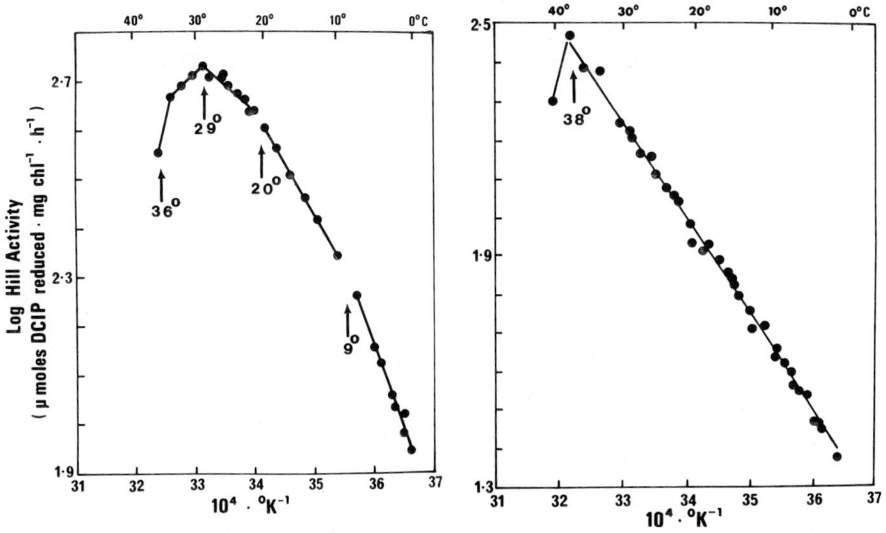

Fig. 3. Arrhenius plot of photoreduction of 2,6-dichlorophenol-indophenol by barley chloroplasts. Left, 74 mM methylamine. Right, no methylamine. Data from Nolan and Smillie[3].

concentration of cations, or by using a lipophilic electron acceptor[3]. Yamamoto and Nishimura[7] have recently reported changes at 8° and 20°C in Ea for proton transport by illuminated spinach chloroplasts.

One explanation for these results is that at temperatures around 9°, 20° and 29°C alterations occur in the microenvironment of plastid membranes, or in the structure of membrane components, which result in a change in the properties of the membrane system. Not only is membrane function in mature chloroplasts affected by these changes, but so is membrane assembly in differentiating chloroplasts. The results given in Fig. 2 suggest that other cellular membranes undergo changes in their properties at these temperatures.

If this explanation is correct then perhaps its most important implication is that such temperature-induced changes in membrane properties provide a direct mechanism by which cellular growth can be regulated by the temperature of the environment.

A study of chloroplast development in a plant which displays a marked difference compared with barley in its growth response to temperature might provide additional support for this hypothesis. For this reason the effect of temperature on chloroplast development was also studied in cucumber, a chilling-sensitive plant.

The effect of temperature on chloroplast development in a chilling-sensitive plant: Most plants indigenous to lowland tropical and sub-tropical climates show little or no growth at low temperatures above freezing, commonly between 0° and $12^{o}C$. Prolonged exposure to such temperatures produces irreversible tissue damage (chilling injury).

Fig. 4 shows the results of two separate experiments to determine the effect of low and high temperatures on chloroplast development in cucumber cotyledons. The results at low temperatures were markedly different from those for barley since there was a definite low cut-off temperature ($9^{o}C$) for net synthesis of chlorophyll in cucumber. Between 9° and $20^{o}C$ chlorophyll production increased rapidly with increasing temperature.

Hypothesis to account for the restricted growth and development of chilling-sensitive dicotyledonous plants at chilling temperatures: A plausible explanation for the difference in rates of growth of chilling-sensitive and chilling-resistant plants at chilling temperatures and the subsequent development of chilling injury in the former group has been advanced by Lyons[4] and Raison[5]. They suggested that a reversible temperature-induced phase change in membrane lipids occurs at chilling temperatures in the membranes of chilling-sensitive plants. This change alters the properties of the membrane and in particular of membrane-associated enzymes, and can be detected by a decrease in fluidity of chloroplast and mitochondrial membranes[5] and by a decrease in Ea of membrane-associated mitochondrial oxidative reactions[6] at chilling temperatures. The

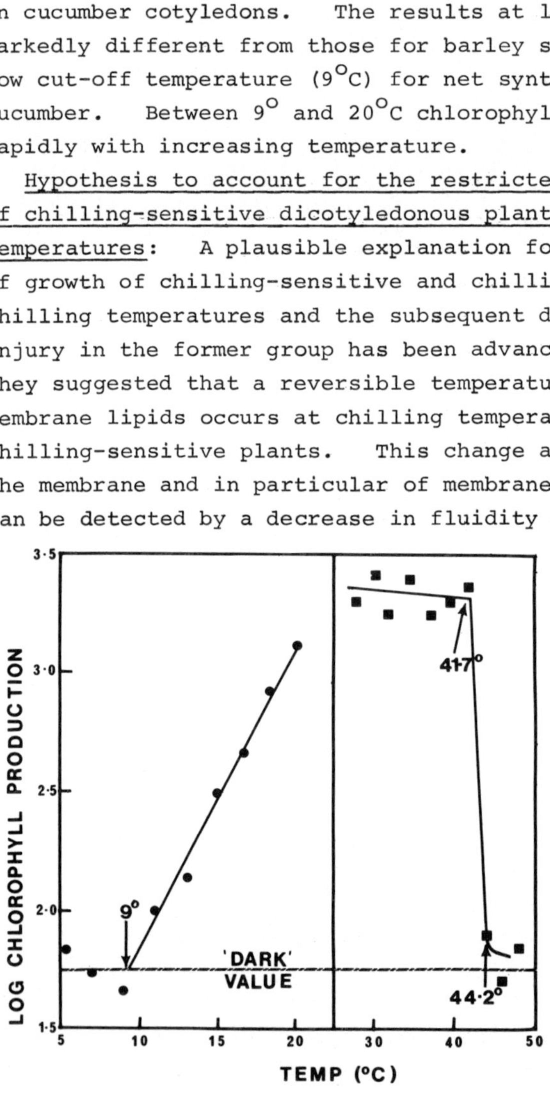

Fig. 4. The effect of temperature on chloroplast development (ng chlorophyll·h^{-1}· cotyledon) in cucumber.

TABLE 1

Q_{10} FOR CHLOROPLAST DEVELOPMENT AND HYPOCOTYL GROWTH IN CUCUMBER

	Q_{10}		
Temperature range	<9°	9° to 22°	>22°
Chlorophyll production	-	18	1.6
Hypocotyl elongation	-	6.3	1.6

depressed growth at chilling temperatures was correlated with the temperature-induced change in structure and function of cellular membranes. Subsequent development of symptoms of chilling injury were envisaged as the outcome of an imbalanced metabolism at chilling temperatures, notably of membrane-associated enzymic pathways relative to soluble enzymic systems.

The data shown in Fig. 3 suggest an alternative explanation for the inability of cucumber to develop at chilling temperatures. The two experiments shown in Fig. 3 indicate there must be at least one major change in slope for chlorophyll synthesis between 20° and 30°C. Other experiments disclosed a major change around 22°C and that further, as in the case of barley, there was a coincidence of the effects of temperature on chloroplast development and seedling growth (Table 1). Below 9°C only a very slight growth rate could be detected. The increase in slope below 22°C of the logarithmic plot either of growth or chlorophyll synthesis (cf. Fig. 3) against temperature is of such a magnitude that a projection of the plot shows that growth and development must almost cease at some temperature well above 0°C. If the change to a high Q_{10} is brought about by a temperature-induced change around 22°C, then it is this change and not a change occurring only at chilling temperatures, that largely determines if reasonable growth rates are still possible at chilling temperatures. By analogy with barley in which temperature-induced changes around 20°C in both chloroplast membrane function and assembly were demonstrated, the changes brought about in cucumber by temperatures below 22°C probably also result from a temperature-induced change in membrane properties.

This hypothesis differs from previous ones proposed in that the critical temperature change takes place at a temperature which can be several degrees above chilling temperatures. Cucumber mitochondria do in fact show a change in Ea for succinate oxidation near 12°C[6], but a membrane change at this temperature could not have much additional effect on developmental processes whose rates have declined almost to zero by this temperature. Barley likewise shows changes around 10° and 20°C but the magnitude of the growth change at 20° is smaller than for cucumber and a projection of the relationship between growth and temperature (Fig. 2) to low temperatures indicates that growth will continue well into the chilling temperature range.

It is concluded that both chilling-sensitive and chilling-resistant plants show temperature-induced changes in the temperature coefficients of developmental processes, that these changes correlate with temperature-induced alterations in the properties of chloroplast and other membranes, and that differences between temperature-growth responses of the two categories of plants are explainable in terms of the magnitude of the changes in temperature coefficient and the temperatures at which these changes occur.

The distinction made between the effects of low temperature on development and growth of chilling-sensitive plants on the one hand and chilling injury on the other needs emphasising since the hypothesis presented above does not attempt to explain the latter. In cucumber a further change in membrane properties at chilling temperatures may result in deleterious effects, for instance a change in membrane permeability[8], that are responsible for or contribute to the onset of chilling injury.

REFERENCES

1. Smillie, R.M., Nielsen, N.C., Henningsen, K.W., and von Wettstein, D. (1974) Proceedings of the IIIrd International Congress of Photosynthesis Research (Avron, M., ed) pp. 1841-1860, Elsevier Scientific Publ. Co., Amsterdam.
2. Boardman, N.K., Anderson, J.M., Björkman, O., Goodchild, D.J., Grimme, L.H., and Thorne, S.W. (1974) Portulgaliae Acta Biol. Ser. A. 14, 213-236.
3. Nolan, W.G. and Smillie, R.M. (1976) Biochim. Biophys. Acta, in press.
4. Lyons, J.M. (1973) Annu. Rev. Plant Physiol. 24, 445-466.
5. Raison, J.K. (1973) Symp. Soc. Exp. Biol. 27, 485-512.
6. Lyons, J.M. and Raison, J.K. (1970) Plant Physiol. 45, 386-389.
7. Yamamoto, Y. and Nishimura, M. (1976) Plant and Cell Physiol. 17, 11-16 and 17-21.
8. Wright, M. and Simon, E.W. (1973). J. Exp. Bot. 24, 400-411.

Genetics and Biogenesis of Chloroplasts and Mitochondria, Th. Bücher et al. eds.

CHLOROPHYLL SYNTHESIS AND THE DEVELOPMENT OF PHOTOSYNTHETIC ACTIVITY

W.T. Griffiths, N.L. Morgan and R.E. Mapleston
Department of Biochemistry,
University of Bristol,
Bristol, U.K.

The formation of chlorophyll in higher plants and the complex changes accompanying the etioplast to chloroplast transition are all well known to be light dependent processes. As yet, however, the only light reaction to be well-characterised in this complex photomorphogenic process is the photoreduction of protochlorophyll(ide) to chlorophyll(ide)-a step in the pathway leading to chlorophyll a formation. Following the almost instantaneous formation of chlorophyll resulting from illumination of dark grown plants, there occurs, over a longer time scale, the transformation of the characteristic prolamellar body membranes of the etioplasts producing the photosynthetically functional membranes of the mature chloroplast. The work to be discussed here deals with some aspects of chlorophyll synthesis by isolated etioplasts and also describes some data relating to the acquisition of photosynthetic function on greening.

Chlorophyll a formation has been extensively studied both at the whole-cell level and by identification of possible intermediates accumulated following the interruption of synthesis genetically in mutants[1] or in normal plants chemically[2,3]. However, cell-free preparations, active in any or all of the reactions involved in chlorophyll formation have only relatively recently produced any real success[4], this undoubtedly resulting from the increased sensitivity achieved by the application of tracer techniques and new innovations in pigment separating and assay methods. As a result, complete sequences depicting the pathway of chlorophyll a formation now appear quite commonly. In contrast, the single light requiring step involved in the pathway has for a long time been extensively studied at the cell-free level following the isolation of a solubilised PChl(ide)-protein complex from etiolated plants - the Protochlorophyll holochrome - this however retaining the ability to photoreduce only a proportion of its pigment component to Chl(ide)[5,6]. Spectrophotometry of whole leaves has also indicated the existence of both photoconvertible and non photoactive PChl(ide) in etiolated plants, these displaying light absorption maxima at 630 and 650 nm respectively[7].

Many of the earlier conclusions about chlorophyll synthesis have recently been confirmed and extended using isolated etioplasts. These have been shown to be

Abbreviations: PChlide-protochlorophyllide; PChl-esterified protochlorophyllide; Chlide-chlorophyllide a; Chl-chlorophyll a; DVP-Mg divinyl phaeoporphyrin a_5.

capable of mimicking whole plants in regenerating photoactive PChl(ide), such that under a flashing light regime all their endogenous pigment is converted to Chl(ide)[8]. Washed membranes, prepared from these etioplasts and containing only a trace of photoactive pigment, could, on supplementation with NADPH, again achieve complete pigment photoconversion in flashing light[9]. Further, the process was shown to involve the successive dark, NADPH, dependant formation of a photoactive complex, P638/652, absorbing light maximally at 638 and 652 nm. The mechanism by which P638/652 is formed from NADPH remains unknown, but it is reasonable to assume that NADPH is the long sought after source of hydrogen for Chl(ide) formation. These membranes have also been shown to catalyse the conversion of exogenously added PChlide to Chlide when flash illuminated in the presence of NADPH - the dark formation of P638/652 being again an obligate intermediate[10]. However, when protochlorophyll (PChl) isolated from etiolated barley was added under similar conditions no Chl(ide) was produced. Recently we hav e directed our attention to other possible precursors which might be transformed into chlorophyll(ide) by the barley etioplast system.

Scheme I summarises current views of the later stages of chlorophyll formation from Mg-protoporphyrin IX and demonstrates that alternate routes have been proposed at two different points viz. at PChlide and at Mg-protoporphyrin. For these studies a continuous assay has been developed for assaying the activity of the NADPH dependent photoactive P638/652 synthesizing system. In the assay the rate of saturating flash induced NADPH dependent Chlide synthesis is noted by dual wavelength spectrophotometry as an increased absorbance at 672 nm relative to 710 nm. The basis of the assay is illustrated in Fig.1. Thus, on initial flashing of etioplasts, Chlide can be seen to be formed from the endogenous P638/652 already present which, however, soon becomes exhausted. Addition of

SCHEME 1

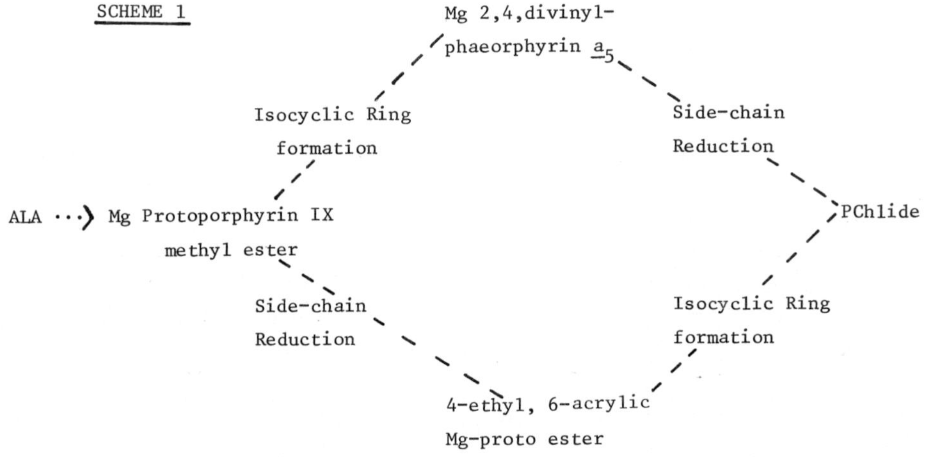

SCHEME 1 - contd.

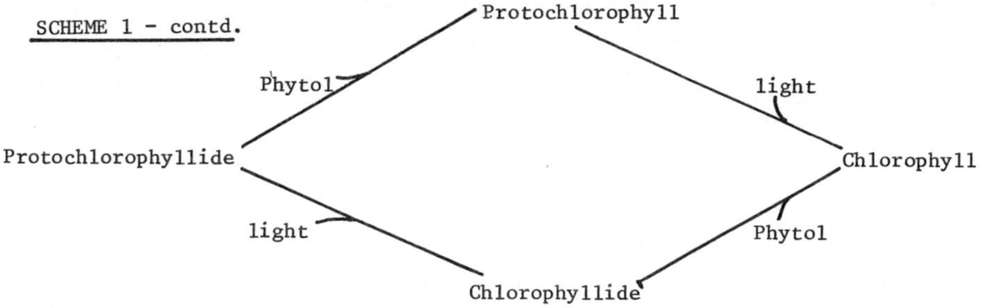

Scheme 1 - Suggested routes for Chlorophyll *a* formation.

NADPH, after first disrupting the etioplasts with Triton, results in more Chlide being formed on flashing from the endogenous non-photoconvertible PChl(ide). After exhaustion of this supply, addition of exogenous PChl(ide) results in a

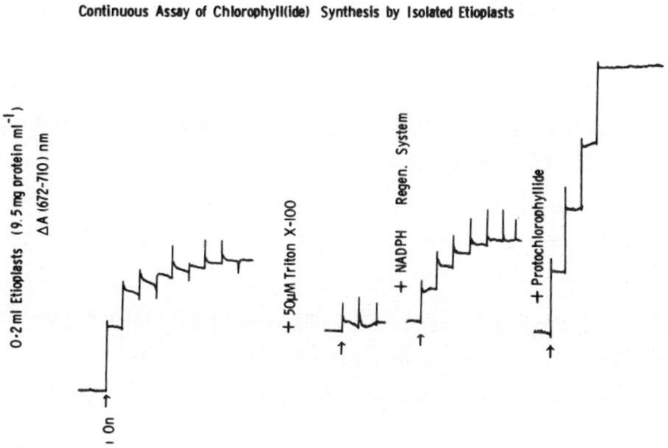

Fig. 1. Continuous assay of chlorophyll(ide) synthesis by isolated etioplasts.
Barley etioplasts were exposed to a short Xe flash every 20 sec. and absorbance changes at 672 nm relative to 710 nm monitored in a dual wavelength spectrophotometer. Illumination was through a $CuSO_4$ filter whilst the photo-multiplier was protected with 2 layers of Wratten 70 geltine filters.

linear rate of Chlide production on flash illumination - a rate which reflects the initial rate of dark P638/652 formation. Returning to the alternatives in Scheme I, from work already mentioned, PChlide is photoconverted prior to phytolation, or at least so in etioplasts isolated from seven day old barley.This is again illustrated in Fig. 2 using the continuous assay on substrate depleted 'green' etioplast membranes, prepared by extensive flashing of etioplasts at 0°C. followed by membrane isolation. Thus, with no added exogenous pigment, no Chlide is formed on flashing in the presence of NADPH. Addition of PChlide results in rapid Chlide formation. PChl in the presence of NADPH however, failed to photoproduce any Chlide, whereas addition of PChlide once more restored Chlide synthesis. Thus, it appears that the esterifying side-chain in PChl

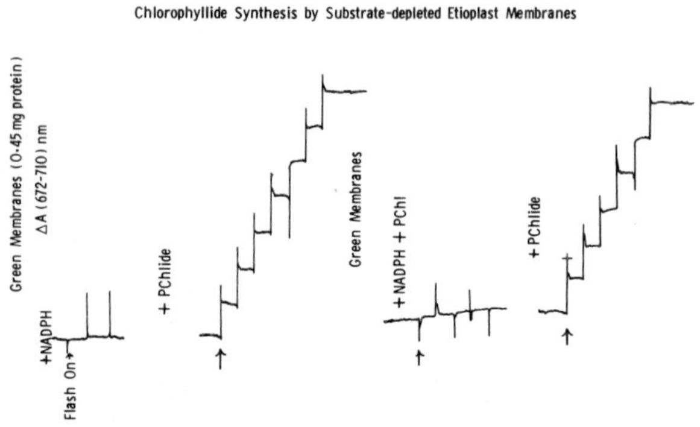

Fig. 2. Chlorophyll(ide) synthesis by substrate-depleted etioplast membranes.

Barley etioplasts were flashed at 0°C. to deplete them of photoconvertible pigments followed by isolation of the resulting 'green' prolamellar body membranes. Chlorophyllide synthesis by these membranes in the presence of added cholate solubilised protochlorophyll or protochlorophyll(ide) was monitored as in Fig. 1.

prevents it from interacting with the system to form an active complex. (While we designate the esterified PChlide we isolate from etiolated barley as protochlorophyll it should be mentioned that despite extensive efforts we have repeatedly failed to detect, by TLC or mass spectrometry, neither phytol nor geranyl-geraniol as one of the products of hydrolysis of our highly purified barley protochlorophyll[11]).

The remaining region of uncertainty in Scheme 1 concerns intermediates between Mg protoporphyrin and PChlide. Thus, Ellsworth and co-workers on the basis of identification of intermediates accummulated by various Chlorella mutants claim that reduction of the 4-vinyl group precedes formation of the iso-cyclic ring at position 6, in a sequence involving the 4-ethyl, 6-acrylate derivative as an

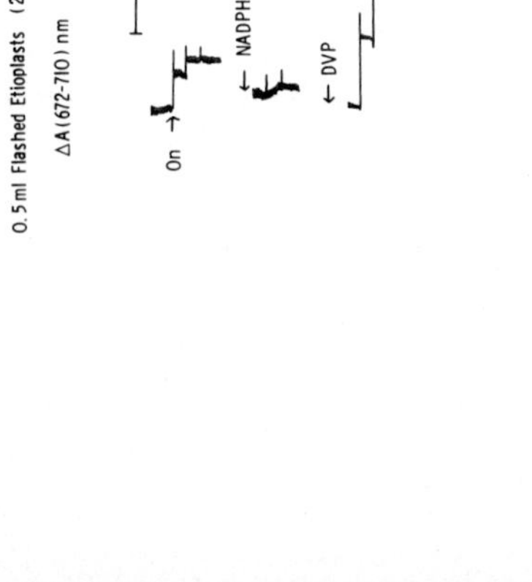

Fig. 3. Mg-divinyl-pheoporphyrin a_5 as a precursor for chlorophyllide formation by substrate depleted etioplasts.

Barley etioplasts (2.8 mg protein) were substrate depleted by pre-flashing at 0°C. Triton X-100 (50µM) was then added and flash induced chlorophyll formation monitored by dual wavelength spectrophotometry following the addition of NADPH and the porphyrin (DVP) substrate.

intermediate. Jones, however[13], has claimed the compound Mg-2,4,divinyl-pheoporphyrin (DVP), which he characterised from various organisms, to be an intermediate in chlorophyll a formation, the process involving firstly the formation of the isocyclic ring followed by reduction of the 4-vinyl group. The barley etioplast membranes have produced evidence in favour of the latter formulationation[14], Fig. 3. Thus, DVP- (a compound excreted by a Rhodoseudomonas mutant V_3, isolated and kindly donated by O.T.G. Jones) was efficiently converted into Chlide in the presence of NADPH, whereas with the ethyl-acrylate derivative featured in Scheme 1 (synthesised and kindly donated by Prof.G.Kenner and Dr K. Smith of Liverpool University) no such conversion was seen, even in the presence of an extensive range of possible co-factors that might be required. With this simplified assay of PChlide activation we have investigated some of the systems' properties, a summary of which are presented in Table 1.

Table 1. PROPERTIES OF THE SYSTEM FORMING THE PHOTOACTIVE PROTOCHLOROPHYLLIDE COMPLEX.

1. Reaction Catalysed:

   P Chlide + NADPH —(ENZYME SYSTEM)→ Photoactive Complex (+ NADP ?) —(light)→ Chlide

2. System completely specific to NADPH (K_m = 35μM) and Protochlorophyllide (K_m = 0.46μM) as substrates
3. Reaction sensitive to -SH reagents e.g. DTNB, NEM but insensitive to Arsenite.
4. pH optimum 7.2; Temperature Optimum 18°C.
5. Enzyme system stable at -20°C. Unstable at room temperature particularly in the presence of chlorophyllide formed on illumination.
6. Reaction sensitive to viscosity changes e.g. completely inhibited by 35% v/v glycerol
7. Reaction assayed in all species so far investigated.
8. System normally membrane bound by solubilised by 50μM Triton X-100 without loss of activity.

The nature of the photoactive complex, P638/652, remains uncharacterised but possible representations could be a ternary (PChlide/protein/NADPH) complex, or alternatively, a binary (reduced protein/PChlide) complex. The system shows a high degree of specificity for both reductant[15] and porphyrin, being completely inactive with the closely related compounds NADH or PChl[11].

The sensitivity of the reaction to DTNB is of interest, the effect of which is demonstrated in Fig. 4. It can be seen that DTNB completely inhibits the

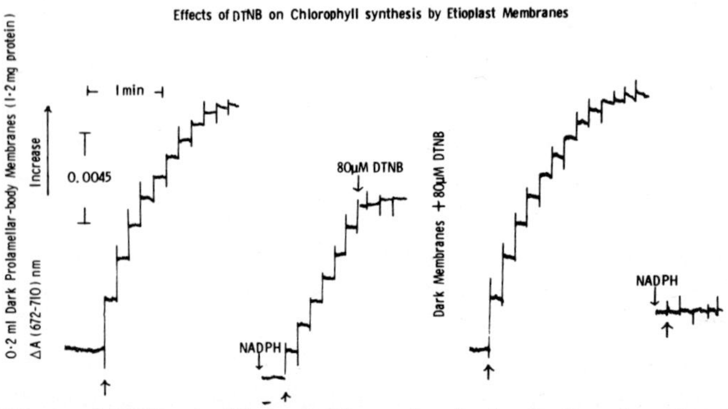

Fig. 4. Effect of DTNB on chlorophyll synthesis by isolated prolamellar-body membranes from barley etioplasts.

Dark prolamellar-body membranes were prepared from barley etioplasts following water lysing. Chlorophyll formation by these membranes from their endogenous porphyrin substrates was monitored by dual wavelength spectrophotometry with the additions indicated.

activation process whilst having no effect on preformed P638/652 present initially in the membrane. This suggests that in the active complex some essential-SH groups are protected, these becoming exposed to DTNB after photo-conversion, thereby inhibiting any further regeneration.

Included amongst species found to contain the PChlide activating system is dark grown spruce (Picea abies) which is capable of extensive dark chlorophyll synthesis[16]. Plastids from dark grown spruce when flash illuminated in the presence of PChlide and NADPH have been found to be active in Chlide formation. However, to date we have failed to detect dark Chlide synthesis by the spruce plastids even when mixed with barley membranes containing photo-convertible P638/652. This probably suggests that a completely different system is involved with the terminal reactions of dark Chlide synthesis.

In keeping with tradition, we have studied the effects of varying plant age and illumination on the activity of the PChlide activating system. With varying age (Fig. 5) the system's activity in 4 to 10 day old barley closely paralleled the capacity of the plants to synthesise chlorophyll, both attaining maximum levels at 6 days. This suggests the possibility that the activating system may function as a coarse control in Chlide formation. On illumination of 6 day old plants however, (Fig. 6) the activity of the enzyme in isolated plastids fell dramatically, being reduced by 70% of the dark level after only 30 min. illumination. This opens up the possibility of a fine control mechanism for the regulation of Chlide synthesis in the light, e.g. Fig. 6 may reflect the formation of an inhibitor (Chlide-?) of the enzyme on illumination. Interestingly, we have evidence[17] that on standing in the dark a reactivation of the

Fig. 5.

Comparison of Chlorophyll synthesis rates in whole plants and isolated prolamellar body membranes in relation to age.

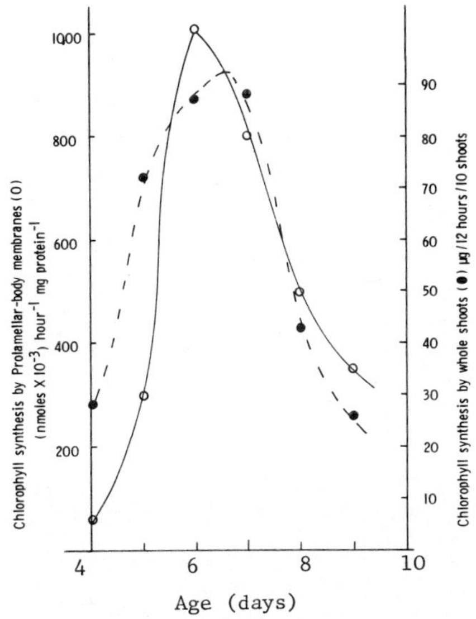

Fig. 6.

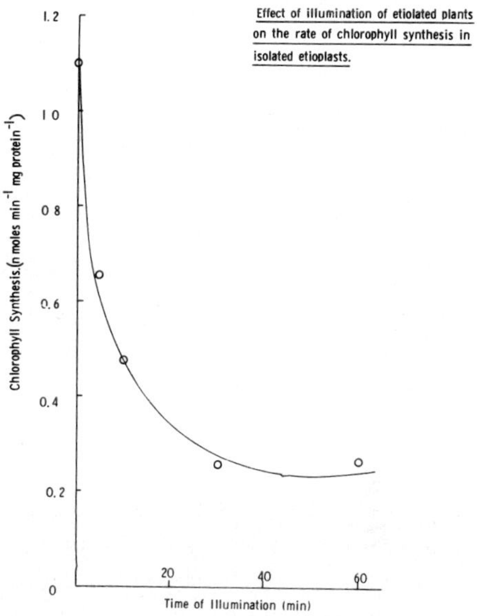

Fig. 5. Comparison of chlorophyll synthesis rates in whole plants and isolated prolamellar-body membranes in relation to age.

Etiolated barley plants of different ages were either (i) illuminated for 12 hours and chlorophyll content estimated, or (ii) used to prepare prolamellar body membranes and the in-vitro rate of chlorophyll synthesis noted, in the presence of saturating levels of NADPH and protochlorophyllide, by dual wavelength spectrophotometry.

Fig. 6. Effect of illumination of etiolated plants on the rates of chlorophyll synthesis assayed in the isolated plastids.

6 day old isolated barley plants were illuminated and etioplasts prepared after varying time intervals and these assayed for rates of chlorophyll synthesis as in Fig. 7.

---

enzyme is seen reflecting, perhaps, the esterification of Chlide, thus lifting the light induced inhibition?

A question often asked in discussions on chloroplast development is whether the etioplast membrane is retained to form the eventual thylakoid membrane of the chloroplast? Thus it has been claimed[18] that following the in-vitro illumination of purified PChlide-holochrome preparations from etiolated barley these acquire photochemical activity characteristic of a functional photo-system I, implying the identity of the holochrome protein with the apo-protein of the system I reaction centre. To some extent in agreement with this, we have immunologically detected the apo-protein of photosystem I in barley etioplasts. This was made possible with a sample of the antiserum to the purified system I prepared and kindly donated by Dr N. Nelson (Haifa). This same antibody was ineffective in inhibiting the PChlide activating enzyme implying at least the lack of identity of this with the P700 apo-protein. Further, in contrast to the work mentioned above[18], we have reported[19] plastid protein synthesis to be required for building a functional photosystem I reaction centre from the apo-protein and chlorophyll.

In conclusion, we believe the work reported here provides some clarification of the terminal reactions involved in chlorophyll formation and also on the early stages of formation of photosynthetically functional units from this chlorophyll.

Acknowledgements. We thank Mrs D. Wiles for skilled technical assistance during the latter part of this work. NLM and REM are in receipt of SRC postgraduate studentships.

REFERENCES.

1. Gough, S. (1972) Biochim. Biophys. Acta. 286, 36.
2. Granick, S. (1961) J. Biol. Chem. 236, 1168.
3. Duggan, J. & Gassman, M. (1974) Plant Physiol. 53, 206.
4. Rebeiz, C.A. & Castelfranco, P.A. (1973) Ann. Rev. Plant Physiol. 24, 129.
5. Smith, J.H.C. & Kupke, D.W. (1956) Nature 178, 751.
6. Boardman, N.K. (1962a) Biochim. Biophys. Acta. 62, 63.
7. Gassman, M. & Bogorad, L. (1967) Plant. Physiol. 42, 781.
8. Griffiths, W.T. (1975) Biochem. J. 146, 17.
9. Griffiths, W.T. (1974) FEBS Lett. 46, 301.
10. Griffiths, W.T. (1975) FEBS Lett. 49, 196.
11. Griffiths, W.T. (1976) In Preparation.
12. Ellsworth, R.K. & Aronoff, S. (1969) Arch. Biochem. Biophys. 130, 374.
13. Jones, O.T.G. (1968) In "Biochemical Society Symposia, No.28". (ed. T.W. Goodwin) p131. Academic Press, London and New York.
14. Griffiths, W.T. & Jones, O.T.G. (1975) FEBS Lett. 50, 355.
15. Griffiths, W.T. (1975) Biochem. J. 152, 623.
16. Oku, T., Hayashi, H. & Tomita, G. (1975) Plant and Cell Physiol. 16, 101.
17. Mapleston, R. & Griffiths, W.T. (1976) Unpublished Observation.
18. Henningsen, K.W., Thorne, S.W. & Boardman, N.K. (1974) Plant Physiol. 53, 419.
19. Morgan, N.L. & Griffiths, W.T. (1975) Biochem. Soc. Trans. 3, 391.

Genetics and Biogenesis of Chloroplasts and Mitochondria, Th. Bücher et al. eds.

GENETIC REGULATION OF CHLOROPHYLL SYNTHESIS ANALYZED WITH DOUBLE MUTANTS IN BARLEY

Albert Kahn[a], Naomi Avivi-Bleiser[a,b] and Diter von Wettstein[a,b]

[a]Institute of Genetics, University of Copenhagen

and

[b]Department of Physiology, Carlsberg Laboratory, Copenhagen, Denmark

INTRODUCTION

Nuclear genes involved in the genetic control of chlorophyll biosynthesis have been identified in barley[1] and Chlamydomonas[2]. In barley, some of these genes have been designated as structural genes and others as regulatory genes. The former give rise to mutants which accumulate the chlorophyll precursors Proto IX* and/or Mg-proto when fed the early metabolite ALA in darkness. The latter yield mutants which accumulate Pchlide constitutively up to 20 times the wild-type level in darkness. The structural genes (*xantha*) are envisaged to code for enzymes catalyzing the conversion of ALA to Pchlide. The regulatory genes (*tigrina*) are viewed to code for repressor(s) interacting with the structural genes for one or several enzymes catalyzing ALA synthesis and for peptides involved in feedback inhibition of ALA synthesis. Double mutants containing pairwise combinations of mutations involved in chlorophyll synthesis have been particularly useful in testing the notion that both repression and feedback inhibition of ALA synthesis play crucial roles in regulating chlorophyll biosynthesis in eukaryotes[1,2].

In Chlamydomonas[2], the mutants br_s-1 and br_c-1 accumulate small amounts of Proto IX, and the level is increased 20-fold to about 70% of the wild-type chlorophyll level, when either mutant is combined with the *r*-1 mutation. This mutant by itself accumulates 40% less chlorophyll than the wild-type and hence is not a constitutive mutant for chlorophyll synthesis. Wang et al.[2] proposed that the *r*-1 gene codes for an ALA-synthesizing enzyme and that the mutation leads to an altered enzyme with lower activity and reduced sensitivity to feedback inhibition

* Abbreviations: ALA, δ-aminolevulinate; Pchlide, protochlorophyllide; Proto IX, protoporphyrin IX; Mg-proto, magnesium protoporphyrin; *tig*, *tigrina*; *xan*, *xantha*

by protoporphyrin or heme.

Mutations in the regulatory genes *tig-d*, *tig-b* and *tig-o* in barley elicit constitutive synthesis of Pchlide in darkness[3,1]. Gough (personal communication) has demonstrated constitutive synthesis of ALA in darkness in these mutants. Double mutants homozygous for *tig-d*^{12} and *xan-l*^{35} accumulate Mg-proto constitutively, and double mutants homozygous for *tig-o*^{34} and *xan-f*^{10} accumulate detectable amounts of Proto IX. Neither of the *xan* mutants accumulate precursors of Pchlide in sufficient quantity to allow detection by the absorption spectrophotometric procedures used[4,1]. These results led to the interpretation that the *tig-d* gene product functions in the repression of the ALA-synthesizing enzyme(s) in darkness, whereas the pleiotropic effects of *tig-b* and *tig-o* mutations made it doubtful whether these genes have simple functions in the repression of ALA-synthesis[1].

We report here on a quantitative analysis of critical double mutant combinations of *tig* and *xan* mutants. The results further support the previous suggestions that chlorophyll biosynthesis in eukaryotes is controlled both by repression and feedback inhibition. The notion that the *tig-d* gene codes for the repressor of a gene for an ALA-synthesizing enzyme is retained. The gene interaction effects between the *xan* mutations and the *tig-b* or *tig-o* mutations are explained most simply by the assumption that the products of the *tig* genes are involved in the feedback control of the ALA-synthesizing system through Proto IX or a closely allied derivative.

MATERIALS AND METHODS

Double mutants: The *xan* + *tig* double mutants of barley (Hordeum vulgare L) were obtained from progenies of F_1 plants heterozygous for the relevant *xan* and *tig* mutations as exemplified for the cross *tig-d*^{12} x *xan-g*^{45} in Table 1. Kernels of F_1 plants germinated in the dark[4], and the 6-day-old dark grown seedlings were sorted into 6 classes after removing apical 1-cm segments of the primary leaves and then illuminating them over-night in a moist chamber. Leaf segments of wild-type seedlings and those heterozygous for the *xan* gene green over-night. Seedlings heterozygous for *tig-d*^{12} and homozygous or heterozygous for the *xan* wild-type allele are recognized by a yellow tip on the greening leaf segment. Seedlings homozygous for *tig-d*^{12} with at least one *xan* wild-type allele are green in darkness (accumulate Pchlide) and bleach in the light. The homozygous *xan-g*^{45} segments remain yellow in the light, whereas *xan-g*^{45} seedlings heterozygous for *tig-d*^{12} bleach. The double mutants in darkness are yellow with a green tinge due to accumulated Pchlide and bleach in the light. The observed frequencies of the six classes agreed well with those expected for a dihybrid segregation (Table 1).

Segregations for the other mutant combinations considered in this paper are listed in Table 2. As the genotypes heterozygous at the *tig* locus and containing at least one *xan* wild-type allele frequently cannot be identified reliably, this class has been pooled with the class containing the wild-type phenotypes. For the same reason, seedlings homozygous for the *xan* mutant allele and heterozygous or homozygous for the *tig* wild-type allele have been combined into one class. Apart from the two exceptions mentioned below and the combinations in which the double mutants were not detected, the segregation ratios found are not considered to deviate significantly from the expected 9:3:3:1 ratio. The double mutants can be distinguished phenotypically as follows: 1) *tig-d*^{12} + *xan-g*^{45}, *tig-d*^{12} + *xan-f*^{26}, *tig-b*^{19} + *xan-g*^{45}, *tig-b*^{19} + *xan-f*^{26} and *tig-o*^{34} + *xan-g*^{45} homozygous seedlings in darkness are yellow with a green tinge due to accumulated Pchlide and bleach in the light. 2) *tig-d*^{12} + *xan-l*^{35}, *tig-b*^{19} + *xan-l*^{35} and *tig-o*^{34} + *xan-l*^{35} homozygotes in darkness are olive green and bleach in the light. 3) Dark-grown *tig-b*^{19} + *xan-g*^{28}, *tig-b*^{19} + *xan-f*^{10}, *tig-o*^{34} + *xan-g*^{28} and *tig-o*^{34} + *xan-f*^{10} homozygotes (occasionally reddish in colour) are recognized by an absorption band at 630 nm characteristic of Proto IX and/or by the absence of a pronounced α-band of a reduced c-type cytochrome (presumably cytochrome f) in the 548 to 555 nm region. Spectra of single *xan* mutants show a pronounced α-band, whereas spectra of *tig* mutants do not allow evaluation in this regard because of Pchlide absorption in this region. The *tig-o*^{34} + *xan-f*^{26} double mutant is off-yellow in darkness and its leaf spectra are characterized by a higher Pchlide absorption than wild-type leaves as well as the absence of a pronounced absorption band of reduced cytochrome. The dark-grown *tig-d*^{12} + *xan-u*^{21} homozygote is recognizable spectrophotometrically by its elevated Pchlide absorption and by a darker orange colour than *xan-u*^{21} after exposure to light.

It should be mentioned that *xan-u*^{21} is a complex mutation and behaves as if it contains besides the *xan* mutation a mutation in the *tig-n* gene. It has not been possible to separate by recombination the postulated components of the *xan-u* mutation. The homozygous "double" mutants obtained in the cross *tig-d*^{12} x *xan-u*^{21} are therefore interpreted to have the following genotype: *tig-d*^{12}/*tig-d*^{12}, *tig-n*^{21}*xan-u*^{21}/*tig-n*^{21}*xan-u*^{21}.

The significant deviation from the expected ratio in the cross *tig-b*^{19} x *xan-g*^{45} is due to a surplus of *xantha* and a deficit of wild-type seedlings (Table 2). The gametic selection for the *xan* gametes is peculiar to this progeny, as studies of other progenies have revealed entirely normal segregation ratios. The progeny of the cross *tig-o*^{34} x *xan-g*^{28} is significantly deficient in the double mutant class and has a large surplus of *tigrina* seedlings. In other progenies similar gametic selection against the *xan-g*^{28} allele and in favour of the *tig-o*^{34} allele was observed.

It has not been possible to identify the double mutants in the two crosses *tig-d*^{12} x *xan-g*^{28} and *tig-d*^{12} x *xan-f*^{10}. The segregation for the latter cross (Table 2) fits well to a 9:3:4 ratio, indicating the double mutant to be indistinguishable from *xan-f*^{10}. In the *tig-d*^{12} x *xan-g*^{28} cross a slightly signi-

TABLE 1

GENETIC ANALYSIS OF CROSS BETWEEN *tigrina-d*^{12} AND *xantha-g*^{45}

F_2 GENOTYPES	3 TTXX TTXx	6 TtXX TtXx	3 ttXX ttXx	1 TTxx	2 Ttxx	1 ttxx
F_2 PHENOTYPES						
In darkness	yellow	yellow	green	yellow	yellow	yellow-green
In light	green	green base	bleaches	yellow	bleaches	bleaches
tig-d^{12}x *xan-g*^{45}	60	101	58	28	38	16

χ^2-test: 0.2 < P < 0.3

TABLE 2

GENETIC ANALYSIS OF CROSSES AMONG *tigrina* AND *xantha* MUTANTS

F_2 GENOTYPES		9 TTXX,TtXx TtXX,TTXx	3 ttXX ttXx	3 TTxx Ttxx	1 ttxx	χ^2-test
F_2 PHENOTYPES						
In darkness		yellow	green	yellow	double	
In light		green	bleaches	yellow	mutant	
tig-d^{12}x *xan-g*^{28}		53	22	12	*	0.025 < P < 0.05
tig-d^{12}x *xan-f*^{26}		41	18	14	4	0.7 < P < 0.8
tig-d^{12}x *xan-f*^{10}		139	40	59	*	0.7 < P < 0.8
tig-d^{12}x *xan-l*^{35}		120	41	32	18	0.3 < P < 0.5
tig-d^{12}x *xan-u*^{21}		160	50	59	10	0.2 < P < 0.3
tig-b^{19}x *xan-g*^{45}		283	107	136	36	0.005 < P < 0.01
	exp.	316.1	105.4	105.4	35.1	
tig-b^{19}x *xan-g*^{28}		146	55	39	8	0.05 < P < 0.1
tig-b^{19}x *xan-f*^{26}		180	41	60	17	0.1 < P < 0.2
tig-b^{19}x *xan-f*^{10}		222	61	72	14	0.1 < P < 0.2
tig-b^{19}x *xan-l*^{35}		96	31	29	10	0.95 < P < 0.98
tig-o^{34}x *xan-g*^{45}		230	70	88	27	0.5 < P < 0.7
tig-o^{34}x *xan-g*^{28}		68	43	30	2	0.0005< P < 0.001
	exp.	80.4	26.8	26.8	8.9	
tig-o^{34}x *xan-f*^{26}		209	55	65	8	0.01 < P < 0.02
tig-o^{34}x *xan-f*^{10}		70	21	32	4	0.1 < P < 0.2
tig-o^{34}x *xan-l*^{35}		169	49	55	11	0.3 < P < 0.5

* Double mutant not distinguished; χ^2-test on 9:3:4 ratio

ficant deficit occurs in the *xantha* class. From progeny testing of *xan-g*^{28} we know that gametic selection against *xan-g*^{28} can take place and this can explain the deficit in the *xantha* class noted here. Probably the *tig-d*^{12} + *xan-g*^{28} double mutant is indistinguishable from *xan-g*^{28}.

Pigment analyses: The leaf material used for the extraction of pigments received only dim green safe-light or light of the spectrophotometer's measuring beam. Accordingly, the extracts contained no significant quantities of chlorophyll(ide). Etiolated primary leaves were ground with a mortar and pestle using 5 ml of ammoniacal aqueous acetone (1 ml of 25% NH_4OH : 9 ml water : 90 ml acetone) per g leaves. After clarification by centrifugation, the extracts were made up to 5.9 ml per g leaves with absolute acetone to replace acetone lost by evaporation during grinding and to bring the acetone concentration to ca. 80% (v/v) after the dilution by water from the leaves. Spectra were made either with a Cary 17 or an Aminco DW-2 spectrophotometer.

Thin layer chromatography and qualitative identification of the pigments by cochromatography with authentic compounds were performed essentially by the method of Gough[4], but using an aged solvent system A which gave separations intermediate between those obtained by Gough with solvent systems A and B. This allowed the resolution of Proto IX and its esters, Mg-proto and its monomethyl ester, and Pchlide on a single thin layer plate. The lower limit of reliable detectability of Proto IX was 10 picomoles applied to the plate, which extrapolates to ca. 0.5 nanomoles per g fresh wt of leaf.

Since the quantitative analyses do not discriminate between the methyl esters and their parent compounds Proto IX and Mg-Proto,respectively, the designations imply nothing about the extent of esterification. For uniformity in spectrophotometric quantitation, one set of quantitative equations was derived and used throughout. Simpler procedures, where adequate, gave similar results. No greater uncertainties or limitations than are common in such analyses were encountered, and problems with pigment breakdown or interfering absorption by irrelevant pigments were minimal.

Three equations with the three unknowns were solved simultaneously:

1) $O.D._{628} = 31.1\,(\text{Pchlide}) + 5.5\,(\text{Proto IX}) + 0.5\,(\text{Mg-proto})$

2) $O.D._{590} = 5.9\,(\text{Pchlide}) + 3.2\,(\text{Proto IX}) + 17.9\,(\text{Mg-proto})$

3) $O.D._{575} = 10.7\,(\text{Pchlide}) + 7.5\,(\text{Proto IX}) + 4.4\,(\text{Mg-proto})$

The constants are millimolar extinction coefficients[4,5], some of which were redetermined in this study. Solution of these equations leads to:

4) $(\text{Pchlide}) = 0.0426\ O.D._{628} + 0.0072\ O.D._{590} - 0.0343\ O.D._{575}$

5) $(\text{Proto IX}) = 0.1962\ O.D._{575} + 0.0466\ O.D._{590} - 0.0587\ O.D._{628}$

6) $(\text{Mg-proto}) = 0.0618\ O.D._{590} + 0.0238\ O.D._{575} - 0.0035\ O.D._{628}$

with the pigment concentrations in μmoles per ml and more figures than are significant in the constants. Further rounding off to two significant figures was performed in the tabulation of results. As expected from the nature of the analysis, small negative or positive quantities of pigments that were not detected on thin layer chromatograms were calculated frequently. These quantities are given as zero.

RESULTS AND DISCUSSION

Quantitative pigment analyses of double mutants: Relevant results of the porphyrin determinations in the wild-type and in single *xan* and *tig* mutants together with their double mutant combinations are presented in Fig. 1 on a fresh weight basis. Since no major or consistent variation in fresh weights among the various genotypes was observed, the diagrams look very similar, when the pigment amount is expressed per seedling (not shown). As evident from Fig. 1a, the double mutants containing *xan-l*^{35} together with either *tig-d*^{12}, *tig-b*^{19} or *tig-o*^{34} in homozygous form have similar patterns of pigment accumulation. They all accumulate Mg-proto and Pchlide in darkness to a level approaching Pchlide accumulation in the respective single *tigrina* mutant and from several to 10-fold higher levels than Pchlide accumulation in the wild-type. The constitutive pigment synthesis in these double mutants indicates that mutation in each of the *tig* genes leads to a relaxation of the control of ALA-synthesis. The accumulation of both Mg-proto and Pchlide is due to the leakiness of the *xan-l*^{35} block in the conversion of Mg-proto to Pchlide. When the block is complete, only Mg-proto accumulates as shown by two newly isolated double mutants induced in the *tig-d*^{12} line and described in the next section of this paper.

The single gene mutant *xan-l*^{35} accumulates only Pchlide in the same quantity as the wild-type. This mutation does not interfere with the regulation of the pathway and its block first becomes apparent when the mutant is provided with ALA, a metabolite beyond the point of regulation. Precise phenocopies of the double mutant by feeding ALA to 7-day-old detached shoots of *xan-l*^{35} for 22 hours in darkness were not obtained (Fig. 1a). A low concentration of ALA (1 mM) elicited only small pigment accumulation, whereas a higher concentration (5 mM) led to little increase in the level of Pchlide and marked accumulation of Mg-proto and Proto IX. In an experiment (not shown) in which shoots were transferred from the ALA medium to buffer, an increase of Pchlide at the expense of previously formed Mg-proto was observed with time. Such successive slow conversion of Mg-proto to Pchlide over several days could account for the high proportion of Pchlide in accumulated pigment of the double mutants. More remarkable is the absence of significant Proto IX accumulation in the pigment patterns of the double mutants. Trace amounts of Proto IX detected on the thin layer chromatograms of extracts from *tig-d*^{12} + *xan-l*^{35} double mutants and *xan-l*^{35} fed 1 mM ALA could have resulted from the loss of Mg

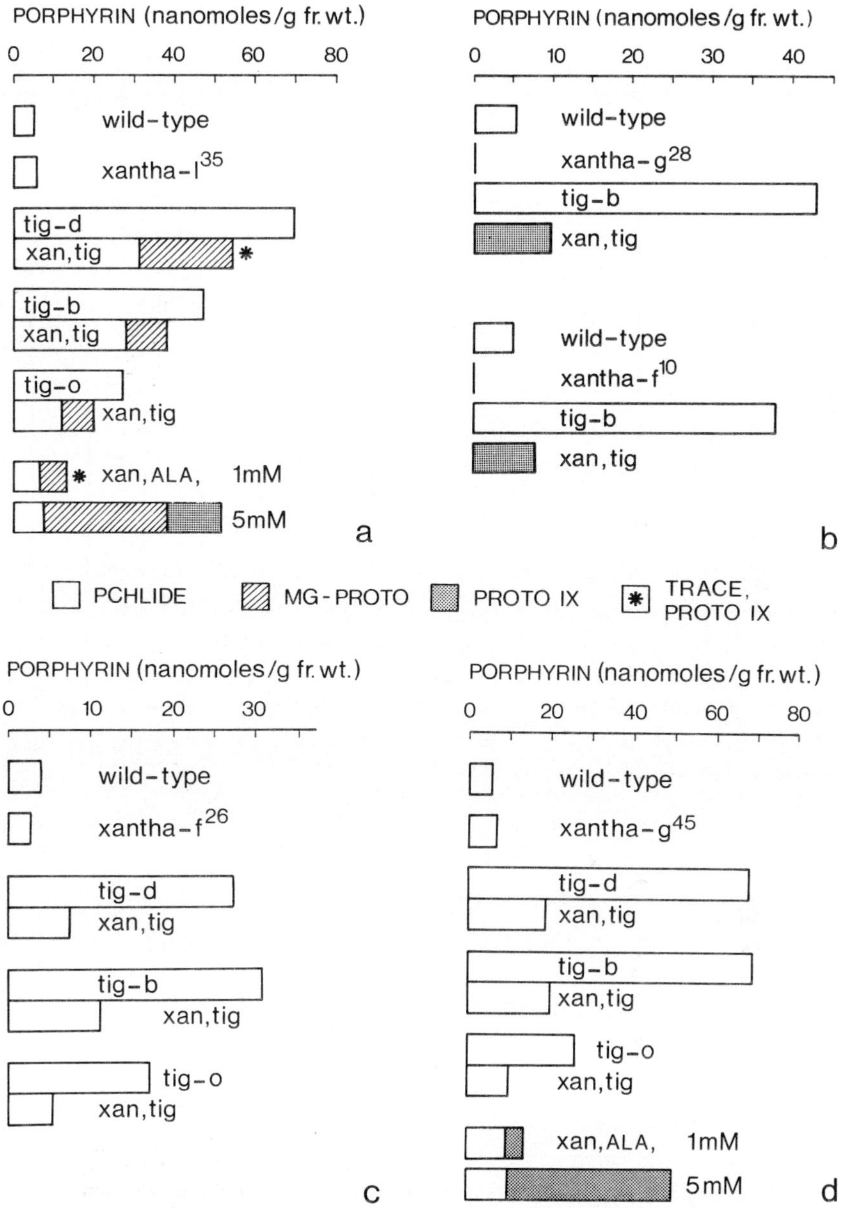

Fig. 1a-d. Amounts of porphyrin precursors of chlorophyll found in wild-type, *xantha* mutants, *tigrina* mutants and *xan* + *tig* double mutants. In a) and d) the amounts of porphyrins formed after feeding *xan-l*^{35} for 22 hr and *xan-g*^{45} for 29 hr with 1 or 5 mM ALA are given for comparison.

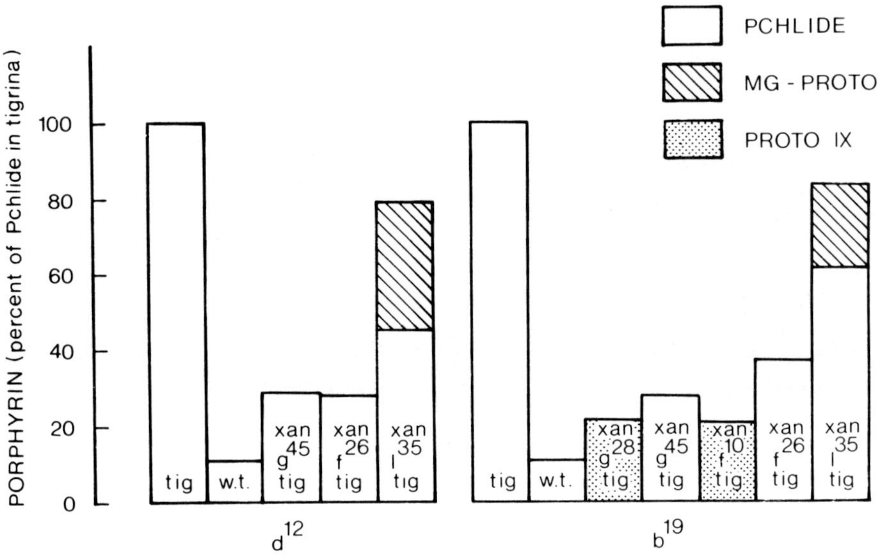

Fig. 2. Relative amounts of porphyrins found in three double mutants containing homozygous *tigrina-d*^{12} and five double mutants containing homozygous *tigrina-b*^{19}. w.t. = wild-type

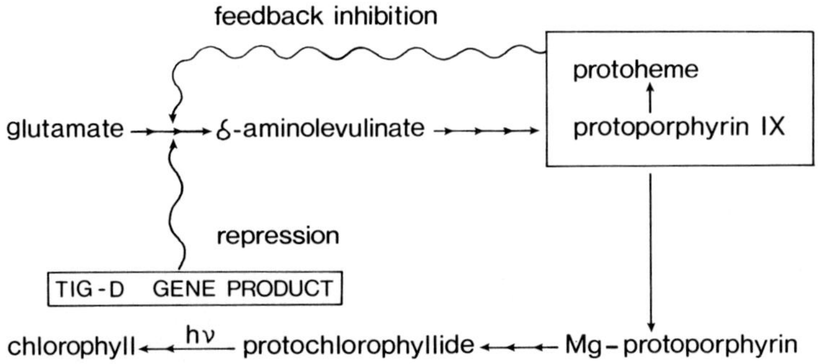

Fig. 3. Model of the genetic regulation of chlorophyll biosynthesis employed in the interpretation of the results with double mutants in barley.

from Mg-proto during extraction or thereafter. The absence of Proto IX in the double mutants could result from its effective conversion into Mg-proto so long as ALA formation does not exceed a certain rate.

The mutants *xan-f*^{10} and *xan-g*^{28} are almost completely blocked at a step between Proto IX and Mg-proto in Pchlide synthesis [4,6]. In these mutants the pathway is regulated since they do not accumulate intermediates. The block first becomes apparant when the mutants are provided with ALA which causes massive accumulation of Proto IX. The relaxed regulation of ALA synthesis caused by mutations in the *tig-d*, *tig-b* and *tig-o* genes should lead to constitutive synthesis and accumulation of Proto IX in double mutants containing a *tig* mutation and the *xan-f*^{10} or *xan-g*^{28} allele. Constitutive production of Proto IX was observed in *tig-o*^{34} + *xan-f*^{10} double mutants[1], and as evident from Fig. 1b, in *tig-b*^{19} + *xan-f*^{10} as well as *tig-b*^{19} + *xan-g*^{28} double mutants. The amounts of Proto IX accumulated are, however, only about 20 per cent of the level of Pchlide accumulation in *tig-b*^{19} and only twice the level of Pchlide in the wild-type (Fig. 2). It can be concluded from these results on the double mutants and the massive pile up of Proto IX in the *xan-f* and *xan-g* mutants on ALA that a block in the conversion of Proto IX to Mg-proto interferes with the relaxed production of ALA caused by the *tig* mutations. Most likely the accumulation of Proto IX *per se* inhibits the production of ALA in the double mutants. In double mutant seedlings of the *tig-o*^{34} + *xan-g*^{28} combinations and in some *tig-o*^{34} + *xan-f*^{10} double mutant seedlings accumulation of Proto IX was not demonstrable. This is taken to mean that the extent of inhibition of the production of ALA in double mutants is subject to variation.

Despite an extensive spectrophotometric search, no *tig-d*^{12}+ *xan-f*^{10} or *tig-d*^{12} + *xan-g*^{28} double mutants containing Proto IX were detected. As mentioned above, these double mutants also could not be distinguished visually and were assumed to be in the *xantha* class. Since we see no reason why these double mutants should be zygotic lethals, we propose that the production of ALA for the chlorophyll pathway is inhibited stringently in the combinations of the *tig-d*^{12} mutation with alleles of *xan-f* or *xan-g* which block the pathway almost completely.

In double mutants homozygous for leaky alleles at the *xan-f* (Fig. 1c) or *xan-g* (Fig. 1d) loci and homozygous for either a *tig-d*, *tig-b* or *tig-o* mutant allele, accumulation of Proto IX is not detectable. These double mutants contain about 30 per cent of the Pchlide present in the relevant *tigrina* mutant and up to 3-fold more Pchlide than wild-type seedlings (Fig. 2). Phenocopies of double mutants could not be produced by feeding two concentrations of ALA to detached *xan-g*^{45} shoots for 29 hours (Fig. 1d). At both concentrations Proto IX accumulated and at the higher concentration this porphyrin amounted to about 85 per cent of the pigment present, the rest being Pchlide. The amount of Pchlide accumulated in the *xan-f*^{26} and *xan-g*^{45} containing double mutants is similar to the amount of Proto IX present in double mutants homozygous for non-leaky *xan-f* or *xan-g* alleles

(Fig. 2). In double mutants containing a leaky block in the conversion of Proto IX to Mg-proto obviously all Proto IX formed can readily be converted into Pchlide.

Relaxed control of porphyrin synthesis can be effected in two ways, namely by defective repression or by defective feedback inhibition of ALA synthesis (Fig.3). Significantly higher capacity for ALA synthesis than in the wild-type has been found in all three dark-grown *tigrina* mutants employing the competitive ALA dehydratase inhibitor levulinate (S. Gough, personal communication). The accumulation of Mg-proto and/or Pchlide does not affect ALA synthesis. On the other hand, accumulation of Proto IX seems to reduce ALA synthesis, limiting the amount of porphyrin formed. Based on observations indicating rapid protoheme turnover, a regulation of ALA synthesis in barley by protoheme has been postulated by Castelfranco and Jones[7].

There are two characteristics that distinguish *tig-d*^{12} from the other *tig* mutants: Mutant *tig-d*^{12} does not affect any other function than the synthesis of Pchlide in the dark, whereas the *tig* mutations in the other loci are highly pleiotropic, e.g. with regard to carotene composition[8]. Accumulation of Proto IX was not found in suspected double mutants involving *tig-d*^{12}, but was found in several instances in those involving *tig-b* or *tig-o*. Assuming that Proto IX is a powerful feedback inhibitor of ALA synthesis, the absence of *tig-d* double mutants accumulating Proto IX can be interpreted simply: ALA synthesis is feedback inhibited as soon as Proto IX starts to accumulate in a *tig-d*^{12} + *xan-f*^{10} double mutant as a consequence of the almost complete block in the conversion of Proto IX. On the other hand, the accumulation of Proto IX in double mutant combinations like *tig-b*^{19} + *xan-g*^{28} or *tig-o*^{34} + *xan-f*^{10} hints that these *tig* mutants are less sensitive to feedback inhibition than wild-type or *tig-d*^{12}.

Isolation of new structural gene mutants in the chlorophyll pathway: The mutant *tig-d*^{12} can be grown to maturity in continuous light, a condition which allows the plant to survive its aberrant regulation of the porphyrin pathway. Thus, seeds homozygous for *tig-d*^{12} can be obtained. When these are treated with the potent mutagen, sodium azide[9], and the M_1 plants are grown in continuous light, many seedlings bearing new mutations which cause chlorophyll deficiency appear in the M_2 generation. Since *tig-d*^{12} causes constitutive synthesis of Pchlide in darkness, mutations in structural genes that cause a block in the pathway of chlorophyll synthesis might be expected to lead to accumulation of precursor porphyrins of Pchlide in darkness.

We screen for new mutations in the chlorophyll pathway in two ways:
1) spike progenies from M_1 plants are grown in natural light-dark cycles until the M_2 seedlings have developed sufficiently for display of the *tig-d*^{12} or a mutant phenotype (shortly after emergence from the vermiculite), whereupon the mutants are marked and all the seedlings transferred to darkness. After 3 to 4

days, mutant leaf segments that developed in darkness and therefore had not previously received light are examined spectrophotometrically for precursor porphyrins. 2) M_1 spike progenies are grown in darkness, where *tig-d*^{12} is green and certain mutants are less green or lack green colour. Sometimes such mutants can be distinguished in dim green safelight, but often they are found during a brief exposure to dim white light. Here too the mutants are analysed spectrophotometrically for precursor porphyrins.

Double mutants of *albina* phenotype with little or no Pchlide accumulation were most frequent. Mutants with *viridis* and *xantha* phenotypes accumulating Pchlide to a level approximating or exceeding that in the wild-type also occurred.

Noteworthy in the present context is that among the 339 new double mutants four accumulated Mg-proto constitutively in darkness, while none accumulated Proto IX detectably. Two of the Mg-proto mutants were completely blocked in Pchlide synthesis, and the other two were leaky and contained Pchlide in addition to Mg-proto, as does the double mutant *xan-l*^{35} + *tig-d*^{12}. Fig. 4 shows the absorption spectrum of a portion of leaf developed in darkness by one of the non-leaky Mg-proto accumulating double mutants. The lack of specific absorption at 630 nm indicates the absence of Pchlide and Proto IX, while the strong absorption bands at 550 and 590 nm are diagnostic for Mg-proto. The other spectrum in Fig. 4 comes from an ammoniacal aqueous acetone extract after treatment with petroleum ether to remove carotenoids. Thin layer chromotography of the extract revealed Mg-proto and Mg-proto monomethylester. No Proto IX or Pchlide was detected. The double mutant accumulated Mg-proto to about half the Pchlide level of *tig-d*^{12} and ca. 7-fold the Pchlide level of the wild-type grown in darkness.

The relative ease with which new Mg-proto accumulating mutants were obtained stands in sharp contrast to the failure to detect Proto IX accumulating mutants. Congruent with the earlier results on double mutants obtained by recombination, the contrast is again explained readily by the proposition that *tig-d*^{12} is derepressed in ALA synthesis, Mg-proto accumulation does not inhibit the enzyme, and Proto IX or a close derivative functions as a powerful feedback inhibitor of ALA synthesis at concentrations that are lower than we normally detect. That some of the new mutants are blocked between Proto IX and Mg-proto was shown by feeding ALA and inducing Proto IX accumulation.

Linkage of genes involved in the genetic control of chlorophyll biosynthesis: Close linkage has been found among the *tig-b*, *tig-n* and *xan-u*^{21} genes. The frequency of crossing over has been analysed extensively in the cross *tig-b*^{23} x *xan-u*^{21}. In four tests comprising 3450 gametes, 55 cross over events were detected giving a linkage value of 1.51 ± 0.14 per cent. Employing a different allele at the *tig-b* locus, a frequency of 0.9 per cent crossing over (6 in 696 gametes) was determined in the *tig-b*^{19} x *xan-u*^{21} combination, whereas no cross over has been detected in the *tig-b*^{24} x *xan-u*^{21} combination (714 gametes tested).

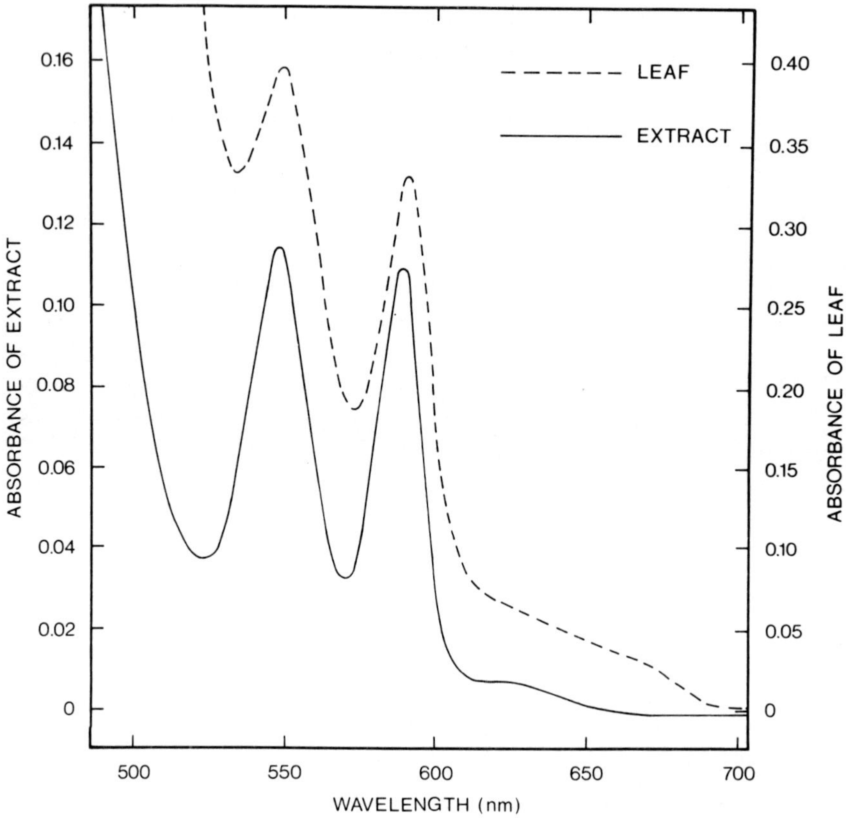

Fig. 4. Absorption spectra of a Mg-protoporphyrin accumulating double mutant in barley.

It was mentioned above that *xan-u*^{21} probably is a double mutant containing, in addition to the *xan-u*^{21} mutation, a mutation in the closely-linked *tig-n* gene. Recombinants between *xan-u*^{21} and *tig-n*^{21} have not been found. In the cross *tig-b*^{19} x *tig-n*^{32} linkage with a cross over frequency of 0.7 per cent (1 in 144 gametes) has been determined.

Another case of linkage became apparent from progeny of the cross *tig-d*^{12} x *tig-o*^{34}, where 1.26 per cent crossing over (9 in 716 gametes) was found. Normal dihybrid segregation ratios and absence of linkage were found in the combinations: *tig-o*^{34} x *tig-b*^{19}, *tig-o*^{34} x *tig-n*^{32}, *tig-o*^{34} x *xan-u*^{21}, *tig-d*^{12} x *tig-b*^{19}, and *tig-d*^{12} x *tig-n*^{32}.

In summary, these studies show the *tig-b*, *tig-n*, and *xan-u* genes to be closely linked and that *tig-d* and *tig-o* also are closely linked, but are separated from the former group. Whereas the linkage intensity can be considered high with regard to the barley chromosome, this need not imply close linkage at the molecular level or significant effects on the genetic control of chlorophyll biosynthesis.

ACKNOWLEDGEMENTS

Support by grants GM 22051 from the U.S. Public Health Service, National Institutes of Health, and 511-555 from the Danish Natural Science Research Council is gratefully acknowledged. We are indebted to Professor Knud W. Henningsen for the use of the Aminco DW-2 spectrophotometer supplied through grant 511-3513 by the Danish Natural Science Research Council and to Mr. Sven Møller for assistance.

REFERENCES

1. Wettstein, D. von, Kahn, A., Nielsen, O.F. and Gough, S. (1974) Science 184, 800-802.
2. Wang, W., Boynton, J.E., Gillham, N.W. and Gough, S. (1975) Cell 6, 75-84.
3. Nielsen, O.F. (1974) Hereditas 76, 269-304.
4. Gough, S. (1972) Biochim. Biophys. Acta 286, 36-54.
5. Nielsen, O.F. and Kahn, A. (1973) Biochim. Biophys. Acta 292, 117-129.
6. Henningsen, K.W., Boynton, J.E., Wettstein, D. von and Boardman, N.K. (1973) in the Biochemistry of Gene Expression in Higher Organisms. pp. 457-478, Australian and New Zealand Book Comp. Sydney.
7. Castelfranco, P.A. and Jones, O.T.G. (1975) Plant Physiol. 55, 485-490.
8. Nielsen, O.F. and Gough, S. (1974) Physiol. Plant. 30, 246-254.
9. Nilan, R.A., Sideris E.G., Kleinhofs, A., Sander C. and Konzak, C.F. (1973) Mutation Res. 17, 142-144.

MITOCHONDRIAL ATPase COMPLEX

Genetics and Biogenesis of Chloroplasts and Mitochondria, Th. Bücher et al. eds.

MUTATION IN SACCHAROMYCES CEREVISIAE MITOCHONDRIAL F_1 LEADING TO AUROVERTIN RESISTANCE

E. Agsteribbe, M. Douglas, E. Ebner,
T.Y. Koh and G. Schatz
Department of Biochemistry,
Biocenter, University of Basel
CH-4056 Basel, Switzerland

INTRODUCTION

The mitochondrial ATPase complex* has received increasing attention in recent years as various laboratories have concentrated on the structural and mechanistic role of the enzyme in mitochondrial energy transfer. Most of the work on the ATPase complex from yeast has dealt with the biosynthesis and genetics of the proteins that bind F_1 to the mitochondrial inner membrane (1,2,3). In the present study, this biochemical genetic approach has been applied to the F_1 molecule itself. We have isolated yeast mutants resistant to aurovertin, an antibiotic which specifically binds to, and inhibits F_1. Griffiths has already briefly mentioned that such mutants can be isolated, but the mutational lesion remained unknown (4). We shall show that a very small fraction of mutants resistant to the inhibitor carries a mutation in one of the structural genes for F_1. These mutants are of interest for the following reasons: (1) They make available mutationally altered F_1 and thus allow one to study the role of individual F_1 subunits in aurovertin binding and, more generally, in energy conservation. (2) They represent one of the very few known structural gene mutants for mitochondrial membrane proteins, thereby opening up many new possibilities for genetic studies on mitochondrial assembly.

RESULTS AND DISCUSSION

The antibiotic aurovertin introduced by Lardy et al. (5) as an inhibitor of various mitochondrial energy-linked reactions was subsequently shown to inhibit the F_1-ATPase (6). In this study yeast

---

*In this communication, the term "ATPase complex" refers to the functional complex consisting of the ATPase F_1, oligomycin sensitivity conferring protein and several hydrophobic proteins.

mutants resistant to aurovertin were generated by ethylmethanesulfonate mutagenesis of a haploid strain [D41; α ura, his 4] whose growth on a non-fermentable carbon source was blocked by 1 μg of aurovertin/ml. Mutants designated aur^R were selected for their ability to grow on a non-fermentable carbon source in the presence of 50 μg aurovertin/ml. Subsequent examination of the ATPase activity** in these colonies showed two mutant classes. Class I aur^R mutants exhibited aurovertin resistant ATPase whereas mutants designated Class II contained aurovertin sensitive ATPase comparable to wild-type. Of approximately 300 aurovertin resistant clones examined in this manner only 2 were of Class I. Therefore mutants of Class II presumably involve lesion(s) which affect the accessibility of F_1 to aurovertin. Sporulation of diploids obtained from the two Class I aur^R mutants and an aurovertin sensitive tester strain and subsequent tetrad analysis showed a 2:2 segregation of the resistance allele indicating the presence of a single nuclear mutation.

The F_1 and the oligomycin sensitive ATPase complex were prepared from wild-type and from one of the Class I aurovertin resistant mutants (aur^R-1). Analysis of the enzyme from both strains on SDS gradient slab gels (Figure 1) show that the 4 subunits*** of F_1 or the 9 subunits of the oligomycin sensitive complex have the same size in the two strains. (Note the small but reproducible differences when yeast F_1 is compared with a similar preparation from beef heart). Enzymatic analysis of the ATPase preparations (Table I) shows that both the F_1 and ATPase complex from the mutant aur^R-1 are only little inhibited by concentrations of aurovertin which inhibit the wild-type enzymes by approximately 80%. It is noteworthy that the mutant F_1 did not show cross resistance to other F_1 inhibitors (Table II) indicating that the site of action of these compounds is different from that of aurovertin.

**Mutant clones (aur^R) were grown to stationary phase in small liquid cultures containing aurovertin, harvested, washed and disrupped mechanically by grinding frozen cells. ATPase activity of the supernatant was assayed by the tetrastat method after unbroken cells and cell debres had been removed.

***In line with observations in other laboratories, the F_1 preparations routinely obtained in this study contained only 4 subunits. The fifth subunit δ is easily detached, but can be observed in preparations of the ATPase complex.

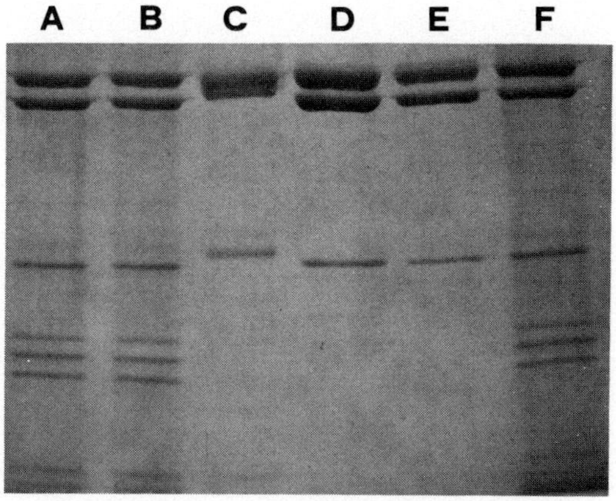

Fig. 1. Slab gel analysis of ATPases from wild-type and an aurovertin resistant mutant.

F_1 was prepared by chloroform extraction of SMP (9) from wild-type D41 and mutant aur^R-1 followed by DEAE cellulose and Ultrogel AcA34 chromatography. ATPase complex was isolated as previously described (10) and further purified on Ultrogel AcA22. Subunit resolution was performed on a 10-15% exponential gradient slab gel (11). A, aur^R-1 complex; B, and F, D41 complex; C, beef-heart F_1 included for comparison; D, aur^R-1 F_1 and E, D41 F_1.

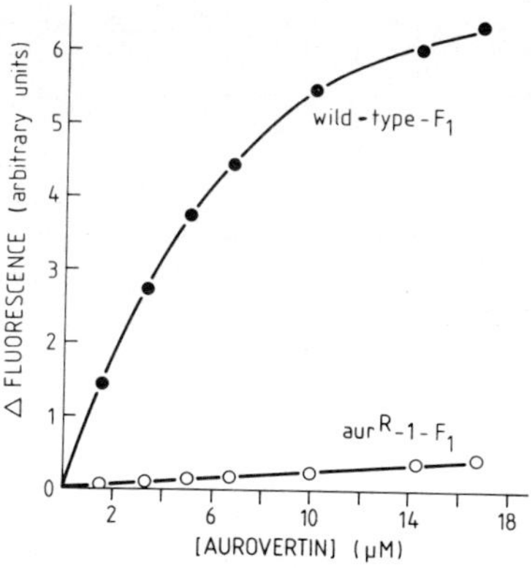

Fig. 2. Mutant F_1 does not bind aurovertin.

Fluorescence (excitation at 368 nm and emission at 470 nm) was measured for the indicated aurovertin additions in the presence of 0.25 μM F_1 from mutant and wild-type (assumed molecular weight = 380,000). The enzyme was suspended in 0.25 M sucrose, 1 mM EDTA, 50 mM Tris SO_4 pH 8.5. The temperature was 23°.

TABLE I

ENZYMATIC ACTIVITY OF ATPase PREPARATIONS FROM MUTANT AUR^R-1 AND WILD-TYPE D-41 S. CEREVISIAE

Enzyme	Strain	Specific ATPase activity (μmole ATP.min^{-1}.mg^{-1} protein)		% Inhibition
		-Aurovertin	+Aurovertin	
F_1	D-41	54.6	10.2	82
F_1	aur^R-1	62.4	51.1	18
ATPase-complex	D-41	32.2	9.0	72
ATPase-complex	aur^R-1	28.1	24.2	14

ATP hydrolysis was determined in a titrator in a medium consisting of 100 mM NaCl, 20 mM ATP, 20 mM $MgSO_4$ adjusted to pH 7.4. The reaction at 30° was initiated by adding enzyme. After a constant rate of ATP hydrolysis had been established, aurovertin was added (where indicated) to a final concentration of 0.6 μg/ml.

TABLE II

INHIBITION OF D41 AND AUR^R-1 F_1-ATPase BY VARIOUS ANTIBIOTICS

Antibiotic	% Inhibition	
	D41-F_1	aur^R-1 F_1
Aurovertin (0.3 μg/ml)	75	17
Efrapeptin (0.5 ng/ml)	50-60	57
Dio 9 (2.5 μg/ml)	81	79
Bathophenanthroline:Fe^{3+} (3:1) (0.25 μM)	74	73

ATPase activities were measured titrimetrically as described in Table I. Efrapeptin was a generous gift from Dr. H. Lardy.

The binding of aurovertin to F_1 is accompanied by a large enhancement of aurovertin fluorescence as shown previously for beef heart F_1 (7). This fluorescence increase is one of the most sensitive methods for detecting the binding of aurovertin to F_1. As shown in Figure 2, titration of 0.25 μM wild-type F_1 with aurovertin caused a marked fluorescence enhancement whereas almost no increase could be detected with 0.25 μM F_1 from the mutant aur^R-1. Subsequent addition of the wild-type F_1 to the mutant enzyme plus aurovertin (not shown) was accompanied by fluorescence enhancement which was proportional to the added wild-type enzyme. This supports the notion that the mutant F_1 does not form a non-fluorescent complex with aurovertin. Identical results were obtained when the oligomycin sensitive complex of the two strains were compared.

Since the "forward" (i.e. ATP-forming) reactions of the ATPase complex in beef-heart submitochondrial particles were reported to be more sensitive to aurovertin inhibition than the ATPase reaction (8) the oligomycin sensitive ATPase complexes from aur^R-1 and wild-type were reconstituted into phospholipid vesicles in order to compare the effect of aurovertin on their ATP-$^{32}P_i$ exchange activities. The representative data shown in Table III indicate that differences in aurovertin-sensitivity between the mutant and wild-type are much greater in this assay than in the ATPase reaction (cf. Table I). This is primarily a result of a greater sensitivity of the ATP-P_i exchange to inhibition by aurovertin in the wild-type complex. Titration experiments (not shown) indicated that the ATPase complex from mutant aur^R-1 was at least 100 times less sensitive to aurovertin inhibition of ATP-$^{32}P_i$ exchange than the wild-type enzyme.

We are currently attempting to identify the F_1 subunit altered by the aur^R-1 mutation. Various approaches are being employed including fingerprint analysis of isolated subunits. Preparations of F_1 made by the $HCCl_3$ extraction procedure lack the delta subunit and usually also the epsilon subunit (cf. D41-F_1 in Figure 1), but still retain ATPase activity and aurovertin sensitivity. The aurovertin-binding site and, by implication, also the mutation must therefore reside in one of the three large subunits. Several lines of evidence narrow the choice to the two largest subunits: (1) When F_1 or ATPase complex are incubated with trypsin, the gamma subunit can be digested away without losing aurovertin sensitivity; at the same time, however, the alpha and beta subunits are considerably reduced and the ATPase activity is lowered. (2) Preliminary micro fingerprint analysis of labeled tryptic peptides of the mutant and wild-type F_1 subunits show differences in the beta subunit peptide pattern.

TABLE III

AUROVERTIN SENSITIVITY OF ATP-$^{32}P_i$ EXCHANGE CATALYSED BY ATPase COMPLEXES FROM WILD-TYPE D41 AND THE AUROVERTIN RESISTANT MUTANT AUR^R-1

Enzyme	ATP-$^{32}P_i$ exchange*	% Inhibition
D41 - wild type		
complete	51.7	0
+ aurovertin	1.9	96
+ CCCP	3.3	94
+ oligomycin	0	100
aur^R-1 mutant		
complete	115.5	0
+ aurovertin	104.3	10
+ CCCP	14.0	88
+ oligomycin	1.7	99

*Activity expressed as nmole/min/mg protein

Inhibitor levels were 10.5 µg/mg and 1.25 µg/mg for aurovertin and oligomycin respectively. CCCP was used at 0.2 mM. For aur^R-1 ATPase vesicles, 85% inhibition of ATP-$^{32}P_i$ exchange was obtained when aurovertin was added at 1.5 mg/mg protein. Sonicated vesicles (5 mg) prepared from purified commercial soy bean phospholipid were mixed with 600 µg of ATPase complex in 0.3 M sucrose, 0.2 mM EDTA, 10 mM Tricine-KOH pH 8.0, 0.7% Na cholate, 1 mM dithiothreitol and incubated at 0° for 30 minutes. ATP-$^{32}P_i$ exchange activity at 37° was determined by Racker's dilution method (12). Time courses of ATP-$^{32}P_i$ exchange were followed up to 20 minutes. Inhibitor preincubations were for 5 min.

In summary, the present data leave little doubt that mutant aur^R-1 carries a mutation in the structural gene for one of the three largest F_1 subunits. While current evidence implicates the beta subunit, further work is required to clearly establish this point.

REFERENCES

1. Tzagoloff, A., Rubin, M. and Sierra, M. (1973) Biochim. Biophys. Acta 301, 71-104.
2. Schatz, G. and Mason, T. (1974) Ann. Rev. Biochem. 43, 51-88.
3. Senior, A. (1973) Biochim. Biophys. Acta 301, 249-277.

4. Griffiths, D.E., Houghton, R.L. and Lancashire, W.E. (1974) in The Biogenesis of Mitochondria (eds. A. Kroon and C. Saccone), Academic Press, pp. 215-223.
5. Lardy, H., Connethy, J. and Johnson, D. (1964) Biochemistry 3, 1961-1968.
6. Robertson, A.M., Beechey, R.B., Holloway, C. and Knight, I.G. (1967) Biochem. J. 104, 54.
7. Chang, T. and Penefsky, H.S. (1973) J. Biol. Chem. 248, 2746-2754.
8. Lee, C.P. and Ernster, L. (1968) Eur. J. Biochem. 3, 391-400.
9. Beechey, B., Hubbard, S., Linnett, P., Mitchell, A. and Munn, E. (1975) Biochem. J. 148, 533-537.
10. Tzagoloff, A. and Meagher, P. (1971) J. Biol. Chem. 246, 7328-7336.
11. Douglas, M. and Butow, R. (1976) Proc. Natl. Acad. Sci. 73, 1083-1086.
12. Racker, E., Chien, T. and Kandrach, A. (1975) FEBS Letters 57, 14-18.

This work was supported by a long-term EMBO fellowship to E. Agsteribbe.

Genetics and Biogenesis of Chloroplasts and Mitochondria, Th. Bücher et al. eds.

MITOCHONDRIALLY ENCODED OLIGOMYCIN-RESISTANT MUTANTS OF *S. CEREVISIAE* : STRUCTURAL INTEGRATION OF ATPase AND PHENOTYPE

Marie Somlo and Jacky Cosson
Centre de Génétique Moléculaire du C.N.R.S.
91190 Gif-sur-Yvette, France

INTRODUCTION.

Each of 4 loci (1,2,3) on mitochondrial DNA are known to confer oligomycin-resistance to *S. cerevisiae* cells grown on oxidizable carbon sources. We have chosen one allele from respectively the OLI 1 and OLI 2 loci, the mitochondrial ATPase (mit ATPase) of which shows partial oligomycin-resistance (4,5,6), and compared their properties to those from the isonuclear wild type, in order to determine how the genetically produced alteration affects various functional and structural properties of the ATPase complex at different integrational levels : in its highly purified form, in isolated mitochondria, in the whole cell.

Table I summarizes the essential data on the ATPase complex of *S. cerevisiae:* Structure and fonction of the mitochondrial ATPase complex of yeast as described by Tzagoloff and coll. (7).

Name	F_1 (8)	OSCP (9)	(7) Membrane factor	ATPase complex
Fonction	Catalyses hydrolysis of ATP; insensitive to oligomycin	Necessary but insufficient to confer oligomycin sensitivity to F_1	Attachment to the membrane ; implicated in sensitivity to oligomycin	Catalyses hydrolysis of ATP ; sensitive to oligomycin ; implicated in ATP synthesis
Molecular weight	360 000	18 000	71 000	468 000
			(+ phospholipids (10))	
synthesized on	cytoribosomes	cytoribosomes	mitoribosomes	cyto- and mitoribosomes
Structure	5 types of subunits	1 subunit	4 types of subunits, one of which (subunit 9) exists in highly polymerized or dissociated form	

MATERIALS AND METHODS

The strains PS 194 (wild type), PS 195 (O^R_{144} locus OLI 2) and PS 211 (O^R_{146} locus OLI 1) were described in (4). The references for all methods are given in Table II.

RESULTS

Table II contains various structural and functional properties which differentiate the mutants at the OLI1 and OLI 2 loci and the isonuclear wild type at different levels of integration of the ATPase.

TABLE II : Altered properties due to mutationally affected ATPase.

Observed property	Wild type	OLI 2	OLI 1	Remarks	Ref.
A. PURIFIED ATPase COMPLEX				Triton extraction of mitochondria followed by glycerol gradient purification	11
1. Loss of activity after Triton extraction	no	yes	yes		6
Reactivation by heat	no	yes	yes	S value close to 1.5 (see A-4)	
2. Arrhenius plots : - ATPase activity Transition temperature	20°	24°	24°		12
- Fluorescence of complex between ATPase, Cl-tetracyclin and ADP Transition temperature	21.7	23.5	23.9		
3. Acrylamide-SDS-gels: densitometric scanning	Both mutants : simultaneous decrease in staining density of the 4 MF bands			For the three strains the same 9 bands of the ATPase complex are seen, plus 2-3 minor contaminants	6 13
4. Analytical ultracentrifugation 11 S 1.5 S 17 S (Apparent S values)	 8% 86 6	 15% 80 5	 15% 80 5	1.5 S = phospholipid-rich-ATPase complex; 11 S = same as for F_1 (14),but not cold-sensitive although oligomycin-resistant. 11 S could be either a cold-resistant F_1 or a lipid-depleted ATPase complex ; 17 S has no ATPase activity.	13
5. Polyacrylamide-SDS-gel analysis of a mixture of ATPase complexes from OLI 2 (^{3}H-leu-labelled) and wild type (^{14}C-leu-labelled)	Simultaneous decrease in label for all MF bands of OLI 2			Growth without antibiotics,labelled cells mixed and coextracted, ATPase copurified. Reverse experiment gave analogous results.	13
B. MITOCHONDRIA				Preparation described in ADP/O and RC ratios identical for the three strains.	15
1. Coupling : effect of oligomycin μg/mg I_{50}	1.4	1.4	28-42	Substrate : ethanol. Oligomycin sensitivity of wild type and OLI 2 was closely similar also with α-ketoglutarate and succinate as substrates, at all inhibitor concentrations,at 2 pH values and with different protein concentrations.	16

TABLE II (Contd)

Observed property	Wild type	OLI 2	OLI 1	Remarks	Ref.
2. Rates of respiration (μl O_2/h/mg protein) in state 3	100	-	75	Substrate:ethanol.The same slight but consistent decrease was found for α-ketoglutarate and citrate, the 3 substrates linked to NADH. No decrease was found for succinate andD-lactate,both unlinked to NADH. Effect of excess Phospholipids (B-5) on NADH-dehydrogenase?	15 16
3. Respiratory driven H^+-translocation	+	n.d.	+	No apparent change in H^+ permeability for OLI 1.	15
4. Effect of DCCD on ATPase (5.10^{-7} M)					17
Inhibition %	25	25	50		
5. Phospholipids					18
mg/mg protein	0.16	0.16	0.21	The 30% mean excess for OLI 1 phospholipids is compatible with the preferential loss of mitochondrial protein (B-6) without accompanying loss of phospholipids.	
6. Yield of mitochondrial protein during preparation	1	1	0.65	The loss appears only after mitochondria have been sedimented from the crude extract. ATPase is preferentially retained (increased specific activity for OLI 1).	15
Buoyant density	1.19	1.19	1.19		
C. WHOLE CELLS					
1. Growth on oxidizable carbon source:generation time (hours)					
-nutrient-rich natural medium 28°	3	3	3	Yeast extract (Difco) 1%,ammonium sulfate 0.12%,éthanol 1%	15 19
-chemically defined suboptimal medium 28°	5	6	7	Yeast Nitrogen Base (Difco 0.7% ethanol 1%.No difference on fermentable substrate (glucose).	
-nutrient-rich natural medium 18°	5	5.5	6	No difference on glucose	12
2. Effect of oligomycin on growth Inhibition threshold (μg oligomycin/ml)	0.5-1	5-10	>100	Nutrient rich natural medium, pH 6.2, 28°	

DISCUSSION

The fundamental observation that the oligomycin-resistance of the mitochondrial ATPase increases in the following order (5,6) : wild type < OLI 2 (O^R_{144}) < OLI 1 (O^R_{146}) implies mutational alteration in the MF. Assuming that this modifies not only the oligomycin-resistance of the ATPase complex but also some of the functions where the ATPase intervenes, we have examined a series of properties at various levels of integration.

Thus, at the cellular level, it was interesting to find out whether the *in vivo* oligomycin-resistance of growth follows the same order as in the case of the isolated mitochondrial ATPase. On the other hand the presence of an alteration in the hydrophobic MF of the mitochondrial ATPase in the cell, under conditions where all the energy supply depends on oxidative phosphorylation, might perturb some features of growth, such as the rate, or the sensitivity to low temperature (cf. (20)).

At the level of isolated mitochondria, one or several functions related to energy production or conservation might be modified. The oligomycin-resistance of the coupling reaction is of special interest compared to the related behaviour of whole cells and of the ATPase reaction. Apart from these functional aspects, the altered MF might well modify the overall membrane architecture to the extent of producing some structural anomaly such as particular instability of the mitochondria from the mutants.

At the level of the isolated ATPase complex, interaction of the enzyme with some of its effectors, as well as the interaction between the hydrophobic MF and the rest of the complex might be altered.

The order of oligomycin-resistance OS < OLI 2 < OLI 1 for the mitochondrial ATPase reaction implies that the wild type differs from the two mutants, and the two mutants from each other. On examining the results of Table II it appears that this relationship does not hold for all properties : while for some of them the two mutants do differ both from the wild type and from each other, for others only one mutant differs from the wild type, and for still others, both mutants differ from the wild type but not from each other. Inspection of those properties which distinguish at least one of the mutants from the wild type shows a different relationship for each level of integration : at the cellular level the three strains differ from each other, and the degree of difference for all examined properties follows the same order as the oligomycin-resistance of mitochondrial ATPase. At the mitochondrial level, in contrast, the mutant at the OLI 2 locus cannot be distinguished from the wild type by either of the examined properties. Knowing that the ATPase in mitochondria from the mutant at the OLI 2 locus is oligomycin-resistant, it is particularly remarkable that the coupling in these mitochondria is oligomycin-sensitive. The discrepany might be explained by a change in conformation occurring between coupling and hydrolytic conditions as detailed below. At the purified complex level, the two mutants differ from the wild type, but not from each other.

Starting from the mitochondrial level, Tzagoloff et al. (21) have shown for mutants at the OLI 1 locus that the proteolipid (subunit 9 synthesized on mitoribosomes) of the MF extracted from whole mitochondria was "spontaneously" converted from the high (45000) to the low (8000) molecular weight form. This observation suggests that subunit 9 is the product of the OLI 1 locus. It is tempting to correlate this property of "spontaneous" dissociation with the specific "instability" of mitochondria from the mutant at the OLI 1 locus during purification (loss of protein, Table II-B-6). Also, significantly, no such "instability" was observed for mitochondria from the mutant at the OLI 2 locus, nor was the high molecular weight form of the proteolipid of the mutant at the OLI 2 locus (or the wild type) dissociated unless deliberately treated by organic solvents (21).

Thus our relationship $\overset{\text{wild type}}{\text{OLI 1} \neq \text{OLI 2}}$ for the properties examined at the mitochondrial level also holds for the observations concerning subunit 9 after extraction from mitochondria (21). It is reasonable to expect that the difference between the two mutants with respect to dissociation properties of subunit 9 will also be observed after extraction of the purified ATPase complex. This expectation is substantiated by our finding that labelled subunit 9 obtained by SDS-treatment of the purified ATPase of the mutant at the OLI 2 locus does not undergo dissociation to the low molecular weight form. Therefore even at the purified ATPase complex level, there would be an intrinsic difference between the MF of the two mutants ; if this is so, the fact that we have found no difference between a series of properties examined at the level of the purified ATPase complex cannot be attributed to the loss during purification of a component responsible for the difference at the mitochondrial level. How can the similarity of our findings at the purified ATPase complex level be accounted for ? The properties we have selected to study the whole purified complex all tended to test how the interaction between the MF and the rest of the complex might be modified by the mutational alteration in the MF: in fact any change in ATPase activity (loss of activity during Triton extraction and heat reactivation (Table II-A-1), change in transition temperature in Arrhenius plots (A-2)) necessarily corresponds to a modified interaction between the MF and the catalytic part of the complex. Likewise any change in cohesiveness of the ATPase complex (partial loss of MF subunits, Table A-3 and 5, increased proportion of the 11S component, A-4) will also be due to such a modified interaction.

Now a physically more or less extended zone of interaction between the MF and the rest of the ATPase complex could account for the apparent identity of the two mutants at the level of the purified complex and for their differences at the two other levels,as described in the following model;

In the purified complex the zone of interaction between the MF and the rest of the complex is enlarged by the presence of Triton (for the effect of Triton, see (6,17). For either mutant, the mutational alteration is expressed as a long range disturbance of interaction which is independent of the starting point of this disturbance. As a result there occurs in the complex from each mutant a displacement from the stable wild type species towards an unstable species in equilibrium with a mixture of F_1 + MF, accounting for the similarity in the properties observed for the ATPase complexes of the two mutants.

At the level of mitochondria, as for the purified ATPase complex, the mutationally modified zone of interaction can account for the fact that the ATPase from the mutants is oligomycin-resistant. However, with regard to the coupling reaction, the "ATP-synthetase" from the mutant at the OLI 2 locus is oligomycin-sensitive. In order to account for this behaviour we assume that in the isolated mitochondria the ATPase is integrated through a much more restricted zone of interaction than in the purified complex. In the case of the mutant at the locus OLI 1, the altered site of the MF is inside this zone of interaction, allowing expression of the mutated phenotype both in the ATPase and the coupling reactions. In the case of the mutant at the locus OLI 2, the altered site is outside of the zone of interaction and the mutated phenotype cannot be expressed in the coupling reaction. In contrast,the conformational state prevailing during the ATPase reaction establishes contact between the mutationally altered part and the zone of interaction, thereby allowing expression of the resistant phenotype.

		Wild type	OLI 1	OLI 2
Coupling	altered site	-	included	excluded
	reaction with oligomycin	O^S	O^R	O^S
ATPase	altered site	-	included	included
	reaction with oligomycin	O^S	O^R	O^R

In the mitochondria *in situ* (whole cells), it might be supposed that the zone of interaction between MF and the rest of the ATPase is large enough to include the mutationally altered site of each mutant (although not in the "Triton-stretched" conformation imagined for the purified complex). Therefore oligomycin-resistance of coupling, and hence of growth, would be operative in the whole cells of the mutant at the OLI 2 locus, although to a slighter extent than for the mutant at the OLI 1 locus.

As mentioned before, the observations of Tzagoloff et al. (21) suggest that subunit 9 is the product of the OLI 1 locus. As regards the OLI 2 locus the product is still unknown. However, one can already exclude the hypothesis that the OLI 1 and OLI 2 loci belong to the same gene : this would be incompatible with the high frequency of recombination between OLI 1 and OLI 2 (1) (subunit 9 corresponds to a short polynucleotide stretch of not more than 300 base pairs) ; also another gene is located between OLI 1 and OLI 2 (22). Therefore either the products of the OLI 1 and OLI 2 loci are two different polypeptide chains, or we are dealing with two duplicated DNA segments which were separated and have evolved differently in the course of time.

ACKNOWLEDGEMENTS. Helpful discussions with Dr. P.P. Slonimski are gratefully acknowledged. This work was supported by a grant of D.G.R.S.T. N° 659 1338.

REFERENCES

1. Avner, P.R., Coen, D., Dujon, B. and Slonimski, P.P. (1973) Mol. Gen. Genet. 125, 9-52.
2. Lancashire, W.E. and Griffiths, D.E. (1975) Eur. J. Biochem. 51, 403-413.
3. Clavilier, L. (1976) Genetics 83, 227-243.
4. Somlo, M., Avner, P.R., Cosson, J., Dujon, B., and Krupa, M. (1974) Eur. J. Biochem. 42, 439-445.
5. Griffiths, D.E., and Houghton, R.L. (1974) Eur. J. Biochem. 46, 157-167.
6. Cosson, J. and Spiridakis, A.(1974) Biochem. Biophys. Res. Commun. 59, 1039-1046.
7. Tzagoloff, A., Rubin, M.S. and Sierra, M.F. (1973) Biochim. Biophys. Acta 301, 71-104.
8. Penefsky, H.S., Pullman, M.E., Datta, A. and Racker,E.(1960) J. Biol. Chem. 235, 3330-3336.
9. MacLennan, D.H. and Tzagoloff, A. (1968) Biochemistry 7, 1603-1610.
10. Bulos, B. and Racker, E. (1968) J. Biol. Chem. 243, 3901-1905.
11. Tzagoloff, A. and Meagher, P. (1971) J. Biol. Chem. 246, 7328-7336.
12. 13. Cosson, J. and Spiridakis, A. Manuscripts in preparation.
14. Catterall, W.A. and Pedersen, P.L. (1974) Biochem. Soc. Spec. Publ. 4, 63-88.
15. Somlo, M., Reid, R.A. and Krupa, M. (1976) Biochem. J. In the press.
16. Somlo, M. Manuscript in preparation.
17. Somlo, M. and Krupa, M. (1974) Biochem. Biophys. Res. Commun. 59, 1165-1171.
18. Kader, J.C., Krupa, M. and Somlo, M. Manuscript in preparation.
19. Michaelis, G. and Somlo, M. (1976) J. Bioenergetics, 8, 93-107.
20. Trembath, M.K., Monk, B.C., Kellerman, G.M., and Linnane, A.W. (1975) Mol. Gen. Genet. 141, 9-22.

21. Tzagoloff, A., Akai, A. and Foury, F. (1976) Febs Letters, 65, 391-395.

22. Slonimski, P.P. and Tzagoloff, A. (1976) Eur. J. Biochem. 61, 27-41.

Genetics and Biogenesis of Chloroplasts and Mitochondria, Th. Bücher et al. eds.

INVESTIGATION OF THE OLIGOMYCIN BINDING PROTEIN IN YEAST MITOCHONDRIAL ATPase

R.S. Criddle, C. Arulanandan, T. Edwards, R. Johnston, S. Scharf, and R. Enns
Department of Biochemistry and Biophysics
University of California
Davis, California, USA

INTRODUCTION

Oligomycin has been shown to be a potent inhibitor of oxidative phosphorylation in mitochondria. The site of action of the antibiotic has further been shown to be at the membrane bound mitochondrial ATPase enzyme which is directly involved in the coupling of oxidation and phosphorylation (for review see 1). When the soluble F_1-ATPase catalytic unit is separated from the membrane fraction, it is no longer inhibited by oligomycin. Sensitivity to inhibitor can be restored, however, by recombination of F_1-ATPase with a membrane fraction[2]. Alternatively, an oligomycin sensitive ATPase (OS-ATPase) can be isolated by Triton X-100 extraction of submitochondrial particles which contains both F_1-ATPase and a "membrane fraction"[3].

These studies have led to the suggestion that the site for binding oligomycin is associated with the membrane fraction of the ATPase enzyme complex[4,5]. Direct identification of an oligomycin binding protein has not been previously reported, although considerable indirect evidence has accumulated. For example, Cattell et al.[6] showed that dicyclohexylcarbodiimide (DCCD) could be bound to a 10,000 molecular weight protein in submitochondrial particles which probably corresponds to subunit 9 of OS-ATPase (using the subunit designations of Tzagoloff and Meagher[7]). DCCD has much the same inhibitory effects on ATPase as does oligomycin. Both the DCCD binding protein and subunit 9 of ATPase can be extracted with chloroform-methanol mixtures[8,9]. This led Sierra and Tzagoloff to suggest that subunit 9 may also be the oligomycin binding protein[10]. Ernster et al.[11] have reached a similar conclusion based on comparative studies of inhibition by the two inhibitors.

RESULTS AND DISCUSSION

To examine the question of the identity of the oligomycin binding protein further and to investigate changes resulting from formation of oligomycin resistant mutants in the yeast mitochondrial ATPase, an affinity labeling technique has been developed. This technique was based on the postulate that one or more of the five keto moieties in the oligomycin molecule may be bound in the vicinity of an amino group on the binding protein and may participate in formation of a Schiff's base complex. Reductive alkylation with ^{3}H-sodium borohydride would then result in formation of a covalent complex between oligomycin and the binding protein with the introduction of a tritium label.

Addition of oligomycin to preparations of OS-ATPase, or to submitochondrial particles, followed by borohydride reduction does cause an oligomycin dependent uptake of ^3H into the OS-ATPase. The label becomes associated with a low molecular weight fraction that migrates with subunit 9 of the ATPase when analyzed by SDS polyacrylamide gel electrophoresis as shown in Figure 1. The ^3H-label appears to be bound to protein, as proteolysis with pronase followed by chymotrypsin reduces the size of the labeled products (Fig. 2). Moreover, the label seems to be covalently bound to the protein as it remains associated with subunit 9 sized material through heating with SDS and urea, solubilization in alkali, and thin layer chromatography on silica gel. The oligomycin affinity labeled protein is soluble in chloroform:methanol (2:1) solutions and therefore also resembles subunit 9 in this property.

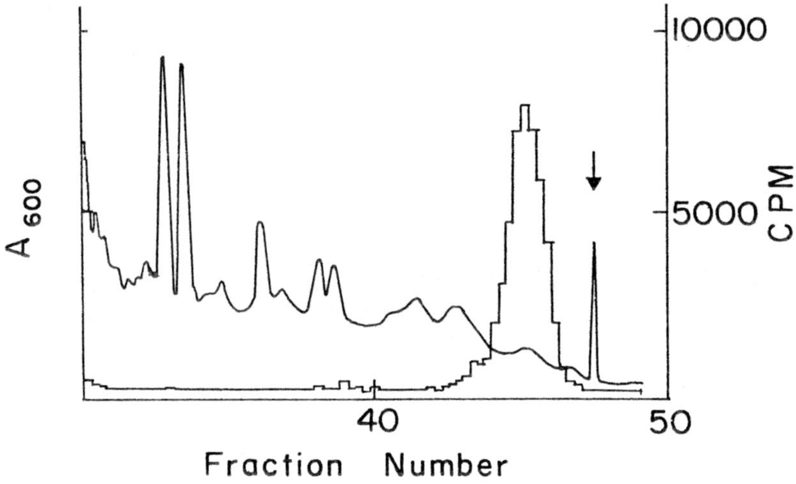

Fig. 1. SDS gel electrophoretic analysis of oligomycin affinity labeled ATPase. Oligomycin sensitive ATPase was purified from the Triton X-100 extract of yeast submitochondrial particles by chromatography through Sepharose 6-B[14]. Oligomycin was added at a concentration of 200 μg/mg protein and then ^3H-sodium borohydride was used for reductive alkylation. Low molecular weight reaction by products were removed by chromatography on Sephadex G-25 and the protein was then extracted with 90% methanol:10% water. The protein was then solubilized in SDS-urea and electrophoresed in 12% gels by the methods of Webber and Osborne. Gel samples were analyzed by staining protein with Coomassie blue or by slicing the gels, dissolving fractions in 10% H_2O_2 and counting in PCS scintillation fluid.

When the reductive alkylation procedure is carried out with submitochondrial particles rather than with isolated OS-ATPase, a portion of the label goes into protein that is solubilized with Triton X-100 but is not associated with ATPase activity (Fig. 3). The proteins of this low molecular weight fraction (peak III of Fig. 3) was also analyzed by gel electrophoresis and again the label was found to migrate only with subunit 9 sized material (Fig. 4). It appears quite probable that a portion of the subunit 9 peptides may not be recovered as part of the OS-ATPase complex, thus raising questions regarding the absolute stoichiometry of this peptide in the enzyme complex.

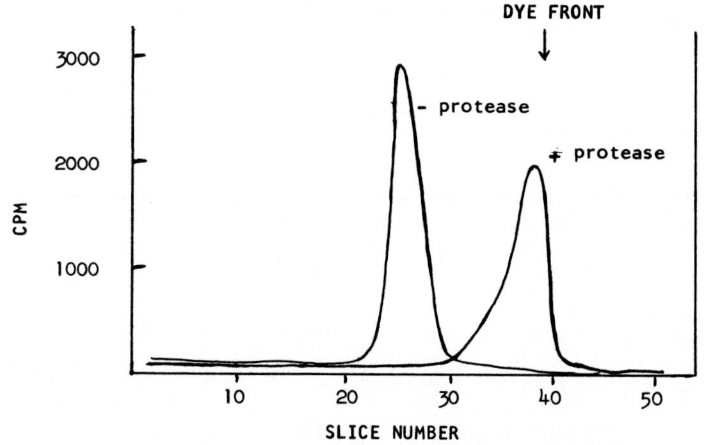

Fig. 2. SDS gel electrophoretic analysis of oligomycin affinity labeled ATPase following degradation with proteases. The protein samples were prepared as in the legend of Fig. 1. The samples were then boiled for 10 min, cooled, and pronase was added (0.1 mg/mg protein). After 90 min incubation, the samples were again boiled, rapidly cooled and chymotrypsin was added (0.1 mg/mg protein). The samples were then electrophoresed in 14% acrylamide SDS gels and analyzed as in Fig. 1.

It is difficult to establish homogeneity of protein preparations with the low molecular weight and solubility properties exhibited by the oligomycin affinity labeled protein products in these studies. The gel electrophoretic analyses establish only that label is associated with material having a relatively uniform size but cannot rule out either the possible association of label with many similar sized peptides or the specific binding of label to only one out of such a group of similar peptides. It is possible, however, to demonstrate specificity of the affinity labeling by analysis of effects on the enzymic activity and by comparing oligomycin sensitive and oligomycin resistant mutant strains of yeast.

The reductive alkylation procedures used for affinity labeling do not inhibit ATPase in the absence of added antibiotic. Furthermore, activity loss is proportional to the amount of oligomycin bound, suggesting specific alkylation of some functional unit of the enzyme complex. Maximum inhibition of OS-ATPase is observed at a level of about 0.5 to 1.5 ^{3}H-incorporated per ATPase molecule.

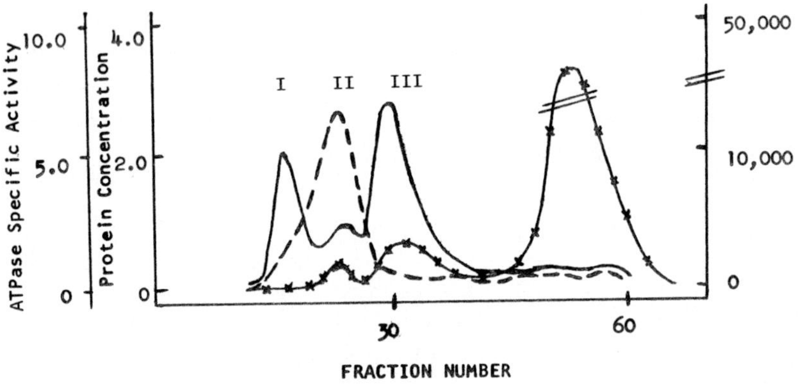

Fig. 3. Separation of the Triton X-100 extraction products of submitochondrial particles. Submitochondrial particles were extracted with 0.25% Triton using the procedure of Tzagoloff and Meagher. The soluble material was concentrated by ultrafiltration and fractionated by chromatography on Sepharose 6-B[14]. ATPase activity was monitored (-----) and compared with the eluted protein (———) and with ^3H-label incorporated into protein by reductive alkylation (-X-X-). The protein peaks were termed peaks I, II, and III based on order of elution.

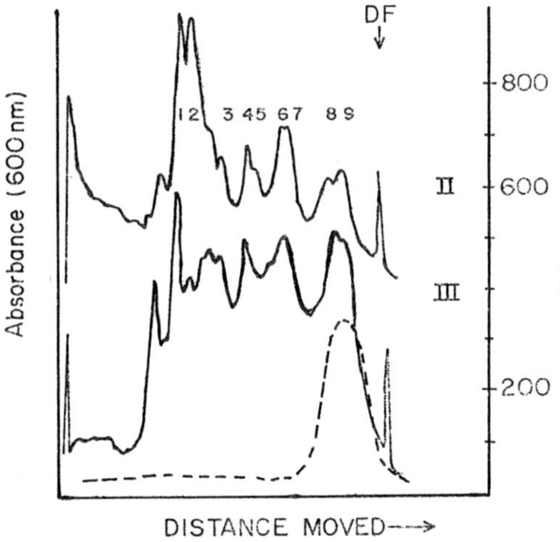

Fig. 4. Separation of low molecular weight components from a Triton X-100 extract of submitochondrial particles by SDS gel electrophoresis 7.5% polyacrylamide gels. For comparison, the distributions of proteins from both peaks II and III (see Fig. 3) are shown and compared with the distribution of radioactivity (-----) obtained in proteins associated with peak III.

A measurement of relative oligomycin binding affinities for different samples of ATPase can be obtained by carrying out the reductive alkylation procedure in the presence of varying amounts of added oligomycin. Figure 5 shows such binding curves for mitochondria from an oligomycin sensitive strain and for two oligomycin resistant mutants derived from this strain. These results indicate that strains resistant to inhibition by oligomycin also require higher levels of oligomycin for the affinity labeling reaction. Both of the mutant strains illustrated in Figure 5 have been shown by genetic analysis to belong to the class B oligomycin resistant mutations defined by Trembath et al[12]. Since the affinity labeling reaction appears to be specific for subunit 9, these results suggest that subunit 9 is the gene product of the class B oligomycin resistance region of the mitochondrial DNA. These studies are in agreement with tentative observations by Groot Obbink et al.[13] that some alteration in aggregation properties of subunit 9 may result from a class B oligomycin resistant mutation.

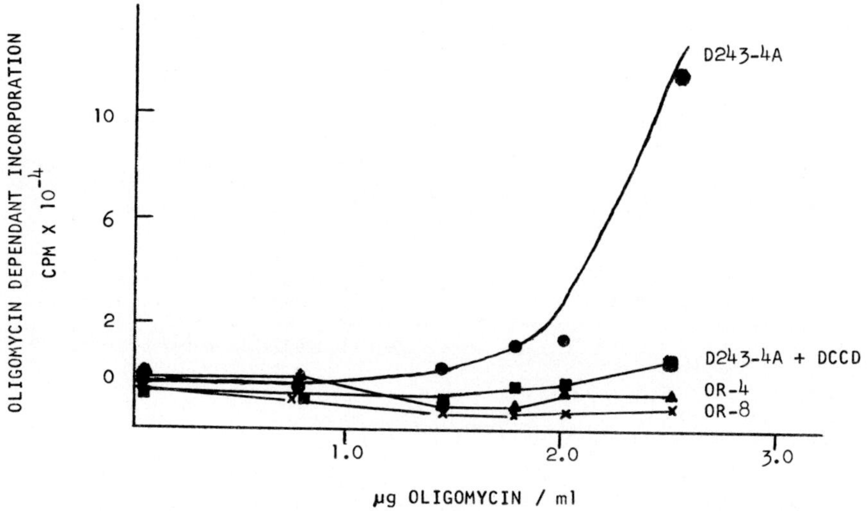

Fig. 5. Oligomycin binding curves for affinity labeling of mitochondrial ATPases using submitochondrial particles from oligomycin sensitive and oligomycin resistant strains of yeasts. Oligomycin at the concentrations indicated was added to 5 mg samples of submitochondrial particles. ^{3}H-sodium borohydride was then added to each sample and reacted for 30 min at zero degrees. Reaction by-products were removed by extraction four times with 90% methanol:water. The protein samples were then solubilized with formic acid and counted in PCS scintillation fluid. The results shown are for strain D243-4A (-●-●-), oligomycin sensitive parent strain of OR-4 (-▲-▲-) and OR-8 (-X-X-). In addition, 0.1 µg/ml of DCCD was added to a preparation of D243-4A prior to the addition of oligomycin and reductive alkylation. The results of this experiment are also shown (-■-■-).

Figure 5 also shows that addition of low levels of DCCD, sufficient to just block ATPase activity, prior to reductive alkylation of OS-ATPase, blocks incorporation of ^{3}H-label into the protein. This again suggests that subunit 9 may be the binding site of both DCCD and oligomycin. This conclusion is strengthened by comparing the effects of DCCD on oligomycin sensitive and oligomycin resistant strains of yeast. Studies using submitochondrial particles from OR-4 and from its sensitive parent strain, D243-4A, are shown in Figure 6. It can be seen that cross-resistance between DCCD and OR-4 is observed.

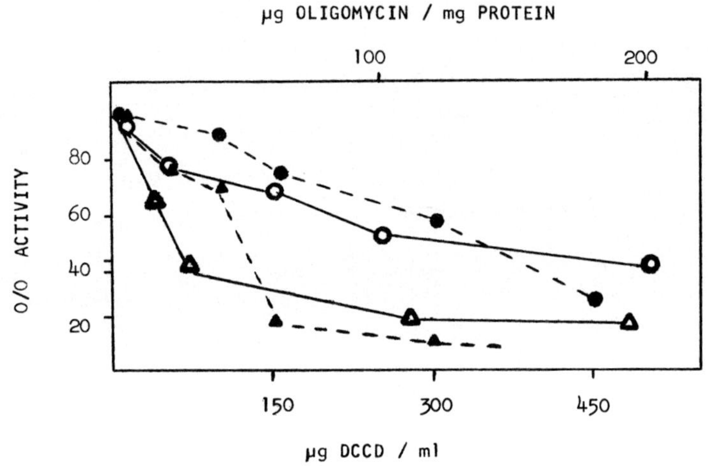

Fig. 6. Sensitivity of oligomycin resistant and oligomycin sensitive strains of yeast to inhibition by DCCD. The ATPase activity of mitochondrial particles from sensitive (D243-4A) (-▲-▲-), and resistant (OR-4) (-●-●-) strains were assayed at the DCCD concentrations indicated. For comparison, sensitivity of these preparations to oligomycin is also shown: OR-4 (-○-○-); D243-4A (-△-△-).

SUMMARY

These experiments have established that reductive alkylation of mitochondrial ATPase causes specific introduction of ^{3}H-label into a protein corresponding in properties to subunit 9 of ATPase. This binding affinity for oligomycin is decreased by mutations which decrease sensitivity of oligomycin inhibition of ATPase activity. The same mutations result in a simultaneous decrease in sensitivity to inhibition by DCCD which is consistent with previous suggestions that DCCD binds to subunit 9. It now appears that both inhibitors not only interact with the same protein but may also overlap in their binding sites.

Since the mutants studied are class B oligomycin resistance mutants, it is tempting to speculate that the changes observed in oligomycin and DCCD binding to subunit 9 are the direct result of mutational changes in the protein of this subunit. Further support for this proposal must await similar binding studies on oligomycin resistant mutants of class A and class C types.

REFERENCES

1. Racker, E. (1970) in Membranes of Mitochondria and Chloroplasts, Van Nostrand Reinhold, New York.
2. Tzagoloff, A. (1969) J. Biol. Chem. 244, 5020.
3. Tzagoloff, A., Rubin, M.S. and Sierra, M.F. (1973) Biochim. Biophys. Acta 301, 71.
4. Tzagoloff, A. (1971) in Current Topics in Membranes and Transport (Bronnet, F. and Kleinzeller, A., eds.), Vol. 2, pp. 157-205, Academic Press, New York.
5. Shannon, C., Enns, R., Wheelis, L., Burchiel, K. and Criddle, R.S. (1973) J. Biol. Chem. 248, 3004.
6. Cattell, K.J., Knight, I.G., Lindop, G.R. and Beechey, R.B. (1970) Biochem. J. 117, 1011.
7. Tzagoloff, A. and Meagher, P. (1971) J. Biol. Chem. 246, 7328.
8. Tzagoloff, A. and Akai, A. (1972) J. Biol. Chem. 247, 6517.
9. Cattell, K.J., Kindop, G.R., Knight, I.G. and Beechey, R.B. (1971) Biochem. J. 125, 169.
10. Sierra, M.F. and Tzagoloff, A. (1973) Proc. Natl. Acad. Sci. USA 70, 3155.
11. Ernster, L., Nordenbrand, K., Chude, O. and Juntti, K. (1974) in Membrane Proteins in Transport and Phosphorylation (Azzone, G.F., Klingenberg, M.E., Quagliariello, E. and Siliprandi, N., eds.), pp. 29-41, North-Holland Publ. Co., Amsterdam.
12. Trembath, M.K., Molloy, P.L., Sriprakash, K.S., Cutting G.J., Linnane, A.W. and Lukins, H.B. (1976) Mol. gen. Genet. 145, 43.
13. Groot Obbink, D.J., Hall, R.M., Linnane, A.W., Lukins, H.B., Monk, B.C., Spithill, T.W. and Trembath, M.K. (1976) in The Genetic Function of Mitochondrial DNA (Kroon, A.M. and Saccone, C., eds.), North-Holland Publ. Co., Amsterdam (in press).
14. Enns, R. (1976) Ph.D. Thesis, University of California, Davis.

Genetics and Biogenesis of Chloroplasts and Mitochondria, Th. Bücher et al. eds.
© 1976 Elsevier/North-Holland Biomedical Press, Amsterdam, The Netherlands.

POLYPEPTIDES ENCODED BY MITOCHONDRIAL GENES IN ASCOMYCETES

H. Küntzel, M.A. Marahiel, D.E. Leister and P. Nelson
Max-Planck-Institut für experimentelle Medizin
Abteilung Chemie, 34 Göttingen, Germany

It is now well established that the three largest subunits of cytochrome oxidase, the apoprotein of cytochrome b and at least two subunits of the ATPase complex are synthesized in ascomycetes on mitochondrial ribosomes (1) and are controlled by mitochondrial genes (2). However, not a single individual polypeptide has yet been identified as a product of one of the various mitochondrial loci.

We have previously reported that most in vivo products of mitochondrial protein synthesis in yeast and Neurospora crassa can be heat-dissociated in SDS buffer to small polypeptides of apparent molecular weights (AMW) 8 to 10 k, if all sulfur-containing amino acid residues are completely oxidized to cysteic acid and methionine sulfone by performic acid (3,4). Since peptide bonds are not hydrolyzed under the condition used (5), the results suggest that the major mt-translation products migrating in SDS gels with AMW greater than 12 k are SDS-stable complexes of small, hydrophobic polypeptides.

Furthermore, we have demonstrated that the majority of the in vitro translation products of an E.coli cell-free system programmed with Neurospora mtDNA or of a submitochondrial system from Neurospora are small polypeptides (AMW 10-12 k), which have a strong tendency to form SDS-stable heavy aggregates at high protein concentration or at low SDS concentration (6,7).

These results suggest that most, if not all, major polypeptides encoded by mitochondrial genes are small and that some of them are recovered from SDS-lysed membranes or enzyme complexes in monomeric form and others as specific, SDS-stable complexes.

Here we report that the smallest ATPase subunit (AMW 6 k) in Aspergillus nidulans is the product of a mitochondrial gene carrying an oligomycin resistance locus. Furthermore, we demonstrate that the largest subunit of yeast cytochrome oxidase is further dissociated by SDS denaturation to polypeptides of AMW ~10 k after complete oxidation of sulfur-containing amino acids.

ALTERED ATPAse SUBUNIT IN AN EXTRANUCLEAR OLI^R MUTANT OF ASPERGILLUS NIDULANS

We have isolated about 100 oligomycin-resistant mutants of Aspergillus. Heterokaryon analysis (N.J. Pieniazek, unpublished) established that 11 of these are extranuclear mutants.

The three most stable mutants were biochemically analyzed. Gradient-purified mitochondria were lysed with Triton X-100, and ATPase activity was measured at pH 6.5 in the presence of increasing concentrations of oligomycin (10).

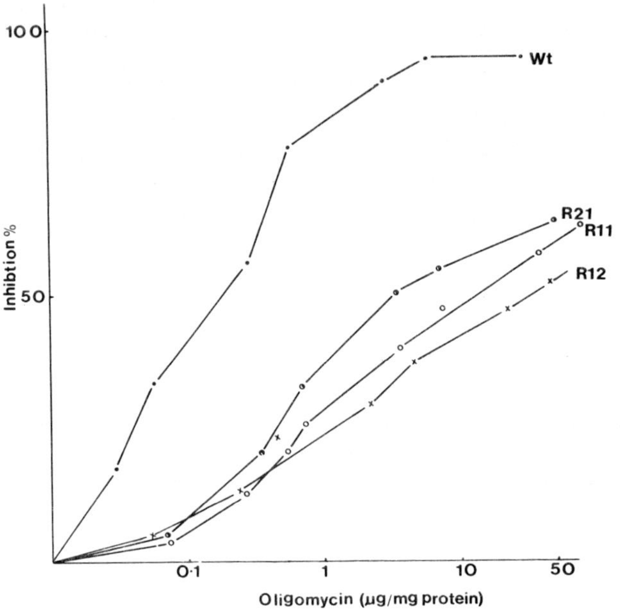

Fig. 1. Oligomycin sensitivity of solubilized mitochondrial ATPase from wild type and three extranuclear oligomycin-resistant mutants of Aspergillus. Submitochondrial particles were suspended to 6 mg/ml protein and lysed by adding 0.3% Triton X-100, 2 mM ATP and 2% methanol. Aliquots of a 100,000 x g supernatant were assayed at pH 6.5 (10).

Fig. 1 shows that the oligomycin concentration required for 50% inhibition of ATPase activity is about 20, 80 and 200 times higher in the mutant extracts R 21, R11 and R 12, respectively, than in the wild type extract.

The ATPase complex of wild type and mutant R 11 was purified to electrophoretic homogeneity (see gel A, Fig. 2) by glycerol gradient centrifugation and analyzed by SDS-urea-PAGE (8).

The dissociated enzyme contains 15 major bands of AMW between 65 k and 5.6 k (gel B, Fig. 2). Four of these bands (AMW 5.6 k, 10 k, 19 k and 40 k) were identified as mitochondrial translation products by in vivo pulse-labelling in the presence of cycloheximide and/or chloramphenicol.

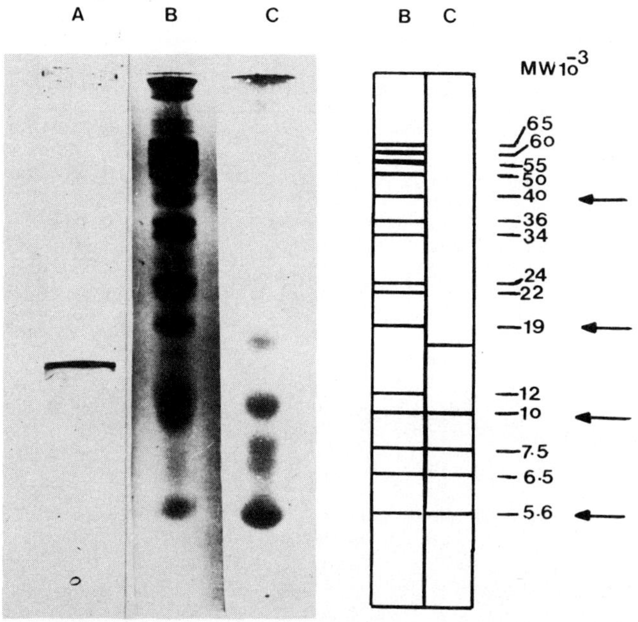

Fig. 2. Polyacrylamide gel electrophoresis of the oligomycin-sensitive ATPase complex from Aspergillus (wild type). A, PAGE (3% acrylamide) of the undissociated enzyme. B, SDS-urea-PAGE (12% acrylamide) of the dissociated enzyme. C, SDS-urea-PAGE of a fraction of mitochondrial proteins extracted with chloroform:methanol (2:1, 0.2% acetic acid). The arrows indicate protein subunits synthesized on mitochondrial ribosomes.

The majority of the in vivo mitochondrial translation products of Aspergillus can be extracted from mitochondria by acidified chloroform:methanol (2:1), and about 20% of the extractable mitochondrial proteins elute from SDS-Sephadex G-200 with an average AMW of 8 k. Two of the major proteins in this fraction are indistinguishable from the intramitochondrially synthesized 5.6 k and 10 k subunits of ATPase in their site of synthesis and their mobility in SDS-urea-PAGE (gel C, Fig. 2).

The subunit patterns of the SDS-dissociated ATPase complexes from wild type and mutant R 11 are identical. However, the 5.6 k protein isolated from wild type cells differs from that of the mutant R 11 in its stronger tendency to form an SDS-stable dimer and larger aggregates upon concentration, and in its lower polarity (see Table 1).

Table 1. Amino acid composition of the smallest ATPase subunit from wild tpye and oligomycin-resistant mutant R 11 of Aspergillus.

Amino acid	Mole percentage		Number of residues	
	wt	R 11	wt	R 11
Asx	6.11	7.47	4	5
Thr	1.59	2.77	1	2
Ser	7.64	10.84	4	6
Glx	6.97	9.19	4	5
Pro	2.93	3.44	2	2
Gly	13.45	16.14	8	9
Ala	11.86	9.64	7	6
Cys	n.d.	n.d.	-	-
Val	8.92	6.80	5	4
Met	5.13	1.64	3	1
Ile	8.01	5.98	5	4
Leu	10.15	8.22	6	5
Tyr	3.55	3.36	2	2
Phe	6.36	4.56	4	3
Lys	3.24	4.41	2	3
His	0.86	1.64	1	1
Arg	3.24	3.89	2	2
Try	n.d.	n.d.	-	-
Minimal m.w.			6331	6229
Polarity			30%	39%

Samples were hydrolysed for 20, 48, 72 and 96 hours at 110° in 5.7 N HCl. Values for Ser and Thr were derived by extrapolation to zero time and those for Val, Ile and Leu are maximal values.

The data of Table 1 clearly indicate that some apolar amino acids (including two of three methionine residues) of the wild type protein have been replaced in the mutant by more polar amino acids (including two serine residues).

We conclude that the smallest subunit of the ATPase complex in Aspergillus is coded by a mitochondrial gene carrying an oligomycin resistance marker and that the oligomycin resistance of the mutant R 11 is caused by the replacement of at leasttwo amino acids in this ATPase subunit.

COMPLEX STRUCTURE OF THE LARGEST SUBUNIT OF YEAST CYTOCHROME OXIDASE

Cytochrome oxidase was isolated from Saccharomyces cerevisiae by using the method of Poyton and Schatz (9). The subunits of the SDS-dissociated enzyme were resolved by polyacrylamide gel electrophoresis in urea-SDS as described by Swank and Munkres (8).
Fig. 3A shows a typical densitogram of a stained gel. The AMW agree well with published data (9) except for subunit 6 which migrates with an AMW of 5.5 k instead of 10 k. Furthermore subunit 1 forms a less diffuse band than in gel systems without urea and the free electrophoretic mobility of the band determined in a Ferguson plot appears to be normal.

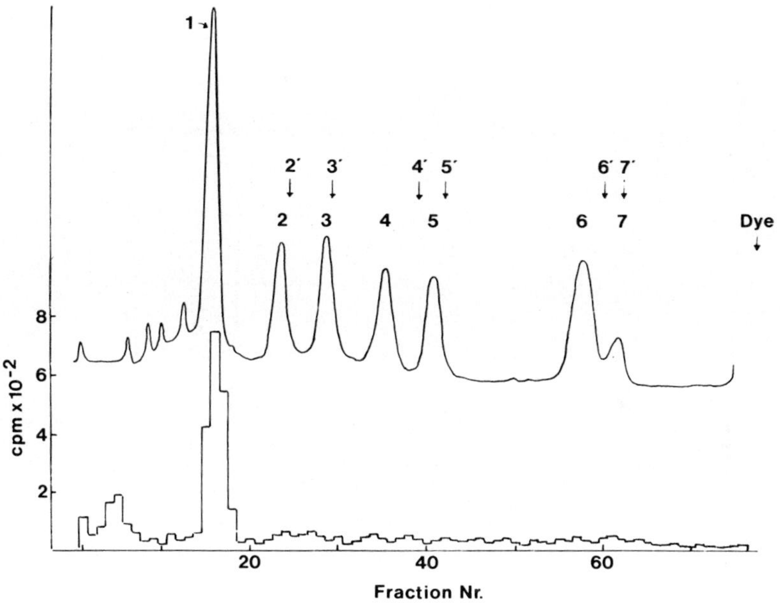

Fig. 3. SDS-urea-PAGE (12% acrylamide) of yeast cytochrome oxidase (upper part) and of re-isolated, [3H]dansylated subunit 1 (lower part). The arrows indicate the relative positions of fluorescent bands after dansylation of the whole enzyme.

Subunit 1 was purified by preparative slab gel electrophoresis essentially as described (9). The protein was [^{3}H]-dansylated at pH 9.6 in a pH-stat and purified by Sephadex G-25 filtration in 0.1% SDS. Aliquots were analyzed by urea-SDS-PAGE (Fig. 3) or by Sephadex G-200 filtration in the presence of 1% SDS (Fig. 4A).

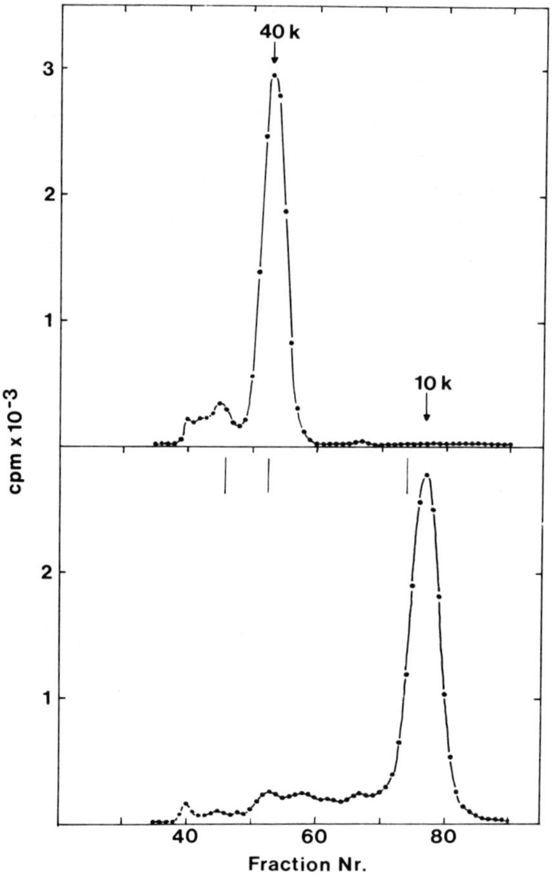

Fig. 4. SDS-Sephadex G-200 filtration of the re-isolated subunit 1 of cytochrome oxidase. Upper part: The protein was [3H]-dansylated, purified by SDS-Sephadex G-25, heat-dissociated in 1% SDS and subjected to Sephadex G-200 filtration in 1% SDS. Lower part: The protein was treated with performic acid at 4° to completely oxidize sulfur-containing amino acids, and then [3H]-dansylated and analyzed as above. The bars indicate the elution positions of bovine serum albumin (69 k), ovalbumin (43 k) and cytochrome c (12 k).

From both experiments it is obvious that the AMW of the subunit has not been altered by dansylation. Another aliquot of re-isolated subunit 1 was oxidized with performic acid at 4° for 12 hours, and after complete removal of performic acid treated with [^{3}H]-danyl-chloride and purified by SDS-G-25 filtration. An aliquot of this sample was diluted with 1% SDS, 50 mM Tris-HCl, 1 mM DTT (pH 7.5), heated in a boiling water bath for 2 min and subjected to G-200 filtration. Fig. 4B shows that the protein elutes as a symmetric peak with an AMW of 10 k. The same protein was analyzed by urea-SDS-PAGE. The gels exhibited a fluorescent and radioactive band of AMW 9 k in addition to poorly resolved larger bands which we attribute to re-aggregation of the oxidized small proteins.

These results suggest that subunit 1 of cytochrome oxidase is a SDS-stable complex of several, probably four, small polypeptides of rather uniform size.

Unfortunately, most of the [3H]-dansyl chloride is incorporated into internal lysyl and tyrosyl residues, and the yield of dansylated N-terminal amino acids is too low to decide whether the small polypeptides differ in their end groups.
However, the recent finding that cytochrome oxidase subunit 1 is controlled in yeast by at least two different mitochondrial genes (2) raises the interesting possibility that different mitochondrially coded small polypeptides contribute to the structure of this subunit.

ACKNOWLEDGEMENT: This study was supported by the Deutsche Forschungsgemeinschaft.

REFERENCES

1. Schatz, G. and Mason, T.L. (1974) Ann.Rev.Biochem. 43, 51-87.
2. Tzagoloff, A., Akai, A., Needleman, R.B. and Zulch, G. (1975) J.Biol.Chem. 250, 8236-8242.
3. Leister, D.E., Rogers, P.J. and Küntzel, H. (1976) in Mitochondrial Genetics, Biogenesis and Bioenergetics (De Gryuter, Berlin), in press.
4. Küntzel, H., Pieniazek, N.J., Pieniazek, D. and Leister, D.E. (1975) Eur.J.Biochem. 54, 567-575.
5. Hirs, C.H.W. (1967) in Methods in Enzymology, Vol. XI, p. 197.

6. Blossey, H.C. and Küntzel, H. (1972) FEBS-Letters 24, 335-338.
7. Küntzel, H. and Blossey, H.C. (1974) Eur.J.Biochem. 47, 165-171.
8. Swank, R.T. and Munkres, K.D. (1971) 462-477.
9. Poyton, R.O. and Schatz, G. (1975) J.Biol.Chem. 250, 752-761.
10. Baginski, E.S., Foa, P.P. and Zak, B. (1967) Clin.Chim.Acta 15, 155-158.

Genetics and Biogenesis of Chloroplasts and Mitochondria, Th. Bücher et al. eds.

CYTOPLASMIC SYNTHESIS OF THE DICYCLOHEXYLCARBODIIMIDE-BINDING PROTEIN IN NEUROSPORA CRASSA

W. Sebald, Th. Graf and G. Wild

Institut für Physiologische Chemie und Physikalische Biochemie der Universität München

1. INTRODUCTION

Polypeptides of low molecular weight which are soluble in neutral chloroform-methanol have been isolated from the ATPase complex of several organisms[1-3]. They have been referred to as "proteolipids", even though the presence of a lipid component could not be established in these proteins up to now. Beechey and coworkers have demonstrated that the proteolipid from beef heart mitochondria reacts with the ATPase inhibitor dicyclohexylcarbodiimide (DCCD)[4]. Sierra and Tzagoloff have found that the proteolipid from yeast ATPase is a major mitochondrial translation product in this organism[2].

The present communication is concerned with the proteolipid present in the ATPase complex from Neurospora crassa. As in beef heart, the protein was found to bind DCCD. In contrast to Saccharomyces cerevisiae, however, the proteolipid appeared to be synthesized on cytoplasmic ribosomes, since it was formed in the presence of chloramphenicol but not in the presence of cycloheximide.

2. METHODS

(^{14}C)DCCD was synthesized from (^{14}C)urea and cyclohexylamine as described[5,6]. The specific ^{14}C radioactivity was 20 000 cpm per nmol DCCD.

For inhibitory and binding studies whole mitochondria (5 mg per ml) were incubated with (^{14}C)DCCD for 4 h at 0 °C in 10 mM Tris-acetate buffer pH 7.4 . 2 µg DCCD per mg mitochondrial protein was applied, if not otherwise indicated.

ATPase complex was isolated from whole mitochondria by immunoprecipitation with antiserum to F_1 as described[7]. The precipitated protein was washed three times with buffer (10 mM Tris-acetate pH 7.4) to remove unbound DCCD. Identical results were obtained when the ATPase protein was washed further with organic solvents as described below for whole mitochondrial protein.

Removal of unbound (^{14}C)DCCD. After incubation with (^{14}C)DCCD, mitochondria were diluted ten-fold with buffer (10 mM Tris-acetate pH 7.4) and sedimented by centrifugation. The sedimented protein was washed three times with chloroform-methanol-diethylether (2 : 1 : 12 v/v/v) at 0 ^{o}C. One ml of the organic solvents was used per 2 - 10 mg protein.

Isolation of the proteolipid. Whole mitochondria or ATPase complex were washed with organic solvents as described above. The proteins were then incubated with neutral chloroform-methanol (2 : 1) for 30 min at room temperature (10 mg protein per ml). The residue obtained after centrifugation was extracted a second time with chloroform-methanol. To the combined extracts four volumes of diethylether were added. The precipitated protein was separated by centrifugation and again dissolved in chloroform-methanol. Unsoluble material was removed. The precipitation was repeated three times.

Mitochondrial translation products were labelled by a pulse of (^{3}H)leucine in the presence of cycloheximide after a transitory incubation of the cells with chloramphenicol. Cytoplasmic translation products were labelled with (^{3}H)leucine in the presence of chloramphenicol. The chloramphenicol was removed and the cells were grown in the absence of the inhibitor for another 60 min before the isolation of the mitochondria. Further details are described in ref. 7.

Gelelectrophoresis in the presence of dodecylsulfate or phenolformic acid was performed as described[8].

3. RESULTS

3.1 Inhibition of mitochondrial ATPase in situ and binding of (^{14}C)-DCCD. Whole mitochondria were incubated for 4 h at 0 °C with different amounts of (^{14}C)DCCD. The radioactive DCCD inhibited ATPase activity halfmaximally at 0.5 - 1 µg per mg mitochondrial protein as shown in Fig. 1. Maximal inhibition was obtained at 2 - 4 µg per mg protein. The same values were measured with cold DCCD. At 2 µg DCCD about 0.35 nmol Inhibitor were bound per mg mitochondrial protein. Below this value DCCD-binding and inhibition of ATPase activity exhibited a similar dependence on DCCD concentration.

About 35% of the mitochondrial DCCD label was recovered in the immunoprecipitated ATPase complex (Fig. 1). In the presence of oligomycin the binding of DCCD was found to be highly reduced. This strongly suggests that the DCCD is bound to the inhibitory site.

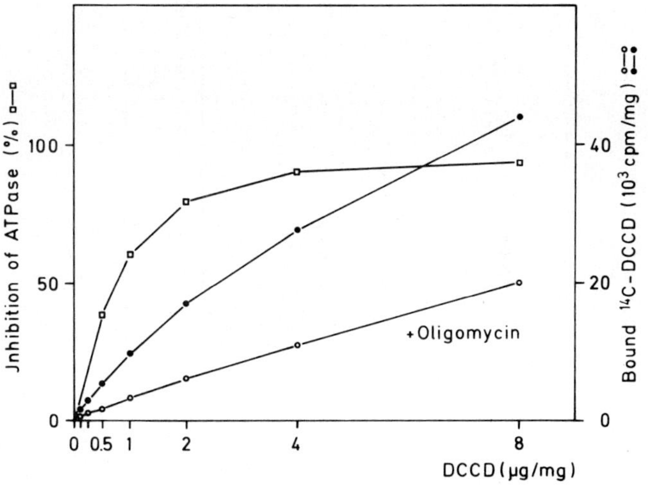

Fig. 1. □—□ Inhibition of mitochondrial ATPase at various (14-C)-DCCD concentrations.

●-● Recovery of bound (14-C)DCCD in isolated ATP-ase complex.

o-o Recovery of bound (14-C)DCCD in isolated ATP-ase complex after incubation of mitochondria with oligomycin (20 µg per mg protein) plus (14-C)DCCD.

3.2 Identification of the DCCD-binding protein. The binding of DCCD was found to be highly selective when mitochondria were incubated with low amounts (up to 2 µg per mg protein) of (^{14}C)DCCD. After gel electrophoresis in the presence of dodecylsulfate only a single DCCD-binding protein was observed. This applies for the ATPase complex (Fig. 2A) but also for whole mitochondrial protein (see Fig. 4A). The bound (^{14}C)DCCD migrated with the smallest ATPase subunit of 8 000 Dalton.

3.3 Extraction of the DCCD-binding protein by neutral chloroform-methanol. Under certain conditions (see methods) the extraction appeared to be highly specific. The same amounts of (^{14}C)DCCD were found in the proteolipid extracted from whole mitochondria and in the proteolipid isolated from the ATPase complex. This suggests that the ATPase subunit is the only proteolipid present in Neurospora mitochondria.

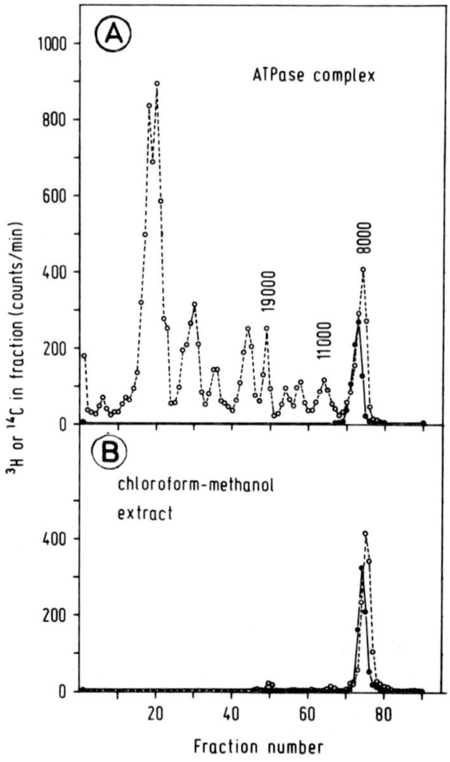

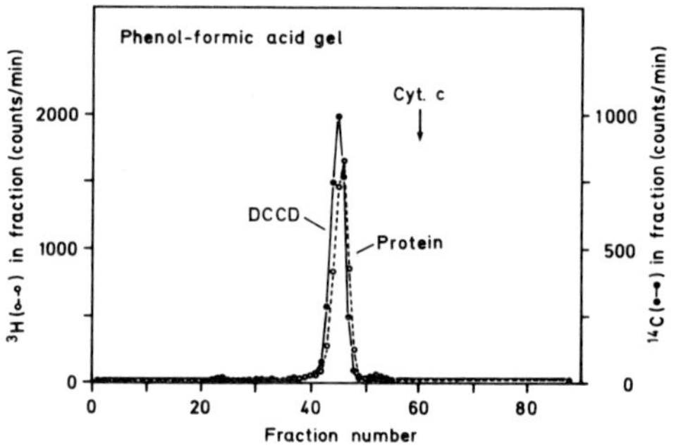

Fig. 2. (left) Dodecylsulfate-gel electrophoresis of (A) ATPase complex and (B) proteolipid isolated from ATPase complex.
●-● (14-C)DCCD label; o-o (3-H) leucine protein label (345 000 cpm per mg protein).

Fig. 3. (above) Phenol-formic acid-gel electrophoresis of proteolipid isolated from whole mitochondrial protein.
●-● (14-C)DCCD label; o-o (3-H) leucine protein label (500 000 cpm per mg protein).

The isolated proteolipid behaved as a homogeneous protein when analysed by different methods: A single band was observed after electrophoresis on dodecylsulfate gels (Fig. 2B) as well as on phenol-formic acid gels (Fig. 3). Gelchromatography in the presence of 80% formic acid results in the elution of one peak (Sebald unpub. result). Tyrosine was found as only N-terminal amino acid. On the other hand in all systems the (^{14}C)DCCD label did not exactly coincide with the protein peak. This could indicate that the electrophoretic and chromatographic behaviour of the proteolipid is changed by the bound inhibitor. However, the possibility cannot be excluded at present that several closely related polypeptides are present in the proteolipid and that only one of them binds the DCCD.
The proteolipid is a major constituent of the Neurospora ATPase complex. It comprises at least 10% of the total enzyme protein[7]. Thus, six or more polypeptides of 8 000 Dalton may be present in one ATPase molecule.

3.4 Site of synthesis. Mitochondrial translation products were pulse-labelled in vivo with (^{3}H)leucine in the presence of cycloheximide. Mitochondria were then reacted with (^{14}C)DCCD. From part of the double-labelled mitochondria the ATPase complex was isolated. The proteolipid was purified from whole mitochondria as well as from ATPase complex. All proteins were analysed by dodecylsulfate-gel electrophoresis (Fig. 4). Neither in whole mitochondrial protein nor in the ATPase complex, any mitochondrial translation products migrated with the DCCD-binding protein. The isolated proteolipids contained the (^{14}C)DCCD but only negligible amounts of mitochondrially synthesized polypeptides. Apparently, the proteolipid was not synthesized in the presence of cycloheximide and this suggested a cytoplasmic site of translation.
Corresponding results were obtained by labelling experiments in the

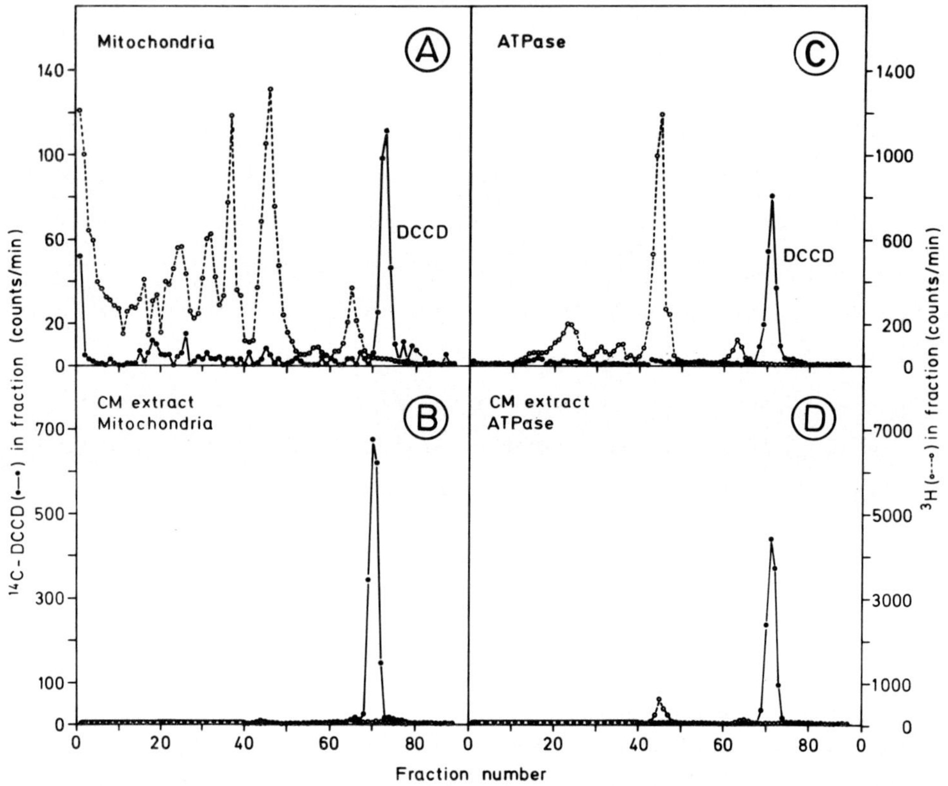

Fig. 4. Distribution of (14-C)DCCD label and cycloheximide-resistant (3-H)leucine label after gel electrophoresis in the presence of dodecylsulfate.
(A) Whole mitochondrial protein; (B) proteolipid isolated from whole mitochondrial protein; (C) ATPase complex; (D) proteolipid isolated from ATPase complex.
●-● (14-C)DCCD label; o-o mitochondrial translation products labelled with (3-H)leucine.

presence of chloramphenicol. All proteins were first labelled homogeneously with (^{14}C)leucine. The cells were then poisoned with chloramphenicol and incubated with (^{3}H)leucine in order to allow a synthesis and labelling of only the cytoplasmic translation products. The double-labelled mitochondria were reacted with cold DCCD. Isolated ATPase complex as well as the proteolipids purified from whole mitochondria and ATPase complex were separated on dodecylsulfate gels as

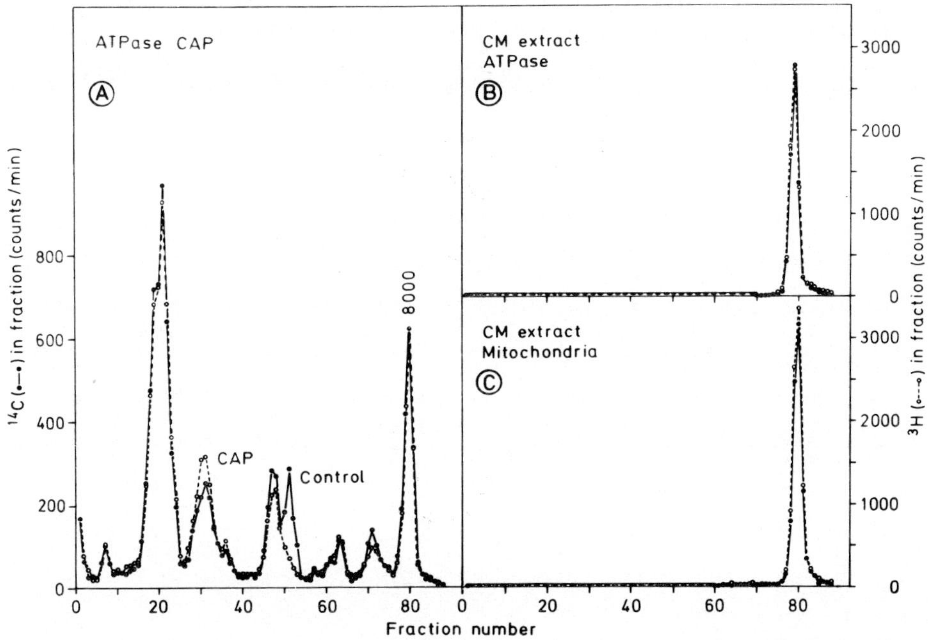

Fig. 5. Distribution of (14-C)leucine control label and chloramphenicol-resistant 3-H label after gel electrophoresis in the presence of dodecylsulfate.
(A) ATPase complex; (B) proteolipid isolated from ATPase complex; (C) proteolipid isolated from whole mitochondrial protein.
●-● (14-C)leucine control label; o-o cytoplasmic translation products labelled with (3-H)leucine. The 3-H/14-C ratio of whole mitochondrial protein was 4.7 .

Fig. 6. Electrophoresis on phenol-formic acid gel of proteolipid isolated from whole mitochondrial protein.
The same protein as described in Fig. 6C was separated.

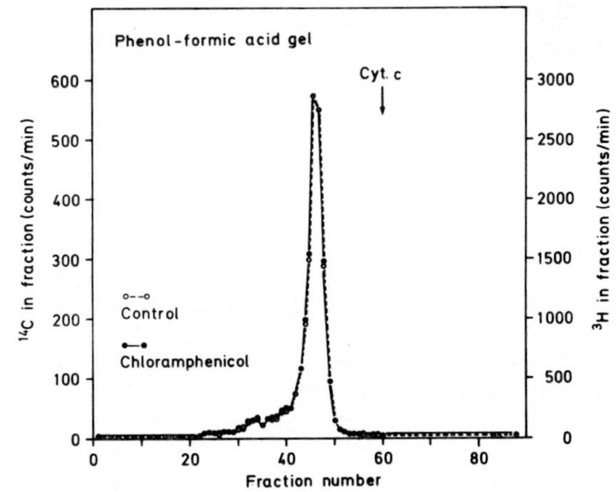

shown in Fig. 5. In the smallest ATPase subunit the chloramphenicol-resistant ^3H label comigrated exactly with the ^{14}C control label. The $^3H/^{14}C$ ratio was found as high as e.g. in the two large subunits of F_1. In the isolated proteolipids too, both labels were found to coincide after electrophoresis on dodecylsulfate gels (Fig. 5 B,C) and phenol-formic acid gels (Fig. 6).

All described results consistently indicate that the DCCD-binding proteolipid from Neurospora ATPase is synthesized outside the mitochondria on cytoplasmic ribosomes. It is still uncertain whether or not the Neurospora protein is identical with the mitochondrially synthesized proteolipid dedected in oligomycin-sensitive ATPase from Saccharomyces cerevisiae[2]. Nevertheless, the possibility may be seriously considered that homologous polypeptides are synthesized in different organisms at different sites of the cell.

4. REFERENCES

1. Stekhoven, F.S., Waitkus, R.F. and Moerkerk, H.Th.B. (1972) Biochemistry 11, 1144-1150
2. Sierra, M.F., Tzagoloff, A. (1973) Proc. Nat. Acad. Sci. U.S.A. 69, 930-932
3. Sebald, W. and Jackl, G. (1975) in "Electron Transfer Chains and Oxidative Phosphorylation" (Quagliariello,E., Papa, S., Palmieri, F., Slater, E.C. and Siliprandi, N., eds) pp. 193-198, North Holland, Amsterdam
4. Cattell, K.J., Lindop, C.R., Knight, I.G. and Beechey, R.B. (1971) Biochem. J. 125, 169-177
5. Amiard, G., Heymes, R. and Vellus, L. (1958) Chem. Abstr. 52, 426
6. Walther, H. (1963) Chem. Abstr. 58, 2382
7. Jackl, G., Sebald, W. (1975) Eur. J. Biochem. 54, 97-106
8. Sebald, W., Machleidt, W. and Otto, J. (1973) Eur. J. Biochem. 38, 311-324

This work was supported by the Deutsche Forschungsgemeinschaft, Sonderforschungsbereich 51, Medizinische Molekularbiologie und Biochemie.

Genetics and Biogenesis of Chloroplasts and Mitochondria, Th. Bücher et al. eds.

BIOCHEMICAL GENETICS OF OXIDATIVE PHOSPHORYLATION AN APPROACH TO THE REACTION MECHANISM

D.E.Griffiths
Department of Molecular Sciences
University of Warwick
Coventry CV4 7AL
U.K.

INTRODUCTION

Biochemical genetic studies [1] of mitochondrial drug and inhibitor resistance have led to the isolation of cytoplasmic mutants which are resistant to triethyl tin, a known inhibitor of oxidative phosphorylation, Pi-ATP exchange activity and OS-ATPase. These mutants, in contrast to specific oligomycin resistant and venturicidin resistant mutants, show cross-resistance to venturicidin, the uncoupling agent "1799", rhodamine 6G and bongkrekic acid [2,3].

Triethyl tin resistance thus may reflect modification of a component, a previously unidentified component of OS-ATPase, which serves as a link between the ATP synthase complex and interacting systems including the respiratory chain complexes. To identify this component a specific affinity label derivative for the trialkyl tin binding site in ATP synthase was developed, dibutyl chloromethyl tin chloride (DBCT), $Bu_2(CH_2Cl).Sn.Cl$. This compound was shown to have the properties of a site directed inhibitor which competes for the triethyl tin binding site [4] and which reacts covalently with a mitochondrial component.

DBCT is a potent inhibitor of energy conservation reactions and inhibits oxidative phosphorylation, ADP-stimulated respiration in coupled mitochondria, P_i-ATP exchange, energy driven reduction driven by ATP and mitochondrial OS-ATPase (ATP synthase) but has no effect on F_1-ATPase. It should be stressed that DBCT has no effect on respiration, including respiration with pyruvate/malate as a substrate. DBCT has all the inhibitory properties of its structural analogues, triethyl and tributyl tin, and appears to be a specific inhibitor of oxidative phosphorylation whose locus of action is a component of the OS-ATPase complex. This is supported by studies with [^3H]-DBCT labelling of yeast sub-mitochondrial particles and analysis by gel electrophoresis of the OS-ATPase complex. Two radioactive bands are seen corresponding to components of approximately molecular weight 46,000 and 8,000. These correspond to the aggregated (hexameric) and monomeric forms of subunit 9, a component of OS-ATPase and a major product of mitochondrial protein synthesis in yeast [5] which appears to be present in excess as compared with ATP synthase.

The properties of the DBCT binding component were investigated as it appeared to provide a means of identifying a component of the mitochondrial inner membrane which may be related to the DCCD binding site in mitochondria which may also be located on subunit 9, a product of mitochondrial biogenesis.

This paper will show that the interaction site of DBCT is not an amino acid residue of subunit 9 but a lipophilic cofactor which has been tentatively identified as lipoic acid. The implications of this finding for the mechanism of oxidative phosphorylation are discussed , as well as for studies of the biogenesis of the OS.ATPase complex.

MATERIALS AND METHODS

Dibutyl chloromethyl tin chloride (DBCT) and [^{3}H]-DBCT were synthesised as described previously[6] and were used as ethanolic solutions. Lipoic acid, dihydrolipoic acid, lipoamide, reduced glutathione, coenzyme A and dithiothreitol were obtained from Sigma Chemical Company. Reduced lipoamide was made by reduction of lipoamide with dithiothreitol. Solutions of dihydrolipoic acid and dihydrolipoamide were made up in buffers containing mM EDTA in order to avoid rapid autoxidation.

The following preparations were used: mitochondria and sub-mitochondrial particles (SMP) and Complex V from beef heart[7] ; yeast OS-ATPase[8].

Binding of [^{3}H]-DBCT to mitochondria, sub-mitochondrial particles and chloroplasts was estimated by addition of excess (^{3}H)-DBCT (10-30 nmole/mg protein) to membrane preparations and washing four times with suspending medium after high speed centrifugation. Free lipoic acid was assayed by bioassay of sterile protein fractions or acid extracts, using a lipoic acid requiring mutant of E.Coli[9]. Autoclaving of extracts in 6N H_2SO_4 was not used as it led to loss of lipoic acid.

RESULTS

DBCT reacts covalently with a single component of yeast and heart mitochondrial membranes and specifically inhibits ATP synthesis and a variety of energy linked reactions driven by ATP. At least 95% of this ^{3}H-DBCT derivative is extractable by chloroform/methanol mixtures which extract mitochondrial proteolipids and on examination by gel electrophoresis appears to be associated with components such as subunit 9, which run near the dye marker front. However, the DBCT derivative is separable from subunit 9 and other chloroform/methanol soluble proteolipids by thin layer chromatography as a fast running lipophilic component, i.e.DBCT reacts with a low molecular weight lipophilic component and not a protein subunit of OS-ATPase.

The DBCT binding component has been shown to be lipoic acid. A second DBCT binding component is present in the chloroform/methanol extract which also contains lipoic acid. This has not been identified but has the properties of a low molecular weight ester of lipoic acid. The evidence that DBCT reacts with lipoic acid is summarised in Table 1.

TABLE 1

EVIDENCE THAT DBCT REACTS WITH LIPOIC ACID

1. DBCT reacts rapidly with authentic lipoic acid to give DBCT-S-lipoate.
2. [^{3}H] DBCT specifically reacts with a single component of the mitochondrial inner membrane.
3. This component is extractable with chloroform/ methanol and is a low molecular weight lipophilic component separable from subunit 9.
4. This component has been identified as DBCT-lipoate by thin layer chromatography in four solvent systems, but the extract also contains another lipoic acid derivative which is suspected to be a lipoic acid ester.
5. After reaction with DBCT, lipoic acid in membranes and OS-ATPase is no longer detectable by bioassay.

The DBCT binding component is present in high concentration in all energy conserving membranes examined at levels in excess of all other components (Table 2). That the DBCT binding component is lipoic acid has been confirmed by bioassay in the samples examined to date (heart SMP and yeast OS ATPase).

TABLE 2

DBCT INTERACTION SITES (LIPOIC ACID) IN BIOENERGETIC MEMBRANES

	nmole/mg protein
Yeast SMP	8 - 9
Heart SMP	9 - 10
Chloroplasts	5 - 6
E.Coli plasma membrane	30 - 40

Membranes treated with excess [^{3}H] - DBCT and washed 4 times with suspension buffer by high speed centrifugation.

Thus the mitochondrial inner membrane contains 8 - 10 nmoles of lipoic acid or a lipoic acid derivative, a previously unknown pool of reducing equivalents. This lipoic acid component of the inner membrane appears to be titrated specifically by DBCT.

Studies with yeast mitochondria and the purified yeast OS-ATPase indicate that the OS-ATPase contains approximately 6 binding sites for triethyl tin and an equivalent number for DBCT [Table 3] , but this accounts for less than 30% of the DBCT interaction sites in the inner membrane. Up to 35% of the DBCT interaction sites remain in the membrane when OS-ATPase is extracted and of the extractable DBCT interaction sites, 30% are not precipitated by an antibody to OS-ATPase and 50-60% of the DBCT interaction sites present in the Triton extract are separable from OS-ATPase by density gradient centrifugation.

TABLE 3

COMPONENTS OF YEAST OS-ATPase (Tzagoloff)

ATPase S.A. = 20 - 28 Pi-ATP exchange = 0 , 10-12 subunits. Molecular weight 450 -500,000.

COMPONENTS	nmole/nmole protein
Triethyl tin* sites	6
DBCT* sites	5 - 6
Lipoic acid (bioassay[9])	> 3
DCCD* sites	(O)

* Binding of radioactive inhibitor.

Bioassay[9] of the lipoic acid content of yeast OS-ATPase shows that it contains amounts of lipoic acid stoichiometric with the enzyme complex. Similar results have been obtained with other OS-ATPase complexes obtained from heart mitochondria by the methods of Hatefi et al[7] and Serrano et al[10]. The demonstration that the OS-ATPase complex contains stoichiometric amounts of lipoic acid and that this lipoic acid content is less than 30% of the total lipoic acid content (DBCT interaction sites) in the inner membrane raises the question as to whether this pool of lipoic acid residues plays a functional role in oxidative phosphorylation and as to whether lipoic acid derivatives are intermediates in this process as found in substrate level phosphorylation.

Evidence that lipoic acid is involved in oxidative phosphorylation is indicated by the specific interaction of DBCT with lipoic acid leading to inhibition and the specific reversal of inhibition by dihydrolipoic acid.

TABLE 4

REVERSAL OF DBCT INHIBITION OF ATPase BY DIHYDROLIPOATE

ATPase activity was estimated at pH 9 in a volume of 1 ml at 30° as described previously[11]. Heart SMP (100-150 µg protein) was added to start the reaction which was stopped by addition of 1.0 ml 10% w/v TCA and inorganic phosphate measured. Inhibited particles were preincubated with dihydrolipoate in Tris-sucrose-EDTA buffer, pH 7.5 for 20 min. Qualitatively similar results are obtained if the inhibitor is added to the incubation mixture containing enzyme followed by dihydrolipoate and the reaction started by ATP addition. Similar results are obtained with yeast SMP and yeast OS-ATPase.

Additions (nmole/mg protein)	ATPase Activity (% of Control)
None *	100
DBCT (0.5 nmole)	5
DBCT (10 nmole)	0
DBCT (10 nmole) + Dihydrolipoate (100 nmole)	100
DBCT (10 nmole)	0
DBCT + lipoic acid (100 nmoles)	0
DBCT + lipoic acid + DTT (40 nmoles)	80
DBCT + lipoamide (100 nmoles)	0
DBCT + lipoamide + DTT (40 nmoles)	0
DBCT + reduced glutathione (60 nmoles)	0
DBCT + Dithiothreitol (DTT) (40 nmoles)	0
DBCT + 2,3, dimercaptopropanol, B.A.L. (100 nmoles)	0

* Specific activity 2.5 µmoles/mg protein/min.

Table 4 shows that inhibition by DBCT of mitochondrial ATPase is specifically reversed by dihydrolipoic acid and by a mixture of lipoic acid and dithiothreitol which generates dihydrolipoic acid. Oxidised lipoic acid, oxidised or reduced lipoamide, reduced glutathione, 2,3 Dimercaptopropanol (B.A.L.) coenzyme A and dithiothreitol were ineffective. Reversal of DBCT inhibition is readily achieved by incubation of inhibited particles or mitochondria with an excess of dihydrolipoate for 20 min at 4° and 10 min at 30°, but the reversal can be demonstrated clearly by addition of dihydrolipoate to the ATPase incubation medium. The mechanism of reversal of DBCT inhibition is under investigation and it is not known if the reaction involves simply replacement of a DBCT-lipoate complex by dihydrolipoate or whether activation and formation of a high energy derivative of lipoate is involved.

Reversal of DBCT Inhibition of Oxidative Phosphorylation by Dihydrolipoate

DBCT is a potent inhibitor of oxidative phosphorylation with pyruvate/malate, succinate and ascorbate-TMPD as substrates (maximal inhibition at 4-5 nmoles/mg protein). The amount of DBCT required is about 10 times that required to inhibit ATPase activity in heart sub-mitochondrial particles and approaches the pool concentration of DBCT interaction sites present in the mitochondrial inner membranes.

As in the case of mitochondrial ATPase inhibition of mitochondrial oxidative phosphorylation at all three sites by DBCT is reversed specifically by dihydrolipoate (Griffiths, D.E. & Hyams, R.L., unpublished results). If appropriate preincubation conditions are used then oxidised lipoic acid is effective due to its prior reduction to dihydrolipoate or conversion to an activated form of lipoic acid. Similar results are obtained with ATP driven energy linked reductions where again DBCT inhibition is specifically reversed by dihydrolipoic acid. As DBCT reacts specifically with a membrane bound species of lipoic acid a role for lipoic acid in oxidative phosphorylation is indicated. This conclusion is fully supported by experiments with 8-methyl lipoic acid, a potent growth inhibitor and competitive inhibitor of lipoic acid requiring reactions. 8-methyl lipoic acid is a potent inhibitor of mitochondrial ATPase and of oxidative phosphorylation in the reduced form but is totally ineffective in the oxidised form , again pointing to an enzyme catalysed intermediary role of lipoic acid.

The role of lipoic acid has been further investigated, particularly with respect to the role of the OS ATPase complex and the question that has been asked is - are there any basic similarities in the mechanism of oxidative phosphorylation and the reactions involved in substrate level phosohorylation?

Table 5 shows that submitochondrial particles inhibited with antimycin and rotenone will catalyse net ATP synthesis with dihydrolipoate as substrate.

TABLE 5

NET SYNTHESIS OF ATP WITH REDUCED LIPOATE BY SMP

	ADDITIONS	μmoles / 20 mins	
		Δ Lip $(SH)_2$	Δ Pi
1.	Lip $(SH)_2$ (2.73 μmoles)	1.17	4.04
2.	Lip $(SH)_2$ + KCN (100 nmoles)	0	0.16
3.	Lip $(SH)_2$ + VEN (2 μg)	0.2	0
4.	Lip $(SH)_2$ + 1799 (5μg)	0.1	0

Glucose Hexokinase trap system. 1.5 ml.30°. SMP (2 mg), inhibited with Antimycin A (2 μg), and rotenone (2 μg), 250 mM Sucrose, 22 mM Glucose, 2mM $MgCl_2$, EDTA 0.5 mM, 2mM ADP, 6.6 mM phosphate, 20mM-Tris-Cl buffer pH 7.3. Hexokinase, 30 units. Incubated aerobically with shaking. Lip $(SH)_2$ (2.73 μmoles), Lip $(SH)_2$ estimated by DTNB (Ellman Reagent).

When net ATP synthesis with a purified OS-ATPase complex (heart complex V[7] and yeast OS-ATPase[8]) was attempted no ATP production was observed. This was probably due to the lack of a component in the reaction system necessary for generation of an active intermediate. By analogy with substrate level phosphorylation where this component is succinate and the generated intermediates are succinyl-S-lipoate and succinyl CoA[12] an investigation into the identity of a lipophilic carboxylic acid which could serve as the intermediate was made. The answer came from an investigation of the nature of the DCCD binding component[13], the other covalent inhibitor of oxidative phosphorylation which has an interaction site on subunit 9 of OS-ATPase. It was demonstrated that DCCD inhibition could be reversed by low concentrations of unsaturated fatty acids and unsaturated fatty acids were therefore tested in the phosphorylation system with isolated OS-ATPase.

Table 6 shows net synthesis of ATP with dihydrolipoate as substrate using an isolated OS-ATPase preparation (Complex V) and demonstrates the requirement for oleic acid and oleoyl CoA. The function of oleoyl CoA may be to generate an oleoyl-ATPase complex intermediate or to generate oleoyl-S-lipoate.

TABLE 6

NET SYNTHESIS OF ATP WITH REDUCED LIPOATE BY OS-ATPase(COMPLEX V)

	ADDITIONS	Total Pi (µmoles)	ΔPi/30 min (µmoles)
1.	Lip $(SH)_2$(2.73 µmoles)	10.04	0.12
2.	Lip $(SH)_2$ + Oleoyl CoA	10.19	0.0
3.	Lip $(SH)_2$ + OlCoA + Oleate	6.95	3.25
4.	Lip $(SH)_2$ + OlCoA + Oleate + OLIGO	9.97	0.23

Glucose-Hexokinase trap system. 30°. Initial P_i = 10.2 µmoles. Shaken in air. Lip$(SH)_2$ 2.73 µmoles; Oleoyl CoA (10 nmoles) Oleate (15 nmoles); Oligomycin (2µg). COMPLEX V, 1mg.

With this demonstration it has become apparent that the terminal reactions of oxidative phosphorylation are not significantly different from those of substrate level phosphorylation. The key intermediates in oxidative phosphorylation are lipoic acid (or a lipoic acid ester) and unsaturated fatty acid (UFA) which are present in the inner membrane. A postulated intermediate, Oleoyl-S-lipoate has been synthesised by reaction of oleoyl chloride with dihydrolipoic acid and has been shown to catalyse net ATP synthesis with sub-mitochondrial particles and with an isolated OS-ATPase (Table 7)

TABLE 7

ATP SYNTHESIS FROM OLEOYL-S-LIPOATE BY COMPLEX V

Glucose Hexokinase trap system 30°. Complex V, 0.5 mg. Lip-X in DMF.

	ADDITIONS	ΔPi/20' (µmoles)	ΔSH/20' (µmoles)
1.	Lip-X (0.7 µmoles)*	1.34	< 0.05
2.	Lip-X (1.4 µmole)	2.68	< 0.1

FULLY SENSITIVE TO 1799 (5µg), OLIGOMYCIN (2µg) DCCD (2µg) AND EFRAPEPTIN (2 µg)

* 1µmole based on weight. Oleoyl-S-lipoate content estimated to be 70%. Contains lipoic and oleic acid as contaminants.

This reaction has been shown to be sensitive to uncoupling agents, Oligomycin, DCCD and to efrapeptin, a F_1-ATPase inhibitor. A puzzling feature of this reaction is that 2 moles of ATP are produced per mole of oleoyl-S-lipoate. As a net increase in dihydrolipoate is not observed it is assumed that further reactions with dihydrolipoate are involved. By analogy with succinic thiokinase a possible intermediate would be oleoylphosphate, which was synthesised and tested (Table 8).

TABLE 8

ATP SYNTHESIS FROM OLEOYL PHOSPHATE BY OS-ATPase (COMPLEX V)

	ADDITIONS	Glucose -6-P (μmoles/20 mins)
1.	Oleoyl-P (0.5 μmole)	0.375
2.	O-P + 1799 (5μg)	0.385
3.	O-P + OLIGO (2μg)*	0.36
4.	O-P + DCCD (2μg)	0.0
5.	O-P + Efrapeptin (2μg)	0.0

Complex V, 0.5 mg, Sucrose-Tris-Mg-EDTA pH 7.5 12μmoles Glucose 0.2 μmoles ADP, 20 units Hexokinase. Total Vol.1.0 ml. Stop PCA. Assay G-6-P.
* Also not sensitive to TET and DBCT.

Apart from the obvious analogies to the succinic thiokinase reaction where enzyme bound succinyl phosphate is an intermediate and added succinyl phosphate generates ATP, the main feature of the results in Table 8 is the inhibitor specificity. The transphosphorylation reaction is insensitive to 1799 and to oligomycin but is sensitive to DCCD and to efrapeptin, a F_1-ATPase inhibitor.

DISCUSSION

The results in Tables 6, 7 and 8 provide full support for the concept that the terminal reactions of oxidative phosphorylation are basically similar to those in substrate level phosphorylation and that they are contained within the isolated OS-ATPase complex.

These findings have major implications, not only for the mechanism of oxidative phosphorylation, but for studies of the biogenesis of the OS-ATPase complex. It is now possible to define intermediate reaction steps and to correlate them with nuclear or mitochondrial (cytoplasmic) coded components of the OS-ATPase. The demonstration of different reaction sites for uncoupling agents, oligomycin, triethyl tin and DCCD fully support previous biochemical genetic investigations in this laboratory which have indicated that these are associated with separate and distinct mitochondrial (cytoplasmic)-coded gene products [1,2,3]. This work also provides a biochemical function for unsaturated fatty acids which are the specific cofactor required in oxidative phosphorylation and provides an explanation for the long standing association of UFA-deficiency with reduced efficiency of oxidative phosphorylation and for the lack of ability of E.coli and yeast mutants which require UFA to grow on oxidisable substrates [14, 15]. The importance of UFA in mitochondrial biogenesis has been emphasised in extensive studies by Linnane and co-workers [16] and a specific biochemical explanation is now available.

With the elucidation of the reaction mechanism of oxidative phosphorylation and the demonstration that the essential reactions are contained within the OS-ATPase complex it is now possible to devise biogenesis experiments to establish the role of components of the OS ATPase. A particular example is the acceptor for the oleoyl group from oleoyl-S-lipoate, i.e. the equivalent of succinyl CoA in substrate level phosphorylation. As there is little evidence for the presence of CoA in OS ATPase, an investigation was made into the possibility that a protein-bound CoA, i.e. an acyl carrier protein, was present. Growth of yeast during glucose derepression in a pantothenate deficient medium in the presence of added ^{14}C-pantothenate leads to incorporation of radioactivity into the mitochondrial inner membrane. 80% of the counts are extractable by chloroform-methanol and 40% of the counts are extractable by Triton X-100. Examination of the Triton extract shows that the radioactivity is associated with a protein band of molecular weight 12,000. The role of this component and its association with the ATPase complex is under investigation.

A combination of the techniques of mitochondrial genetics, mitochondrial biogenesis and mitochondrial biochemistry has thus given important information as to the mechanism of oxidative phosphorylation and the structure and function of the OS-ATPase complex. Many features of this system still remain to be elucidated including the functional role of subunit 9, the major product of mitochondrial protein synthesis. Evidence for a functional association of UFA and lipoic acid with subunit 9 suggests that specific binding sites for these cofactors are present on subunit 9 and the possibility that it may serve as a mobile " carrier" for these components in the mitochondrial inner membrane is being investigated.

This work was supported by research grants from the Science Research Council and NATO Grant 572. I wish to thank my colleagues, Kelvin Cain, Robert Hyams and Michael Partis for their support.

REFERENCES

1. Griffiths, D.E. (1972) in Mitochondria : Biogenesis and Bioenergetics (S.G. van der Bergh, P.Borst and E.C.Slater eds.) pp 95 - 104, North Holland, Amsterdam.
2. Lancashire, W.E. and Griffiths, D.E.(1975) Eur.J.Biochem. 51 , 403-413.
3. Lancashire, W.E. and Griffiths, D.E.(1975) Eur. J.Biochem. 51,377-392.
4. Cain, K and Griffiths, D.E. (unpublished results).
5. Tzagoloff, A., Rubin, M.S. and Sierra, M.F. (1973) Biochim. Biophys.Acta, 301, 71 - 104.
6. Cain, K., Partis, M.D. and Griffiths, D.E.(1975) International Symposium on Electron Transfer Chains and Oxidative Phosphorylation Abstract Vol.p.47, University of Bari, Bari.
7. Hatefi, Y., Stiggall, D.L., Galante, Y., and Hanstein, W.G. (1974), Biochim. Biophys. Res.Commun. 61 , 313-321.
8. Tzagoloff, A. and Meagher, P. (1971) J.Biol.Chem. 248 , 7328-7336.
9. Herbert, A.A. and Guest, J.R. (1970) Meth.Enzymol.18A, 269-272
10. Serrano, R., Kanner, B.I. and Racker, E. (1976), J.Biol.Chem. 251 ,2453-2461.
11. Griffiths, D.E. and Houghton, R.L. (1974), Eur. J.Biochem. 46 ,157-167.
12. Reed, L.J. and Cox, D.J. (1966) Ann. Rev. Biochem. 35, 57-84.
13. Cattell, K.J., Lindop, C.R., Knight, I.G. and Beechey, R.B.(1971) Biochem.J. 125, 169-177.
14. Silbert, D.F. and Vagelos (1967) Proc. Nat.Acad. Sci. U.S. 58 , 1579-1586.
15. Resnick, M.A. and Mortimer, R.K. (1966), J.Bacteriol 92 , 597-600.
16. Haslam, J.M., Proudlock, J.W. and Linnane, A.W. (1971) J.Bioenergetics 2 , 351-370.

MITOCHONDRIAL RESPIRATORY COMPLEXES

Genetics and Biogenesis of Chloroplasts and Mitochondria, Th. Bücher et al. eds.

ANALYSIS OF THE POLYPEPTIDE CHAINS OF CYTOCHROME OXIDASE FROM BEEF HEART MITOCHONDRIA

G. Buse and G.J. Steffens
Fachgebiet Molekulare Biologie der Proteine
Abteilung Physiologische Chemie, RWTH Aachen
Melatener Straße 211
D-5100 Aachen

INTRODUCTION

As compared to cytochrome c and hemoglobin, the functionally adjacent proteins of electron and oxygen transport, cytochrome c-O_2-oxidoreductase |EC 1.9.3.1| is presently not well known from its molecular structure. Though it is now well documented that the enzyme complex in the inner mitochondrial membrane is composed of cytoplasmically and mitochondrially synthesized polypeptides, functional, e.g. heme or copper carrying subunits have not been prepared nor has the character of membraneous peptides in the complex been established.

Our present studies with the beef heart enzyme are undertaken in order to contribute to the chemical structure of Warburg's "Atmungsferment". Our sequence work concentrates mainly on the cytoplasmic 14000 Dalton peptide and the mitochondrial 28000 (21000) Dalton peptide contained in the enzyme complex.

MATERIALS AND METHODS

Cytochrome oxidase from beef heart mitochondria is prepared according to Fowler[1] and Yonetani[2], using a solubilisation procedure with successive use of desoxycholate, cholate and Tween 80. The final purification is achieved by repeated fractionations with ammoniumsulfate. Details of this procedure are described elsewhere[3].

The preparation ends up with an enzyme with the generally obtained spectral and activity characteristics and with the heme, copper and lipid contents as given in table 1.

Preparative isolation of the polypeptides is achieved using two different separation parameters, first gel permeation chromatography on Bio-Gel P-60 (-400 mesh) columns (5 x 100 cm) in 2 % sodiumdodecylsulfate (SDS) and second different Craig distribution systems[4]. Purity of the polypeptides is checked with SDS polyacrylamide gel electrophoresis with 15 % acrylamide.

Peptide separations were done - mainly on BrCN-peptides - first with various Bio-Gel P-6 columns in 10 % acetic acid and second preparative paper chromatography in the system n-butanol-acetic acid-water 4 : 1 : 5.

Amino acid analyses were performed according to Spackman et al.[5] on a Beckman Multichrom. Sequences and endgroups have been obtained with a Beckman Sequencer using the Quadrol protein program[6] and a modified peptide program of Crewther and Inglis[7] which allows sequencing peptides of different length and hydrophobicity without prior chemical modification.

RESULTS AND DISCUSSION

Figure 1 shows a preparative run of cytochrome oxidase polypeptides on Bio-Gel P-60 columns with seven fractions separated. There are six (I - VII) containing seven polypeptides in nearly 1 : 1 ratio with apparent molecular weights 36000, 28000, 19000, 14000, (12500, 11000) and 9500. Fraction VIII is a non-stoichiometric mixture of polypeptides (∼5000 Dalton) and contains the three N-terminal amino acids isoleucine, serine and phenylalanine. Polypeptide II with a column derived apparent mol. weight of 28000 runs in SDS gel electrophoresis together with polypeptide III with an apparent mol. weight of 19000 (Figure 2) and thus may have been overlooked in other fractionations of the beef heart cytochrome oxidase peptide chains. The polypeptides contained in fraction V - VII are again not separated in the same order as in the analytical SDS electrophoresis gels.

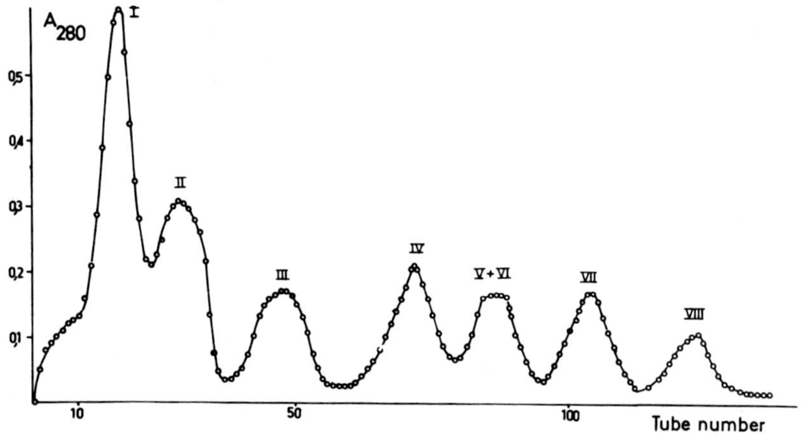

Fig. 1: Preparative gel filtration on Bio-Gel P-60 (-400 mesh) in the presence of 2 % SDS.

Table 1

Characteristic components of cytochrome oxidase	
components	yield per mg protein
heme a	10 - 11 nmol
iron	10 - 11 nmol
copper	12 - 13 nmol
phosphorus	2 µg
phospholipid	0,05 mg

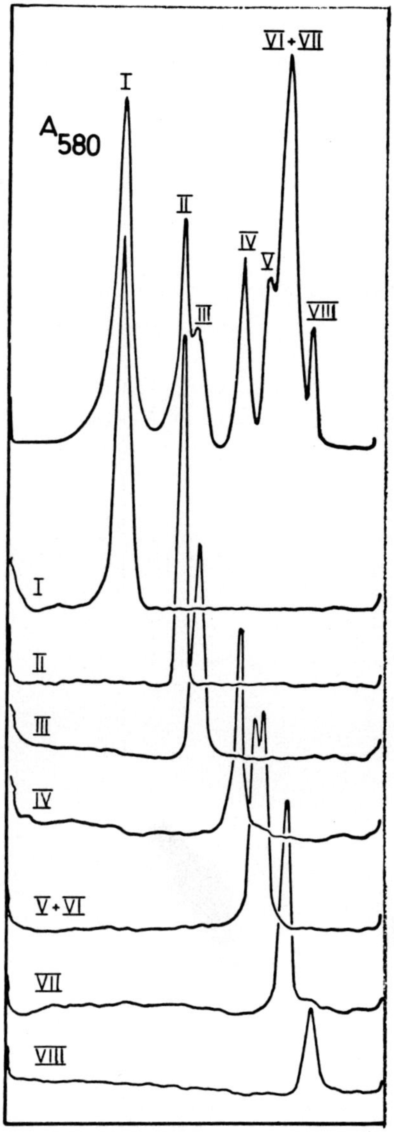

Fig. 2: Densitometric tracings of cytochrome oxidase (top) and the isolated polypeptides (I to VIII) analyzed by analytical SDS polyacrylamide gel electrophoresis (15 % acrylamide, stained with Coomassie Blue R-250).

Table 2

Amino acid composition of cytochrome oxidase and of peptide components I - VIII

Amino acid mol/100 mol	Cytochrome oxidase	I	II	III	IV	V+VI	VII	VIII
Lysine	4.88	2.60	3.34	3.83	10.30	6.24	7.05	7.9
Histidine	3.90	3.75	3.46	4.91	2.65	2.68	3.00	2.8
Arginine	4.32	2.63	3.12	3.58	4.41	5.62	7.07	3.8
Aspartic acid	7.38	7.25	7.28	6.55	9.28	10.92	9.38	7.3
Threonine	7.20	7.71	7.80	7.91	4.62	5.45	5.71	6.0
Serine	6.97	6.69	9.78	7.34	7.76	4.78	5.44	6.3
Glutamic acid	7.48	4.70	9.71	8.47	11.63	11.93	9.28	8.1
Proline	5.47	5.91	6.25	5.31	5.24	7.45	5.60	7,8
Glycine	7.82	9.26	5.05	8.80	5.64	7.71	8.33	8,0
Alanine	8.11	8.05	4.56	8.39	9.39	8.85	10.83	9.2
Cysteic acid[+]	0.72	n.d.	0.74	n.d.	0.37	n.d.	n.d.	n.d.
Valine	5.97	6.19	4.79	5.71	6.38	6.45	5.35	5.4
Methionine	4.48	5.77	7.08	3.42	2.52	1.35	2.10	1.8
Isoleucine	4.62	5.83	5.00	4.14	3.41	4.89	3.78	4.1
Leucine	11.13	12.19	14.00	11.16	8.13	9.29	7.56	11.3
Tyrosine	4.00	3.91	5.06	3.79	4.43	3.40	3.69	4.0
Phenylalanine	6.06	7.51	3.70	6.63	4.45	2.99	5.75	6.4
Total	100.52	99.95	100.72	99.94	100.39	100.00	99.92	100.2

[+] after performic acid oxidation

Amino acid analyses of the different polypeptide fractions are given in table 2. They indicate two groups of polypeptides: I - III (especially I and II), probably mitochondrial, with low lysine and arginine, more histidine and an extremely high content of leucine and methionine, and the polypeptides IV - VII, probably cytoplasmic, with a composition close to that of normal enzymes[8].

Where there are free N-terminal amino acids in this group the N-terminal sequences seem to indicate their homology:

		1 2 3 4 5 6 7	
IV	14000	Ala-His-Gly-Ser-Val-Val-Lys-	N-terminal sequences of
V	12500	Ala-His-Gly-Ser-His-Glu-Thr-	IV and VII were obtain-
VI	11000	Ala-Ser-Gly-Gly-Gly-Val-Pro-	ed with 20 % yields
VII	9500	Ala-Ser-Ala-Ala-Lys-Gly-Asp-	only.

The large polypeptides I and II have blocked N-termini. Under deformylating conditions[9] the blocking group can be split off releasing methionine as the N-terminal residue in both cases. Thus formylmethionine should be at the start of these polypeptides, again indicating their mitochondrial origin.

Our sequence studies - mainly with polypeptide II (28000 Dalton) - show the terminal sequence:

1 2 3 4 5 6 7 8 9 10
Formyl-Met-Ala-Tyr-Pro-Met-Gln-Leu-Gly-Phe-Gln-

for this chain, which might be recognized among sequences of the mitochondrial DNA, even if they are obtained from other species. Further preliminary sequences of BrCN-peptides derived from the polypeptide II are given in the poster presentation. Among them there is a sequence containing most of the tyrosine and tryptophan residues of this chain:

-His-Gln-Trp-Tyr-Trp-Ser-Tyr-Glu-Tyr-Thr-X-Tyr-Glu-

In this respect it is totally different from other sequences obtained so far from this polypeptide. This leads to our suggestion that polypeptide II is involved in electron transfer through membraneous structures.

Though our attempts to crystallize the cytochrome oxidase complex have not been successful up to now, the sequence results, when completed, may give information on structural and functional aspects of the terminal oxidase. This may be the more so since promising attempts to derive secondary and tertiary structure information from the primary structures of proteins have been made recently[10].

REFERENCES

1. Fowler, L.R., Richardson, S.H. and Hatefi, Y. (1962) Biochem. Biophys. Acta, 64, 170-173.
2. Yonetani, T. (1961), J. Biol. Chem. 236, 1680-1688.
3. Steffens, G. and Buse, G. (1976) Hoppe-Seyler's Z. Physiol. Chem., in press
4. Hill, R.J., Konigsberg, W., Guidotti, G. and Craig, L.C. (1963) in Biochem. Prep. Vol. X, 55-66.
5. Spackman, D.H., Stein, W.H. and Moore, S. (1958) Anal. Chem. 30, 1190-1206.
6. Edman, P. (1975) in Molecular Biology, Biochemistry and Biophysics (Needleman, S.B. ed.) Vol. VIII, pp. 214-226, Springer Verlag, Berlin.
7. Crewther, W.G. and Inglis, A.S. (1975) Anal. Biochem. 68, 572-585.
8. Dayhoff, M.O. and Hunt, L.T. (1972) in Atlas of Protein Sequence and Structure (Dayhoff, M.O. ed.) pp. 355-359.
9. Sheehan, J.G. and Yang, D.H. (1958) J. Amer. Chem. Soc. 80, 1154-1158.
10. Chou, P.Y. and Fasman, G.D. (1974) Biochemistry 13, 211-221.

ACKNOWLEDGEMENT

This work was supported by the Sonderforschungsbereich 160 "Biologische Membranen".

Genetics and Biogenesis of Chloroplasts and Mitochondria, Th. Bücher et al. eds.

PARTIAL SEQUENCE OF A CHLOROFORM-METHANOL SOLUBLE POLYPEPTIDE FROM NEUROSPORA MITOCHONDRIAL MEMBRANES

Werner MACHLEIDT, Rainer MICHEL, Walter NEUPERT and Elmar WACHTER
Institut für Physiologische Chemie, Physikalische Biochemie
und Zellbiologie der Universität München
Goethestrasse 33, D 8000 München 2, Germany.

INTRODUCTION

A chloroform-methanol soluble polypeptide with a molecular weight of 8 500 (HP 8 500) has been isolated from the mitochondrial membrane of the nuclear mutant cni-1 of Neurospora crassa[1]. Labeling studies in the presence of cycloheximide resp. chloramphenicol have shown that this polypeptide is translated on mitochondrial ribosomes. It has an unusual amino acid composition with a very high proportion of nonpolar residues (Table 1). The polarity calculated according to the definition of Capaldi and Vanderkooi[2] is 0.25.

Although the functional role of the polypeptide remains to be solved, studies on its primary structure have been undertaken for the following reasons:

1) The extremely hydrophobic character of the HP is suggestive of a close connection with the lipid bilayer of mitochondrial membranes. Amino acid sequence will provide information on the structural basis of this interaction.

2) The HP is used as a model for the application of solid-phase methods[3-5] to amino acid sequence analysis of hydrophobic proteins.

RESULTS

1. Cyanogen bromide fragments

HP 8 500 dissolved in 80% formic acid was cleaved with cyanogen bromide (500-fold molar excess over methionine). The resulting fragments were separated by gel chromatography on Bio-Gel P-30 in 80% formic acid (Fig. 1). Incomplete cleavage of at least two of the three methionyl bonds leads to overlapping fragments in addition to the theoretically expected peptides (Table 1, Fig. 2). Peptides C (18 residues) and D (9 residues) were obtained in pure form; E_1 (4 residues) and E_2 (9 residues) were resolved by rechromatography on Bio-Gel P-2.

Peptides C, D, and E_2 were sequenced by automated solid-phase Edman degradation on aminopropyl glass after attachment via their

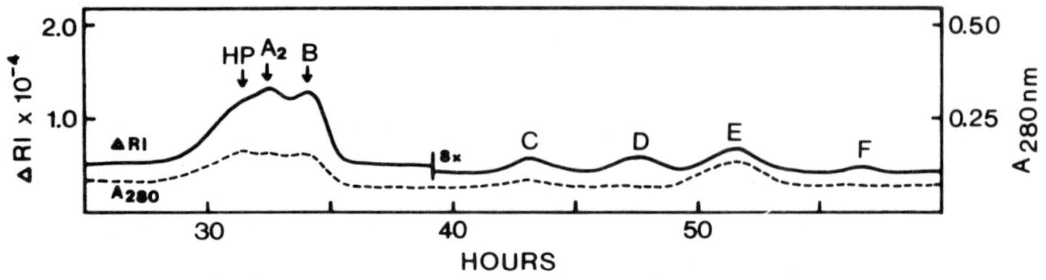

Fig. 1. Chromatography of the cyanogen bromide fragments on 0.8 x 150 cm Bio-Gel P-30 in 80% formic acid (1.0 ml/hr). For the detection of peptides without 280 nm absorption a differential refractometer was used (ΔRI).

TABLE 1

AMINO ACID COMPOSITION OF HP 8 500 AND ITS CYANOGEN BROMIDE FRAGMENTS

The composition of HP 8 500 (from ref. 1) is corrected for destruction and incomplete hydrolysis. Compositions of peptides are from 24 hrs hydrolysis without any correction. Amino acids signed (+) were identified but not quantitated. N-terminal amino acids were determined by dansylation.

Amino acid	HP 8 500	B	D	E_1	E_2
Lys	2 (2.07)		1 (1.11)	1 (1.14)	
Arg	2 (1.93)	2 (2.26)			
Asp	4 (4.13)	3 (3.34)	1 (0.97)		
Thr	3 (2.53)	1 (1.26)		1 (0.90)	
Ser	5 (5.00)	2 (2.34)	1 (0.94)		2 (2.09)
Homoserine		1 (+)	1 (+)		1 (+)
Glu	5 (5.44)	2 (2.46)	1 (0.89)		2 (1.96)
Pro	2 (1.63)	1 (1.26)			
Gly	11 (11.05)	10 (10.23)	1 (1.08)		
Ala	14 (14.23)	11 (10.58)		1 (1.33)	2 (1.96)
Val	6 (6.10)	3 (2.95)	2 (1.76)		
Met	3 (3.21)				
Ile	6 (6.09)	5 (4.53)			1 (1.08)
Leu	11 (11.06)	9 (8.90)	1 (1.25)		
Tyr	2 (2.05)	1 (1.24)			1 (+)
Phe	6 (5.86)	5 (5.23)		1 (0.96)	
Total	82	56	9	4	9
N-terminal amino acid	Tyr	Gly	Val	Ala	Tyr

C-terminal homoserine lactone residues[4,5]. The amino acid phenylthiohydantoins from the degradations were identified by GLC and chemical ionization mass spectrometry (for details see ref. 6). 35 nmoles of peptide C resp. 45 nmoles of peptide D were sufficient for sequence determination. The C-terminal tetrapeptide E_1 was sequenced both by solid-phase degradation (after C-terminal coupling with a carbodiimide[5]) and manually by the dansyl-Edman procedure. From the cyanogen bromide fragments a partial sequence for HP was deduced (Fig. 2) which was confirmed and extended by solid-phase degradation of the intact polypeptide.

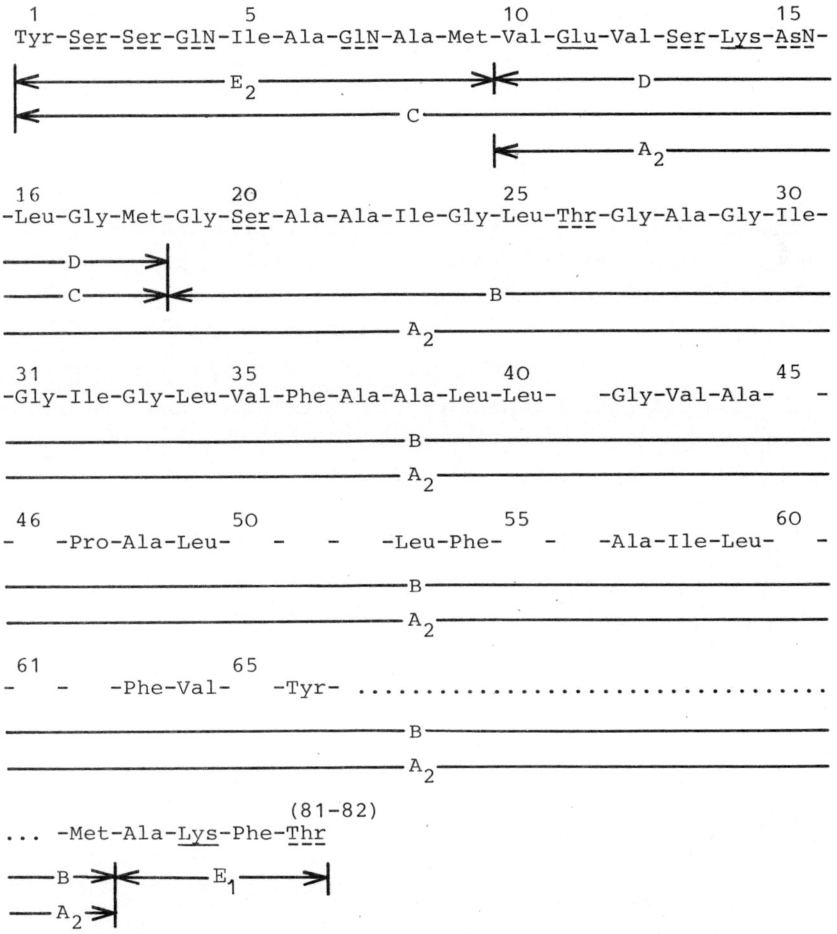

Fig. 2. Partial sequence of HP 8 500. Xxx = polar residues without charge; Xxx = charged residues.

2. Automated solid-phase degradation of the intact polypeptide

Intact HP 8 500 was coupled to p-phenylene diisothiocyanate activated aminopropyl glass (DITC-glass[5]) via its two lysine residues (positions 14 and "79") in more than 90% yield. The covalently bound HP (200 nmoles) was degraded over 80 steps. The amino acid phenylthiohydantoins were determined by GLC and by computerized mass spectrometry (A. Schwab et al., unpublished) in electron impact mode. The second tyrosine was found in step 66. Residues 1 - 40 and several of the later ones were unequivocally identified (Fig. 2).

DISCUSSION

In the partial sequence so far obtained (Fig. 2), polar residues seem to occur preferentially in the N-terminal and C-terminal regions, whereas the midpart contains a long series of nonpolar residues. At least this hydrophobic segment might be located within the lipid bilayer of the membrane.

We hope to complete the amino acid sequence of HP by quantitated solid-phase degradations of the intact polypeptide as well as of the large cyanogen bromide fragment B. In the case that these degradations should leave gaps, appropriate means for a further fragmentation of the polypeptide will have to be found.

REFERENCES

1. Michel, R., Liebl, A., Machleidt, W., Otto, J. and Neupert, W. (1975) Hoppe-Seyler's Z. Physiol. Chem. 356, 1595-1604.
2. Capaldi, R.A. and Vanderkooi, G. (1972) Proc. Nat. Acad. Sci. U.S.A. 69, 930-932.
3. Laursen, R.A. (1971) Eur. J. Biochem. 20, 89-102.
4. Machleidt, W., Hofner, H. and Wachter, E. (1975) in Solid-Phase Methods in Protein Sequence Analysis (Laursen, R.A., ed.), pp. 17-30, Pierce Chem. Co., Rockford, Ill.
5. Wachter, E., Hofner, H. and Machleidt, W. (1975) in ref. 4, pp. 31-46.
6. Wunderer, G., Fritz, H., Wachter, E. and Machleidt, W. (1976) Eur. J. Biochem., in press.

ACKNOWLEDGEMENTS

The authors are grateful to Prof. Th. Bücher for generous support and wish to thank Miss K. Deppner, Miss H. Kummer, Mrs. G. Behrens and Mr. A. Liebl for their excellent technical assistance.

This investigation was supported by the Deutsche Forschungsgemeinschaft, Sonderforschungsbereich 51, "Medizinische Molekularbiologie und Biochemie", München.

Genetics and Biogenesis of Chloroplasts and Mitochondria, Th. Bücher et al. eds.

IDENTIFICATION OF AN ASSEMBLY INTERMEDIATE OF CYTOCHROME OXIDASE IN NEUROSPORA CRASSA.

Sigurd WERNER and Maria NEUNER-WILD

Institut für Physiologische Chemie, Physikalische Biochemie und Zellbiologie der Universität München

Cytochrome oxidase is a supramolecular enzyme complex located in the inner membrane of mitochondria. The oxidase consists of a set of different components, viz. polypeptides, heme a, copper and phospholipids (1-4). Close cooperation between two genetic systems and two compartments of the cell is necessary for the biosynthesis of the oxidase (1,5-7). The complex mechanism of enzyme formation involves a) synthesis of different components in cytoplasm and in mitochondria, b) transport and integration of the enzyme precursors, c) organisation of precursor molecules in an assembly process, and d) arrangement of the final cytochrome complex within the membrane layer. Co-ordination of these individual processes is required for the formation of a functional cytochrome oxidase.

We have focused our attention on the assembly process of the polypeptide subunits of the enzyme. The oxidase contains at least 7 distinct polypeptides, with apparent molecular weights of 1, 41000; 2, 28000; 3, 21000; 4, 16000; 5, 14000; 6, 12000; 7, 10000; subunits 1, 2 and 3 are translated on mitochondrial ribosomes and subunits 4-7 are synthesized on cytoplasmic ribosomes (4,6).

In an initial study of the assembly process we determined pool sizes and half-lives of individual subunit precursors and later succeeded in isolating a precursor polypeptide of subunit 3 (8,9). Currently we are trying to isolate intermediate complexes of the enzyme assembly. Isolation and comparison of such intermediates would provide the basis for determining the assembly sequence of individual components. In the present paper we describe a system for the identification of intermediate complexes of cytochrome oxidase.

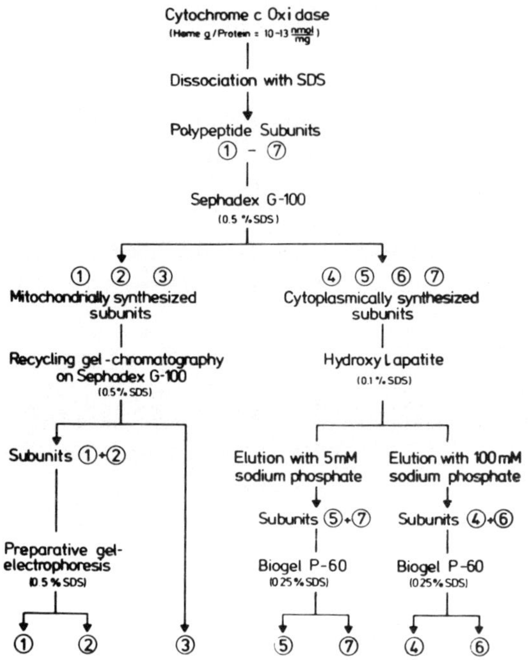

Fig. 1. Purification scheme for 7 polypeptide subunits of Neurospora cytochrome oxidase.

Our experimental approach was as follows: 1) Isolation of individual polypeptide subunits of cytochrome oxidase 2) Preparation of specific antibodies to these polypeptides 3) Pulse-chase labelling of mitochondrial proteins in vivo with radioactive amino acids 4) Precipitation of intermediates from isolated mitochondria by the specific antibodies.

The Neurospora enzyme was prepared from mitochondria by a modified procedure of the ammoniumsulfat fractionation technique described by Tzagoloff (10).

The oxidase complex was dissociated with SDS (0,4g SDS per g protein) and the subunits obtained were isolated following the purification scheme given in Fig. 1. Subunit 1 could be separated from subunit 2 only by means of preparative gel-electrophoresis. All other subunits were isolated by column chromatography which allows

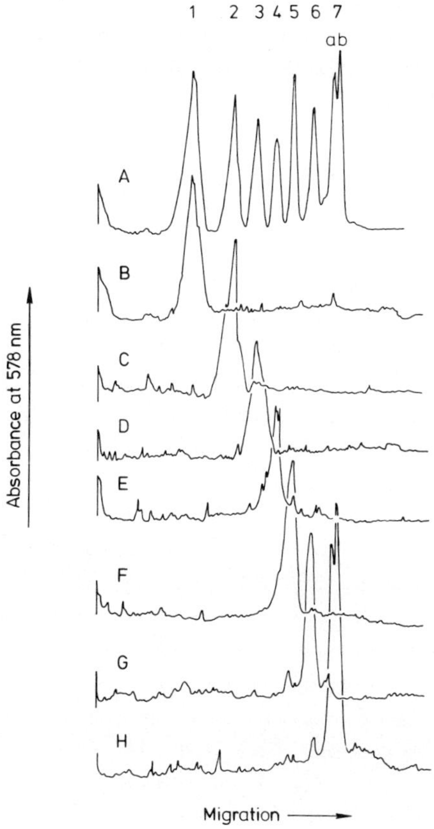

Fig. 2. Densitometric traces of cytochrome oxidase and isolated subunits. Gel-electrophoresis was in the presence of dodecylsulfat. Gels were stained with coomassie brillant blue and recorded at 578 nm. (A) cytochrome oxidase, (B) subunit 1, (C) subunit 2, (D) subunit 3, (E) subunit 4, (F) subunit 5, (G) subunit 6, (H) subunit 7.

large scale preparation. Yields of individual polypeptides varied from 35% to 75% of the applied material. The final preparations were checked for purity by SDS gel-electrophoresis. The densitometric traces of the final subunit preparations are shown in Fig. 2.

The polypeptides were injected into rabbits (for immunisation scheme see Ref. 9) and the γ-globulin fractions were isolated from the hyperimmune sera.

In the following we describe two sets of experiments, which were carried out with wild-type cells using an antibody to subunit 1 and an antibody to subunit 5.

To an exponentially growing cell culture ^{14}C-leucine was added and the culture was aerated for a further 3 hours. After that period one can assume that all radioactive amino acids have been incorporated into stable cell proteins. Then ^{3}H-leucine was given to the culture and one minute later a chase of unlabelled leucine was added. At different times after pulse-labelling samples of the culture were taken, cells were harvested and mitochondria were prepared. The mitochondria were solubilized and treated first with a subunit-specific immunoglobulin, produced in rabbit, and then with a second antibody, an antirabbit immunoglobulin from sheep (9). The resulting immunoprecipitates were dissolved in SDS and analysed by SDS gel-electrophoresis.

Fig. 3A shows the radioactivity pattern obtained with an antibody to subunit 1. In this experiment mitochondria were used which were prepared from cells harvested 4 min after addition of ^{3}H-leucine. Four ^{3}H-labelled bands with apparent molecular weights of 46000, 41000, 14000 and 12000 were observed. The smaller 3 bands coincide exactly with marker subunits 1, 5 and 6 of cytochrome oxidase. This is remarkable since the antibody does not crossreact with isolated subunits 5 or 6. Moreover, these three polypeptides show the characteristics of precursors: a) the ^{3}H/^{14}C ratios calculated for these polypeptides (see Fig. 3A) exceed by far the ratio of 3.9 obtained for the average cell protein; b) most of the ^{3}H-radioactivity found in the polypeptide bands with molecular weights of 41000, 14000 and 12000 is no longer detectable in an immunoprecipitate obtained with mitochondria, which were prepared from cells harvested 90 min after addition of ^{3}H-leucine (see Fig. 3B). The ^{3}H-label in these polypeptides is clearly seen to be chased by the cold leucine. The remaining radioactive protein of the immunoprecipitate displays ^{3}H/^{14}C ratios close to the ratio found for the average cell protein. Thus, a group of three precursor proteins is detectable with an antibody to subunit 1 using a pulse-chase technique for labelling mitochondrial protein.

The precipitated polypeptide with a molecular weight of 46000, however, does not meet the above listed criteria for a precursor protein. The ^{3}H/^{14}C ratio of this polypeptide does not exceed the level of an average cell protein during the pulse-chase experiment.

In the second set of experiments mitochondria were used, which had been labelled and prepared as described above. However, in this case they were treated with an antibody to subunit 5. The applied γ-globulin to subunit 5 showed a slight crossreaction with isolated subunit 6 in the Ouchterlony test. Other subunits than 5 and 6 did

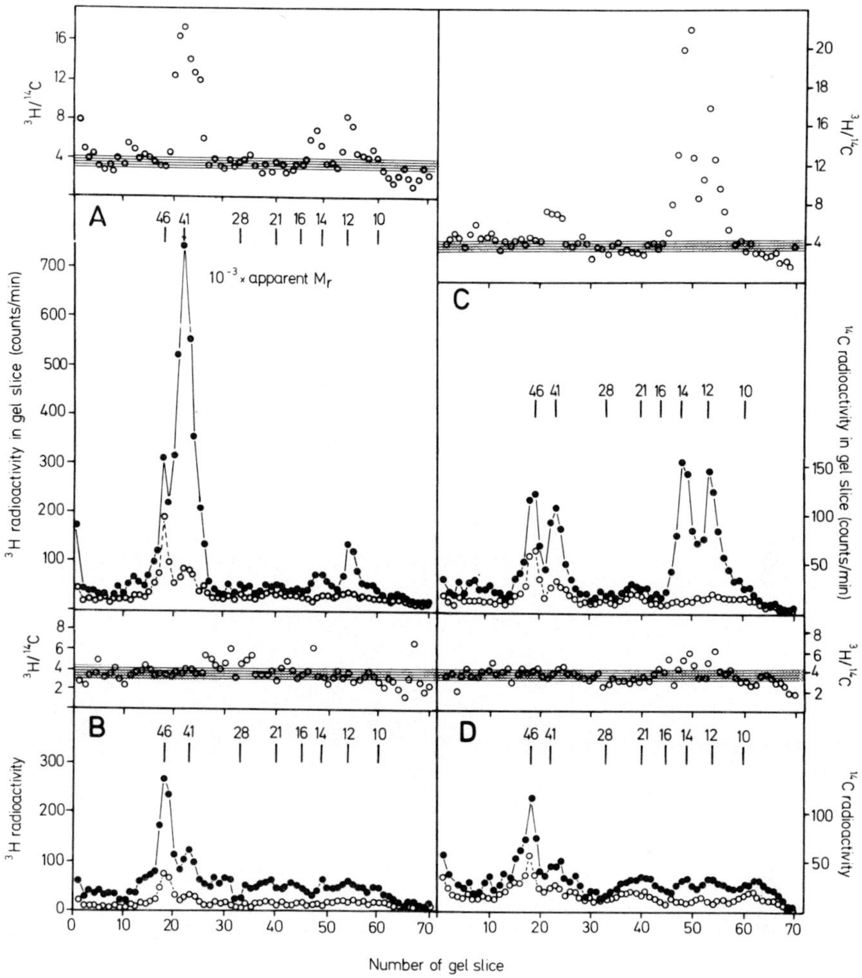

Fig. 3. Gel-electrophoretic analysis of immunoprecipitates obtained with an antibody to subunit 1 (A and B) and an antibody to subunit 5 (C and D). Cells were incubated for 3 hrs with ^{14}C-leucine. Then ^3H-leucine was added. One min later the culture was chased with unlabelled leucine. At different times after addition of ^3H-leucine, samples of the culture were harvested. Mitochondria were prepared, solubilized in 100mM phosphate buffer-1% Triton X-100 and treated successively with the indicated anti-subunit γ-globulin from rabbit and anti-rabbit γ-globulin from sheep. Two of the precipitates obtained with each subunit specific antibody are shown: (A) and (C) 4 min, and (B) and (D) 90 min after addition of ^3H-leucine. $^3H/^{14}C$ ratios of all gel slices were calculated and plotted above the radioactivity patterns. The ratio of total membrane protein was 3.9. ●—● ^3H, o--o ^{14}C radioactivity, o $^3H/^{14}C$ ratio.

not react. Fig. 3C and D shows the gel electrophoretic analysis of the immunoprecipitates. Again three precursor proteins can be identified, which coincide on the gel with marker subunits 1, 5 and 6. Furthermore a 46000 molecular weight polypeptide was observed. It did not show precursor kinetics and seems to be identical to that one precipitated with the antibody to subunit 1.

The labelling patterns obtained from these two sets of experiments can be interpreted as follows: The specific antibody to subunit 1 precipitates a pre-assembly form of subunit 1, predominantly.

Moreover, the antibody recognizes a small portion of subunit 1 to which either subunit 5 or subunit 6, or possibly both subunits 5 and 6 are attached. Ninety min after pulse-labelling these precursors can no longer be detected with the subunit specific antibody, since they are integrated in a final cytochrome oxidase complex. An analogous situation is observed with the antibody to subunit 5.

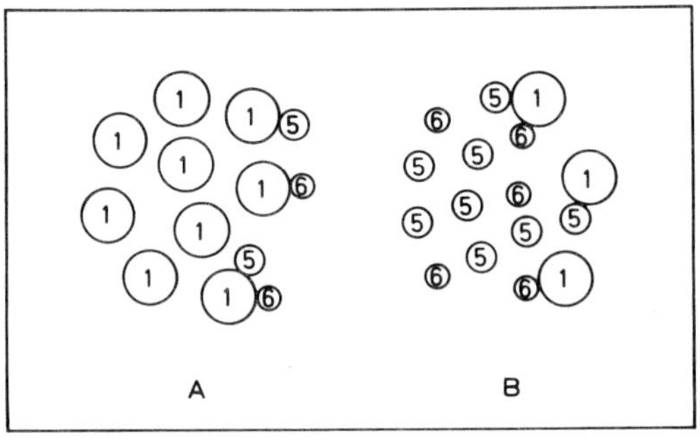

Fig. 4. Proposed conformation of precursors found in the immunoprecipitates obtained with an antibody to subunit 1 (A) and an antibody to subunit 5 (B). Numbers indicate the subunits corresponding to the precursors.

The coprecipitated subunit 1 should be bound to subunit 5, since no crossreaction between the antibody to subunit 5 and isolated subunit 1 could be detected. In Fig. 4 we present a schematic description of the precursors found in the immunoprecipitates.

If this interpretation is valid we would expect that these intermediate complexes would be labile and hence readily dissoziatable. The effect of different salt concentrations on the dissoziation of these complexes is shown in the following experiment.

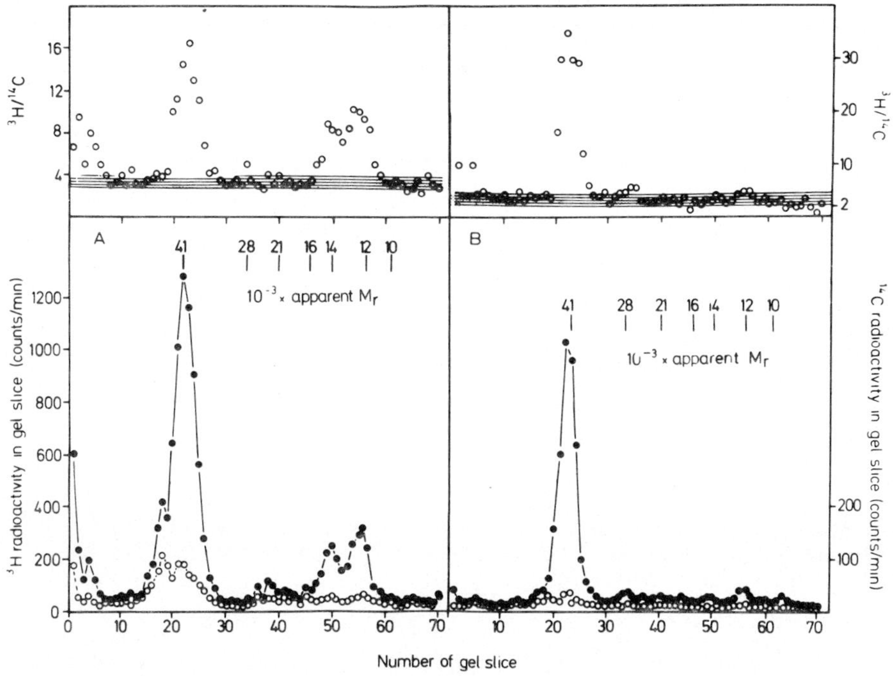

Fig. 5. Gel-electrophoretic analysis of immunoprecipitates obtained with an antibody to subunit 1. Cells were labelled as described in legend to Fig. 3. Mitochondria were prepared from cells which were harvested 4 min after pulse labelling. Mitochondria were solubilized with 100 mM phosphate buffer-1% Triton X-100 containing either 0,3M KCL (A) or 2M KCL (B). Then the mixtures were treated with antibodies (see Fig. 3). $^3H/^{14}C$ ratios of gel slices are given above the radioactivity patterns. The $^3H/^{14}C$ ratio of total membrane protein was 2.8. ●—● ^3H, ○- - -○ ^{14}C radioactivity, ○ $^3H/^{14}C$ ratio.

Mitochondria which were prepared from cells, harvested 4 min after pulse-labelling were solubilized and treated with either 0,3M KCL or 2M KCL before the antibody to subunit 1 and the anti-rabbit γ-globulin were added. These salt concentrations do not influence the immunoprecipitation.

The gel-electrophoretic patterns of the precipitates reveal that in case of 0,3M KCL (Fig. 5A) the same precursor triplet is precipitated as already demonstrated in Fig. 3A; however, in case of 2M KCL (Fig. 5B) subunit 1 is present, but only minimal amounts of subunits 5 and 6.

The status of the 46000 molecular weight polypeptide remains an enigma. It does not have the characteristics of a precursor and it is not found in control precipitates obtained with non-specific rabbit γ-globulin. Further experiments are necessary before any conclusions on the role of this polypeptide can be reached.
The data presented here are of a preliminous nature. We believe however that the experimental approach described can be extremely useful in the study of the assembly process of supramolecular enzyme complexes.

Acknowledgment: This investigation was supported by the Deutsche Forschungsgemeinschaft.

References

1. Weiss,H., Sebald,W. and Bücher,Th. (1971) Eur. J. Biochem. 22, 19-26.
2. Mason,T.L., Poyton,R.O., Wharton,D.C. and Schatz,G. (1973) J. Biol. Chem. 248, 1346-1354.
3. Rubin,M.S. and Tzagoloff,A. (1973) J. Biol. Chem. 248, 4269-4274.
4. Sebald,W., Machleidt,W. and Otto,J. (1973) Eur. J. Biochem. 38, 311-324.
5. Sebald,W., Weiss,H. and Jackl,G. (1972) Eur. J. Biochem. 30, 413-417.
6. Mason,T.L. and Schatz,G. (1973) J. Biol. Chem. 248, 1355-1360.
7. Rubin,M.S. and Tzagoloff,A. (1973) J. Biol. Chem. 248, 4275-4279.
8. Weiss,H., Schwab,A.J., and Werner,S. (1975) in Membrane Biogenesis (Tzagoloff,A., ed) pp. 125-153 Plenum Press, New York.
9. Werner,S. (1974) Eur. J. Biochem. 43, 39-48.
10. Rubin,M.S., and Tzagoloff,A. (1973) J. Biol. Chem. 248, 4269-4274.

Genetics and Biogenesis of Chloroplasts and Mitochondria, Th. Bücher et al. eds.

THE ASSEMBLY OF CYTOCHROME C OXIDASE FROM SACCHAROMYCES CEREVISIAE

R. O. Poyton and E. McKemmie

Department of Microbiology,
University of Connecticut Health Center,
Farmington, Connecticut

INTRODUCTION

The finding that cytochrome c oxidase from a variety of eucaryotes is a transmembranous oligomeric enzyme composed of subunits synthesized on both cytoplasmic and mitochondrial ribosomes[1] raises interesting questions concerning both the structure of the holoenzyme as it sits in the inner mitochondrial membrane and its assembly *in vivo*.

Attempts to determine the topography of cytochrome c oxidase subunits in the membrane have relied on a variety of approaches including the use of enzyme and subunit-specific antisera[2-4], surface labelling with covalent probes[5-6], an analysis of the chemistry of isolated subunits[7-8], and reconstitution from isolated subunits[9]. Although results from these studies have not yet made possible a definitive model for the structure of the holoenzyme they have afforded a first glimpse. Indeed, these studies have made clear that cytochrome c oxidase from yeast is composed of stoichiometric amounts of each of seven subunits[4], that the three mitochondrially-made subunits (I-III) are rather hydrophobic polypeptides[4,7,8] which are either partially or completely buried within the lipid bilayer of the inner mitochondrial membrane[6], and that at least three (IV-VI) of the cytoplasmically-made subunits (IV-VII) are rather hydrophilic polypeptides[4,7,8] localized at or near the surface of the membrane[6]. In addition, recent studies with cytochrome c oxidase from beef heart have suggested that the hydrophilic subunits may be distributed asymmetrically on both sides of the membrane[10].

While the above model is clearly only a "rough approximation" and leaves many important questions regarding the functional organization of the enzyme unanswered it does provide a useful starting point for formulating questions pertinent to the *in vivo* assembly of the holoenzyme. In particular, this model raises the following questions:

1. To what extent and by what mechanisms are the synthesis of mitochondrial and cytoplasmic subunits coordinated; and is the synthesis of subunits coupled directly to the assembly of a functional holoenzyme?
2. What are the intermediate steps (e.g., post-translational processing, subunit-subunit interactions) in the assembly of the holoenzyme?
3. Does assembly take place entirely within the milieu of the membrane or are the cytoplasmic subunits assembled into proenzymes in the cytoplasm?
4. If the cytoplasmic subunits are indeed localized asymmetrically on both sides of the membrane[10] by what mechanism does the cell manage to build an asymmetric oligomeric membrane enzyme?

In this paper we will first review our experiments on the coordination between the synthesis of cytoplasmic and mitochondrial subunits and then consider some recent progress on the assembly of the holoenzyme.

RESULTS AND DISCUSSION

1. Coordination of the synthesis of mitochondrial and cytoplasmic subunits.

Previous studies concerned with the interplay between mitochondrial and cytoplasmic protein synthesis have been done *in vivo* with cells incubated in the presence of drugs which selectively inhibit one or the other protein synthetic system. While these studies[1] have convincingly demonstrated what appears to be a one-way control exerted by cytoplasmic proteins on mitochondrial protein synthesis they have not permitted an identification of the regulatory proteins, an assessment of their mode of action, or a determination of the number of proteins involved. In order to approach these questions and to determine the identity of cytoplasmic proteins which regulate the synthesis of subunits I,II, and III of cytochrome *c* oxidase we have used isolated yeast mitochondria in an *in vitro* system which has already been shown to synthesize identifiable polypeptides[11] (including subunits I,II, and III of cytochrome *c* oxidase) and to be subject to some of the same physiological controls which are operative *in vivo*[12]. With this system a direct demonstration of the importance of cytoplasmic proteins became possible when we observed (1) that a post-mitochondrial supernatant fraction from yeast cells could restore protein synthesis in isolated mitochondria which had stopped synthesizing protein, and (2) that the stimulatory activity was heat and trypsin sensitive, RNase resistant, and non-dialyzable[13,14].

The availability of this *in vitro* assay has allowed us to determine if the cytoplasmic subunits (IV-VII) of cytochrome *c* oxidase are among the stimulatory proteins and have any role in regulating the synthesis of the mitochondrial subunits, I-III. We have taken two approaches. In the first, the post-mitochondrial supernatant was replaced by a mixture of purified cytoplasmic subunits, then the degree of stimulation of both total and cytochrome *c* oxidase protein synthesis was determined. In the second, the post-mitochondrial supernatant was titrated with an antiserum to holocytochrome *c* oxidase, in order to remove all cross-reacting material, and the loss of stimulatory activity determined. Although the first approach proved unsuccessful the second approach has provided direct evidence that at least one cytoplasmic protein which is antigenically related to at least two (subunits IV and VI) of the cytoplasmic

Table 1. Removal of stimulatory activity for cytochrome *c* oxidase synthesis by immunotitration.

	Stimulation(cpm/mg mitochondrial protein) of incorporation into:	
Cell fraction	Total protein	Cytochrome c oxidas
Post-mitochondrial supernatant	24,150	2,510
Post-mitochondrial supernatant titrated with:		
Anti holoenzyme serum	20,690	160
Anti-subunit IV serum	21,750	250
Anti-subunit VI serum	19,960	300

Stimulatory activity for the *in vitro* incorporation of (^{3}H)leucine into total protein and cytochrome *c* oxidase protein was determined as in Ref. 14. A post-mitochondrial supernatant[14] prepared from cells grown in a mixture of (^{14}C) labelled amino acids was divided into four aliquots. Three aliquots were treated with the indicated antiserum to remove all cross-reacting material. An equivalent amount of control serum was added to the remaining "untreated" aliquot.

subunits acts as a specific regulator of the synthesis of the mitochondrial subunits, I-III. As seen in Table 1, post-mitochondrial supernatants which have been titrated with antisera to the holoenzyme or to either of two cytoplasmic subunits, IV and VI, lose the ability to stimulate the synthesis of subunits I, II, and III (Fig. 1) but still retain the ability to stimulate total mitochondrial protein synthesis[14]. Supernatants titrated with antisera against the other two cytoplasmically-made subunits, V + VII, failed to give any cross-reacting material and hence could not be used in these studies.

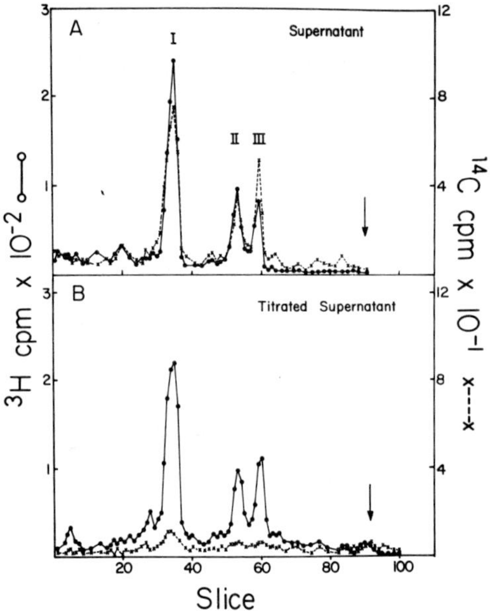

Fig. 1. Polypeptides immunoprecipitated by anti-holocytochrome c oxidase serum from isolated mitochondria stimulated with: (A) a post-mitochondrial supernatant; and (B) a titrated post-mitochondrial supernatant. Mitochondria suspended in the protein-synthesizing mixture were divided into three aliquots. One aliquot was labelled with (^{3}H)leucine for 45 min. while the other two aliquots received an equivalent amount of unlabelled L-leucine. At 45 min. the first aliquot was processed for immunoprecipitation[14] while the other two aliquots were labelled for 30 min. with (^{14}C)leucine in the presence of either a post-mitochondrial supernatant or a titrated post-mitochondrial supernatant[14]. After 30 min., these two aliquots were processed for immunoprecipitation. A portion of the ^{3}H labelled immunoprecipitate was mixed with each ^{14}C labelled immunoprecipitate and co-electrophoresed on SDS polyacrylamide gels.

2. Assembly of Cytochrome c Oxidase

Our analysis of the assembly of cytochrome c oxidase received an important stimulus from the studies described above when upon examining the nature of the stimulatory material removed by immunotitration we observed that immunoprecipitates from all three titrations listed in Table 1 gave essentially identical results. On SDS polyacrylamide gels each exhibited a single major peak of radioactivity at a molecular weight of 55,000 daltons (55K), and no peaks in the 5,000-15,000 dalton region of the gel where the cytoplasmically-made subunits would be expected to migrate. By examining the immunological properties of this 55K polypeptide with subunit-specific antisera and comparing tryptic peptides prepared from it with those prepared from a mixture of sub-units IV - VII we have observed that: (1) the 55K polypeptide cross reacts with antisera to the holoenzyme and to subunits IV and VI but not with antisera specific for subunits I,II,V or VII, (2) the 55K polypeptide gives a line of identity in its cross reaction with subunits IV and VI, and (3) the tryptic peptides prepared from the 55K polypeptide are identical to those prepared from the mixture of subunits IV - VII with the exception of 4-6 peptides released from the 55K polypeptides but not from the mixture of subunits[4]. These results clearly indicate that the 55K polypeptide is a cytoplasmic precursor to all four of the cytoplasmic subunits (IV - VII) of cytochrome c oxidase. Since the cumulative molecular weight of subunits IV - VII is 45,000 daltons[8] the most feasible interpretation of the tryptic mapping results is that the extra peptides from the 55,000 dalton polypeptide are derived from the 10,000 dalton piece which is not converted into any of the four subunits.

The finding that the cytoplasmic subunits are part of a higher molecular weight precursor together with the observation that this precursor, and not the individual subunits, is present in the cytoplasm suggests that the precursor is processed only after its incorporation into the mitochondrion. To confirm this we prepared immunoprecipitates from the post-mitochondrial supernatant of cells grown in the presence of (^{3}H)leucine and from solubilized submitochondrial membrane of cells grown in the presence of (^{14}C)leucine, and compared them on SDS polyacrylamide gels. Immunoprecipitates prepared from submitochondrial membranes contain a peak which co-electrophoreses with the 55K precursor (Fig. 2) and which is not made in the presence of cycloheximide (Fig. 3). Since this peak also coelectrophoreses with the 55K precursor on isoelectric focusing gels we conclude that it is identical to the 55K cytoplasmic precursor.

The existence of a precursor to the cytoplasmic subunits in the membrane together with the fact that the mitochondrial subunits are hydrophobic makes it very likely that the assembly of the protein component of cytochrome c oxidase

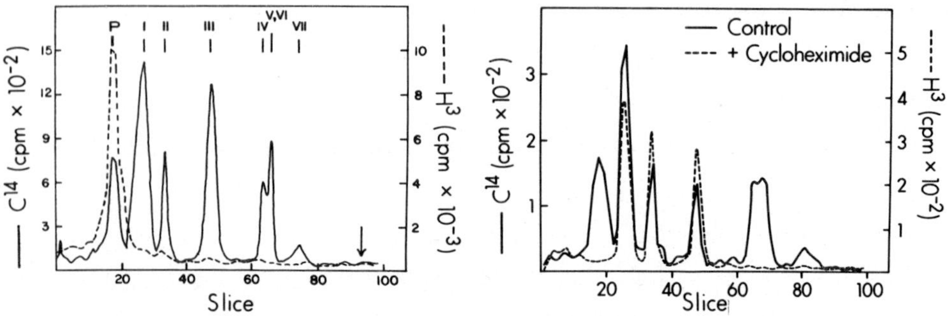

Fig. 2. Co-electrophoresis of immunoprecipitates prepared from a ^3H labelled post-mitochondrial supernatant and ^{14}C labelled submitochondrial membranes on SDS polyacrylamide gels. Immunoprecipitation from the supernatant was carried out in the presence of 1% Triton X-100, 10mM KPO_4, pH7.4, 0.15M KCl, cold carrier cytochrome c oxidase, and 0.5mM tosyl lysine chloromethyl ketone. The membranes were solubilized in cholate-KCl[11] but not fractionated prior to immunoprecipitation. P = precursor.

Fig. 3. SDS gel electrophoretic analysis of immunoprecipitates from submitochrondial membranes isolated either from cells grown for 6 generations in the presence of ^{14}C leucine (——) or from cells labelled for 60 min in the presence of cycloheximide (---).

takes place entirely within the milieu of the membrane. As a first step toward analyzing the assembly process we have followed the kinetics of appearance of the precursor and each of the subunits in the membrane. Cells were grown for 4 generations to mid-exponential phase with [^{14}C]leucine to label stable cellular protein. At this time all of the free amino acid had been incorporated into cellular protein and L-[^3H]leucine was added. Aliquots were then taken at the times indicated, and processed immediately for the isolation of mitochondria. The mitochondrial were converted to submitochondrial particles which were subsequently solubilized in cholate-KCL[11] and subjected to immunoprecipitation with an antiserum to the holoenzyme which cross reacted with all seven cytochrome c oxidase subunits. Under our experimental conditions half of the [^3H]leucine added to the culture is incorporated into protein in less than 1 min and there is no detectable turnover of total cellular or cytochrome c oxidase protein during the tenure of the experiment. Therefore, the $^3H/^{14}C$ ratio corresponds to the ^3H specific radioactivity and can be used to determine the kinetics of appearance of polypeptides in the membrane.

The $^3H/^{14}C$ ratio in the average membrane protein increases very rapidly, is half-maximal by 1.5 min., and reaches a plateau by 20 min. (Fig. 4). The labelling of cytochrome c oxidase protein also exhibits an initial rapid phase but, in addition, exhibits at least one slower phase. Since the kinetics of labelling of cytochrome c oxidase immunoprecipitates is the resultant of the labelling of seven subunits and the 55K precursor to the cytoplasmic subunits, and since the precursor appears to be required for the synthesis of subunits I-III, an unambiguous interpretation of the differences in labelling of the average membrane protein and cytochrome c oxidase protein is, at the moment, impossible. Analysis of the kinetics of labelling of the individual polypeptides in each cytochrome c oxidase immunoprecipitate (Fig. 5) reveals that the precursor is labelled very rapidly (as rapidly as the average membrane protein) followed shortly thereafter by subunits VII and III. Since there are no detectable pools of the cytoplasmic subunits in the unintegrated state either in the cytoplasm, the mitochondrial matrix, or the mitochondrial membrane[4] and since the 55K precursor accounts for about 0.3% of the total cellular protein the most likely interpretation of these results is that the 55K precursor pool turns over rapidly, the 55K precursor is incorporated into the membrane first,

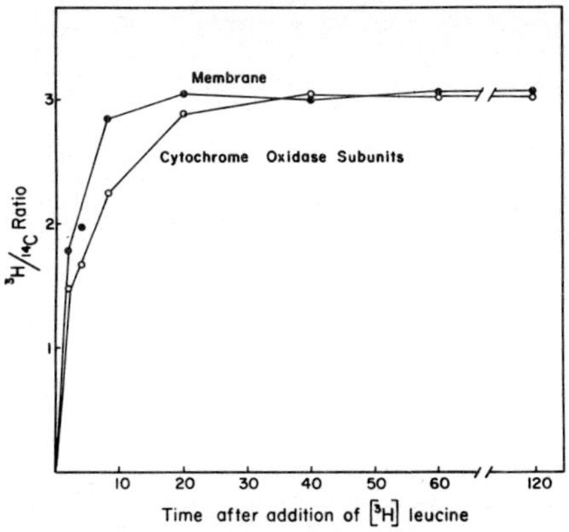

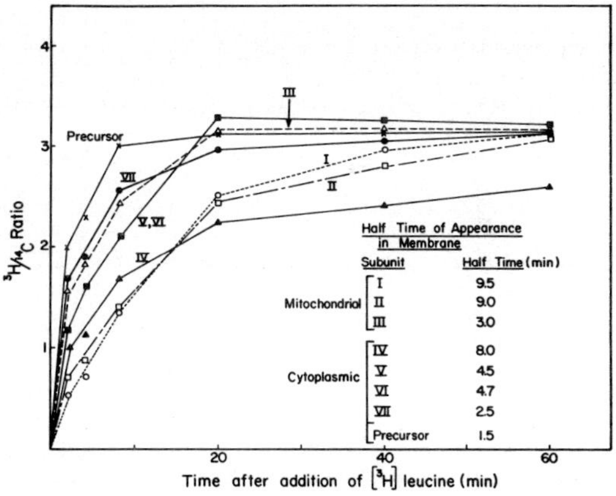

Fig. 4. Kinetics of labelling of total mitochondrial membrane protein (TCA precipitable) and cytochrome c oxidase (immunoprecipitable) protein.

Fig. 5. Kinetics of labelling of the cytoplasmic precursor and the seven subunits of cytochrome c oxidase. Immunoprecipitates prepared in the experiment shown in Fig. 4 were resolved on SDS polyacrylamide and isoelectric focusing gels (for subunits V and VI). The half times of appearance are defined as the time at which the $^3H/^{14}C$ ratio is half-maximal.

and that the time of appearance of subunits IV, V, VI, and VII is related to the order in which they are cleaved from the 55K precursor. Experiments investigating the intermediate steps in this processing pathway support this interpretation and will be detailed elsewhere[4]. The kinetics of labelling of the mitochondrial subunits reveals that subunits I and II follow essentially the same time course and appear in the membrane long after subunit III. At the moment at least two interpretations of these results are possible. Either the 55K precursor differentially stimulates the synthesis of subunit III before it effects the synthesis of subunits I and II, or subunits I and II are at first translated into polypeptides which are not recognized by antiserum to the holoenzyme and are later processed into recognizable subunits[15].

CONCLUSIONS

While these experiments are clearly only a beginning they have given us a preliminary look at the regulatory role played by a specific cytoplasmic protein in the synthesis of three mitochondrial translation products and, at the same time, have allowed us to reach the following conclusions regarding the assembly of yeast cytochrome c oxidase: (1) the cytoplasmically-made subunits are present in the cytoplasm as part of a 55K precursor which is processed after entering the membrane, (2) the assembly of the holoenzyme is most likely initiated when this precursor (or a polypeptide derived from it) enters the membrane and stimulates the synthesis of one or all of the mitochondrially-made subunits, and (3) the assembly process take place within the milieu of the membrane.

REFERENCES

1. Schatz, G. and Mason, T.L. 1974. Annu. Rev. Biochem. 43, 51.
2. Poyton, R.O. and Schatz, G. 1975. J. Biol. Chem. 250, 762.
3. Hackenbrock, C.R. and K.M. Hammon. 1975. J. Biol. Chem. 250, 9185.
4. Poyton, R.O. and McKemmie, E. in preparation.
5. Kornblatt et al. 1974. Proc. Can. Fed. Biol. Soc. 17, 279.
6. Eytan, G.D. and Schatz, G. 1975. J. Biol. Chem. 250, 767.
7. Sebald et al. 1973. Eur. J. Biochem. 38, 311.
8. Poyton, R.O. and Schatz, G. 1975. J. Biol. Chem. 250, 752.
9. G-Nascimento, C. and Poyton, R.O. unpublished.
10. Eytan et al. 1975. J. Biol. Chem. 250, 8598.
11. Poyton, R.O. and Groot, G.S.P. 1975. Proc. Natl. Acad. Sci. USA. 72, 172.
12. Groot, G.S.P. and Poyton, R.O. 1975. Nature 255, 238.
13. Poyton, R.O. and Kavanagh, J. 1975. in Electron Transfer Chains and Oxidative Phosphorylation (ed. E. Quagliariello et al.) p 75.
14. Poyton, R.O. and Kavanagh, J. 1976. Proc. Natl. Acad. Sci. USA. in press.
15. Werner, S. 1974. Eur. J. Biochem. 43, 39.

Genetics and Biogenesis of Chloroplasts and Mitochondria, Th. Bücher et al. eds.

STRUCTURE AND BIOSYNTHESIS OF CYTOCHROME C OXIDASE

F. Cabral, J. Saltzgaber, W. Birchmeier,
D. Deters, T. Frey, C. Kohler and G. Schatz
Department of Biochemistry
Biocenter, University of Basel
CH-4056 Basel, Switzerland

INTRODUCTION

This article summarizes our recent efforts to understand the structure and the assembly of the mitochondrial enzyme cytochrome c oxidase in the yeast Saccharomyces cerevisiae. Cytochrome c oxidase is tightly associated with the mitochondrial inner membrane and consists of seven distinct polypeptide subunits. Three of these (MW: I, 40K; II, 33K; III, 22K) are made on mitochondrial ribosomes and four (MW: IV, 14K; V, 12K; VI, 12K and VII, 4.5K) are made on cytoplasmic ribosomes (1-3). The biogenesis of cytochrome oxidase appears to mirror many of the events which control the biogenesis of the mitochondrial inner membrane as a whole.

RESULTS AND DISCUSSION

Chemical properties of the isolated subunits: The amino acid composition of six of the seven subunits and the N- and C-terminal amino acids of the four cytoplasmically-made subunits have been determined (4,5). In collaboration with A. Tsugita we are currently analyzing the amino acid sequence of the four cytoplasmically-made subunits. Initially, this work was greatly hindered by the microheterogeneity of the subunits isolated from commercial bakers' yeast. It is not clear at present whether this results from heterogeneity of the cell population or the known aneuploidy of these commercial strains. In order to circumvent this problem, we had to isolate the subunits from 30 kg batches of the haploid laboratory strain D273-10B (ATCC 25657).

The isolation of the mitochondrially-made subunits in amounts sufficient for sequence studies proved to be a much more difficult task since these subunits are very hydrophobic and tend to aggregate even in the presence of urea, guanidine hydrochloride or SDS. We are now isolating these subunits by preparative SDS-polyacrylamide gel electrophoresis on 13 x 10 x 2.4 cm rectangular 14% polyacrylamide gel slabs. Effective separation requires careful control of several seemingly minor experimental variables. Up to 15 mg of partially purified cytochrome oxidase (containing 10-12 nmole heme a/mg protein) can be adequately resolved on each slab gel, four of which are usually run simultaneously. The gels are then superficially stained, the desired protein bands are cut out and the subunits are eluted electrophoretically. If necessary, they can be purified further by a second electrophoresis on 13 x 10 x 0.6 cm 20% polyacrylamide gel slabs (6). Each of the three mitochondrially-made subunits is now available in sufficient amounts for detailed chemical studies.

Involvement of a mitochondrially-made subunit in cytochrome c binding: In order to identify possible substrate-binding subunit(s) of yeast cytochrome *c* oxidase, the purified enzyme was reacted with yeast iso-1-cytochrome *c* whose single free cysteinyl residue at position 107 had been activated with 5,5'-dithiobis(2-nitrobenzoate) (7). The resulting cytochrome *c* derivative appeared to function as an "affinity label" of cytochrome *c* oxidase since it rapidly inactivated the enzyme. Inactivation was competitively prevented by underivatized cytochrome *c*. When the "affinity-labeled" oxidase was analyzed by SDS-polyacrylamide gel electrophoresis, a new protein band of apparent molecular weight 38,000 was observed and the relative amount of subunit III was diminished. This new band disappeared almost completely upon treatment of the cross-linked samples with 2-mercaptoethanol. At the same time, subunit III was restored to its normal level (Figs. 1,2). This indicated that the new band consisted of smaller components held together via a disulfide bridge. Similar results were obtained when underivatized iso-1-cytochrome *c* was cross-linked to the oxidase via disulfide-bridge formation in the presence of o-phenanthroline and Cu^{++} (8).

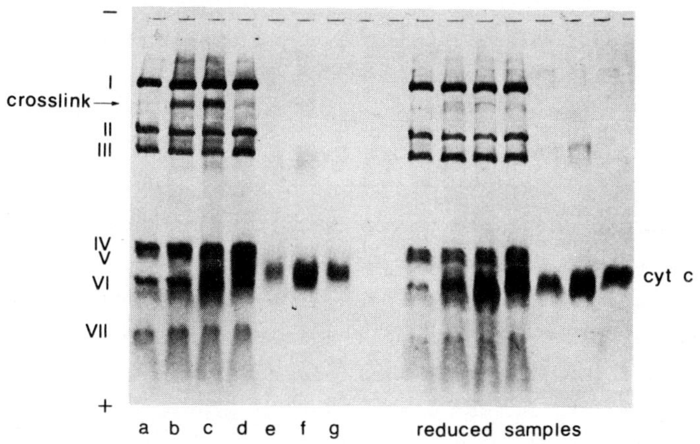

Fig. 1. Cross-linking of cytochrome c oxidase to thionitrobenzoate-cytochrome c. The samples were subjected to electrophoresis on 10/20% polyacrylamide gel slabs (6) and the polypeptide bands were stained with Coomassie Blue. (a) cytochrome c oxidase alone (b) cytochrome c oxidase plus 1 equivalent of modified yeast cytochrome c (c) cytochrome c oxidase plus 2.5 equivalents of modified yeast cytochrome c (d) cytochrome c oxidase plus 1 equivalent of (thiol-free) horse-heart cytochrome c that had been treated with 5,5'-dithiobis(2-nitrobenzoate) in the same way as the yeast cytochrome c. Samples (e)-(g) correspond to (b)-(d) except that cytochrome c oxidase was omitted. The unlettered samples on the right correspond to (a)-(g) except that they were treated with 0.3 M 2-mercaptoethanol before electrophoresis.

More direct evidence for the composition of the cross-linked band was obtained by two-dimensional gel electrophoresis (9). In the first dimension, the cross-linked band was resolved from the other cytochrome c oxidase subunits by SDS-polyacrylamide gel electrophoresis in the absence of added thiol compounds. In the second dimension, it was separated into its component polypeptides by electrophoresis in the presence of excess mercaptoethanol. The cleavage products of the 38,000 dalton band were subunit III and cytochrome c. These data indicate that subunit III is at the very least situated close to one of the cytochrome c binding sites on cytochrome c oxidase and almost certainly on the same side of the mitochondrial inner membrane as cytochrome c itself, i.e. on the outer side.

Structure of cytochrome c oxidase as determined by electron microscopy: Vanderkooi *et al*. have found that extraction of beef-heart mitochondria with increasing concentrations of nonionic detergents yields almost pure cytochrome c oxidase membranes in which the enzyme is organized in a crystalline array (9a). Following their procedure we have obtained electron micrographs of crystalline cytochrome c oxidase membranes which exhibit greater regularity than previously observed and have subjected these micrographs to optical diffraction and optical filtration (10). The reconstructed image reveals slightly oblong (50Å x 85Å) units arranged on a two-dimensional lattice. Significantly, each unit clearly reveals a fairly consistent substructure which had not been observed previously (Fig. 3). SDS-polyacrylamide gel electrophoresis of the preparations revealed essentially only the typical subunit structure of bovine cytochrome c oxidase with minimal contamination by other polypeptides. We are currently attempting to equate the different substructures with individual subunits and to study unstained specimens by low-dose electron microscopy (11). These studies are intended to refine our earlier experiments employing radioactive "surface labels" (12,13) and could ultimately provide clear-cut information on the three-dimensional architecture of the enzyme.

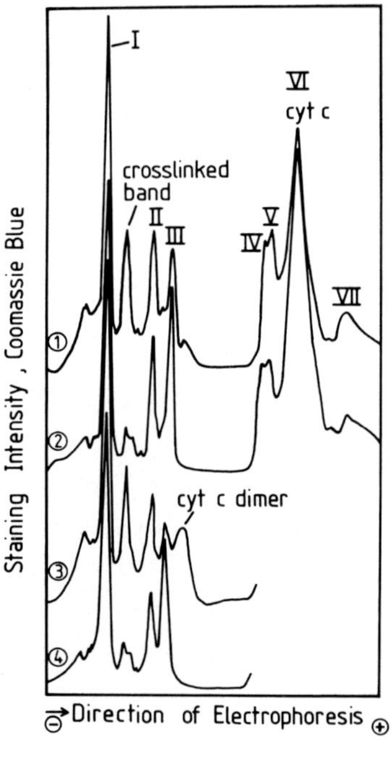

Fig. 2. Scans of the Coomassie Blue stain on four gel strips selected from those shown in Fig. 1 and from a similar gel of samples cross-linked with o-phenanthroline and Cu^{++}. (1) is (c) of Fig. 1 (cytochrome c oxidase crosslinked with a 2.5 molar excess of thionitrobenzoate-cytochrome c). (2) is the same sample but treated with 2-mercaptoethanol following the initial incubation. (3) cytochrome c oxidase crosslinked with a twofold molar excess of unmodified cytochrome c and Cu^{++}-phenanthroline (4) sample (3) incubated with 2-mercaptoethanol before electrophoresis.

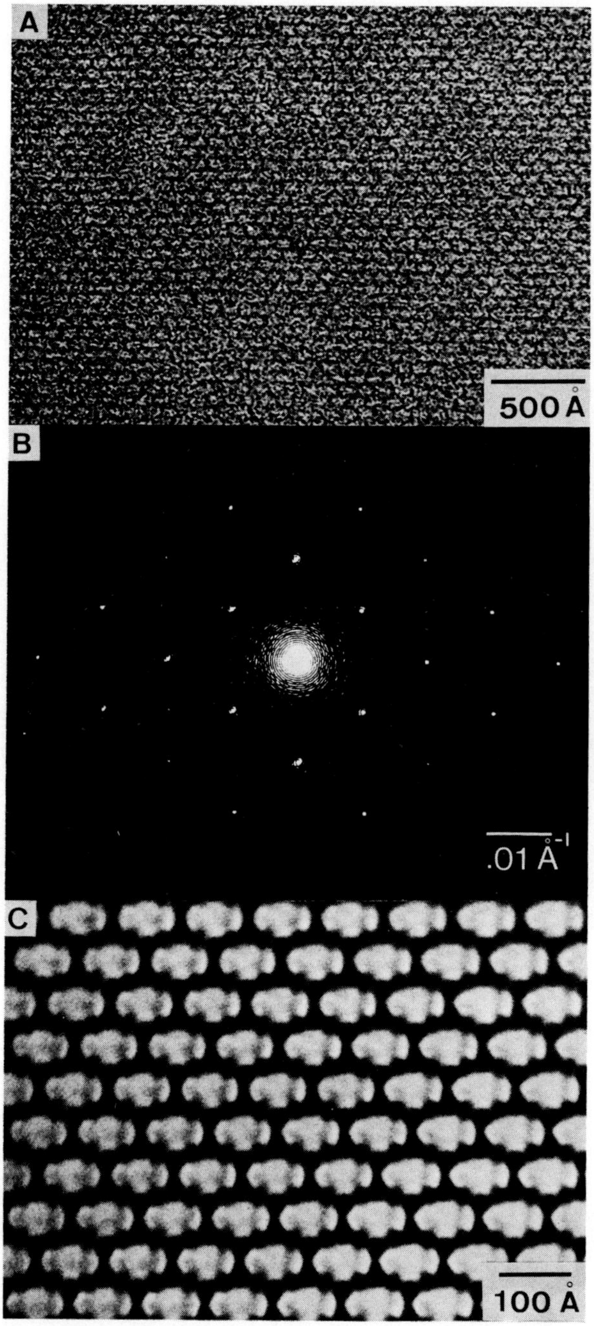

Fig. 3. Image reconstruction of electron micrographs of crystalline cytochrome c oxidase from bovine heart.

The crystalline enzyme was prepared essentially according to Sun *et al.* (14) at pH 7.2, negatively stained with 1% uranyl acetate pH 4.5 and examined in a Philipps 300 electron microscope. Optical diffraction and optical filtration were performed as described in ref. 10.

Top: Unprocessed image
Middle: Optical diffraction pattern
Bottom: Optically filtered image.

Effect of heme on the formation of cytochrome c oxidase: Heme has been reported to control the formation of yeast mitochondria (15). Additionally, heme precursors have been suggested to function as repressors of cytochrome c oxidase formation (16). Earlier work (17) with the heme-deficient yeast mutant GL1-38 (18) has shown that the cytochrome c_1 apoprotein is not inserted into the mitochondrial membrane in the absence of heme. This contrasted with the observation (19) that lack of heme does not prevent accumulation of the cytochrome apoproteins in the cytoplasmic membrane of a heme-deficient E. coli mutant.

In view of these conflicting results we have used the heme-deficient yeast mutant GL1-38 (18) to study the effect of heme on the incorporation of the individual cytochrome oxidase subunits into the mitochondrial organelles. The defect in this mutant has been traced to a lack of δ-aminolevulinic acid synthetase (18). When the mutant is grown on a glucose medium (17) in the absence of δ-aminolevulinic acid, it lacks any absorption peaks in the visible region (Fig. 4C). However, if the medium is supplemented with δ-aminolevulinic acid, the cytochrome spectrum of the mutant (Fig. 4B) is essentially identical to that of the wild-type (Fig. 4A). In agreement with these spectral data, mitochondria isolated from the mutant grown in the absence of the heme-precursor lack significant activities of cytochrome c oxidase, succinate-cytochrome c reductase and cyanide-sensitive respiration. They also lack P_i-ATP exchange activity. Oligomycin-sensitive ATPase and mitochondrial protein synthesis are still present, although with reduced activity. Growth in the presence of δ-aminolevulinic acid restores all these activities to wild-type levels.

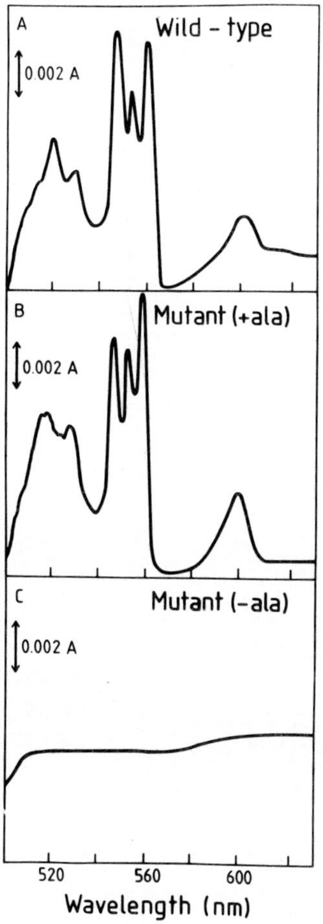

Fig. 4. Low-temperature spectra of mitochondria from wild-type yeast cells (strain X-2180) and the δ-aminolevulinic acid synthetase-deficient mutant GL1-38. Reduced-oxidized difference spectra were recorded at liquid nitrogen-temperature in an Aminco DW-2 recording spectrophotometer at the following protein concentrations.

Wild-type:	12.0 mg/ml
Mutant (+ala):	14.7 mg/ml
Mutant (-ala):	14.0 mg/ml

In order to check for the presence of mitochondrially-made cytochrome c oxidase subunits, mitochondria prepared from cells labeled with ^{35}S-methionine in the presence of cycloheximide were analyzed by electrophoresis on exponential polyacrylamide gels in the presence of SDS. Autoradiography of the dried gels was used to visualize the protein bands labeled in the presence of the inhibitor (20). This method reveals the mitochondrially-made subunits of cytochrome c oxidase and of cytochrome b without the need for immunoprecipitation. In the mitochondria from the mutant grown in the absence of δ-aminolevulinic acid, subunit I was undetectable whereas subunit II was present (Fig. 5). Subunit III appeared to be present as well, although this conclusion is weakened by the fact that subunit III is accompanied by another mitochondrial translation product of very similar mobility (cf. Fig. 6 below). The presence of subunit III will thus have to be corroborated by an independent method. Growth of the mutant in the presence of δ-aminolevulinic acid restored the normal mitochondrial translation pattern.

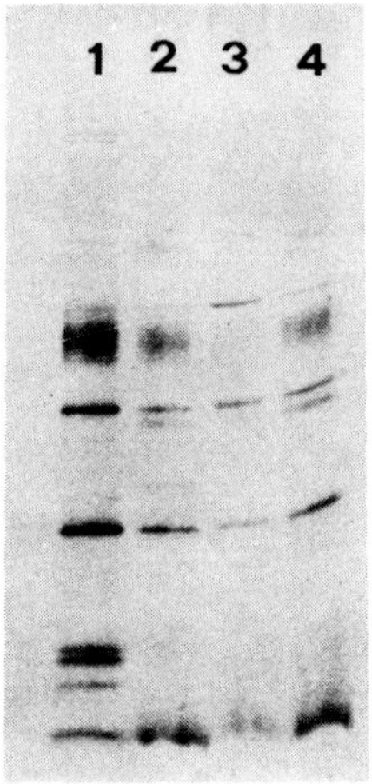

Fig. 5. Heme-deficient yeast cells fail to accumulate the mitochondrially-synthesized cytochrome c oxidase subunit I.

1 Cytochrome c oxidase immunoprecipitated from wild-type (X-2180) mitochondria uniformly labeled with $^{35}SO_4^{--}$.

2 Mitochondria from wild-type cells labeled with ^{35}S-methionine in the presence of cycloheximide.

3 Mitochondria from the heme-deficient mutant GL1-38 grown without added δ-aminolevulinic acid and labeled as in 2.

4 Same as 3 except that the mutant was grown with added δ-aminolevulinic acid.

The presence of subunits in the mutant was also investigated by testing serial dilutions of the isolated mitochondria against constant amounts of subunit-specific antisera by double-diffusion in agar (21). According to this semiquantitative method, the mutant mitochondria appear to lack subunits V and VII whereas subunits II, IV and VI are present in essentially normal amounts (Table I).

TABLE I

TWO OF THE FOUR CYTOPLASMICALLY-MADE CYTOCHROME c OXIDASE SUBUNITS ARE NOT DETECTABLE IN MITOCHONDRIA FROM THE HEME-DEFICIENT YEAST MUTANT GL1-38

Subunit No.	% of wild-type level detected in mutant mitochondria grown	
	-ala	+ala
II	32	100
IV	100	100
V	<4	49
VI	100	100
VII	<3	40

Double diffusion was performed according to Ouchterlony in the presence of 1% Triton X-100. Mitochondria treated with 2% SDS were allowed to diffuse against 20 µl of antiserum.

We conclude that heme affects the accumulation of both mitochondrially-made as well as cytoplasmically-made cytochrome c oxidase subunits. The subunits remaining in the heme-deficient mitochondria are apparently not firmly attached to each other since an antiserum directed primarily against the cytoplasmically-made subunits fails to precipitate the mitochondrially-made subunits (not shown).

Analysis of cytochrome c oxidase-less yeast mutants: Several years ago, Ebner *et al.* (21,22) isolated numerous nuclearly-inherited yeast mutants lacking cytochrome c oxidase activity and analyzed several of these by immunoprecipitation with antisera directed against holocytochrome c oxidase. Within the resolution limits of the methods then available, all mutants appeared to contain a normal set of cytoplasmically-made cytochrome c oxidase subunits but lacked at least one of the mitochondrially-made subunits. We have attempted to refine the analysis of these mutants by introducing the following methods:

(1) Concentrated polyacrylamide step gels (10/20%) offering excellent resolution of small polypeptides (6).

(2) Exponential acrylamide gradient gels (10-15%) affording high resolution of mitochondrial translation products (20).

(3) Two-dimensional separation methods combining isoelectric focusing in the presence of urea and nonionic detergent with SDS-polyacrylamide gel electrophoresis (6,23,24).

These methods have so far yielded the following results (cf. Fig. 6):

(1) The pet 494-1 mutation leads to a loss of subunit III. (Earlier results had only shown that this subunit was no longer integrated with the other subunits - cf. ref. 21). While all four cytoplasmic subunits are still present, the level of subunit IV is lowered.

(2) Mutant pet E11-1 has lost subunits I and II but has retained subunit III as well as all four cytoplasmically-made subunits.

(3) The residual cytochrome c oxidase subunits in the mutants are no longer firmly bound to each other since a given antiserum precipitates only those subunits against which it is directed and other residual subunits do not co-precipitate (not shown).

(4) In line with (3), subunits IV and VI can be readily detached from pet 494-1 and pet E11-1 mitochondria by sonic oscillation whereas they remain firmly bound to wild-type mitochondria (cf. Fig. 9 in the following section).

(5) Thus far, none of the mutants could be shown to contain a cytoplasmically-made cytochrome c oxidase subunit with an altered isoelectric point (cf. below). Our hypothesis (21) that some of these mutants may carry mutations in one of these subunits thus remains without experimental support.

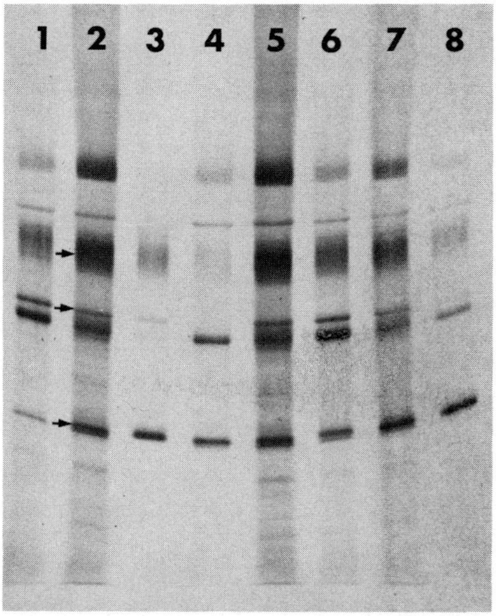

Fig. 6. High resolution analysis (20) of mitochondrial translation products in nuclear cytochrome c oxidase-less yeast mutants.

1 pet 494-1 mitochondria labeled with $^{35}SO_4^{--}$ in vivo in the presence of cycloheximide
2 and 5 Wild-type mitochondria labeled as in 1
3 Cytochrome c oxidase immunoprecipitated from a mitochondrial lysate of wild-type cells grown in $^{35}SO_4^{--}$
4 pet E11-1 mitochondria labeled as in 1
6 Mixture of 1 and 3
7 Mixture of 2 and 3
8 Mixture of 4 and 3

The arrows mark the position of the three mitochondrially-made cytochrome c oxidase subunits. In wild-type mitochondria, subunit III is barely separated from a slightly faster moving mitochondrial translation product which is still present in mutant pet 494-1. These two closely adjacent bands can be differentiated by the mixing experiments shown in slots 6-8.

Towards a two-dimensional polypeptide map of the mitochondrial inner membrane: We are convinced that progress in elucidating mitochondrial membrane assembly is now critically dependent on the ability to screen the polypeptide patterns of a large number of samples and to detect missense mutations in individual membrane proteins without having to isolate these proteins. This requirement is best met by two-dimensional separation techniques which separate polypeptides both by charge as well as by size. The method employed in our laboratory is as follows:

(1) Mitochondria or submitochondrial fractions are first dissociated in urea - Triton X-100, or in SDS followed by urea - Triton X-100 and subjected to isoelectric focusing in the presence of urea and Nonidet P40 in cylindrical polyacrylamide gels.

(2) The isoelectric focusing gels are then placed on top of gel slabs containing an exponential (10-15%) gradient of polyacrylamide and electrophoresed in the presence of SDS.

(3) The gel slabs are stained with Coomassie Blue; if radioactive material is being used, the gels are subsequently dried and radioautographed.

(4) For identifying polypeptide spots, unlabeled mitochondria are mixed with uniformly labeled cytochrome c oxidase, cytochrome c oxidase subunits, or F_1-subunits and carried through steps (1) to (3). When the radioautograms are superimposed upon the stained gel slabs, individual polypeptides can be readily identified.

Figures 7-9 illustrate this approach. Since the intensities of the individual spots are so different, not all of the spots visible on the original gel or radioautogram may be discernible on these reproductions. Nevertheless, it can be seen that nine important polypeptides of the mitochondrial inner membrane have already been "placed on the map".

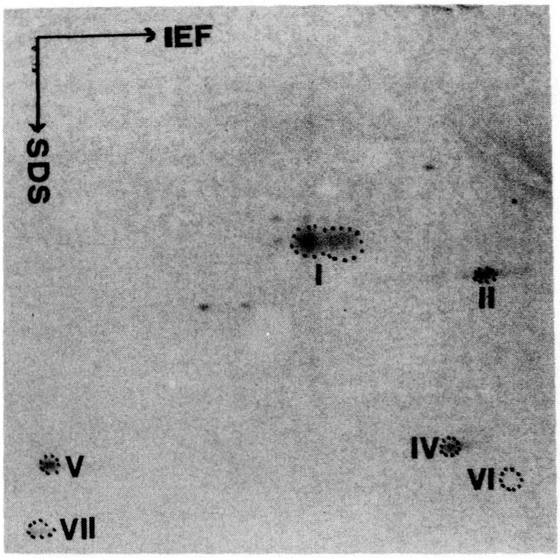

Fig. 7. Two-dimensional analysis of radioactive immunoprecipitated cytochrome *c* oxidase subunits.

Wild-type yeast cells (D273-10B) were grown in the presence of $^{35}SO_4^{--}$. The mitochondria were isolated, extracted with cholate/KCl (21) and immunoprecipitated with an antiserum directed mainly against subunits V and VII. The immunoprecipitate was analyzed as described in the text. The Roman numerals identify the individual cytochrome *c* oxidase subunits. (Subunit III cannot be clearly displayed in this system).

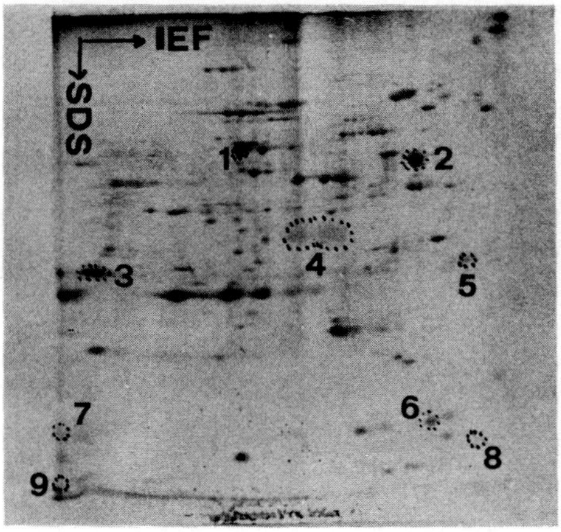

Fig. 8. A rudimentary two-dimensional "map" of the mitochondrial inner membrane.

Wild-type yeast cells were uniformly labeled as in Fig. 7 and converted to spheroplasts. The mitochondria were isolated in the presence of the protease inhibitor phenylmethyl sulfonyl fluoride and the mitochondria were isolated and purified on a Metrizamide density gradient. They were then converted to submitochondrial particles by sonication in the presence of the protease inhibitor and the submitochondrial particles were isolated by centrifugation and analyzed as in Fig. 7. See text for further details.

1 α subunit of F_1
2 β subunit of F_1
3 γ subunit of F_1
4 cytochrome *c* oxidase subunit I
5 cytochrome *c* oxidase subunit II
6 cytochrome *c* oxidase subunit IV
7 cytochrome *c* oxidase subunit V
8 cytochrome *c* oxidase subunit VI
9 cytochrome *c* oxidase subunit VII

As discussed in the text, some of the minor spots may not be visible in this reproduction.

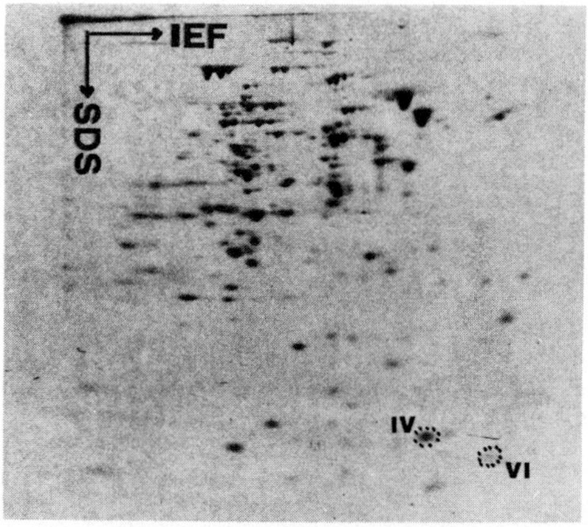

Fig. 9. Two-dimensional analysis of a sonic extract of uniformly labeled mitochondria from the cytochrome c oxidaseless mutant pet 494-1. The experimental details were as described in Fig. 8 except that the centrifuged sonic extract was analyzed. Cytochrome c oxidase subunits IV and VI, which had been released into solution, are marked with Roman numerals.

Since so many functionally-defined mitochondrial polypeptides can be purified, we should soon have a usable polypeptide map of the mitochondrial inner membrane. Significantly, this method is also capable of resolving the total mitochondrial translation products into more than a dozen discrete spots (not shown). This opens new possibilities for identifying the primary structural lesions in some of the mitochondrially-inherited mutants lacking various components of oxidative phosphorylation (25,26,27).

ACKNOWLEDGEMENTS

These studies were supported by grants 3.2350.74, 3.399.74 and 3.433.74 from the Swiss National Science Foundation. We wish to thank Urs Müller and Henri Homberger for excellent technical assistance.

REFERENCES

1. Tzagoloff, A., Rubin, M.S. and Sierra, M. (1973) Biochim. Biophys. Acta 301, 71-104.
2. Mason, T., Ebner, E., Poyton, R.O., Saltzgaber, J., Wharton, D.C., Mennucci, L. and Schatz, G. (1972) in Mitochondria: Biogenesis and Bioenergetics (van den Bergh, S.G. *et al.*, editors), North Holland, Amsterdam, p. 53-70.
3. Schwab, A.J., Sebald, W. and Weiss, H. (1972) Europ. J. Biochem. 30, 511.
4. Poyton, R.O. and Schatz, G. (1975) J. Biol. Chem. 250, 752-761.
5. Birchmeier, W., Tsugita, A. and Schatz, G., in preparation.
6. Cabral, F. and Schatz, G. (1976) in Methods Enzymol. (Fleischer, S. and Packer, L., editors), Academic Press, New York, in press.
7. Birchmeier, W., Kohler, C. and Schatz, G. (1976) submitted for publication.
8. Kobashi, K. and Horecker, B.L. (1967) Arch. Biochem. Biophys. 121, 178-186.
9. Wang, K. and Richards, F.M. (1974) J. Biol. Chem. 249, 8005-8018.
9a. Vanderkooi, G., Senior, A.E., Capaldi, R.A. and Hayashi, H. (1972) Biochim. Biophys. Acta 274, 38-48.
10. Bijlenga, R.K.L., Aebi, U. and Kellenberger, E. (1976) J. Mol. Biol. 103, 469-498.
11. Unwin, P.N.T. and Henderson, R. (1975) J. Mol. Biol. 94, 425-440.
12. Eytan, G.D. and Schatz, G. (1975) J. Biol. Chem. 250, 767-774.
13. Eytan, G.D., Carroll, R.C., Schatz, G. and Racker, E. (1975) J. Biol. Chem. 250, 8598-8603.
14. Sun, F.F., Prezbindowski, K.S., Crane, F.L. and Jacobs, E.E. (1968) Biochim. Biophys. Acta 153, 804-818.
15. Sugimura, T., Okabe, K., Nagao, M. and Gunge, N. (1966) Biochim. Biophys. Acta 115, 267-275.
16. Charalampous, F. (1974) J. Biol. Chem. 249, 1014-1021.
17. Ross, E. and Schatz, G. (1976) J. Biol. Chem. 251, 1997-2004.
18. Gollub, E.G., Trocha, P., Liu, P.K. and Sprinson, D.B. (1974) Biochem. Biophys. Res. Commun. 56, 471-477.
19. Haddock, B.A. and Schairer, H.U. (1973) Europ. J. Biochem. 36, 34-45.

20. Douglas, M.G. and Butow, R.A. (1976) Proc. Natl. Acad. Sci. USA 73, 1083-1086.
21. Ebner, E., Mason, T.L. and Schatz, G. (1973) J. Biol. Chem. 248, 5369-5378.
22. Ebner, E., Mennucci, L. and Schatz, G. (1973) J. Biol. Chem. 248, 5360-5368.
23. O'Farrell, P.H. (1975) J. Biol. Chem. 250, 4007-4021.
24. Ames, G.F.-L. and Nikaido, K. (1976) Biochemistry 15, 616-623.
25. Tzagoloff, A., Akai, A., Needleman, R.B. and Zulch, G. (1975) J. Biol. Chem. 250, 8236-8242.
26. Louie, A., Boerner, P., Biron, C.A. and Mason, T.L. (1975) Federat. Proc. 34 (3), 1556.
27. Storm, E.M. and Marmur, J. (1975) Biochem. Biophys. Res. Commun. 64, 752-759.

Genetics and Biogenesis of Chloroplasts and Mitochondria, Th. Bücher et al. eds.

COORDINATION OF MITOCHONDRIAL AND CYTOPLASMIC PROTEIN SYNTHESIS IN NEUROSPORA CRASSA

W. Neupert and A. v. Rücker
Institut für Physiologische Chemie, Physikalische Biochemie und Zellbiologie
der Universität München
Germany

INTRODUCTION

The formation of some multi-component mitochondrial enzyme complexes such as cytochrome oxidase requires the cooperation of cytoplasmic and mitochondrial translation. Products from both systems are integrated to form functional enzyme complexes with definite subunit stoichiometry. Control of the coordination of these two distinct systems may be exerted at many levels, so that a general study of the coordination presents many difficulties. Recently, a detailed review on these problems was presented by SCHATZ and MASON (1).

In the present investigation we have limited ourselves to one facet of the interaction between cytoplasmic and mitochondrial translation, viz. the synthesis and membrane integration of mitochondrial translation products under conditions when cytoplasmic translation does not operate. We have linked the general aspects of translation to the synthesis and assembly of specific mitochondrial products, namely the subunits of cytochrome oxidase.

RESULTS

1. Amino acid incorporation into mitochondrial membrane proteins of cells with blocked cytoplasmic translation

Cells were prelabelled with ^{14}C-leucine to obtain homogeneous labelling of total cellular protein. At 16 hrs growth (log phase) cycloheximide (CHI) was added. After the times indicated in Fig. 1, aliquots were withdrawn, mitochondria prepared and the ^{3}H/^{14}C-ratios of the mitochondrial membrane protein determined. Labeling with ^{3}H-leucine was performed at a very low added leucine concn. (0.3 nmol/mg cell protein), a procedure usually employed to effectively label mitochondrial translation products.

When leucine was added 2.5 min after addition of CHI, incorporation proceeds for about 10-15 min and then stops. When the cells were preincubated for 62.5 min, incorporation of ^{3}H-labelled leucine still takes place and furthermore continues for a longer time span. The level of incorporation is lower, however, than in the cells exposed to CHI for a shorter time.

This observation leads to suggest that the incorporation of ^{3}H-leucine in the presence of CHI is not a true measure of the rate of mitochondrial protein

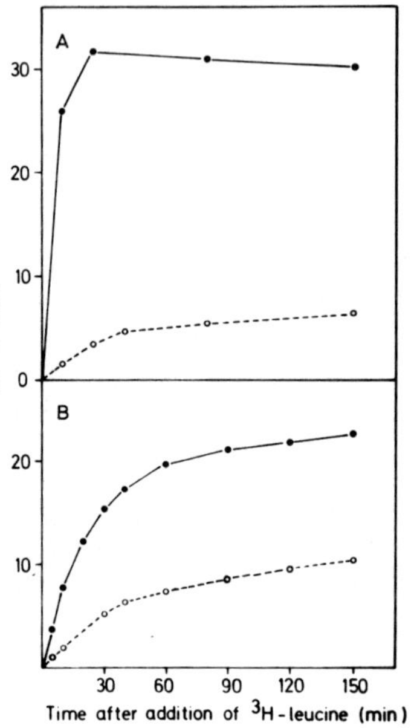

Fig. 1. Time course of 3H-leucine incorporation into mitochondria in whole cells as a function of time of incubation in cyclohex-imide (0.1 mg/ml).
Cells were prelabelled with 14C-leucine. Incorporation was carr-ied out with (A) low and (B) high concn. of added leucine (see text)
-●——●- 2.5 min CHI preincub.;
-o----o- 62.5 min CHI preincub.

synthesis (see also ref. (2)). This lack of validity may be related to changes in pool size and to metabolism of intracellular leucine.

To obviate these effects, leucine was added at concns. sufficiently high to offset these changes. At the concn. added (500 nmol/ mg cell protein), leucine uptake still takes place and within 10 min about 75% of the added radioactivity was accumulated in the cells. At this high added leucine concn. incorporation proceeds in both the short and long time CHI preincubation. Incorporation even continues at 150 min after addition of ^{3}H-leucine. When the initial rate of the 62.5 min preincubation sample is compared with the rate observed at 60 min with the 2.5 min preincubated sample, it can be seen that the rates are not equal. The longer preincubated sample shows a rate faster by a factor of six than the 2.5 min sample. Therefore, even at this high added leucine concn. the incorporation does not truely reflect the real rate of protein synthesis. This discrepancy between real and apparent rates can be explained by the following observations:
(a) The intracellular concns. of amino acids were found to strongly increase after exposing cells to CHI. Leucine concn. increases by a factor of about 10 during the first 90 min. (b) It was found that leucine is rapidly metabolized, a large part of the ^{3}H-radioactivity appearing in a-ketoisocaproic acid.

The effect of preincubation of cells with CHI on the protein synthesis of isolated mitochondria was measured and the results are presented in Fig. 2. The mitochondria from cells kept for 1 hr in the presence of CHI displayed a rate of incorporation which was very similar to that of mitochondria from cells treated for 2.5 min with CHI (or to that of mitochondria from untreated cells). Even after treatment of cells for 4 hrs the rate of incorporation was not reduced by more than about 30%. Similar observations were made by Sebald et al. (3).

It appears therefore that the mitochondrial protein synthesizing system is not essentially linked to the cytoplasmic system as suggested by some investigators (see ref. (1)).

2. Translational activity of mitochondrial ribosomes in cycloheximide blocked cells

In view of the unexpected lack of tight coupling of cytoplasmic and

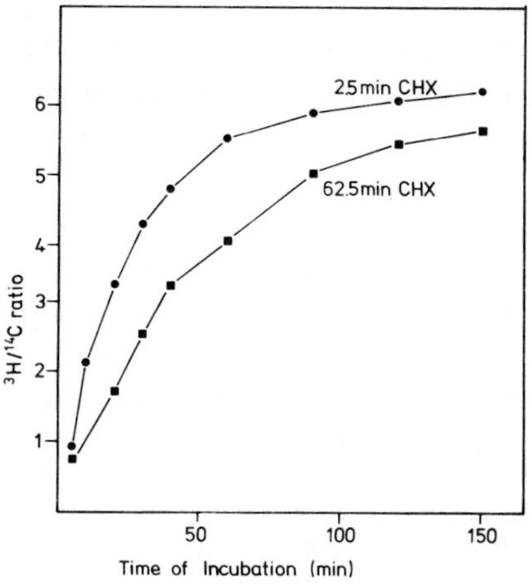

Fig. 2. Time course of 3H-leucine incorporation into isolated mitochondria from 14C-prelabelled cells and the effect of preincubation of cells in cycloheximide on this incorporation.

For experimental details see ref.(8).

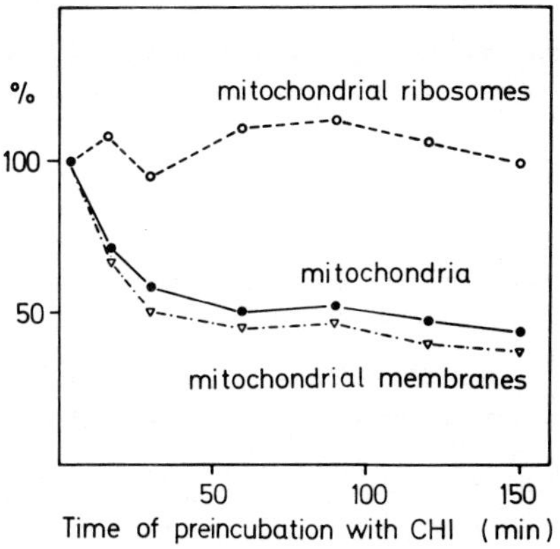

Fig. 3. The effect of cycloheximide preincubation on the *in vivo* incorporation of leucine into different mitochondrial fractions.
Whole cells were pulse labelled with 3H-leucine for 5 min at the indicated times. All incorporation measurements were corrected for the intracellular leucine pool. The 100% value represents incorporation after 2.5 min preincub.

mitochondrial protein synthesis, the rate of translation on mitochondrial ribosomes was measured by a different approach. Cells were preincubated for different time periods in the presence of CHI and then pulsed with ^{3}H-leucine for 5 min. To aliquots of these samples a chase of unlabelled leucine was given for 10 min. Mitochondrial ribosomes were then isolated from the pulsed cells. Mitochondria and mitochondrial membranes were isolated from the chased cells. ^{3}H-radioactivity was measured in these fractions and the actual rates of leucine incorporation were determined by correcting for the intracellular leucine concns. at the different times. The result is shown in Fig. 3. The amount of leucine in nascent polypeptide chains is found to be constant at all times of CHI preincubation. The amount of leucine incorporated into mitochondrial membrane protein however decreases during the first 30 min to about 50% compared to the control. (Preincubation for 2.5 min is taken as a control because the addition of CHI is necessary to assess specifically radioactivity in mitochondrial translation products). After 30 min preincubation incorporation stays constant at a level of about 40%.

To measure the velocity of replacement of labelled nascent chains on mitochondrial ribosomes by unlabelled ones, i.e. to determine the rate of elongation, cells were pulse labelled in the presence of CHI and then exposed to a leucine chase of increasing times. The rate of elongation was found to be the same, irrespective of whether the cells were preincubated with CHI for 2.5 min or for 62.5 min before the pulse. Furthermore, when ribosomes from corresponding mitochondria were analysed by gradient centrifugation, the same distribution of radioactive nascent chains was seen among mono- and polymeric ribosomes, more than 80% of the ^{3}H-leucine being associated with polymers.

These observations firmly support the conclusion that mitochondrial translation continues in the absence of cytoplasmic translation. They also suggest that at least part of the polypeptides formed, are integrated into the mitochondrial membrane. They further suggest that part of the translation products are degraded since completed mitochondrial translation products were neither detected in the mitochondrial matrix nor in the postmitochondrial supernatant.

3. Translation products formed in mitochondria in cycloheximide blocked cells

Cells homogeneously labelled with ^{14}C-leucine were exposed to CHI for 2.5 min, then a pulse of ^{3}H-leucine was given for 2.5 min, followed by a 60 min chase. This experiment was made in an attempt to catch the two systems as close to equilibrium as possible. Mitochondria were isolated and subjected to SDS gel electrophoresis (Fig. 4). ^{14}C-labelled total mitochondrial proteins are mainly found at apparent MWs between 60,000 and 10,000. Welldefined mitochondrial translation products are displayed by the ^{3}H-label, which can be attributed to cytochrome oxidase, cytochrome b and OS-ATPase. It should be especially noted that the proportion of mitochondrial translation products with low apparent MWs (10,000 and less) is extremely low.

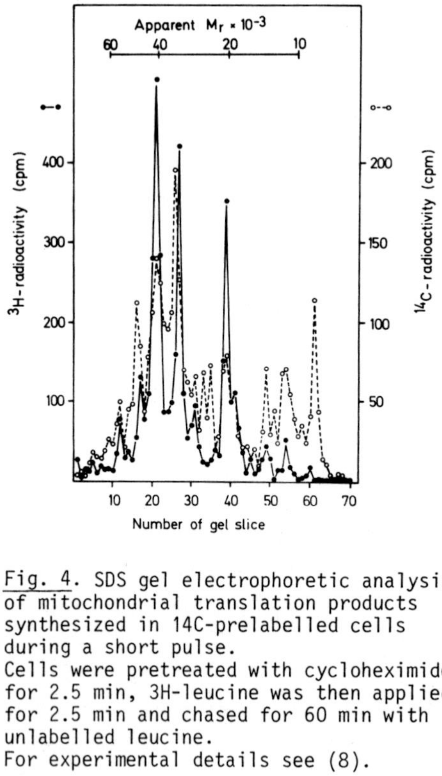

Fig. 4. SDS gel electrophoretic analysis of mitochondrial translation products synthesized in 14C-prelabelled cells during a short pulse.
Cells were pretreated with cycloheximide for 2.5 min, 3H-leucine was then applied for 2.5 min and chased for 60 min with unlabelled leucine.
For experimental details see (8).

In a further experiment, cells were preincubated with CHI for 2.5 min, 62.5 min and 122.5 min, respectively, then ^{3}H-leucine was incorporated for 45 min and a chase of unlabelled leucine was given for further 10 min. The SDS gel electrophoretic analysis of the isolated mitochondria is presented in Fig. 5. Clearly, the same products are formed in cells, in which cytoplasmic translation was blocked for 120 min compared to those blocked only for 2.5 min. However, the translation products with apparent MWs of more than 20,000 are present in lower proportion in cells treated with

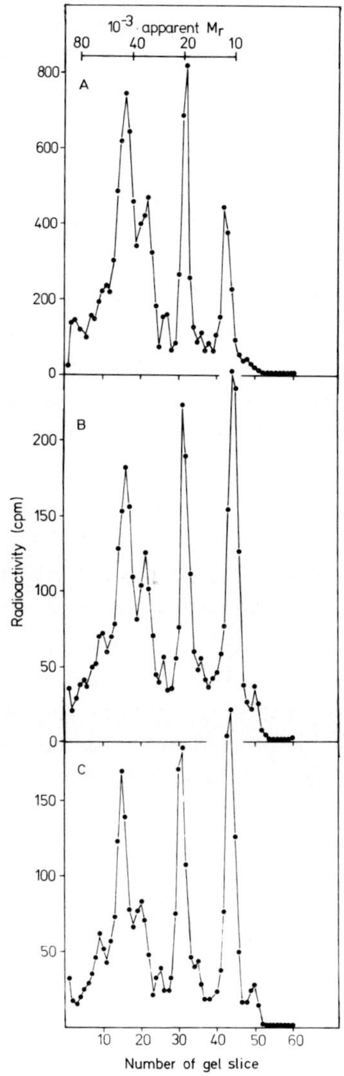

Fig. 5. SDS gel electrophoretic analysis of mitochondrial translation products synthesized after different times of pretreatment of cells with cycloheximide.
Cells were incubated with cycloheximide for (A) 2.5 min, (B) 62.5 min and (C) 122.5 min.3H-leucine was added to the cells and after further 45 min incubation a chase of unlabelled leucine was given for 10 min.

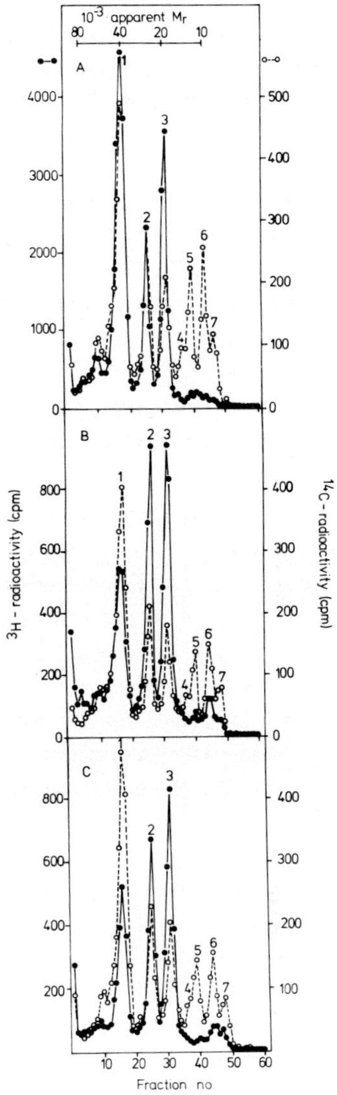

Fig. 6. SDS gel electrophoretic analysis of cytochrome oxidase immunoprecipitated from the total homogenate of cells which were prelabelled with 14C-leucine. Incubation of cells with cycloheximide and labelling with 3H-leucine was carried out as described in legend to Fig. 5. Samples A, B and C correspond to samples A, B and C in Fig. 5.

cycloheximide for longer time. Furthermore, translation products are found in increasing amounts in the low apparent MW region (11,000 and less) after longer preincubation of cells with CHI (cf. Fig. 4).

These observations suggest that original translation products are degraded to low MW components and that part of the degradation products are retained in the membrane. When mitochondrial translation products are formed only during a short pulse period immediately after blocking cytoplasmic translation, this degradation appears to be negligible. It is tempting to suggest that under these conditions cytoplasmically synthesized partner proteins are available in such amounts that the newly formed mitochondrial partners can be assembled to functional complexes and thereby preserved in the mitochondrial membrane. After longer exposure of the cells to CHI, the pools of cytoplasmic partner proteins are exhausted, the mitochondrially translated components are continued to be synthesized but only part of them can be integrated into the membrane. Another part may be degraded probably before being integrated.

4. Synthesis and assembly of cytochrome oxidase components in cells with blocked cytoplasmic translation

We next studied the influence of the availability of cytoplasmically synthesized subunits of cytochrome oxidase on the production and insertion of the mitochondrially synthesized subunits. ^{14}C-homogeneously labelled cells were exposed to ^{3}H-leucine in the presence of CHI as described in Fig. 5. After incubation, the cells were homogenized and extracted with a buffer containing Triton X-100. Cytochrome oxidase was precipitated from these extracts with an antibody against the whole enzyme. SDS gel electrophoretic analysis of the immunoprecipitates is shown in Fig. 6.

The ^{14}C-radioactivity on the gels displays the seven subunits of the enzyme. After short CHI treatment, the three mitochondrially synthesized subunits (1-3) are found to be labelled with ^{3}H in about the same ratio, at which they are present in the ^{14}C-profile, subunit 3 being somewhat higher. After 62.5 min and 122.5 min preincubation of cells still considerable synthesis of subunits 1-3 occurs. The proportions of the three subunits however change. Subunits 1 and 2 decrease with respect to subunit 3. This is in close agreement with the findings with whole mitochondrial membranes (cf. Fig. 5). In the samples preincubated with CHI for longer times, lower MW material is precipitated which may represent breakdown products of subunits 1-3 which are recognized by the antibody.

These data show that immunoprecipitation of cytochrome oxidase from total cell homogenate appears to capture not only fully assembled enzyme but also precursor polypeptides. However, if immunoprecipitation was carried out with isolated mitochondria, a quite different picture was observed. The labelling pattern in this case resembles that of the chemically isolated complex (4), showing that after short CHI exposure only small amounts of ^{3}H-labelled subunits 1 and 2 are found (see also (5)), whereas after 62.5 min and 122.5 min CHI none of these subunits were detected. This is in agreement with the data of Schwab et al. (4), that the pool of cytoplasmically formed subunits allows assembly of total enzyme only for a few min. Subunit 3 may get into the immunoprecipitate by an exchange process or by specific attachment to a complete enzyme complex.

DISCUSSION

The data presented show that mitochondrial translation in Neurospora continues even when cytoplasmic translation is shut off for periods as long as 2 hrs, i.e. one half of a doubling period of the cells. We find no evidence for a mechanism which regulates mitochondrial translation in such a way that synthesis of mitochondrial translation products is stopped, when these products cannot any longer become assembled to functional membrane complexes. With respect to the question, what then happens to these products, the experimental data presented would suggest that (a) part of these products are taken up into the membrane and remain there, and (b) another part of the translation products is degraded to lower MW components which partly remain in the mitochondrial membrane. Such a mechanism would require the existence of specific mitochondrial proteinases, since it was observed that preexisting mitochondrial proteins are not degraded when the cells are kept in the presence of CHI. Based on experiments which were carried out on isolated mitochondria to investigate the formation of cytochrome oxidase components in a reconstituted system, we have arrived at a similar conclusion, viz. that mitochondrial translation products are formed in the absence of cytoplasmic partner proteins but are subject to proteolytic breakdown, probably since they cannot be assembled to functional complexes (6).

Cycloheximide treatment of cells may result in a situation analogous to that in some nuclear mutants. In such mutants defective assembly of cytochrome oxidase or other complexes may result from the inabilty of cytoplasmic protein synthesis to provide certain partner proteins (for review see (1)). Specific degradation of mitochondrially synthesized partner proteins may occur in these mutants in the same way as in CHI treated cells.

Further studies of the regulatory processes on the level of transcription and translation seem desireable (e.g. (7)), however our limited knowledge of these reactions imposes severe experimental limitations.

ACKNOWLEDGEMENTS

We wish to thank Dr. Sigurd Werner for kindly providing us with antibodies against holo-cytochrome oxidase from Neurospora crassa. This work was supported by the Deutsche Forschungsgemeinschaft Schwerpunktprogramm "Biochemie der Morphogenese".

REFERENCES

(1) Schatz, G. and Mason, L. (1974) Annu. Rev. Biochem. 43, 51
(2) Schwab, A. (1972) Ph.D. Thesis, University Munich
(3) Sebald, W., Schwab, A. and Bücher, Th. in "Inhibitors - Tools in Cell Research" Bücher, Th. and Sies, H.,eds., Springer, Heidelberg, Berlin, New York (1969) p. 140
(4) Schwab, A., Sebald, W. and Weiss, H. (1972) Eur. J. Biochem. 30, 511
(5) Werner, S., Schwab, A. and Neupert, W. (1974) Eur. J. Biochem. 49, 607
(6) Rücker, A.v., Neupert, W. and Werner, S. in "Genetics, Biogenesis and Bioenergetics of Mitochondria", Bandlow, W. et al. eds., DeGruyter, Berlin (1976) in press
(7) Ibrahim, N.G. and Beattie, D.S. (1976) J. Biol. Chem. 251, 108
(8) Rücker, A.v., Werner, S. and Neupert, W. (1974) FEBS Lett. 47, 390

Genetics and Biogenesis of Chloroplasts and Mitochondria, Th. Bücher et al. eds.

THE USE OF DOUBLE MUTANT STRAINS CONTAINING BOTH HEAT- AND COLD-SENSITIVE MUTATIONS IN STUDIES OF MITOCHONDRIAL BIOGENESIS

Thomas Mason, Paula Boerner and Christine Biron*
Department of Biochemistry
University of Massachusetts
Amherst, Massachusetts 01002, USA

INTRODUCTION

We have been exploring new techniques for selectively arresting assembly of a fully functional mitochondrial inner membrane in yeast cells so that assembly events might be more readily subjected to detailed biochemical analysis. The conceptual basis for one approach we are using has been presented by Jarvik and Botstein (1). It relies on the ability to construct double mutant cells containing both cold-sensitive (<u>cs</u>)[†] and heat-sensitive (<u>hs</u>) mutations which conditionally and reversibly block membrane assembly at 18° and 37°, respectively. In practice shift experiments between two different nonpermissive temperatures can be employed with the double mutant strains in a manner similar to the sequential use of chloramphenicol (CAP) and cycloheximide (CHI) so common to studies of organelle biogenesis (3,4,5,6). By monitoring assembly during shift experiments, it is possible to distinguish a process which requires concomitant events or a fixed-order sequence from one which may proceed randomly. Furthermore it may be possible to order events in a pathway.

$$I_1 \xrightarrow{cs} I_2 \rightarrow I_3 \xrightarrow{hs} P$$

<u>Growth</u>

18° $[I_1]$

37° $[I_3]$

<u>Shift</u>

18° → 37° $[I_1] \rightarrow \;\; \rightarrow [I_3]$

37° → 18° $[I_3] \rightarrow P$

Fig. 1 Hypothetical scheme for the analysis of assembly pathways in <u>cs</u> - <u>hs</u> double mutant strains (see ref. 1 for detailed discussion).

---

*Present address: Department of Bacteriology and Immunology, University of North Carolina School of Medicine, Chapel Hill.

[†]Terminology and abbreviations: TS, temperature-sensitive phenotype, a general class referring to cold-sensitive and/or heat-sensitive phenotypes; CS, cold-sensitive phenotype; HS, heat-sensitive phenotype; <u>ts</u>, <u>cs</u>, and <u>hs</u>, the corresponding genetic determinants. Adopted from (2).

For example, consider the simple case of the linear pathway depicted in Fig. 1. During growth at 18° the cold-sensitive block is imposed and intermediate I_1 may accumulate. When the cells are shifted to 37° this precursor may be converted to I_3 but due to the imposition of the heat-sensitive block at 37° no product (P) will appear. If the heat-sensitive block is imposed first by growth at 37°, accumulated I_3 may be converted to P upon a shift to 18°. In this method of analysis, when either the 18° → 37° or the 37° → 18° shift (but not both!) is productive, then a temporal order of expression for a given pair of genes can be inferred.

In this paper work with two double mutant strains will be presented in order to demonstrate the feasibility of this approach for the study of membrane biogenesis. Part of this work has been reported in preliminary form (7,8).

RESULTS

Genetic description of TS strains

The three mutants described in this paper were obtained by mutagenesis of a wild type strain of Saccharomyces cerevisiae originally derived from S288C by Dr. G. Fink, Cornell University, and here designated as A6 (a his 7-1229 trp 5 [rho^+]). These strains were first isolated as antibiotic-resistant mutants which were subsequently found to be TS for growth on non-fermentable substrates. Two of the mutants, cs23 and hs69, were concluded to be cytoplasmically inherited because the TS phenotype was not restored in diploids formed when the mutants were mated to rho° tester strains, exhibited non-Mendelian segregation upon meiosis, and was observed in mitotic segregants of crosses to rho^+ testers. In addition the genetic character responsible for the mutant phenotype was lost when these strains were converted to neutral petites (presumably rho°) by growth in the presence of ethidium bromide. The third mutant, cs65, exhibited a 2:2 meiotic segregation pattern characteristic of a single nuclear mutation.

The transmission frequencies of the cs23 and hs69 markers in crosses to ω^+ and ω^- tester strains are given in Table I. The high polarity of transmission of the E^R marker in crosses 1 and 2 and the C^S marker in cross 4 suggests that E23-5A and OL69-3B carry the ω^+ marker. Because the cs23 marker also showed high transmission frequencies in heteropolar crosses, this mutation can be tentatively localized in or near the ribosomal RNA region of mitochondrial DNA (9). The hs69 marker does not show polarity in either heteropolar or homopolar crosses and is therefore unlinked to ω.

Although detailed recombination analysis of the cytoplasmic markers will be published elsewhere, some of the conclusions drawn from these studies are summarized below:

1) The E^R and cs23 markers in E23 strains show bias in the frequency of recombination with the O_1^R locus.

TABLE I

TRANSMISSION OF MARKERS IN HOMOPOLAR AND HETEROPOLAR CROSSES

Cross			% of total diploids TS	C^S	E^R	O^S	Total colonies
1. E23-5A ($\alpha\ \omega^+\ C^SE^RO^S\ \underline{cs}^-$)	X	41-6/161* ($\underline{a}\ \omega^-\ C^RE^SO_1^R\ \underline{cs}^+$)	86.2	92.0	85.9	57.5	1415
2. E23-5A ($\alpha\ \omega^+\ C^SE^RO^S\underline{cs}^-$)	X	55R5-3c/321** ($\underline{a}\ \omega^-\ C^R_{321}E^SO^S\ \underline{cs}^+$)	89.9	92.2	91.0	——	1430
3. E23-4D ($\underline{a}\ \omega^+\ C^SE^RO^S\ \underline{cs}^-$)	X	DPI-1B/7** ($\alpha\ \omega^+\ C^SE^SO_1^R\ \underline{cs}^+$)	27.8	——	33.3	30.5	1345
			TS	C^S	E^S	O^R	
4. OL69-3B ($\alpha\ \omega^+\ C^SE^SO^R\ \underline{hs}^-$)	X	41-6/161 ($\underline{a}\ \omega^-\ C^RE^SO_1^R\ hs^+$)	32.9	96.4	——	75.3	1350
5. OL69-1B ($\breve{\alpha}\ \omega^+\ C^SE^SO^R\ \underline{hs}^-$)	X	IL216-1A ($\underline{a}\ \omega^-\ C^SE^R_{221}O^S\ \underline{hs}^+$)	40.7	——	48.6	55.9	1133

*Strain provided by Dr. R. Butow, University of Texas Health Center, Dallas

**Strain from Yeast Stock Center, University of California, Berkeley

2) The cs23 mutation is expressed in erythromycin-sensitive mitotic segregants.

3) The O^R marker present in OL69-1B is not linked to O_1^R (also see cross 2, Table I).

4) hs69 O^S mitotic segregants are phenotypically heat-sensitive, i.e. heat-sensitivity is not dependent on the O^R marker.

5) Based on restoration tests with several of the mit$^-$ mutants recently described by Slonimski and Tzagoloff (10), the hs69 mutation lies in the OXI3 region of mitochondrial DNA.

The reversion frequencies for these strains were less than 1 revertant/10^8 cells. In fact we have been unable to obtain either spontaneous or induced revertants of cs23 and hs69. These strains were stable with respect to conversion to the rho$^-$ state even after 8-10 generations of growth at nonpermissive temperatures.

Properties of mutant and wild type cells after growth at 18° and 37°

Mutant strains grown at their respective nonpermissive temperatures had respiration rates which were less than 10% of wild type values and when the respiratory deficient cells were shifted to permissive conditions, respiratory adaptation occurred which was sensitive to either CHI or CAP. The respiratory capacity of these strains did not decay within five hours after a shift to the restrictive temperature provided there was no net cell growth. These observations indicate that the enzymes of the electron transfer chain are not heat or cold labile *in vivo* and suggest that these mutations are affecting steps in the assembly process.

A survey of the enzyme activities of mitochondria isolated from respiratory deficient cells (Table II) revealed severely reduced levels of cytochrome *c* oxidase activity and cytochrome aa_3. The *cs*65 and *cs*23 mutations are somewhat pleiotropic as indicated by lowered NADH- and succinate-cytochrome *c* reductase activities and partial deficiencies in cytochromes *b* and c_1.

These strains retain mitochondrial protein synthesis at nonpermissive temperatures. The mitochondrial translation products were labeled *in vivo* at 18° and 37° using $[^{35}S]$-$SO_4^{=}$ according to the procedure outlined by Douglas and Butow (14). The labeled polypeptides were analyzed by slab gel electrophoresis in the presence of sodium dodecyl sulfate (14). Figure 2 shows that bands B, C and E of the normal labeling pattern are composed at least in part of the three large subunits of cytochrome oxidase. It is clear that *cs*23 and *hs*69 exhibit relative deficiencies of label associated with band B in cells grown and labeled at nonpermissive temperatures. In addition a relative excess of label migrated with band D in *cs*23 strains labeled at 18°. Based on molecular weight estimates, band D is probably associated with the apoprotein of cytochrome *b* (15). Thus, the gel profiles of *cs*23 and *hs*69 are consistant with the lack of cytochrome oxidase in these strains. The gel pattern for *hs*69 labeled at 37° is also consistant with that observed by Tzagoloff *et al*. (16) for *mit*$^-$ mutants in the OXI3 region. The nuclear mutant *cs*65 had patterns of mitochondrial products similar to the wild type at either 18° or 37° (data not shown).

Double mutant strains in the study of membrane assembly

Two *cs*-*hs* double mutant strains have been constructed from the three strains described above. A diploid *cs*23*hs*69 was obtained as a mitotic segregant from a cross between the two haploid parents. A haploid *cs*65*hs*69 strain was obtained as a spore from the cross α *cs*65[rho°] to *a* *hs*69[rho$^+$]. These strains were respiration deficient when grown at either 18° or 37°. The pattern of mitochondrial translation products of *cs*23*hs*69 labeled at 18° and 37° corresponded to those observed in the parental strains labeled under nonpermissive conditions (see Fig. 2).

The respiratory rates of the double mutants during temperature-shift experiments are shown in Fig. 3. When cells grown at 18° were shifted to 37°, there was no significant respiratory adaptation during a period of time when the cold-sensitive mutants would have regained full activity. However, in the reciprocal experiment cells grown at 37° showed a 10-fold increase in respiration during the incu-

TABLE II

PROPERTIES OF DEFICIENT MITOCHONDRIA

Cells were grown at either 18° or 37° on 0.3% glucose growth medium (11) and harvested at early stationary phase. Mitochondria were isolated from spheroplast lysates (12). Enzyme assays and chemical determinations were performed at 25° as described earlier (13).

	cs65 at 18°	cs23 at 18°	hs69 at 37°
Enzymes	(% of wild type specific activity)		
NADH-cyto. c reductase	24.2	34.0	68.1
Succinate - cyto. c reductase	32.6	28.4	67.0
Cytochrome c oxidase	2.8	1.6	2.1
Oligomycin-sensitive ATPase	35.9	19.3	66.3
Cytochromes present			
b	±	±	+
c_1	±	±	+
c	+	+	+
aa_3	-	-	-

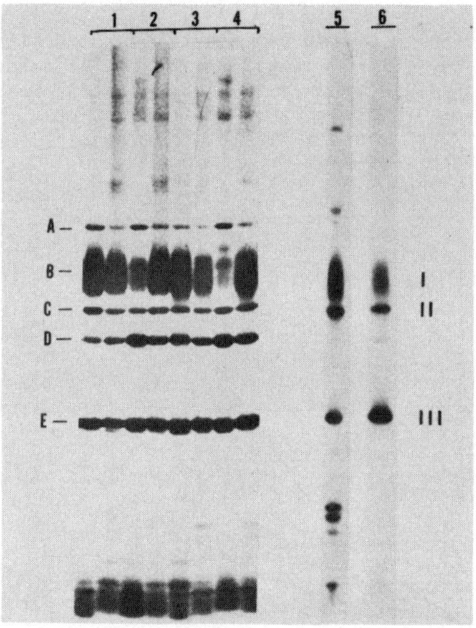

Fig. 2 Mitochondrial translation products labeled in mutant strains at 18° and 37°. Cells for gels 1-4 were obtained as mitotic segregants from the cross E23(cs⁻) X OL69(hs⁻). The cells were grown at either 18° or 37° and then labeled at the temperature of growth with $[^{35}S]$-$SO_4^=$ plus CHI. Gel 1: cs^+hs^+; Gel 2: cs^-hs^+; Gel 3: cs^+hs^-; Gel 4: cs^-hs^-. For reference, cytochrome oxidase immunoprecipitates from mitochondrial extracts are represented in gels 5 and 6. Gel 5: cyto. ox. from wild type cells grown in the presence of $[^{35}S]$-$SO_4^=$; Gel 6: cyto. ox. from wild type labeled in the presence of CHI. For each strain, the 18° and 37° samples are on the left and right, respectively.

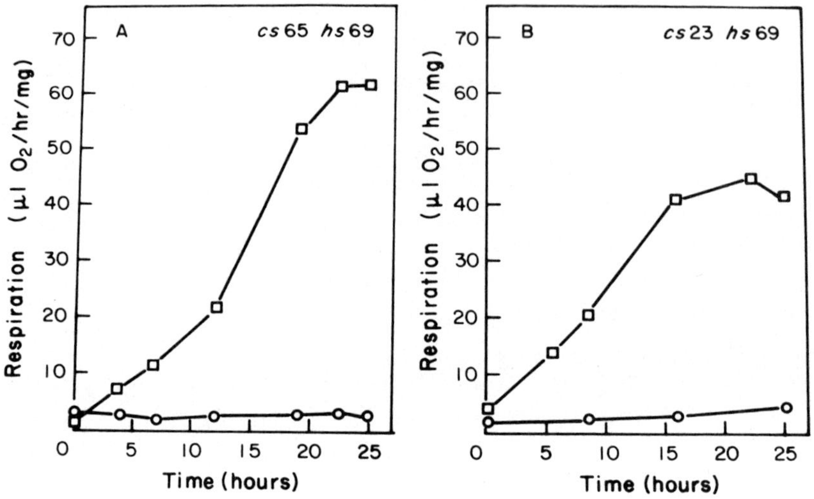

Fig. 3. Respiratory adaptation of double mutants in temperature-shift experiments. Cells were grown at 18° and 37° on 0.3% growth medium (11), harvested at early stationary phase, washed, resuspended in 0.1% glucose, 1% ethanol growth medium and incubated at the appropriate temperatures. The cyanide-sensitive respiration rates of whole cells were determined polarographically at 25° (11). Shifts were 18°→37° (o——o) and 37°→18° (□——□).

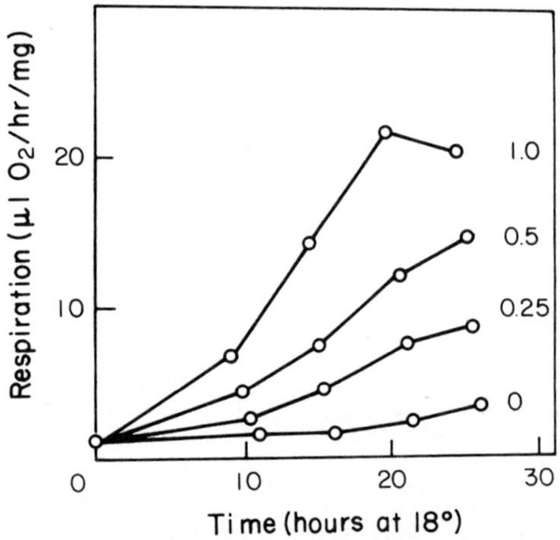

Fig. 4. "Double shift" experiments (18°→37°→18°) with cs23hs69. Cells grown at 18° (0.3% growth medium) were incubated at 37° for 0, 0.25, 0.5 and 1.0 hour (0.1% glucose, 1% ethanol growth medium) and then returned to 18°. There was no increase in whole cell respiration during the incubation at 37°.

bation at 18°. In control experiments, cells grown at either 18° or 37° were capable of temperature adaptation at 28°. These results suggest that the assembly events monitored here must follow an ordered sequence of events. Furthermore, because the shift is productive with these strains only when performed from 37° to 18°, the blocks imposed by cs23 and cs65 must presumably be expressed prior to the heat-sensitive block in the assembly process.

If an ordered pathway exists, it should be possible to bypass the block imposed at 18° in cs23hs69 by short-term incubations at 37°. This possibility was tested in "double shift" experiments. Cells grown at 18° were shifted to 37° for varying periods of time, returned to 18° and monitored for respiratory adaptation (Fig. 4). Clearly the adaptation at 18° is determined by a time-dependent process that occurs at 37° even though there is no increase in respiratory capacity during the 37° incubation. Inhibitor studies have shown that the process occurring at 37° is completely sensitive to CAP and insensitive to CHI (Table III). This suggests an essential mitochondrial translation product (perhaps the largest subunit of cytochrome oxidase) must be made during the incubation at 37°. However, on the basis of labeling experiments in the presence of CHI, the products made after the shift to 37° appear to be identical to those made before the shift (Boerner, P. unpublished). Most notably there was no obvious increase in the synthesis of material in Band B until the cells were allowed to adapt at 18°. Therefore, the question remains; what is the nature of the CAP-sensitive product made at 37° which allows productive assembly when the cells are returned to 18°?

TABLE III

EFFECT OF INHIBITORS OF PROTEIN SYNTHESIS IN "DOUBLE SHIFT" EXPERIMENTS

Cells were treated as described in Fig. 4 except that inhibitors were present during the 60 min incubation at 37°. Inhibitors were removed by extensive washing prior to the incubation at 18°. QO_2 = μl O_2/hr/mg dry wt. at 25°

Inhibitors present during 37° incubation	QO_2 after 15 hours incubation 18°	28°
None	27.0	57.6
+ CAP (4 mg/ml)	3.1	49.0
+ CHI (100μg/ml)	25.3	52.4

CONCLUDING REMARKS

At this time we know little of how these mutations perturb membrane assembly or of the molecular events occurring during the shift experiments. However, the evidence available so far points to some very interesting lines of investigation.

The genetic behavior of cs23 strongly suggests the mutation might be expressed at the level of the mitochondrial ribosome; perhaps through an altered 21S rRNA. Indeed, preliminary experiments have shown a deficiency of mature 50S ribosomal subunits in mitochondria isolated from cs23 cells grown at 18° (T. Mason,

G.S.P. Groot and N.G. Clark, unpublished observations). If the primary defect in this mutant is ribosomal, it will be of interest to understand why there is a selective effect on the mitochondrial translation products labeled in the presence of CHI (see Fig. 2). It will also be important to identify the hs69 gene product and to understand its role in the assembly of cytochrome oxidase.

The experiments with double mutant cells have demonstrated the feasibility of this approach in the study of membrane assembly. It should be possible to process many different mutations in this manner and to position them on an order-of-function map for the *in vivo* assembly of the mitochondrial inner membrane.

ACKNOWLEDGEMENTS

This study was supported in part by grant GM-20909 from the United States National Institutes of Health. We thank Marilyn Breitbart and Susan Moynihan for expert technical assistance.

REFERENCES

1. Jarvik, J. and Botstein, D. (1973) Proc. Nat. Acad. Sci. USA 70, 2046.
2. Jarvik, J. and Botstein, D. (1975) Proc. Nat. Acad. Sci. USA 72, 2738.
3. Rouslin, W. and Schatz, G. (1969) Biochem. Biophys. Res. Commun. 37, 1002.
4. Tzagoloff, A. (1971) J. Biol. Chem. 246, 3050.
5. Ibrahim, N.G., Stuchell, R.N. and Beattie, D.S. (1973) Eur. J. Biochem. 36, 519.
6. Eytan, G. and Ohad, I. (1970) J. Biol. Chem. 245, 4297.
7. Biron, C.A. and Mason, T.L. (1975) Abstracts of the Annual Meeting of the American Soc. for Microbiology, 183.
8. Boerner, P., Breitbart, M. and Mason, T. (1976) Fed. Proc. 35, 1406 abs.
9. Sanders, J.P.M., Heyting, C., DiFranco, A., Borst, P. and Slonimski, P.P. (1976) The Genetic Function of Mitochondrial DNA, (A.M. Kroon and C. Saccone, ed.) North Holland; Amsterdam, in Press.
10. Slonimski, P.P. and Tzagoloff, A. (1976) Eur. J. Biochem. 61, 27.
11. Mason, T.L. Poyton, R.O., Wharton, D.C. and Schatz, G. (1973) J. Biol. Chem. 248, 1346.
12. Schatz, G. and Kovac, L. (1974) Methods Enzymol. 31, 627.
13. Ebner, E., Menucci, L. and Schatz, G. (1973) J. Biol. Chem. 248, 5360.
14. Douglas, M. and Butow, R.A. (1976) Proc. Nat. Acad. Sci. USA 73, 1083.
15. Lin, L. and Beattie, D.S. (1976) Fed. Proc. 35, 1406 abs.
16. Tzagoloff, A., Akai, A., Needleman, R.B. and Zulch, G. (1975) J. Biol. Chem. 250, 8236.

Genetics and Biogenesis of Chloroplasts and Mitochondria, Th. Bücher et al. eds.

COLD SENSITIVITY OF MITOCHONDRIAL BIOGENESIS IN A NUCLEAR MUTANT OF NEUROSPORA CRASSA

Rose KIENTSCH and Sigurd WERNER
Institut für Physiologische Chemie, Physikalische Biochemie und Zellbiologie der Universität München

Cni-1, a chromosomal mutant of Neurospora crassa, is characterized by cyanide-insensitive respiration and a deficiency of cytochromes aa_3 and b (1). We have demonstrated that mitochondria from cni-1 mutant undergo dramatic changes during the growth cycle of the culture (2). Cells from early log phase cultures grown at 30° C were deficient in cytochromes aa_3 and b, showed an altered pattern of mitochondrial translation products and had a cyanide-insensitive respiration. Mitochondria from late log and stationary phase cells, however, contained all cytochromes, synthesized the normal complement of mitochondrial translation products and had high levels of cyanide-sensitive respiration. Recently we have found growth conditions for cni-1 which allow a stable expression of either the wild type or the mutant phenotype during the whole growth cycle.

Cells were grown in Vogel's minimal medium with 2% sucrose as carbon source at two temperatures: 12° C and 37° C. At both temperatures cni-1 cells showed growth rates comparable to those obtained with wild type strain.

When the mutant was grown at 12° C no cytochromes aa_3 and b could be detected spectrophotometrically throughout the growth cycle of the culture. The cytochrome c oxidase activities which could be tested remained at an extreme low level (see Table). An elevated level of cytochrome c related to that of the wild type was found in cni-1 mitochondria.

The analysis of cells prepared from 37° C cultures showed clearly that all cytochromes were present in the mitochondria. Data listed in the table display that the specific contents of cytochromes aa_3 and b remained relatively constant in the various growth phases of the hyphae. The specific content of cytochrome aa_3 of cni-1 mitochondria was about 50% of that measured in wild type mitochondria. The cytochrome b contents obtained with both strains grown at 37° C differed only slightly. Concerning the cytochrome c content of cni-1 mitochondria a decrease was observed during the increase of cell mass, but less marked than that found in 12° C cultures.

Table I: Determination of Various Parameters in cni-1 Mutant and Wild Type

	cni-1, 12° C			wild type, 12° C			cni-1, 37° C			wild type, 37° C		
Growth phase	A	B	C	A	B	C	A	B	C	A	B	C
Cytochrome oxidase activity $\frac{\mu\text{Mole Cyt. c}}{\text{min x g protein}}$	0.4	0.4	0.4	5	6	7	2	6	8	12	20	23
Cytochrome content µMole/g protein												
aa_3	not measurable			0.15	0.15	0.15	0.1	0.1	0.1	0.2	0.2	0.2
b	not measurable			0.45	0.5	0.4	0.6	0.5	0.5	0.6	0.6	0.6
$c + c_1$	2.1	1.4	1.2	0.8	0.7	0.7	1.6	1.4	1.2	1.2	1.1	0.95
RNA content µg/mg protein	15	29	42	38	37	44	48	75	82	50	69	80
Contribution of mitochondrial protein synthesis (% of total cellular synthesis)		2.5			11			11			14	
Ratio $\frac{\text{large subunit}}{\text{small subunit}}$		13.3			2.1			2.9			2.1	

Cells were grown at the indicated temperatures and samples of the cultures were harvested at early log phase (A), mid-log phase (B), and late log/stationary phase (C). Mitochondria were prepared and analysed for cytochrome oxidase activity (8), cytochrome contents (7), RNA-contents (9), and mitochondrial protein synthesis (7). The ratios of large to small ribosomal subunits were calculated from density gradient profiles. The gradients were loaded with dissociated ribosomes as described in legend to Fig. 2

The data on the cytochromes emphasize again that the mutant is equipped with a regulatory system controlling the synthesis of cytochromes aa_3 and b. The formation of cytochromes is triggered when the cells are grown at higher temperature.

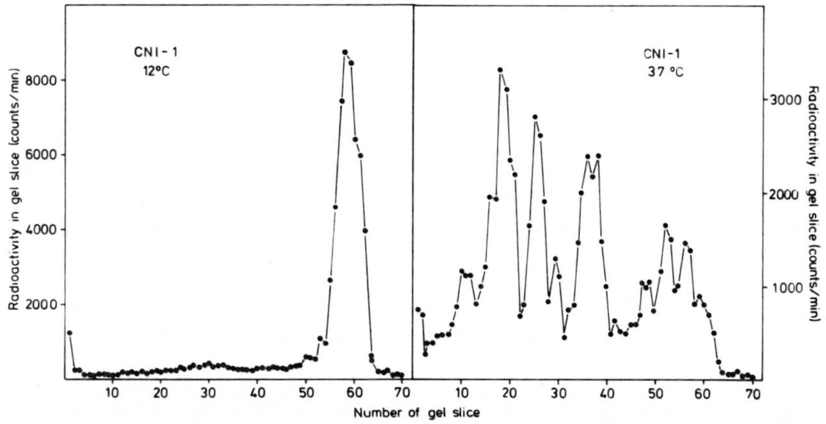

Fig. 1:
Electrophoretic analysis of mitochondrial translation products in cni-1 mutant grown at 12° C and 37° C, respectively.
Samples of the cultures were withdrawn at mid-log phase, adapted at 25° C for 30 min and treated with cycloheximide; 2.5 min later radioactive leucine was added to these samples and incubated for an additional 60 min at 25° C. Mitochondrial membranes were prepared and subjected to SDS gel electrophoresis.

The cycloheximide resistant incorporation patterns of radioactive aminoacids into cell proteins observed with the mutant (Fig. 1) suggest that mitochondrial translation products are involved in that regulation process. Mitochondria of cni-1 cells grown at 12° C synthesize exclusively small molecular weight polypeptides, whereas the normal complement of mitochondrial translation products is obtained from 37° C cultures. The absolute amount of mitochondrial polypeptides which were synthesized in mutant cells from 12° C cultures was only 20% of that formed in wild type cells grown at the same temperature. This does not result immediately from the incorporation rates in the presence of cycloheximide. In the case of cni-1 it was found that most of the radioactive leucine was incorporated into the lipid fraction and not into polypeptides (3). In wild type, however, only 5 - 10% of the specific label was found in the fraction. The corrected values for mitochondrial protein synthesis measured in cells at both temperatures are given in the table.

How can this severely reduced mitochondrial protein synthesis in cni-1 cells grown at low temperature be explained? Possible regulation processes may occur both on a transcriptional and on a translational level. The relative low contents of mitochondrial RNA found in the mutant grown at 12° C (see Table) directed further investigation to the translation machinery.

Mitochondrial ribosomes were prepared from 12° C and 37° C cultures and analysed on sucrose density gradients. The profile obtained from the 37° C culture of cni-1 was similar to the standard profile observed with wild type ribosomes (4). However, in the profile recorded with ribosomes prepared from cultures of cni-1 grown at 12° C no small ribosomal subunit and no polysomes could be detected. Only a small portion of monosomes and large ribosomal subunit were present in approximately equal amounts. In further experiments

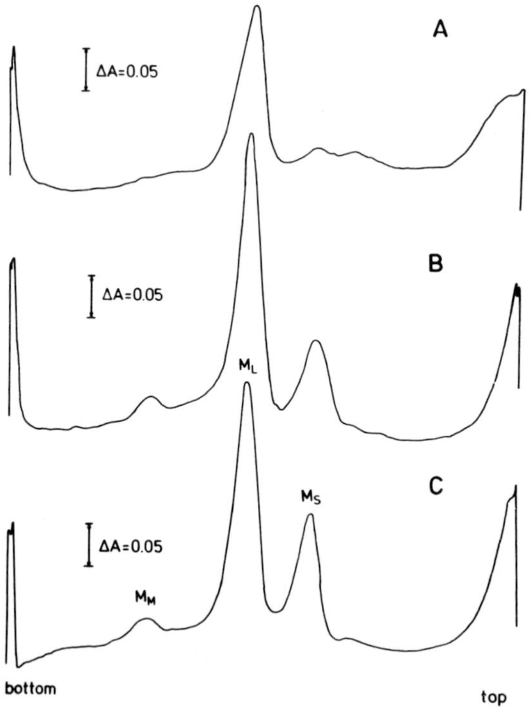

Fig. 2: Sedimentation profiles of mitochondrial ribosomal subunits prepared from cni-1 mutant grown at 12° C (A), cni-1 mutant grown at 37° C (B), and wild type grown at 12° C (C). Mitochondria were prepared and lysed in AMT buffer (100 mM NH_4Cl, 10 mM $MgCl_2$, 10 mM TRIS-HCl, 0.5 mg heparin/ml, pH 7.5) containing 2% Triton X-100. Ribosomes were isolated as described by Michel and Neupert (4). The ribosomal pellets were resuspended in AMT buffer without heparin. The suspension was frozen, thawed twice and layered on a convex sucrose density gradient (from 10-35%). Centrifugations were carried out at 40000 rpm for 240 min (Spinco SW 41 rotor) at 4° C. Gradients were analysed in a flow cell by recording the absorbancy at 260 nm.

M_M = monomer, M_L = large ribosomal subunit, M_S = small ribosomal subunit mitochondrial ribosomes were first dissociated before loading on a sucrose density gradient. In Fig. 2 the profiles obtained after centrifugation are presented. They clearly show that cni-1 mutant grown at 12° C contains defective mitochondrial ribosomes. Minimal amounts of the small ribosomal subunit are recorded. The ratios of large to small subunit obtained from the various cultures can be found in the table.

A similar phenomenon was previously observed with the cytoplasmic mutant mi-1 (poky). This mutant was shown to be also deficient in small ribosomal subunits of mitochondria during the exponential growth phase (5, 6). The authors initially explained the ribosomal deficiency in the extrachromosomal mutant with an altered mitochondrial RNA. In the case of the chromosomal mutant cni-1 we are tempted to speculate that the underlying mutation involves a nuclear coded ribosomal protein.

Irrespective of whether the primary defect in the cell can be located or not, we would like to use that mutant as a tool for our current studies on the assembly process of cytochrome oxidase. In a preliminary investigation we tried to determine what polypeptide subunits of cytochrome oxidase are present in the mutant, when it is grown under the conditions described. An immunological technique was used for the analysis of the subunit composition (7). Fig. 3 shows that only cytoplasmatically synthesized polypeptides (subunits 5, 6 and 7) are detectable in mitochondria isolated from 12° C cultures. The lack of mitochondrially synthesized polypeptides of the oxidase with molecular weights of 41000, 28000 and 21000 (subunits 1, 2 and 3) fits well with the deficiency of high molecular weight translation products (see Fig. 1) and the presence of defective mitochondrial ribosomes. When cni-1 mutant is grown at the permissive temperature the cells contain the complete set of seven polypeptides, which are assembled to a functional enzyme complex as judged by the high enzymatic activity obtained with the mitochondria (see Table). The assembly of cytochrome oxidase may be due to the availability of the mitochondrially synthesized components of the enzyme.

Thus the temperature sensitivity of that mutant provides a control to switch on and off the formation of the oxidase. This effect may be profitably exploited in the study of the complex assembly process of cytochrome oxidase.

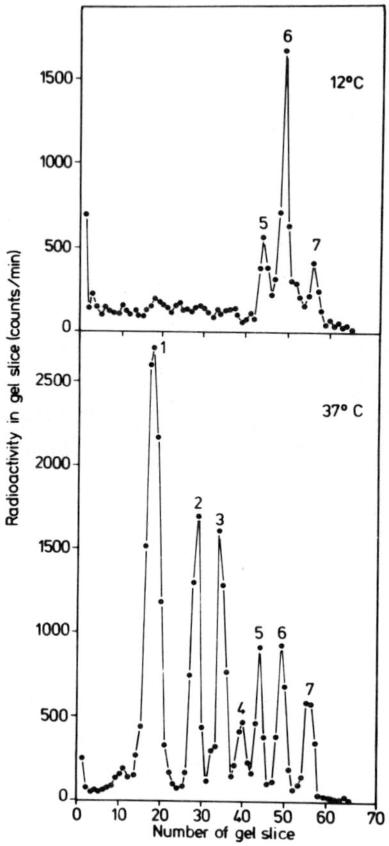

Fig. 3:
Electrophoretic analysis of polypeptide components of cytochrome oxidase in cni-1 mutant grown at 12°C and 37°C, respectively. Cultures were grown to mid-log phase, adapted at 25°C for 30 min, then labelled continuously with ^{3}H-leucine. Mitochondria were isolated from these cells and mixed with unlabelled wild type mitochondria. The suspensions were solubilized with Triton X-100 and treated with an antibody to cytochrome oxidase. The immunoprecipitates were subjected to SDS gel electrophoresis. Numbers indicate the enzyme subunits.

Acknowledgement:
This investigation was supported by the Deutsche Forschungsgemeinschaft.

References

1. Edwards, D.L. and Kwiecinski, F. (1973) J. of Bact. 116, 610-618
2. Klein, J.L. and Edwards, D.L. (1975) J. Biol. Chem. 250, 5852-5858
3. Michel, R., Liebl, A., Machleidt, W., Otto, J., and Neupert, W. (1975) Hoppe-Seyler's Z. Physiol. Chem. 356, 1595-1604
4. Michel, R. and Neupert, W. (1973) Eur. J. Biochem. 36, 53-67
5. Neupert,W.,Massinger, P., and Pfaller, A. (1971) in Autonomy and Biogenesis of Mitochondria and Chloroplasts (Boardman, N.K., Linnane, A.W., and Smillie, R.M. eds) pp. 328-338, North-Holland, Amsterdam-London
6. Rifkin, M.R. and Luck,D.J.L. (1971) Proc. Natl Acad. Sci. USA 68, 287-290
7. Werner, S.,Schwab, A.J., and Neupert, W. (1974) Eur. J. Biochem. 49, 607-617
8. Werner, S. and Neupert,W. (1972) Eur. J. Biochem. 25, 379-396
9. Lusk, J.E., Williams, R.J.P., and Kennedy, E.P. (1968) J. Biol. Chem. 243, 2618-2624

Genetics and Biogenesis of Chloroplasts and Mitochondria, Th. Bücher et al. eds.

ISOLATION AND CHARACTERIZATION OF A CYTOCHROME OXIDASE DEFICIENT MUTANT OF NEUROSPORA CRASSA

Klaus BRUCKMOOSER and Sigurd WERNER

Institut für Physiologische Chemie, Physikalische Biochemie und Zellbiologie der Universität München

Only a few procedures for enrichment of respiratory deficient mutants of Neurospora crassa are known (1). The methods in use depend on the selective removal or killing of growing cells relative to non-growing mutants.

The technique described here is based on the same principle using salicyl hydroxamate as selecting agent.

The rationale of the selection results from the observation that Neurospora possesses a branched respiratory system consisting of the standard cytochrome chain and the alternative oxidase. The first pathway is inhibited by cyanide, the second by salicyl hydroxamate (for review see ref. 2). The efficiency of the alternative oxidase in producing ATP and supporting growth is much less than that of the cytochrome chain (3). Thus mutants deficient in cytochromes should be unable to grow in the presence of salicyl hydroxamate. Fig. 1 shows the effect of different concentrations of salicyl hydroxamate on the mycelial growth of two mutants deficient in cytochromes aa_3 and b (cni-1 and sas-1), and on wild type cells.

Selection of mutants

Cells (WT 74A) were mutagenized by treatment with 2-aminopurine in the presence of cycloheximide, essentially as described by Rosenberg et al. (4). Following mutagenesis cells were thoroughly washed and suspended in growth medium (Vogel's minimal medium plus 2% sucrose) containing 4 mM salicyl hydroxamate. The culture was shaken at 25°C for 35 hours. Then the filtrate was transferred to fresh medium supplemented with salicyl hydroxamate. This procedure was repeated 4-5 times until no visible growth could be detected. At the end of the incubation non-growing cells were collected and plated on sorbose medium (1). Colonies which grew on the plates were overlaid

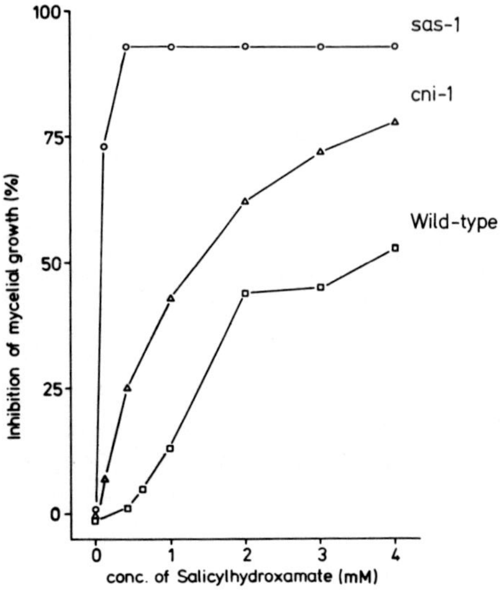

Fig. 1: Selective growth inhibition of wild type, cni-1 and sas-1 by salicyl hydroxamate. Cultures were grown for 35 hours at 25°C in a gyratory shaker on Vogel's minimal medium plus 2% sucrose. Medium was supplemented with different amounts of salicyl hydroxamate. Inhibition of cell growth was determined on a dry weight basis.

with a medium containing 0.25% 2,3,5-triphenyl-2H-tetrazolium (1,5). Colonies which did not reduce the dye were isolated and transferred to slants. The selected cells were analysed for cytochromes by means of a hand spectroscope which was mounted on a microscope tube. For that purpose mycelium was suspended in water, dithionite was added, and cells were transferred to a microscope slide. The material was frozen in liquid nitrogen and the absorption bands were observed at low magnification. A number of such salicyl hydroxamate sensitive (sas) mutants deficient in cytochromes aa_3 and b could be isolated.

A preliminary characterization of one mutant (sas-1) is presented. The mutant grows well in Vogel's minimal medium plus 2% sucrose without further supplement. The doubling time of cell mass at 25°C is approx. 5 hours.

<u>Genetic characterization of sas-1 mutant</u>

In order to determine the genetic nature of the sas-1 mutation, the mutant was crossed to an adenine auxotroph (ad-1), which includes the normal complement of cytochromes. sas-1 was used as the conidial parent. An analysis of 30 ascospores showed that all cells contained the full complement of cytochromes. These preliminary data suggest that sas-1 is an extranuclear mutant. This result has yet to be verified by tetrad analysis.

<u>Biochemical characterization of sas-1 mutant</u>

1) Respiration of whole mycelium

Fig. 2 shows the oxygen electrode traces of the respiration of mycelium harvested from mid-log phase cultures of sas-1 and wild type. Addition of KCN has no effect on the respiration rate of the sas-1 culture. However, respiration was strongly inhibited after application of salicyl hydroxamate. The delayed onset of inhibition after addition of salicyl hydroxamate is probably due to the difficult translocation of the inhibitor under these conditions. The

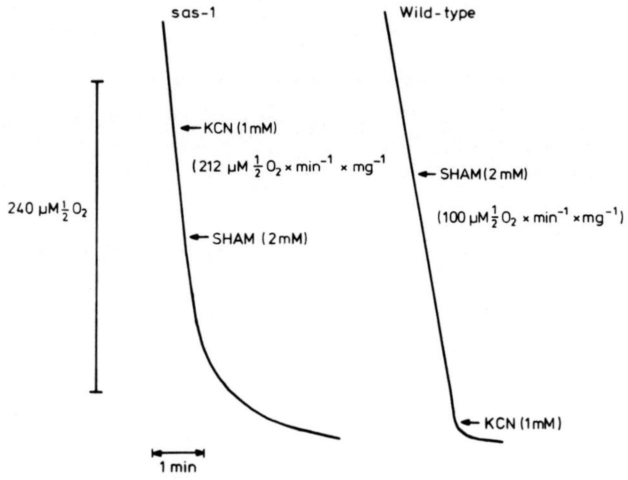

Fig. 2: Oxygen electrode traces of the respiration of mycelium from sas-1 and wild type. Cultures were grown at 25°C in Vogel's minimal medium containing 2% sucrose. Respiration measurements were made in the same medium. Numbers in paranthesis give the rates of oxygen utilisation. At the points indicated potassium cyanide (1 mM) or salicyl hydroxamate (SHAM) (2 mM) was added.

inhibition of the respiration occured while the cells were still in an aerobic state. This observation with sas-1 is in marked contrast to the effects of salicyl hydroxamate and cyanide on wild type cultures.

2) Contents of cytochromes

Room-temperature absorption spectra (reduced minus oxidized) of sas-1 mitochondria are shown in Fig. 3. No cytochrome aa_3(605 nm

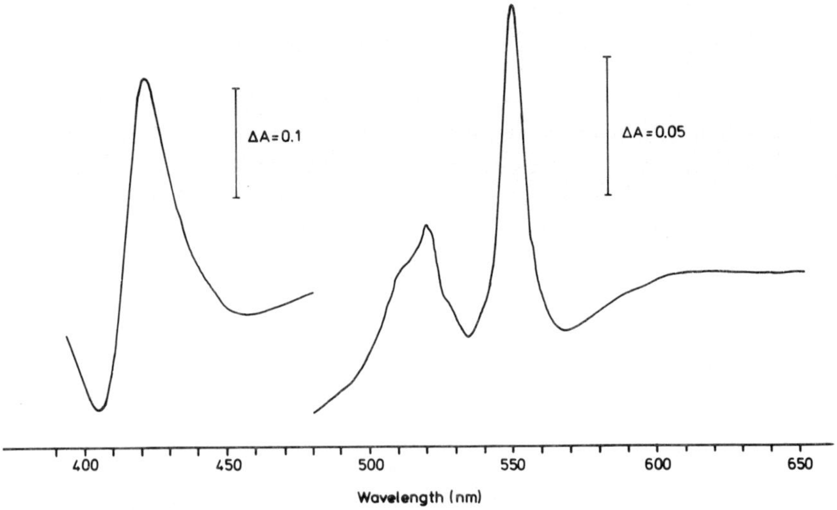

Fig. 3: Room-temperature absorption spectra of the cytochromes in sas-1 mitochondria. The test sample was reduced with dithionite. The reference sample was left in oxidized state. Protein concentration of the mitochondrial suspension was 4.0 mg/ml.

and 442 nm, respectively) and no b-type cytochromes (558 nm and 428 nm, respectively) were detected with the applied technique. The high absorption bands at 550 nm and 420 nm reflect the presence of cytochrome c. The specific content of cytochrome c ($c+c_1$) in sas-1 mitochondria was 1.3 nmol/mg protein. A similar value was obtained with wild type mitochondria (6).

A low-temperature spectrum of submitochondrial particles from sas-1 is given in Fig. 4. It clearly shows that the mutant contains cytochrome c_1 (552 nm). Surprisingly one b-type cytochrome(556 nm) is present. However, no band at 562 nm is detectable under these conditions. The absorption band at 546 nm (6) is due to cytochrome c, which has not been extracted.

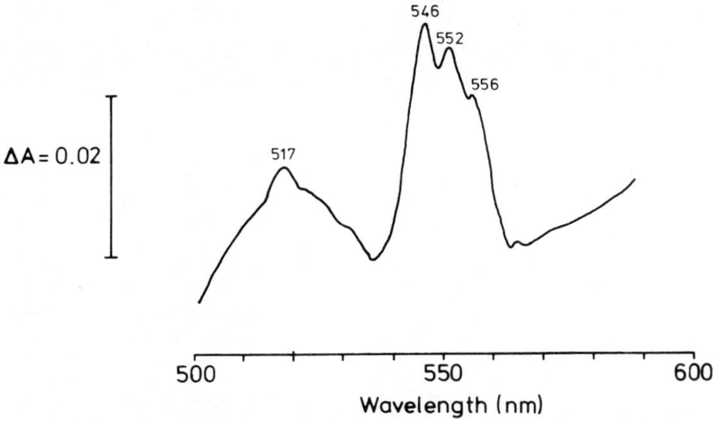

Fig. 4: Low-temperature absorption spectrum of submitochondrial particles from sas-1 mutant. Difference spectrum was performed in liquid N_2 with 0.5 M sucrose in the reaction medium. Test sample was reduced by dithionite; reference sample was kept aerobic.

3) Mitochondrial translation products

Fig. 5 displays the cycloheximide-resistant labelling pattern of mitochondrial membranes from sas-1 mutant. Both high and low molecular weight polypeptides are present. However, compared to wild type membranes, the synthesis of high molecular weight polypeptides is significantly reduced (7).

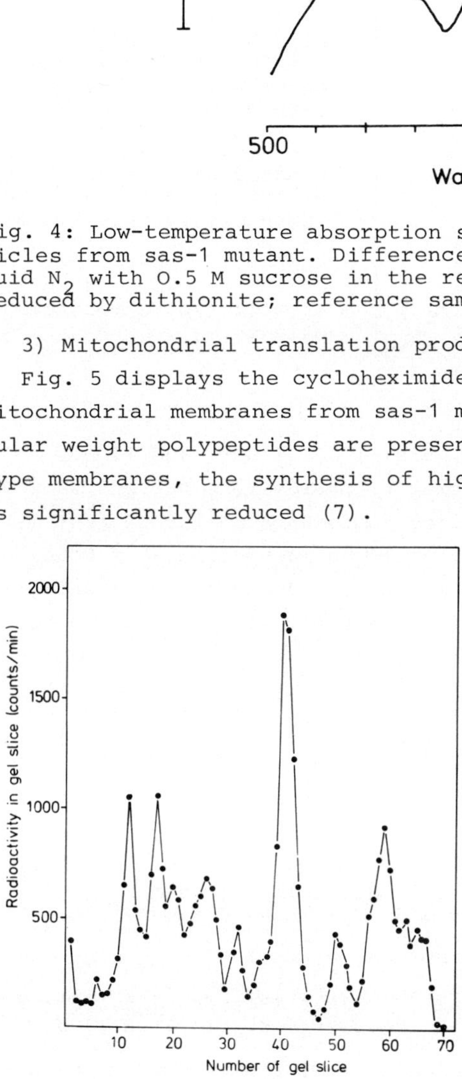

Fig. 5: Gel-electrophoretic analysis of mitochondrial translation products in sas-1 mutant. Cells were labelled in the presence of cycloheximide with ^3H-leucine. Mitochondrial membranes were prepared and subjected to dodecylsulphate-gel electrophoresis.

4) Analysis of polypeptide components of cytochrome oxidase

An immunological technique was used to isolate polypeptide components of cytochrome oxidase (7). The gel electrophoretic pattern obtained with the immunoprecipitate is shown in Fig. 6. Components

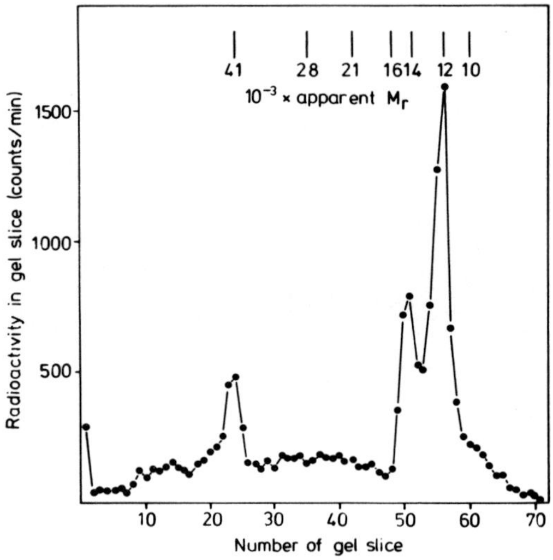

Fig. 6: Gel electrophoretic analysis of polypeptide components of cytochrome oxidase in sas-1 mutant. Cells were labelled continously with ^{3}H-leucine. Mitochondria were isolated from the cells and mixed with unlabelled wild type mitochondria. The suspension was solubilized with triton X-100 and treated with an antibody to cytochrome oxidase. The immunoprecipitate was subjected to SDS gel electrophoresis. The gel was calibrated with subunits of wild type cytochrome oxidase.

5,6 and 7 (apparent molecular weights 14000, 12000, and 10000), which are synthesized on cytoplasmic ribosomes were present. Concerning the mitochondrially synthesized components only component 1 (apparent molecular weight 41000) could be identified. A further analysis to detect the missing components 2,3 and 4, using subunit specific antibodies, is necessary to decide whether or not these components are synthesized in sas-1 mutant.

Acknowledgment: The authors thank G. von Jagow for recording the low-temperature spectra and M. L. Teles Grilo for valuable help in the genetic analysis. This investigation was supported by the Deutsche Forschungsgemeinschaft.

References:

1. Edwards, D.L., Kwiecinsky, F., and Horstmann, J. (1973) J. of Bact. 114, 164-168
2. Henry, M.-F. and Nyns, E.J. (1975) Sub-cell. Biochem. 4, 1-65
3. Slayman, C.W., Rees, D.C., Orchard, P.P., and Slayman, C.L. (1975) J. Biol. Chem. 250, 396-408
4. Rosenberg, E., Mora, C., and Edwards, D.L. (1976) Genetics 83, 11-24
5. Ogur, M., John, R.S., and Nagai, S. (1957) Science 125, 928- 929
6. Jagow, G. von, Weiss, H., and Klingenberg, M. (1973) Eur. J. Biochem. 33, 140-157
7. Werner, S., Schwab, A.J., and Neupert, W. (1974) Eur. J. Biochem. 49, 607-617

Genetics and Biogenesis of Chloroplasts and Mitochondria, Th. Bücher et al. eds.

SUBUNIT STRUCTURE AND ARRANGEMENT OF MITOCHONDRIAL CYTOCHROME B

Hanns Weiss and Barbara Ziganke

European Molecular Biology Laboratory
P.O.Box 10 22 09, 69 Heidelberg
Federal Republic of Germany

INTRODUCTION

In previous papers we have reported a cytochrome b preparation from Neurospora crassa[1-3] mitochondria which is soluble in the presence of deoxycholate binding 0.3 g detergent per g protein. From the sedimentation equilibrium of this preparation the molecular weight 51 000 was determined for pure cytochrome b by applying a buoyant density factor calculated for the protein-detergent complex[4]. From the specific heme content, however, the molecular weight of only one half, namely 27 000 was derived. By hydroxylapatite chromatography in the presence of dodecylsulfate the protein moiety of the preparation was separated into two parts which both showed upon dodecyl sulfate gel electrophoresis, the molecular weight 28 000. Both parts were labelled by in vivo incorporation of [^{3}H]-leucine in the presence of cycloheximide but were not labelled by incorporation of [^{3}H]-leucine in the presence of chloramphenicol.

From these data we suggested that mitochondrial cytochrome b is a dimeric heme protein consisting of two different heme binding subunits which both are translated on mitochondrial ribosomes.

In the present paper we describe some recent experiments which were performed in order to answer the following questions: a) are the two cytochrome b subunits identical polypeptide chains or not; b) is this dimeric cytochrome b the molecular basis of the two functionally different cytochrome b species of the mitochondrial respiratory chain; and, c) how is this cytochrome b arranged in the inner mitochondrial membrane?

METHODS

Purification of Neurospora crassa cytochrome b: The purification procedure involved chromatography on oleyl-polymethacrylic acid, chromatography on DEAE-cellulose and gel filtration on Sephadex G-100, all steps were performed in the presence of bile acids and salt. The final yield was 5%. The specific heme content was enriched 40 times to 35 µmol/g protein. The preparation showed the characteristic absorption spectra of cytochrome b[1-3].

Separation of the cytochrome b subunits: 1 mg cytochrome b was dialysed against water for 6-8 h and lyophilized. Then it was dissolved in 0.7 ml 5% Na dodecyl sulfate, 5% mercaptoethanol and 0.1 M Tris-acetate pH 8 and subjected to gel filtration on a 0.9 x 60 cm Sephadex G-150 column equilibrated with 0.15% dodecyl sulfate, 0.01 M Na phosphate pH 6.4, and 2 mM dithioerytritol. About 80% of the protein appeared in 3-4 ml eluate at an apparent molecular weight of 28 000. It was well separated from some aggregated protein, from free heme, bile acids and the excess of dodecyl sulfate. It was applied on a 0.6 x 8.5 cm hydroxylapatite (Bio-Gel HT) column equilibrated with 0.15% Na dodecyl sulfate, 0.01 M Na phosphate pH 6.4 and 2 mM dithioerytritol. After washing the column with 5 ml of the same solution the protein was eluted with a linear 150 ml gradient ranging from 0.25 M to 0.60 M Na phosphate pH 6,4 containing additionally 0.15% Na dodecyl sulfate and 2 mM dithioerytritol[5]. The flow rate was 4 ml/h, the temperature 28°. The protein was assayed by continous monitoring of the eluates at 206 nm with an Uvicord III, the eluate of a dummy column passing the reference cell (Fig. 1).

Amino acid analysis: The amino acid composition was determined after acid hydrolysis (24h, 48h, 72h) with a Biotronik analyser. Methionine was determined as methionine sulfone and cysteine as cysteic acid after performic acid oxidation. The content of tryptophan was estimated spectrophotometrically[6].

C-terminal amino acid analysis: Samples of 0.1 mg cytochrome b protein in 20 µl 1.5% Na dodecyl sulfate, 0.2 M Na hydrogen carbonate pH 8.5 were incubated with a) 7 µg carboxypeptidase A, b) 4 units carboxypeptidase B or c) 7 µg carboxypeptidase A plus 4 units carboxypeptidase B[7].

Cyanogen bromide cleavage of the cytochrome b subunits and peptide analysis by gel filtration: 30 µg cytochrome b was dissolved in 100 µl 90% formic acid. 3 mg cyanogen bromide was added, the tube was flushed with nitrogen and sealed. For separation of the fragments the reaction mixture was applied 27 h later on a 0.6 x 250 cm Bio-Gel P 30 column in 85% formic acid. At a hydrostatic pressure of 120 cm the flow rate was 0.5 ml/h, the temperature was 25°C.

Iodination of mitochondria: Enzymatic iodination was performed as described by Boxer[8] in a medium containing 5 mg/ml mitochondrial protein, 0.44 M sucrose, 0.02 M Tris-acetate pH 7.7, 0.06 µCi Na $[^{125}J]$ /ml (approx. 10 mCi/µg J) and 50 µg/ml lactoperoxidase. The reaction was initiated by H_2O_2 which was added in five aliquots at intervals of 2 min each sufficient to bring the H_2O_2 concentration to 25 µM. Then the reaction was stopped by the addition of 2 mM KJ and 1% mercaptoethanol. The mitochondria were washed 5 times with a solution containing 0.44 M sucrose, 0.02 M Tris-acetate pH 7.7, 2 mM KJ and 1% mercaptoethanol. They were then sonficated in 0.3 M KCl, 0.02 M Tris-acetate pH 7.7, and 2 mM KJ and the resulting membranes were washed 3 times in that solution. Before measurement of radioactivity the mitochondrial membrane protein and the mitochondrial matrix protein was washed with 5% trichloro-acetic acid and water.

RESULTS AND DISCUSSION

Separation of the cytochrome b subunits by hydroxylapatite chromatography in the presence of dodecyl sulfate: At a low Na phosphate concentration in the presence of Na dodecyl sulfate and dithioerytritol the total protein of the cytochrome b preparation was bound to hydroxylapatite. Afterward, using a linear gradient of Na phosphate, 70-80% of the protein was eluted as two distinct peaks which are referred as subunit I and II (Fig. 1).

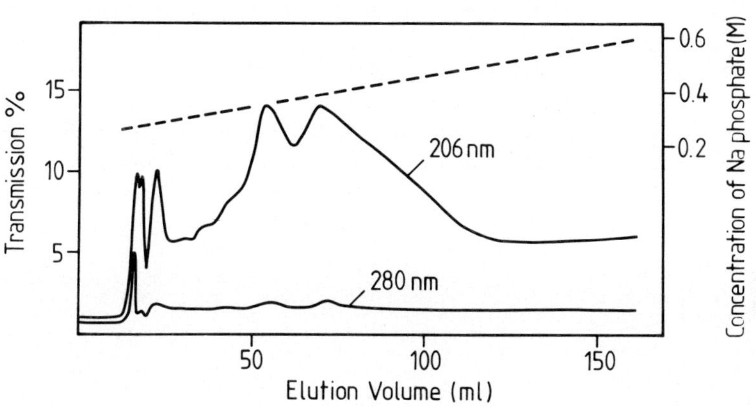

Fig. 1. Hydroxylapatite chromatography in dodecyl sulfate of the cytochrome b apoprotein.

Amino acid composition of the cytochrome b subunits: The amino acid composition of the two cytochrome b subunits showed a high content of unpolar amino acids and a low content of basic amino acids. Their compositions were very similar. Differences were found essentially only in their lysine and cysteine contents. Whether these differencies, however, really exist or whether they are caused by lysine and cysteine rich contaminations still present in the preparation of subunit I cannot yet be decided with certainty (Table 1).

Endgroup analysis: Attempts to determine the N-terminal amino acids of the cytochrome b subunits by means of the dansyl chloride-method have not given any result, nor could the protein be attacked by the solid-phase Edman degradation*[9]. This suggests that the N-termini of both subunits are masked.

The C-terminal amino acids of the two subunits were determined by means of enzymatic hydrolysis[7] (Table 2). No amino acid was released by the action of car-

* The solid-phase Edman degradation was kindly assayed by Dr. E.Wachter,München.

boxypeptidase A alone. Carboxypeptidase B released lysine from each of the two subunits. The combined action of carboxypeptidases A and B produced a release of lysine, tyrosine and leucine. From these data we suggest that both subunits have the C-terminal sequence lys-tyr-leu.

TABLE 1

AMINO ACID COMPOSITION OF THE CYTOCHROME B SUBUNITS

Amino acid	subunit I	subunit II
	MOL mol %	
Aspartic acid	8.6	8.4
Threonine	4.7	4.8
Serine	7.9	8.3
Glutamic acid	7.4	6.2
Proline	4.7	4.7
Glycine	7.1	7.5
Alanine	7.2	7.3
Cysteine	1.2	0.7
Valine	6.6	6.7
Methionine	2.1	2.1
Isoleucine	8.0	8.3
Leucine	12.0	12.8
Tyrosine	4.0	4.1
Phenylalanine	5.5	6.5
Lysine	3.8	2.7
Histidine	2.4	2.5
Arginine	4.1	3.6
Tryptophan	2.7	2.9

TABLE 2

RELEASE OF AMINO ACIDS FROM THE CYTOCHROME B SUBUNITS BY CARBOXYPEPTIDASES

Amino	Carboxypeptidase					
acid	A		B		A + B	
released	subunit I	subunit II	subunit I	subunit II	subunit I	subunit II
	nmol amino acid per nmol protein					
Lysine	0.0	0.0	0.4	0.6	0.6	0.7
Tyrosine	0.0	0.0	0.1	0.1	0.6	0.6
Leucine	0.0	0.0	0.1	0.1	0.4	0.5

<u>Peptide pattern resulting from cyanogen bromide cleavage of the cytochrome b subunits:</u> When the two subunits were cleaved with cyanogen bromide and then subjected to gel filtration on Bio-Gel P 30 in 85% formic acid elution patterns were obtained which were indistinguishable. Fig. 2A shows the pattern obtained after cleavage of total cytochrome b labelled by in vivo incorporation of [^{3}H] -leucine. Fig. 2B shows the patterns obtained after cleavage of a mixture of subunit I from [^{3}H]-leucine labelled cytochrome b and subunit II from [^{14}C] -leucine labelled cytochrome b. As can be seen cyanogen bromide cleavage of the two subunits yielded indistinguishable fragment patterns indicating that both subunits might have the same primary structure. Unfortunately, real finger prints from the two subunits are not yet available because firstly the peptide fragments obtained by cyanogen bromide cleavage are insoluble in the solvents usually applied for two dimensional peptide-mappings and, secondly, the subunits proved to be very resistent to digestion by trypsin or chymotrypsin.

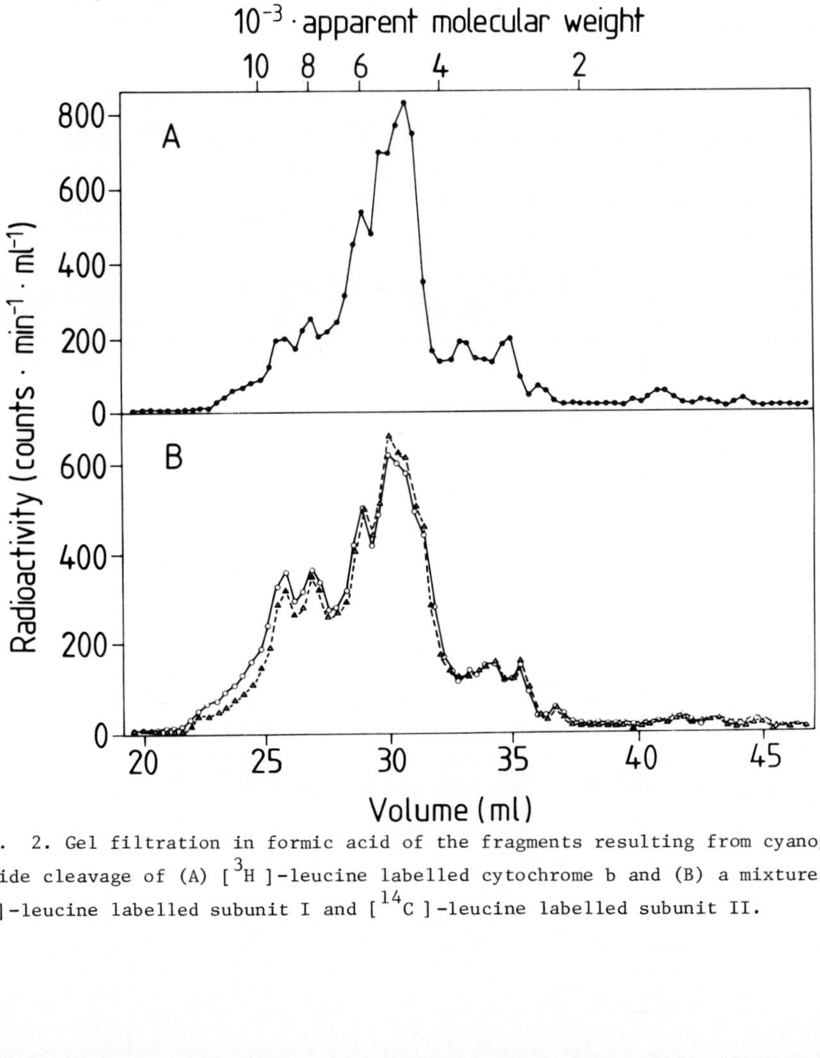

Fig. 2. Gel filtration in formic acid of the fragments resulting from cyanogen bromide cleavage of (A) [^{3}H]-leucine labelled cytochrome b and (B) a mixture of [^{3}H]-leucine labelled subunit I and [^{14}C]-leucine labelled subunit II.

From the data available so far, namely that the two cytochrome b subunits could be separated by hydroxylapatite chromatography but were indistinguishable on the basis of the gel filtration pattern obtained after cleavage with cyanogen bromide, the following speculation might be drawn: both subunits are specified by the same (mitochondrial) gene but are modified after their translation on mitochondrial ribosomes.

The dimeric cytochrome b preparation, the isolated form of the cytochrome b_Tb_K-pair: Two functionally different cytochrome b species (b_T and b_K) form part of the ubiquinone cytochrome c reductase[10]. They differ with respect to the wavelength-position of their light absorption maxima, their redox potentials and the redox kinetics. Upon isolation from the proper membrane surrounding, however, the two species became indistinguishable[3]. For example, with Neurospora crassa, the low temperature absorption spectrum of intact mitochondria shows two cytochrome b bands in the α-region, at 556 nm and at 563 nm, whilst only one band at 559 nm is seen in the low temperature absorption spectrum of purified cytochrome b. Titration of cytochrome b in intact Neurospora mitochondria, using a combined potentiometric-spectrophotometric technique, revealed two redox components with midpoint-potentials $E_{m,7}$=-40 mV and $E_{m,7}$=+60 mV. Titration of purified cytochrome b, however, revealed only one component with $E_{m,7}$=-56 mV. Since the purified cytochrome b can be obtained at a yield of only a few per cent it was not yet possible to decide whether it represents the cytochrome b_Tb_K-pair or only one of these two species.

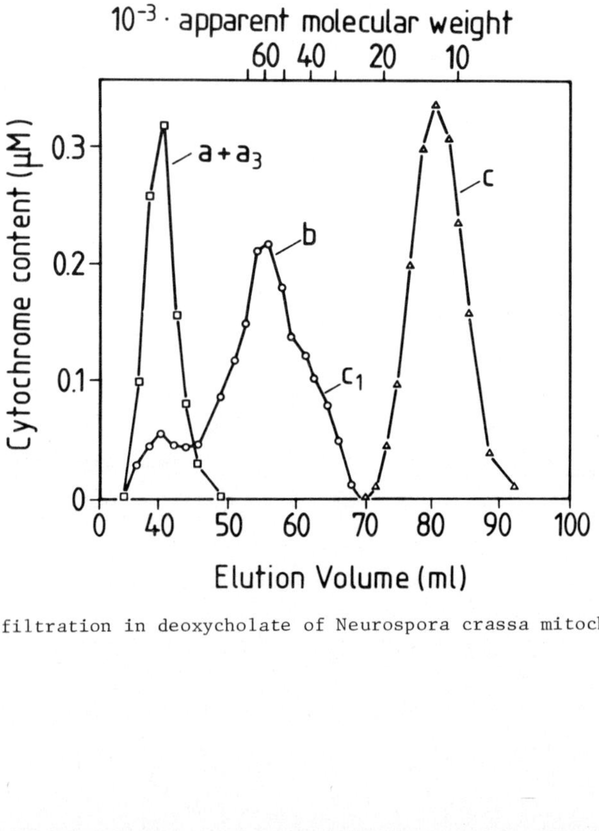

Fig. 3. Gel filtration in deoxycholate of Neurospora crassa mitochondria

When whole Neurospora mitochondria were solubilized with deoxycholate (1 mg/mg protein at 1.0 M KCl and 0.05 M Tris-acetate pH 9) and then subjected to gel filtration on Sephadex G-100 (in a solution containing 0.5% K deoxycholate, 0.2 M KCl and 0.05 M Tris-acetate pH 9) 60-80% of the cytochrome b present in the starting mitochondria was eluted at an apparent molecular weight of 58 000 (Fig. 3). This value was identical to that which was obtained for purified cytochrome b by gel filtration in the deoxycholate and salt solution. The low temperature absorption spectrum of this crude cytochrome b preparation was also indistinguishable from that of purufied cytochrome b. From these results we suggest that the purified dimeric cytochrome b represents the molecular basis of the two cytochrome species (b_T and b_K) involved in the mitochondrial respiratory chain.

Location of cytochrome b in the inner mitochondrial membrane: Intact mitochondria (and sonicated mitochondria) were subjected to lactoperoxidase catalysed radioiodation in order to label those proteins which are exposed to the cytoplasmic and (or) the mitochondrial matrix sites. Then the mitochondrial membranes, the soluble mitochondrial matrix protein, and purified cytochrome b were investigated for radioactivity.

Consistent with surface labelling, the specific radioactivity which was incorporated into the mitochondrial membrane protein was much higher than that which was incorporated into the mitochondrial matrix proteins. The membrane protein cytochrome b, however, was almost completely inaccessible (sonification of the mitochondria prior to iodination did not significantly alter this small accessibility). These data suggest that the bulk of cytochrome b is localized in the interior of the membrane (Table 3). A distinction between the two cytochrome b subunits was not possible by this procedure as no distinct radioactivity profile pattern was obtained when they were chromatographically separated. This might be due to a much higher specific radioactivity of some polypeptides which were present as contaminations.

TABLE 3

ENZYMATIC IODINATION OF NEUROSPORA CRASSA MITOCHONDRIA

Preparation	specific [^{125}J]-radioactivity of protein
	10^{-3} x cpm/mg
Mitochondrial membranes	128
Mitochondrial matrix	32
Cytochrome b	21

ACKNOWLEDGEMENT

We wish to thank Brigitte Juchs and Annette Krebs for excellent technical assistence and Gisela Pouvachy for writing the manuscript.

We are greatly indebted to Sir John Kendrew, the Director-General of EMBL, for his support and interest in this project.

REFERENCES

1. Weiss, H. and Ziganke, B. (1974), Eur. J. Biochem. 41, 63-71
2. Weiss, H., Ziganke, G. and Kolb, H.J. (1975) in Electron Transfer Chains and Oxidative Phosphorylation (Quagliariello, E. et al. eds.) pp. 15-22, North Holland, Amsterdam
3. Weiss, H., Biochim.Biophys.Acta, submitted
4. Tanford, C., Nozaki, Y., Reynolds, J.A. and Makono, S. (1974) Biochemistry, 13, 4452-4459
5. Moss, B. and Rosenblum, E.N. (1972), J.Biol.Chem. 247, 3656-3661
6. Bencze, W.L. and Schmid, K. (1957) Anal.Chem. 29, 1193-1196
7. Ambler, R.P. (1972) in Methods in Enzymology, Vol. XXV, (Hirs, C.H.W. and Timaskoff, S.N.eds.) pp. 143-154, Academic Press, New York
8. Boxer, D.H. (1975) FEBS Lett. 59, 149-152
9. Wachter, E., Machleidt, W., Hofner, H. and Otto, J. (1973) FEBS Lett.35, 97-102
10. Wikström, M.R.F. (1973) Biochim.Biophys.Acta 301, 155-193

Genetics and Biogenesis of Chloroplasts and Mitochondria, Th. Bücher et al. eds.
© 1976 Elsevier/North-Holland Biomedical Press, Amsterdam, The Netherlands.

THE $\underline{bc}_1$-COMPLEX FROM BEEF HEART PREPARED BY ADSORPTION CHROMATOGRAPHY IN TRITON X-100

G. von Jagow, W.D. Engel, P. Riccio and H. Schägger

Institut für Physiologische Chemie und Physikalische Biochemie der Universität München

Starting point of our studies was the aim to purify a $\underline{bc}_1$-complex which may be useful for protein chemistry as well as for reconstitution experiments. We are especially interested in the components, the mechanism of the electron transfer and energy conservation in the $\underline{bc}_1$-region. The isolated complex may be highly pure and depleted of phospholipids and ubiquinone.

First we have developed a new preparation method for the $\underline{bc}_1$-complex with protection of the $\underline{b}$ cytochromes by antimycin. After elaboration of this procedure we were successful in preparing the complex without antimycin; under these conditions the $\underline{b}$ cytochromes had to be protected by reduction with dithionite.

The isolation was performed with Triton X-100 as detergent. The purification was carried out on hydroxyapatite.

In literature (1,2) the preparation of the complex pursues a common scheme consisting in a series of ammonium sulfate precipitations under dispersion of the protein by bile salts. This procedure needs more time and much more preparative expense than the method described here.

With the combination of NaCl and Triton X-100 about 85% of the complex could be extracted from submitochondrial particles. The dissolved proteins were applied to a hydroxyapatite column.

Fig. 1. gives a typical elution pattern of such a chromatographic procedure. With the first buffer, containing a high concentration of phosphate and 0.5% Triton X-100, about 70% of the total protein including cytochrome $\underline{aa}_3$ and a small part of denatured cytochrome $\underline{b}$ came off.

The first buffer was changed after elution of the main part of the cytochrome oxidase.

Subsequently, as indicated by arrow No. two, a so called wash buffer was applied to decrease the salt and detergent concentration before elution of the complex. After about one hour the buffer was changed again as indicated by arrow No. three.

The complex was eluted by means of 8 mM citrate in the presence of 0.2% Triton X-100

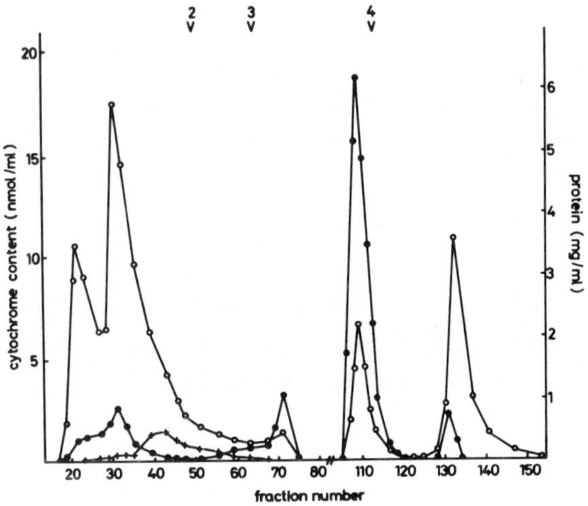

Fig. 1. The elution profile of the antimycin-loaded bc_1-complex
The chromatography was performed by stepwise elution. The buffer changes are indicated by the arrows and described in the text. o—o curve of protein concentration, ●—● curve of cytochrome b, +—+ curve of cytochrome aa_3. The fraction volume amounted to 2.4 ml

The whole isolation procedure lasts for about 10 hours. For protein balance the remaining proteins can then be eluted quantitatively by a high citrate concentration as indicated by arrow No. four.

Table I

Contents of chain components of the two preparations

component	antimycin-complex µmole/g prot.	mol.ratio	dithionite-complex µmol/g prot.	mol.ratio
cyt. b	9.0	≡2.0	6.5	≡2.0
cyt. c_1	5.5	1.2	4.9	1.5
ubiquinone	<0.1	<0.02	<0.1	<0.02

The complex contains per g of protein 9 µmoles cytochrome b and 5.5 µmoles cytochrome c_1 as calculated from absorbance spectra (Table 1.). No ubiquinone could be detected in the complex with a lower limit of detection under the conditions used of 0.2 µmoles/g protein. In the second column of Table I the cytochrome contents are compared on a molecular basis. In contrast to literature the cytochrome $\underline{c}_1$ content is higher than one molecule per two molecules cytochrome $\underline{b}$. This may be due to an overestimation of the $\underline{c}_1$ content which results from a definite interference of the cytochromes $\underline{b}$ at 553 nm.

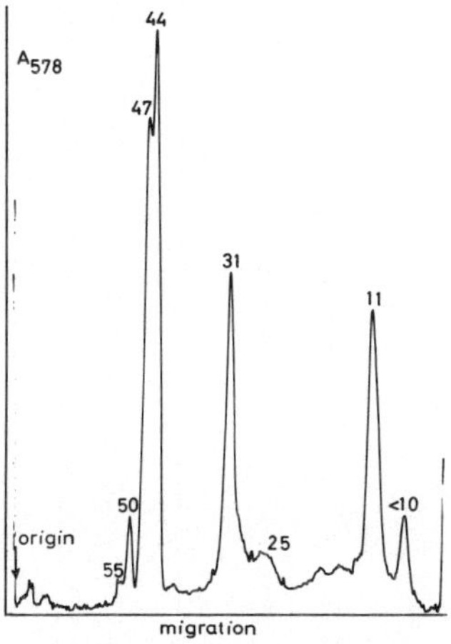

Fig. 2. The polypeptide pattern of the antimycin-complemented bc_1-complex The gel consisted of 8% acrylamide and the samples run in a buffer which contained 0.1 M Tris-HCl, 0.5% SDS, pH 8. The samples were preincubated in 2% SDS, 2 mM Dithioerythrol at 50° C for 20 min. The gels were stained in 0.25% Coomassie Brilliant Blue, destained in 10% acetic acid and scanned at 578 nm.

The SDS-gel electrophoresis (Fig. 2.) shows five main bands. In the high molecular weight region there are three bands with 50, 47 and 44 kdalton. In the intermediate molecular weight region appears one polypeptide with 31 kdalton and a small shoulder at 25 kdalton.

Finally in the low molecular weight region two bands with 11 and less than 10 kdalton are visible.

The polypeptide pattern resembles that of the $\underline{bc}_1$-complex of S. cerevisiae as prepared by Katan and Groot (3).

The 50 kdalton peptide may be attributed to one major subunit of the F_1-ATPase. The 31 kdalton band should represent the cytochromes $\underline{b}$ and $\underline{c}_1$. Assuming for both a molecular weight of 31.000 and the estimated high heme contents of our complex these cytochromes should account for about 40% of the total protein of the complex. The result of this simple calculation is completely inconsistent with the small area of the 31 kdalton band of the scan. This inconsistency holds more or less for all published polypeptide patterns of the varioust species. It is under investigation whether part of cytochromes $\underline{b}$ remains specifically at the origin (4) and/or an unusual low staining index mimics a too low content of the 31 kdalton proteins (5).

Table II

Enrichment and yield of the cytochrome *b* of the two different *bc*$_1$-complexes during preparation

preparation	fraction	total protein (mg)[+]	total heme *b* (nmol)[o]	heme *b* per protein (nmol/mg)	purification fold	yield (%)
antimycin-complex	submitochondrial particles	375	397	1.06	1.0	100
	Triton X-100 extract of SMP	228	338	1.48	1.4	85
	hydroxyapatite chromatography	20	182	9.0	8.5	46
dithionite-complex	submitochondrial particles	451	455	1.01	1.0	100
	Triton X-100 extract of SMP	306	317	1.04	1.0	70
	hydroxyapatite chromatography	22.5	145	6.44	6.2	32

[+]The protein contents were determined by a modified Lowry or Biuret method respectively.

[o]The cytochrome *b* contents were measured at 562-575nm with a Cary 118 spectrophotometer.

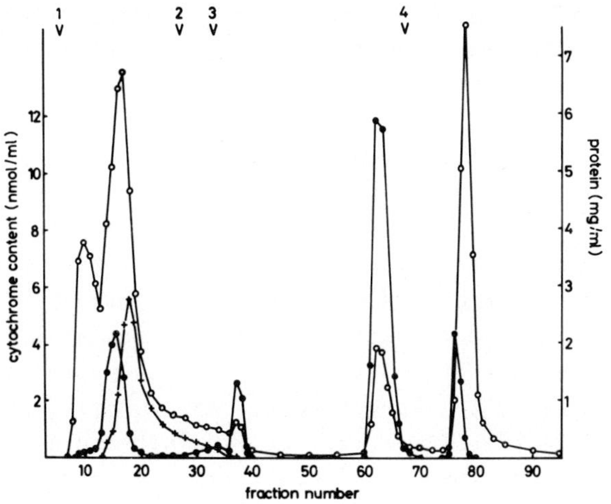

Fig. 3. The eluate profile of the dithionite-stabilized bc_1-complex. The isolation procedure is similar to that of the antimycin-complex with the exception that the composition of the solubilisation and elution buffers was changed. o—o curve of protein, ●—● curve of cytochromes b, +—+ curve of cytochrome aa_3. The fraction volume amounted to 4.4 ml.

Fig. 3. shows an elution pattern of a dithionite-stabilized complex. It is similar to the former one. Again the main part of the proteins is eluted by the first buffer. The relative high amount of cytochrome oxidase is caused by modified solubilisation conditions.

It was possible to lower the detergent concentration drastically to 0.02% Triton X-100 in the second and third buffers which may be of importance for functional studies.

The bc_1-complex is eluted in similar way as in the first chromatography. The yield of the four best fractions amounted to 46% of the applied b cytochromes.

The dithionite-complex has a lower b content than the antimycin-complex, i.e. only 6-7 µmoles/g protein (Table I). It also contains no detectable ubiquinone.

The polypeptide pattern looks very similar to that of the antimycin-complex.

Although antimycin is not present, the dithionite-complex has nearly no ubiquinol or duroquinol cytochrome c reductase activity. The electron flows were tested with 50 µM ubiquinol-3 or duroquinol as electron donors and 50 µM cytochrome c as electron acceptor.

Table II compars the values of enrichment and yield of the cytochrome b contents of the bc_1-complex during the various preparation steps. Oviously the stabilisation of the cytochromes b by antimycin is more effective than that by dithionite as indicated by a higher enrichment and yield in the former case. But also under the latter conditions the yield of the hemes b after chromatography still amounts to about 50%.

Summary The antimycin-loaded bc_1-complex of beef heart prepared in Triton X-100 by a one-step procedure on hydroxyapatite may be a valuable starting material for preparation and study of the distinct polypeptide subunits of this complex. The adaption of this preparation procedure to Neurospora crassa and yeast which is under elaboration may be a tool for further biosynthetic studies.

Reconstitution experiments of the dithionite-stabilized bc_1-complex with phospholipids, liposoms and ubiquinone will show whether this depleted system can be restored in its enzymatic activity.

Acknowledgement The authors are grateful to Prof. M. Klingenberg for his helpful suggestions and criticism. We thank Dr. B. Lang, Genetische Institut der Universität München, for kindly supplying us with ubiquinol-3. This work was supported by a grant from the Deutsche Forschungsgemeinschaft.

Literature

1) Hatefi, Y., Haavik, A.G. and Griffiths, D.E. (1962) J. Biol. Chem. 237, 1681-1685
2) Rieske, J.S., Hansen, R.E. and Zaugg, W.S. (1964) J. Biol. Chem. 239, 3017-3022
3) Katan, M.B., Pool, L. and Groot, G.S. (1976) Eur. J. Biochem. 65, 95-105
4) Katan, M.B. and Groot, G.S., personal communication
5) Weiss, H. (1976) Biochim. Biophys. Acta, in press

Genetics and Biogenesis of Chloroplasts and Mitochondria, Th. Bücher et al. eds.
© 1976 Elsevier/North-Holland Biomedical Press, Amsterdam, The Netherlands.

COMPLEX III OF YEAST: SUBUNIT COMPOSITION AND BIOSYNTHESIS

Martijn B. KATAN and Gert S.P. GROOT

Section of Molecular Biology and Medical Enzymology, Laboratory of Biochemistry, University of Amsterdam, Amsterdam, The Netherlands.

INTRODUCTION

In recent years, much attention has been devoted to the elucidation of the biosynthesis of the enzyme complexes of the mitochondrial inner membrane, in particular to those complexes in whose biosynthesis both the nucleo-cell-sap and the mitochondrial genetic system are involved. For example, cytochrome *c* oxidase and the oligomycin-sensitive ATPase complex from both *Saccharomyces cerevisiae*[1,2] and *Neurospora crassa*[3,4] have been isolated and characterized with respect to their subunit composition and the site of biosynthesis of these subunits. Much less is known about Complex III (cytochrome bc_1 complex) although it is clear from studies of rho$^-$ yeast mutants[5] - in which there is no functional DNA - and of wild type yeast grown in the presence of inhibitors of mitochondrial protein synthesis[6] that also in this case both the nucleo-cell-sap and the mitochondrial genetic system contribute to its formation. In agreement with this, Weiss and Ziganke[7] have shown that in *N. crassa* the polypeptide(s) carrying the heme *b* is synthesized on mitochondrial ribosomes, whereas in yeast cytochrome c_1 is synthesized on cytoplasmic ribosomes[8].

We have recently described the isolation of Complex III from baker's yeast[9] using a method adapted from Tzagoloff[10]. In this paper we report on our attempts to establish the subunit composition, the stoichiometry of the subunits and their site of biosynthesis.

RESULTS AND DISCUSSION

Subunit composition and stoichiometry

From submitochondrial particles solubilized with bile salts in the presence of antimycin, a spectrally pure Complex III preparation can be obtained which contains 6-7 μmol cytochrome *b* and 3-3.5 μmol cytochrome c_1 per g protein. Analysis of this preparation on polyacrylamide gels in the presence of sodium dodecylsulphate (SDS) shows 6 bands with molecular weights of 44 000, 40 000, 32 000, 17 000, 14 000 and 11 000 (Fig. 1). We have shows earlier[9] that the band at M_r 32 000 is double and contains both the heme-carrying polypeptide

Abbreviations: SDS, sodium dodecylsulphate; pI, isoelectric point; n.d., not determined.

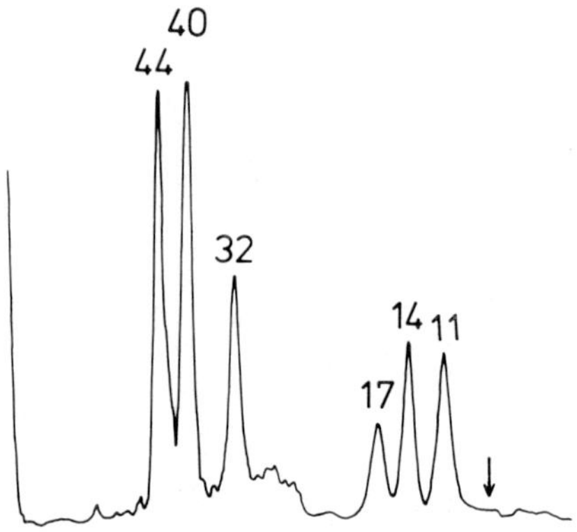

Fig. 1. Electrophoresis of cytochrome *bc*$_1$ complex on a 12% polyacrylamide gel in the presence of SDS. After the electrophoresis the gel was stained with Coomassie blue and scanned at 550 nm[9]. The arrow indicates the dye front.

of cytochrome c_1 and the major polypeptide associated with cytochrome *b*, so that the total number of different polypeptides in the complex is 7.

The stoichiometry of these polypeptides is of interest for two reasons. First, if, as we have argued earlier[9], all 7 polypeptides are associated with each other *in vivo*, then they should be present in simple molar ratio's. Second, the stoichiometry is important in current models of energy transducing at the b-c_1 site; also on the protein chemistry level Weiss and Ziganke[11] has shown that cytochrome *b* from *Neurospora crassa* can be isolated in the form of a dimer and they have suggested that it consists of two polypeptides of slightly different amino acid composition.

We have analysed the stoichiometry in several different ways[9] and found quantitative amino acid analysis after separation of the polypeptides on SDS-polyacrylamide gels the most satisfactory.

The results of such an analysis are shown in Table 1.

Several conclusions can be drawn:

1. Only the polypeptide associated with cytochrome *b* shows the apolar character that is typical for membrane proteins[12] and which is also found for other mitochondrial translation products[2, 3, 11, 13]. The other 6 polypeptides all have polarities in the same range as those of soluble proteins[12].

2. The ratio's of acidic to basic amino acid residues are rather similar for all polypeptides except for the 17 k dalton polypeptide

TABLE 1

PROPERTIES OF THE POLYPEPTIDES OF THE CYTOCHROME bc_1 COMPLEX OF YEAST

The stoichiometry determination was performed in duplicate. Gel strips were cut from slab gels stained with Coomassie blue and combined with 10 ml of 6 N HCl containing 1% mercaptoacetic acid, 0.2 μmol of norleucine and 0.2 μmol of α-amino β-guanidino propionic acid. After evacuation, sealed ampoules were heated at 110°C for 24 h, the precipitate of polyacrylic acid was removed by centrifugation and the hydrolysates were dried down by rotary evaporation. Residues were taken up in 0.2 M citrate buffer, pH 2.1 and analysed on a Beckman Multichrom M. Recoveries varied from 85% to 95%; blank slices from the same gels yielded negligeable amount of amino acid. Duplicate values for corresponding slices were generally within 10% of each other.

Isoelectric focusing was performed in a Triton X 100-urea medium followed by SDS gel electrophoresis in the second dimension, essentially as described by O'Farrel[17]. pH gradients were determined non-destructively with a micro-electrode; the pI-values found in this way were essentially independent of the manner of pre-treatment of the sample (plus or minus boiling in SDS + dithiothreitol) and of the amount of sample applied to the focusing gel. Application of the sample at the acidic instead of alkaline side yielded similar patterns.

Amino acid	44 k	40 k	c_1 (1)	*b* 32 k	17 k	14 k	11 k
	mol/100 mol						
Asx	11.1	10.8	10.2	8.6	17.8	11.4	8.8
Thr	5.0	4.8	4.7	5.3	4.0	4.5	4.0
Ser	6.1	6.9	5.5	5.7	3.4	9.4	7.4
Glx	10.3	9.8	10.7	6.6	28.6	16.4	13.4
Pro	3.7	4.2	9.3	4.9	2.9	5.1	5.9
Gly	9.4	11.2	9.3	6.5	6.0	4.4	10.6
Ala	10.2	8.1	11.2	7.6	6.0	9.8	7.6
Cys	n.d.	n.d.	n.d.	n.d.	n.d.	n.d.	n.d.
Val	6.6	8.3	5.9	9.4	4.4	4.0	5.5
Met	1.5	1.2	1.9	2.6	n.d.	n.d.	n.d.
Ile	6.2	4.2	3.4	8.1	1.9	6.0	4.9
Leu	11.0	8.8	7.5	11.5	6.2	9.5	6.9
Tyr	1.7	3.3	2.2	3.4	2.6	3.0	4.6
Phe	4.1	3.5	3.7	8.4	3.3	2.9	4.8
Trp	n.d.	n.d.	n.d.	n.d.	n.d.	n.d.	n.d.
Lys	7.2	8.1	4.1	4.4	6.6	6.6	8.0
His	1.7	2.1	3.4	2.9	3.8	2.8	3.4
Arg	4.3	4.8	5.3	3.9	2.9	4.6	4.6
Polarity	45.7%	47.3%	44.7%	37.4%	66.9%	61.0%	49.4%
Molar ratio, Coomassie blue	2.0	2.0	1.3		1.1	2.3	2.5
Molar ratio, amino acid analysis	2.0	2.0	1.6		1.0	2.2	2.2
pI	6.4 (2)	6.1 (3)	n.d.	n.d.	4.7	5.9	5.5

(1) Based on data from ref. 18.

(2) Plus a satellite at 6.25.

(3) Plus satellites at 5.9 and 5.7

which contains large amounts of glutamic acid (or glutamine) and aspartic acid (or asparagine). These results were partially confirmed by isoelectric focusing of the complex followed by SDS electrophoresis in the second dimension. As shown in Table 1, the isoelectric points (pI) of the 44, 40, 14 and 11 k dalton polypeptides lie closely together between 6.4 and 5.5, while the 17 k dalton polypeptide has a pI of 4.7. Unfortunately we have not yet been able to detect the 32 k dalton polypeptides in our isoelectric focusing system.

3. The molar ratio's calculated on the basis of Coomassie blue staining of SDS gels agree rather well with those calculated from the amino acid analysis.

4. The most likely stoichiometry is 2:2:3:1:2:2 for the 44, 40, 32, 17, 14 and 11 k dalton polypeptides respectively. The number of copies proposed for the 32 000 dalton polypeptides deserves some comment. The preparation used for the amino acid analyses contained some material that did not enter the gels but remained on top. The amino acid composition of this material was quite hydrophobic and resembled that of the 32 000 dalton cytochrome *b* polypeptide (not shown). Of the 3 copies of 32 000 dalton polypeptides given above, almost half is contributed by this aggregated cytochrome *b*-like material. This does of course decrease the accuracy of the determination. On the other hand, 3 copies is what would be expected from spectral data, because spectrally the complex contains 2 moles of cytochrome *b* per mole of cytochrome c_1[9].

If our figures are correct then the total number of polypeptides per unit of bc_1 complex would come to 12, with a total M_r of 331 000. It is worth noting that King et al.[14] have tentatively suggested a total number of at least 18 polypeptides for beef heart Complex III.

We are currently attempting to find conditions under which all components enter the gel and under which cytochrome *b* and cytochrome c_1 can be separated, so as to increase the reliability of future determinations of the stoichiometry.

Biogenesis

When Complex III is isolated from cells which have been labelled in the presence of cycloheximide, and is subsequently analysed on SDS polyacrylamide gels, only one labelled band is found, with a mobility corresponding to an M_r of 32 000. The specific radioactivity of the complex is rather low under these conditions.

It has been found that preincubation of the cells in the presence

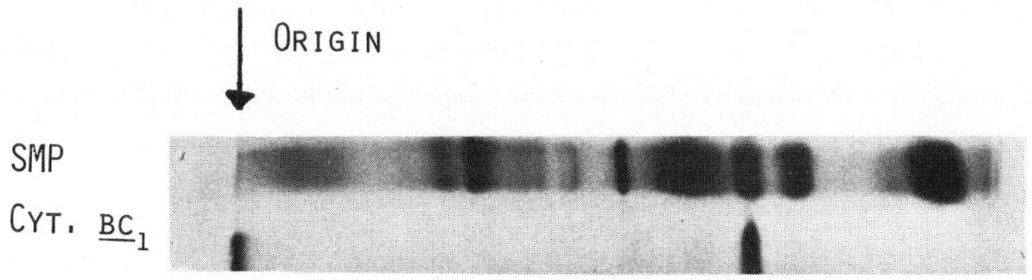

Fig. 2. Electrophoresis of submitochondrial particles and cytochrome bc_1 complex, isolated from yeast cells labelled in the presence of cycloheximide after a transient preincubation with chloramphenicol. The preparations were electrophoresed side by side on a 12% polyacrylamide slab gel, which was then stained, dried and autoradiographed[16].

of chloramphenicol before labelling in the presence of cycloheximide increases the specific radioactivity of certain mitochondrial complexes[15]. We have used such a preincubation and found that the radioactivity of Complex III indeed increases about five-fold. Analysis of such preparations on gels shows that again only one band is labelled with a mobility corresponding to an M_r or 32 000 (Fig. 2). At the same time, the relative amount of radioactivity in the corresponding band of submitochondrial particles is also increased.

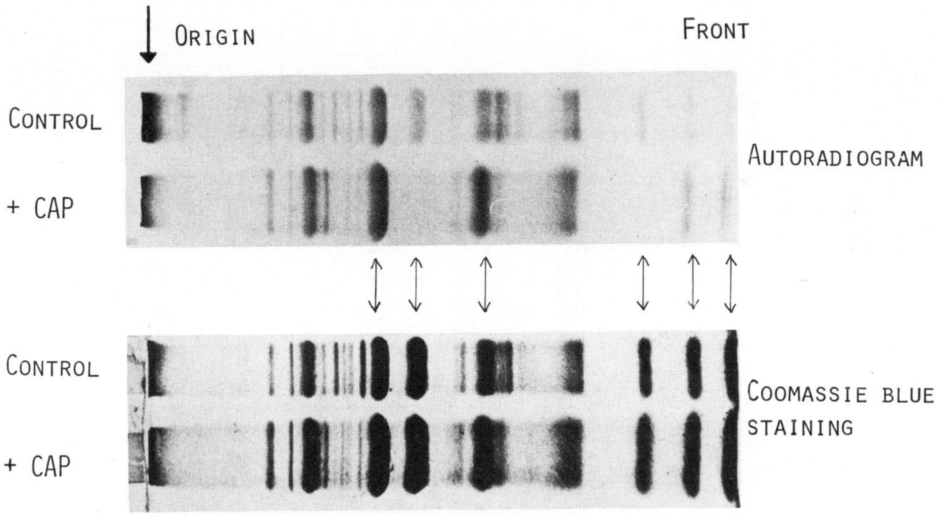

Fig. 3. Electrophoresis of cytochrome bc_1 complex isolated from cells labelled in the presence (+ CAP) or absence (control) of chloramphenicol. The preparations were electrophoresed as described in Fig. 2, stained, dried and autoradiographed.

These results suggest that one polypeptide of 32 000 dalton - most likely cytochrome *b*, as in *Neurospora crassa* - is synthesized on mitochondrial ribosomes. One would therefore expect that the other polypeptides should be labelled in the presence of chloramphenicol. Fig. 3 shows such an experiment. It is evident that no label is incorporated into the polypeptides with M_r 40 000 and 17 000. This surprising result is not due to some unspecific side effect of chloramphenicol, because it can be reproduced using acriflavin (not shown). Also the control (Fig. 3) shows that these polypeptides can incorporate [^{35}S]methionine in the absence of inhibitor.

The lack of incorporation of radioactivity into these polypeptides might be caused by the fact that during the labelling period only very small amounts of new polypeptides are added to the bulk of existing protein, and if some of these polypeptides fail to be properly integrated then they would not be copurified with Complex III in our isolation procedure. In order to detect also such non-integrated polypeptides we have treated an extract of submitochondrial particles from cells labelled in the presence of chloramphenicol with an antiserum against Complex III. Fig. 4 shows that this procedure

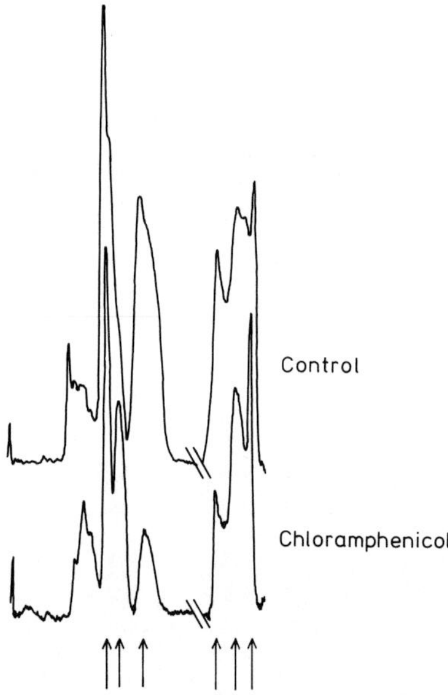

Fig. 4. Radioactivity pattern of cytochrome bc_1 complex, isolated by immunoprecipitation from cells labelled in the absence (control) or presence (chloramphenicol) of chloramphenicol. The autoradiograms obtained as described in Fig. 2 were scanned using the Gilford Linear Transport system. The arrows indicate the position of the polypeptides of Complex III.

leads to the precipitation of the complete cytochrome bc_1 complex from control cells, and that immunoprecipitation from chloramphenicol cells does yield the 40 000 and 17 000 dalton polypeptides in a labelled form.

We therefore conclude that of Complex III of yeast one polypeptide - most likely cytochrome *b* - is synthesized on mitochondrial ribosomes. The other are synthesized on cytoplasmic ribosomes, but under our labelling conditions two -with M_r 40 000 and 17 000- are not integrated into holo Complex III although they are present in the total membrane fraction.

Studies to untangle the apparently complex mode of assembly of Complex III in yeast are at present in progress.

ACKNOWLEDGEMENTS

We thank Mr. P. Severin for performing the iso-electric focusing experiments, Mr. W.J.M. Pluyms and Mr. J. Post for help and advice on amino acid analyses, Prof. Dr. P. Borst for helpful comments and criticism and Mrs. N. van Harten-Loosbroek for expert technical assistance. Part of this work was supported by a grant to P. Borst from the Netherlands Foundation for Chemical Research (S.O.N.) with financial aid from the Netherlands Organization for the Advancement of Pure Research (Z.W.O.).

REFERENCES

1. Mason, T.L., Poyton, R.O., Wharton, D.C. and Schatz, G. (1973) J. Biol. Chem. 248, 1346-1354.
2. Tzagoloff, A. (1973) Biochim. Biophys. Acta 301, 71-104.
3. Sebald, W., Machleidt, W. and Otto, J. (1973) Eur. J. Biochem. 38, 311-324.
4. Jackl, G. and Sebald, W. (1975) Eur. J. Biochem. 54, 97-106.
5. Ephrussi, B. and Slonimski, P.P. (1955) Nature 176, 1207-1208.
6. Clark-Walker, G.D. and Linnane, A.W. (1967) J. Cell Biol. 34, 1-14.
7. Weiss, H. and Ziganke, B. (1974) Eur. J. Biochem. 41, 63-71.
8. Ross, E. and Schatz, G. (1976) J. Biol. Chem. 251, 1997-2004.
9. Katan, M.B., Pool, L. and Groot, G.S.P. (1976) Eur. J. Biochem. 65, 95-105.
10. Tzagoloff, A. (1969) J. Biol. Chem. 244, 5020-5026.
11. Weiss, H. and Ziganke, B. (1975) in Electron Transfer Chains and Oxidative Phosphorylation (Quagliariello, E., Papa, S., Palmieri, F., Slater, E.C. and Siliprandi, N., eds), pp. 15-22, North-Holland, Amsterdam.
12. Capaldi, R.A. and VanderKooi, G. (1972) Proc. Natl. Acad. Sci. U.S. 69, 930-932.

13. Poyton, R.O. and Schatz, G. (1975) J. Biol. Chem. 250, 752-761.

14. King, T.E., Yu, C.A., Yu, L. and Chiang, Y.L. (1975) in Electron Transfer Chains and Oxidative Phosphorylation (Quagliariello, E., Papa, S., Palmieri, F., Slater, E.C. and Siliprandi, N., eds), pp. 105-118, North-Holland, Amsterdam

15. Tzagoloff, A. (1971) J. Biol. Chem. 246, 3050-3056.

16. Katan, M.B. and Groot, G.S.P. (1975) in Electron Transfer Chains and Oxidative Phosphorylation (Quagliariello, E., Papa, S., Palmieri, F., Slater, E.C. and Siliprandi, N., eds), pp. 127-132, North-Holland, Amsterdam.

17. O'Farrel, P.H. (1975) J. Biol. Chem. 250, 4007-4021.

18. Ross, E. and Schatz, G. (1976) J. Biol. Chem. 251, 1991-1996.

Genetics and Biogenesis of Chloroplasts and Mitochondria, Th. Bücher et al. eds.

PURIFICATION AND BIOGENESIS OF CYTOCHROME B IN BAKERS' YEAST

Leu-Fen Hou Lin and Diana S. Beattie
Department of Biochemistry
Mount Sinai School of Medicine
of the City University of New York
New York, N. Y. 10029

SUMMARY

Cytochrome b has been purified from mitochondria of Bakers' yeast. The final preparation is spectrally pure and shows only one band with a molecular weight of 28,000 daltons upon polyacrylamide gel electrophoresis in the presence of sodium dodecyl sulfate. The reduced minus oxidized difference spectra of cytochrome b revealed absorption bands at 562, 530 and 427 nm at room temperature. Yeast cells were labeled with ^{3}H-leucine in the presence or absence of cycloheximide. Electrophoretic analysis of the immunoprecipitates prepared from Triton-solubilized mitochondria indicated that label was incorporated into cytochrome b in the presence or absence of cycloheximide suggesting that cytochrome b is translated on mitochondrial ribosomes in yeast.

INTRODUCTION

Previous studies [1,2] on the control of mitochondrial protein synthesis in yeast had indicated that the mitochondrial synthesis of proteins in the cytochrome b-c_1 region of the respiratory chain is controlled by proteins synthesized in the cytoplasm. A study of the biogenesis of the cytochrome b-c_1 complex in yeast is further warranted in view of the considerable knowledge now available about the biogenesis and assembly of both the oligomycin-sensitive ATPase complex [3] and cytochrome oxidase [4,5] in yeast; however, no comparable information is available for the cytochrome b-c_1 complex.

In an initial attempt to study the biogenesis and assembly of the cytochrome b-c_1 complex, we have concentrated our efforts on a relatively well defined component, cytochrome b. In this study we have identified the intracellular site of cytochrome b synthesis by immunological techniques using a purified preparation of cytochrome b.

MATERIALS AND METHODS

L-[4,5-^{3}H]-leucine (54 Ci mmol^{-1}) was obtained from Amersham/Searle, and L-leucine from J.T. Baker Chemical Co. Commercially grown bakers' yeast (Fleischmann's yeast) were obtained from Standard Brands Inc. New York, N.Y. Enzyme grade ammonium sulfate was obtained from Schwarz/Mann; sodium cholate, sodium deoxycholate and sodium dodecyl sulfate (SDS) were purchased from Sigma; acrylamide and N N'-methylene-bisacrylamide from Eastman; and Tween 80 from Fisher.

Cytochrome b was purified from commercial bakers' yeast by a procedure involving solubilization in deoxycholic and cholic acids, ammonium sulfate fractionation, proteolytic digestion and sucrose gradient centrifugation in the presence of Tween 80 [6]. The heme content of cytochrome b was measured at 562-577 nm using a millimolar extinction coefficient of 25.6 [7] for the reduced minus oxidized difference spectrum. Absorption spectra were obtained with a Cary model 15 at room temperature. SDS-gel electrophoresis in 10% acrylamide gels was performed according to Weber and Osborn [8] except that half of the amount of ammonium persulfate was used. The molecular weight of the polypeptide band in the gel was estimated using the molecular weight of cytochrome c (11.7 x 10^{3}), myoglobin (17 x 10^{3}), carbonic anhydrase (29 x 10^{3}), ovalbumin (45 x 10^{3}) and bovine serum albumin (68 x 10^{3}) as standards.

Preparation of antiserum to cytochrome b - Purified yeast cytochrome b in 0.02 N Na-P_i buffer and 0.5% Tween 80 was homogenized with an equal volume of Freund's complete adjuvant. A rabbit was injected with a total of 5.4 mg of protein in three equal doses, subcutaneously into one footpad, in the back and in the neck region. Each injection was one week apart. After 2.5 weeks, the rabbit received a booster injection of 1 mg of protein intraveneously. Samples of blood were collected weekly after the booster injection. Control serum was collected from the same rabbit prior to the immunization.

Labeling of yeast cells in vivo - Yeast were grown aerobically in 5% glucose growth medium [9] to early stationary phase. The cells were harvested by centrifugation, and then washed twice with water and once with 0.05 M Na-P_i buffer, pH 7.4, containing 0.1% glucose. The cells were suspended to a final concentration of 250 mg/ml in the above buffer. Where indicated, cycloheximide

(100 µg/ml) was added and the culture incubated for 20 min prior to addition of 50 µCi per ml [^{3}H]leucine. After incubation for 30 min at 30°C, 10 mM unlabeled leucine was added and the incubation continued for another 4 hours. The cells were then harvested and mitochondria prepared as described previously [9].

Immunological techniques - The ^{3}H-labeled mitochondria were suspended in 5 mM Tris-acetate, pH 7.5, at a protein concentration of 4 mg per ml. To this suspension was added a solution of 10% Triton X-100 (w/v) to a final concentration of 0.5%. After 30 min, the mixture was centrifuged at 35,000 x g for 10 min. To the supernatant was added ½ volume of antiserum and incubated overnight at 4°. The mixture was centrifuged at 12,000 x g for 10 min. The supernatant was incubated for another 20 hrs. at 6° and then centrifuged as above. The second precipitate was washed 3 times with 5 mM Tris-acetate, pH 7.5, containing 0.5% Triton. The washed precipitate was depolymerized by incubation at 70° for 20 min with 100 µl of 0.02 N Na-P_i buffer, pH 7.0, containing 1% SDS, 1% mercaptoethanol and 10% glycerol, and then analyzed by SDS gel electrophoresis. The procedures for slicing and counting of gels have been described previously [10]. A parallel experiment was performed in which control serum was used instead of antiserum.

TABLE 1

PURIFICATION OF CYTOCHROME B FROM COMMERCIAL BAKERS' YEAST

Fraction	Total protein	Heme b	Yield
	mg	nmoles/mg protein	%
Submitochondrial particles	9,000	0.365	(100)
Low deoxycholate precipitate	4,872	0.606	90
High deoxycholate extract	1,789	0.930	51
$(NH_4)_2SO_4$ fractionation	550	1.914	32
Cholate-$(NH_4)_2SO_4$ fractionation	282	3.174	27
b-c_1 separation	147	4.385	20
Protease-$(NH_4)_2SO_4$ fractionation	45	9.658	13
Centrifugation	29	13.731	12
Sucrose gradient centrifugation	1.7	2.832	-

RESULTS

Purification of yeast cytochrome b - The purification of cytochrome b from yeast submitochondrial particles is summarized in Table I. Cytochrome b was relatively unstable and large losses in the heme content encountered during the final stages of the purification procedure. An indication of the purity of the final preparation as compared to various fractions from the purification scheme is clear by the gel patterns obtained after electrophoresis in polyacrylamide gels containing SDS (Fig. 1). The purified cytochrome b preparation revealed only one polypeptide band which has a molecular weight of 28,000 daltons.

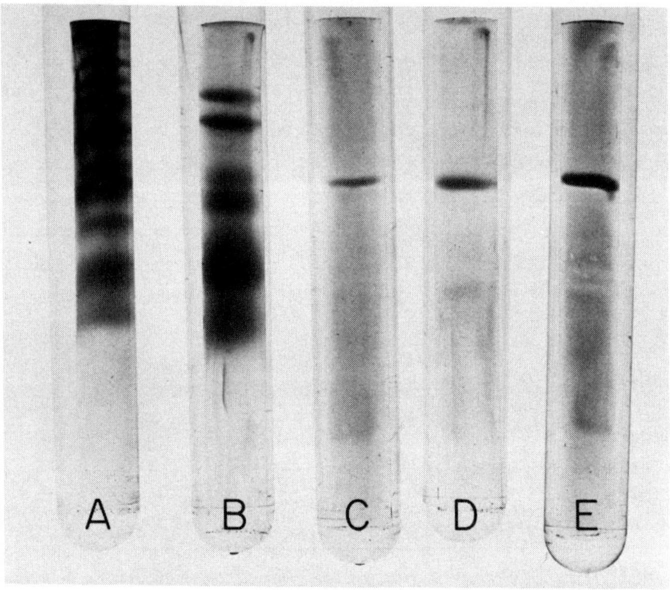

Fig. 1. SDS gel electrophoresis of various fractions obtained during purification of cytochrome b, gel A, high deoxycholate extract; gel B, protease-$(NH_4)_2SO_4$ fractionation; gel C,D, and E, 50 µg, 100 µg and 150 µg respectively of purified cytochrome b.

Spectral analysis - Cytochrome b prepared by the procedure described above exhibited α,β and γ bands at 562, 530 and 427 nm in difference spectra (Fig. 2). No contamination by the other cytochromes could be detected spectrophotometrically. The pyridine hemochromogen spectrum revealed only the presence of protoheme.

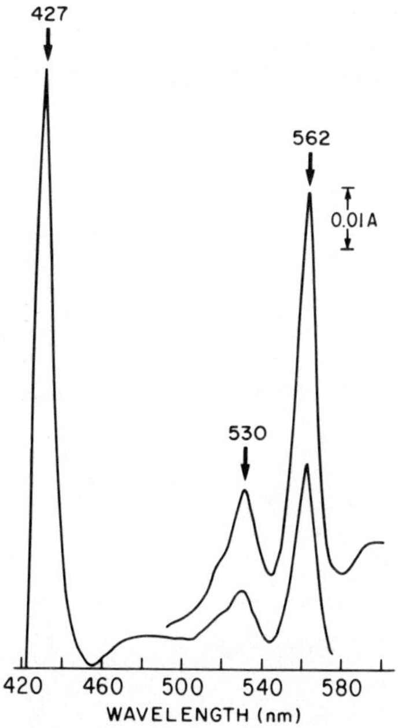

Fig. 2. Difference spectra of yeast cytochrome b at room temperature. Sample used was the protease-$(NH_4)_2SO_4$ fraction. Protein concentration was 0.25 mg ml^{-1} in the upper curve and 0.12 mg ml^{-1} in the lower curve. $Na_2S_2O_4$ reduced versus air oxidized difference spectrum was recorded.

Biosynthesis of cytochrome b - In order to identify the site of synthesis of cytochrome b, an antiserum against the purified protein was prepared. Yeast were labeled in vivo under nongrowing conditions with radioactive leucine in the absence or in the presence of the cytoribosomal inhibitor cycloheximide. After incubation, the mitochondria were isolated, extracted with Triton, and then subjected to immunoprecipitation with anticytochrome b antiserum as described under "Materials and Methods". The radioactive polypeptides present in the immunoprecipitates were analyzed by SDS-gel electrophoresis (Fig. 3). In the absence of inhibitor, the 28,000 dalton-polypeptide is the predominant band in the immunoprecipitates (Fig. 3A). Labeling of this fraction was not blocked by the presence of cycloheximide during the labeling of yeast cells (Fig. 3B). The relative amount of

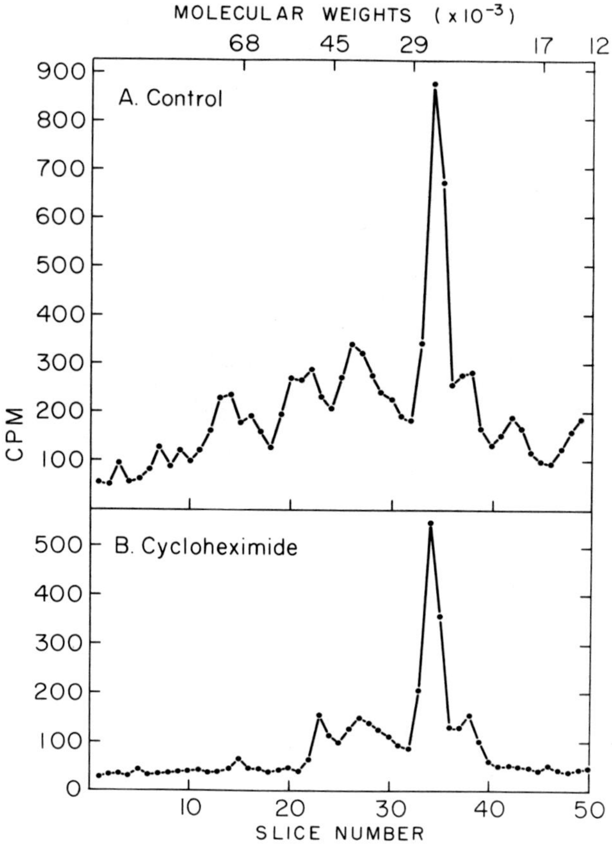

Fig. 3. Electrophoretic analysis of cytochrome b immunoprecipitated from labeled yeast mitochondria. See "Materials and Methods" for details.

radioactivity incorporated into cytochrome b was calculated by integration of the area under the protein peak. As shown in Table II, more than 80% of [^{3}H]leucine was still incorporated into the cytochrome b region in the presence of cycloheximide as compared to the control without inhibitor; however, under these conditions, cycloheximide inhibited more than 80% of the incorporation of [^{3}H]leucine into the whole mitochondrial membrane. Furthermore, labeling of cytochrome b was completely inhibited by chloramphenicol or erythromycin.

TABLE II

EFFECT OF CYCLOHEXIMIDE ON THE [^{3}H]LEUCINE INCORPORATION BY INTACT YEAST CELLS

See "Materials & Methods" for details

Inhibitor	Mitochondria		^{3}H-Labeled cytochrome b in the immunoprecipitate
	Specific ^{3}H activity cpm/mg	%	%
none	1.18×10^6	100	100
cycloheximide	0.22×10^6	18.6	82

DISCUSSION

Spectrally pure cytochrome *b* from yeast mitochondria has been described only once previously (11) but was not further characterized especially with regard to the subunit composition of the preparation. The procedure described in the present investigation yields a cytochrome *b* preparation which has a molecular weight of 28,000 daltons. The purity of the cytochrome *b* preparations is established by its absorption spectrum; by a single polypeptide band upon SDS-gel electrophoresis, and also by a single precipittine line upon double diffusion test using cytochrome *b* and anti-cytochrome *b* antiserum (not shown). However, since the procedure involved protease treatment, the molecular weight of 28,000 daltons of the heme *b*-containing polypeptide might be an underestimation for native cytochrome *b*. Despite the observation that when yeast cells were labelled *in vivo*, the immunoprecipitated mitochondrial membranes revealed the presence of only one radioactive polypeptide band of 28,000 daltons, we cannot rule out the possibility that endogeneous proteases in yeast convert native cytochrome *b* to a protein with an apparent molecular weight of 28,000 daltons.

During the preparation of this manuscript, Katan et al. [12] have published an entirely different procedure for the preparation of cytochrome *b* from a cytochrome *b*-c_1 complex of yeast. This preparation of cytochrome *b* consists of a major band at 32,000 daltons, two minor bands at 14,000 and 11,000 daltons, and traces of visible contaminants at higher molecular weight regions. Also, the preparation is not free from contamination of cytochrome c_1 as judged from the absorption spectrum [12].

The use of purified enzymes and immunological techniques have proven fruitful in the identification of several products of mitochondrial protein synthesis, e.g., 3 subunits of cytochrome oxidase [4,5] and the 4 subunits of oligomycin-sensitive ATPase [3]. To this list we now can add cytochrome b, which previously was shown to be a product of mitochondrial protein synthesis in *Neurospora*[13].

ACKNOWLEDGEMENTS

This work was supported, in part, by NIH Grants HD-04007 and NSF Grant GB-39896. Dr. Lin is a fellow on Training Grant GM-07036.

REFERENCES

1. Lin, L. F. H., Kim, I. C. and Beattie, D. S. (1974) Arch. Biochem. Biophys. 160, 458-464.
2. Beattie, D. S., Lin, L. F. H. and Stuchell, R. N. (1974) in The Biogenesis of Mitochondria (Kroom, A. M. and Saccone, C., eds), pp. 465-475, Academic Press, New York.
3. Tzagoloff, A. and Meagher, P. (1972) J. Biol. Chem. 247, 594-603.
4. Mason, T.L. and Schatz, G. (1973) J. Biol. Chem. 248, 1355-1360.
5. Rubin, M. S. and Tzagoloff, A. (1973) J. Biol. Chem. 248, 4275-4279.
6. Lin, L. F. H. and Beattie, D. S. (1976) Submitted for publication.
7. Berden, J. and Slater E. (1970) Biochim. Biophys. Acta 216, 237-249.
8. Weber, K. and Osborn, M. (1969) J. Biol. Chem. 244, 4406-4412.
9. Kim, I. C. and Beattie, D. S. (1973) Eur. J. Biochem. 36, 509-518.
10. Ibrahim, N. G., Stuchell, R. N. and Beattie, D. S. (1973) Eur. J. Biochem. 36, 519-527.
11. Tzagoloff, A. (1969) J. Biol. Chem. 244, 5020-5026.
12. Katan, M. B., Pool, L. and Groot, G. S. P. (1976) Eur. J. Biochem. 65, 95-105.
13. Weiss, H., Schwab, A. J. and Werner, S. (1975) in Membrane Biogenesis (Tzagoloff, A., ed.), pp. 125-153, Plenum Press, New York and London.

Genetics and Biogenesis of Chloroplasts and Mitochondria, Th. Bücher et al. eds.

BIOSYNTHESIS OF CYTOCHROME C IN THE HONEY BEE, APIS MELLIFERA

M. Osanai and H. Rembold
Max-Planck-Institut für Biochemie
D-8033 Martinsried bei München

Concentration of cytochrome c in the honey bee is very low in larval and early pupal stages followed by a rapid increase in the newly hatched adult. It corresponds to a fortyfold concentration in comparison with the early pupal stage. This fact suggests that biosynthesis of cytochrome c is very active in the late pupal stages of honey bee (Fig. 1). The active synthesis of the respiratory pigment is also indicated by both active incorporation of radioactive iron and of 5-aminolaevulinate (ALA) into heme c, the prosthetic group of cytochrome c, and of amino acids into the protein moiety (procytochrome c). Especially, the incorporation rate of ALA into cytochrome c is very high.

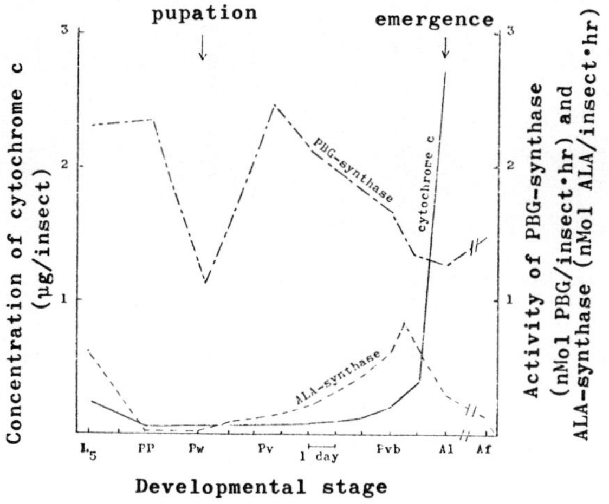

Fig. 1. Relation between cyt. c synthesis and activities of ALA- and PBG-synthase during development of honey bee, A. mellifica

ALA-Synthase, forming ALA from glycine and acetyl CoA, must be a reaction-determining enzyme for heme synthesis, corresponding to the very active cytochrome c biosynthesis. This was confirmed by the following reasons (Fig. 1):

1. the enzyme exhibits its maximal activity directly before the

active cytochrome c synthesis.
2. Compared with porphobilinogen-synthase, one of the other enzymes in the heme-synthesizing pathway, the activity of ALA-synthase is low throughout all developmental stages.
3. Corresponding to external changes, ALA-synthase changes its enzyme activity. This originates in quantitative changes of ribosomal ALA-synthase protein synthesis.

In spite of these facts, no close regulation mechanism could be confirmed between both synthesizing systems for the heme and the protein moiety of cytochrome c.

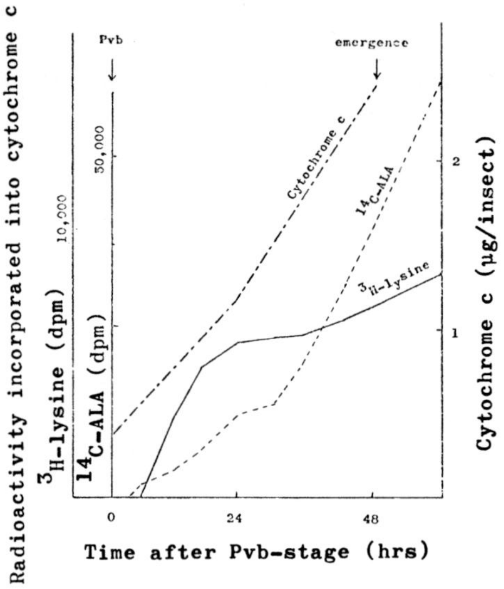

Fig. 2. Incorporation of ^{14}C-ALA and ^{3}H-lysine into cytochrome c during development of honey bee

Incorporation of labelled ALA into cytochrome c begins a little earlier than that of amino acids (Fig. 2). An active heme synthesis seems to precede the synthesis of procytochrome c. Incorporation rate of ALA coincides with rate of cytochrome c synthesis in the late pupal stages. However, rate of lysine incorporation is different. At the beginning of cytochrome c synthesis, incorporation of lysine is rather high, but then decreases 18 hours after injection into stage Pvb (Fig. 2). A similarly low rate of incorporation was also recognized after injection of another radioactive amino acid, i.e., valine.

In vitro synthesis of cytochrome c was further investigated with homogenates from thorax flight muscle of the honey bee. Content of cytochrome c in thorax reaches about 3/4 of that in the whole body. Different from results obtained in vivo, both incorporations of ALA and valine run parallel even after two hours incubation, although a short lag time was observed during incorporation of the amino acid (Fig. 3).

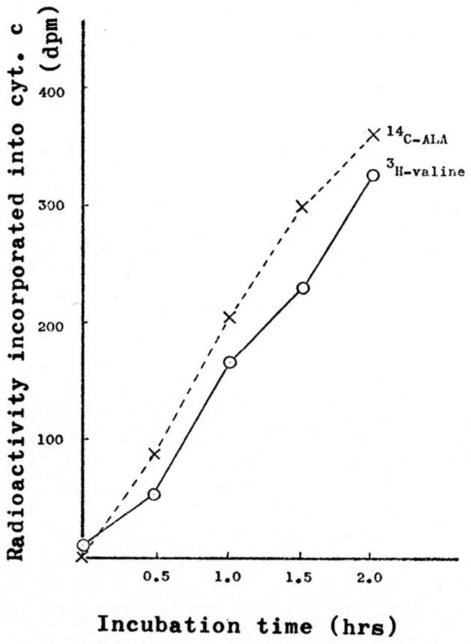

Fig. 3. Incorporations of ^{14}C-ALA and ^{3}H-valine into cytochrome c by homogenate of flight muscle of honey bee as a function of time

Since the flight muscle homogenate has capacity enough to synthesize both heme and protein moiety of cytochrome c, the decrease of incorporation rate in vivo may be interpreted by a dilution of the added radioactive amino acid in the amino acid pool of the honey bee.

After demonstration of a cytochrome c synthesis in vitro, the different cell particles were isolated by fractional centrifugation. Cytochrome c, labelled either through radioactive ALA or valine, was almost exclusively found in the cytosol (Tab. 1). These results might demonstrate that not only the protein moiety of cytochrome c is synthesized in ribosomes, but also insertion of the prosthetic group into the protein moiety is carried out in the same organelles. Consequently, holocytochrome c is already synthesized in ribosomes and penetrates through the cytosol into the mitochondrion. Free cytochrome c must then settle on the cristae. The isotonic wash solution showed a slight radioactivity from both the labelled precursors. They must have resulted from cytochrome c, which has not yet been firmly combined with mitochondria.

Localization of cytochrome c synthesized in vitro

cell fraction	protein contents (mg)	radioactivity in cyt. c (dpm)			specific activity (dpm/mg protein)	
		^{3}H-valine	^{14}C-ALA	^{3}H-valine/^{14}C-ALA	^{3}H-valine	^{14}C-ALA
homogenate	16.20	412	170	2.4	25.4	10.5
mitochondria (sarcosomes)	12.14	0	0	-	0	0
washing of mitochondria	1.32	17	8	2.2	12.9	6.1
microsomes	0.62	0	0	-	0	0
cytosol	3.53	451	174	2.6	127.8	49.3
total	17.61	468	182			
recovery	108.8%	113.6%	107.0%			

Tab. 1. Incorporations of ^{14}C-ALA and ^{3}H-valine into cytochrome c of different cell-fractions after in vitro synthesis

ACKNOWLEDGEMENT

These studies were supported by a grant, BCT 86, by Bundesministerium für Forschung und Technologie.

CHLOROPLAST DNA:
GENES AND MOLECULES

Genetics and Biogenesis of Chloroplasts and Mitochondria, Th. Bücher et al. eds.

THE CIRCULAR DIPLOID MODEL OF CHLOROPLAST DNA IN CHLAMYDOMONAS

Ruth Sager
Sidney Farber Cancer Institute
Boston, Massachusetts 02115
U.S.A.

The classical methods of genetic analysis based upon examining progeny of crosses for recombination events are powerful tools in revealing the structural organization as well as the genetic composition of DNA. The newly developing physical and chemical methods of DNA mapping complement recombination analysis, but do not replace it.

Genetic analysis can be used for at least three distinct purposes: (1) To establish gene identity, order, and distance by mapping; (2) To examine molecular organization and distribution of DNA molecules at cell division; and (3) To analyze patterns of molecular exchange events leading to recombination. As an example of the power of genetic analysis, this paper will summarize that aspect of the work of my laboratory over the past several years which was devoted to the genetic analysis of chloroplast DNA in Chlamydomonas.[1-5;7-11]

Early developments in our work which made the recent analysis possible include: the discovery of non-Mendelian inheritance of streptomycin resistance in *Chlamydomonas*[1] and subsequently the same pattern of inheritance of many additional genes[2]; the discovery that streptomycin is a potent mutagen for non-Mendelian genes in *Chlamydomonas*[3]; and the discovery that maternal inheritance (which helps in identifying non-Mendelian genes, but blocks their recombination) could be converted to biparental inheritance by UV-pretreatment of *mt+* gametes before mating.[4] The lines of evidence establishing that the non-Mendelian genes that we have described are located in chloroplast DNA have been discussed elsewhere.[5,6]

The life cycle of *Chlamydomonas* is exceedingly advantageous for genetic analysis.[7] Haploid gametes of opposite mating type fuse, two by two, to form zygotes which, in our experiments, are the only diploid stage in the life cycle. We induce them to undergo meiosis synchronously about 5-7 days after zygote formation, with the production of 4 zoospores, representing the 4 meiotic products. The zoospores, which are haploid for the nuclear genome, can be isolated at this time, and allowed to form zoospore colonies for analysis; or they can be separated after one or two further doublings, to provide small pedigrees of 8 or 16 cells per zygote (Fig. 1), each of which is then allowed to form a colony for further study. The results to be reported here are based entirely on pedigree analysis.

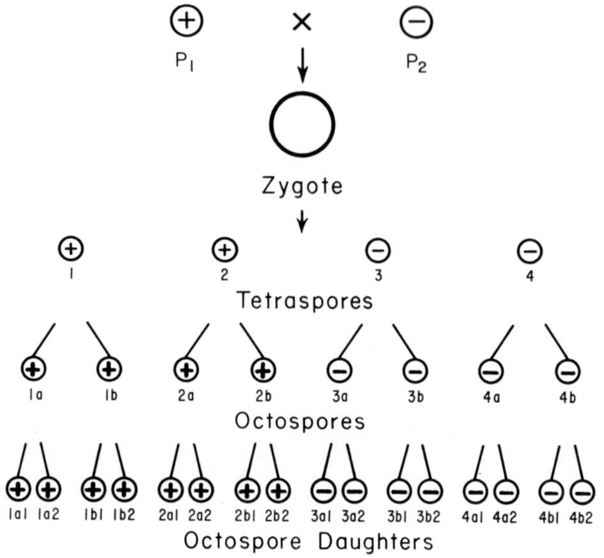

Figure 1. Procedure for pedigree analysis. After germination, zoospores are allowed to undergo one mitotic doubling and then the eight cells (octospores) are transferred to a fresh Petri plate and respread. After one further doubling, each pair of octospore daughters is separated and allowed to form colonies. The sixteen colonies, derived from the first two doublings of each zoospore are then classified for all segregating markers.

In order to distinguish recombination events occurring among nuclear genes from those involving chloroplast genes, we have incorporated three pairs of unlinked nuclear genes into all crosses. They serve to distinguish the four meiotic products unambiguously, and to monitor the germination process so that we can establish in each experiment whether lethality, differential growth rates of progeny, or other non-random events are perturbing our data. The genetic behavior of chloroplast genes is then assessed on this nuclear background, which serves as a powerful control.

The results to be summarized in this paper[8,9,10] have established the following regularities of behavior of chloroplast DNA in Chlamydomonas, and have led us to propose the circular diploid model:

(1) Genetic recombination occurs between the two parental chloroplast genomes in the progeny of biparental zygotes (in which maternal inheritance is inhibited[11]).

(2) Recombination occurs principally in vegetative growth of zoospore progeny clones (Fig. 1), rarely in zygotes where nuclear gene recombination occurs during meiosis. Thus recombination of nuclear and of chloroplast genes are separated both in time and in space.

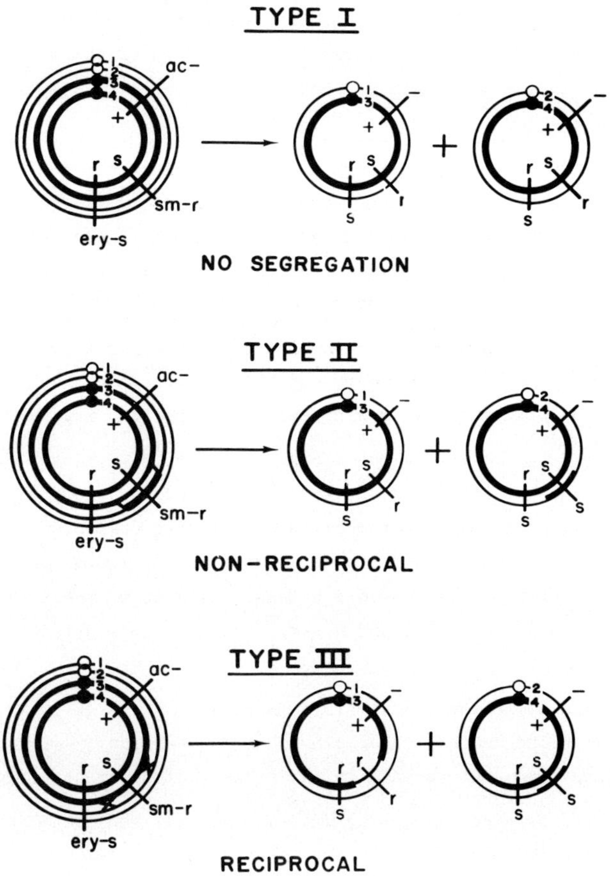

Figure 2. The chloroplast genome is shown on the left in circular form at a four-strand stage after replication and before cell division; and at the right after cell division. Each line represents a double-stranded DNA. Type I, the commonest pattern, resembles mitotic segregation of nuclear chromosomes. All three markers shown remain heterozygous (het). Type II depicts a gene conversion-like event at the *sm* locus, produces one *het* daughter and one homozygous *sm-s*. Type III depicts a reciprocal event giving rise to one pure *sm-r* and one pure *sm-s* daughter.

(3) In vegetative cells, recombination occurs at a 4-strand stage, as inferred from genetic analysis of zoospore pedigrees (Fig. 1). The results are equivalent to results obtained by mitotic crossing-over of nuclear chromosomes.[12]

(4) Two kinds of recombination events occur, reciprocal and non-reciprocal (Fig. 2). Alleles segregate 1:1 in reciprocal recombination, and 1:1 on the average in non-reciprocal recombination (see below).

(5) As can be seen in Fig. 2, allelic segregation, in which a cell that was heterozygous for a pair of alleles, gives rise to daughter cells that are homozygous, is the result of recombination.

(6) Reciprocal recombination occurs with a different but constant probability per doubling for each gene; and thereby provides evidence for the existence of a centromere-like attachment point that governs the distribution of chloroplast DNAs to daughter chloroplast and to daughter cells at cell division (Fig. 2). Reciprocal recombination frequencies provide the basis for mapping the relative distances of each gene from this attachment point (ap).

(7) Non-reciprocal recombination resembles gene conversion in yeast, but occurs with the previously unheard-of frequency of 40% per gene per doubling. Each gene shows the same probability of non-reciprocal recombination, regardless of map position.

(8) The simultaneous conversion of two or more genes at the same doubling provides a means of mapping genes on the basis of relative co-conversion frequencies.

(9) Recovery of pairs of daughter cells after one and two mitotic doublings, i.e. pedigree analysis (Fig. 1), permits strand analysis of recombination events, establishing that both reciprocal and non-reciprocal events often occur at the same doubling.

(10) The results of pedigree analysis, especially points (3), (4), and (9) provide the critical evidence that chloroplast DNA is genetically diploid, with one chloroplast genome from each parent, in the progeny of biparental zygotes.

(11) The results of genetic mapping by two methods: gene-attachment-point distances, and co-conversion frequencies, taken together, demonstrate that the chloroplast DNA is genetically circular.

The principal lines of evidence in support of the circular diploid model of chloroplast DNA come from pedigree analysis of 12 crosses, in each of which 4 to 7 pairs of chloroplast alleles were segregating.[8,9,10] These lines of evidence include the establishment of 1:1 ratios as the regular pattern of allelic segregation, with deviations that are localized in time and space[8]; the establishment of a unique order of genes by co-conversion mapping in a linear array that is consistent from cross to cross despite large differences in overall recombination frequencies[9]; the development and use of a novel mapping procedure for evaluating gene-attachment-point distances (see below)[10]; the demonstration that the UV irradiation pretreatment employed to obtain biparental zygotes by inhibiting maternal inheritance does not significantly distort 1:1 segregation ratios or recombination frequencies[8,9]; and the demonstration that the two different linear maps obtained by gene-attachment-point mapping and by co-conversion mapping can be reconciled by circularization.[10]

Allelic segregation ratios: The patterns of allelic segregation in pedigrees are critically important in the diploid model, since deviations from 1:1 suggest the presence of multiple genomic copies. We compared the segregation ratios of genes located all around the linkage map, and furthermore were able, by pedigree analysis, to distinguish segregation events occurring in the zygote from those occurring during vegetative multiplication of zoospore clones.[8] We did find some deviations from 1:1 but found they were specific to particular genes, to particular alleles, and to particular times in the life cycle. Thus, the genes ac1, ac2, and tm1, located adjacent to the attachment point ap, showed little if any deviation from 1:1 in the relatively rare zygotic events as well as in the more frequent recombination events in vegetative clones. On the other hand, the genes located in the region furthest from the attachment point showed considerable deviations from 1:1, both in the zygote and in vegetative cells. In the zygote, the deviations were towards an excess of alleles of maternal origin. This finding cannot be explained as excess copies of the chloroplast genome of maternal origin because ac and tm genes, present in the same cells at the same time, do not show a parallel excess. We have interpreted this result as preferential enzymatic degradation in zygotes of paternal chloroplast DNA in the region not protected by proximity to the attachment point.[8] In vegetative cells, however, deviations from 1:1 are allele-specific, with the mutant allele usually favored over the wild-type. Here too, genes further from the attachment point are more affected than those close to ap. We have postulated a different enzyme, capable of attacking and repairing heteroduplex regions of chloroplast DNA, as effector of these events in vegetative cells.[8]

Thus, the detailed evaluation of allelic segregation ratios has illuminated two fundamental features of chloroplast genetics: (1) the diploidy of the chloroplast genome; and (2) the different enzymatic bases of recombination-repair of chloroplast DNA in zygotes and in vegetative cells. Recently we have isolated and characterized a site-specific endonuclease from vegetative cells of Chlamydomonas, which may be a part of the chloroplast recombination-repair system.[13,14]

Evidence that the gene-conversion-like process of non-reciprocal recombination occurs not only with a high probability per gene per doubling, but also for long distances corresponding to one, two, or more genes, comes from our studies of co-conversion frequencies.[9] The relative frequencies with which two or more genes co-convert has been used directly, as shown in Figure 3, to establish a unique linear array of the five key genes: ac, sm2, ery, spc, and tm1, each present in all 12 crosses used in this analysis. The map is based on the single assumption that the closer two genes are to each other physically, the higher will be the frequency of their co-conversion. The frequencies of single gene conversion are the same for all genes whatever their map location: about 40% per gene per doubling, whereas only the highest co-conversion frequencies (Figure 3A) ap-

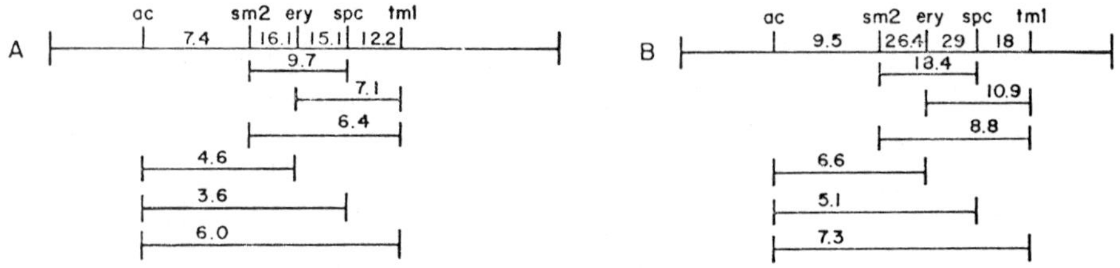

Figure 3. Maps of chloroplast markers.

A. Linear arrangement of 5 key markers based upon frequencies of Type II co-segregation.

B. Linear arrangement of 5 key markers based upon frequencies of Type II, Type III and zygote co-segregation.

proach the random expectations of 16% for two unrelated events. Thus, co-conversion events show high positive interference. For this and other reasons, we consider it premature to develop a metric for map distance at this time. The numbers shown in Figure 3 simply provide a basis for unambiguous assignment of gene order.

A different mapping method is shown in Figure 4 and Table 1.[10] Zygotes that are heterozygous for several chloroplast genes, are germinated synchronously, and the emerging zoospores are grown in liquid, and sampled at frequent intervals to determine the rate of segregation of each allelic pair from the heterozygous state to homozygosity. The results plotted as in Figure 4 show that each gene pair segregates at a different and reproducible rate. This rate is a function of the

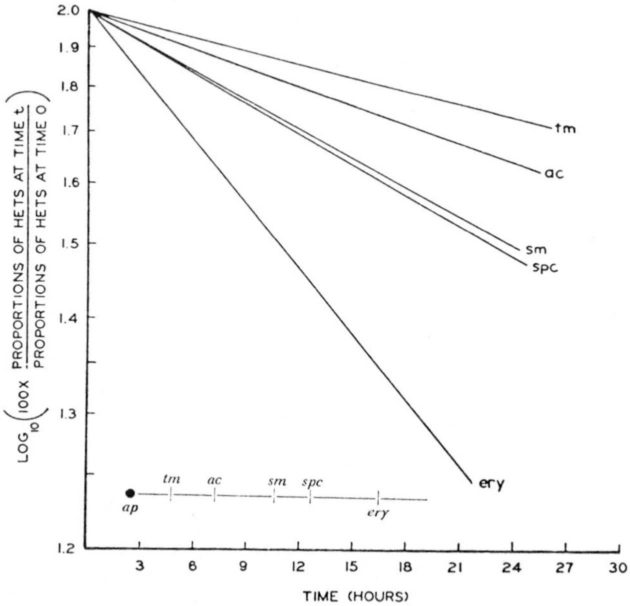

Figure 4. Segregation rates of principal genes.

TABLE 1.

Summary slopes and variability assessment for principal markers

Gene	Slope ± variability	Number of crosses
tml	-0.01204 ± .0021	15
ac	-0.01709 ± .0084	13
sm2	-0.02459 ± .0027	13
spc	-0.02513 ± .0138	14
ery	-0.04512 ± .0060	10

frequency of recombination between the gene and the centromere-like attachment point, postulated as shown in Figure 2; and thus provides a means to evaluate gene-attachment point distances. A statistical method was developed to estimate these rates, using data from many different experiments and different crosses.[10] Map distances are taken as proportional to rate (i.e. slope).

This method does not distinguish between one-armed or two-armed linear or circular maps. Thus, the map derived from the pooled data in Figure 4 needs to be evaluated in conjunction with the co-conversion map shown in Figure 3. Taken together, they provide unambiguous evidence of the genetic circularity of the Chlamydomonas chloroplast genome, as shown in Figure 5.

RELATIVE MAP POSITIONS OF 15 CHLOROPLAST GENES

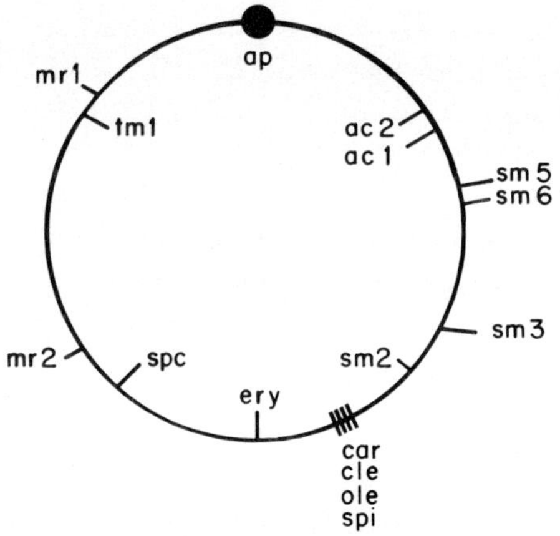

Figure 5. Based upon data in references 2,8,9,10.

Thus, the methods of classical genetic analysis alone have provided us with the means to establish certain fundamental properties of chloroplast DNA in Chlamydomonas: that after meiosis, vegetative cells contain two genetically circular copies of chloroplast DNA, that these molecules pair and recombine in every doubling, that they are distributed with mitotic-like regularity to daughter cells at each cell division, that this distribution is governed by a centromere-like attachment point, and that both reciprocal and non-reciprocal modes of recombination occur, with different net frequencies. From allelic segregation ratios and frequencies, it is further inferred that recombination events occurring in zygotes and in vegetative cells are effected by different enzymes.

These results need to be incorporated into a more comprehensive understanding of the molecular organization of chloroplast DNA. At present, the genetic data are in apparent conflict with physical evidence from reannealing kinetic analysis, which has revealed a major genomic component of about 10^8 daltons in size.[15,16] Since Chlamydomonas contains about 4 X 10^9 daltons of chloroplast DNA, the 10^8 dalton component may be present in as much as 40 copies, rather than the two copies identified by genetic analysis. The genetic and physical evidence could be reconciled if a slowly annealing component, corresponding to single copy DNA and not previously noted, were present; this possibility has not yet been adequately evaluated.

The multi-copy model of chloroplast DNA proposed by Gillham et al.[17] is at variance with our diploid model. Gillham et al. found excess copies of chloroplast alleles of maternal origin among the progeny of crosses, and interpreted the maternal excess as evidence of multiple copies. As discussed in detail elsewhere[5], in our view, Gillham et al. have misinterpreted their data for two principal reasons. First, they did not do pedigree analysis nor did they use nuclear markers to monitor their results, and as a consequence could not identify lethality of zoospores or other events which could have distorted allelic ratios; and second, they used only three linked markers, all located in the region furthest from the attachment point, where maternal bias does occur, as we pointed out above. In crosses we have made between Gillham's strains and ours, we confirmed their results for markers in the region furthest from the attachment point, but not for linked markers present in the same crosses located near the attachment point. Thus maternal bias does not result from the presence of multiple copies of the chloroplast genome, since different linked genes show different extents of maternal bias in the same crosses. In conclusion, genetic analysis of chloroplast DNA has provided a framework for further investigations of the molecular organization of this genome and its interactions with the nuclear and mitochondrial genomes. In addition, chloroplast DNA represents a unique and valuable system for the investigation of the molecular and enzymatic bases of recombination.[13,14]

REFERENCES

1. Sager, R. (1954) Proc. Nat. Acad. Sci. USA 40: 356-363.
2. Sager, R. and Ramanis, Z. (1970) Proc. Nat. Acad. Sci. USA 65: 593-600.
3. Sager, R. (1962) Proc. Nat. Acad. Sci. USA 48: 2018-2026.
4. Sager, R. and Ramanis, Z. (1967) Proc. Nat. Acad. Sci. USA 58: 931-937.

5. Sager, R. (1976) in Advances in Genetics, Vol. 19 (ed. E. Caspari), Academic Press, New York (in press).

6. Gillham, N.(1974) in Ann. Rev. of Genetics, Vol. 8 (ed. H.L. Roman, A. Campbell, and L.M. Sandler), Annual Reviews, Inc., Palo Alto, California.

7. Sager, R. (1972) Cytoplasmic Genes and Organelles, Academic Press, New York.

8. Sager, R. and Ramanis, Z. (1976) Genetics 83: 303-321.

9. Sager, R. and Ramanis, Z. (1976) Genetics 83: 323-340.

10. Singer, B., Sager, R., and Ramanis, Z. (1976) Genetics 83: 341-354.

11. Sager, R. and Ramanis, Z. (1973) Theor. and Appl. Genet. 43: 101-108.

12. Pontecorvo, G. (1958) Trends in Genetic Analysis, Columbia University Press, New York.

13. Burton, W., Roberts, R., Myers, P., and Sager, R. (1976) Fed. Proc. (in press).

14. Burton, W., Roberts, R., Myers, P., and Sager, R., manuscript in preparation.

15. Wells, R. and Sager, R. (1971) J. Mol. Biol. 58: 611-622.

16. Bastia, D., Chiang, K., Swift, H., and Siersma, P. (1971) Proc. Nat. Acad. Sci. USA 68: 1157-1161.

17. Gillham, N., Boynton, J., and Lee, R. (1974) Genetics 78: 439-457.

Supported by grants from the National Institutes of Health and American Cancer Society.

Genetics and Biogenesis of Chloroplasts and Mitochondria, Th. Bücher et al. eds.
© 1976 Elsevier/North-Holland Biomedical Press, Amsterdam, The Netherlands.

ON THE SEARCH FOR A MOLECULAR MECHANISM OF CYTOPLASMIC INHERITANCE: PAST CONTROVERSY, PRESENT PROGRESS AND FUTURE OUTLOOK

K. S. Chiang
Department of Biophysics and Theoretical Biology
University of Chicago
Chicago, Illinois 60637
U.S.A.

Following the demonstration that chloroplast (Chl-) DNA replication is semiconservative and is temporally separated from the nuclear DNA replication in the vegetative cell cycle of _Chlamydomonas_,[1] it has become evident that organelle DNA molecules themselves may be responsible for cytoplasmic inheritance, since the physical preservation and thus informational continuity of the organelle DNA in eukaryotes is maintained in much the same manner as that of the chromosomal (nuclear) DNA through the mitotic divisions and/or vegetative multiplication. This result also implies that the marked difference between the inheritance patterns of cytoplasmic and nuclear genes may well be based on the differences in the physical behavior of organelle and nuclear DNA molecules during meiosis and zygotic development.

The initial efforts in the search for a molecular mechanism of cytoplasmic inheritance were perhaps overly influenced by the conventional belief that the paternal contribution must be physically excluded.[2,3,4,5] Disappearance of paternal chloroplast DNA (chl-DNA) in zygotes of _Chlamydomonas_ has been reported to be the molecular basis for maternal inheritance.[4,5] Simple logic dictates, however, that the molecular mechanism of cytoplasmic inheritance in any organism cannot be ascertained simply by following the fate of the considerably reiterated (e.g. 25 to a few hundred) parental organelle DNA molecules in the zygotes alone. This is because in determining the genetic continuity of any DNA species, irrespective of its origin, not only the preservation or disappearance of parental DNA molecules, but also the newly replicated daughter molecules must be considered, since two consecutive semi-conservative replications of a single parental molecule would be sufficient to preserve the physical and thus the genetic continuity of a given DNA species in succeeding generations. In consequence, even if all the 25 to 50 paternal chl-DNA molecules[6,7,8] _completely_ "disappeared" in the zygotic development [which is, of course, not the case (see below)], it would still be incorrect to conclude that the genetic continuity of the paternal chl-DNA must be disrupted in the ensuing vegetative divisions of zoospores, unless it can be demonstrated that further replication of the paternal organelle DNA molecules did not occur before the complete "disappearance." This is also not the case, as the further replication of parental DNA was detected only a few hours after zygote formation (see below). Due to the considerable reiteration of organelle DNA molecules, the experimental determination of whether a very small number (e.g. one or two) of parental DNA molecules is conserved and whether any of these molecules is further replicated is certainly beyond the resolution and detecting sensitivity of the analytical CsCl gradient centrifugation technique employed.[4,5]

In addition, the interpretation of our previous experimental results[2,3] and that of others[4] was made without knowledge of an apparently compartmentalized chl-DNA precursor pool complication that was discovered much later. If stable isotope ^{15}N, which we first employed to elucidate the vegetative cell chl-DNA replication mode,[1] or radioactive Ade, was used to label vegetative cells, a labeled chl-DNA pool appeared during the ensuing zygotic development, even though we had ascertained before interpreting the data that no labeled DNA precursor pool was involved in the synthesis of a nuclear DNA component in the zygotic maturation period.[2] Perhaps largely due to this pool complication, the behavior of chl-DNA during the zygotic development in *Chlamydomonas* was not unequivocably resolved as conflicting results on the fate of parental chl-DNA were reported.[2,4] More recently we have been able to eliminate the labeled DNA precursor pool complication by employing thymidine (dThd) and 5-bromodeoxyuridine (BrdUrd) as specific radioactive and density labels, respectively, for chl-DNA replication.[9,10] This development has dramatically changed the entire prospect for the search for an unequivocal molecular mechanism of cytoplasmic inheritance. Some of our recent progress on the elucidation of the physical behavior of chl-DNA molecules during the sexual reproductive cycle in *Chlamydomonas reinhardtii* will be summarized below.

RECENT PROGRESS

1. The indiscriminate partial conservation of both parental chl-DNAs: As shown in Fig. 1, approximately 90 to 95% of ^{3}H-dThd activity incorporated into the chl-DNA during gametogenesis of either male or female gametes was no longer detectable at the mature zygote or zygospore stage under our experimental conditions. No further reduction of this gametic chl-DNA was observed in the ensuing zoospore stage.[11] This unbiased partial conservation of both parental chl-DNA molecules was also observed if chl-DNA was labeled by ^{14}C-or ^{3}H-dThd during the vegetative growth stage. These results also substantiated our earlier suspicion that a labeled precursor pool(s) of previously employed isotope ^{15}N and/or non-specific radioactive ^{3}H-or ^{14}C-Ade was manifested in zygotic development.[2,3] This complication misled us in our previous interpretation on the extent of this unbiased conservation of both parental chl-DNA molecules.[2]

Since neither the female nor the male parental gametic chl-DNA was preferentially conserved in the zygotes, these results on the fate of gametic chl-DNA during zygotic development do not by themselves constitute any direct determinant for the mechanism of maternal inheritance in *Chlamydomonas*.

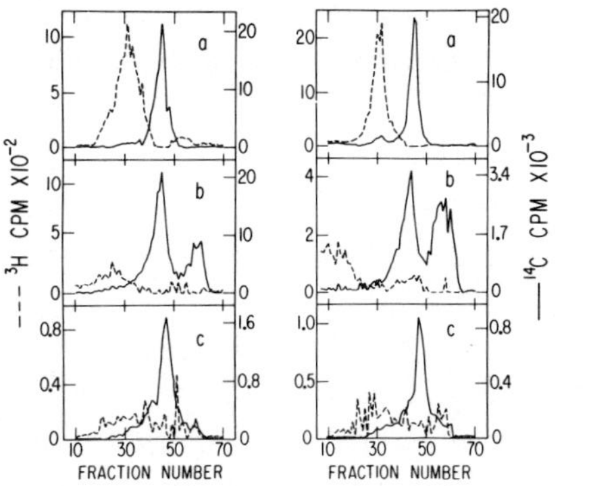

Fig. 1. The banding profiles of whole cell DNA in CsCl equilibrium density gradient.
Left panel: The male gametogenic culture was labeled with ^{3}H-dThd and ^{14}C-Ade. After chasing, the labeled gametes were mixed with an equal number of unlabeled female gametes. The homogeneous population of young zygotes thus obtained were allowed to mature and germinate. (a) Equal numbers of male and female gametes; (b) Mature zygotes; and (c) Zoospores.
Right panel: Same as the left panel except that the female gametes were labeled with ^{3}H-dThd and ^{14}C-Ade, while the male gametes were unlabeled.

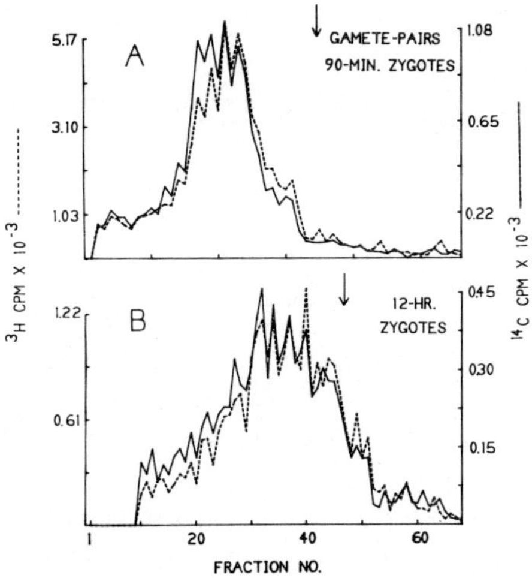

Fig. 2. The whole cell DNA banding profile in CsCl equilibrium density gradient.
A. Gamete-pairs: A homogeneous population of freshly formed zygotes (gamete-pairs) obtained 90 min. after mixing equal numbers of ^{3}H-dThd labeled male non-Mendelian mutants ESS_p-14[15] and ^{14}C-dThd labeled 137c female gametes.
B. Developing young Zygotes: The zygotic population in A was allowed to mature for 10-1/2 hr. in the presence of 1×10^{-4}M 5-bromodeoxyuridine (BrdUrd).
The arrow in (A) and (B) marks the nuclear DNA peak position in the gradient.

2. Structural alterations of gametic chl-DNA: The CsCl equilibrium banding pattern of ^{3}H-dThd labeled chl-DNA was altered after gametogenesis.[11] Unlike the sharp and smooth chl-DNA peak of the vegetative cells, the female as well as the male gametic chl-DNA banded with an unevenly serrated peak with characteristic multiple zig-zag submaxima over a considerably broadened density range, indicating that the chl-DNA samples isolated from gametes were either excised or weakened at selected rather than random sites. This peculiar banding pattern is more pronounced in the female gametes than the male gametes; however, upon aging of the unmated gametic populations, the slight initial differences became less obvious as the peculiar banding characteristics became more pronounced.[12] This apparent "restriction" of both parental chl-DNA was augmented rapidly and markedly immediately after fusion of gametes (see Fig. 2a and 4a). This banding pattern of

apparently restricted maternal and paternal chl-DNA continued and became even more pronounced for at least another 10-1/2 hr. (see Fig. 2b) while replication of both parental chl-DNA and recombination also occurred (see below). Furthermore, the structure of the newly replicated chl-DNA in the early zygotic maturation stage was also altered or restricted in a similar manner (see Fig. 5), perhaps in preparation for the extensive recombination observed between these newly replicated and pre-existing parental chl-DNA molecules (see Fig. 3 and text below).

3. Replication of parental chl-DNA: As shown in Fig. 3, in a BrdUrd supplemented developing zygote culture, the appearance of a *de novo* synthesized ^{3}H-dThd labeled major DNA peak at a density of 1.719-1.720 g/ml (see Fig. 3, Fraction 40-41), which is equal to that of the BrdUrd labeled vegetative cell chl-DNA obtained after prolonged incubation,[10] indicates that at least two rounds of semiconservative replication of chl-DNA have been completed 10 hr. after zygote formation. Neither parental chl-DNA was favored in this early zygotic chl-DNA replication as both maternal and ^{3}H-dThd labeled paternal chl-DNA molecules co-migrated coordinatedly toward the heavier density regions (see Fig. 2b). The replication of both parental chl-DNA as detected by the BrdUrd and dThd incorporation occurred as early as 5-1/2 hr. after zygote formation under our experimental conditions.[12] However, the earliest time of such replication has not yet been determined specifically. As a control experiment this rather extensive *de novo* replication of parental chl-DNA was also demonstrated in the absence of BrdUrd during the early zygotic maturation period (see Fig. 5).

In contrast to the chl-DNA replication in vegetative cells, this early zygotic chl-DNA replication was characteristically asynchronous and accompanied by extensive recombination events (see below). The *de novo* replicated heavy (BrdUrd and ^{3}H-dThd incorporated into both complementary polynucleotide chains) DNA peak

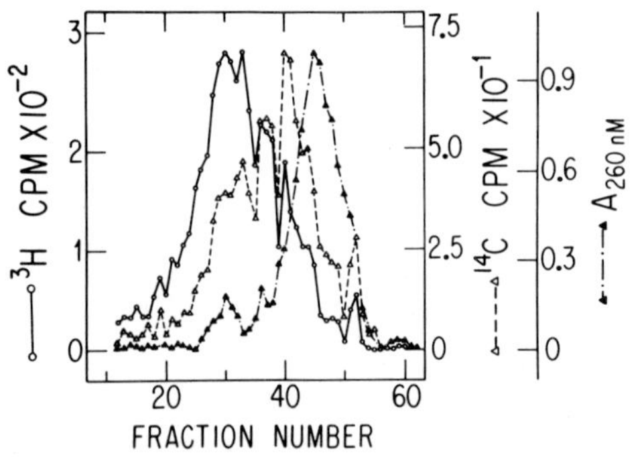

Fig. 3. The whole cell DNA banding profile in CsCl equilibrium density gradient. A homogeneous population of freshly formed zygotes obtained 60 min. after mixing equal numbers of ^{14}C-dThd labeled 137C female gametes and unlabeled 137C male gametes was incubated in the presence of 3H-dThd and BrdUrd for 10 hours. Unlabeled C. reinhardtii vegetative cell DNA was included in the gradient as a density marker for both Chl-DNA (fraction 30) and nuclear DNA (fraction 43).

(fraction 40-41 in Fig. 3) was found to co-exist with hybrid chl-DNA molecules (fraction 36-38) carrying both the maternal ^{14}C-dThd and the de novo synthesized ^{3}H-dThd as early as 5-1/2 hr. after zygote formation. This observation may suggest that some of the parental chl-DNA is incapable of further replication after the initial round of semiconservative replication.

4. High Frequency Recombination of chl-DNA: Using a differential vegetative labeling scheme with ^{15}N, ^{3}H-Ade for one parent and ^{14}N, ^{14}C-Ade for the other, we first observed an apparent extensive recombination between the two parental chl-DNA molecules in the early zygotic maturation[3] (between 12-24 hr. after zygote formation). The unexpected high frequency of such recombinations was previously considered to be due largely to the precursor pools of ^{15}N and labeled Ade. However, the high frequency of chl-DNA recombinations in the early zygotic maturation period has now been confirmed in the absence of complications due to pools. Considerable maternal ^{3}H-dThd labeled chl-DNA appeared in the minor DNA peaks (fraction 40 and higher in Fig. 3) with a density much higher than that of the hybrid peak (i.e., fraction 36-38), while an even larger amount of the de novo synthesized ^{14}C-dThd labeled chl-DNA appeared in the light ^{3}H-dThd labeled maternal DNA region (fraction 34 and lower). This result indicates that physical exchanges of DNA fragments between parental and the newly replicated chl-DNA must have occurred at a high frequency. Fig. 4 shows that paternal chl-DNA also participated in this type of recombinational event just as the maternal chl-DNA shown in Fig. 3. This physical exchange, which accompanied the parental chl-DNA replications (see

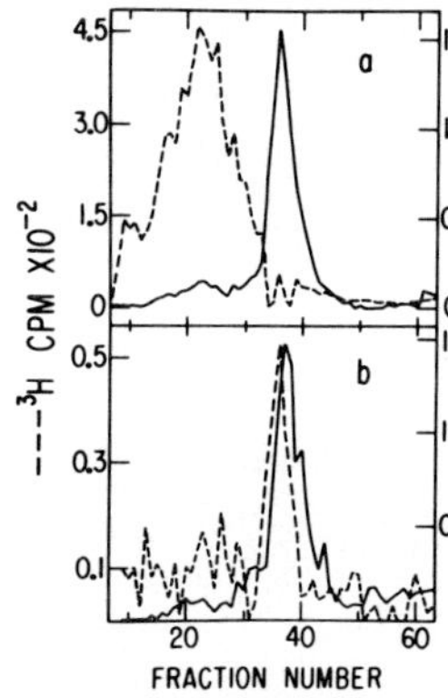

Fig. 4. The whole cell DNA banding profiles in CsCl equilibrium density gradient.
A. Gamete-pairs: A homogeneous population of freshly formed zygotes obtained 120 min. after mixing equal numbers of ^{3}H-dThd labeled ESS_p-14^{15} and unlabeled $nESS_p$-4^{15} gametes.
B. Mature Zygotes: Same as A after a 2-day light and 5-day dark maturation period.
In both panels, ^{14}C-Ade labeled whole-cell DNA has been added as a density marker.

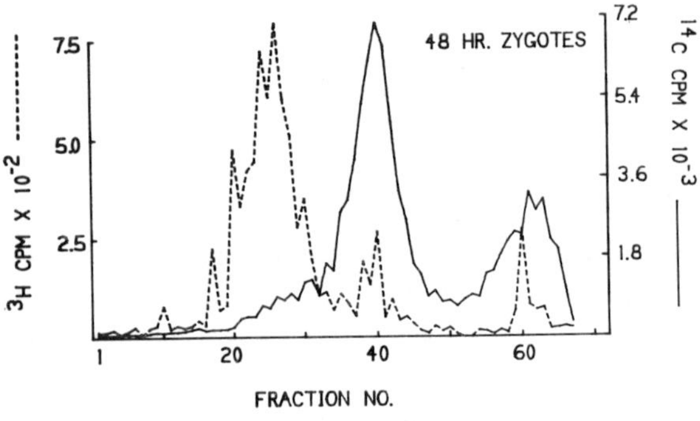

Fig. 5. The whole cell DNA banding profile in CsCl equilibrium density gradient. A homogeneous population of freshly formed zygotes obtained 90 min. after mixing equal numbers of unlabeled 137C male and 137C female gametes was allowed to mature in the presence of ^{3}H-dThd for 2 days under illumination. ^{14}C-Ade labeled whole cell DNA was included in the gradient as a density marker. Both gametic populations were obtained under special gametogenic conditions in order to block ^{3}H-dThd incorporation into a nuclear DNA component (i.e., M-band) which appeared approximately 30-36 hr. after mating under normal conditions.[2,3,16]

Fig. 3), was observed as early as 5-1/2 hr. after zygote formation.

5. Kinetics and Nature of chl-DNA degradations: Despite the fact that considerable replication and recombination of parental chl-DNA occurred during early zygotic development as outlined above, approximately 90-95% of the parental chl-DNA was degraded before the zygotes reached maturation (see Fig. 1). The degradation of parental chl-DNA commenced soon after zygote formation in the midst of the replication and recombinational events described above. The kinetics of such degradations are significantly faster for the paternal chl-DNA. As shown in Fig. 2, the ratio of the total amount of ^{3}H-dThd labeled paternal chl-DNA to the ^{14}C-dThd labeled maternal chl-DNA decreased more than 75% in 10-1/2 hr. after zygote formation. This faster degradation of paternal chl-DNA apparently occurred randomly as the "restricted" banding did not change significantly after such degradation (compare Fig. 2b with 2a). Although the de novo synthesized chl-DNA molecules in the early zygotic development were also degraded by the zygospore stage, these newly replicated and/or recombined chl-DNA molecules appeared to have survived better than the original paternal molecules. As seen in Fig. 4, in the presence of BrdUrd during zygotic development most of the remaining ^{3}H-dThd labeled paternal chl-DNA appeared at a density much higher than the original paternal chl-DNA. A detailed characterization of the kinetics and nature of the zygotic chl-DNA is currently under investigation in our laboratory.

FUTURE OUTLOOK

The elucidation of this complex and intriguing chl-DNA behavior in zygotic development under normal experimental conditions has had a pronounced impact on our ongoing search for a molecular mechanism of cytoplasmic inheritance. Since replication and recombination of both parental chl-DNAs occur long before their substantial but never complete degradation, the critical issue to be resolved here, therefore, is not so much the disappearance or degradation of parental chl-DNA molecules per se but whether any differential protective mechanism(s), either structural, biochemical or temporal, is selectively accorded the daughter (or parental) chl-DNA molecules of either parental origin, and whether any advantage or disadvantage in the kinetics of degradation versus further replication/recombination is selectively accorded to one of the two parental chl-DNA populations. I believe that no unequivocal molecular mechanism(s) of cytoplasmic inheritance can be elucidated before the above questions can be answered by direct experimental means.

A serious but often neglected dilemma exists in the search for a mechanism of cytoplasmic inheritance through the study of organelle DNA behavior. Only one or a very few out of a population of considerably reiterated organelle DNA molecules, completely homogeneous or not, is required to transmit and preserve the organelle genetic specificity. The physical behavior of this particular genetically competent molecule(s) may not necessarily be identical to that of the overwhelming reiterated majority, whose behavior is what we usually detect under normal experimental conditions. To ascertain the possible behavioral difference of a single organelle DNA molecule in a population of 50 to a few hundred similar, if not identical, molecules is, of course, not easily achievable.

We will attempt to overcome this difficulty by employing special experimental conditions, such as UV irradiation[13,14], so that the normal maternal transmission mechanism can be variably attenuated. Under such special conditions, we hope to ascertain whether any deviation in the normal physical behavior of organelle DNA molecules and any possible protective mechanism(s) to be elucidated can be correlated with the alteration of cytoplasmic gene inheritance patterns. Despite its obvious indispensibility, this type of experiment was not performed in a previous study on the mechanism of cytoplasmic inheritance. The absence of this important control experiment may in part have been responsible for an erroneous and unjustified claim that the disappearance of paternal chl-DNA in Chlamydomonas zygotes is the molecular basis for maternal inheritance.[4] Assuming that UV irradiation will not introduce serious unexpected technical difficulties, such as rapid and quantitative fusion of gametes, we may at last have reached a final stage in our search for an unequivocal molecular mechanism of cytoplasmic inheritance.

ACKNOWLEDGEMENTS

This work was supported by grants from the U.S. Institutes of Health (GM-21955, HD-05804, CA-14599, and HD-07110). The author wishes to acknowledge with gratitude the assistance of Dr. David O. Swinton, Ms. Eva Eves, and Mrs. Tina Su Chiang in the preparation of this manuscript.

REFERENCES

1. Chiang, K.-S., and Sueoka, N. (1967) Proc. Natl. Acad. Sci. U.S.A. 57, 1506-1513.
2. Chiang, K.-S. (1968) Proc. Natl. Acad. Sci. U.S.A. 60, 194-200.
3. Chiang, K.-S. (1971) in Autonomy and Biogenesis of Mitochondria and Chloroplasts, Boardman, N.K., Linnane, A.W., and Smillie, R.M., Eds. North-Holland, Amsterdam, 235-249.
4. Sager, R., and Lane, D. (1972) Proc. Natl. Acad. Sci. U.S.A. 69, 2410-2413.
5. Sager, R. (1972) Cytoplasmic Genes and Organelles, Academic Press: New York.
6. Bastia, D., Siersma, P., Swift, H., and Chiang K.-S. (1970) J. Cell Biol. 47, 15a.
7. Bastia, D., Chiang, K.-S., Swift, H., and Siersma, P. (1971) Proc. Natl. Acad. Sci. U.S.A. 68, 1157-1161.
8. Wells, R., and Sager, R. (1971) J. Mol. Biol. 58, 611-622.
9. Swinton, D., and Hanawalt, P. (1972) J. Cell Biol. 54, 592-597.
10. Chiang, K.-S., Eves, E., and Swinton, D. (1975) Developmental Biol. 42, 53-63.
11. Chiang, K.-S. (1975) in Les cycles cellulaires et leur blocage chez plusieurs protistes. Editions du centre national de la recherche scientifique, Paris, 147-158.
12. Chiang, K.-S., Manuscript in preparation.
13. Sager, R., and Ramanis, Z. (1967) Proc. Natl. Acad. Sci. U.S.A. 58, 931-937.
14. Gillham, N., Boynton, J., and Lee, R. (1974) Genetics 78, 439-457.
15. Chu-Der, O., and Chiang, K.-S. (1975) Proc. Natl. Acad. Sci. U.S.A. 71, 153-157.
16. Sueoka, N., Chiang, K.-S., Kates, J. (1967) J. Mol. Biol. 25, 47-66.

Genetics and Biogenesis of Chloroplasts and Mitochondria, Th. Bücher et al. eds.

TRANSMISSION, SEGREGATION AND RECOMBINATION OF CHLOROPLAST GENES IN CHLAMYDOMONAS

John E. Boynton, Nicholas W. Gillham, Elizabeth H. Harris, Constance L. Tingle, Karen Van Winkle-Swift, and G.M.W. Adams, Duke University, Durham, NC 27706

Chlamydomonas reinhardtii, a eukaryotic unicellular green alga with a single chloroplast and several mitochondria, is an exemplary organism for investigation of organelle genetics[1-4]. Both chloroplast and mitochondrial DNAs have been characterized physically and appear to be present in many copies per cell. The chloroplast genome contains sufficient unique sequence DNA to code for several hundred proteins of average molecular weight, whereas the mitochondrial DNA has less than 10% of this coding capacity. Phenotypes of chloroplast mutants include resistance to or dependence on antibacterial antibiotics, in many cases attributable to direct effects on chloroplast ribosomes[5], as well as temperature sensitivity and requirement for acetate[1,3,4]. With very few exceptions[3,6,14], the allelic relationships of chloroplast gene mutants with similar phenotypes are not known.

Mutations in *C. reinhardtii* show one of three distinct patterns of inheritance. Nuclear gene mutations, which fall into 16 linkage groups[7], regularly segregate 2:2 at meiosis in reciprocal crosses and do not segregate somatically Putative mitochondrial gene mutations[8,9] are transmitted by either parent to the meiotic progeny in a non-Mendelian fashion. Chloroplast gene mutations are usually transmitted uniparentally by the maternal (*mt*$^+$) parent[1-4,10]; the physical basis for this uniparental transmission is not completely understood. A few zygotes transmit chloroplast genes from both parents (biparental zygotes) or, rarely, only from the *mt*$^-$ parent (paternal zygotes). The frequency of these exceptional zygotes can be greatly increased by UV irradiation of *mt*$^+$ gametes prior to mating[11,12]. Segregation and recombination of chloroplast genes have been studied among the progeny of biparental zygotes by pedigree analysis,[3,13,14] by analysis of successive generations of zoospore progeny growing in liquid culture[3,15] and by zygote clone analysis[1,12,14,16]. The relative merits of different methods for mapping chloroplast genes are discussed by Adams *et al*[1]. In the zygote clone analysis method used in our laboratory, 64 progeny clones are selected randomly from 50-100 biparental zygotes. Recombination frequencies are calculated as (recombinant progeny)/(total progeny from all biparental zygotes), and in this respect our method resembles the random diploid method used for mitochondrial genes in yeast[17]. The few remaining heteroplasmic progeny are excluded from calculations.

Using pedigree analysis, Sager and Ramanis[3,13,14] have reported a 1:1 ratio of parental chloroplast alleles among the progeny of biparental zygotes, and have interpreted this as evidence for the presence of only two copies of the chloroplast genes per vegetative cell. Recently they have also found the same 1:1 ratio in zygote clones irrespective of the UV dose used to increase the frequency of biparental zygotes[14]. In contrast, we have found a biased output in zygote clones favoring chloroplast genes from the maternal parent[12] (Fig. 1). With increasing UV dose transmission of paternal chloroplast genes also increases, and the average ratio of maternal to paternal genes in a population of biparental zygotes approaches 1:1. However, individual biparental zygotes vary widely in output of paternal and maternal chloroplast genomes, as defined by allelic ratios, and are not normally distributed around the mean of 0.5 which presumably represents equal allelic input. The extreme classes are as frequent as those with a ratio of 0.5.

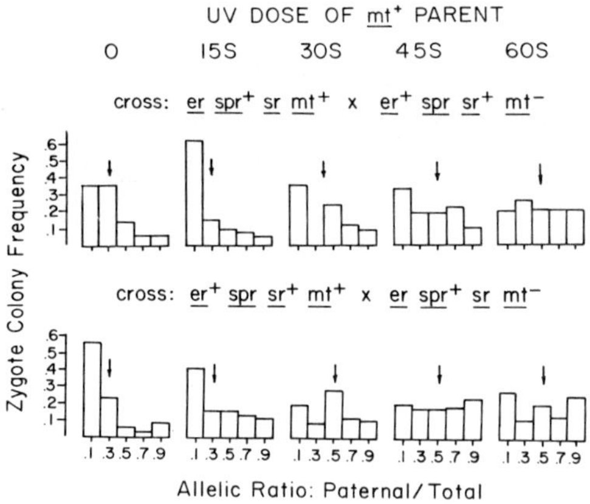

FIGURE 1. Allelic ratios among biparental zygotes in reciprocal crosses. The mt+ parent was treated with various doses (seconds) of UV prior to mating. Zygote clones were grouped into classes based on the fraction of progeny in each clone carrying the alleles from the paternal parent. From Gillham et al. 1974[12]

er = er-u-37

spr = spr-u-1-6-2

sr = sr-u-2-60

We have also examined the meiotic progeny of complete tetrads produced from biparental zygotes (Table 1). Among 1012 meiotic products, 50% were homoplasmic at the end of the second meiotic division for the three chloroplast markers studied. The ratio of products homoplasmic for maternal and paternal chloroplast markers approximates 3:1, and is clearly not consistent with Sager's two-copy model. Assuming that the entire chloroplast genome is transmitted en bloc, including the region delimited by these three markers, then tetrads in which the meiotic products show a predominance of maternal chloroplast genomes comprise 70% of the total. Thus our results indicate that these biparental zygotes must have contained more than two copies of the chloroplast genome prior to meiosis, with the bias favoring the maternal parent. If each meiotic product receives at least two copies of the chloroplast genome, then in the most extreme cases one has to assume a minimum of 7 maternal:1 paternal genome (21% of observed tetrads) and 1 maternal:7 paternal (2% of observed tetrads). We do not yet know whether our stocks continue to show a biased segregation of chloroplast genes during the first two mitotic divisions following meiosis. Neither the consistent maternal bias seen in the progeny of biparental zygotes (Fig. 1) nor that seen among the meiotic products (Table 1) can be accounted for by differential growth of specific genotypes.

TABLE 1. Inheritance of chloroplast genomes in meiotic products of biparental zygotes. Analysis is based on 1012 meiotic products from 253 complete tetrads formed by biparental zygotes in the crosses shown in Tables 2 and 3. A biparental zygote is here defined as one in which identifiable chloroplast markers (er-u-37, spr-u-1-6-2, sr-u-2-60) from both parents can still be scored at the end of meiosis. Each meiotic product is presumed to have a minimum of two copies of the chloroplast genome.

Cross: M/M mt^+ X P/P mt^-, where M/M indicates maternal genotype, P/P indicates paternal genotype, and M/P indicates biparental genotype (carrying chloroplast markers from both parents).

CHLOROPLAST GENOTYPES OF THE FOUR MEIOTIC PRODUCTS			MINIMUM NUMBER OF CHLOROPLAST GENOMES IN ZYGOTE AFTER REPLICATION		OBSERVED FREQUENCY OF TETRAD TYPE	
					NUMBER	%
Zygotes with maternal bias						
3 M/M	1 M/P	----	7 M	1 P	52	21
2 M/M	2 M/P	----	6 M	2 P	60	24
3 M/M	----	1 P/P	6 M	2 P	0	0
1 M/M	3 M/P	----	5 M	3 P	51	20
2 M/M	1 M/P	1 P/P	5 M	3 P	13	5
Zygotes with no bias						
1 M/M	2 M/P	1 P/P	4 M	4 P	27	11
----	4 M/P	----	4 M	4 P	15	6
2 M/M	----	2 P/P	4 M	4 P	3	1
Zygotes with paternal bias						
----	3 M/P	1 P/P	3 M	5 P	12	5
1 M/M	1 M/P	2 P/P	3 M	5 P	3	1
----	2 M/P	2 P/P	2 M	6 P	11	4
1 M/M	----	3 P/P	2 M	6 P	1	0.5
----	1 M/P	3 P/P	1 M	7 P	5	2

By application of the Visconti-Delbrück[18] theory of phage genetics, the relative input ratio of maternal and paternal chloroplast genomes in individual biparental zygotes should influence the frequency of recombination of chloroplast genes[1,12]. The same analysis may be applied to mitochondrial genes in yeast[19]. Data from reciprocal crosses involving three chloroplast markers with different phenotypes support this hypothesis (Table 2): Progeny of the most paternally and most maternally skewed biparental zygotes show substantially less recombination than progeny from biparental zygotes with more nearly equal allelic ratios. Since biparental zygotes with nearly equal maternal and paternal ratios of chloroplast genes are relatively infrequent, one cannot realistically use only progeny from these zygotes for mapping chloroplast genes. Instead our approach is to keep the extent of maternal bias constant by using standardized conditions for gametogenesis, UV irradiation and mating.

TABLE 2. Recombination analysis of the progeny of biparental zygote clones arising from reciprocal crosses. Data are pooled into 5 allelic ratio classes from a series of crosses in which the mt+ parent received from 0 to 60 s UV prior to mating.

(Cross: er-u-37 spr-u-1-6-2+ sr-u-2-60 X er-u-37+ spr-u-1-6-2 sr-u-2-60+)

PATERNAL ALLELIC RATIO OF BIPARENTAL ZYGOTE	TOTAL SAMPLED		% RECOMBINATION		
	BIPARENTAL ZYGOTES	PROGENY	er--sr	er--spr	spr--sr
0.01 to 0.20	249	15,791	2.9	2.2	1.4
0.21 to 0.40	79	4,961	9.8	7.5	4.4
0.41 to 0.60	60	3,820	7.1	5.6	3.5
0.61 to 0.80	41	2,586	6.6	6.3	2.0
0.81 to 0.99	47	2,975	4.1	3.8	1.6
overall	476	30,133	5.0	4.0	2.2

To demonstrate that the parental combination of chloroplast markers has no intrinsic effect on recombination of chloroplast genes, we have made all possible reciprocal crosses of stocks carrying markers for resistance to streptomycin, erythromycin, and spectinomycin (Table 3). From among the progeny of the initial pair of reciprocal crosses, we selected recombinant genotypes which served as parents for the five subsequent crosses. The seven crosses not only produced the same map order for the three chloroplast markers involved, but also yielded remarkably similar recombination frequencies for all three intervals.

TABLE 3. Effect of marker combination on recombination of the chloroplast genes er-u-37, spr-u-1-6-2, and sr-u-2-60 analyzed by the zygote clone method. In each cross, mt+ gametes were subjected to 15 s UV irradiation prior to mating. Approximately 3200 progeny from 50 biparental zygotes were analyzed in each cross.

CROSS				% RECOMBINATION		
	mt^+	X	mt^-	er--sr	er--spr	spr--sr
1.	er spr^+ sr	X	er^+ spr sr^+	4.3	4.0	0.6
2.	er^+ spr sr^+	X	er spr^+ sr	5.0	4.8	1.5
3.	er^+ spr^+ sr	X	er spr sr^+	5.2	4.2	1.5
4.	er spr sr^+	X	er^+ spr^+ sr	6.0	5.7	1.3
5.	er^+ spr sr	X	er spr^+ sr^+	4.1	3.8	2.2
6.	er spr^+ sr^+	X	er^+ spr sr	4.8	3.9	2.7
7.	er^+ spr^+ sr^+	X	er spr sr	3.7	2.8	2.3
	overall			4.7	4.2	1.7

Thus although the zygote clone analysis protocol is potentially subject to problems with differential growth of certain genotypes[1], such problems do not seriously affect our ability to map chloroplast genes. We feel that since a large number of progeny can be quickly and accurately scored by the zygote clone method, and since segregation is almost entirely complete at the time of analysis (less than 1% of total progeny heteroplasmic for any marker), this method is valid for chloroplast gene mapping in Chlamydomonas. In mapping by pedigree analysis[3,13,14] one faces the problem that many chloroplast genes are still heteroplasmic at the time the analysis is done.

To establish a map of the chloroplast genes for antibiotic resistance that affect chloroplast ribosomes, we first tested the known chloroplast markers with similar antibiotic resistance phenotypes for allelism. Somatic segregation of chloroplast genes in diploids as well as haploids prevents the use of the classical dominance/complementation test used in vegetative diploids to determine whether one or more gene functions are involved. Instead allelism must be judged entirely on the ability of two antibiotic resistant mutants with similar phenotypes to produce recombinant progeny which are antibiotic sensitive[6]. Table 4 illustrates two streptomycin resistant mutants which recombine at an appreciable frequency (1.6%) and are not alleles (cross 1), in contrast to two which show less than 0.03% recombination and appear to be alleles (cross 2).

TABLE 4. Results from typical crosses made to test chloroplast mutants with similar phenotypes for allelism. The mt^+ gametes received 15 s UV prior to mating.

	NUMBER OF PROGENY								% RECOMBINATION	
CROSS	er + sr	+ nr sr	er nr sr	+ + +	+ + sr	er + +	+ nr +	er nr +	sr-sr*	er-nr
1	3353	863	16	0	194	0	25	10	1.6	4.9
2	1576	4853	96	0	272	0	0	0	<0.029	5.4

Cross 1: sr-u-2-60 er-u-37 mt^+ X sr-u-sm3 nr-u-2-1 (spr-u-1-27-3) mt^-

Cross 2: sr-u-sm5 nr-u-2-1 mt^+ X sr-u-sm-3a er-u-1a mt^-

*(number of streptomycin sensitive progeny X 2 / total progeny) X 100

Among mutants from our laboratory and from Sager and Bogorad, we have now unequivocally demonstrated 4 discrete loci for streptomycin resistance and 2 for erythromycin resistance in the chloroplast genome (Fig. 2)[6]. At least 4 independently isolated mutant alleles have been identified at one streptomycin locus (Table 5). Similarly, all three chloroplast mutants resistant to spectinomycin and the mutant resistant to neamine appear to be alleles at a single locus[6,20].

Eight different mapping crosses were analyzed by the zygote clone method to establish the order and relative position of these seven loci on the chloroplast genome. The frequencies of parental and recombinant progeny appear in Table 6. In 7 of the 8 crosses, a bias exists favoring the maternal genotype, and in some instances reciprocal recombinant types occur in vastly unequal frequencies. The genotypes er spr nr sr and spr nr sr appear to be at a selective disadvantage whether they are recombinant (Table 6) or parental (Table 7). Nonetheless these

FIGURE 2. Tentative map of chloroplast ribosome genes in Chlamydomonas reinhardtii based on zygote clone analysis of allele test and mapping crosses. Parentheses indicate preliminary data from a small number of progeny.

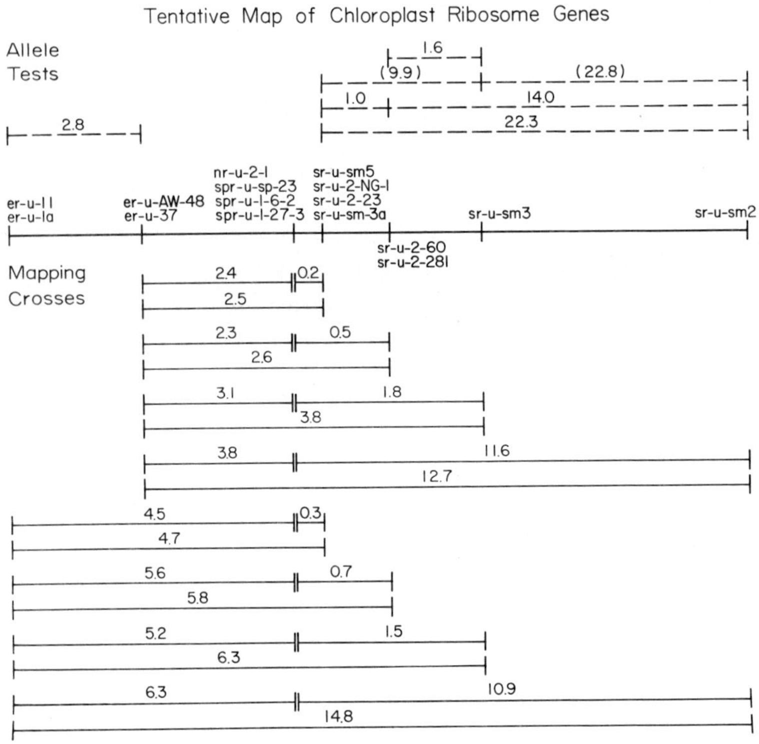

TABLE 5. Recombination analysis of chloroplast gene mutants conferring antibiotic resistance on chloroplast ribosomes: Allelism at one streptomycin locus. Recombination of each mutant with sr-u-2-60 is shown for comparison.

NUMBER OF STREPTOMYCIN-SENSITIVE RECOMBINANTS/ TOTAL PROGENY

	sr-u-sm-3a	sr-u-2-23	sr-u-2-NG-1	sr-u-sm5	sr-u-2-60
sr-u-sm-3a	0/2884	0/3175	0/3163	0/6158	16/4252
sr-u-2-23		1*/2928	0/3127	0/1280	12/3129
sr-u-2-NG-1			0/3175	0/640	30/2805
sr-u-sm5				--	18/6400
sr-u-2-60					0/3182

*possibly either a wild type contaminant or a back-mutation

TABLE 6. Results of mapping crosses between mutants at 7 different chloroplast gene loci, using the zygote clone method. Possible recombinant classes where no progeny were observed are not listed. spr = spr-u-1-27-3 and nr = nr-u-2-1.

CROSS		NUMBER OF PROGENY								
		PARENTAL		RECOMBINANT						
mt$^+$(15 s UV) X mt$^-$		+	er	er	+	er	+	+	er	+
		+	nr	+	nr	+	nr	+	nr	+
		+	spr	+	spr	+	spr	+	spr	spr
		sr	+	sr	+	+	sr	+	sr	+
sr-u-sm-3a	X er-u-37 nr spr	3921	2878	116	155	1	2	0	0	0
sr-u-2-60	X er-u-37 nr spr	5439	755	64	76	9	0	21	2	0
sr-u-sm3	X er-u-37 nr spr	2891	2904	40	115	24	9	62	12	0
sr-u-sm2	X er-u-37 nr spr	3257	2208	72	84	85	1	649	1	0
sr-u-sm-3a	X er-u-1a nr spr	4063	2017	88	197	2	0	17	0	0
sr-u-2-60	X er-u-1a nr spr	4056	1505	157	158	14	3	26	1	1
sr-u-sm3	X er-u-1a nr spr	3707	2128	36	275	7	3	74	7	0
sr-u-sm2	X er-u-1a nr spr	2089	1058	82	108	46	0	360	0	1

crosses have yielded a highly consistent map order and remarkably similar map distances for the 7 loci involved (Fig. 2). Comparison of our map with the linear map of Sager and Ramanis[13], which is based on recombination frequencies observed in pedigrees, gives the identical gene order for four markers which these maps have in common, i.e. er-u-11/er-u-1a (ery), nr-u-2-1 (nea), sr-u-sm3 (sm3) and sr-u-sm2 (sm2). However, this order does not agree well with the circular map subsequently proposed by Sager on the basis of data derived primarily from co-segregation analysis and marker segregation in liquid culture[3,14,15,21], unless the circle is broken between ery and sm2. We are presently attempting to resolve these discrepancies by additional allele test and mapping crosses with the mutants which are at variance. The linkage and arrangement of the chloroplast genes which are known to affect chloroplast ribosomes in Chlamydomonas is also of interest in view of the clustering of ribosomal genes observed in bacteria[22].

In the case of yeast mitochondrial genetics most segregation and recombination analysis has been carried out on vegetative diploid progeny from mitotic zygotes[23]. To allow more direct comparison between chloroplast genetics of Chlamydomonas and mitochondrial genetics of yeast, we have also studied the transmission, segregation, and recombination of chloroplast genes in vegetative diploids (mitotic zygotes) of Chlamydomonas[24]. Gillham[16,25] found that vegetative diploids isolated using the method of Ebersold[26] transmit chloroplast genes from both parents at a higher frequency than is normally observed for meiotic zygotes. Van Winkle-Swift has shown that this high biparental transmission of chloroplast genes depends upon the newly formed mitotic zygote dividing somatically soon after gamete fusion. Isolation of mitotic zygotes under conditions which delay the first somatic division following mating, e.g. incubation in darkness (Fig. 3), leads to a pronounced drop in the frequency of biparental diploids, and in the most extreme instances results in diploid populations exhibiting predominantly maternal inheritance[24,27]. Delayed division apparently allows the chloroplast genomes of the $\underline{mt}^-$ parent to be lost systematically, or at least not transmitted to the diploid progeny. While routine procedures for selection of mitotic zygotes promote the rapid onset of somatic division, meiotic zygotes are not usually induced to divide until 6-7 days after mating. We are presently investigating whether premature germination of meiotic zygotes increases the transmission of chloroplast genes from the paternal parent.

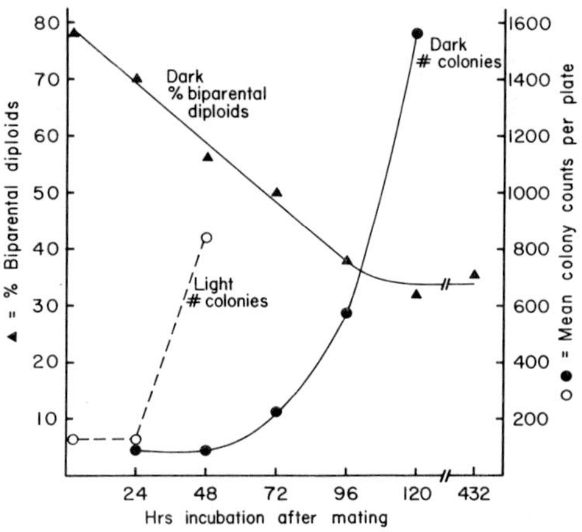

FIGURE 3. Decline in frequency of biparental diploids (mitotic zygotes carrying chloroplast gene markers from both maternal and paternal parents) when first mitotic division is delayed by incubation in darkness. Respreading experiments with plates incubated in light compared to plates incubated in dark indicate time of first division.

We have also compared recombination of four chloroplast gene markers in two crosses involving vegetative diploids with a cross involving meiotic zygotes (Table 7). The gene order observed in all three crosses is the same as found previously (Table 4, Fig. 2). The somewhat higher recombination frequencies observed for all intervals in these crosses compared to those in Table 2 may be due to the additional presence of the nr-u-2-1 marker in the quadruple antibiotic resistant stocks used as one parent in each of these crosses. As discussed above, stocks carrying nr in combination with sr and spr grow slowly, possibly because all three mutations alter the same chloroplast ribosomal subunit and in combination impair its function in the absence of antibiotics. This preferentially lowers the frequency of one parental class (Table 7), so that the overall frequency of recombinants appears higher.

TABLE 7. Comparison of recombination of the chloroplast genes sr-u-2-60, nr-u-2-1, spr-u-1-27-3 and er-u-37 in zygote clones and vegetative diploid clones. Cross 3 was not scored for spectinomycin resistance.

	NUMBER OF PROGENY												
	PARENTAL		RECOMBINANT										
											% RECOMBINATION		
CROSS	+ + + +	er nr spr sr	er + + +	+ nr spr sr	+ + + sr	er nr spr +	er + + sr	+ nr spr +	er + spr +	+ + spr sr	er-sr	er-nr	nr-sr
1	5,406	241	208	21	252	103	89	77	1	0	9.1	6.2	8.2
2	5,930	6	97	5	249	33	46	33	0	1	6.0	2.8	5.6
3	10,467	191	277	35	131	85	55	50	--	--	4.7	3.7	2.8

Cross 1: wild type mt^{+} (15 s UV) X er nr spr sr mt^{-} (zygote clones)

Cross 2: wild type arg-7 mt^{+} X er nr spr sr arg-2 mt^{-} (vegetative diploid clones)

Cross 3: er nr spr sr arg-2 mt^{+} X wild type arg-7 mt^{-} (vegetative diploid clones)

In conclusion, *Chlamydomonas* is at present the only photosynthetic organism in which one can map chloroplast genes by formal genetic methods. Although certain discrepancies in map order exist between Sager's laboratory and our own, we concur in the demonstration of a single chloroplast linkage group. The majority of the chloroplast mutations studied to date confer antibiotic resistance on chloroplast ribosomes and are thought to code for specific ribosomal proteins[5]. Even after the known chloroplast mutants and the cistrons coding for chloroplast ribosomal RNA are accounted for, there remains in the chloroplast genome sufficient information to code for a significant number of additional proteins. Isolation and characterization of a broad spectrum of acetate-requiring (non-photosynthetic) mutants in the chloroplast genome should enable us to determine which other photosynthetic components are coded by chloroplast genes and how these genes map with respect to the chloroplast ribosomal genes described here.

REFERENCES

1. Adams, G.M.W., K.P. Van Winkle-Swift, N.W. Gillham & J.E. Boynton, 1976, in *The Genetics of Algae*, R.A. Lewin, ed. Botanical Monographs Vol. 12, Blackwell Scientific, Oxford, pp. 69-118.
2. Gillham, N.W., 1974, *Ann. Rev. Genet.* *8*: 347-391.
3. Sager, R., 1972, *Cytoplasmic Genes and Organelles*, Academic Press, New York, 405 pp.
4. Sager, R., 1974, in *Algal Physiology and Biochemistry*, W.D.P. Stewart, ed., Botanical Monographs Vol. 10, U. California Press, Berkeley, pp. 314-345.
5. Gillham, N.W., J.E. Boynton, E.H. Harris, S.B. Fox, & P.L. Bolen, this volume.
6. Conde, M.F., J.E. Boynton, N.W. Gillham, E.H. Harris, C.L. Tingle & W.L. Wang, 1975, *Molec. gen. Genet.* *140*: 183-220.
7. Hastings, P.J., E.E. Levine, E. Cosbey, M.O. Hudock, N.W. Gillham, S. J. Surzycki, R. Loppes & R.P. Levine, 1965, *Microb. Genet. Bull.* *23*: 17-19.
8. Alexander, N.J., N.W. Gillham & J.E. Boynton, 1974, *Molec. gen. Genet.* 112: 225-228.

9. Wiseman, A., N.W. Gillham & J.E. Boynton, 1975, Genetics 80: s84-s85.

10. Sager, R., 1954, Proc. Nat. Acad. Sci. USA 40:356-362.

11. Sager, R. & Z. Ramanis, 1967, Proc. Nat. Acad. Sci. USA 58: 931-937.

12. Gillham, N.W., J.E. Boynton & R.W. Lee, 1974, Genetics 78: 439-457.

13. Sager, R. & Z. Ramanis, 1970, Proc. Nat. Acad. Sci. USA 65: 593-600.

14. Sager, R. & Z. Ramanis, 1976, Genetics 83: 303-321.

15. Sager, R. & Z. Ramanis, 1976, Genetics 83: 323-340.

16. Gillham, N.W., 1969, Amer. Naturalist 103: 355-388.

17. Coen, D., J. Deutsch, P. Netter, E. Petrochilo & P.P. Slonimski, 1970, Symp. Soc. Exp. Biol. 24: 449-496.

18. Visconti, N. & M. Delbrück, 1953, Genetics 38: 5-33.

19. Dujon, B., P.P. Slonimski & L. Weill, 1974, Genetics 78: 415-437.

20. Boynton, J.E., E.H. Harris, C.L. Tingle, S.B. Fox & N.W. Gillham, 1975, J. Cell Biol. 67: 41a.

21. Singer, B., R. Sager & Z. Ramanis, 1976, Genetics 83:341-354.

22. Jaskunas, S.R., M. Nomura & J. Davies, 1974, in Ribosomes, ed. M. Nomura, A. Tissieres & P. Lengyel, Cold Spring Harbor Laboratory, New York, pp. 333-368.

23. Birky, C.W., Jr., 1976, in Genetics and Biogenesis of Mitochondria and Chloroplasts, ed. C.W. Birky, Jr., P.S. Perlman & T.J. Byers, Ohio State University Press, Columbus, pp. 182-224.

24. Van Winkle-Swift, K.P., 1976, The transmission, segregation and recombination of chloroplast genes in diploid strains of *Chlamydomonas reinhardtii*, Ph.D. thesis, Duke University.

25. Gillham, N.W., 1963, Nature 200: 294.

26. Ebersold, W.T., 1963, Genetics 48: 888.

27. Van Winkle-Swift, K.P., 1976, Genetics, abstract in press.

ACKNOWLEDGEMENTS

We thank Sue Fox for her able assistance. This work was supported by grants NSF-GB22769 and NIH-GM19427, and by NIH RCDA awards GM70453 to J.E.B. and GM70437 to N.W.G.

Genetics and Biogenesis of Chloroplasts and Mitochondria, Th. Bücher et al. eds.

A UNIPARENTAL MUTANT OF CHLAMYDOMONAS REINHARDTII WITH A VARIANT THYLAKOID MEMBRANE POLYPEPTIDE

Nam-Hai Chua
The Rockefeller University
New York, N.Y.10021,U.S.A.

INTRODUCTION

Recent studies on the sites of translation of thylakoid membrane polypeptides in *C. reinhardtii* show that at least 9 polypeptides are made inside the chloroplast (Chua and Gillham, manuscript in preparation). If the sites of transcription are in the same compartment as the sites of translation, these polypeptides would be expected to be coded for by chloroplast DNA. One approach to defining the functions of chloroplast DNA in thylakoid membrane biogenesis is to isolate uniparental mutants which have altered thylakoid membrane phenotypes. In this paper, I describe preliminary characterization of a spontaneous thylakoid membrane mutant which makes a variant polypeptide slightly larger than its wild type counterpart. Genetic analysis shows that the mutation is inherited in a uniparental manner and therefore presumably localized in chloroplast DNA.

MATERIALS AND METHODS

The wild type strain (137c) and two arginine auxotrophs, arg 2 and arg 7, were cultured under standard conditions (1, 2). Crosses, maturation of zygotes, tetrad analysis, and construction of diploids were performed according to established techniques (3-5). Thylakoid membranes were purified and membrane polypeptides analysed as described previously (1, 2).

RESULTS

In the course of verifying the thylakoid membrane polypeptide profiles of our arg 2 and arg 7 stocks used for the construction of diploids it was found that our mt^- arg 7 strain synthesizes a variant polypeptide in place of polypeptide 5 (apparent molecular weight = 50,000). The variant, designated polypeptide 5', is larger than the wild type counterpart by about 1,000 daltons. Since the mt^+ arg 7 strain is identical to wild type in its membrane polypeptide profile the mt^- arg 7 strain must contain, in addition to arg 7, a mutation which results in the synthesis of polypeptide 5' instead of polypeptide 5. This novel mutation which arose spontaneously is designated thm-u-1 (thm stands for thylakoid membrane and u for

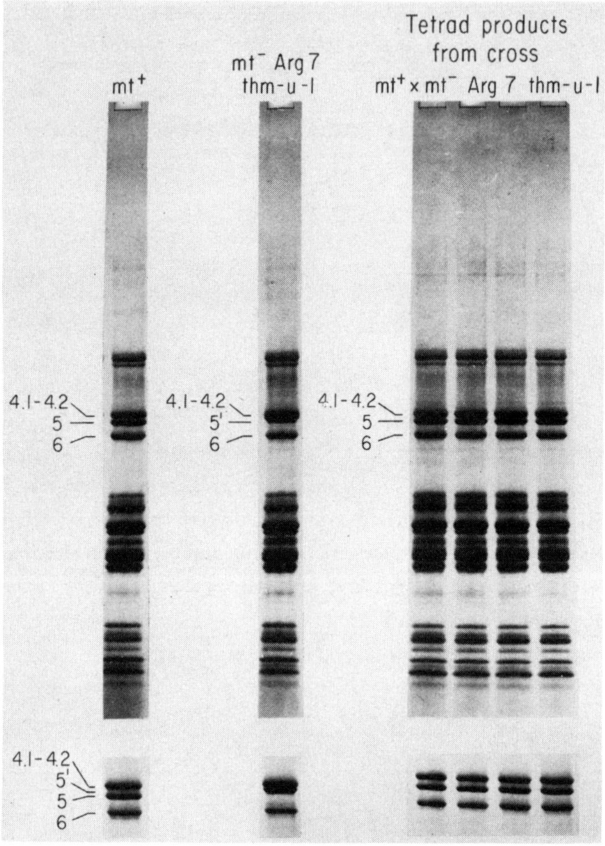

Fig. 1 Thylakoid membrane phenotypes of tetrad products from the cross mt^+ x mt^- arg 7 thm-u-1. In the lower panel, the gel was overrun to increase the resolution of polypeptides 5 and 5'. Heated, SDS-solubilized membrane sample containing 20 µg chlorophyll was loaded in each slot. For sample preparation and conditions of electrophoresis, see (1) and (2).

---

uniparental).

The new mutant mt^- arg 7 thm-u-1 was crossed to mt^+ and 10 complete tetrads were isolated. Since the thm-u-1 mutation does not result in an impairment of photosynthetic capacity, all tetrads were scored by their thylakoid membrane phenotypes. Fig. 1 shows that no thm-u-1 genes were transmitted to the progeny in the cross mt^+ x mt^- arg 7 thm-u-1 suggesting a uniparental mode of inheritance. In the same cross the nuclear markers, arg 7 and mt, segregated 2:2.

A mt^+ strain carrying both arg 7 and thm-u-1 was recovered from a biparental zygote obtained by UV irradiation of the mt^+ parent prior to mating (6). Thylakoid membrane phenotypes of the meiotic progeny derived from the biparental zygote are shown in Fig. 2. Fig. 3 shows the typical membrane polypeptide phenotypes of tetrads from the reciprocal cross mt^+ arg 7 thm-u-1 x mt^-. Among 16 complete tetrads analyzed thm-u-1 was found to segregate 4:0 whereas arg 7 and mt segregated 2:2, as expected of nuclear genes. These experiments confirm that thm-u-1 behaves as a uniparental gene.

So far, the results show that the thm-u-1 mutation resembles a large number of antibiotic-resistant mutations investigated by Sager (7) and Gillham (8, 9) in that it is inherited in a uniparental manner. Since mitochondrial mutations in C. reinhardtii are inherited in a non-Mendelian but biparental manner (cf. 9), thm-u-1 is probably located in chloroplast DNA. To compare further thm-u-1 and other uniparental, antibiotic-resistant mutations the transmittance of thm-u-1 genes in heterozygous zoospores in biparental diploids was investigated. Fig. 2 shows that in the heterozygous haploid uniparental genes segregate during mitosis giving rise to clones which make either polypeptide 5 or polypeptide 5'. Gillham (8, 9) has reported that a large proportion of diploids carry uniparental genes from both parents. During mitotic division of the biparental diploids segregation of the gene pair occurs. Gillham's observations with sr-2, a uniparental mutation which confers resistance to streptomycin, are essentially confirmed by similar experiments with thm-u-1. Among 75 diploids arising from the reciprocal crosses, mt^+ arg 7 thm-u-1 x mt^- arg 2 and mt^- arg 7 thm-u-1 x mt^+ arg 2, approximately 60% are biparentals, 38% are maternals and the remainder are paternals. As expected of uniparental genes, the biparental diploids segregate thm-u-1 and its allele during mitotic division (Fig. 4). However, a small percentage of the biparental diploids are persistent in that the segregation of the gene pair occurs slowly (diploid #20; Fig. 4).

The mutant thm-u-1 does not require acetate for growth and its growth rate at 25° is similar to that of wild type. Thylakoid membranes prepared from thm-u-1 are able to carry out all photochemical reactions at rates comparable to those of wild type. Thus, the replacement of polypeptide 5 by polypeptide 5' does not have any apparent effect on the photosynthetic capacities of the membranes. Inhibitor studies show that polypeptide 5', like its counterpart in

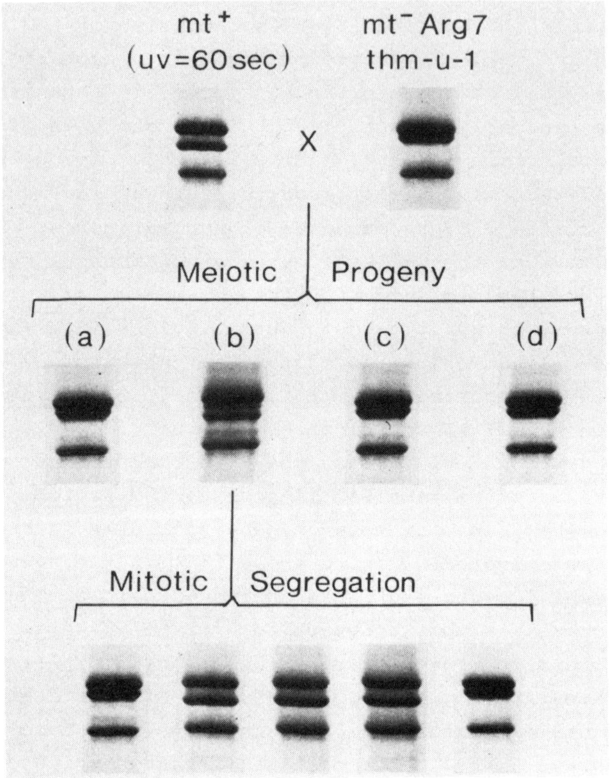

Fig. 2. Effect of UV irradiation on the transmittance of thm-u-1 and mitotic segregation of heterozygous haploid. mt^+ gametes were irradiated with UV (3×10^3 ergs/cm^2-sec) for 60 sec prior to mating. Among the four meiotic progeny three have the thm-u-1 gene (a, c, and d) whereas the remainder (b) has both thm-u-1 and its allele. During mitosis the heterozygous zoospore (b) gives rise to haploid clones synthesizing either polypeptide 5 or polypeptide 5'. Only the gel region encompassing polypeptides 4.1 and 4.2, 5, and 6 is shown. Heated, SDS-solubilized membrane sample containing 20 µg chlorophyll was used for each slot.

---

wild type, is made inside the chloroplast.

DISCUSSION

This paper presents preliminary characterization of a mutant, thm-u-1, which has a novel membrane polypeptide phenotype. Three lines of evidence show that thm-u-1 is a uniparental gene mutation:

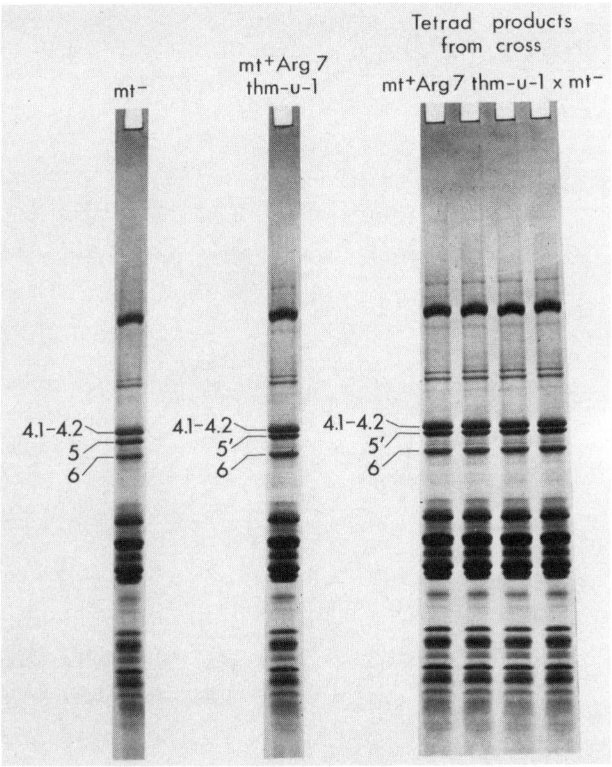

Fig. 3. Thylakoid membrane phenotypes of tetrad products from cross mt^+ arg 7 thm-u-1 x mt^-. Non-heated, SDS-solubilized membrane sample containing 20 µg chlorophyll was used in each slot.

---

(1) Thm-u-1 is transmitted to all meiotic progeny when carried by the maternal parent but not transmitted at all when carried by the paternal parent; (2) In heterozygous haploids derived from biparental zygotes, thm-u-1 and its allele segregate out during mitosis; and (3) Mitotic segregation of thm-u-1 and its allele also occurs in biparental diploids. Thm-u-1 is the first uniparental gene known to affect thylakoid membrane polypeptide in C. reinhardtii. The mutation results in the synthesis of a variant thylakoid membrane polypeptide which is ~1,000 daltons larger than its wild type counterpart. The chain extension, either at the amino or carboxy

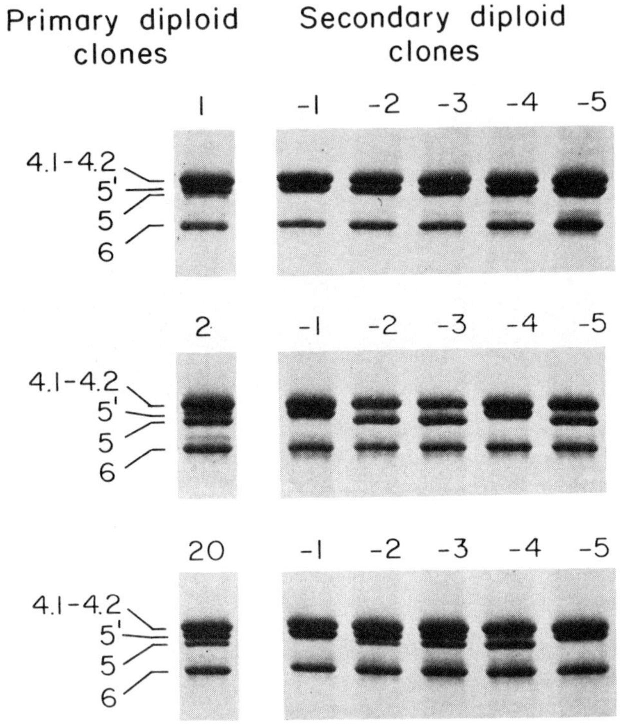

Fig. 4. Mitotic segregation of thm-u-1 and its alleles in biparental diploids. Biparental diploids (#1, #2, #20) were obtained from cross mt$^-$ arg 7 thm-u x mt$^+$ arg 2 and cloned on agar plates. Five secondary diploid clones were picked from each and their thylakoid membranes were analyzed together with those of the primary diploids. Only the gel region between polypeptide 4.1 to polypeptide 6 is shown. Load was equivalent to 20 µg chlorophyll.

---

terminus, does not prevent the assembly of the variant polypeptide into the membranes and also does not seem to impair the photosynthetic capacities of the thylakoids. In *Antirrhinum majus* and *Pelargonium zonale*, plastome mutants which are defective in photosynthetic electron transport and have altered membrane polypeptide profiles have been described previously (10, 11). The plastome mutations, like the uniparental mutations in *C. reinhardtii*, are presumably localized in chloroplast DNA. Work is in progress to determine whether thm-u-1 is a structural or regulatory gene.

Acknowledgement

I thank Drs. D. Luck, N. Gillham, M. Adams, and G. Schmidt for fruitful discussions. This work was supported in part by NIH Grant GM-21060.

REFERENCES

1. Chua, N.-H., and Bennoun, P. (1975) Proc. Nat. Acad. Sci. U.S.A. 72, 2175.

2. Chua, N.-H., et al. (1975) J. Cell Biol. 67, 361.

3. Levine, R.P. and Ebersold, W.T. (1960) Ann. Rev. Microbiol. 14, 197.

4. Ebersold, W.T. (1967) Science (Wash., D.C.) 157, 447.

5. Gillham, N.W. (1963) Nature (Lond.) 200, 294.

6. Sager, R., and Ramanis, Z. (1967) Proc. Nat. Acad. Sci. U.S.A. 58, 931.

7. Sager, R. (1972) Cytoplasmic genes and organelles. Academic Press, N.Y.

8. Gillham, N.W. (1969) Am. Naturalist 103, 355.

9. Gillham, N.W. (1974) Ann. Rev. Genet. 8, 347.

10. Hermann, F. (1971) FEBS Letters 19, 267.

11. Hermann, F. (1972) Exp. Cell Res. 70, 452.

Genetics and Biogenesis of Chloroplasts and Mitochondria, Th. Bücher et al. eds.
© 1976 Elsevier/North-Holland Biomedical Press, Amsterdam, The Netherlands.

PLASTID DISTRIBUTION AND PLASTID COMPETITION IN HIGHER PLANTS AND THE INDUCTION OF PLASTOM MUTATIONS BY NITROSO-UREA-COMPOUNDS

Rudolf Hagemann
Department of Genetics, Section Biosciences
Martin-Luther-University, Domplatz 1
DDR - 402 H a l l e / S., German Dem. Rep.

SUMMARY

1. Electron microscopical studies in angiosperms reveal that during the development of male and female gametophytes and gametes resp., an unequal plastid distribution takes place. During pollen formation the first pollen mitosis is unequal: The vegetative cell always contains plastids. By contrast, the generative cell of many species does not contain any plastids (Gossypium, Mirabilis), or, very infrequently, only single plastids (Lycopersicon, Beta; Antirrhinum). However, generative cells of Pelargonium and Oenothera do contain numerous plastids.

2. The genetic results (uniparental maternal or biparental plastid inheritance) are in full agreement with the cytological findings. There is no proof for the occurrence of plastid destruction by restriction mechanisms in zygotes as the basis of uniparental plastid inheritance.

3. In the period of fertilization and embryogenesis competition between genetically different types of plastids (e.g. green and white) takes place in Oenothera and Pelargonium. The pattern of sorting-out of genetically different plastids and the segregation in F_1 depend upon the input of maternal and paternal plastids into the zygote and the plastid competition during embryogenesis.

4. In Pelargonium, during embryogenesis green wild-type plastids were found to be superior when competing with mutant yellow and white plastids (e.g. green 'Trautlieb' > yellow 'Mrs.Pollock' > white 'Mrs.Parker' and 'Flower of Spring').

5. By contrast, in the course of the formation and growth of several foliage leaves the same mutant plastids of Pelargonium are no longer inferior at all. During this later developmental stage the mutant plastids of 'Mrs.Pollock' and 'Flower of Spring' and the cells carrying these plastids have the same (or even a slightly better) multiplication and growth rate resp. as the green plastids and cells. Thus, competition processes between genetically different types of plastids can be different in successive developmental stages.

6. Our experiments with Antirrhinum majus and, to a lesser extent, with Pelargonium zonale show that N-Nitroso-N-methyl-urea (NMU) and N-Nitroso-N-ethyl-urea (NEU) are potent chemical mutagens for the induction of plastom mutations. The proof that the induced variegation in M_1 plants is due to plastom mutations, is based on three criteria (a fine chequered variegation pattern, the demonstration of mixed cells and the non-Mendelian type of inheritance).

INTRODUCTION

The problems of the genetics and biogenesis of chloroplasts in higher plants have been studied since 1909. There are two modes of plastid transmission in angiosperms: In most species a uniparental maternal inheritance; in a minority of species a biparental inheritance[13,27,36,44]. This is due to characteristic cytological mechanisms.

RESULTS AND DISCUSSION

Unequal plastid distribution and plastid competition in angiosperms

Higher plants contain numerous plastids per cell. Thus, the plastid distribution during cell divisions and processes of competition between genetically different plastid types can be studied in different stages of plant ontogeny. These processes can best be understood if we consider separately three different ontogenetic stages: (1) development of gametophytes and gametes, (2) zygote formation and embryogenesis, and (3) the ontogenetic development of the plant after the seedling stage. - Electron microscopical studies have led to the discovery of unequal plastid distribution during meiosis and mitoses in both male and female gametophytes and gametes.

Unequal plastid distribution during the formation of the generative cell of the pollen of angiosperms

In all angiosperms the undivided microspores, formed by meiosis, and the vegetative cells of the pollen contain plastids, mitochondria, dictyosomes etc. Presence of plastids in these cells has convincingly been demonstrated by electron microscopy[10,23-25,30,31] and by anther culture[33]. The decisive event with regard to plastid distribution is the first pollen mitosis resulting in a vegetative cell (which later grows out into the pollen tube) and a generative cell. According to the plastid content of the generative cell the species of angiosperms fall into three different groups:

Group I: Very careful electron microscopical studies have revealed that many species do not contain any plastids in their generative cells (and thus also not in the sperm cells, formed by division of the generative cells), e.g. Gossypium[23,24], Capsella[39], Mirabilis [30,31] and several other species[25,31].

Group II: Very infrequently the generative cells of some species possess single plastids (whereas most cells do not), e.g. Beta[21] and Lycopersicon (S.Krahnert unpubl.). According to genetic[11] and electron microscopical studies (R.Knoth: Plate I), Antirrhinum majus belongs to this group; as an exception transmission of paternal plastids occurs in 0.03% of hybrid seedlings.

Apparently there is no difference in principle between groups I and II. The mechanism that excludes plastids from the generative cell during first pollen mitosis does not work as efficiently in some species (group II) as in others (group I).

In striking contrast to groups I and II is the behaviour of Group III: In some species the generative cells regularly contain numerous plastids. Many plastids have been found in the generative cells of Pelargonium zonale[30,31] (and R.Knoth: Plate I), Oenothera hookeri[10], Castilleia wightii[28] and Lobelia exinus[9].

The genetic results and the electron microscopical data are in full agreement. The uniparental, purely maternal inheritance of plastid differences in many angiosperm species is due to the exclusion of the paternal plastids from the generative cells (and consequently from the sperm cells) by unequal plastid distribution during first pollen mitosis and thus already before fertilization. This exclusion is usually perfect (Gossypium, Mirabilis); in single cases there are exceptions which cause the occurrence of trace paternal plastids in some crosses (Antirrhinum). So far, there is no indication and no proof for the occurrence of destruction of paternal plastids by restriction mechanisms in zygotes as the basis of uniparental plastid inheritance.

Only in a relatively small group of species (group III), including the genetically well-studied genera Pelargonium and Oenothera, the generative cells and sperm cells contain numerous plastids, and there is a regular transmission of paternal plastids into the zygote; the well-known biparental type of plastid inheritance in these species is the direct consequence of the cytological facts.

It should be emphasized that unequal plastid distribution was also found during female gametophyte development. In Epilobium palustre[3] and in some orchids[12] the plastids are unequally distributed during meiotic divisions: Only the megaspore that develops into the

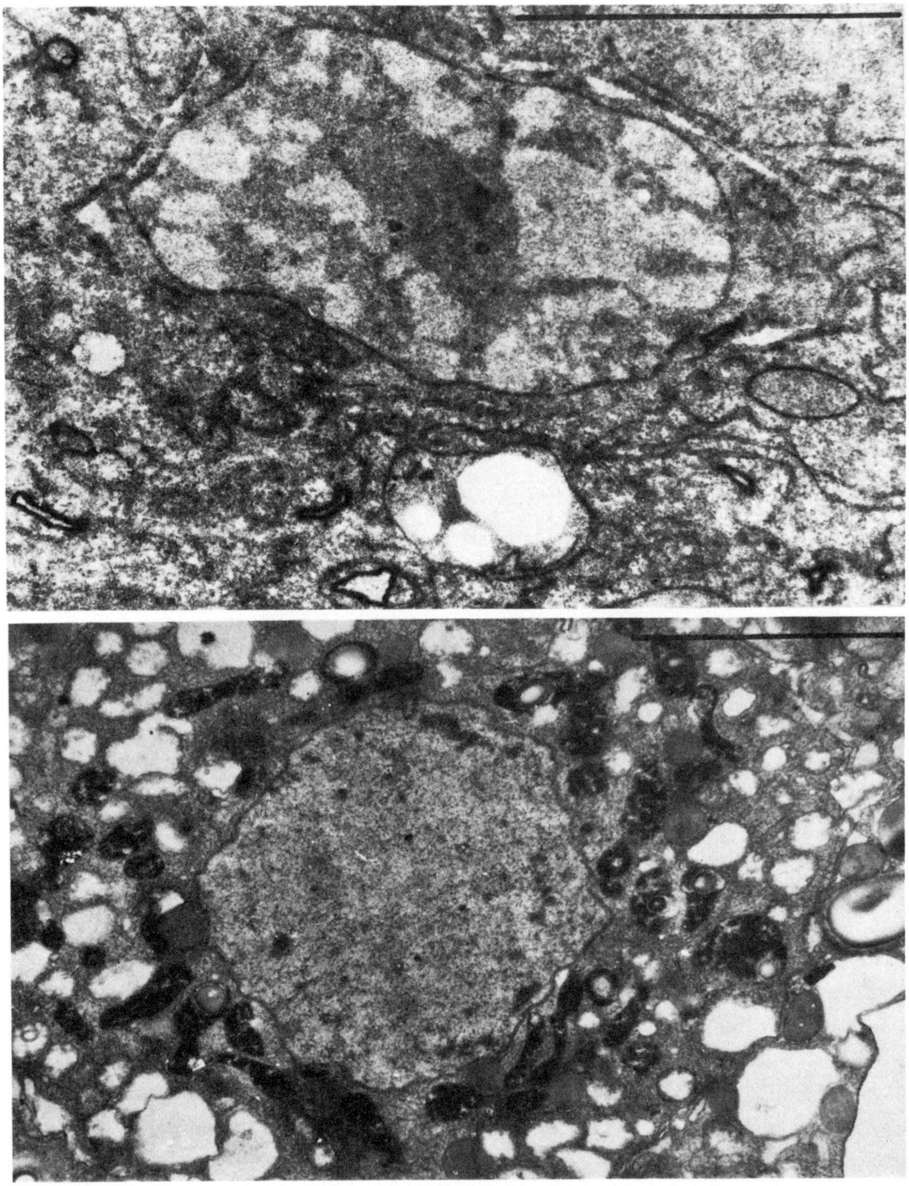

Plate I. Electron micrographs of pollen cells. In Antirrhinum majus (1) the generative cell (GC) contains some mitochondria, but no plastids; in the vegetative cell there are many plastids with starch grains. In Pelargonium zonale (2) the generative cell contains many plastids, frequently without, but sometimes with small starch grains; in the vegetative cell there are many plastids with big starch grains. The bar indicates 5 µm in (1) and (2).

embryo sac acquires a high number of plastids, whereas the other three cells, that degenerate later, contain almost no plastids.

Plastid competition during embryogenesis

In the genera Oenothera and Pelargonium interesting results concerning the distribution and competition of genetically different plastids have been elaborated after crossing plants with green and yellow or white plastids.

In Oenothera the pattern of distribution and sorting-out of genetically different plastids and the frequency of green, variegated and white seedlings in F_1 depend upon (a) input of maternal and paternal plastids into the zygote[32] and (b) plastid competition during embryogenesis; the different multiplication rates of plastids of different species are mainly determined by the plastom, but there is also a minor influence of the genomes[37, 38,40,41] (reviews[13,14,27,36,44]). It should, however, be mentioned that differences in the 'success in competition' or 'multiplication rate' do not necessarily imply a higher speed of multiplication of a particular plastid type compared with another; they can likewise be due to different times of onset of multiplication of genetically different plastid types during interphase.

In our experiments with Pelargonium[4,13,17,19] during embryogenesis the green wild-type plastids have been superior when competing with mutant yellow and white plastids; the success in plastid competition was: green 'Trautlieb'$>$ yellow 'Mrs.Pollock'$>$ white 'Mrs. Parker' and 'Flower of Spring'[17] (compare [42-44]).

Competition during the formation and growth of foliage leaves

In the foliage leaves a completely different situation was found with regard to competition of different plastids and cells containing them. During this later developmental stage the mutant plastids of 'Mrs.Pollock' and 'Flower of Spring' and the cells carrying these plastids have the same (or even a slightly higher) multiplication and growth rate resp. as the green plastids and cells[17]. Thus, competition processes between genetically different types of plastids can be different in successive developmental stages.

Experimental induction of plastom mutations in higher plants by N-Nitroso-N-methyl-urea (NMU) and N-Nitroso-N-ethyl-urea (NEU)

An important presupposition for genetic, electron microscopical and biochemical studies on plastid (= plastom) mutants of higher plants, as e.g. done in our Department of Genetics in Halle[4-6,13-17,19,20,28], is a sufficient number of different plastom mutants. In this respect, however, geneticists working with higher plants have had problems in the past. (1) For a long time spontaneous plastom mutants have been the only source in many species. The frequency of variegated plants, whose variegation is due to spontaneous plastom mutations, is for several angiosperms within the range of 0.02 and 0.3%[13,29]. We studied such spontaneous mutants in Antirrhinum and Pelargonium. (2) A second way to obtain plastom mutants is by nuclear gene induction. In several genera, e.g. Hordeum, Zea, Arabidopsis, Epilobium and Oenothera, nuclear mutant genes are known which induce plastom mutations[13,27,44]. In the paper of BÖRNER et al. in this Symposium the work of our research group with the barley mutant albostrians is reported. (3) The third way of obtaining plastom mutants in higher plants is the experimental induction by mutagenic physical and chemical agents. However, for a long time many attempts failed or gave unsatisfactory results. But now there is a number of reports about the induction of plastom mutations by N-Nitroso-N-methyl-urea (NMU) in species which so far have played no role in plastid genetics: the sunflower Helianthus annuus[2,3,29] and the ornamental plant Saintpaulia ionontha[18,26,34,35].

In our mutation experiments we treated seeds of Antirrhinum majus, and seeds and shoot cuttings of Pelargonium zonale with NMU and NEU. Our research group, including my coworkers M.Scholze and F.Lieberwirth, was able to induce many different plastom mutants in both objects. NMU and NEU are very efficient carcinogens[12,22] and widely used for cancer induction. Obviously NMU and NEU induce many nuclear gene and chromosome mutations[12,22], too. The recessive gene mutations are detected in M_2. We focussed our attention on variegated M_1 plants. Part of them are certainly dominant nuclear mutations or chromosome aberrations, but part of them are plastom mutations. The percentage of variegated M_1 plants is surprisingly high after NMU and NEU treatment in A.majus (Table 1). We found that about a twofold dose of NEU was as efficient as NMU. In mutation experiments with durum wheat NMU was found to be more toxic than NEU, whereas the mutagenic efficiency of NMU was much higher[7]. The plastom mutations among the M_1 plants with chlorophyll deficiencies were found by using 3 criteria:

(1) the fine chequered variegation pattern, which is typical for sorting-out of genetically different types of plastids
(2) the cytological demonstration (by light and electron microscopy) in these plants of mixed cells containing in the same cell

Table 1

Induction of variegated plants of Antirrhinum majus by treatment with N-Nitroso-N-methyl-urea (NMU) and N-Nitroso-N-ethyl-urea (NEU)

Mutagen treatment	Concentration (mM)	Total number of seedlings	Number of variegated seedlings	Percentage of variegated seedlings
	4	150	4	2.7
Presoaking	5	112	4	3.6
6 hours	6	128	3	2.3
	7	149	8	5.4
2 hours	8	155	8	5.2
NMU	9	119	10	8.4
	10	96	21	21.9
	11	28	14	50.0
Experiment	12	39	23	59.0
of	13	27	13	48.1
M.SCHOLZE	14	24	18	75.0
	4	174	1	0.57
Presoaking	5	103	0	0
6 hours	6	134	7	5.2
	7	138	10	7.2
2,5 hours	8	81	7	8.6
NMU	9	50	11	22.0
	10	120	24	20.0
	11	108	34	31.5
	12	86	31	36.0
	13	82	24	29.3
	14	91	17	18.7
Experiment	15	68	28	41.2
of F.	16	42	7	16.7
LIEBERWIRTH	17	18	17	94.4
	4	153	1	0.65
Presoaking	7	190	2	1.05
6 hours	10	127	4	3.15
3 hours	13	156	14	9.0
NEU	16	99	10	10.1
	19	117	21	18.0
	22	109	21	19.3
Experiment	25	78	14	18.0
of F.	28	32	8	25.0
LIEBERWIRTH	31	24	10	41.7

two different types of plastids (normal green and white or yellow mutant plastids)[13,14] and

(3) the proof of the non-Mendelian, uniparental type of inheritance of these characters in Antirrhinum.

In our opinion, these results are very promising. We have now compounds at hand which enable us to induce experimentally plastom mutations in the plant species of our choice, and we are convinced that these Nitroso-urea-compounds will be applicable for a wide range of higher plants.

REFERENCES

1. Bednara,J. and Rodkiewicz,B. (1974) Bull.Acad.Polon.Sci. Cl.II, 22, 847-850
2. Beletski,Y.D. et al. (1969) Dokl.Akad.Nauk SSSR 186, 1425-1426
3. Beletski,Y.D. and Razoriteleva,E.K. (1972) Genetika (Moskva) 8, 17-22
4. Börner,T. et al. (1972) Theoret.Applied Genetics 42, 3-11
5. Börner,T. et al. (1973) FEBS Letters 37, 117-119
6. Börner,T. et al. (1975) Biochem.Physiol.Pfl. 168, 185-193
7. Desai,R.M. and Bhatia,C.R. (1975) Mutation Res. 27, 119-121
8. Corti,E.F. and Cecchi,A.F. (1970) Caryologia 23, 715
9. Dexheimer,J. (1965) Compt.Rend.Acad.Sci. (Paris) 260,6963-6965
10. Diers,L. (1963) Z.f.Naturforsch. 18b, 562-566
11. Diers,L. (1971) Molec.Gen.Genetics 113, 150-153
12. Fishbein,L. et al. (1970) Chemical Mutagens, Environmental effects on biological systems, Academic Press, New York, London
13. Hagemann,R. (1964) Plasmatische Vererbung. VEB G.Fischer, Jena
14. Hagemann,R. (1965) in Genetics Today, Proc.XI. Internat.Congr. Genetics, Vol.3, pp.613-625, Pergamon Press, Oxford, New York
15. Hagemann,R..et al. (1975) in Genetic Aspects of Photosynthesis (Nasyrov,Y.E. and Sestak,Z., eds.) pp.115-118, Junk, The Hague
16. Hagemann,R..and Scholz,F. (1962) Züchter 32, 50-59
17. Hagemann,R. and Scholze,M. (1974) Biol.Zentralbl. 93, 625-648
18. Hentrich,W. and Beger,B. (1974) Arch.Züchtungsf..4, 29-43
19. Herrmann,F. and Hagemann,R. (1971) Biochem.Physiol.Pfl. 162, 39o-409
20. Herrmann,F. et al. (1974) Biochem.Physiol.Pfl. 165, 393-400
21. Hoefert,L.L. (1969) Protoplasma 68, 237-240
22. Hollaender,A. (ed.) (1971,1973) Chemical Mutagens, Principles and methods for their detection, Vol. 1,2,3, Plenum Press, New York, London
23. Jensen,W.A. (1972) The embryo sac and fertilization in angiosperms, Univ.Hawaii, H.L.Lyon Arboret.Lecture No.3, 1-32
24. Jensen,W.A. and Fisher,D.B. (1968) Protoplasma 65, 277-286
25. Jensen,W.A. et al. (1974) Botan.Gaz. 135, 210-218
26. Jungnickel,F. (1977) Biol.Zentralbl. 96 (in press)
27. Kirk,J.T.O. and Tilney-Bassett,R.A.E. (1967) The Plastids, W.H.Freeman & Comp., London and San Francisco
28. Knoth,R. (1975) Biol.Zentralbl. 94, 681-694
29. Kutzelnigg,H. and Stubbe,W. (1974) Sub-Cell.Biochem..3, 73-89
30. Lombardo,G..and Gerola,F.M. (1968) Planta 82, 105-110
31. Lombardo,G. and Gerola,F.M. (1968) Giorn.Bot.Ital. 102,353-380
32. Meyer,B. and Stubbe,W..(1974) Ber.Dtsch.Bot.Ges. 87, 29-38
33. Nilsson-Tillgren,T. and v.Wettstein-Knowles,P. (1970) Nature 227, 1265-1266
34. Pohlheim,F. (1974) Biol.Zentralbl. 93, 141-148
35. Pohlheim,F. and Beger,B. (1974) Biol.Rundschau 12, 204-206
36. Sager,R. (1972) Cytoplasmic Genes and Organelles, Academic Press, New York and London
37. Schötz,F. (1954) Planta 43, 182-240
38. Schötz,F. (1974) Biol.Zentralbl. 93, 41-64
39. Schulz.R. and Jensen,W.A. (1968) Am.J.Bot.55,541-552,807-819
40. Stubbe,W. (1960) Z.f.Bot. 48, 191-218
41. Stubbe,W. (1963) Ber.Dtsch.Bot.Ges. 76, 154-167
42. Tilney-Bassett,R.A.E. (1970) Genet.Res. (Camb.) 16, 49-61
43. Tilney-Bassett,R.A.E. (1974) Heredity 33, 353-360
44. Tilney-Bassett,R.A.E. (1975) in Genetics and Biogenesis of Mitochondria and Chloroplasts (Birky,C.W. et al., eds.) pp. 268-308, Ohio State Univ. Press, Columbus

Genetics and Biogenesis of Chloroplasts and Mitochondria, Th. Bücher et al. eds.

STRUCTURAL AND FUNCTIONAL ANALYSIS OF EUGLENA GRACILIS CHLOROPLAST DNA

E. Stutz, E.J. Crouse, L. Graf, B. Jenni and H. Kopecka*)
Laboratoire de Biochimie, Université de Neuchâtel
CH-2000 Neuchâtel

*) Laboratoire de Génétique Moléculaire,
Institut de Biologie Moléculaire,
Paris 5e, France

The unicellular alga Euglena gracilis contains between 1.2 to 12 x 10^{-15} g of double stranded DNA per chloroplast (1,2). This DNA occurs in circular form (3) having a circumference of 44 µm equivalent to a molecular weight of 9.2 x 10^7. The overall G+C content is very low (25%), and several major segment classes having a G+C content in the range of 25%, 33% and 41%, respectively, have been postulated from melting studies and from CsCl equilibrium density experiments (4,5). About two thirds of the chloroplast genome are composed of segments having a low G+C content (22-26%) while one third is composed of segments with a G+C content in the range of 31-41% (6). However, these reports were based on analyses of randomly degraded DNA and therefore represent approximate values.

SOME CHARACTERISTICS OF Hae III-CHLOROPLAST DNA FRAGMENTS

A more precise analysis of the compositional heterogeneity of chloroplast DNA became possible with the introduction of site-specific restriction endonucleases. In a recent study (Kopecka, Crouse and Stutz, submitted for publication), we have shown that Euglena chloroplast DNA yields numerous fragments using Hind II, Hind III, Hpa I, Hpa II, Hae III and Eco RI (enzyme nomenclature according to Smith and Nathans, ref. 7). Sma I enzyme did not measurably degrade the chloroplast DNA sample which had an average starting molecular weight of 15-20 x 10^6. Total chloroplast DNA, restricted by Hae III enzyme, yielded 36 bands (Table 1) which were resolved under a variety of gel conditions. Some of these bands were considered to be composed of 2 to 3 either very similar or identical DNA fragments, representing a total of 51 to 52 fragments. The total molecular weight amounted to 87 x 10^6 or 95% of the size of one 44 µm circular DNA (Table 2).

A similar study was made with sheared chloroplast DNA, pre-fractionated on a preparative CsCl density gradient and analyzed as shown in Fig. 1. Sample I contains mainly (ca. 60%) the 1.701 g/cc chloroplast DNA (41% G+C)

TABLE 1

MOLECULAR WEIGHTS OF Hae III RESTRICTION DNA FRAGMENTS FROM EUGLENA CHLOROPLAST DNA

Band number	MW of DNA $x\ 10^{-6}$	Fragment repetitions	MW of DNA per band $x\ 10^{-6}$
1	5.7	3	17.1
2	5.3	2	10.6
3	4.3	1	4.3
4	3.9	1	3.9
5	3.7	2	7.4
6	3.2	2	6.4
7	2.4	3	7.2
8	2.2	1	2.2
9	2.1	1	2.1
10	1.9	1	1.9
11	1.8	1-2	1.8-3.6
12	1.6	1	1.6
13	1.4	1	1.4
14	1.2	1	1.2
15	1.1	1	1.1
16	0.88	1	0.88
17	0.82	1	0.82
18	0.81	1	0.81
19	0.80	3	2.40
20	0.78	1	0.78
21	0.73	2	1.46
22	0.72	1	0.72
23	0.69	1	0.69
24	0.68	1	0.68
25	0.64	2	1.28
26	0.59	1	0.59
27	0.51	1	0.51
28	0.50	1	0.50
29	0.38	1	0.38
30	0.36	1	0.36
31	0.29	3	0.87
32	0.28	1	0.28
33	0.20	2	0.40
34	0.19	2	0.38
35	0.16	1	0.16
36	0.15	1	0.15

while sample II contains almost exclusively the 1.685 g/cc chloroplast DNA (25% G+C). These two samples gave qualitatively and quantitatively different banding patterns with all enzymes tested. Schematic banding patterns for total chloroplast DNA, sample I and sample II, obtained with the Hae III enzyme, are shown in Fig. 2 (three left side columns). Bands marked with a square indicate fragments that hybridized to rRNA. Thus, some sequences within the circular chloroplast DNA are very similar or identical (repeated) and such sequences contain the rRNA genes among other information.

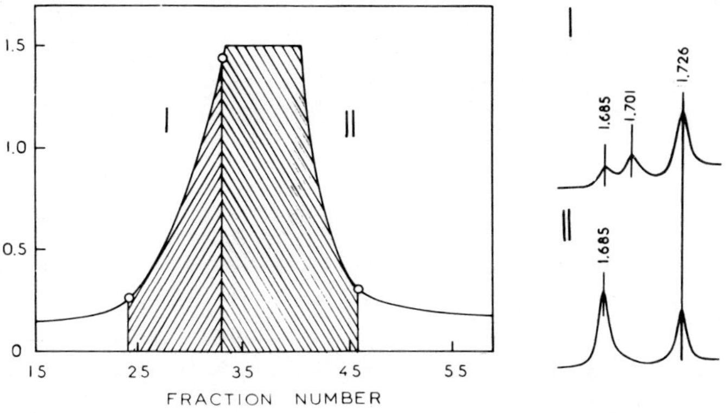

Fig. 1. CsCl density gradients of Euglena chloroplast DNA.
20 A_{260} units of chloroplast DNA were sheared to an average molecular weight of 4-6 x 10^6 and equilibrated in a 10 ml neutral CsCl density gradient at 33 krpm for 64 h, 20°C, in a Beckman type 40 rotor. Samples I and II were fractionated and combined as indicated. 1-2 µg of the samples I and II were centrifuged in a Beckman analytical centrifuge equipped with a scanner, at 44 krpm, for 20 h, 25°C. Marker: Pseudomonas aeruginosa DNA (1.726 g/cc).

TABLE 2

SOME PROPERTIES OF TOTAL CHLOROPLAST DNA
COMPARED TO THE G+C-RICH (I) AND THE A+T-RICH (II) SUBFRACTIONS

Sample	Hybridization with rRNA %	Number of fragments generated by Hae III digestion	Sum of MW of fragments x 10^{-6}
Total	5.1	51-52	87
I	9.1	38	36
II	2.1	38-40	56

SOME CHARACTERISTICS OF Eco RI-CHLOROPLAST DNA

A similar study was made using the Eco RI endonuclease to fragment total chloroplast DNA and samples I and II. In Table 3, the DNA fragments are ordered according to molecular weights which range from approximately 15 x 10^6 to 0.5 x 10^6. A similar result was recently reported by Millenz and Hershberger (personal communication). The bands E, K, N were estimated to be multiple ones. The sum of the molecular weights of all fragments was close to 9 x 10^7. For samples I and II, we obtained banding patterns in 1% agarose as schematically presented in Fig. 2 (right side patterns). Sample I yielded almost exclusively the multiple bands E, K, N which must originate from segments with an average buoyant density of 1.701 g/cc (see density profile in Fig. 1).

Eco RI DNA fragments were transferred to filter strips and hybridized against 16S+23S ^{125}I-rRNA. The extent of hybridization was monitored by autoradiography. In Fig. 3, the banding pattern is shown along with two autoradiograms and the Eco RI digested λ-phage DNA (marker). The DNA in the

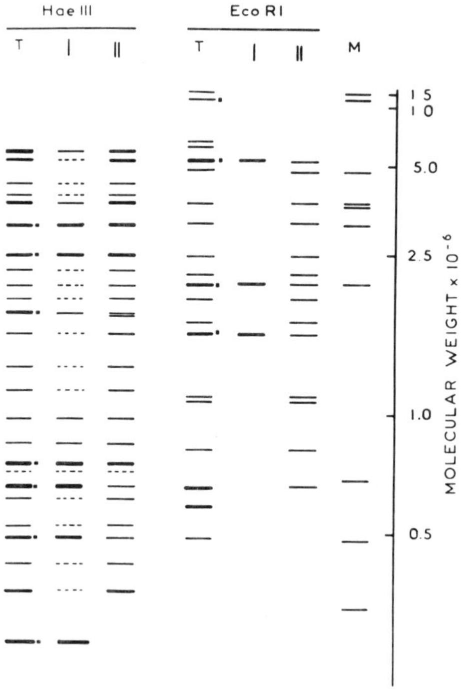

Fig. 2. Scheme of banding patterns of restricted chloroplast DNA analyzed on a 1% agarose gel. Total chloroplast DNA (T) and sample I (I) and sample II (II) were digested with the Hae III (left side column) or Eco RI enzyme (right side column) and the fragments were resolved in 1% agarose gels, 4°C, 30 mA, 16 h. Marker (M): phage λ DNA-Eco RI plus SV40 DNA - Hind II + III. Bands hybridizing to chloroplast rRNA are marked with a square.

Heavy solid lines symbolized bands which stained more strongly with ethidium bromide than the average band (light lines) and were considered to be due to more than one similar or identical fragment. Dashed lines indicate faint bands.

TABLE 3

APPROXIMATE MOLECULAR WEIGHTS OF

Eco RI RESTRICTION DNA FRAGMENTS FROM EUGLENA CHLOROPLAST DNA

Band	MW of DNA $\times 10^{-6}$	Fragment repetitions	MW of DNA per band $\times 10^{-6}$
A	15.0	1	15.0
B	13.0	1	13.0
C	7.5	1	7.5
D	6.8	1	6.8
E	5.2	2-3	10.4-15.6
F	4.7	1	4.7
G	3.6	1	3.6
H	3.1	1	3.1
I	2.4	1	2.4
J	2.2	1	2.2
K	2.1	3	6.3
L	1.9	1	1.9
M	1.7	1	1.7
N	1.6	3	4.8
O	1.1	1	1.1
P	1.0	1	1.0
Q	0.81	1	0.81
R S	0.65 (ave.)	2	1.30
T U	0.57 (ave.)	2	1.14
V	0.48	1	0.48

multiple bands E, K and N strongly hybridized to 16S+23S rRNA (Table 3 and Fig. 3). A faint band is seen (autoradiogram) in the upper part, suggesting that band B (13×10^6) also carries rRNA related DNA sequences. From filter hybridization experiments with total chloroplast DNA, we have concluded that a 44 μm circular DNA contains three rRNA genes (Table 2: 5.1% hybridization) requiring approximately 9.6×10^6 daltons of double stranded DNA. According to the results of Table 3, the size of the DNA fragments containing rRNA genes could be up to one third of the total genome; however, the stoichiometry of the multiple bands E, K, N is not exactly established. Nevertheless, this result strongly suggests that the three rRNA genes are considerably spaced within the DNA circle, an observation also made for spinach chloro-

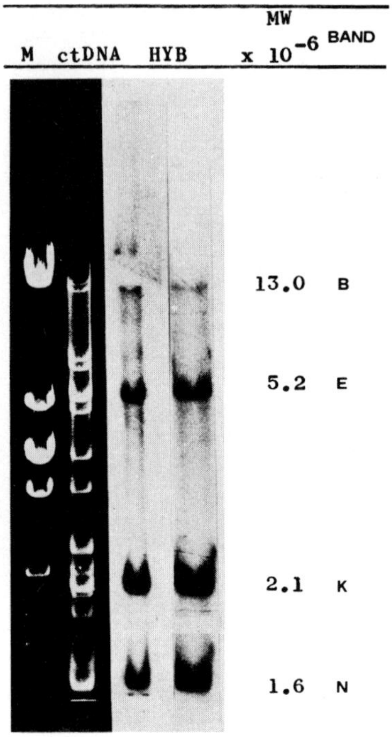

Fig. 3. Hybridization of chloroplast rRNA to Eco RI generated fragments of chloroplast DNA. 5 μg of DNA were digested with Eco RI, 1 h, 37°C, in 0.01 M Tris-HCl pH 7.5, 0.05 M NaCl, 0.01 M $MgCl_2$, 0.01 M 2-mercaptoethanol. The fragments were separated on 1% agarose gels (16 x 22 cm) at 20 mA, 20 h, 4°C using 0.04 M Tris-acetate, 0.02 M sodium acetate pH 7.8 as running buffer. DNA was transferred to Millipore HAWP 30 4FO filter strips according to Southern (9). The filter strips were incubated in ^{125}I-rRNA (approximately 1 x 10^6 cpm/μg RNA, 10 μg RNA/1 μg DNA) in 0.30 M NaCl and 0.03 M sodium citrate, 16 h, 65°C. The filter strips were exposed for 3-7 days using a Typon X-ray film, at 20°C. The two left panels show the ethidium bromide stained band patterns of λ phage DNA (M) and chloroplast DNA (ctDNA). The middle panel shows the corresponding autoradiogram. The respective gel was cut between the two top bands and a blank gel was inserted before transferring the DNA to the filter strip. The right panel shows the autoradiogram from an identical filter strip obtained from an uncut gel. Numbers at the right indicate molecular weights of DNA fragments.

TABLE 4

SOME PROPERTIES OF Eco RI FRAGMENTS OF EUGLENA CHLOROPLAST DNA

BAND	MW x10^{-6}	PROFILE neutral CsCl	d g/cc	G+C mole%	HYB rRNA
A B	15 13	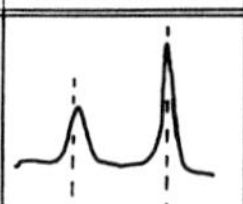	1.687	27	+
E F	5.2 4.7	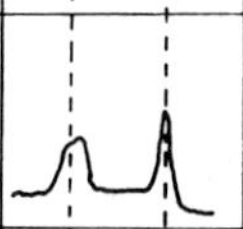	1.691 1.685	31 25	+
H	3.1	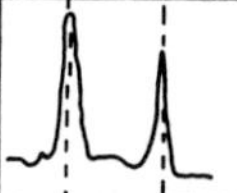	1.686	26	−
K L	2.1 1.9	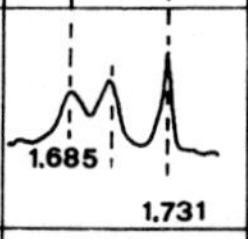	1.702 1.684	42 24	+
N	1.6			(41)	+

plast DNA (8). Similar hybridization experiments with Euglena chloroplast DNA, digested with Hind II endonuclease, yielded mainly two bands having molecular weights of approximately 3.7 and 8.1 x 10^6, which strongly hybridized with rRNA (unpublished results).

In previous studies (4,5), we have suggested that Euglena chloroplast DNA consists of segments varying in G+C content. This should be apparent in buoyant density analysis of DNA restriction fragments. In Table 4, we show the density profiles of Eco RI generated DNA fragments isolated out of five gel regions (the fifth profile is lacking for technical reasons). M. lysodeikticus DNA (1.731 g/cc) was used as the density marker (see Fig. 1 for centrifugation details and Fig. 3 for gel details). Although the top region contained two bands of low G+C content, the autoradiogram in Fig. 3 shows that band B faintly hybridized with rRNA. Band B probably carries part of a rRNA gene (terminal sequence), but should an entire rRNA cistron be located on this fragment, it might differ from the other two rRNA cistrons in its base sequence by lacking Eco RI specific site(s). The second gel region (bands E, F) gave a broad density profile with values in the range of 1.685 and 1.691 g/cc. From the autoradiogram, we concluded that band E carries the rRNA sequences and we tentatively assigned it a G+C content of 31%. Band E seems to be a multiple band which could be further subdivided into two types of fragments. The DNA of the third region gave a uniform density profile corresponding to 26% G+C. The DNA in this gel region did not hybridize to rRNA. The DNA in the fourth gel region gave a clear bimodal density profile, with densities corresponding to 42 and 24% G+C. All indications suggest that the band K (42% G+C) is the only one hybridizing to rRNA within that gel region. The DNA of the fifth gel region strongly hybridized to rRNA. The density scan is missing, but from electron microscopy studies, we know that this fragment is uniform in size and shows melting characteristics expected for a G+C-rich fragment, e.g. 41% (unpublished results).

These preliminary results give further evidence for the presence of DNA segments which are clearly distinct in G+C content from the bulk DNA. The future physico-chemical characterization of restricted DNA fragments should allow to map these segments on the circular DNA.

ACKNOWLEDGMENTS

This research was supported by Fonds National Suisse de la Recherche Scientifique, grant number 3.0300.73 and Fonds Dr. Säuberli to E.S. Some of the λ-phage DNA and Eco RI endonuclease were kind gifts of Dr. Hedgpeth, Université de Genève. We are grateful to Prof. A. Tissière, Université de

Genève, for the use of the analytical ultracentrifuge and to Prof. G. Bernardi, Université de Paris VII, for the liberal use of his research facilities.

REFERENCES

1 Edelman, M., Cowan, C.A., Epstein, H.T. and Schiff, J.A. (1964) Proc. Natl Acad. Sci. U.S.A. 52, 1214-1219

2 Scott, N.S. (1973) J. Mol. Biol. 81, 327-336

3 Manning, J.E. and Richards, O.C. (1972) Biochim. Biophys. Acta 259, 285-296

4 Stutz, E. and Vandrey, J.P. (1971) FEBS Lett. 17, 277-280

5 Crouse, E.J., Vandrey, J.P. and Stutz, E. (1974) in Proc. IIIrd Intern. Congress on Photosynthesis, Israel (Avron, M., ed.) pp. 1775-1786, Elsevier Publ. Co., Amsterdam

6 Slavik, N.S. and Hershberger, C.L. (1976) J. Mol. Biol. 103, 563-581

7 Smith, H. and Nathans, D. (1973) J. Mol. Biol. 81, 419-423

8 Herrmann, R., Hobom, G. and Bohnert, H. (1976) in Proc. NATO Course on Nucleic Acids and Protein Synthesis in Plants, Strasbourg, in press

9 Southern, E.M. (1975) J. Mol. Biol. 98, 503-517

Genetics and Biogenesis of Chloroplasts and Mitochondria, Th. Bücher et al. eds.

RESTRICTION ENDONUCLEASE MAP OF CHLOROPLAST DNA FROM *Euglena gracilis*

Patrick W. Gray and Richard B. Hallick
Department of Chemistry
University of Colorado
Boulder, Colorado 80309
U.S.A.

INTRODUCTION

In the course of our studies on the RNA transcripts from *Euglena* chloroplast DNA it became apparent that a detailed physical map of the chloroplast genome would be invaluable in studying the control of RNA synthesis. The technique of restriction nuclease mapping was chosen for this analysis. Each cleavage site of a site-specific nuclease on a defined DNA such as *Euglena* chloroplast DNA can provide a physical marker that can be related to biochemically defined chromosome loci. In this report we describe the cleavage site map of *Euglena* chloroplast produced by restriction endonucleases *Sal*I and *BamH*I.

MATERIALS AND METHODS

Preparation of Chloroplast DNA: For chloroplast DNA isolation, *Euglena gracilis* Klebs, Strain Z Pringsheim cells were grown to stationary phase on a photoautotrophic medium, and chloroplasts were isolated as previously described (1). Covalently closed, superhelical *Euglena* chloroplast DNA was isolated by centrifugation of chloroplast lysates in CsCl-ethidium bromide (2).

Restriction Endonucleases: Restriction endonuclease *Sal*I was isolated from *Streptomyces albus* G by the method of Thomas and colleagues. (C. A. Thomas, Jr., personal communication). Restriction endonuclease *BamH*I was purified from *Bacillis amyloliquefaciens H* (3).

Agarose Gel Electrophoresis: Separation and analysis of restriction endonuclease fragments was achieved by electrophoresis in 0.7 percent agarose slab gels, 14.5 cm x 12.5 cm x 0.3 cm thick. Following electrophoresis, DNA cleavage products were visualized by ethidium bromide fluorescence during ultraviolet irradiation (4).

RESULTS AND DISCUSSION

SalI Digestion of Chloroplast DNA: The fragment pattern resulting from the limit digestion of chloroplast DNA with *Sal*I endonuclease is shown in Figure 1, lanes 1 and 4. Only three fragments are produced. Estimates of the molecular weight of these fragments are as follows: *Sal* A, 80 kbp (kilo basepair); *Sal* B, 56 kbp; *Sal* C, 4.4 kbp. We have assumed a length of 139 kbp (9.2 x 10^7 daltons) for the intact chloroplast genome (5). There is some uncertainty in the length estimates of the two large fragments. Very large fragments have limited mobility in 0.7% agarose gels, making accurate molecular weight estimates difficult. For three fragments from a circular chromosome, there is only a single fragment order possible (Figure 2).

BamHI Digestion of Chloroplast DNA: When *Euglena* chloroplast DNA is treated with *BamH*I endonuclease, six fragments are produced. Only five of these are resolved on 0.7% agarose gels (Figure 1, lane 2). The smallest two fragments co-electrophorese. Estimates of the lengths of the six fragments are as follows: *Bam* A, 64 kbp; *Bam* B, 41 kbp; *Bam* C, 15 kbp; *Bam* D, 7.3 kbp; *Bam* E, 5.9 kbp; *Bam* F, 5.9 kbp. The fragment pattern resulting from partial digestion of chloroplast DNA with *BamH*I is shown in Figure 1, lane 3. Two partial fragments are apparent with electrophoretic mobilities between fragments *Bam* C and *Bam* D, and lengths of approximately 13.2 kbp and 12 kbp. We can assign these partial digestion products the structures *Bam* D-*Bam* E or *Bam* F, and *Bam* E-*Bam* F respectively. This leads to the conclusion that the linkage group *Bam* D-(*Bam* E-*Bam* F) must exist. Likewise, the four partial fragments (Figure 1, lane 3) with lengths greater than *Bam* C and less than *Bam* B can only result from a linkage between *Bam* C and the *Bam* D-(*Bam* E-*Bam* F) group. The remaining two fragments, *Bam* A and *Bam* B must therefore be linked together, and also linked to both ends of the *Bam* C-(*Bam* D-*Bam* E, *Bam* F) group.

BamHI-SalI Cleavage Map of Euglena Chloroplast DNA: The fragment pattern resulting from a double digestion of chloroplast DNA with *BamH*I and *Sal*I is shown in Figure 1, lane 5. Nine fragments, designated BS 1-9, are produced as expected for cleavage of a circular DNA at three *Sal*I and six *BamH*I sites. The fragments *Bam* A, D, E, F and *Sal* C have the same electrophoretic mobilities as fragments BS 1, 5, 6, 7 and 8, respectively. Fragment *Bam* B is cleaved by *Sal*I into the following three fragments: BS 2, 26 kbp; BS 3, 13 kbp; and BS 8 (*Sal* C); 4.4 kbp. *Sal* C is the internal fragment. Fragment *Bam* C is cleaved by *Sal*I into the following two fragments: BS 4, 13 kbp; and BS 9, 2.6 kbp.

Fig. 1. Analysis of *Euglena* chloroplast DNA by restriction endonuclease *BamH*I or *Sal*I cleavage and electrophoresis on 0.7% agarose gels. 1. *Sal*I limit digest. 2. *BamH*I limit digest. 3. *BamH*I partial digest. 4. *Sal*I limit digest. 5. *Sal*I and *BamH*I double digest.

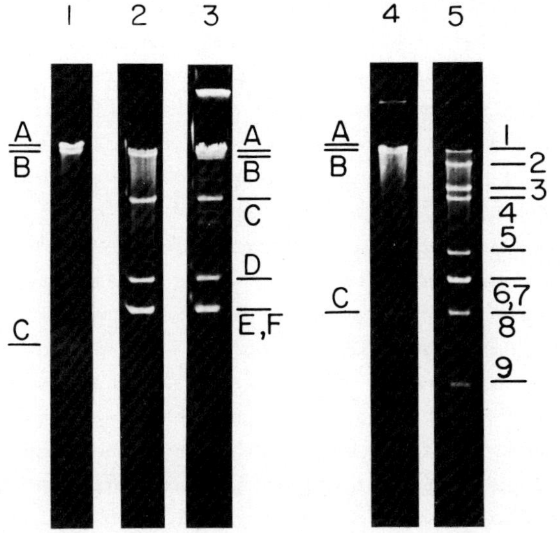

Fig. 2. *BamH*I and *Sal*I restriction endonuclease cleavage map of *Euglena* chloroplast DNA. The numbers refer to the *BamH*I-*Sal*I double digestion fragments.

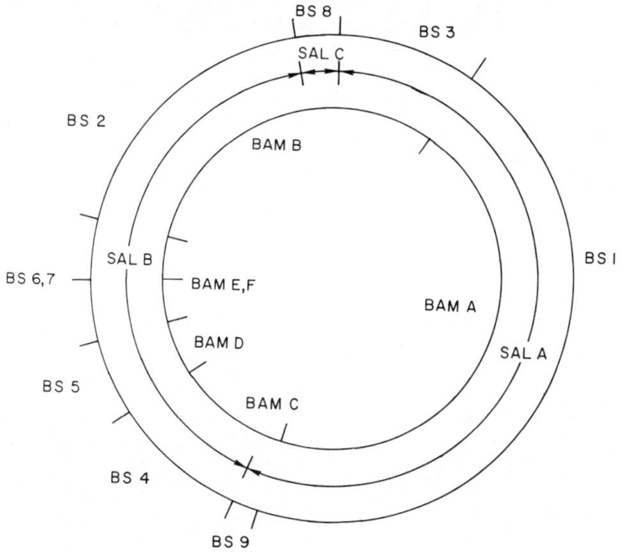

From the data on *BamH*I fragment linkage groups deduced from partial digests, the above results on *BamH*I-*Sal*I double digests can be reduced to only two possible sets of linkage groups. In the first possibility *Sal* B contains the fragments BS 4-(5,6,7)-2, and *Sal* A contains BS 3-1-9. Alternatively *Sal* B is composed of BS 4-2, and *Sal* A contains BS 3-1-(5,6,7)-9. Only the first possibility is consistent with the estimated lengths of fragments *Sal* A and *Sal* B. This leads to the *BamH*I-*Sal*I cleavage site map illustrated in Figure 2. It is noted that there is unresolved ambiguity in the order of fragments BS 6-7 (*Bam* E-F), however fragment BS 5 (*Bam* D) is ordered in fragment *Sal* B as follows: BS 4-5-(6,7)-2. This is based on an experiment (not shown) in which *Sal*I digested chloroplast DNA was treated for varying lengths of time with *E. coli* exonuclease III. Subsequent digestion with *BamH*I showed that BS 6-7 were more resistant to exonuclease III than BS 5.

Work is currently in progress to achieve more detailed cleavage maps of *Euglena* chloroplast DNA by utilizing other restriction endonucleases. In results not shown, we have resolved 21 cleavage fragments produced by digestion of chloroplast DNA with *EcoR*I restriction endonuclease. With the exception of a few unresolved fragments, these cleavage products are present in stoichiometric amounts. These observations, along with the *BamH*I and *Sal*I data support the conclusion that *Euglena* chloroplast DNA is a homogenous population of molecules of uniform contour length. This work was supported by grant GM 21351 from the National Institutes of Health and a grant to the University of Colorado from the American Cancer Society.

REFERENCES

1. Hallick, R. B., Lipper, C., Richards, O. C., and Rutter, W. J. (1976) Biochemistry 15, in press.
2. Chelm, B. K., Hoben, P. J., and Hallick, R. B., submitted for publication.
3. Wilson, G. A., and Young, F. E. (1975) J. Mol. Biol. 97, 123.
4. Helling, R. B., Goodman, H. M., and Boyer, H. W. (1974) J. Virol. 14, 1235.
5. Manning, J. E., and Richards, O. C. (1972) Biochim. Biophys. Acta 259, 285.

Genetics and Biogenesis of Chloroplasts and Mitochondria, Th. Bücher et al. eds.

THE LOCATION OF rRNA GENES ON THE RESTRICTION ENDONUCLEASE MAP OF THE SPINACIA OLERACEA CHLOROPLAST DNA

Reinhold G. Herrmann, Hans-Jürgen Bohnert, Albert Driesel and Gerd Hobom

Institut für Botanik I der Universität
4 Düsseldorf, FRG
Institut für Biologie III der Universität
78 Freiburg i. Br., FRG

The total plastid-located genetic information is contained in one circular DNA molecule[1] which has been called a "chromosome". This paper presents a physical map of the Spinacia oleracea plastid chromosome constructed with the aid of restriction endonucleases and summarizes the current status of attempts to localize the constituents of the high-molecular-weight precursor for plastid ribosomal RNA[2,3] on this map. This precursor contains the 23S, 16S, 5S, 4.5S[4] and about 30% spacer RNA.

ENDONUCLEASE DIGESTION OF CHLOROPLAST DNA

The methods for preparation of unbroken plastids from Spinacia tissue, extraction and analysis of DNA have been described in detail in previous reports[5,6]. The DNA was routinely checked by electron microscopy and generally contained between 60 and 75% circular molecules. No DNA fibrils were seen that exceeded the contour length of circular dimers, thus indicating the absence of major contamination by nuclear DNA[5].

Chloroplast DNA was digested by various restriction enzymes, and the resulting mixtures of DNA fragments were analyzed along with appropriate marker series on slab gels containing concentrations of 0.6 to 3% agarose. The following enzymes were used for this purpose: SalI, SmaI or XmaI, HindIII, HindII, BamI, Bsu, EcoRI and EcoRII. Complete digestion with the individual restriction endonucleases produced a unique series of fragment bands representing stoichiometric amounts of individual DNA fragments. In several instances two or even three fragments of identical or nearly identical size were present per band. No undigested material or submolar fragment bands could be detected in any of the digestions, thus suggesting a unique constitution for Spinacia chloroplast DNA[1]. Several of these digestion patterns are displayed in Fig. 1.

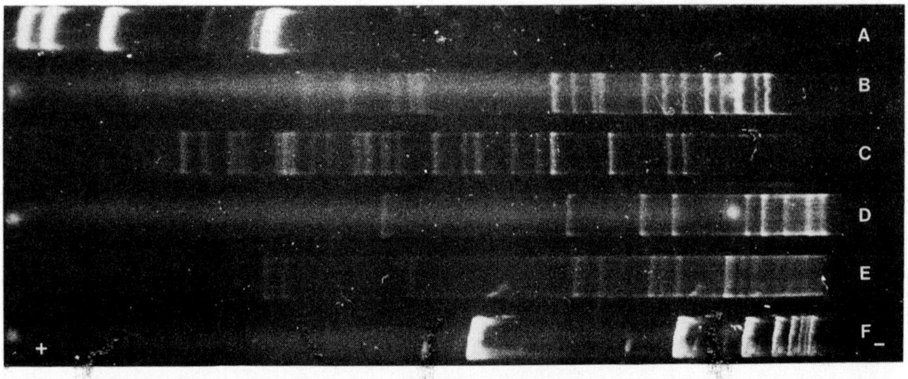

Fig. 1. Restriction fragment patterns of chloroplast DNA from Spinacia. Slot B-E: HindIII, EcoRI, SalI and SmaI pattern (1% gels). The flanking marker series in slot F was obtained by partial EcoRI digestion of an octamer plasmid λdv 021-DNA resulting in regular oligomers of 2.07 megadalton. The marker series in slot A originated from a HindII + HindIII digestion of plasmid λdv 1-DNA which gives rise to fragments of 1.15, 1.10, 0.81, 0.72 and 0.67 megadaltons together with several smaller ones not visible here[7].

Due to their limited complexity three of the patterns were selected for further studies, those resulting from SalI, SmaI (XmaI) and HindIII. These patterns contained 12, 20 and 40 fragements, respectively (excluding very small fragments of less than 60 base pairs which may have remained undetected).

Two of the SalI fragments migrated as a single (two-molar) band at the position of 6.8 megadalton. Similarly, the presence of four two-molar bands and of one three-molar band reduced the number of bands observed in the SmaI pattern to 14. Thirty-one bands were detected in the HindIII series. From preparative gel electrophoreses all fragments or groups of fragments larger than one megadalton were isolated from the agarose gels by centrifugation on KI/ethidium bromide gradients[8], followed by extraction of the dye, and dialysis against 0.01 M Tris/HCl, pH 8.

The DNA fragments resulting from the two- and three-molar bands obtained with the three different endonucleases were analyzed for their complexity by digestion with several of the other enzymes listed above. It was concluded that the bands contained identical fragments present twice per plastid chromosome, if they were digested exclusively into two-molar subfragments by at least three enyzmes, and different fragments of similar size, if they gave rise to one-molar subfragments with one or more of the enzymes.

In this way, the only two-molar band in the Sal digestion pattern (6.8 megadalton) was shown to contain identical fragments. The Sma series contained five such two-molar bands (5.1, 3.4, 2.5, 1.2 and 1.15 megadalton), and the HindIII series at least three (5.8, 4.7 and 2.6 megadalton) with the possibility of an additional four among the series of smaller yet unresolved fragments. In only three instances was there a coincidence in size of different fragments.

These results suggested a long stretch or stretches of sequence repetition within the Spinacia chloroplast DNA. These segments apparently also gave rise to a number of two-molar bands present in digestion patterns obtained with other endonucleases, but these were not analyzed in more detail.

PHYSICAL MAP OF CHLOROPLAST DNA

In order to be able to construct a physical map of the chloroplast DNA from restriction fragments, the isolated fragments in all three series were hydrolysed individually in turn with one or both of the other enzymes. The resulting subfragment patterns were analyzed again by agarose gel electrophoresis with double digests and a triple digest of total chloroplast DNA serving as a marker series. Identical sizes of e.g. Sal/Sma fragments obtained first from a Sal fragment redigested by the Sma enzyme, and again obtained by Sal digestion of a Sma fragment were taken to represent the overlapping link between the two respective fragments of both series. While the larger of the redigestion fragments could be easily analyzed in this manner, it was not possible to detect the resulting small linking fragments. In order to be able to analyze these small fragments, redigestion mixtures of fragments were terminally labelled with α-^{32}P-rGTP by terminal transferase from calf thymus in the presence of 1 mM Co^{++}, which had been reported to result in terminal addition of 2-3 moles of rGMP[9]. The mixtures were then separated by electrophoresis, the gels dried and analyzed by autoradiography. In this way a circular map of Spinacia chloroplast DNA could be constructed which is given in Fig. 2. The third series of HindIII fragments was used for confirmation of the Sal/Sma map, and allowed a greater resolution in the serial ordering of the fragments present in double and triple digests (well over 100).

The most prominent feature of the physical map of the chloroplast chromosome besides being circular is a large inverted repeat of at least 14 megadalton. This inverted repeat, separated by 14

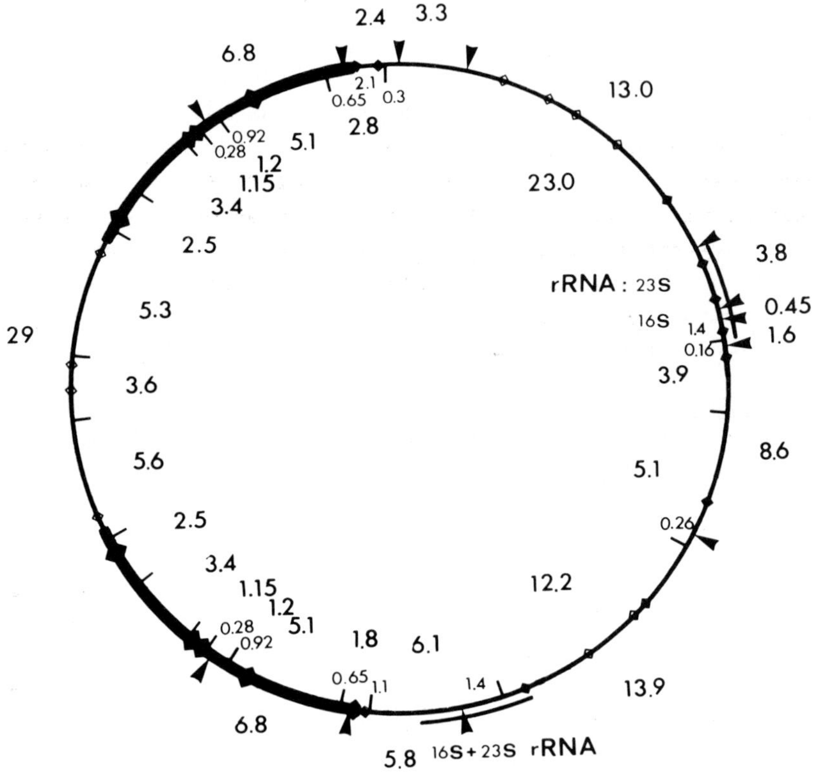

Fig. 2. Restriction endonuclease map of the *Spinacia oleracea* chloroplast DNA (SalI/SmaI/HindIII). Fragment sizes are given in megadaltons, that of the Sal digestion on the outer, that of the Sma digestion on the inner side of the circle. Terminal fragments resulting from the Sal/Sma digestion are given in small numbers. The inverted repeat is indicated by thick lines. ▼ = Sal cut, | = Sma cut, ◆ = HindIII cut, ◇ = HindIII cut, serial order arbitrary.

megadalton of unique DNA contains all the two-molar fragments listed above for the different series, and may have to be increased somewhat in size, if parts of the adjacent unique fragments can be shown to contain identical sequences. The nature of this repeat is unknown.

The minor ambiguities remaining in the ordering of SmaI and HindIII fragments result from the large inverted repeat which is digested by these enzymes into several fragments. Therefore, no definite connection between the large part of the unique DNA (50 megadalton) and the small part of the unique DNA (14 megadalton) could be obtained in these series, and the latter (Sma 5.6, 3.5 and 5.3 megadalton) could be equally as well be the inverse of what is given in the map.

Some additional ambiguities in the HindIII pattern arise from the presence of several HindIII cuts within the larger segments of the Sal/Sma map (Sal 13.9 and 13 megadalton). The resulting subfragments could not be positioned with respect to each other, due to the lack of end fragments.

LOCALIZATION OF THE rRNA CISTRONS

The distribution of the discrete rRNA components of the two rRNA cistrons per circular chromosome[1] along the physical map was studied by hybridization to individual DNA fragments.

a. Purification of rRNA species

In order to obtain reliable hybridization data, the plastid rRNA species were isolated from ribosomes (Figs. 3 and 4) or from their subunits and in exceptional cases (Fig. 5) from unbroken EDTA-treated (3 mM) plastids.

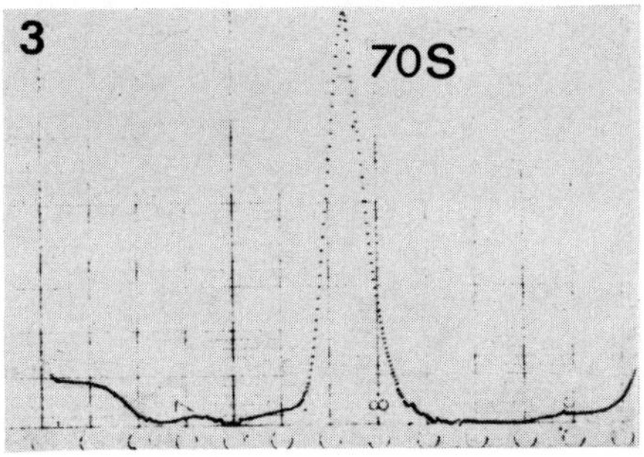

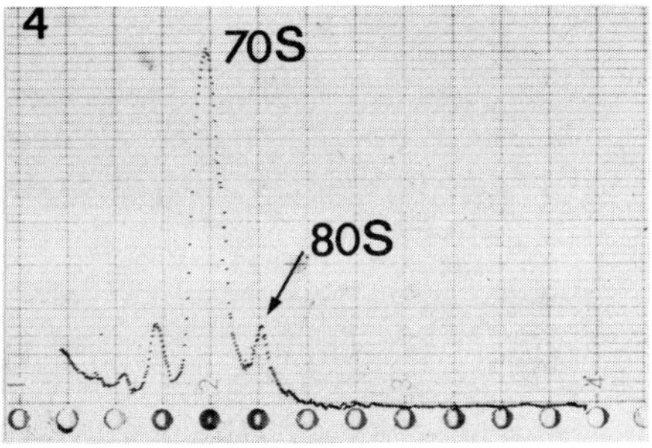

Figs. 3 and 4. Fractionation by sucrose density gradient centrifugation of a ribosome preparation from Spinacia plastids (Fig. 3) and of an artificial mixture of plastid ribosomes, their subunits and cytosolic ribosomes (Fig. 4). To prepare plastid ribosomes, EDTA-treated organelles were washed once in the presence of 10 mM $MgCl_2$ and ruptured by hypotonic dilution with 15 volumes of 10 mM $MgCl_2$, 30 mM Tris/HCl, pH 7.6. The free monomer fraction, representing about 90% of the total ribosomes, was isolated by centrifuging the 21,000 x g supernatant of the lysate at 105,000 x g for 2-3 hours at 4 ^{o}C. After resuspension, the ribosomes were purified by velocity sedimentation through a 8% (4.5 ml) to 28% (7.5 ml) sucrose gradient in the SW 40.1 rotor at 36,000 rpm for 2-3 hours at 2 ^{o}C. No contamination by cytosolic ribosomes which separate under these conditions (Fig. 4) was detectable.

The quality and purity of the 23S and 16S rRNA was finally analyzed by electrophoresis in 2.4% (Fig. 6a-c), that of the 5S and 4.5S species in 10% acrylamide gels. The purified RNA species were radioactively labelled by iodination[10].

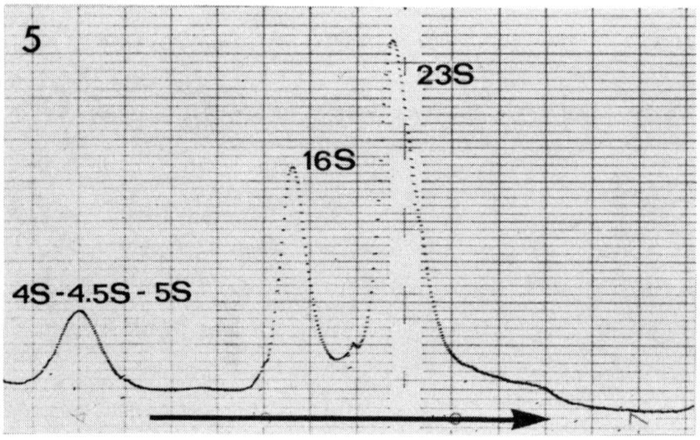

Fig. 5. Purification of the 23S, 16S and low-molecular weight RNA components from plastid fractions. The RNA was extracted by standard procedures[2,3] and fractionated on a 0.4 to 1.3 M linear sucrose gradient in 100 mM NaCl, 1 mM EDTA, 0.2% sodium dodecylsulfate and 20 mM Tris/HCl, pH 7.6 at 120,000 x g for 24 h at 4 °C in the SW 40.1 rotor.

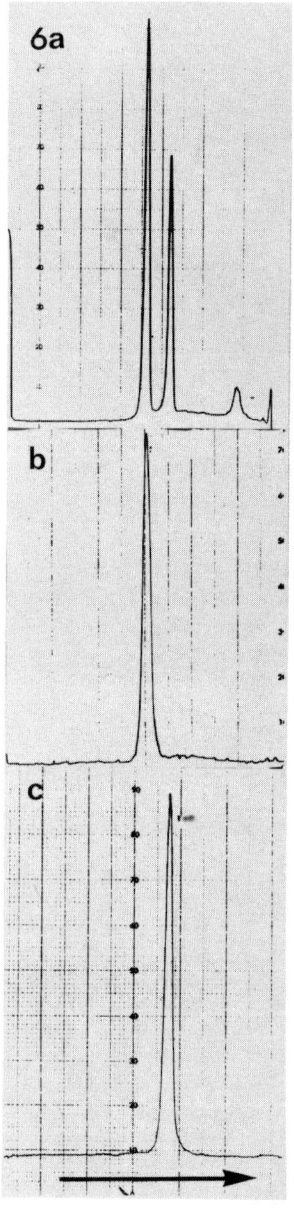

b. Location of the 23S and 16S rRNA

In order to associate the ribosomal RNA species with specific endonuclease fragments, SalI and HindIII digests of 4 µg chloroplast DNA were separated on agarose gels. The fragments were transferred to strips of nitrocellulose filters according to Southern (ref. 11). After incubation with iodinated RNA (20 µg RNA/ml, 1 x SSC, 64 °C for 2-8 hours), the strips were cut into 1 mm slices and the slices counted for radioactivity.

Fig. 6a-c. Electrophoresis in acrylamide gels of (a) RNA from plastid ribosomes of *Spinacia oleracea* and of (b) 23S and (c) 16S rRNA purified by sucrose gradient centrifugation.

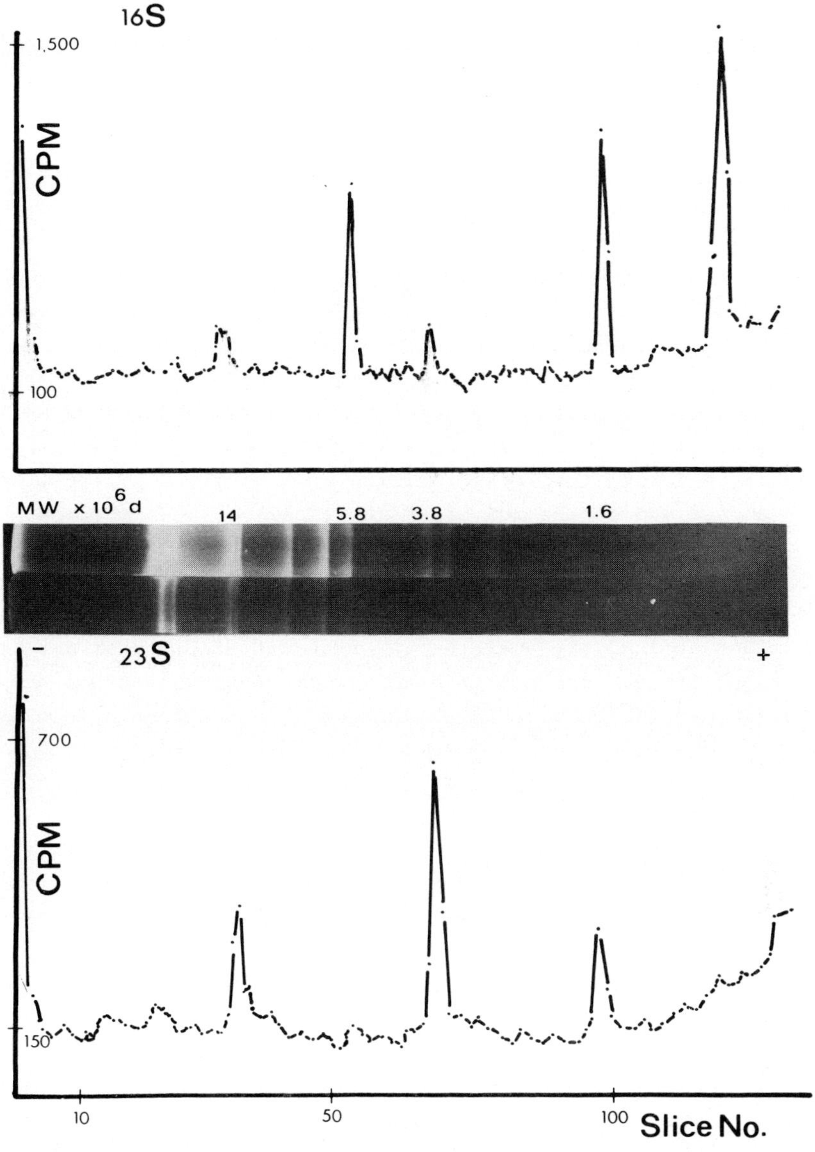

Fig. 7. Hybridization of ^{125}I-labelled 23S (bottom) and 16S (top) rRNA to Sal fragments transferred from agarose gels to nitrocellulose strips. A large excess of cold 23S RNA was added to the radioactive 16S component and vice versa to compete out any contamination. After hybridization, the strips were washed in 1 x SSC, treated with RNase, dried, cut and counted.

The resulting radioactivity profiles show that the genes for 23S and 16S RNA in each of the two rRNA cistrons occupy adjacent regions of the chromosome (Figs. 7 and 2). Additional evidence comes from hybridization with HindIII gels. (Fig. 2 shows that this enzyme cuts the Sal fragments of 5.8 and 13.9 megadaltons overlapping, whereas the other rDNA region is degraded more intensively. The radioactivity profiles obtained are consistent with this pattern.) This finding is in excellent agreement with results obtained in our laboratory from studies on transcription in the isolated organelle [2,3]. These results indicated that the 23S and 16S rRNA are transcribed as a single unit and should therefore be physically linked in the *Spinacia* plastid chromosome. Preliminary hybridization experiments with the RNA fraction migrating with a molecular weight of 0.45 megadalton yielded radioactivity in the 14 and 3.8 megadalton Sal fragments suggesting that some component of this fraction (assumed to be transcribed spacer of the rRNA cistron[2,3]) is either situated at the end of the 23S or, more likely, between the 16S and 23S rDNA sequences. This RNA fraction must, however, contain additional RNA species since hybridization also occurred with other fragments.

Our data also indicate that the two rRNA cistrons do not map in the same vicinity but are approximately 20 megadaltons apart, and that the 16S and 23S components are arranged in the same directional sense along the molecule. Essentially similar findings with regard to the map distance of the two rRNA cistrons have been reported, using *Zea mays* chloroplast DNA[12].

Another interesting point emerging from our results is that the two rRNA cistrons localized outside the large inverted repeat are different. Only unique, non-repetitive fragments have been observed throughout these areas with SalI, SmaI, HindIII and BamI. In the EcoRI gels, hybridization was obtained for two- as well as for one-molar bands, indicating some similarity between the two different cistrons. While most of the endonucleolytic cuts could still be assigned to spacer sequences, at least one of the HindIII cuts (in the Sal 1.6 megadalton fragment), present in one of the cistrons but absent in the other, appears to be within the 16S rRNA sequence itself.

Acknowledgements: We are grateful to Miss Barbara Schiller, Miss Christine Michalowski and to Miss Monika Streubel for expert technical assistance. We also thank Dr. Ch. Leaver for grammatical correction of the manuscript. This investigation was supported by the Deutsche Forschungsgemeinschaft.

REFERENCES

1. Hobom, G., Bohnert, H.-J., Driesel, A. and Herrmann, R.G. (1976) Nucleic Acids and Protein Synthesis in Plants. Plenum Press, eds. L. Bogorad and J.H. Weil, in press.
2. Bohnert, H.-J., Schmitt, J.M. and Herrmann, R.G. (1974) Portugal. Acta Biol. 14, 71-90.
3. Bohnert, H.-J., Driesel, A. and Herrmann, R.G.(1976) this Volume.
4. Driesel, A. (1975) Diplomarbeit, Universität Köln, 43 p.
5. Herrmann, R.G., Bohnert, H.-J., Kowallik, K.V. and Schmitt, J.M. (1975) Biochim. Biophys. Acta 378, 305-317.
6. Schmitt, J.M. and Herrmann, R.G. (1976) Methods in Cell Biology, Vol 15, ed. D. Prescott, in press.
7. Streek, R.E. and Hobom, G. (1975) Eur. J. Biochem. (1975), 57, 595-606.
8. Blin, H., Gabain, A.v. and Bujard, H. (1975) FEBS Letters 53, 84-86.
9. Roychoudhury, R., Jay, E. and Wu, R. (1976) Nucleic Acid Res. 3, 863-878.
10. Sriprakash, K.S., Choo, K.B., Nagley, P. and Linnane, A.W. (1976) Biochem. Biophys. Res. Commun. 69, 85-91.
11. Southern, E.M. (1975) J. Mol. Biol. 98, 503-517.
12. Bedbrook, J.R. and Bogorad, L. (1976) this Volume.

Genetics and Biogenesis of Chloroplasts and Mitochondria, Th. Bücher et al. eds.

ANALYSIS OF THE CODING CAPACITY OF EcoRI RESTRICTION FRAGMENTS OF SPINACH CHLOROPLAST DNA

Paul R. Whitfeld, Bentley A. Atchison, Warwick Bottomley
and Christopher J. Leaver
Division of Plant Industry, CSIRO, P.O. Box 1600,
Canberra City A.C.T. 2601, Australia

INTRODUCTION

As part of an overall programme directed towards ascertaining the various coding functions of chloroplast DNA, we have been using specific restriction endonucleases to fragment chloroplast DNA in the hope that the segments produced could be more easily characterized than the intact molecule. In this paper we describe our initial experiments on the cleavage of spinach chloroplast DNA by EcoRI endonuclease which cleaves at the sequence GAATTC[1]. The DNA fragments produced range in size from 3.7×10^6 to less than 0.2×10^6 daltons and are quite heterogeneous in terms of their GC-content. The results of experiments involving membrane filter hybridizations and Cot curve analyses of chloroplast RNA with the DNA fragments have permitted the allocation of coding functions in respect of both stable and messenger RNA species to a number of the DNA bands. Evidence is also provided that the sequence of ribosomal RNA genes in the chloroplast DNA is probably 16S-23S-5S.

MATERIALS AND METHODS

Methods for the isolation of spinach chloroplasts and the preparation of chloroplast DNA and total RNA have been described elsewhere[2,3]. 23S and 16S RNA were extracted from chloroplast ribosomal subunits which has been processed twice through sucrose gradients. Low molecular weight RNAs were obtained by fractionation of total chloroplast RNA on a sucrose gradient, and subsequently separated into 5S- and 4S-RNA fractions by sieving on a column of Sephadex G-100.

^{125}I-labelling of RNA fractions was carried out as described by Robertson <u>et al.</u>[4]. ^{125}I-RNA was purified by gel filtration, CF11-cellulose chromatography and finally by phenol extraction.

EcoRI endonuclease was purified from *E. coli* strain RY13 by ammonium sulphate precipitation and DEAE-cellulose column chromatography[5]. Digestion of chloroplast DNA was usually carried out in a volume of 30 µl in 20 mM $MgCl_2$, 10 mM Tris HCl, pH 7.8, at 35° for 1 hr. Electrophoresis on 1% agarose, staining and photography were done as previously detailed[2].

Labelling of chloroplast DNA with *E. coli* DNA polymerase 1, its hybridization to excess RNA, and the analysis of the DNA-RNA hybrid by S1 nuclease digestion followed published procedures[6].

RESULTS

Digestion of spinach chloroplast DNA with EcoRI endonuclease and electrophoresis of the products on 1% agarose results in a pattern such as is shown in Fig. 4. Over 25 bands can be detected on the gel; however, for most of our initial studies we have elected to group some bands which migrate very close together and ignore some of the minor low molecular weight bands. The 14 regions we shall be dealing with are specified in the figure.

The size of the DNA fragments was determined by comparison of their electrophoretic mobilities with those of fragments produced by EcoRI cleavage of λ and ϕ80 phage DNAs, and also by electron micrograph measurements of the DNA eluted from the various bands on the gel. The fragments range in molecular weight from 3.7×10^6 to less than 0.2×10^6 (Table 1.).

As may be noted from the relative intensity of staining of the various bands with ethidium bromide there appears to be more than a single fragment of DNA in many of the bands. For instance, bands 3, 4 and 11 stain more intensely than bands 1 and 2. Multiplicity in the various bands was estimated by restricting ^{32}P-labelled DNA from chloroplasts and determining the amount of radioactivity associated with each band. The results are shown in Table 1 (column 3) and it is

TABLE 1

SIZE, MULTIPLICITY AND CODING FUNCTION OF THE FRAGMENTS FROM EcoRI DIGESTED CHLOROPLAST DNA

Band No.	Mol. wt. ($\times 10^{-6}$)	Estimated Multiplicity	Total Mol. wt. of DNA in band ($\times 10^{-6}$)	Cistrons present	
				Stable RNA	mRNA
1	3.7	1	3.7	-	++
2	3.5	1	3.5	-	-
3	2.7	2	5.4	16S, 23S	+
4	2.0	6	12.0	16S	++
5	1.7	1	1.7	-	+
6	1.5	3	4.5	-	++
7	1.3	3	3.9	-	+
8	1.1	5	5.5	-	+
9	0.9	5	4.5	-	+
10	0.8	2	1.6	-	+
11	0.7	10	7.0	4S, 5S, 23S	-
12	0.57	3	1.7	-	+
13	0.45	8	3.6	-	++
14	0.20	3	0.6	-	-
Total	21.12	Total	59.2		

apparent that all bands except 1, 2 and 5 are multiple. Although the multiplicity in some instances is due to our grouping of bands, in other instances (e.g. band 3), multiple-fragment bands always behave as a single molecular weight species despite our attempts to split them into individual components by prolonged electrophoresis. Whether this means that the component fragments are completely identical or only of the same size is not known.

Calculation of the total molecular weight represented by the DNA content of each of the specified bands (i.e. M.Wt. x multiplicity) and summation of all the bands results in a figure of approximately 60×10^6 (Table 1). This value is low compared to the known molecular weight (89×10^6) of spinach chloroplast DNA[7] and probably is due to slight inaccuracies in the estimated molecular weights and multiplicities of the DNA fragments.

<u>CsCl fractionation of EcoRI restricted chloroplast DNA</u>: When EcoRI fragmented DNA was centrifuged to equilibrium in a CsCl buoyant density gradient the absorbance profile showed a broad peak, skewed on the higher density side (Fig. 1).

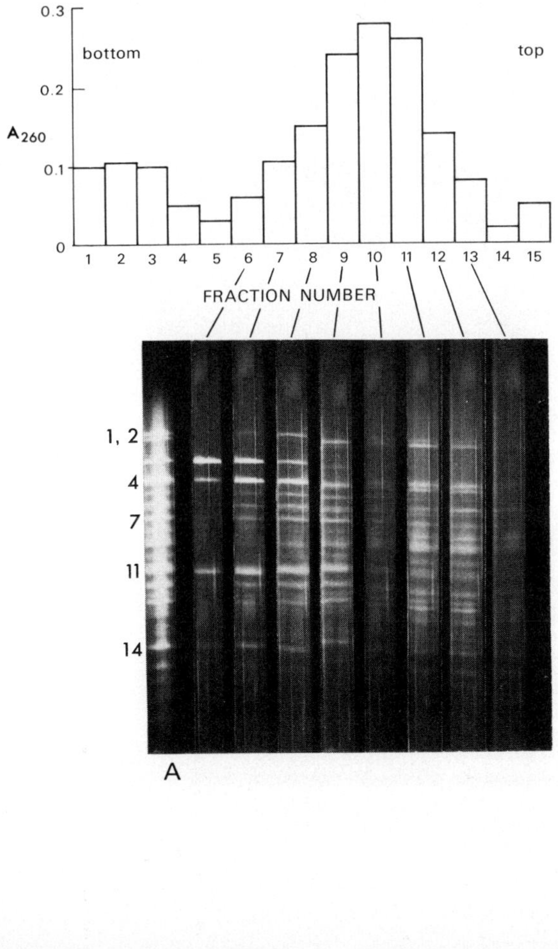

Fig. 1. Electrophoretic analysis of fractions from CsCl density centrifugation of EcoRI-digested chloroplast DNA. 30μg chloroplast DNA were digested with EcoRI, adjusted to ρ=1.700 g/cc with CsCl and centrifuged 65 hr at 40,000 r.p.m. DNA recovered from the gradient fractions was electrophoresed on 1% agarose and stained with ethidium bromide.
Top panel: A_{260} of fractions from gradient. Absorbance in fractions 1 to 3 was due to contaminating RNA.
Bottom panel: Electrophoretic patterns of DNA from major fractions. "A" is a gel of restricted, unfractionated DNA. Numbers refer to band number in electrophoresis pattern.

Intact chloroplast DNA gives a much sharper and more symmetrical peak (not shown). Electrophoretic analysis of the DNA recovered from the gradient fractions yielded the patterns shown in the lower half of Fig. 1. It can be seen that the buoyant densities of the various fragments differ quite markedly and that partial fractionation of some of the bands is possible. This could be particularly useful in the case of those bands which contain a multiplicity of identical-size fragments (e.g. bands 4 and 11). It is to be expected that the cistrons coding for chloroplast ribosomal RNA will have a higher GC-content than the rest of the DNA[8]. Thus DNA fragments occurring in bands 3, 4 and 11 should be considered as possibly arising from the ribosomal RNA cistrons. That this is in fact so will be demonstrated below. There do not appear to be any fragments of very high AT-content which could reflect the existence of AT-rich regions in the chloroplast DNA. Such regions are known to be a feature of certain mitochondrial DNAs[9].

Filter hybridization of chloroplast RNA to DNA fragments: EcoRI digested chloroplast DNA was electrophoresed on agarose and the fragments were recovered from regions of the gel as specified in Fig. 2. The DNA was denatured, bound to nitrocellulose filters and then hybridized with total spinach chloroplast RNA which had been labelled with ^{125}Iodine. As can be seen in Fig. 2 the RNA hybridized to a significant extent to DNA from all regions of the gel. However, hybridization to band 4 was clearly greater than to any other band and, because total chloroplast RNA consists mainly of ribosomal RNA, it can be concluded that part at least of the DNA present in band 4 is involved in coding for ribosomal RNA.

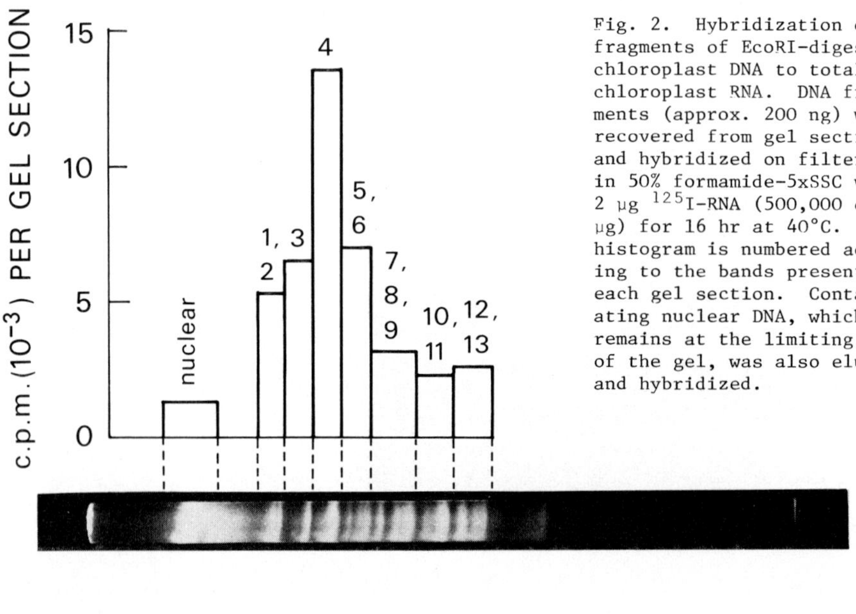

Fig. 2. Hybridization of fragments of EcoRI-digested chloroplast DNA to total chloroplast RNA. DNA fragments (approx. 200 ng) were recovered from gel sections[10] and hybridized on filters[11] in 50% formamide-5xSSC with 2 μg ^{125}I-RNA (500,000 cpm/μg) for 16 hr at 40°C. The histogram is numbered according to the bands present in each gel section. Contaminating nuclear DNA, which remains at the limiting size of the gel, was also eluted and hybridized.

Hybridization of total chloroplast ^{125}I- RNA to the DNA from all the regions of the gel may be attributed partly to the presence of messenger RNA species in the chloroplast RNA preparation since the conditions used (Cot = 0.3) were such as to allow mRNA to hybridize. However, it may also have been partly due to a non-specific binding of ^{125}I-RNA to some contaminant eluted from the agarose along with the DNA. In view of the unsatisfactory nature of this result we turned to more definitive forms of analysis.

Cot curve analysis of chloroplast DNA: If DNA of high specific activity is hybridized to a vast excess of RNA and the rate and amount of hybrid formation are measured, one can determine the fraction of DNA involved in coding for the various RNA species, including messenger RNA-species. When whole chloroplast DNA, which had been labelled by incubating it with DNA polymerase I, dATP, dGTP, ^{3}H-dCTP and ^{3}H-TTP, was hybridized to total chloroplast RNA significant hybridization was not detected at Cot values less than 10^{-2} (Fig. 3). Above that value, however, there was a gradual increase in the amount of hybrid formed so that at a Cot of 10^{2} approximately 40% of the DNA had been hybridized. Hybridization at Cot values greater than 10^{-2} is due mainly to the presence of non-ribosomal RNAs. On theoretical grounds[12] one can calculate that a messenger RNA species of molecular weight 500,000, which is present at a concentration 1% that of ribosomal RNA in the total chloroplast RNA preparation, will hybridize with a Cot of about 0.5. Messenger RNA species present in smaller amounts will hybridize at correspondingly higher Cot values. Hybridization of the ribosomal RNAs is expected to occur at a Cot of 5×10^{-3} under the conditions used. Failure to detect a break in the curve in that region is probably due to the fact that the ribosomal RNA will hybridize to less than 4% of the chloroplast DNA[13]. This analysis indicates that RNA transcripts of at least 40% of the chloroplast DNA are present in the chloroplast, although most of them are there in very low concentration relative to the ribosomal RNA.

Cot curve analysis of chloroplast DNA fragments: Chloroplast DNA, labelled as above, was digested with EcoRI enzyme, electrophoresed on 1% agarose, and the bands eluted and hybridized to total chloroplast RNA. The Cot curves from these analyses are shown in Fig. 4 and it can be seen that band 4 has a

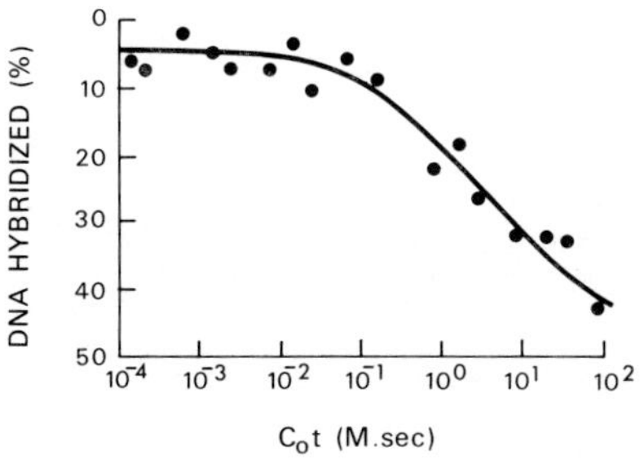

Fig. 3. Cot curve analysis of labelled chloroplast DNA hybridizing to total chloroplast RNA. 25 ng ^{3}H-DNA (800,000 cpm/μg) were hybridized to a range of chloroplast RNA concentrations in 0.18 M NaCl, 10 mM Tris-HCl, pH 7, 1 mM EDTA, 0.05% SDS, in a final volume of 30 μl for 4 hr at 60°C. Unhybridized DNA was digested with S1 nuclease and the DNA-RNA hybrid was precipitated with cold TCA, washed and counted in a scintillation spectrometer.

component which hybridizes to ribosomal RNA. The Cot curve for band 3 also shows a break at 5×10^{-3}, and, to a lesser degree so do those for bands 5 and 13. Bands 1, 5, 6 and 13 show considerable hybridization at Cot values between 10^{-1} and 10^{0} suggesting that the DNA contained in them might code for messenger RNAs present in relatively high concentrations in the chloroplast. Bands 2 and 11 do not appear to hybridize at all to the RNA under these conditions but it should be noted that the analysis was carried out to a Cot value of 10^{0} only.

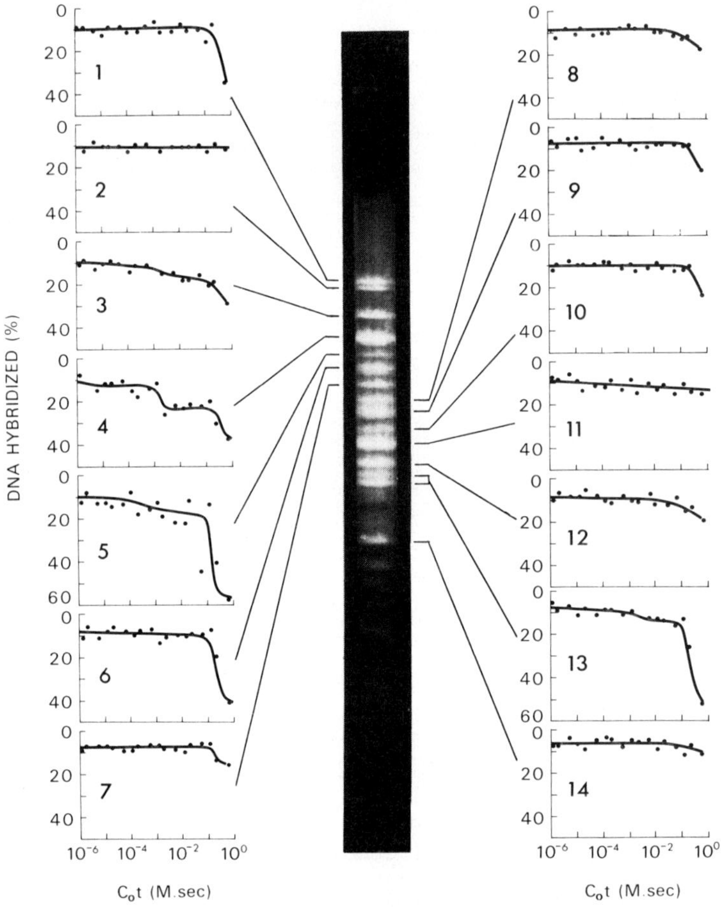

Fig. 4. Cot curve analysis of total chloroplast RNA hybridizing to labelled fragments derived from chloroplast DNA by digestion with EcoRI endoculease. Chloroplast DNA was labelled with ^{3}H-dCTP and ^{3}H-TTP using E. coli DNA polymerase 1 and then digested with EcoRI endonuclease. DNA fragments were recovered from the agarose gel and hybridized to chloroplast RNA as detailed for Fig. 3.

In situ hybridization of stable chloroplast RNA species to DNA fragments: A method has been described recently[14] for transferring DNA fragments directly from an agarose gel to a cellulose nitrate membrane without disturbing the geographical relationship of the fragments to each other. Normal RNA-DNA hybridization procedures are then applied and the DNA bands to which the labelled RNA hybridizes are located by autoradiography. We have examined the hybridization of 23S, 16S, 5S and 4S RNAs, individually and together, to EcoRI digested chloroplast DNA and the results are shown in Fig. 5. The 23S RNA hybridized mainly to band 3 but also significantly to band 11; 16S RNA hybridized mainly to band 4 but also slightly to band 3. Chloroplast 5S RNA hybridized to band 11 only, while transfer RNA hybridized to a number of bands (4,11,14). A mixture of 23S, 16S and 5S RNAs hybridized to bands 3, 4 and 11 as would be expected from the patterns obtained for the individual species.

DISCUSSION

From the data presented in this paper a number of conclusions can be drawn concerning the distribution of various coding functions among the fragments produced by EcoRI cleavage of spinach chloroplast DNA. These are summarised in Table 1. Allocation of messenger RNA cistrons to the various bands follows from the Cot analyses and bands 1, 5, 6 and 13 are clearly involved in coding for messenger RNA species present in significant amounts in the chloroplast. The

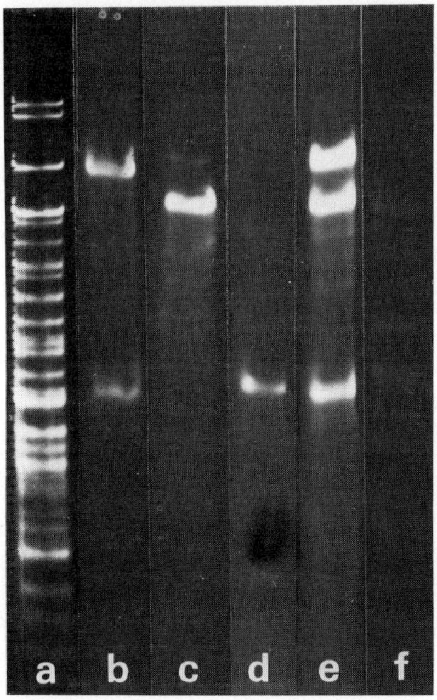

Fig. 5. Hybridization of chloroplast ribosomal and transfer RNAs to EcoRI-digested chloroplast DNA by the Southern technique[14]. Restricted chloroplast DNA was electrophoresed on a 1% agarose slab gel (4 μg DNA/slot), stained with ethidium bromide, photographed and each channel then cut out. The DNA fragments were alkali-denatured in the gel, neutralized and transferred to a cellulose nitrate membrane exactly as described by Southern[14]. Membranes were hybridized with ^{125}I-RNA (2 x 10^6 cpm/μg) for 16 hr at 45° in 50% formamide, 4 x SSC, 10 mM Tris-HCl, pH 7, 1 mM KI, 0.01% SDS. After washing, ribonuclease-treatment and further washing, the membranes were dried and autoradiographed. a Photograph of stained gel. b to f autoradiographs of membranes hybridized to b, 23S RNA; c, 16S RNA; d, 5S RNA; e, 23S, 16S and 5S RNAs; f, 4S RNA.

messenger RNA for the large subunit of Fraction I protein is probably present in greater amounts in the chloroplast than any other messenger species and it is thus likely that the cistron for the large subunit of this protein is located in one of these fragments.

Allocation of the cistrons for the stable RNA species is based mainly on the *in situ* hybridization data but it is supported by the Cot analyses and partly by the filter hybridization results. The fact that the.Cot curve analyses implicate bands 5 and 13 as also being involved in coding for ribosomal RNAs, but direct *in situ* hybridization does not, may be due to the fact that, in the former experiments total RNA from chloroplasts was used, whereas in the latter experiments highly purified 23S, 16S and 5S RNA fractions were used. Total chloroplast RNA could contain significant amounts of RNA fragments which are lost during the processing of the primary ribosomal RNA transcript to the mature 23S, 16S and 5S species.

Finally, the *in situ* hybridization data allow us to order the sequence of the three ribosomal RNA genes within the chloroplast DNA molecule. Assuming that the small fraction of 16S RNA which hybridizes to band 3 hybridizes to the same DNA fragment as 23S RNA does, and that the small fraction of 23S RNA which hybridizes to band 11 hybridizes to the same DNA fragment as 5S RNA does in that band, then one must conclude that the 23S RNA cistron lies between the 16S and 5S RNA cistrons. The assumptions are not unrealistic in the light of the observation that a component fragment of each of bands 3, 4 and 11 is rich in GC (Figure 1). A sequence of 16S - 23S - 5S would be analogous to the sequence of ribosomal RNA genes in bacteria[15].

REFERENCES

1 Hedgpeth, J., Goodman, H.M. and Boyer, H.W. (1972). Proc. Nat. Acad. Sci. U.S.A. 69, 3448-3452.

2 Atchison, B.A., Whitfeld, P.R. and Bottomley, W. (1976). Submitted for publication.

3 Spencer, D. and Whitfeld, P.R. (1967). Arch. Biochem. Biophys. 121, 336-345.

4 Robertson, H.D., Dickson, E., Model, P. and Prensky, W. (1973). Proc. Nat. Acad. Sci. U.S.A., 70, 3260-3264.

5 Hamer, D.H. and Thomas, C.A. (1975). Chromosoma, 49, 243-267.

6 Kemp, D.J. (1975). Nature, 254, 573-577.

7 Manning, J.E., Wolstenholme, D.R. and Richards, O.C. (1972). J. Cell Biol. 53, 594-601.

8 Ingle, J., Possingham, J.V., Wells, R., Leaver, C.J. and Loening, W.E. (1970). Symp. Soc. Exp. Biol. 24, 303-325.

9 Prunell, A. and Bernardi, G. (1974). J. Mol. Biol. 86, 825-841.

10 Thuring, R.W.J., Anders, J.P.M. and Borst, P. (1975). Anal. Biochem. 61, 213-220.

11 Gillespie, D. and Spiegelman, S. (1965). J. Mol. Biol. 12, 829-842.

12 Britten, R.J. & Kohne, D.E. (1968). Science, 161, 529-540.

13 Thomas, J.R. and Tewari, K.K. (1974). Biochim. Biophys. Acta 361, 73-83.

14 Southern, E.M. (1975). J. Mol. Biol. 98, 503-517.

15 Doolittle, W.F. and Pace, N.R. (1970). Nature, 228, 125-129.

Genetics and Biogenesis of Chloroplasts and Mitochondria, Th. Bücher et al. eds.

PHYSICAL AND TRANSCRIPTIONAL MAPPING OF ZEA MAYS CHLOROPLAST DNA

John R. Bedbrook and Lawrence Bogorad
The Biological Laboratories, Harvard University
16 Divinity Ave., Cambridge, Mass. 02138,USA

INTRODUCTION

The closed circular DNA molecules from chloroplasts of higher plants have been well characterized by electron microscopy (1). We present here a map of restriction enzyme recognition sites on circular DNA from Zea mays chloroplasts. The region of the map complementary to the chloroplast ribosomal genes has been identified. Chloroplast DNA (ct DNA) restriction enzyme fragments have been cloned in E.coli. A fine structure map of the ribosomal genes from Zea mays ct DNA has been made using the cloned DNA fragment complementary to ribosomal RNA.

METHODS AND MATERIALS

The isolation and assay of restriction enzymes used in this work have been described previously (2). RNA hybridization to ct DNA restriction enzyme fragments (3) and gel electrophoresis techniques have been presented (4,5). Molecular cloning of DNA fragments in E.coli was carried out essentially as described previously (6,7).

Experimental design for restriction site mapping. Ct DNA is digested with three restriction endonucleases Sal 1, Bam 1 and $EcoR_1$. DNA fragments produced are separated by agarose gel electrophoresis and visualized under UV light after staining with ethidium bromide. The relative stoichiometry of DNA fragments is determined by quantitative fluorimetry. The sequence complexity of size classes of ct DNA fragments produced by one enzyme and present in more than 1 copy per genome is determined by size analysis of the digestion products produced by other enzymes.

Sal 1 gives 8 size classes and 10 different sequences when used to digest maize ct DNA. Bam 1 or $EcoR_1$ fragments from ct DNA which contain recognition sequences for Sal 1 are used to determine which Sal 1 fragments are adjacent. Bam 1 and $EcoR_1$ fragments with Sal 1 sites are termed Bam 1* or $EcoR_1$* fragments. 32P-phosphate labelled copy RNA (cRNA) made using Bam 1* or $EcoR_1$* as templates is hybridized to the fractionated Sal 1 fragments after they have been transferred to Millipore filter strips (3). Autoradiographs of the hybridized strips and photographs of the fractionated Sal 1 fragments are aligned. The Sal 1 fragments overlapped by the Bam 1* or $EcoR_1$* fragments are determined by inspection.

RESULTS

Size and stoichiometry of ct DNA fragments produced by endonucleases Sal 1, Bam 1 and $EcoR_1$. Fig. 1 shows agarose gel electrophoresis of ct DNA terminally

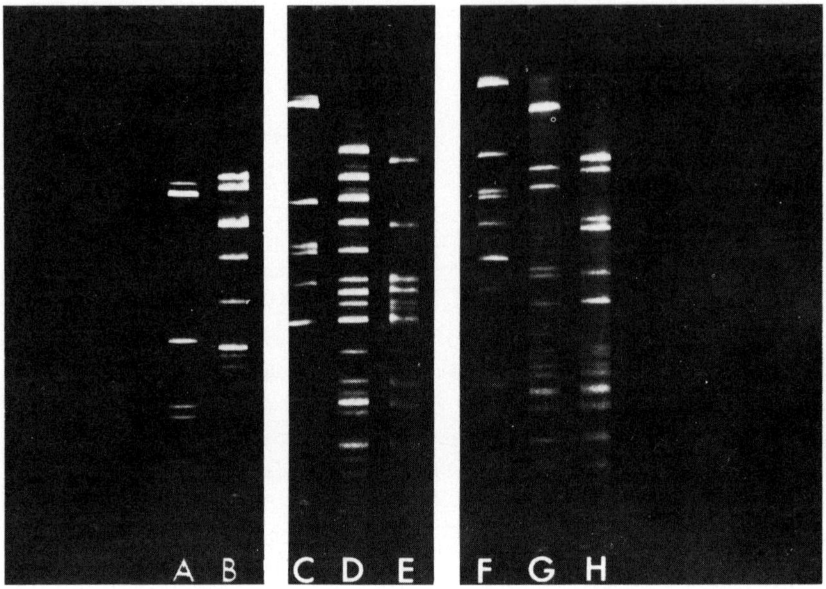

Figure 1 Gel electrophoresis of chloroplast DNA fragments produced by Sal 1, Bam 1 and $EcoR_1$. A.Lambda phage DNA/$EcoR_1$ (0.75% gel); B. Chloroplast DNA/Sal 1 (0.75% gel); C. Lambda phage DNA/$EcoR_1$ (0.85% gel); D. Chloroplast DNA/Bam 1 (0.85% gel); E. Chloroplast DNA/Bam 1 and Sal 1 (0.85% gel); F. Lambda phage DNA/$EcoR_1$ (1% gel); G. Chloroplast DNA/$EcoR_1$ (1% gel); H. Chloroplast DNA/$EcoR_1$ and Sal 1 (1% gel).

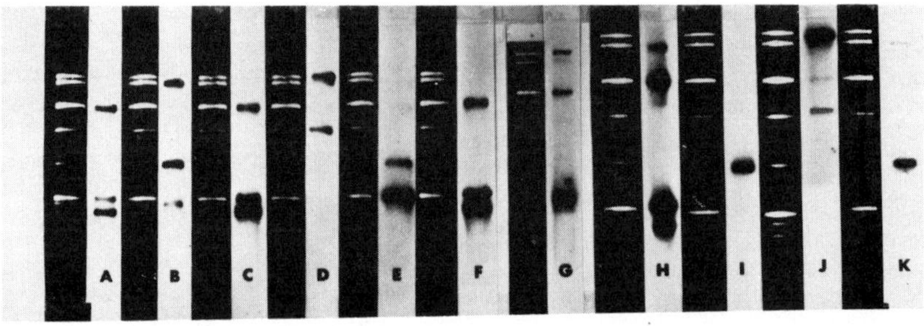

Figure 2 Homology between Bam 1* and $EcoR_1$* fragments with Sal 1 fragments of chloroplast DNA. ^{32}P-phosphate labelled cRNA from Bam 1*or $EcoR_1$* fragments was hybridized to the fractionated Sal 1 fragments. The gels and autoradiographs of hybridizations to the Sal 1 fragments were aligned and photographed. A. cRNA, Bam 1 fragment hybridized to Sal 1 fragments,B. cRNA from Bam 1 fragment 2, C. cRNA from Bam 1 fragment 4, D. cRNA from Bam 1 fragment 5, E. cRNA from Bam 1 fragment 12, F. cRNA from Bam 1 fragments 13 and 13', G. cRNA from Bam 1 fragment 3, H. cRNA from $EcoR_1$ fragment a, I. cRNA from $EcoR_1$ fragment c, J. cRNA from $EcoR_1$ fragment d, K. cRNA from $EcoR_1$ fragment g.

digested with Sal 1 (Fig. 1B) Bam 1 (Fig. 1D) and $EcoR_1$ (Fig. 1G). Figs. 1A, C and F are Lambda phage DNA digested with $EcoR_1$ (8) plus phage DNA digested with Hae III (5,9) as size markers. Table 1 gives the name, size, stoichiometry and sequence complexity of ct CNA fragments seen in Fig. 1. Ct DNA Sal 1 fragments (Fig. 1B) sum to the molecular weight determined for this genome by electron microscopy (1). The Bam 1 and $EcoR_1$ fragments shown in Fig. 1 represent approximately 80% and 60% of the genome respectively. Many more size classes of ct DNA fragments produced by these latter two enzymes can be shown by electrophoresis on higher percentage gels. The gels of Fig. 1 give optimal resolution of Bam 1 and $EcoR_1$ fragments used to map the Sal 1 fragments.

TABLE I

Sal 1 fragments			Bam 1 fragments[1]			$EcoR_1$ fragments[1]		
Name	Size[2]	Copies[3]	Name	Size	Copies	Name	Size	Copies
A	25	1	1	15.0	1	a	12.5	2
B	21	1	2	11.0	1	b	6.3	1
C	16	1	3	9.7	1	c	5.6	1
C'	16	1	4	7.1	1	d	2.85	1
D	12.7	1	5	6.0	1	e	2.8	1
E	9	1	6	4.9	1	f	2.2	1
F	6.7	2	7	4.6	2	g	2.0	1
F'	6.7	1	8	4.3	1	h	1.9	1
G	6.3	0.8	9	3.9	2	i	1.85	1
H	6.1	1	10	3.2	1	j	1.8	2
			11	2.5	1	k	1.7	2
			12	2.4	1	l	1.65	3
			13	2.3	2	m	1.6	2
			13'	2.3	1	n	1.5	3
			14	2.1	1	o	1.25	-
			15	2.0	2	p	1.2	-
			16	1.8	1			
			17	1.5	2			

Size and stoichiometry of chloroplast DNA fragments produced by three restriction endonucleases.
[1]Only those fragments shown in Fig.1 are tabulated.
[2]Size in kilobases
[3]Number of copies of fragments per chloroplast genome

Hybridization of cRNA from Bam 1* and $EcoR_1$*fragments to Sal 1 fragments. Comparison of Figs. 1D and E or Figs. 1G and H show that Bam 1 fragments 1, 2, 3, 4, 5 and 12 and $EcoR_1$ fragments a, c, d and g are further digested by Sal 1. Sal 1 digestion of purified fragments showed that Bam 1 fragments 7 and 13' also contained Sal 1 recognition sequences. The position of Sal 1 sites in these fragments has been determined (2). $EcoR_1$ fragment a and Bam 1 fragment 3 have two Sal 1 recogniton sites and Sal 1 fragments H and C,and B and C' have $EcoR_1$ and Bam 1 fragments in common. Hybridization of cRNA from the Bam 1* and $EcoR_1$* fragments is given in Fig. 2. Hybridization of Bam 1* fragment 7 is not shown in Fig 2. CRNA from fragment 7 hybridizes to Sal 1 fragments F, H and C'. Distinguishing hybridization to C or C' and F or F' was done by hybridizing cRNA from Bam 1* or $EcoR_1$* fragments to isolated C+C' digested with Bam 1 and F+F' digested with $EcoR_1$ after the products had been fractionated by agarose gel electrophoresis.

Order of the Sal 1 fragments. Fig 2 shows that cRNA from $EcoR_1$ fragment a hybridizes Sal 1 fragments B, C, F and H. Hybridization of fragment a to isolated C+C' digested with Bam 1 showed that C+C' were hybridized by cRNA from a. These data along with the hybridization of cRNA from Bam 1 fragments 1, 4, 7 and 13' indicated that the two copies of a extend from H or C through F into B on the one hand and through F into C' on the other. The two copies of a are in an inverted orientation. The rationale for the inverted repeat and for the remainder of the Sal 1 fragment ordering has been given (2). Fig. 2 shows cRNA from Bam 1 fragment 2 hybridizes to Sal 1 fragment B and E as well as to G. Since G does not contain a Bam 1 site and Bam 1 fragment 2 does not contain G within it we assume that 2 overlaps the Sal 1 site between B and E and contains a sequence in common with Sal 1 fragment G. B joining E is confirmed by hybridization of cRNA from $EcoR_1$ fragment d to B and E. CRNA from Bam 1 fragment 3 hybridizes A, C'and G. G is within Bam 1 fragment 3. Therefore Bam 1 fragment 3 extends from C' through G into A. Hybridization of cRNA from Bam 1 fragment 5 and $EcoR_1$ fragment g imply Sal 1 fragments A and D are contiguous. Since this DNA is a covalently closed circle and since the Sal 1 fragments shown in Fig. 1 represent the total genome we assume that Sal 1 fragments D and F' are joined. Fig. 3 gives the map of Sal 1, Bam 1 and $EcoR_1$ restriction sites on chloroplast DNA.

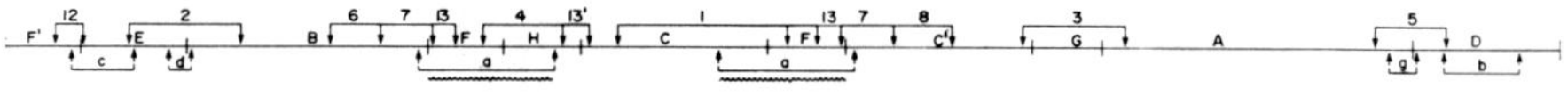

Figure 3 Location of Sal 1, Bam 1 and $EcoR_1$ recognition sites in Zea mays chloroplast DNA. Sal 1 fragments and recognition sites are shown in the central horizontal line by vertical lines. Bam 1 fragments and recognition sites are indicated on the upper horizontal line (arrows with solid tails). $EcoR_1$ fragments and recognition sites are shown on the lower line by arrows with dashed tails. General region of ribosomal RNA genes is within the squiggled horizontal line.

Location and Fine structure mapping of the ribosomal genes of Zea mays chloroplasts. P^{32}-phosphate labelled total chloroplast ribosomal RNA hybridizes to $EcoR_1$ fragment a and to no other $EcoR_1$ fragments. We cloned $EcoR_1$ fragment a in E.coli using the tetracycline resistance plasmid pMB9. Large quantities of a were prepared and the position of recognition sites for the enzymes Sal 1, Bam 1 and Hind III were determined. Purified P^{32}-phosphate labelled 16 and 23S ribosomal RNA were hybridized to the Sal 1, Bam 1 and Hind III digestion products of fragment a after the products had been fractionated by gel electrophoresis. The relationship between restriction enzyme recognition sites and the ribosomal RNAs is given in Fig.4.

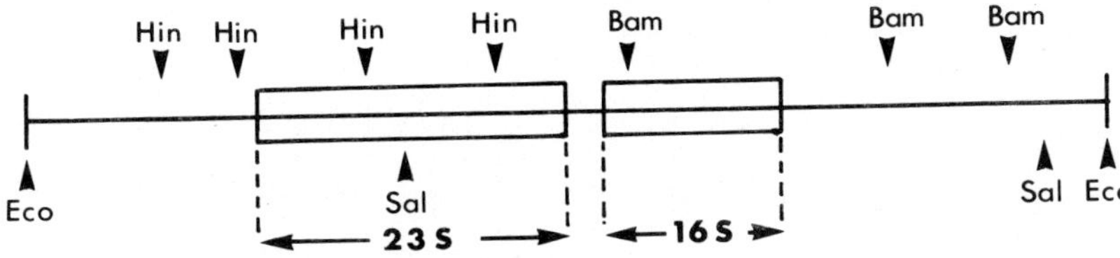

Figure 4 Fine structure map of the ribosomal genes of *Zea mays* chloroplast DNA. The location of the RNAs in the map is approximate because of the relatively large size of the restriction fragments relative to the RNAs. The location given is based on quantitative hybridization to the various restriction fragments.

DISCUSSION

The experiments reported here show that restriction enzyme recognition sites in circular ct DNA can be ascribed a discrete order. The fragment map of ct DNA shown in Fig. 3 contains approximately 15% of the genome length in an inverted repetition. We have no knowledge of a role for the inverted repeat in ct DNA except that it contains the chloroplast ribosomal genes. Fine structure mapping shows that the ribosomal genes for the 16S and 23S RNA are closely linked as in E.coli (10,11) and are represented twice per genome. The two copies of the ribosomal genes are spaced by a non-homologous DNA sequence representing more than 10% of the genome length.

Acknowledgements. Supported in part by GM20470-NIGMS and in part by the Maria Moors Cabot Foundation of Harvard University.

REFERENCES

1. Kolodner,R.D. & Tewari, K.K.(1975) Biochem.Biophys.Acta 402,372-390.
2. Bedbrook, J.R. & Bogorad, L.(1976) Proc.Nat.Acad.Sci.USA (In press).
3. Southern,E.M.(1975) J.Mol.Biol.98, 503-517.
4. Hayward,G.S. & Smith,M.H.(1972) J.Mol.Biol. 63, 383-395.
5. Hamer,D.H. and Thomas,C.A.(1976) Chromosoma 49, 243-267.
6. Cohen,S.N., Chang, A.C.Y., Boyer,H.W. & Helling,R.B.(1973) Proc.Nat.Acad.Sci. USA 70, 3240-3244.
7. Tanaka,T. & Weisblum,B. (1975) J.Bacteriol.121, 354-362.
8. Thomas,M. & Davis,R.W.(1975) J.Mol.Biol. 91, 315-328.
9. Allet, B. & Solem,R. (1974) J.Mol.Biol. 85, 475-484.
10. Lund,E., Dahlberg, J.E., Lindahl, L., Jaskunas, S.R., Dennis, P.P. and Nomura, M. (1976) Cell,7, 165-177.
11. Yamamoto, M., Lindahl, L. and Nomura,M. (1976) Cell, 7, 179-190.

Genetics and Biogenesis of Chloroplasts and Mitochondria, Th. Bücher et al. eds.

STUDIES WITH CHLOROPLAST DNA-PLASMID HYBRIDS FROM CHLAMYDOMONAS REINHARDI

Jean-David Rochaix
Département de Biologie Moléculaire
Université de Genève
1211 Geneva 4, Switzerland

INTRODUCTION

The chloroplast DNA of C. reinhardi has a kinetic complexity of the order of 200 x 10^6 daltons (1,2). It contains the genes for the 16 S and 23 S chloroplast ribosomal RNAs (3) and presumably some tRNA genes. What the remaining of the chloroplast DNA codes for is still largely unknown. Theoretically it could code for over 100 proteins with an average molecular weight of 30,000 daltons. We have attempted to develop a general method for identifying proteins coded for by chloroplast DNA. The idea is to fractionate the chloroplast DNA into specific parts, to clone and amplify these fragments, to use them as a template for in vitro protein synthesis and finally to identify the in vitro products with chloroplast polypeptides. This report presents preliminary results obtained by this method.

RESULTS

Plate I.A shows an agarose gel electrophoresis of chloroplast DNA fragments of C. reinhardi obtained after digesting the DNA with various restriction enzymes. It can be seen that EcoRI generates at least 35 different fragments (slot 3), Bgl II produces 13 (slot 4), Bam I produces 17 (slot 5), Hind III produces over 60 (slot 6) and Sal I cuts the DNA only at a small number of sites. Besides the smear at the top of the gel, five bands can be recognized in the Sal I digest (slot 7). The presence of DNA fragments in nonstoichiometric amounts among the Bgl II, Bam I and Sal I digestion products is not yet understood. These bands could represent partial digests, nuclear contaminants, or more interestingly, they might indicate some heterogeneity in the chloroplast DNA sequences. The size of the EcoRI fragments has been estimated both by electron microscopy and by comparison with DNA fragments of known molecular weight. The fragments shown in slot 3 of Plate I.A range in size from 0.58 to 12.3 x 10^6 daltons. If one sums up the molecular weights of the fragments, taking into account that some of them comigrate in the gel, one obtains a molecular weight of 120 ± 5 x 10^6 daltons, a value which is significantly lower than the one estimated by reassociation kinetics (1,2).

The EcoRI chloroplast DNA fragments have been inserted into the pCRI plasmid (constructed by J. Carbon) and cloned according to published procedures (4). A detailed description will be published elsewhere. The pCRI plasmid contains the

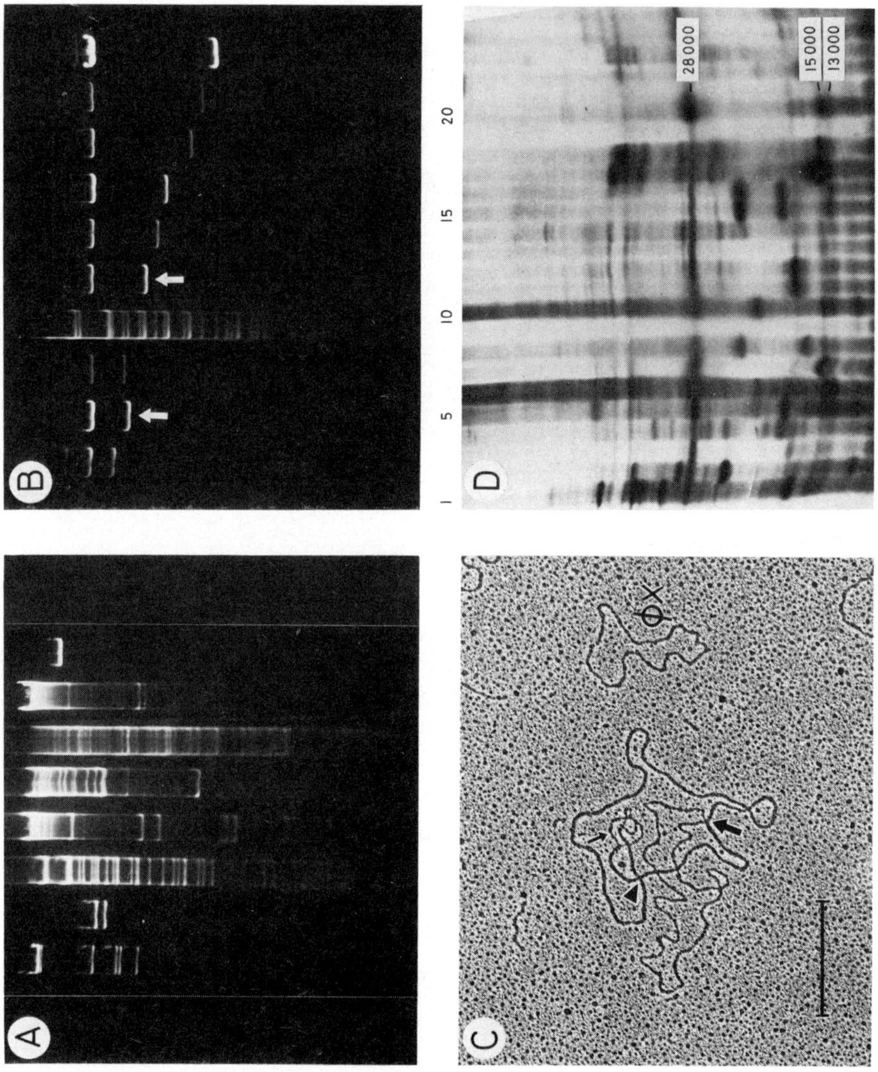
A
B
C
D
ΦX
28000
15000
13000

ColEI plasmid and a DNA fragment coding for kanamycin resistance. It has a molecular weight of 8.7 x 10^6 daltons and contains a single EcoRI site. Hybrid chloroplast DNA-plasmid molecules are recognized by the fact that after EcoRI digestion at least two fragments are visible upon agarose gel electrophoresis (Plate I.B). The upper bands in the gel correspond to linear pCRI DNA, the lower bands represent chloroplast DNA fragments. It can be seen that each of the chloroplast DNA fragments of the hybrids migrates at the same position in the gel as one DNA fragment from a total chloroplast DNA -EcoRI digest shown in slot 4 of Plate I.B. Of the 35 chloroplast DNA EcoRI fragments, 26 have now been inserted into plasmids.

In order to determine which chloroplast DNA EcoRI fragment contains the chloroplast ribosomal RNA genes, 28 hybrids were denatured, loaded on nitrocellulose filters and challenged with ^{3}H-labelled E. coli ribosomal RNA. This RNA is known to crosshybridize with chloroplast DNA (unpublished observations). Two hybrids, (pCP51 and pCP19) were positive in this test. This was confirmed by repeating the hybridization with purified ^{32}P-labelled chloroplast 23 S and 16 S ribosomal RNA. The results are shown in Table I. The two hybrids can be seen in slots 2 and 5 of Plate I.B; the corresponding chloroplast DNA fragments with molecular weights of 5.4 and 4.5 x 10^6 daltons are indicated by arrows. The densities of these fragments in CsCl, as determined by analytical ultracentrifugation, are 1.696 and 1.697 g/cc respectively. Plate I.C shows a heteroduplex formed between pCRI and a hybrid containing a 6.4 x 10^6 daltons chloroplast DNA fragment. The latter appears as a single-stranded DNA loop whose site of insertion is marked with a thick arrow. (For further details see Plate legend).

The chloroplast DNA-plasmid hybrids have been used in an in vitro E. coli coupled transcription-translation system (6). Plate I.D shows an autoradiogram of in vitro synthesized polypeptides from various hybrids fractionaed on an SDS polyacrylamide gel (slots 1-19). The polypeptides were labelled with ^{35}S-methionine.

Plate I. (A). Agarose gel electrophoresis of chloroplast DNA of C. reinhardi digested with EcoRI (slot 3), BglII (slot 4), Bam I (slot 5), Hind III (slot 6) and Sal I (slot 7). Slot 1 contains λ DNA digested with EcoRI, slot 2 contains PML2 DNA (identical to pCRI except that it contains two EcoRI sites) digested with EcoRI and slot 8 contains linear pCRI DNA. These DNA fragments were used as size markers. (B) Agarose gel electrophoresis of chloroplast DNA-pCRI hybrids digested with EcoRI; slot 4 contains total chloroplast DNA digested with EcoRI; see text for further details. (C) Heteroduplex between pCRI and a chloroplast DNA-plasmid hybrid molecule. The thick arrow indicates the insertion site of the chloroplast DNA fragment into the plasmid. The thin arrow indicates two small single-stranded DNA loops at the end of two double stranded DNA stems originating from the same site on the plasmid marked by ∇. This loop-stem structure represents the DNA region coding for kanamycin resistance flanked by inverted sequences (5). The single stranded circle on the right represents a ϕX 174 molecule. The scale is 0.5 μ. (D) Autoradiogram of in vitro synthesized ^{35}-S labelled proteins from various chloroplast DNA-pCRI hybrids electrophoresed on a 12.5% SDS-polyacrylamide gel.

TABLE I

HYBRIDIZATIONS WITH CHLOROPLAST RIBOSOMAL RNA

DNA	μg DNA loaded/filter	cpm/filter
Chloroplast DNA	2.0	1010
pCP 51	5.3	1214
pCP 19	3.2	1178
pCP 27	4.6	74
pCRI	2.9	47

The specific activity of the ^{32}P-labelled chloroplast ribosomal RNA was 74000 cpm/μg. All filters were hybridized in the same vial in the presence of 3.2 μg of RNA. Hybridization reactions were performed in 4 x SSC, 30% formamide at 37°C. pCP 27 is a chloroplast DNA-plasmid hybrid.

Those specified by the pCRI plasmid alone (shown in slot 20) consist of a major polypeptide of 28,000 daltons and of minor polypeptides ranging in size from 13,000 to 15,000 daltons. This gel allows one to distinguish easily between the polypeptides originating from pCRI (visible in all of the slots) and those synthesized from the chloroplast DNA. The latter polypeptides range in size from 12,000 to 45,000 daltons. Some of the hybrids produce identical products, e.g. those of slots 11 and 12 and those of slots 15 and 16. These hybrids were isolated independently. The fact that some of the hybrids are much more active in *in vitro* protein synthesis than the pCRI plasmid suggests that transcription starts at chloroplast DNA "promoters". However, whether these bear a relationship to true chloroplast promoters remains an open question. We are attempting to identify some of the *in vitro* polypeptides with specific chloroplast proteins by comparative electrophoresis and fingerprinting and also by immunological means.

ACKNOWLEDGMENTS

I thank B. Allet for help with the *in vitro* coupled system and for Hind III, F. Rougeon for EcoRI and ligase, T. Bickel for BglII and Bam I and J. van Dillewijn for technical assistance. This work was supported by grant 3.285.74 from the Fonds National Suisse pour la Recherche Scientifique.

REFERENCES

1. Bastia, D. et al. (1971) Proc. Nat. Acad. Sci. U.S. 68, 1157-1161.
2, Wells, R. and Sager, R. (1971) J. Mol. Biol. 58, 611-622.
3. Bastia, D. et al. (1971) 11th Ann. Meeting Americ. Soc. Cell Biol. Abst. 41
4. Morrow, J. et al. (1974) Proc. Nat. Acad. Sci. U.S. 71, 1743-1747
5. Berg D. et al. (1975) Proc. Nat. Acad. Sci. U.S. 72, 3628-3632
6. Zubay, G. et al. (1970) in The Lactose Operon, ed. by Beckwith, J.R. and Zipser, D. Cold Spring Harbor, p. 375-392.

Genetics and Biogenesis of Chloroplasts and Mitochondria, Th. Bücher et al. eds.

REPLICATION OF CIRCULAR CHLOROPLAST DNA

K. K. Tewari, Richard D. Kolodner and William Dobkin
Department of Molecular Biology and Biochemistry
University of California Irvine
Irvine, California 92664

INTRODUCTION

The chloroplast (ct) DNAs from a number of higher plants have been obtained as covalently closed circular DNA molecules and their molecular weights have been found to range from 85×10^6 to 96×10^6.[1] For any given ctDNA, there is an excellent agreement between the molecular weights obtained by equilibrium sedimentation, electron microscopic measurements, and renaturation kinetics, which suggests that the sequence of a single circular ctDNA molecule represents the entire information content of the ctDNA[2]. The denaturation mapping studies on pea ctDNA also support the conclusion that ctDNA molecules are essentially homogeneous[3] in base sequences. In order to study the replication of circular ctDNA, we have examined the pea and corn ctDNA molecules for replicative intermediates. The data have shown that pea and corn ctDNA molecules replicate by the introduction of two displacement (D) loops which are hydrogen bonded to the opposite parental strands.[4] These D-loops expand towards each other forming Cairns type of replicative intermediate when the two D-loops have elongated past each other. Cairns replicative intermediates have been found to elongate bidirectionally. A unique observation has been made in the replication of ctDNA. The ctDNA has also been found to replicate by the rolling circle model of replication.[5] The initiation site for the rolling circle replication has been found to be at or near the termination site of the Cairns round of replication.

The isolated chloroplasts from higher plants are known to synthesize DNA in the presence of four deoxynucleoside triphosphates (dNTP). The polymerized product has been shown to resemble ctDNA by its density in CsCl gradients, renaturation property, and hybridization with the ctDNA.[6] We have succeeded in removing the endogenous DNA from the solubilized DNA polymerase activity of chloroplasts by passing it through a DEAE cellulose column. The solubilized fraction has been found to replicate DNA _in vitro_. Further purification of this fraction on DEAE and phosphocellulose columns has shown that there are at least three proteins which synthesize DNA. Our long term goal is to identify and purify the proteins involved in the formation of replicative intermediates observed _in vivo_, and eventually to reconstitute the faithful replication of ctDNA _in vitro_.

RESULTS AND DISCUSSION

Displacement loops in ctDNA: The ctDNA was isolated from 10 day old pea or corn leaves, fractionated in cesium chloride-ethidium bromide (CsCl-EtBr) gradients, and collected into three fractions; the lower band of closed circular DNA containing 20-25% of the total ctDNA, the upper band of nicked and fragmented circular DNA consisting of 65-70% of total ctDNA, and the area between the two bands (middle band) which contained 1% of the total ctDNA. The closed circular DNA (lower band DNA) from pea was prepared for electron microscopy using the formamide technique.[4] Pea ctDNA molecules containing one or two D-loops, or up to seven denaturation (Den-) loops, or a combination of both structures were observed (Fig. 1). The D-loops could be distinguished from the Den-loops because one side of the D-loop appeared to be double stranded while the other side appeared to be single stranded because of its thinner and kinkier appearance. In addition, the D-loops observed in the pea ctDNA molecule exhibited branch migration. The frequency of molecules containing D-loops was not affected by RNase but brief treatment of ctDNA with alkali released the displacing strand, suggesting that the displacing strand is DNA and hydrogen bonded to the parental molecule. In addition to D-loops, the closed circular molecules contained Den-loops. The Den-loops were located at 10 distinct regions on a circular map that was the size of pea ctDNA. This map closely corresponds to the denaturation maps that were constructed using nicked circular pea ctDNA. The two D-loops which were positively identified are located at two adjacent sites and the positions of the two D-loop sites were different from the positions of the Den-loop sites on the ctDNA. The smallest size of D-loops was 820 base pairs and the average distance between the outside edges of the two D-loops was 7.15 kilo base pairs. The inner distance between the two D-loops was highly variable, indicating that the two D-loops expand towards each other. Small "Cairns type" of replicative forked structure ranging in size from 7.2 to 10.6 kilobases have been observed in closed circular pea ctDNA. These Cairns structures map at the position of the two D-loops and probably result when the two displacing strands expand past each other in the opposite strands. This confirms that the two displacing strands are located on the opposite parental strands of the pea ctDNA molecule. The two D-loops in corn ctDNA are also located at two adjacent sites. The size of the corn ctDNA D-loop corresponds to 860 base pairs and the outer distance between the two D-loops was 7.06 kilobase pairs.

The frequency of D-loops in the total ctDNA of pea was analyzed. For this analysis, all of the D-loops that were larger than the small size class of D-loops were classified as expanding D-loops. In the lower band of the pea ctDNA, the frequency of D-loops was greater than the frequency of expanding D-loops. The frequency of expanding D-loops was higher than the frequency of small D-loops in the middle band. The correlation between D-loop expansion and a higher banding position in the CsCl-EtBr density fradient is an indication of replication on a closed circular template accompanied by nicking and closing cycle. Similar results were obtained with corn ctDNA.

D-loops and Den-loops were also observed in circular dimers and catenated dimers of ctDNA in pea and corn. The relative positions of the D-loops and Den-loops in a given monomer length segment of the two catenated molecules and the three circular dimers of pea ctDNA correspond to the positions of D-loops and Den-loops that were mapped in the circular monomers.

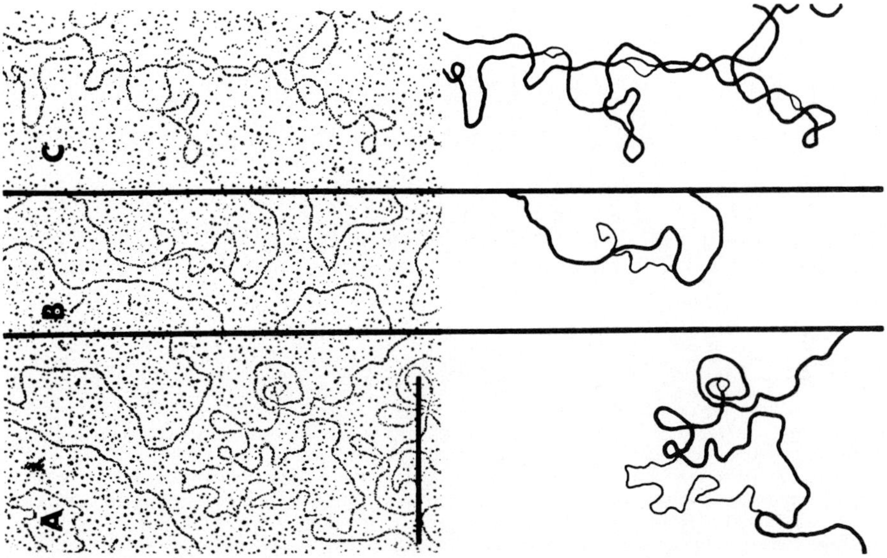

Fig. 1. High magnification electron micrographs of ctDNA molecules. A) a molecule containing one D-loop and one Den-loop from the pea ctDNA, B) a branch migrating D-loop from corn ctDNA. C) a pea ctDNA molecule containing three Den-loops. The thin and thick lines in drawing represent the single stranded and double stranded regions, respectively. The bar indicates 1 μm.

Cairn's replicative intermediates: Cairns type of replicative forked structures were found in the DNA from lower, middle, and upper bands (Fig. 2). In pea ctDNA, three percent of the circular molecules in the lower band were Cairns structures and the extent of their replication ranged from 5.2% to 8.2%. The Cairns replicative intermediates made up an average of 5.7% of the circular DNA in the middle band and the extent of their replication ranged from 7% to 50%. The Cairns structure consisted of 2.9% of the circular ctDNA molecules in the upper band and the extent of their replication ranged from 38% to 87%. The finding of Cairns replicative intermediates in the lower and middle bands of the CsCl-EtBr density gradients, and the correlation between higher banding position and larger amount of replication again suggests that replication takes place on a covalently closed circular template, and is accompanied by nicking and closing cycles. In corn ctDNA, Cairns replicative intermediates made up 4.5% of the total circular ctDNA molecules and the extent of their replication ranged from 9.4% to 70%.

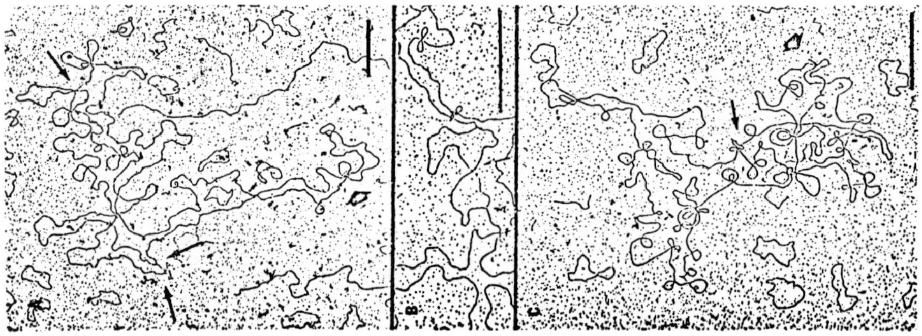

Fig. 2. Cairns replicative intermediates. A) a molecule that is 37% replicated and contains a single strand region (2 small arrows) at one growing fork and a denatured region (large arrow) in the unreplicated portion of the molecule. B) a portion of a molecule that is 5.2% replicated and contains a single strand region (2 small arrows) at one growing fork. C) a Cairns replicative forked molecule showing branch migration at one of the replicative forks (large arrow). The molecules were mounted for electron microscopy by the formamide technique, the spreading solution contained 50% formamide. Single stranded and double stranded OX DNA molecules were used as internal standards. The bars indicate 1 μm.

Rolling circle replicative intermediates: In the pea ctDNA preparations circular molecules with an attached double stranded tail accounted for 4.9% of the circular molecules in the upper band, and 2.8% of the circular molecules in the lower band (Fig. 3). There were no circular molecules with tails in the lower band of pea ctDNA. The lengths of tails in the pea ctDNA ranged from 1.5% to 124% of the length of the attached monomer length circular molecule. In corn ctDNA, 11% of the total circular molecules had tails which ranged in length from 2% to 140% of the attached monomer length circular molecule. The finding of tails that were longer than the attached monomer length circular molecule eliminated the possibility that the tails arose by breakage of a Cairns forked structure at a replicative fork.

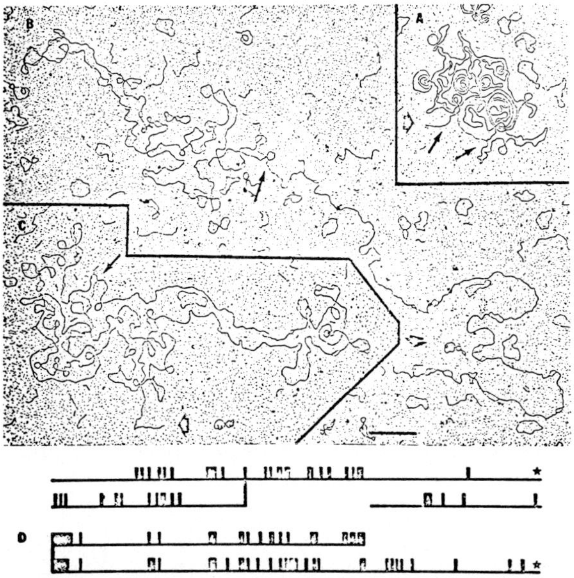

Fig. 3. Rolling circle molecules. A) a molecule with a tail that is 19% of the length of the attached circle. There is a single stranded region at the growing fork that is 3,500 bases long (indicated by arrows). B,C) partially denatured molecules with tails that illustrate the two different denatured patterns at the tip of the tails. D) graphical representation of molecules B and C. The boxes indicate the denatured regions, the * indicates the circular part of the molecule that has been linearized. The unmarked free end is the tip of each tail. The bars indicate 1 μm.

Single stranded regions in the replicative intermediates: The rolling circles and the Cairns forked structure in both pea and corn ctDNA showed that 77% to 81% of the growing forks in these molecules contained single stranded regions which were variable and ranged in length from 250 bases to 7,600 bases. In the Cairns structure, single stranded regions were observed at both of the replicative forks and were found to be trans with respect to each other suggesting that the Cairns structure grows bidirectionally. In both the rolling circles and the Cairns structures, the single stranded regions were located on only one daughter strand at each replicative fork.

Denaturation mapping of the replicative intermediates: The partially denatured Cairns replicative intermediates were ordered with respect to their position in the denaturation map of pea ctDNA that had an average denaturation of 22%.[3] These partially replicated molecules had a pattern that would be expected if they all had initiated replication at the same place and that both replicative forks were being extended at the same rate. The initiation site of these Cairns replicative intermediates lies at the same point in the denaturation map where D-loops were shown to map. The termination site for the Cairns round of replication appears to lie at a site 180° around the pea ctDNA molecule from the initiation site.

The denaturation mapping of rolling circles has shown that there were two types of rolling circle molecules. The tip of the tail of these two types of rolling circle molecules map at the same site but the sequence of two types of tails extend in opposite directions from the tip.

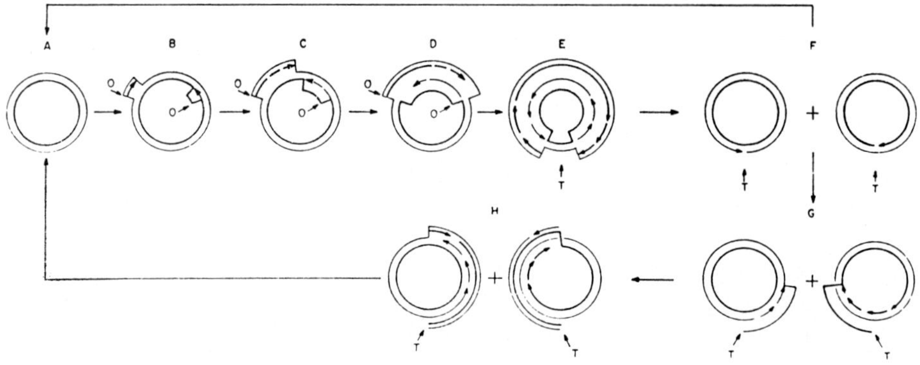

Fig. 4. A model for the replication of ctDNA. A) closed circular parental molecule. B) D-loop containing molecule. C) expanded D-loop containing molecule. D,E) Cairns type of replicative intermediate. F) nicked progeny molecules. G,H) rolling circles. The thin and thick lines mark the opposite strands of a molecule. The lines with the arrows are the daughter strands. The "O" indicates the positions of the two origins of D-loop synthesis which are 5.2% of pea ctDNA apart. The "T" indicates the terminus of the Cairns round of replication which is 180° opposed to the origins of D-loop synthesis.

A model for the replication of chloroplast DNA: Based upon the above results, a model for the replication of pea and corn ctDNA is presented in Fig. 4. The ctDNA replication is initiated by the formation of two displacement loops whose displacing strands are complementary to the opposite parental strands of ctDNA (Fig. 4b). The two displacing strands expand toward each other and initiate the formation of Cairns replicative forked structures (Fig. 4). The small Cairns forked structures (Fig. 4e) expand bidirectionally until termination takes place at a site that is 180° around the circular ctDNA molecule from the initiation site. Separation of the daughter molecules takes place yielding two circular molecules that each have a single strand break or small gap at the same site located in opposite daughter strands (Fig. 4f). In ctDNA, the nicked circles could be sealed to close the circles or the 3'OH of each nicked progeny molecule could be extended by a DNA polymerase molecule. This would displace a single stranded tail (Fig. 4g) from the molecule and this tail could be filled in by discontinuous duplex synthesis to yield a molecule with a double stranded tail (Fig. 4h). The tails might then be converted to circular molecules by an intrastrand recombination event. If both progeny from a Cairns round of replication initiated rolling circle synthesis, two types of rolling circles would be formed. In each case, the tip of the tail would map at the same site but the sequence of the two types of tails would extend in the opposite direction from this site and this is the result that was observed in the denaturation mapping experiments with pea ctDNA rolling circles.

DNA synthesis in vitro: The above experiments on replication have been carried out with in vivo replicative intermediates. In order to further understand the replication process, we must resolve and reconstitute each function outside the cell, and understand its regulation inside the cell. As a first step we have studied DNA synthesizing activity in the isolated chloroplasts. Sucrose density gradient purified pea chloroplasts were found to carry out a very active DNA synthesis (DS) in the presence of dNTP utilizing endogenous ctDNA. The lysis of chloroplasts with triton, deoxycholate and EDTA failed to solubilize any significant amount of DS activity from chloroplasts. The DS activity was completely solubilized when chloroplasts were treated with 0.5% Lubrol.

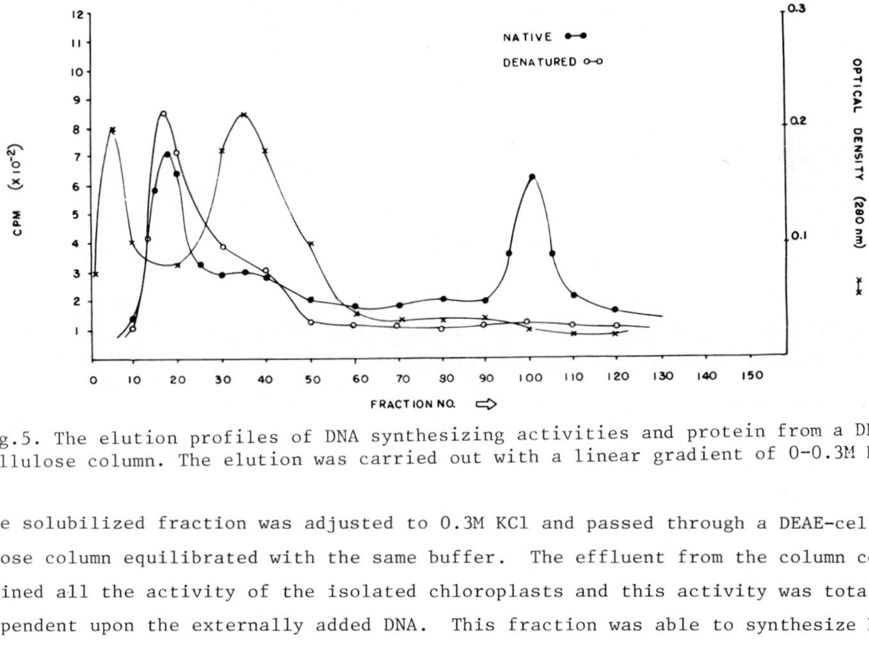

Fig.5. The elution profiles of DNA synthesizing activities and protein from a DEAE cellulose column. The elution was carried out with a linear gradient of 0-0.3M KCl.

The solubilized fraction was adjusted to 0.3M KCl and passed through a DEAE-cellulose column equilibrated with the same buffer. The effluent from the column contained all the activity of the isolated chloroplasts and this activity was totaly dependent upon the externally added DNA. This fraction was able to synthesize DNA to the extent of 1-2% of the template DNA and was active against both double stranded and single stranded DNA. Using single stranded OX174 and G4 DNA as a template, the in vitro synthesized product was found to be very close to the molecular size of the template. This fraction was allowed to synthesize DNA in the presence of ctDNA by substituting bromodeoxyuridine triphosphate for thymidine triphosphate in the DS reaction system. The isolated porduct had a buoyant density expected of a DNA whose one entire strand contains bromodeoxyuridine instead of thymidine. The DNA containing bromodeoxyuridine in both strands was found in only 1% of the synthesized molecules.

This fraction was further fractionated in a DEAE-cellulose column by using 0-0.3M linear KCl gradient. The data are shown in Fig. 5. DS activity was found at three different fractions. The first two peaks eluted at approximately 0.07 and 0.13M KCl. These two peaks were found to be active in the presence of both double stranded and single stranded DNA. The third peak eluted at 0.25M KCl and was active only on the double stranded DNA as a template. The above three different fractions were further purified on a phosphocellulose column. The activity from peak I eluted at 0.3M KCl from phosphocellulose column. The activity found in peak II was found to elute without any salt. The activity corresponding to peak II eluted with 0.4M KCl.

The experiments are currently in progress to completely purify the three DS activities. Using these purified proteins and the native ctDNA, we propose to study replication *in vitro*. It is clearly recognized by the authors that the complete replication of ctDNA might require proteins other than these three fractions. The knowledge of the *in vivo* replicative intermediates will help us to identify all the proteins necessary for the replication of ctDNA.

REFERENCES

1. Kolodner, R. and Tewari, K. K. (1975) *Biochim. Biophys. Acta*, 402, 372.
2. Kolodner, R., Tewari, K. K., and Warner, R. C. (1976) *Biochim. Biophys. Acta*, in press.
3. Kolodner, R. and Tewari, K. K. (1975) *J. Biol. Chem.* *250*, 4888.
4. Kolodner, R. and Tewari, K. K. (1975) *J. Biol. Chem.* *250*, 8840.
5. Kolodner, R. and Tewari, K. K. (1975) *Nature*, *256*, 708.
6. Tewari, K. K. (1971) *Ann. Rev. Plant Physiol.* *22*, 141.

ACKNOWLEDGMENTS

The authors are thankful to the National Science Foundation for the financial support.

Genetics and Biogenesis of Chloroplasts and Mitochondria, Th. Bücher et al. eds.

STUDIES OF THE GROWTH AND REPLICATION OF SPINACH CHLOROPLASTS AND OF THE LOCATION AND SEGREGATION OF THEIR DNA

J. V. Possingham & R. J. Rose
CSIRO Division of Horticultural Research
P.O.Box 350, Adelaide, Australia 5001.

SUMMARY

Chloroplast growth and replication and the location of chloroplast DNA (c-DNA) have been studied in cultured leaf disks of spinach. Interference contrast microscopy observations made of disks grown at low and high light intensity and in darkness establish that light has differential effects on chloroplast growth and replication. Observations of disks transferred from dark and low light to high light intensity show that all chloroplasts in the one cell can divide. Studies involving light and electron microscope autoradiography indicate that the DNA of spinach chloroplasts is associated with grana lamellae and with the nearby clear areas containing fibrillar material. It is suggested that the association of c-DNA molecules with the lamellar system of chloroplasts ensures an orderly segregation of c-DNA to daughter chloroplasts during their binary fission by constriction division.

INTRODUCTION

Light is known to be necessary for the maximum growth and replication of higher plant chloroplasts(1 2). Limited chloroplast replication and extensive multiplication of both proplastids and etioplasts can however occur in the dark(3 4). In spinach relatively high light intensities (4-6 mW cm^{-2}) are required for both maximal cell growth and chloroplast replication, but blue, green and red light as well as white light have all been found to be effective(5 6). In low intensity light (tenth normal) replication is inhibited but chloroplast growth continues to take place(6). The appearance of spinach disks grown in high and low light intensities and in darkness are described and as well the changes which occur when disks grown in low light intensity and darkness are transferred to high light intensity conditions. In addition light and electron microscope autoradiography have been used to study the location of the DNA in the replicating chloroplasts of cultured leaf disks of spinach(7 8 9).

METHODS

The techniques used for growing spinach plants, culturing leaf disks and measuring chloroplast number per cell are as previously described(2). The methods for light and electron microscope autoradiography are as described before(8 9).

EXPERIMENTAL AND RESULTS

Light Effects on Chloroplast Growth and Replication - Disks taken from young leaves of spinach plants grown in nutrient culture for 14 days in a growth cabinet were cultured for 5 days either in low or high intensity white (0.6 or 6.0 mW cm^{-2}) light or in darkness. The appearance of living sections cut from disks with a sledge microtome and viewed by interference contrast (Nomarski) light microscopy are shown in Figs.1-3. The cells of disks grown in the dark contain a relatively few small dumbbell shaped plastids. Those grown in low light contain a reduced number of large dumbbell shaped chloroplasts relative to those grown in high light intensity. The appearance of disks grown initially in darkness or low intensity light after transfer to high intensity light for 12 hours is given in Figs. 4 & 5. The plastids of disks grown in darkness or in low intensity light begin to replicate within 12 hours of transferring them to high intensity light.

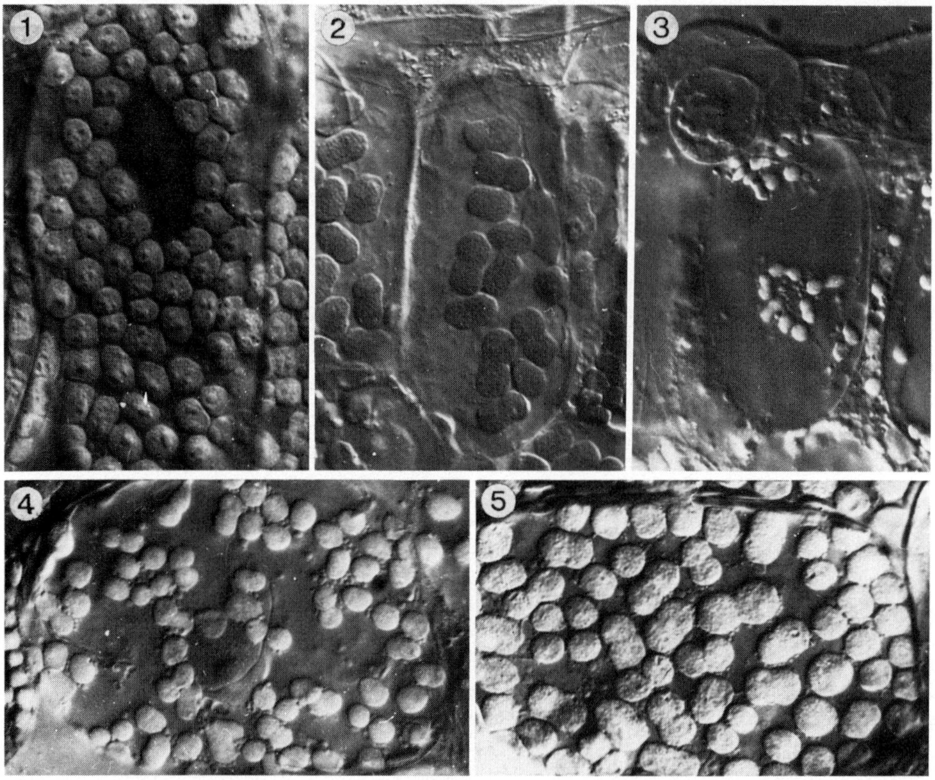

Figures 1, 2 & 3. Sections of disks grown for 5 days either in high or low light intensity (6 or 0.6 mW cm^{-2}) or in the dark, viewed by interference contrast light microscopy. Figures 4 & 5. Sections of disks grown 5 days in darkness or low light intensity and then for 12 hours in high light intensity. Magns. X1,000.

In the case of dark grown disks there is concomitant growth and replication but in the case of disks pregrown in low intensity light replication is associated with a reduction in chloroplast size. In Figs. 4 & 5 virtually all the chloroplasts are either dumbbells or grouped as pairs. The pairs could have arisen by binary fission. These figures suggest that given the right conditions all chloroplasts in young cells divide at about the same time.

Intra Chloroplast location of c-DNA - Spinach leaf disks were cultured on a nutrient agar medium containing ^{3}H-thymidine ([^{3}H]TdR), 50 μCi (6-^{3}H thymidine 27 Ci/mM Radiochemical Centre, Amersham, UK) per 20 ml. Labelling periods of 24 hours in the light were usual after which the disks were fixed, embedded and sectioned as for normal electron microscopy. Ultra thin sections were used for E.M. autoradiography. Light microscope autoradiographs were prepared from some of the disks. Fig. 6 shows the spiral labelling patterns which occur in some autoradiographs. These were frequently observed over individual chloroplasts but were not related to any particular treatments. Figs. 7 & 8 show the type of incorporation patterns obtained in E.M. autoradiographs. Examination of over 100

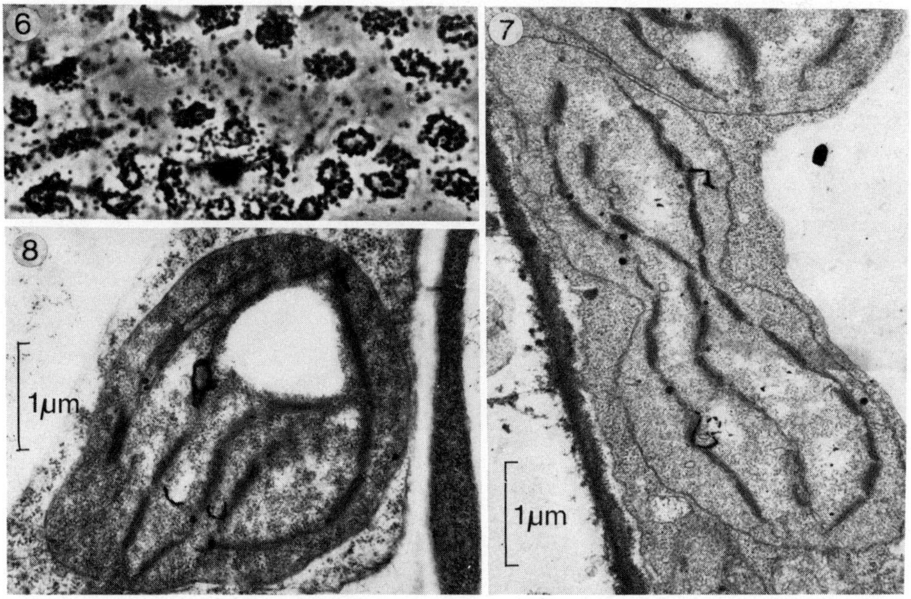

Figure 6. Light microscope autoradiograph of a cell showing helical grain patterns over the chloroplasts. Autoradiograph exposed 13 days. Magn. X 900.
Electron microscope autoradiographs of chloroplasts:-
Figure 7. From a disk excised and incubated in [^{3}H]TdR for 24 h in the light. Autoradiograph exposed 15 weeks. Magn. X 21,000.
Figure 8. From a disk grown in darkness for 5 days, transferred to the light for 1 day and then incubated in [^{3}H]TdR for 24 h in the light. Silver grain on either side of constriction. Autoradiograph exposed 10 weeks. Magn. X 21,000.

chloroplasts from locations in background free areas were made. Less than 5% of grains occurred over the periphery of chloroplasts in association with the chloroplast envelope. Over 70% of the grains occurred over the clear areas containing fibrils (about 3 nm) or in association with grana.

DISCUSSION

The growth and the replication of higher plant chloroplasts are influenced by variations in the intensity of visible light. The nature of the light receptor system for these events in chloroplasts has not been established, however the high intensity requirement for replication and the inhibition of this process by the phosphorylation inhibitor DCMU suggests it could be chlorophyll[1 12]. Difficulties in interpretation arise because some plastid division occurs in dark grown disks and because at low light intensities replication but not chloroplast growth is inhibited. It is relevant here that it has been established that [^{3}H]TdR is readily incorporated into the c-DNA of disks grown in the dark and into those grown in low light intensity[8 10].

The E.M. autoradiography observations indicate that in spinach chloroplasts there is an association of c-DNA with the fibrillar area of low electron density and with the chloroplast lamellar system particularly with grana lamellae. An interpretation of the light microscope autoradiographs showing helical labelling patterns is that they represent incorporation along the chloroplast lamellar system and evidence is available which suggests that the lamellar system of higher plant chloroplasts may be arranged as a continuous helix[11]. Our data suggests a segregation mechanism for c-DNA based on its association with the chloroplast lamellar system.

REFERENCES

1 Boasson, R., Bonner, J.J. and Laetsch, W.M. (1972) Pl. Physiol. 49, 97-101.

2 Possingham, J.V. and Smith, J.W. (1972) J. exp. Bot. 23, 1050-59.

3 Cran, D.G. and Possingham, J.V. (1966) Protoplasma 74, 345-56.

4 Sprey, B. (1970) Planta (Berl.) 78,115-133.

5 Possingham, J.V. (1973) J. exp. Bot. 24, 1247-60.

6 Possingham, J.V., Cran, D.G., Rose, R.J. & Loveys, B.R. (1975) J. exp. Bot. 26, 33-42.

7 Rose, R.J., Cran, D.G. and Possingham, J.V. (1974) Nature 251, 641-2.

8 (1975) J. Cell Sci. 17, 27-41.

9 Rose, R.J. and Possingham, J.V. (1976) J. Cell Sci. 20, 341-55.

10 Possingham, J.V. and Rose, R.J. (1976) Proc.Roy.Soc.Ser.B.193, 295-305.

11 Wildman, S.G., Jope, C. and Atchison, B.A. (1974) Pl. Physiol. 54, 231-7.

12 Possingham, J.V. (1976) J. Microscopie (In Press)

MITOCHONDRIAL DNA: GENES AND MOLECULES

Genetics and Biogenesis of Chloroplasts and Mitochondria, Th. Bücher et al. eds.

MECHANISMS AND RULES FOR TRANSMISSION, RECOMBINATION AND SEGREGATION OF MITOCHONDRIAL GENES IN SACCHAROMYCES CEREVISIAE

B. Dujon and P.P. Slonimski
Centre de Génétique Moléculaire du C.N.R.S.
F-91190, Gif-sur-Yvette, France

INTRODUCTION

In the last few years the understanding of the genetic properties of mitochondrial genes in Saccharomyces cerevisiae has rapidly increased. This was due particularly to the isolation of mitochondrial antibiotic resistant mutations and their use as genetic markers in crosses (1-23). The analysis of the results of a great number of multifactorial mitochondrial crosses involving different combinations of the antibiotic resistant markers have allowed us to describe the characteristic properties of mitochondrial genes and to developp a general model for recombination and segregation of these genes. The bases of this model, together with some examples of the analysis of experimental data it permits, have been published two years ago (24). Since that time we have applied our model to the analysis of many new experimental results obtained in the Gif laboratory (16,21) as well as results obtained in other laboratories (25) and the main ideas of our model have been confirmed. It is our feeling that, in view of the very rapid progresses in isolation of new mitochondrial mutations (either conferring antibiotic resistance (26-30) or affecting mitochondrial inner membrane enzyme complexes (31-34)), a summary of the main conclusions about the mechanisms and the genetic rules governing the hereditary transmission of mitochondrial genes would be useful in a symposium devoted to the genetics and biogenesis of organelles.

The purpose of this article will be therefore (1) to summarize the main conclusions on the mechanisms involved in the hereditary transmission of mitochondrial genes drawn from the analysis of the available data, (2) to propose a small number of simple genetic rules, specific for mitochondrial genes, and to discuss their applications to newly discovered mitochondrial mutations. The complete presentation of the model is to be found in (35).

A MULTIPLE COPY GENETIC SYSTEM

The first idea proposed to account for the genetic properties of mitochondrial genes is to consider that the mitochondrial genetic information is carried by a multiple copy genetic system i.e. that in each cell there are several copies of the mitDNA molecules genetically competent and identical. This idea, proposed first few years ago (24), has been now generally accepted by different authors (15,18,22). It is also interesting to notice that a similar idea has been proposed for the chloroplastic genetic system of Chlamydomonas (36) suggesting the generality of multiple copy systems for organelle genetics.

The second idea (which is in fact directly associated with the first one) is that the number of genetically active copies can differ from cell to cell in a given strain. Furthermore, it can differ from strain to strain and this second variation is genetically determined (6) probably by a series of nuclear genes showing cumulative effects. Finally, this number can be modified experimentally by external conditions (15,16,25,37). No direct measure of the number of genetically active copies per cell is possible until now. The only experimental possibility to determine this number is to make a cross and to measure the transmission of mitochondrial markers in the progeny. From the relative transmission of the alleles of the two parents it is possible to estimate the relative number of genetically active molecules of each parent. This is called the input.

The most direct experimental evidence argueing in favor of these ideas is that the transmissions of mitochondrial markers among the progenies of the different crosses between different strains varies in a continuous manner from almost 0 to almost 1 (see Fig. 1). Similarly a continuous variation is also found after experimental treatments such as UV irradiation of one parent prior to crossing (16). The continuity of the distribution indicates that the number of genetically active copies per cell is relatively large since, if it were not so, a stepwise distribution of transmissions would be expected by variation of the number of copies. The continuity observed represents in fact the overall result of many discrete classes differing from each other by very small steps. Furthermore, the fact that the transmission of alleles at different loci (genetically linked or unlinked to each other) varies in a coordinated way argues strongly in favor of the idea that variations of transmission are representative of variations of number of genetically active copies of the mitochondrial genome.

A direct corollary of the hypothesis of a multiple copy genetic system is that the genetic exchanges (pairing and recombination events) will obviously take place between the different mitDNA molecules forming the intracellular pool. Two main questions can be asked about these processes. The first concerns the mechanisms of pairing of the different copies in the pool, the second the efficiency of genetic exchanges within the pool i.e. the average number of pairing and recombination events that each molecule undergoes in the cell.

MULTIPLE NON SYNCHRONOUS ROUNDS OF RANDOM PAIRINGS LEADING TO RECOMBINATION

The model postulates that pairings and recombination events between the different mitDNA molecules of a cell occur at random i.e. the intracellular population of mitochondrial DNA molecules is regarded as panmictic. The experimental evidences arguing in favor of random pairing processes are based on the low limit of frequencies of recombinants between the distant markers (6,10,22) and on the ratio

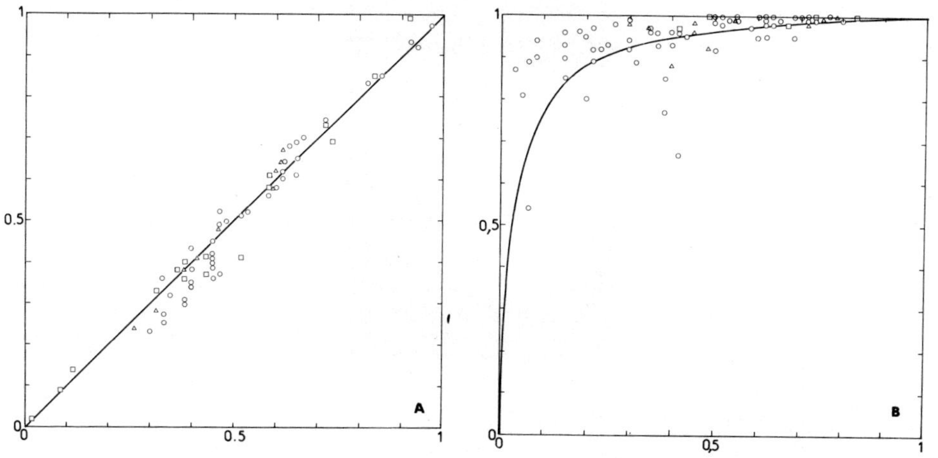

Fig.1. Example of the relationships between transmission of alleles at different genetic loci. These figures represent the transmission of the allele of a given parent at the RIB1 locus (ordinate) as a function of the transmission of the alleles of the same parent at the OLI1 or PAR1 loci (abscissae) respectively for non polar crosses (Fig. 1A) and for polar crosses (Fig. 1B). Each point represent the result of a single cross. Experimental data compiled from (6,10,13,16,21). Theoretical curves are drawn from predictions of our model (24).

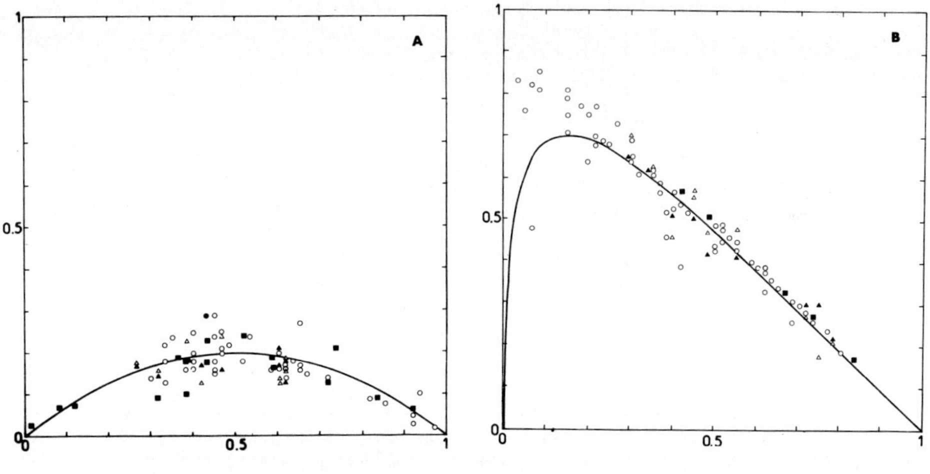

Fig. 2. Example of the relationships between frequency of recombinants and transmission of alleles. These figures represent the total frequency of recombinants between the RIB1 and OLI1 loci or between the RIB1 and PAR1 loci (ordinate) as a function of the transmission of the alleles of a given parent at the OLI1 or PAR1 loci (abscissae) respectively for non polar crosses (Fig. 2A) and for polar crosses (Fig. 2B). Other legend like for Fig. 1,

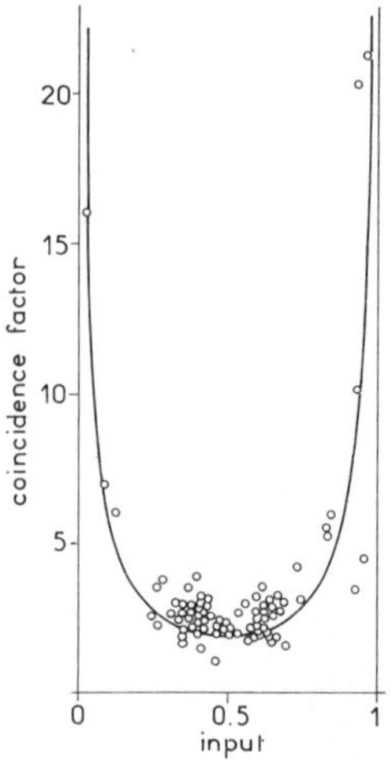

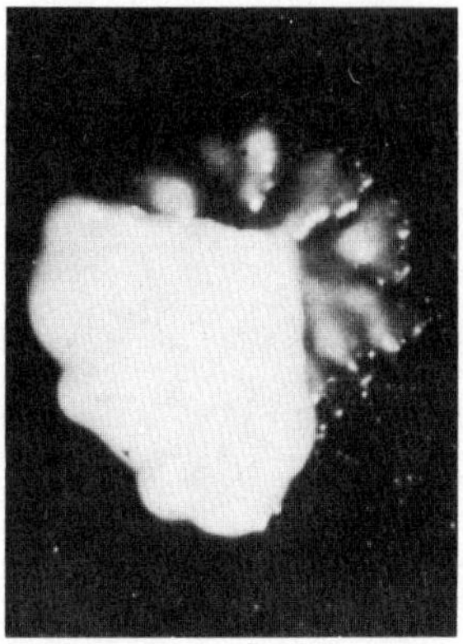

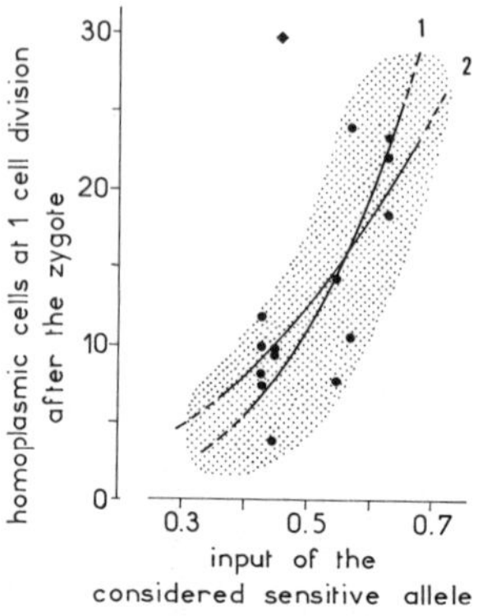

Fig. 3 Fig. 4 Fig. 5

Fig. 3. Relationship between the coincidence factor and the input.
The coincidence is expressed as the ratio of the frequency of double recombinants observed, versus products of frequencies of the two single recombinants. The input is calculated from the transmission of alleles issued from the a parent. For calculation of the coincidence the gene order PCEO has been taken in agreement to (32,49). The line is drawn according to the predictions of the model (for details of calculation see (16)). Data compiled from (6,10,16,21).

Fig. 4. Replica plating of a colony issued from an heteroplasmic cell on an antibiotic containing medium. A sector is composed of resistant cells growing in presence of the antibiotic, while the other part of the colony is composed of sensitive cells which are inhibited. Segregation between resistant and sensitive cells has occured during the growth of the colony issued from the heteroplasmic cell.

Fig. 5. Relationship between the kinetics of appearance of homoplasmic cells for a given allele and the input of that allele. The input of the considered allele (either C^S or E^S or O^S) is estimated from transmission of alleles in crosses (three different three point non polar crosses have been used in this study). The frequency of cells homoplasmic for the considered allele (either C^S or E^S or O^S) at one cellular generation after the zygote is calculated from linear interpolations or extrapolations of the closest results to the exact time of one generation. The curves are theoretical and represents the relationships expected in the simplified hypothesis of non exhaustive random sampling of segregating units assuming that either zygotes and buds can be homoplasmic (1) or only buds (2). The curves are drawn with the best estimate of the number of segregating unit (respectively 3.2 and 2.0 for 1 and 2).
The shadowed part visualizes tne probable limits of the experimental relationship found.
These data have been obtained in collaboration with L. Weill.
◆This point correspond to a parental strain containing a high proportion of rho^- cells and cannot be taken into account.

of frequencies of double recombinants found versus the one expected, which is always greater than 2 (see Fig. 3).

The model postulates also that non-synchronous multiple events of pairing take place in each intracellular population and that the number of pairings undergone by each molecule is randomly distributed around an average number of mating rounds for the total population.

The most direct evidence in favor of multiple mating rounds is based on the transmission of alleles of the polar region in polar crosses (6,10,22) and on the fact that the total frequencies of recombinants in polar crosses can largely exceed the upper limit found for the non polar crosses (see Fig. 2). These arguments cannot be discussed in details here but it is important to mention that the complete set of experimental results fit with an average number of mating rounds close to 4 (see 35) i.e. that there is in average several mating rounds and not one.

HIGHLY EFFICIENT NON RECIPROCAL RECOMBINATION EVENTS OCCUR BETWEEN mitDNA MOLECULES

The existence of a specific mitochondrial locus, called ω, responsible for the polarity of recombination in the RIB1 - RIB2 - RIB3 region of the mitDNA (called polar region and closely linked to ω) has been demonstrated previously (4,6,10, 13,16,22). It has been shown also that in crosses between strains carrying the same allele of ω, similar frequency of the two reciprocal recombinants (absence of polarity) was found for all pairs of markers considered. Absence of polarity has also been found between markers of the non polar regions in $\omega^+ \times \omega^-$ crosses.

Similar frequencies of recombinants among the total progeny of a cross can be the final result of either a series of reciprocal elementary acts of recombination or a series of non reciprocal recombination acts statistically equilibrating each other in the population. In such a system the absence of polarity offers therefore little information about the molecular recombination events themselves.

On the contrary the occurrence of a high polarity of recombination found between markers of RIB1 - RIB2 - RIB3 region in $\omega^+ \times \omega^-$ crosses suggests the occurrence of non reciprocal events of recombination.

Thus gene conversion appears as a major mechanism for genetic recombination between the mitDNA molecules. The ω locus itself acts as an initiator of a dissymetrical gene conversion process (for details see 24, 35) i.e. at each pairing the alleles carried by the ω^- molecule are converted into copies of the alleles carried by the ω^+ molecule. Reiteration of this process by multiple rounds of pairings is necessary and sufficient to account for the polarity of recombination observed.

Recent results on mutants of the ω locus (21) indicate that the different alleles of ω can be regarded as macrolesions i.e. differ by many base pairs and the physical analysis of this region is now in progress in our laboratory (38,39).

A RANDOM SEGREGATION OF A SMALL NUMBER OF SEGREGATING UNITS

One of the characteristic property of mitochondrial genes is the fact that they segregate during the vegetative phase of growth in such a way that a heteroplasmic(*) cell gives rise to a clone composed of different homoplasmic cells in a few cell divisions. Three main questions about the segregation process are important to consider. The first one is to determine whether or not segregation of the mitochondrial DNA molecules occurs at random. The second one is to estimate the number of segregating units entering the bud. The third one is to determine the relationship between the segregating units and the mitochondrial DNA molecules themselves. All these parameters will obviously determine the kinetics of appearance and the distribution of different homoplasmic cells from heteroplasmic ones. Previous observations based on the estimation of frequency of pure sensitive clones issued from the first buds of zygotes suggested that segregation occurs at random and that the number of segregating units is rather small in comparison to the total number of mitDNA molecules (24). Other experiments based on the kinetics of formation of homoplasmic cells during vegetative divisions of zygotes (3,12,44) as well as on the pedigree analysis of first cell divisions (3,40,41,42,45) could also be interpreted as suggesting random segregation. However, the existence of cells appearing with a low frequency (c.a.10^{-4}) in a small fraction of zygotic clones was taken as an evidence indicating that the number of mitDNA molecules entering the bud cannot be very small (43).

We have recently reexamined this question by the analysis of frequencies of homoplasmic cells during the first divisions of zygotes in crosses showing differently biased inputs. The details of the results will be presented elsewhere (46). It suffices here to mention that : 1) a relatively high proportion of cells are already homoplasmic for one allele after the first division of the zygote only, 2) this frequency depends of the input of the allele considered as shown in Fig. 5. The frequency is high for alleles carried by the parent which has the high input. These results argue in favor of the idea that segregation occurs at random since the probability for a bud to be homoplasmic for a given allele depends on

(*) By definition, a homoplasmic cell is a cell in which all mitDNA molecules carry the same alleles while in a heteroplasmic cell different molecules carry different alleles. However, this definition is not directly amenable to the experiment. We prefer rather an operational definition : if a cell gives rise, under non selective conditions, to a colony composed of only one cell type (at the limit of the frequency of spontaneous mutants arised during the growth of the colony) this cell will be called homoplasmic. If, on the contrary, a cell gives rise to a colony containing more than one cell types with a frequency higher than the spontaneous mutants, this cell will be called heteroplasmic. An actual example of such a colony is given on figure 4. This operational definition is valid in case of mitochondrial genes since it has been shown, by the Luria Delbrück fluctuation test, that the variance of spontaneous mutants is low (see 21).

the input of this allele i.e. on the relative number of molecules carrying this allele versus the number of molecules carrying the other allele. Calculations based on a simplified model of non exhaustive random sampling of segregating units during cell division (see Fig. 5) indicate that the number of independant segregating units entering the bud is small in comparison to the total number of mitDNA molecules. The question therefore remains as to determine whether one segregating unit corresponds to one mitDNA molecule or, in average, to several mitDNA molecules segregating together. In the first hypothesis the distribution of frequencies of different alleles in the progeny of the buds is expected to show discrete classes while a continuous distribution is expected in the second hypothesis. The intra and interclonal distributions observed in our laboratory (3,46) as well as in others (40-43) and the existence of cell type appearing with a very low frequency argue in favor of the second hypothesis.

It can be concluded therefore that a small number of segregating units is segregated at random during cell divisions, each segregating unit being composed, in average, of several mitDNA molecules. The process of "packaging" of several mitDNA molecules to constitute a segregating unit is not understood. A plausible hypothesis is to consider that one segregating unit would represent mitDNA molecules involved in or attached to one site of the organelle like it is suggested by ultrastructural observations (47).

GENETIC RULES FOR TRANSMISSION AND RECOMBINATION OF MITOCHONDRIAL MARKERS

Mitochondrial mutations have been classically characterized by four criteria : 1) segregation of the two alleles at a given locus during vegetative cell divisions of a heteroplasmic zygote clone, 2) 4:0 or 0:4 segregation of the two alleles during meiosis, 3) elimination of the allele in all rho^{0} mutants and 4) elimination in some of the rho^{-} mutants but not in all of them.

These criteria are, of course, still valid. However, the good agreement between the experimental results and the quantitative predictions of the model allows us to propose four rules (*) which constitute additional positive criteria for characterizing the mitochondrial genes. They can be used to determine whether a given mutation is carried on the mitDNA molecule and eventually to determine its location on the molecule.

(*) These rules are applicable only under conditions where there is no selective advantage for any of the cell types during the vegetative growth of the population issued from the cross.

1st rule. The transmissions of alleles at different loci, issued from the same parent, are always correlated. The correlation is, in most cases (absence of polarity of recombination) a linear one (see Fig. 1A as an example). In some cases (presence of polarity of recombination) a distortion is found, the correlation is no longer linear but shows a characteristic form which can be predicted from considerations of the model (see Fig. 1B as an example).

2nd rule. There is an upper limit of the recombinant frequency between two markers. This limit is close to 20-25 % (see Fig. 2A). This upper limit can be exceeded only in cases when the two reciprocal recombinants appear with statistically different frequencies (polarity of recombination). In cases of high polarity the maximal frequency of recombinants may be as high as 80 % i.e. there appears much more recombinant than parental types (see Fig. 2B as an example).

3nd rule. There exists a relationship between the frequency of recombinants between two given loci and the transmissions of alleles at any locus. This relationship is a parabolic function as long as the two reciprocal recombinants appear with equal frequencies (see Fig. 2A) while it is a dissymetrical function when a polarity of recombination occurs. The shape of the dissymetrical function can be predicted from the model (see Fig. 2B).

4th rule. In three point crosses showing no polarity of recombination a high positive coincidence is always found i.e. the frequency of double recombinants observed is at least 2 to 3 times higher than that expected from the product of frequencies of the two single recombinants. Furthermore, the coincidence factor is dependant upon the input and therefore a relationship, as shown in Fig. 3, exists between the coincidence factor and the transmission of alleles.

However it is important to stress that the coincidence factor for a single given cross is often difficult to estimate since the fluctuations due to sampling errors on numbers of colonies scored are relatively large in comparison to the expected frequencies of recombinants (notice the scattering of experimental data on Fig.3).

The four rules proposed define a physical and genetic linkage group. Therefore, although the frequencies of recombinants found between many pairs of markers presently known are close to the upper limit of recombinant frequencies and thus these markers cannot be mapped relatively to each other by standard methods, it is clear that all mutations which fullfill these 4 rules are carried on the same mitDNA molecule. This is obviously the case for the mutations at the loci RIB1, RIB2, RIB3, OLI1, OLI2 and PAR1 which have been used to establish these rules. It is also the case for a cytoplasmically inherited temperature sensitive mutation (48,49) and for newly found mutations conferring resistance against antibiotics such as venturicidin (20), antimycin A (28,29) or oligomycin (50). These rules have been found to apply also for the mutations expressed to a deficiency in cytochrome oxidase, cytochrome b and/or ATPase although the reciprocity in recombination frequencies has not been yet demonstrated in mit^- mutants (32,33).

CONCLUDING REMARKS

The mechanisms involved in the hereditary transmission of mitochondrial genetic information, namely a multiple copy genetic system with multiple rounds of mating, a high efficiency of genetic recombination and a random and rapid segregation, have obviously several consequences. Among those, two of them are particularly important for the geneticist.

The first one is the fact that most of the genetic markers presently known behave as genetically unlinked to each other even though for some pairs of markers the physical distances on the mitDNA molecule must be relatively short. This is, of course, an important drawback to built up the genetic map of the mitDNA by simple recombination tests and therefore more sophisticated methods have to be developped (32,49,51,52).

The second consequence is the fact that true genetic complementation test cannot be applied to define the functional units on the mitochondrial genome since: 1-wild type recombinant molecules are always formed rapidly and give rise to phenotypes indistinguishable from the expected complementation ; 2- heteroplasmic cells (necessary for complementation test) are never stable but segregate rapidly different homoplasmic cell lines.

For these reasons mutants totally or partially deficient in the mitochondrial genetic recombination would be usefull for the geneticist as well as mutants deficient in the segregation process.

Acknowledgements. We thank our colleagues at Gif (M.Bolotin-Fukuhara,D.Coen, J.Deutsch,G.Dujardin,P.Netter,E.Petrochilo and L.Weill), Dr. B.Backhaus and Dr. Z.Kotylak for communicating to us their unpublished results and for their fruitful discussions.

This work has been supported by a grant D.G.R.S.T. No 75 70748.

REFERENCES

1. Linnane, A.W., Saunders, G.W., Gingold, E.B., Lukins, H.B. (1968) Proc. Natl Acad. Sci. 59, 903-910.
2. Thomas, D., Wilkie, D. (1968) Biochem. Biophys. Res. Commun. 30, 368-372.
3. Coen, D., Deutsch, J., Netter, P., Petrochilo, E., Slonimski, P.P. (1970) Symp. Soc. Exp. Biol. n° 24, Cambridge Univ. Press, pp. 449-496.
4. Bolotin, M., Coen, D., Deutsch, J., Dujon, B., Netter, P., Petrochilo, E., Slonimski, P.P. (1971) Ann. Inst. Pasteur 69, 215-239.
5. Kleese, R.A., Grotbeck, R.C., Snyder, J.R. (1972) J. of Bacter. 112, 1023-1025.
6. Avner, P., Coen, D., Dujon, B., Slonimski, P.P. (1973) Mol. Gen. Genet. 125, 9-52.
7. Howell, N., Trembath, M.K., Linnane, A.W., Lukins, H.B. (1973) Mol. Gen. Genet. 122, 37-51.
8. Rank, G.M. (1973) Heredity 30, 265-271.

9. Wakabayashi, K., Kamei, S. (1973) FEBS Letters 33, 263-265.
10. Wolf, K., Dujon, B., Slonimski, P.P. (1973) Mol. Gen. Genet. 125, 53-90.
11. Callen, D.F. (1974) Mol. Gen. Genet. 128, 321-329.
12. Callen, D.F. (1974) Mol. Gen. Genet. 134, 49-63.
13. Netter, P., Petrochilo, E., Slonimski, P.P., Bolotin-Fukuhara, M., Coen, D., Deutsch, J., Dujon, B. (1974) Genetics 78, 1063-1100.
14. Suda, K., Uchida, A. (1974) Mol. Gen. Genet. 128, 331-339.
15. Birky, C.W. (1975) Genetics 80, 695-709.
16. Dujon, B., Kruszewska, A., Slonimski, P.P., Bolotin-Fukuhara M., Coen, D., Deutsch, J., Netter, P., Weill, L. (1975) Mol. Gen. Genetics 137, 29-72.
17. Forster, J.L., Kleese, R.A. (1975) Mol. Gen. Genet. 139, 341-355.
18. Gunge, N. (1975) Mol. Gen. Genet. 139, 189-202.
19. Griffiths, D.E., Houghton, R.I., Lancashire, W.E., Meadows, P.A. (1975) Eur. J. Biochem. 51, 393-402.
20. Lancashire, W.E., Griffiths, D.E. (1975) Eur. J. Biochem.51, 403-413.
21. Dujon, B., Bolotin-Fukuhara, M., Coen, D., Deutsch, J., Netter, P., Slonimski, P.P., Weill, L. (1976) Mol. Gen. Genet. 143, 131-165.
22. Gunge, N. (1976) Mol. Gen. Genet. 146, 5-16.
23. Trembath, M.K., Molloy, P.L., Sriprakash, K.S., Cutting, G.S., Linnane, A.W. Lukins, H.B. (1976) Mol. Gen. Genet. 145, 43-52.
24. Dujon, B., Slonimski, P.P., Weill, L. (1974) Genetics 78, 415-437.
25. Backhaus et al. (in prep.)
26. Subik, J. (1975) Febs Letters 59, 273-276.
27. Trembath, M.K., Monk, B.C., Kellernan, G.M., Linnane, A.W. (1975) Mol. Gen. Genetics 141, 9-22.
28. Michaelis, G; et al. This symposium.
29. Lang et al. This Symposium.
30. Colson A.M. et al. This Symposium
31. Tzagoloff, A., Akai, A., Needleman, R.B., Zulch, G. (1975) J. Biol. Chem. 250, 8236-8242.
32. Slonimski, P.P., Tzagoloff, A. (1976) Eur. J. Biochem. 61, 27-41.
33. Kotylak, Z., Slonimski, P.P. (1976) Proc. of the Bari Symposium (in press)
34. Tzagoloff, A. et al. This Symposium.
35. Dujon, B. (1976) Doctoral Thesis. Université P. et M. Curie Paris
36. Gillham, N.W., Boynton, J.E., Lee, R.W. (1974) Genetics 78, 439-457.
37. Goldthwaite, C.D., Cryer, D.R., Marmur, J. (1974) Mol. Gen. Genet. 133,87-104.
38. Dujon, B., Michel, F. (1976) Proc. of the Bari Symposium (in press).
39. Lazowska, J., Jacq, C., Cebrat, S., Slonimski, P.P. (1976) Proc. of the Bari Symposium (in press).
40. Lukins, H.B., Tate, J.R., Saunders, G.W., Linnane, A.W. (1973) Mol. Gen. Genet. 120, 17-25.

41. Wilkie, D., Thomas, D. (1973) Genetics 73, 367-377.
42. Callen, D. (1974) Mol. Gen. Genet. 134, 65-76.
43. Birky, C.W. (1975) Genetics 141, 41-58.
44. Uchida, A., Suda, K. (1976) Mol. Gen. Genet. 145, 159-163.
45. Aufderheide, K., Johnson, R.G. (1976) Mol. Gen. Genet. 144, 289-299.
46. Weill, L. et al. (In prep.)
47. Yotsuyanagi, Y. (1966) Compt. Rend. Acad. Sci. 262, 1348-1351.
48. Handwerker, A., Schweyen, R., Wolf, K., Kaudewitz, F. (1973) J. of Bact. 113, 1307-1310.
49. Schweyen, R., Steyrer, U., Kaudewitz, F., Dujón, B., Slonimski, P.P. (1976) Mol. gen. Genet. In press
50. Clavilier, L. (1976) Genetics 83, 227-243.
51. Linnane, A.W., Lukins, H.B., Molloy, P.L., Nagley, P., Rytka, J., Sriprakash,K. Trembath, M.K. (1976) Proc. Natl Acad. Sci. 73, 2082-2085.
52. Sanders, J.P.M., Heyting, C., Di Franco, A., Borst,P., Slonimski, P.P. (1976) Proc. of the Bari Symposium. In press.

Genetics and Biogenesis of Chloroplasts and Mitochondria, Th. Bücher et al. eds.

CONFIRMATIONS AND EXCEPTIONS TO THE PHAGE ANALOGY MODEL: INPUT BIAS, BUD POSITION EFFECTS, ZYGOTE HETEROGENEITY, AND UNIPARENTAL INHERITANCE

P. S. Perlman, C. W. Birky, Jr., C. A. Demko† and R. L. Strausberg
†Department of Genetics and Developmental Biology Program
The Ohio State University, Columbus, Ohio 43210

INTRODUCTION

Several years ago Dujon, Slonimski and Weill[1] made a quantitative application of the Visconti-Delbruck model for phage crosses to mitochondrial gene inheritance in yeast (*Saccharomyces cerevisiae*). On the strength of that analogy new insights to events in populations of yeast zygotes were achieved. First, outputs of yeast crosses reflect the input ratios of genetically functional mitochondrial genomes; this helps in understanding the common finding that quantitative transmission levels vary widely among crosses and it follows that different haploid strains may have different inputs. Second, multiple rounds of random recombination (mating) occur; this helps to explain the recombination data as well as the phenomenon of polarity. In their analysis it was assumed that a panmictic pool of genomes is established quickly in the zygote and that recombination is non-reciprocal.

Several lines of investigation in our laboratory have been aimed at evaluating their assumptions, providing further experimental support for their conclusions and assessing whether any constraints must be placed on their phage analogy model of mitochondrial genetics. In this communication we present a brief summary of our results.

BIAS IN MITOCHONDRIA CROSSES

Bias has been defined as the asymmetric recovery of mitochondrial alleles from the two parents of a cross[1,2]. This phenomenon has been observed by many workers in the field (reviewed in ref. 3) and in a few cases has been correlated with nuclear genetic factors[4,5,6]. Dujon *et al*. interpreted biased outputs as being a consequence of a biased input; that is, in biased crosses one parent contributed more genetically functional genomes than did the other. Based on the phage analogy it follows that there is a correlation between the recombination frequency and the bias of a cross, with unbiased crosses showing maximum recombination. Evidence supporting this was obtained by Dujon *et al*. where they altered bias by UV irradiation of one parent prior to mating and observed the predicted changes in bias and recombination.

In their presentation of the phage analogy model, Dujon *et al*. defined "input" in terms of the number of genetically competent mtDNA molecules. We have attempted to determine the extent to which there is a simple relationship (or identity) between input and the amount of mtDNA per haploid parent cell. We have used several experimental protocols that are known to alter the amount of mtDNA per cell prior to mating and then analyzed the consequences of this alteration of DNA input on the genetic output. Cycloheximide (cyclo) treatment of yeast cells is known to allow mtDNA to accumulate in non-dividing cells[7]; in Table 1A are data illustrating that this treatment results in a coordinate increase in transmission of both markers from the treated parent. A systematic analysis of this protocol using many different crosses has demonstrated that it is possible to induce bias within a constant nuclear genetic background by increasing the mtDNA input of one parent[8]. A second precedure having a similar (and more striking) effect on output is the treatment of the a mating type parent with purified α-factor (generously provided by Dr. Mike Douglas) (Table 1B). Simultaneous incubation with hydroxyrea (an inhibitor of DNA synthesis under these conditions) and α-factor inhibits bias. In our studies we have encountered several strains that do not respond to one or the other of these treatments; in the case of several non-responders to cyclo treatment we have shown that they do not replicate mtDNA in the presence of the drug.

Two other approaches to this problem not requiring treatment of cells with exogenous substances were also examined. First we used strain rho5, a mutant that has roughly twice as much mtDNA per cell as does its parent strain, N123[6]. The mutant transmits its markers roughly twice as much as does N123 when both are crossed to the same tester (Table 1C). Second, we studied the effects of nuclear ploidy on output; it is known that diploids have twice as much mtDNA as do isogenic haploids[10]. When an isogenic haploid and diploid are crossed to a common tester strain the output varied in the two crosses roughly to the expected extent (Table 1D). Several other workers have reported similar results[11,12].

These data, supported by much more extensive results to be presented elsewhere provide rather direct evidence that input is strongly dependent on the amount of mtDNA present in each parent before mating and lend credence to the notion that input and mtDNA amount are closely related. It is very improbable that models requiring one or a few master molecules per cell and independent of the actual amount of mtDNA can be consistent with these data. Our use of α-factor to alter input led us to examine events in "normal" matings; consistent with the findings of other workers[9,13] we found that in many crosses one or both parents undergo a round of mtDNA replication while the cells are arrested in G1 prior to mating. Thus it is possible that the ratio of genomes in newly formed zygotes may differ from the ratio in the two parent cultures before mating. This situation frustrates attempts to quantitate mtDNA amounts in individual crosses. In view of our analysis of pre-mating synthesis of mtDNA[8] it is possible that a portion of zygote clone heterogeneity (see below) may be explained if one assumes that not all cells in a culture undergo the same amount of pre-mating replication; this possibility can explain the breadth of the frequency distributions but by itself cannot explain the existence of the extreme classes found. Clearly even if all cells in a culture have the same amount of mtDNA before mating, considerable input variations may arise during the mating reaction.

TABLE 1

GENETIC ANALYSIS OF BIAS

Percent Transmission

Cross				chl-s	ery-s	olil-s	N
A.	-ssss	x	-rsrs				
	N123	x	1-1/1	41.5 ± 4.5	--	44.3 ± 4.8	397
	cy*N123	x	1-1/1	54.8 ± 5.1	--	55.6 ± 5.1	365
B.	N123	x	1-1/1	55.5 ± 4.8	--	63.7 ± 4.7	402
	α†N123	x	1-1/1	83.7 ± 3.9	--	86.9 ± 3.6	344
	α†HUˣN123	x	1-1/1	66.8 ± 4.5	--	69.0 ± 4.5	413
C.	N123	x	1-1/1	44.2	--	48.7	3568
	rho5	x	1-1/1	69.1	--	69.4	5016
D.	+rrss	x	+sssr				
	2-36	x	4810 hap	43.2 ± 3.3	43.4 ± 3.3	--	852
	2-36	x	4810 dip	75.8 ± 3.3	75.0 ± 3.3	--	645

* Treated with cyclo (50 μg/ml) for 3 hrs prior to mating.

† Treated with purified α-factor for 90 min prior to mating.

x Treated with hydroxyurea (25 mg/ml) for 90 min prior to mating.

Genotypes of strains used are as follows: N123, a his; 1-1/1, α ade; 2-36, α trp; 4810 hap, a lys; 4810 dip, a/a lys/lys. Mitochondrial genotypes are abbreviated by listing the allelic forms in the order ω, chl, ery, olil, par.

ZYGOTE CLONES VS. POPULATIONS OF ZYGOTES

The phage analogy model explicitly deals with populations of zygotes; if it is to provide useful insights to the primary events of mitochondrial transmission genetics then it is necessary that most individual zygotes closely resemble the population mean in genetic output. Most laboratories in this field have reported clone analyses and the general observation is that there is considerable genetic heterogeneity among clones[14]. In our laboratory we have undertaken extensive clonal analysis of many crosses; we have also observed zygote clone heterogeneity and in addition find that many crosses yield two, three or more distinguishable classes of zygotes[14]. Figure 1A illustrates that these classes using the cross of D6 X 6-2/7. It is very difficult to interpret these results assuming that all parent cells have the same input or even that a range of inputs exists within each parent culture: if the latter were true then a single broad peak should result. In most crosses there is a class of zygotes with outputs resembling the mean of the total population but in some crosses that class is relatively rare and the more extreme classes are frequent. Thus it is clear that the phage analogy model cannot be extended to individual zygotes without modification.

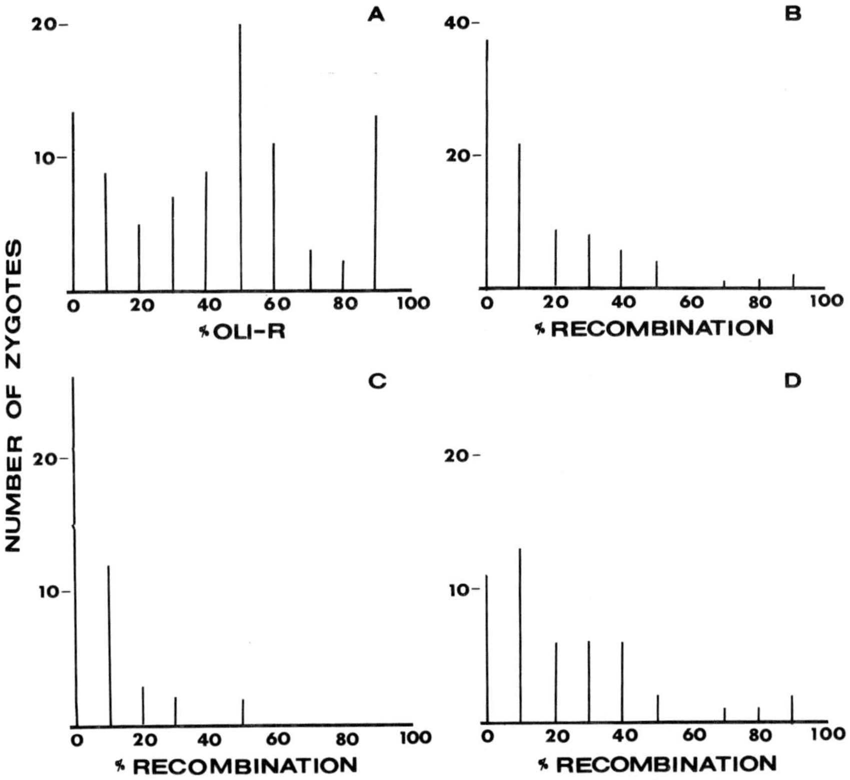

Figure 1. Zygote clone analyses: D6 x 6-2/7. Frequency distribution for the transmission of the oli-r allele (derived from 6-2/7)(A) and recombinants for total zygotes (B), zygotes with first end buds (C) and zygotes with first central buds (D) . The mean value is denoted in each panel by an arrow. The complete genotype in this cross is presented in Table 2.

THE PANMICTIC MATING POOL

Further zygote clone analyses revealed that in most crosses there is also great heterogeneity in recombination frequencies among clones. As is shown in Figure 1B, C, D many zygotes have few or no recombinants; clearly, then panmixis is not actually achieved in all members of the population of zygotes.

TABLE 2

EFFECT OF BUD POSITION ON RECOMBINATION

Cross			First Bud	Percent Recombination			N
				Bud	Residual Zygote	Total*	
1. 4810	x	6-2/5	End	2.5	21.7	12.1	28
+sssr	x	+rsrs	Center	49.8	45.0	47.4	34
a trp	x	α ade					
2. N123	x	41-6/19	End	5.7	8.0	6.8	36
-ssss	x	-rsrr[x]	Center	26.5	21.2	23.8	8
a his	x	α ade lys					
3. D6	x	6-2/7	End	3.0	11.2	7.1	11
+srss	x	+rsrs	Center	22.2	28.8	25.5	14
α arg met	x	a ade					
4. D6	x	6-2/7	End	--	--	11.9	34
			Center	--	--	27.4	34

In each experiment cells were mated in liquid until unbudded zygotes were observed. Then unbudded zygotes were isolated on agar slabs by micromanipulation. For each zygote the first bud position was noted; in experiments 1-3 first buds were separated from the zygote and cells were incubated until colonies were formed. Each colony was then analyzed for genotypic composition and the recombination frequencies for all six recombinant genotypes combined. In experiment 4 first buds were not removed. * Zygote clone values from partial pedigrees were estimated by averaging the values from the bud and the residual zygote. [x]The par locus was not scored in this experiment.

We considered the hypothesis that the mixing of organelles (a likely physical requirement for panmixis) is rather slow and that segregation of mitochondrial genomes during the first round of budding might separate the two genotypes and thus prohibit recombination. Direct examination of zygotes[15] showed that many produce their first bud from one end of the zygote rather than from the fusion neck. We examined recombination levels in zygotes whose first bud positions were known and found that recombination in end-budded zygotes is substantially lower than in zygotes with first central buds (summarized in Table 2). We next performed partial pedigree studies in which the genotypes of progeny of first end and central buds were compared to those of the zygote that issued each bud. In Table 3 are presented a sample of the results of these partial genotypes. Many first end buds yield clones pure for one parental genotype and virtually all such clones are nearly pure. The zygotes that produced those end buds are often greatly enriched for the other parental genotype. That is, first end buds and their zygotes of origin are often very dissimilar. On the other hand, first central buds are usually heteroplasmic as are their zygotes of origin. We conclude that the sample of mitochondrial genophores entering a first bud is a significant fraction (probably 25% or more) of the total present in the unbudded zygote, and that panmixis has not usually been achieved until at least after the first bud has formed. Also it is necessary that the number of genetically functional molecules be fairly large; a more quantitative consideration of this subject will be presented elsewhere.

TABLE 3

PARTIAL PEDIGREE ANALYSIS: 4810 X 6-2/5

First Bud	Residual Zygote			First Bud		
	%rsr	%srs	%recomb.	%rsr	%srs	%recomb.
End	0	100	0	100	0	0
End	91.2	0	8.8	0	92.6	7.4
End	49.6	0	50.4	97.7	0	2.3
End	32.5	27.4	40.1	100.0	0	0
Center	25.8	0	74.2	0	0	100
Center	27.9	23	49.1	100	0	0
Center	0	10.5	89.5	8.6	4.3	87.1
Center	45.9	29.3	24.8	0	100	0

Genotypes of the cross are shown in Table 2. The percent recombination is the total of all recombinant genotypes present in the clone studied. These data are some of the partial pedigrees summarized in cross 1 of Table 2. First buds with both parental types are very rare in all crosses studied; those found were always from the central position. In each cross studied many patterns of segregation between residual zygote and bud were observed and these data illustrate some of those patterns observed.

These observations place real constraints on the phage analogy model and especially on the estimation of the number of actual rounds of mating. There may be some crosses where these constraints are negligible but we have not yet encountered any. Further consideration of the consequences of slow cytoplasmic mixing and first bud position on the genetic composition of zygote clones leads to the conclusion that these factors can cause a zygote with an initially equal input from each parent to yield progeny that fail to reflect the input accurately. This constraint applies whenever the fraction of genomes entering a bud is less than fifty percent and is especially severe for those zygotes where the fraction is small.

UNIPARENTAL INHERITANCE

It has been generally noted that the inheritance of mitochondrial genes in yeast is essentially biparental. Our characterization of multiple transmissional classes of zygote clones and in particular those with greater than 90% transmission from one parent[14] has led to the discovery of uniparental zygotes in a number of crosses. Zygotes formed by mating two wild-type cells but which yield zygotic clones containing alleles from only one parent are termed uniparental (UP). The simplest way to detect such clones is to perform crosses of the type ant-r x ant-s and to test for pure ant-s zygote clone by plating or replica-plating an aliquot (10^4 -10^5 cells) of each clone onto a plate containing the antibiotic[16].

Most crosses selected at random yield a small number of UP clones (Table 4). Since it is possible to alter the input ratio experimentally we analyzed whether there is a correlation between bias and the fraction of UP zygotes. A tentative correlation is seen in Table 4 where an isogenic haploid and diploid were mated to the same tester and where N123 and its mutant derivative rho5 (showing increased bias) were compared. In each case as the bias in favor of the ant-s allele increased, so too did the fraction of UP zygotes. We have performed a more systematic analysis of the N123/rho5 system (data not shown); in this experiment both strains were treated with one or more of the bias-inducing treatments and the fraction of UP zygotes measured. There is a systematic response to increased bias; at very high output ratios (chl-s/chl-r) up to 40% of all zygotic clones were uniparental for chl-s. Thus in UP clones the minor parental type, present initially, is lost from the progeny of the zygote; we stress that the loss of chl-r transmission is not a consequence of there being petites in the chl-r culture.

POSSIBLE MECHANISMS FOR UNIPARENTAL INHERITANCE

Two general mechanisms to account for this loss of transmission can be considered: (i) failure to replicate, or (ii) transformation of ant-r to ant-s by gene conversion. These two mechanisms make distinguishable predictions: in the first case whole genomes should become uniparental while in the second distant markers should become uniparental independently. We have evaluated these predictions by analyzing uniparental inheritance in multi-factor crosses[16]; several examples of the many crosses examined in this way are shown in Table 5. It is clear that the most frequent class of UP clones has all three markers uniparental.

Further analysis of clones pure for only one or two markers shows that they are invariably nearly pure for the other markers. These data do not totally eliminate gene conversion as a factor in uniparental inheritance but clearly limit it to a minor role.

TABLE 4

UNIPARENTAL INHERITANCE OF MITOCHONDRIAL MARKERS

Cross			Percent Transmission			Percent UP Zygotes		
			chl-s	ery-s	oli-s	chl-s	ery-s	oli-s
2-36	x	4810 hap	61.3	62.4	--	1.5	2.2	--
2-36	x	4810 dip	81.6	82	--	12.7	11.9	--
N123	x	ID4/1	38.8	41.2	44.8	3.1	2.7	2.7
αN123	x	ID4/1	56.3	57.3	59.5	12.0	9.9	9.0
rho5	x	ID4/1	61.7	61	61.8	8.4	7.1	8.4
αrho5	x	ID4/1	74.9	75.6	75.6	34.7	35.2	32.6

The genotype of ID4/1 is α ade (nuclear), -rrrr (mitochondrial). Genotypes of the other strains are presented in Tables 1 and 2. Percent transmission is the output of random diploid progeny of many zygote clones. These data illustrate that markers become UP coordinately and that the incidence of UP zygotes is correlated with bias.

TABLE 5

FREQUENCIES OF ZYGOTES UNIPARENTAL AT ONE OR MORE LOCI

		Uniparental Marker											
		None	chl	ery	oli	chl ery	chl oli	ery oli	chl ery oli	Σchl	Σery	Σoli	N
rho5	I	.514	.142	.125	.116	.035	.032	.028	.008	.217	.195	.184	
x	O	.740	.033	.007	.028	.036	.003	.008	.145	.217	.195	.184	609
ID4/1	C	.801	0	0	0	0	0	0	.199	.199	.199	.199	
rho5*	I	.224	.160	.137	.138	.098	.099	.084	.060	.417	.379	.383	
x	O	.497	.044	.007	.063	.071	.017	.016	.285	.417	.379	.383	758
ID4/1	C	.608	0	0	0	0	0	0	.392	.392	.392	.392	

The crosses noted were performed and zygote clones scored for being UP at each locus. In each case all three markers become UP essentially coordinately although some clones are uniparental for less than all three. * Treated with α-factor before mating.

I predicted values if each locus becomes UP independently
O observed values
C predicted values if all loci always become UP coincidentally

Further insights to the mechanism of UP inheritance come from our analysis of zygote bud position[15]. If recombination is required for the elimination of the minor alleles in a cross then as shown previously clones with first central buds should have a higher probability of being UP than would clones with first end buds; as will be reported in detail elsewhere, this prediction was not met. We conclude that the decision to become UP occurs soon after zygote formation and clearly before the first bud is issued.

It appears that UP inheritance occurs as a result of some process that interferes with the replication of mitochondrial genomes from the minority parent. Two general models are being considered. First, there may be processes that randomly eliminate genomes from the zygote; these include outright destruction of a fraction of the total mtDNA or competition for a limited amount of a factor necessary for replication. Second, there may be processes that can discriminate between genomes from each of the two parents and so select specific molecules for elimination; these include a restriction/modification-like process or alternatively preferential binding of replication factors by one parental set of genomes and would indicate that the genomic input ratio does not uniquely define the genetic output. So far we cannot distinguish between these two models, but preliminary studies of the haploid/diploid and N123/rho5 systems indicate bias amplification.

We note an increasing number of parallels between the yeast mitochondrial and the Chlamydomonas chloroplast[17] systems, especially the algal zygotes which undergo mitosis rather than meiosis[18]. The similarities include similar frequencies of uniparental zygotes in some crosses, multiple zygote classes, and zygotes uniparental for markers from parents of either mating type. Moreover, haploid X diploid crosses have been done in both systems with analogous results pointing to effects of genetic input on bias and uniparental inheritance.

ACKNOWLEDGEMENTS

These studies were supported by Grant 19607 from NIH-NIGMS to C. W. Birky and P. S. Perlman. The highly competent technical assistance of Mary Townsend is acknowledged with pleasure and gratitude. Robert L. Strausberg was a predoctoral trainee of PHS Training Grant HD00431 from HICHD; his present address is Department of Biochemistry, University of Texas Health Science Center at Dallas, Dallas, Texas.

REFERENCES

1. Dujon, B., Slonimski, P. and Weill, L. (1974) Genetics 78, 415-437.
2. Wolf, K., Dujon, B. and Slonimski, P. P. (1973) Molec. Gen. Genet. 125, 53-90.
3. Birky, C. W., Jr. (1975) in, Genetics and Biogenesis of Mitochondria and Chloroplasts (Birky, C. W., Jr., Perlman, P. S. and Byers, T. J., eds) Ohio

State University Press, Columbus, Ohio, 182-224.
4. Callen, D. (1974) Molec. Gen. Genet. 128, 321-329.
5. Waxman, M. F. (1975) Molec. Gen. Genet. 141, 285-290.
6. Moustacchi, E., Perlman, P. S. and Mahler, H. R., submitted for publication.
7. Grossman, L., Goldring, E. S. and Marmur, J. (1969) J. Mol. Biol. 46, 367-376.
8. Demko, C. A. and Perlman, P. S., in preparation.
9. Cryer, D. R., Goldthwaite, C. D., Zinker, S., Lam, K.-B., Storm, E., Hirshberg , R., Blamire, J., Finkelstein, D. B. and Marmur, J. (1973) Cold Spring Harbor Symp. Quant. Biol. 38, 17-29.
10. Grimes, G. W., Mahler, H. R. and Perlman, P. S. (1974) J. Cell Biol. 61, 565-574.
11. Gunge, N. (1975) Molec. Gen. Genet. 139, 189-202.
12. Goldthwaite, C. D., Cryer, D. R. and Marmur, J. (1974) Molec. Gen. Genet. 133, 77-104.
13. Sena, E., personal communication.
14. Birky, C. W., Jr. (1975) Molec. Gen. Genet. 141, 41-58.
15. Strausberg, R. L. and Perlman, P. S., in preparation.
16. Birky, C W., Jr., Demko, C. A., Perlman, P. S. and Strausberg, R. L., submitted for publication.
17. Gillham, N. W. (1974) Ann. Rev. Genet. 8, 347-391.
18. Van Winkle-Swift, K. V. (1976) Ph.D. Thesis, Duke University.

Genetics and Biogenesis of Chloroplasts and Mitochondria, Th. Bücher et al. eds.
© 1976 Elsevier/North-Holland Biomedical Press, Amsterdam, The Netherlands.

ON HOMOZYGOTIZATION OF MITOCHONDRIAL MUTATIONS IN SACCHAROMYCES CEREVISIAE

A. Putrament, R. Polakowska, H. Baranowska, A. Ejchart

Institute of Biochemistry and Biophysics,
Polish Academy of Sciences, Warsaw, Poland

Manganese-treated cells were resuspended in glucose medium and samples were plated on ERY dishes and for colony counts immediately after mutagenesis and also after 3-12 h incubation. If upon division of an E^R cell both daughter cells were E^R, doublings of E^R cells would occur at the same rate as those of the entire cell population. We found, however, that 5-6 doublings of the entire cell population were accompanied by 1.5-2.2 doublings of the E^R mutants (Figure). Sometimes the absolute numbers of cells capable of forming colonies on ERY medium decreased after 3-6 h growth in glucose medium. This was not accompanied by a measurable increase of rho^- mutants. The most substantial increase of E^R cells was found when cell population approached stationary phase.

Frequencies of E^R colonies (per 10^6 rho^+ cells) immediately after mutagenesis and after incubation in glucose medium

Time	Exp. 1	2	2+ERY	3	3+ERY
0	369	659	659	328	328
3 h	115	335	310	110	141
6 h	15	122	45	52	31
9 h	22	52	25	27	23
12 h	60	38	37	13	21

These data suggest that after mutagenesis the cells capable of forming colonies on ERY medium carry only very few E^R mitDNA molecules. During first 3-6 h growth in the glucose medium "dilution" of the E^R mitDNA can occur in some cells, so that both daughter cells lose capacity of forming colonies on the ERY medium. At the end of log phase there is intense mitDNA synthesis (5,10) which leads to an enrichment of E^R mitDNA number per cell. Such cells can form colonies on ERY medium, although they may still carry E^S mitDNA. This conclusion is in agreement with the earlier data on intracellular selection

Figure

Interdependence between doublings of total cell population and doublings of E^R mutants during growth in glucose medium

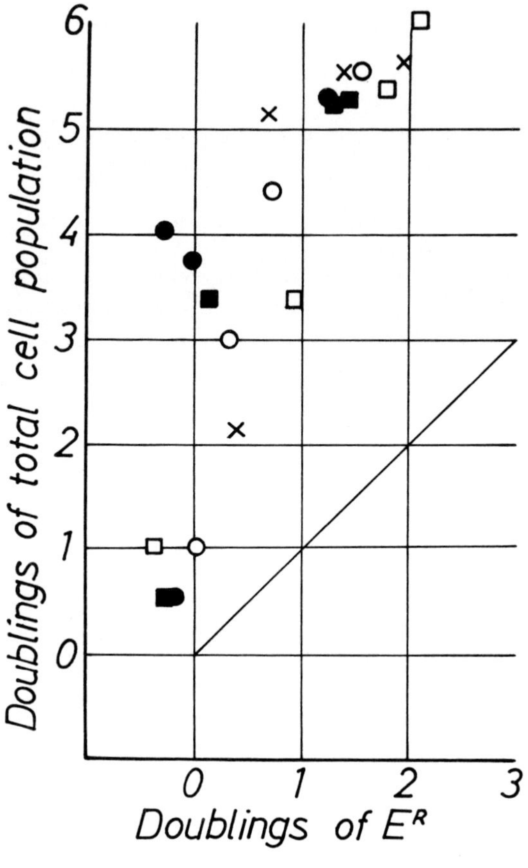

Log phase cells of strain 197/2d were incubated for 6 h in YEP-glucose medium + 10 mM $MnSO_4$. The Mn medium was removed, the cells resuspended in YEP-glu medium with (● ■) or without (x o □) erythromycin (4 mg/ml). First samples were plated immediately afterwards and next after 3-12 h incubation at 30°C on a shaker. Prior to plating each cell sample was sonicated (2x15'') to separate the cells. Diluted cell aliquots were plated on differential medium for cell counts and estimation of % rho^- cells. Each experiment was carried out in duplicate.

of Ant^R mitDNA (1-4).

Our data indicate that cells homozygous for an E^R mutation cannot occur earlier than after 3-4 cell divisions, and probably not until much later. But the mit$^-$ mutants induced with EMS (8) could undergo 1-2 cell divisions before being plated on differential medium. In strain op$_1$, which does not tolerate rho$^-$ mutations (6,9), after manganese mutagenesis mit$^-$ mutations occur with high frequency (P.P. Slonimski, personal communication). We suppose that mit$^-$ mutants are selected from potential rho$^-$ cells rather than beeing segregated from rho$^+$.

In Mendelian populations the equilibrium between two alleles A and a is maintained either by equal frequencies of A a mutations, or by selective advantage of one of the alleles. In S. cerevisiae rho$^-$ mutants do not revert: rho$^+$ rho$^-$ and mit$^-$ rho$^-$. Therefore to survive mit$^-$ must have a selective advantage over rho$^-$. We estimated % rho$^-$ in crosses mit$^-$ M15-190 and M15-233 with mit$^-$ M7-40 (see 7) as follows:

mit$^+$ recombinants were produced by 63.5% of zygotic clones. Assume that (a) each mit$^-$xmit$^-$ zygotic clone produces mit$^+$ recombinants; (b) in the 3 strains % rho$^-$ was identical; (c) the fraction of mit$^-$ cells in population = p, rho$^-$ = q, p+q = 1. Then probabilities of mit$^-$ x mit$^-$, mit$^-$ x rho$^-$ and rho$^-$ x rho$^-$ zygotes are $(p + q)^2 = p^2 + 2pq + q^2$. The estimated $p^2 = 0.635$. Therefore p = 0.79, and q = 1-0.79 = 0.21. So the frequency of rho$^-$ cells in mit$^-$ strains tested was not higher than 21%, although the strains were not purified and re-tested for their ability to carry out mitochondrial protein synthesis for at least 25 cell generations.

Acknowledgements. This work was supported by the Polish Academy of Sciences within Project 0.9.7.2., and partly by the PHS Grant No. 05-001-0.
We are grateful to Professor W. Gajewski and Dr. A. Paszewski for critical discussions.

Literature cited

1. Adoutte, A., Beisson, J. Nature (London) 235, 393 (1972)
2. Beale, G.H., Knowles, J., Tait, A., Nature (London) 235, 396 (1972)
3. Birky, C.W., Genetics 74, 421 (1973)

4. Dujon, B., Bolotin-Fukuhara, M., Coen, D., Deutsch, J., Netter, P., Slonimski, P.P., Weill, L., Molec. gen. Genet. 143, 131 (1976)

5. Goldthwaite, C.D., Cryer, D.R., Marmur, J. Molec. gen. Genet. 133 87 (1974)

6. Kovácová, V., Irmlerová, J., Kovac, L., Biochim. biophys. Acta 162, 157 (1968)

7. Slonimski, P.P., Tzagoloff, A., Eur. J. Biochem. 61, 27 (1976)

8. Tzagoloff, A., Akai, A., Neddlman, R.B., J. Bacteriol. 122, 826 (1975)

9. Ulaszewski, S., Lachowicz, T.M., Molec. gen. Genet. 131, 69 (1974)

10. Williamson, D., H. In: Control of Organelle Development., Cambr. Univ. Press 1970, pp. 247-276

Genetics and Biogenesis of Chloroplasts and Mitochondria, Th. Bücher et al. eds.

GENETIC DETERMINATION OF MITOCHONDRIAL CYTOCHROME b

Alexander Tzagoloff, Françoise Foury and Anna Akai
The Public Health Research Institute
of the City of New York, Inc.
New York, N.Y. 10016

INTRODUCTION

During the past several years we have been studying mutations in mitochondrial DNA of Saccharomyces cerevisiae whose primary effects are on the respiratory chain and oxidative phosphorylation[1]. This class of mutants has been designated as mit^- mutants and the corresponding genes as mit genes[2,3]. A number of genetic loci have been found whose constituent gene or genes are involved in the biosynthesis of cytochrome oxidase[2], of the oligomycin-sensitive ATPase[4] and of cytochrome b, a catalytic subunit of coenzyme QH_2-cytochrome c reductase[5].

In this communication, we will describe some studies of mit^- mutants which lack spectral cytochrome b and exhibit a deficiency in coenzyme QH_2-cytochrome c reductase. Two types of mutants were used in the study. The first lacks specifically cytochrome b and the second is pleiotropic and shows the absence of both cytochrome b and of cytochrome oxidase. Both types of mutants have been previously described[1]. The genetic locus associated with the cytochrome b deficiency was named COB and was localized in the vicinity of the oligomycin resistance locus OLI 1[3]. In the present study, a larger number of strains were examined by a more detailed dissection of the COB locus. Using ρ^- clones which were isolated specifically for their ability to restore the cytochrome b deficient strains, two genetic loci concerned with the specification of cytochrome b have been found. These loci, representing at least two different genes have been called COB 1 and COB 2. Both loci have been ascertained to be localized between the two oligomycin resistance loci, OLI 1 and OLI 2, the order being OLI 2 - COB 2 - COB 1 - OLI 1. Although the functions of the two COB loci have not been defined, indirect evidence strongly suggests that COB 2 contains the structural gene of the cytochrome b apoprotein.

BIOCHEMICAL AND GENETIC PROPERTIES OF CYTOCHROME b DEFICIENT STRAINS

Thirteen mutants of S. cerevisiae have been isolated which lack coenzyme QH_2-cytochrome c reductase activity but retain from somewhat reduced to normal levels of NADH-coenzyme Q reductase, cytochrome oxidase and oligomycin-sensitive ATPase (Table I). A common feature of all the mutants is the absence of the visible absorption bands of cytochrome b (Fig. 1). This phenotype suggests that the enzymatic deficiency is due to a lesion in the cytochrome b component of the respiratory complex.

TABLE I

ENZYMATIC PHENOTYPES OF SPECIFIC AND PLEIOTROPIC CYTOCHROME b MUTANTS

Strain	Enzymatic Deficiency	Specific activity (μmoles/min/mg) NADH-c	CoQH$_2$-c	Cyt.Ox.	ATPase	Cytochr. a	b	c
D273-10B	Wild type	0.35	0.38	0.55	4.45	+	+	+
M6-200	CoQH$_2$-c	0.02	<0.01	0.53	3.43	+	-	+
M10-152	CoQH$_2$-c	0.03	<0.01	0.25	2.50	+	-	+
M9-228	CoQH$_2$-c and Cyt.ox.	0.01	N.D.	<0.01	1.56	-	-	+
M8-130	CoQH$_2$-c and Cyt.ox.	0.02	N.D.	<0.01	1.96	-	-	+

Data taken from Tzagoloff et al (1975).

Another commonly observed phenotype shows a deficiency in cytochrome oxidase and coenzyme QH$_2$-cytochrome c reductase. These strains all show an absence of the visible absorption bands of cytochrome b and of cytochromes a and a_3 (Fig. 1).

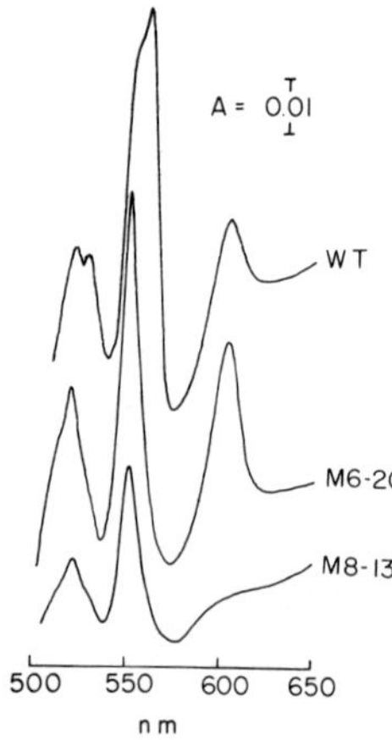

Fig. 1. Absorption spectra of cytochromes in cytochrome b (M6-200) and doubly deficient (M8-130) strains of S. cerevisiae.

All the mutants used in the studies have been ascertained to show mitotic segregation of the enzymatic deficiency when mated to a respiratory competent haploid strain. In addition, the meiotic segregants obtained by tetrad analysis show non-Mendelian segregation ratios. These findings satisfy the usual criteria for cytoplasmic mutations. As will be shown in the subsequent sections, the cytochrome b mutations are localized in mitochondrial DNA.

TWO GENETIC LOCI AFFECTING CYTOCHROME b BIOSYNTHESIS

In previous studies with mit$^-$ mutants defective in cytochrome oxidase, it was reported that mutations could be assigned to different genetic loci based on a simple qualitative test in which two mutant strains are crossed and the diploids scored for growth on glycerol[2]. When the test is carried out by patching the mating partners, the diploids replicated on glycerol either 1) show confluent growth, 2) grow as single colonies or 3) fail to show any growth in the patch. Crosses which exhibit confluent growth, usually

TABLE II

PAIRWISE CROSSES OF CYTOCHROME b MUTANTS

	COB 1								COB 2				
	M6-200	M7-40	M8-53	M8-181	M13-101	M15-207	M17-231	M24-241	M10-152	M17-162	M18-68	M33-119	M21-71
M6-200	-	P	P	P	P	P	-	P	+	+	+	+	+
M7-40		-	-	+	-	P	P	P	+	+	+	+	+
M8-53			-	+	-	P	P	-	+	+	+	+	+
M8-181				-	+	P	+	P	+	+	+	+	+
M13-101					-	-	P	-	+	+	+	+	+
M15-207						-	P	P	+	+	+	+	+
M17-231							-	P	+	+	+	+	+
M24-241								-	+	P	+	+	+
M10-152									-	-	+	+	+
M17-162										-	P	-	-
M18-68											-	-	P
M33-119												-	P
M21-71													-

The symbols P indicate papillae growth, +, confluent growth and -, absence of growth on glycerol.

correspond to mutants in which the mutations recombine with high frequency and belong to different loci. On the other hand, mutants which recombine with low frequencies and belong to the same locus, usually give a negative result or exhibit papillae type of growth. The results of the pairwise crosses of the specific cytochrome b mutants are shown in Table II. Based on these results the cytochrome b mutations have been assigned to two different loci designated as COB 1 and COB 2. Crosses of COB 1 by COB 2 mutants result in high yields of respiratory competent diploids, while crosses involving mutants belonging to the same locus yield low numbers of glycerol positive recombinants.

The assignment of the cytochrome b mutations to the two COB loci was confirmed by quantitating the frequency of recombinants observed in the various crosses. The results of some representative crosses are summarized in Table III. In all the intra-locus crosses, the yields of recombinants was less than 1% and in most cases was less than 0.1%. The recombination values in the inter-locus crosses were significantly higher and ranged from 8-15%. The high values found in the crosses of COB 1 by COB 2 mutants strongly imply that there are at least two genes necessary for the biosynthesis of functional cytochrome b

TABLE III

FREQUENCY OF WILD TYPE RECOMBINANTS

Cross	Total No. colonies scored	Total No. recombinants	Percent recombination
COB 1 x COB 1			
M8-53 x M15-207	69,600	5	0.01
M6-200 x M7-40	30,400	100	0.67
COB 2 x COB 2			
M17-162 x M10-152	88,100	22	0.05
M21-71 x M10-152	100,000	299	0.58
COB 1 x COB 2			
M8-53 x M17-162	1199	80	12.5
M15-207 x M21-71	1317	80	11.6

The mutants were mated in 10% glucose medium for six hours. The diploids were prototrophically selected in liquid minimal medium containing 10% glucose for 20-30 generations,plated for single colonies on minimal glucose and replicated on glycerol. Growth on glycerol was scored after 2 days. The values reported have been corrected for ρ^- cells which were recognized as small colonies. For the intra-locus crosses in which the recombinants were rare, duplicate platings with 100 times as many cells were done. For example in the cross M17-162 x M10-152, 22 glycerol positive colonies were scored out of a total of 88,100 cells plated on glycerol.

LOCALIZATION OF COB 1 AND COB 2

We have recently used ρ^- clones to localize mit$^-$ mutations in the ATPase locus, PHO 1[4]. Because of the close linkage of the ATPase mutations to the OLI 2 locus, the methods used to localize other mit$^-$ loci[2] proved impractical for these strains. The same problem arose in trying to map the COB mutations which showed some linkage to the OLI 1 locus and we therefore employed a deletion mapping method with ρ^- clones to localize the cytochrome b muations.

Approximately 124 stable tertiary ρ^- clones were isolated from the respiratory competent wild type *S. cerevisiae* D22. These clones were isolated for their ability to restore glycerol growth when mated to two representative mutants with lesions in COB 1 (M17-231) and COB 2 (M17-162). The purified clones were tested for the retention of the antibiotic resistance loci, OLI 1, OLI 2 and RIB 3 by gene rescue analysis. Only these three markers were analyzed since other work indicated that the COB loci were either between RIB 3 and OLI 1 or between OLI 1 and OLI 2.

Out of the 124 clones, 106 were found to be non-discriminatory and to restore all the cytochrome b mutations. Twenty clones, however, discriminated to different extents between COB 1 and COB 2 mutations. These results are presented in Fig. 2. Of the twenty

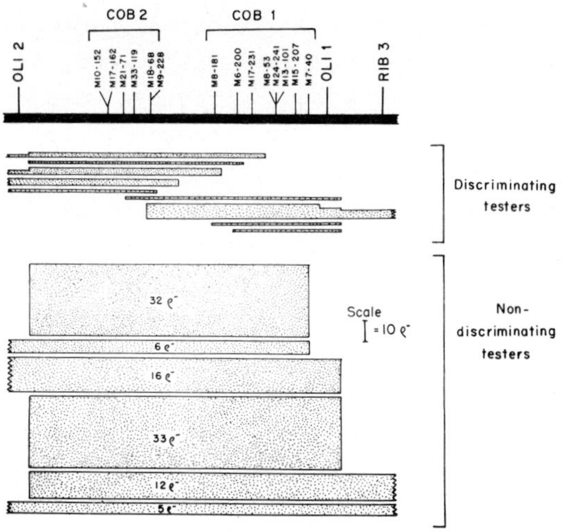

Fig. 2. Restoration of cytochrome b mutations by ρ^- clones. The retention of the different alleles by the ρ^- clones is indicated by an overlap of the stippled areas denoting the segments of mitochondrial DNA retained in the clones and the dark line showing the positions of the mit$^-$ and antibiotic resistance alleles. The number of ρ^- clones is indicated by the width of the stippled areas.

discriminating ρ^- clones, 10 clones restored most of the COB 1 mutations but only a few of the COB 2 mutations. None of these clones retained the OLI 2 marker, but a substantial number had the OLI 1 marker (8 clones). Four of the clones also retained the RIB 3 marker. The other 10 clones restored most of the COB 2 but not the COB 1 mutants. These clones did not have the OLI 1 marker, but 7 of them had the OLI 2 marker. The COB 1 and COB 2 mutations were therefore placed between OLI 1 and OLI 2 in the order OLI 2-COB 2-COB 1-OLI 1.

The results of the restoration tests with the ρ^- clones also indicated that the two COB loci were more closely linked to OLI 1 than to OLI 2. Out of the 124 clones, 74 clones retained OLI 1 and only 34 retained OLI 2. A linkage of the COB 1 mutations to OLI 1 was also supported by the data on the frequency of recombination with the OLI 1 resistance allele. In crosses of mutants in COB 1 with D22/A16 (and OLI 1 resistant strain), the recombination frequencies ranged from 4-8%; values of 15-20% were found when the OLI 2 resistant strain D22/A15 was used.

By combining the results of the pairwise crosses and the deletion mapping with the ρ^- clones, the cytochrome b mutations were arranged in a linear sequence as shown in the top part of Fig. 2.

DOUBLY DEFICIENT STRAINS

In addition to the specific cytochrome b mutants, twenty strains exhibiting a double enzymatic deficiency in cytochrome oxidase and coenzyme QH_2-cytochrome c reductase were also examined. Genetic analyses were done on these strains in order to determine whether the phenotype is due to separate mutations in the OXI and COB loci or whether another locus exerting a dual control of both enzymes is involved. Several lines of evidence indicated that in all the strains, the absence of the enzymatic activities was due to multiple mutations in the known cytochrome oxidase and cytochrome b loci[2,5]. 1) When the strains with an ox^-b^- phenotype are mated to a respiratory competent haploid (ox^+b^+), mitotic segregants with a single enzymatic deficiency of the type ox^+b^- and ox^-b^+ are found.[6] 2) Similarly, ox^-b^+ and ox^+b^- recombinants are found among the meiotic spore progeny in crosses of the mutants to a respiratory competent strain. 3) By using mit^- testers with defined lesions in the different OXI and COB loci, most of the doubly deficient strains have been shown to have mutations in OXI 3 and COB 2. A few of the strains have mutations in OXI 1 and COB 1 loci[6].

IDENTITY OF THE GENES IN THE COB LOCI

Although the genetic evidence suggests that there exist a least two separate loci which are necessary for the biosynthesis of functional cytochrome b, we still do not know whether these loci represent only two or more than two genes.

Weiss and coworkers[7] have obtained evidence that there are two distinct cytochrome b species that are synthesized by Neurospora crassa mitochondria. Since it now appears that the structural genes of cytochrome oxidase[8] and of the ATPase[9] are encoded in the DNA of mitochondria, it may be anticipated that the genome also contains the structural gene(s) of cytochrome b. Two of the mutants in the COB 2 locus, have been found to have a modified form of the cytochrome b apoprotein. Labeling experiments of the mutants in the presence of the cytoribosomal inhibitor cycloheximide and gel electrophoretic separation of the mitochondrially synthesized proteins, have shown that the cytochrome b apoprotein which in wild type cells migrates with a molecular weight of 32,000 daltons[10] appears in the low molecular weight region of the gel[5]. This finding provides some support for the notion that the structural gene of the cytochrome b apoprotein is included in the COB 2 locus.

CURRENT STATUS OF THE MITOCHONDRIAL GENOME

The mit⁻ mutants used in our studies were isolated from two different screenings, each involving approximately 4,000-5,000 glycerol negative mutants. All the mutants that were isolated in the second screening could be placed in genetic loci that were already represented by the mutants obtained in the first screening. Our inability to isolate mutations in new loci despite the large number of strains examined suggests that the mitochondrial genome is probably close to saturated with respect to mit⁻ genes.

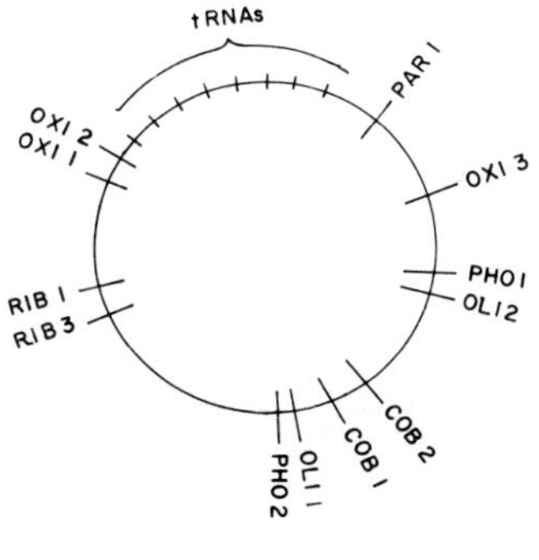

Fig. 3. Location of mit⁻ and antibiotic resistance loci.

A map of the genome of *S. cerevisiae* showing the localization of the mit$^-$ relative to the antibiotic resistance markers is presented in Fig. 3. This map is based on studies of the specific cytochrome oxidase[2], cytochrome *b*[5] and ATPase[4] mutants.

ACKNOWLEDGMENTS

This research was supported by Grants 1RO1-HL 13003 and GM 18868 from the National Institutes of Health, United States Public Health Service.

REFERENCES

1. Tzagoloff, A., Akai, A, Needleman, R.B. and Zulch, G. (1975) J. Biol. Chem. 250, 8236-8242.
2. Slonimski, P.P. and Tzagoloff, A. (1976) Eur. J. Biochem. 61, 27-41.
3. Tzagoloff, A. (1975) in"Electron transfer chains and oxidative phosphorylation" (E. Quagliariello, S. Papa, F. Palmieri, E.C. Slater, N. Siliprandi, eds) North Holland Press, N.Y. pp. 185-192.
4. Foury, F. and Tzagoloff, A. (1976) Eur. J. Biochem. in press.
5. Tzagoloff, A., Foury, F. and Akai, A.(1976) submitted for publication to Mol. Gen. Genet.
6. Foury, F. and Tzagoloff, A. (1976) submitted for publication to Mol. Gen Genet.
7. Weiss, H. and Ziganke (1975) in "Electron transfer chains and oxidative phosphorylation" (E. Quagliariello, S. Papa, F. Palmieri, E.C. Slater, N. Siliprandi, eds) North Holland Press, N.Y. pp. 15-22.
8. Padmanaban, G., Hendler, F., Patzer, J., Ryan, R. and Rabinowitz, M. (1975) Proc. Nat. Acad. Sci. U.S.A. 72, 4293-4297.
9. Tzagoloff, A., Akai, A. and Foury, F. (1976) FEBS Lett. 65, 391-395.
10. Katan, M.B. and Groot, G.S.P. (1975) in "Electron transfer chains and oxidative phosphorylation" (E. Quagliariello, S. Papa, F. Palmieri, E.C. Slater, N. Siliprandi, eds) North Holland Press, N.Y. pp.127-132.

Genetics and Biogenesis of Chloroplasts and Mitochondria, Th. Bücher et al. eds.

THE ISOLATION AND SIMULTANEOUS PHYSICAL MAPPING OF MITOCHONDRIAL MUTATIONS AFFECTING RESPIRATORY COMPLEXES

J. Rytka, Karen J. English, R.M. Hall, Anthony W. Linnane and H.B. Lukins
Department of Biochemistry, Monash University,
Clayton, Victoria, 3168, Australia.

INTRODUCTION

During almost a decade of study, mutations of mitochondrial DNA (mtDNA) conferring resistance to a variety of antibiotics have been employed for the characterisation of mitochondrial genetics in yeast. Thus, antibiotic resistant markers have been used to identify several loci on the mitochondrial genome, for the characterisation of mitochondrial gene recombination, and for the establishment of the first basic map of yeast mtDNA. However, the application of resistance mutants to the identification of genes determining specific mitochondrial gene products has proved unrewarding, although current research is establishing correlations between certain oligomycin[1-3] and antimycin[1,2] resistant mutants and components of the mitochondrial ATPase and coenzyme QH_2-cytochrome *c* reductase complex (complex III) respectively.

The identification of new mitochondrial genes, particularly the structural genes determining protein components of the mitochondrial membranes known to be synthesised on mitochondrial ribosomes, is of immediate importance for the development of our understanding of the mitochondrial genome and its function. A major step towards achieving these objectives has been the isolation of *mit*⁻ mutants, which are unable to utilize non-fermentable substrates for growth and which are distinguished from petite (*rho*⁻) mutants by the retention of mitochondrial protein synthesising activity[4-6]. Among mutations of this type are those that affect one or more of the three multicomponent membrane complexes believed to contain mitochondrially synthesised components, namely cytochrome oxidase, the cytochrome *b* complex and oligomycin-sensitive ATPase[7].

The procedure developed by Tzagoloff *et al.* (1975)[4-6] for the isolation of *mit*⁻ mutants was based on a biochemical screening of many thousands of isolates, unable to grow on glycerol, for the retention of mitochondrial protein synthesising activity, followed by a cross to a *rho-o* strain (lacking mtDNA) to distinguish the mitochondrial mutations from the bulk of mutants carrying nuclear mutations. For the isolation of *mit*⁻ mutants in this laboratory, we have concentrated on a genetic screening procedure based on the use of a collection of *rho*⁻ strains carrying genetically and physically defined segments of the mitochondrial genome. The advantages of this procedure are that, at the one time, we select only mutations of mitochondrial origin and position these within physically defined limits on the map of the mitochondrial genome. In this paper we describe our selection-mapping procedure, some associated phenomena relating to the stability of *mit*⁻ mutations and their effects on mitochondrial recombination.

PROCEDURE FOR SELECTION OF *MIT*⁻ MUTANTS

A procedure for the isolation of mitochondrial mutations of the *mit*⁻ class, suitable for genetic and biochemical analysis, must meet several requirements. A procedure should: (i) yield only mutants arising by single genetic lesions in mitochondrial DNA; (ii) eliminate petite (*rho*⁻) deletion mutants; (iii) eliminate mutants of a similar phenotype but arising by mutation of nuclear genes; (iv) yield mutants retaining activity for mitochondrial protein synthesis. Each of these conditions is satisfied by our procedure which selects *mit*⁻ mutants on the basis of the generation of respiratory competent diploids (*rho*⁺) in crosses to a number of stable petite (*rho*⁻) tester strains.

Each of the petites in our collection of some 15 tester strains has been characterised genetically for retention of antibiotic resistance loci and physically, in terms of the fractional length of the mtDNA genome retained, by DNA-DNA hybridization procedures[8-13]. The petites are overlapping and cumulatively encompass all the sequences of mtDNA. This library of tester strains is used to screen mutant isolates detected by inability to grow on a non-fermentable substrate such as ethanol. Isolates not able to utilize ethanol can include *mit*⁻ mutants, *rho*⁻ mutants and mutants defective in mitochondrial function as a result of nuclear mutations, but the crosses to the *rho*⁻ strains distinguishes between these three classes. Crosses of any individual *mit*⁻ isolate to the tester petites yield respiratory competent (*rho*⁺) diploids by recombination only in the case of those petites retaining the normal genome segment which overlaps and corresponds to the site of lesion in the genome of the *mit*⁻ isolate. Mutant isolates of the *rho*⁻ type do not yield *rho*⁺ recombinants in crosses to any of the tester petites, while any mitochondrial defective mutants resulting from mutation of nuclear genes yield *rho*⁺ diploids by complementation in crosses to all of the tester strains. A *rho-o* (mtDNA less) tester strain is also included in the screen to assist in the differentiation of nuclear mutations from those involving mutation of mtDNA.

An outline of the procedure for isolation of *mit*⁻ mutants is given in Figure 1. We routinely employ mutagenesis with Mn^{++} ions[14], although *mit*⁻ mutants have also been obtained following treatment with either ethylmethane-sulphonate or 2-aminopurine. The mutagenized culture is then plated on 2% galactose medium (containing 0.1% glucose) and small clones are selected after 3 days incubation. Our choice of galactose in the mutant selection medium was based on the report of Puglisi and Algeri (1971)[15] that the integrity of the mitochondrial protein

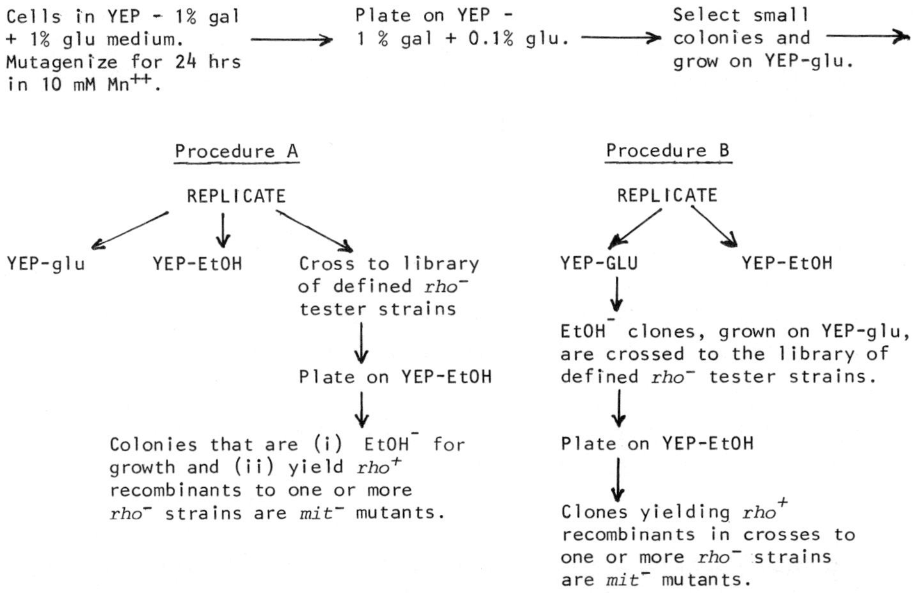

Figure 1. Procedure for the isolation and simultaneous physical mapping of *mit*$^-$ mutations. The alternative procedures A or B, which both follow the initial isolation of small clones on galactose-glucose medium, yield *mit*$^-$ mutants at different frequencies and with different characteristics as discussed in the text. YEP, yeast extract plus peptone; gal, galactose; glu, glucose; EtOH, ethanol.

synthesizing system is required for the induction of the enzymes for galactose utilization. *rho*$^-$ strains lacking mitochondrial protein synthesis are unable to ferment galactose for growth. Hence, galactose should select against *rho*$^-$ mutants when an inducible strain is employed. While it is our experience that *mit*$^-$ mutants are obtained from platings on galactose-glucose medium at a higher frequency than from plating on medium containing glucose alone, we also find that the great majority of *mit*$^-$ mutants obtained by the procedure are subsequently unable to grow on galactose, even though they may retain a high activity for mitochondrial protein synthesis.

Following selection of small colonies from the galactose (+ glucose) plates, the subsequent steps taken in screening these isolates by crossing to the library of defined *rho*$^-$ testers, can noticeably affect the frequency and stability of the *mit*$^-$ mutants obtained. If the isolates, after growth on glucose, are directly crossed to the *rho*$^-$ strains (Fig. 1, procedure A), then as many as 15-20% of the isolates are obtained as *mit*$^-$ mutants on the basis of retention of mitochondrial protein synthesis and capacity to generate *rho*$^+$ recombinants in crosses to one or more *rho*$^-$ testers. However, after subculture and retesting these mutants, a majority are found to be unstable, having either reverted to *rho*$^+$ or having lost the capacity to form *rho*$^+$ recombinants in crosses. Many of this latter type have lost mitochondrial protein synthesis and are *rho*$^-$. Others retain protein synthetic activity and by this criterion are still *mit*$^-$. Their inability to form *rho*$^+$ recombinants probably indicates the presence of more than one lesion in their mtDNA, and for this reason they are of limited use for further genetic or biochemical study. The instability in these freshly-arisen *mit*$^-$ isolates, particularly in regard to the capacity for formation of *rho*$^+$ recombinants in crosses, is similar to the behaviour reported by Clark-Walker and Miklos (1975)[16] for freshly-arisen spontaneous petite isolates.

The problems associated with instability in *mit*$^-$ isolates, as just described, are overcome to a large degree by an additional subculture which first screens the small colonies selected from the galactose (+ glucose) plates for clones that are unable to utilize EtOH for growth (see procedure B, Fig. 1). This additional step eliminates clones showing high reversion frequencies to *rho*$^+$, as well as the high proportion of clones which yield *rho*$^+$ recombinants in initial crosses to *rho*$^-$ but lose this property upon subculture. Only about 1% of the EtOH$^-$ clones yield *rho*$^+$ recombinants when tested with the *rho*$^-$ strains, but these isolates are basically stable mutants which retain their genetic characteristics and capacity for mitochondrial protein synthesis and are therefore suitable for detailed biochemical analysis. Only *mit*$^-$ mutants of this type are considered below.

DETERMINATION OF PHYSICAL MAP LOCATION OF *MIT*$^-$ MUTATIONS

The screening of mutant isolates by crosses to the collection of *rho*$^-$ tester strains simultaneously determines the physical map location of the *mit*$^-$ mutations. The principles involved in mapping with the library of petites have been detailed previously in regard to the mapping of several oligomycin resistant mutants[9,11,12]. Briefly, the location of a particular *mit*$^-$ mutation simply involves the recognition of the segment of the genome which is retained in common by all the petites which yield *rho*$^+$ recombinants in crosses to the *mit*$^-$ mutant. For example, consider the results for the two *mit*$^-$ mutants 41.2 and 37.16 as given in Figure 2. These mutants give *rho*$^+$ recombinants in crosses to all *rho*$^-$ strains retaining the antimycin A resistance locus, *ana*1, and further, the formation of recombinants with petite Y9 but not with petite Y6 limits these mutations within the segment 22-25 units. These two mutants, and others, define a region *cyb*1 in which all mutations affect complex III of the mitochondrial membrane as seen by specific elimination or changes in the spectra of cytochrome *b* and by deficiencies in the activity of coenzyme QH_2-cytochrome *c* reductase[1,2].

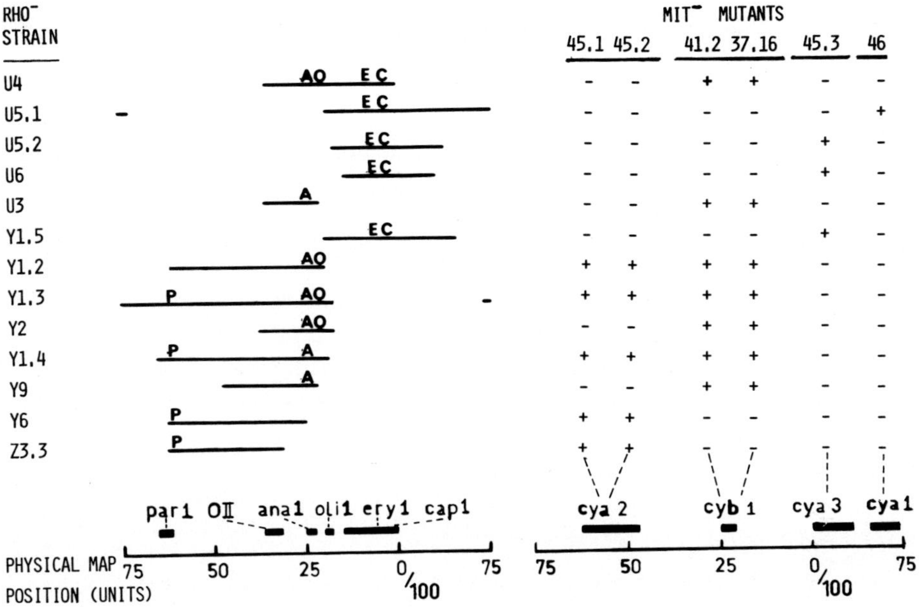

Figure 2. The determination of the physical map location of *mit*⁻ mutations by analysis of crosses to the library of defined *rho*⁻ tester strains. The length and segment of the mtDNA genome retained in each petite is shown on the left side of the figure. The results of crosses of six *mit*⁻ mutants to the *rho*⁻ strains is shown on the right, a "+" indicating the generation of *rho*⁺ recombinants. The physical map location of four loci *cya*1, *cya*2, *cya*3 and *cyb*1, indicated by the *mit*⁻ mutations, are show as bars above the scale of map units on the right and can be compared with the location of antibiotic resistance loci shown on the left. *cya*1, *cya*2 and *cya*3 are loci specifically affecting cytochrome oxidase activity, while *cyb*1 affects the cytochrome *b* complex, complex III (see text).

Also shown in Figure 2 is the mapping data for 4 *mit*⁻ mutants that are examples of mutants specifically defective in cytochrome aa_3 content and cytochrome oxidase activity[1,2]. Mapping of these mutants demonstrated three regions, designated *cya*1, *cya*2 and *cya*3, concerned with cytochrome oxidase synthesis. Based on our present set of physically defined petites, these regions are located within the segments 76-84, 47-63 and 89-100 units respectively. Our loci *cya*3, *cya*1 and *cya*2 probably correspond, respectively, to the loci *OXI*1, *OXI*2 and *OXI*3 described by Slonimski and Tzagoloff (1976)[7]; *OXI*1 and *OXI*2 being broadly located in the region between the loci for chloramphenicol and paramomycin resistance and *OXI*3 in the region between the oligomycin resistance locus *OLI*3 (O_{II}) and the paramomycin resistance locus.

ANOMALOUS RECOMBINATION ASSOCIATED WITH *MIT*$^-$ MUTANTS

The analysis of recombination in intercrosses of *mit*$^-$ mutants (*mit*$^-$ x *mit*$^-$) can be used to complement the mapping of the mutations by *mit*$^-$ x *rho*$^-$ crosses as just described. Pairwise crosses of all the mutants mapped into the four regions *cya*1, *cya*2, *cya*3 and *cyb*1 have been performed and *rho*$^+$ recombinants are generated whenever a mutant from one locus is crossed to a representative mutant at another locus. In addition, *rho*$^+$ recombinants are also generated in several cases involving two mutations mapped within a single region, notably within the regions *cyb*1 and *cya*2.

Recombination of the latter type, between mutations that give a common biochemical phenotype and which are located within a short region, are of particular interest. Either the recombination indicates more than one gene in the region associated with the production of the particular phenotype, or else intragenic recombination if the region contains a single gene. The extent of recombination expected for intragenic mitochondrial recombination in yeast has yet to be clearly established; recombination presumed to be intragenic has been observed only in the case of a few oligomycin resistance mutations which appear to affect a single protein component of the ATPase complex[1,9].

As a means to studying the above questions we are examining recombination between mutations which map within 3 map units in the region *cyb*1. From a consideration of the molecular weight and base composition of yeast mtDNA it can be estimated that 3 units could encode about 2, or perhaps 3 proteins of 20,000 daltons. Table 1 shows the recombination frequencies between three mutants (*ana*1, 17.62.2 and 37.16) within this region and recombination of each of these with the nearby locus *oli*1, 3 to 7 units removed from *cyb*1. The recombination frequency

TABLE 1

ANOMALOUS RECOMBINATION FREQUENCIES

IN CROSSES INVOLVING *MIT*$^-$ MUTATIONS IN THE *cyb*1 REGION

	*ana*1 (*rho*$^+$)	*oli*1 (*rho*$^+$)	17.62.2 (*mit*$^-$)
*ana*1 (*rho*$^+$)	-	6*	
17.62.2 (*mit*$^-$)	>8†	>23	-
37.16 (*mit*$^-$)	>17	>31	>9††

* In *rho*$^+$ x *rho*$^+$ crosses, the recombination frequency is the sum of the reciprocal recombinants expressed as percent of total diploid progeny from the cross.

† In *mit*$^-$ x *rho*$^+$ crosses, the recombination frequency is the number of *rho*$^+$ recombinants expressed as percent of total diploid progeny. The reciprocal *mit*$^-$ recombinant cannot be scored.

†† In *mit*$^-$ x *mit*$^-$ crosses, the frequency is the number of *rho*$^+$ recombinants as percent of total diploids. The reciprocal recombinant cannot be scored.

between *ana*1 and *oli*1 in *rho*$^+$ and *rho*$^+$ crosses can be taken as a measure of "normal" recombination frequencies in this part of the genome, the value of 6% approximating the physical separation of these markers. By contrast, recombination frequencies in crosses involving a *mit*$^-$ mutant are abnormally high, both for recombination between markers within *cyb*1 and for recombination with the *oli*1 locus. Indeed, some of the frequencies exceed 25%, the maximum value observed for unlinked markers, and this is entirely inconsistent with the known physical linkage of these mutations.

It is clear that recombination analysis in crosses involving *mit*$^-$ mutations in the *cyb*1 region is of little value for distinguishing inter- or intra- genic recombination and the number of probable genes in this region. We are presently examining whether *mit*$^-$ mutations in other regions of the genome, such as the *cya*2 region, similarly effect an enhanced rate of recombination as observed in the *cyb*1 region. The high recombination rate between the linked mutations in the *cyb*1 region calls to mind the high recombination rate observed in polar crosses in the *cap*1 - *ery*1 region of the genome under the influence of the alleles ω^+/ω^-. Suggestions have been made[17,18] that ω^+ is a partial deletion of ω^-. Since some *mit*$^-$ mutations may be small deletions, it might be speculated that introduction of such a *mit*$^-$ mutation in a particular region of the genome could have consequences upon recombination in that region analogous to the effects of the ω alleles in the *cap*1 - *ery*1 region.

REFERENCES

1. Groot Obbink, D.J., Hall, R.M., Linnane, A.W., Lukins, H.B., Monk, B.C., Spithill, T.W. and Trembath, M.K. (1976). In, The Genetic Function of Mitochondrial DNA (Kroon, A.M. and Saccone, L. eds.), North-Holland, Amsterdam. pp 163-173.
2. Cobon, G., Groot Obbink, D.J., Hall, R.M., Maxwell, R., Murphy, M., Rytka, J. and Linnane, A.W. (1976). In, Genetics and Biogenesis of Chloroplasts and Mitochondria (Th. Bücher, W. Neupert, W. Sebald and S. Werner, eds.), North-Holland, Amsterdam (this volume).
3. Tzagoloff, A., Akai, A., Foury, F. (1976). FEBS Letters, 65, 391-395.
4. Tzagoloff, A., Akai, A. and Needleman, R.B. (1975). J. Bacteriol. 122, 826-831.
5. Tzagoloff, A., Akai, A. and Needleman, R.B. (1975). Proc. Nat. Acad. Sci. USA, 73, 2054-2057.
6. Tzagoloff, A., Akai, A., Needleman, R.B. and Zulch, G. (1975). J. Biol. Chem. 250, 8236-8242.
7. Slonimski, P.P. and Tzagoloff, A. (1976). Eur. J. Biochem. 61, 27-41.
8. Molloy, P.L., Linnane, A.W. and Lukins, H.B. (1975). J. Bacteriol. 122, 7-18.
9. Trembath, M.K., Molloy, P.L., Sriprakash, K.S., Cutting, G.J., Linnane, A.W. and Lukins, H.B. (1976). Molec. Gen. Genet. 145, 43-52.

10. Sriprakash, K.S., Molloy, P.L., Nagley, P., Lukins, H.B. and Linnane, A.W. (1976). J. Mol. Biol. 104, 485-503.

11. Linnane, A.W., Lukins, H.B., Nagley, P., Molloy, P.L., Rytka, J., Sriprakash, K.S. and Trembath, M.K. (1976). Proc. Nat. Acad. Sci. USA, 73 2082-2085.

12. Nagley, P., Sriprakash, K.S., Rytka, J., Choo, K.B., Trembath, M.K., Lukins, H.B. and Linnane, A.W. (1976). In, The Genetic Function of Mitochondrial DNA (Kroon, A.M. and Saccone, L., eds.), North-Holland, Amsterdam pp 231-242.

13. Nagley, P., Sriprakash, K.S. and Linnane, A.W. (1976). In, Advances in Microbial Physiology (Rose, A.H. and Tempest, D.W., eds.), Academic Press, London (in press).

14. Putrament, A., Baranowska, H. and Prazmo, W. (1973). Molec. Gen. Genet. 126, 357-366.

15. Puglisi, P.P. and Algeri, A. (1971). Molec. Gen. Genet. 110, 110-117.

16. Clark-Walker, G.D. and Miklos, G.L.G. (1975). Proc. Nat. Acad. Sci. USA, 72, 372-375.

17. Perlman, P.S. and Birky, L.W. (1974). Proc. Nat. Acad. Sci. USA, 71, 4612-4616.

18. Dujon, B., Bolotin-Fukuhara, M., Coen, D., Deutsch, J., Netter, P., Slonimski, P.P. and Weill, L. (1976). Molec. Gen. Genet. 43, 131-165.

Genetics and Biogenesis of Chloroplasts and Mitochondria, Th. Bücher et al. eds.

GENETIC ANALYSIS OF MITOCHONDRIAL POLYMORPHIC PROTEINS IN YEAST

M. G. Douglas, R. L. Strausberg, P. S. Perlman* and R. A. Butow
Department of Biochemistry, University of Texas Health Science Center at Dallas, Dallas, Texas USA and *Department of Genetics, The Ohio State University, Columbus, Ohio, USA

Rapid progress is being made through genetic and biochemical investigations toward the goal of defining all of the genes and gene products of mtDNA. Genetic studies using bakers' yeast, Saccharomyces cerevisiae, have been especially productive; there are now nine antibiotic resistance loci[1,2], at least five mit loci[3], two rRNA-specifying regions[4,5], more than ten tRNA loci[6] and a few others that are incompletely mapped[7,8]. Biochemical studies have identified at least eight polypeptide components of the mitochondrial inner membrane that are made on mitoribosomes and each has been shown to be a component of a specific enzyme of the oxidation and phosphorylation machinery[9]. However, little progress has been made in identifying the products of the genetically-defined loci. Among the drug resistance loci only the product of the ery (rib3) locus has been established[4]. Even though the four oli-r loci all affect the mitochondrial ATPase it is not yet known which of its four mitochondrially synthesized polypeptides is encoded by each locus. Analysis of mit- mutants so far has not settled this matter either.

THE INITIAL OBSERVATION: POLYMORPHIC PROTEINS

Douglas and Butow have developed a protocol that allows the high resolution analysis of $^{35}SO_4$-labelled products of mitochondrial protein synthesis on SDS acrylamide gradient slab gels[10,11]. Using a wild-type strain a reproducible pattern of at least 20 bands is obtained. All of these bands meet the usual criteria for mitochondrial translation products but it is likely that some of them are aggregates or precursors of lower molecular weight components or are artifactual cleavage products of larger components. When two different strains (41-6/19 and 55R5-3C/221-2 --- see Table 1 for their complete genotypes) were analyzed simultaneously most bands were identical; however, three differences were evident and these are illustrated in Figure 1 by comparing gels slots 1d and 2a. Variant 1 (var1) appears in two different forms, one migrating faster (var1-f) than the other (var1-s), with an apparent molecular weight difference of 3000 daltons. Var2 and var3 are present in one strain (var2+, var3+) and absent in the other (var2-, var3-). The actual changes in the polypeptides responsible for their altered electrophoretic mobility are not yet established; however, the mobility differences are reproducible and (as shown below) are inherited as are known mitochondrial markers.

TABLE 1

PATTERNS OF INHERITANCE FOR VARIANT POLYPEPTIDES

Cross	Nuclear Marker Segregation	†Variant Polypeptide Segregation
41-6/19 ρ^+ x 55R5-3C/221-2 ρ^+ Diploids	-	mixed
41-6/19 ρ^o x 55R5-3C/221-2 ρ^+ Diploids	-	100% ρ^+ parental
41-6/19 ρ^+ x 55R5-3C/221-2 ρ^o	-	100% ρ^+ parental
41-6/19 ρ^+ x 55R5-3C/221-2 ρ^+ *Tetrads	2:2	4:0

* Diploids displaying parental forms of var species were sporulated and complete tetrads analyzed for var polypeptides as described in the legends to Fig 1. Nuclear markers were scored in the usual fashion.

† Variant polypeptide segregation was scored as varl-f or varl-s.

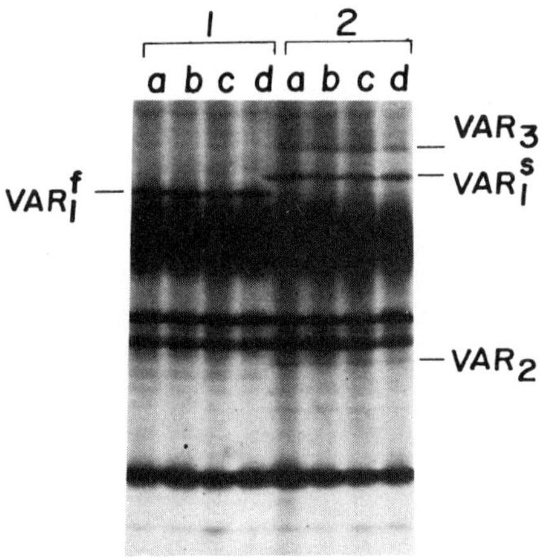

Figure 1: Meiotic segregants of variant proteins. Two purified diploid clones from the cross summarized in Tables 2 and 3 showing the parental arrangements of variants (varl-f var2- var3-, diploid 1; varl-s var2+ var3+ diploid 2) were sporulated and the tetrads dissected. One complete tetrad from each is shown. Each spore clone was labelled with $^{35}SO_4^{-2}$ in the presence of cycloheximide. Mitochondria were isolated, run on an SDS polyacrylamide slab gel and the bands visualized by autoradiography. Complete experimental details have been published elsewhere[10,11]. Only a portion of the gel is displayed.

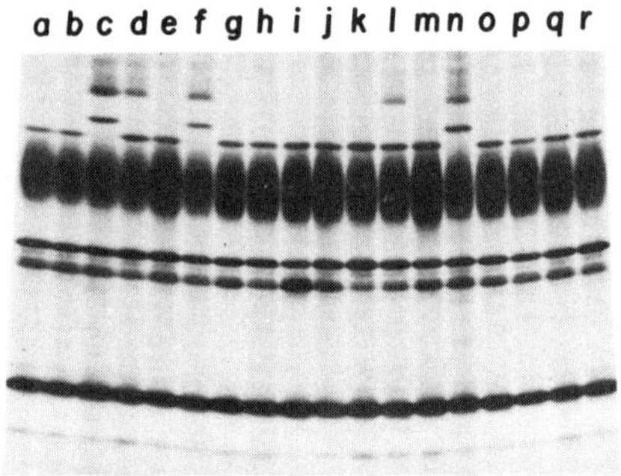

Figure 2: Mitotic segregation of variant proteins. Electrophoretic profiles of eighteen progeny of the cross summarized in Tables 2 and 3 are shown. Several recombinants between varl and var3 are present in this sample (slots d and l). Only a portion of the gel is displayed.

PRELIMINARY GENETIC CHARACTERIZATION

Three criteria met by all mitochondrial genetic markers have been applied to the protein variants; 1. mitotic segregation (Figure 2); 2. lack of meiotic segregation (Figure 1); 3. loss of transmission when converted to the rho-o (petite mutant lacking mtDNA) state (data not shown). As is summarized in Table 1, all three criteria are met by the three variants. For var1, we have studied several hundred diploid segregants and find that each segregant makes one form or the other; we have never observed the absence or presence of both in a progeny cell. These data are sufficient to prove that the variants are due to cytoplasmic genetic factors. It is hoped that the relevant genetic loci are the structural genes for the affected proteins; however, it is possible that the cytoplasmic gene encodes a protease that modifies the molecular weight of a mitochondrial protein product after translation.

COORDINATE TRANSMISSION AND RECOMBINATIONAL MAPPING

We extended these studies by performing a complete progeny analysis of the seven-factor mitochondrial cross outlined in Table 2. The alleles of the four antibiotic resistance/sensitivity markers (chl, ery, oli1 and par) were scored for a large number of diploid progeny while the variants were analyzed on a smaller (random) sample of progeny. Table 2 shows that the variants are transmitted coordinately with the drug markers. In view of the strong evidence that the drug markers are on mtDNA, this result argues that the variants are also located on that cytoplasmic genome. This observation is especially pertinent in view of the isolation by Lancashire and Griffiths of a cytoplasmic mutant (ven-r70) that fails to show coordinate transmission with the mitochondrial markers chl and ery in a homopolar cross; they concluded that ven-r70 resides on a separate cytoplasmic genome[12].

Table 3 summarizes the classes of recombinants obtained in this experiment. Var1 and var2 are very closely-linked and possible allelic but are formally unlinked to the four drug markers and to var3 as well. Var3 is closely linked to oli1; in a repeat of this cross in which 100 progeny were scored for all seven traits, two recombinants between var3 and oli1 were found so it appears that var3 and oli1 are in two adjacent genes or else they are mutations at different sites within a single gene.

DELETION MAPPING OF VARIANTS

It has been shown for drug markers that petite deletion mapping is capable of detecting weak linkage usually missed in recombinational mapping experiments [5,13]. Therefore we undertook such studies to confirm and extend the above data.

Petites of known mitochondrial genotype were isolated (after ethidium bromide induction) from strain 41-6/19 and they were mated to strain

TABLE 2

MARKER TRANSMISSION IN DIPLOIDS FROM THE CROSS

ρ^+ chl-r ery-s oli1-r par-r var1-f var2- var3- x

ρ^+ chl-s ery-r oli1-s par-s var1-s var2+ var3+

	Percent Transmission						
Total Number Scored	var1-s	var2+	var3+	chl-s	ery-r	oli1-s	par-s
99	43.4	43.4	46.4				
626				40.5	40.5	45.3	43.9

TABLE 3

RECOMBINATION ANALYSIS BETWEEN *VAR POLYPEPTIDES AND †ANTIBIOTIC RESISTANCE MARKERS

	Total Recombinants %						
	chl	ery	oli1	par	var1	var2	var3
chl	-						
ery	11.2	-					
oli1	20.9	19.5	-				
par	20.9	21.1	19.0	-			
var1	14.1	15.1	19.1	17.2	-		
var2	14.1	15.1	19.1	17.2	0	-	
var3	19.1	21.2	0	17.2	19.1	19.1	-

* N = 99

† N = 626

55R5-3C/221-2. For each petite studied, respiratory sufficient (rho+) diploid progeny containing the markers contributed by the petite parent (that is, recombinant progeny) were analyzed for the expression of var1-f and var3- (alleles of the parent strain 41-6/19). If either of these variant alleles is found, then this demonstrates that the parent petite retains the relevant structural gene.

The results of this study are presented in Table 4. All petites retaining oli1-r also retain var3- and transmit it to all oli1-r progeny; petites lacking oli1-r failed to transmit that phenotype. This demonstration of coretention of var3- and oli1-r coupled with these additional data showing little or no recombination between var3- and oli1-r confirms the close linkage between them shown in Table 3. Var1-f was retained by some oli1-r petites but not by others. Petites with retained regions flanking oli-r on either side were capable of

TABLE 4

DELETION MAPPING OF VAR1 AND VAR3

Petites were isolated from strain 41-6/19 following mutagenesis by ethidium bromide. Mutagenized clones were subcloned and 100 subclones from each were tested for transmission of chl-r, oli1-r, and par-r by the replica-cross procedure of Coen, et al.[15]; the absence of a marker is denoted by a zero in the table. Drug sensitive alleles were detected by testing for segregation of sensitive cells among the progeny of the cross. Each of the petites used here was subcloned once again and recharacterized before it was used for deletion mapping. Purified petites were then crossed to 55R5-3C/221-2 and wild-type diploids retaining all drug resistance markers from the petite were selected on drug-containing solid media. Two methods were used to measure var transmission: 1) for each of 15 of the petites, an average of 8 recombinant progeny were analyzed individually on gels; 2) for each of the remaining 16 petites, total recombinant progeny were analyzed as a mixture. In all cases clearcut results were obtained on the gels.

Petite Genotype					Number of Isolates Tested	Transmission of Variants	
chl	ery	oli1	oli2	par		var1	var3
0	0	0	0	R	3	-	-
R	S	R	0	0	8	+	+
R	S	R	0	0	1	-	+
0	0	R	0	0	3	-	+
0	0	R	0	0	4	+	+
0	0	R	S	0	2	+	+
0	0	R	?	R	4	+	+
0	0	R	?	R	6	-	+

retaining var1-f. Therefore, var1-f is near oli1-r but it could not be determined from these data whether it is in the ery-oli1 or oli1-oli2 segment of the mitochondrial genome. In another experiment segregants of an oli1-r oli2-s petite of genotype oli1-o oli2-s were found to have lost var1-f; thus the nonidentity of var1 and oli2, implicit from the other data, was shown explicitly.

THE LOCATION OF VAR1

The preceeding sections demonstrate that var1 is far enough from all four drug markers to be formally unlinked; deletion mapping revealed linkage to oli1 but could not distinguish between the ery-oli1 or oli1-oli2 segments of the genome. We have resolved this problem by repeating the cross shown in Table 3 but this time focusing our attention on those progeny that are recombinant. Here the donor of var1-f was a petite derivative of strain 41-6/19; this experiment has also been done using the wild-type parent with similar results. The petite of genotype rho- chl-r ery-s oli1-r par-o var1-f was crossed to strain 55R5-3C/221-2 and wild-type progeny of each of the six genotypes shown in Table

TABLE 5

THE MAP POSITION OF varl: ANALYSIS OF SELECTED GENOTYPES FROM THE CROSS

rho- chl-r ery-s olil-r par-o varl-f x rho+ chl-s ery-r olil-s par-s varl-s

Genotype			Minimum Number of Crossovers Required				Variant Forms Found		
			ery-olil		olil-par				
chl	ery	oli	varl-f	varl-s	varl-f	varl-s	varl-f	varl-s	N
s	r	s	2	0	2	0	-	+	44
r	s	r	2	4	2	2	+	-	8
s	s	r	2	4	2	2	+	-	3
r	r	s	4	2	4	2	-	+	5
r	s	s	2	2	4	2	+	+	13
s	r	r	2	2	2	2	+	+	34

5 were isolated. All progeny of like genotypes were mixed and each mixture was analyzed for the variant alleles.

As is shown in Table 5, the two map orders (ery-varl-olil, olil-varl-oli2) can be distinguished quite readily by comparing the data to the predicted values (based on the number of recombinational exchanges required to yield each recombinant genotype). The number of exchanges noted assumes that the mitochondrial genomes (petite and wild-type) are circular and requires that only rho+ progeny are included in the analysis. The key point is whether or not both varl-f and varl-s are expressed in progeny of genotype srr and rss or rrs and ssr. These data allow the assignment of varl-f to the genome segment between ery and olil. This method can be relatively insensitive if the variant is very close to one of the drug markers; however, the available evidence argues that varl is located roughly in the middle of that region. These results provide strong confirmation of our deletion mapping results; this is important because deletion mapping can yield ambiguous results if double-deleted or rearranged petites were included in that analysis.

THE FUNCTION OF varl and var3 POLYPEPTIDES

While the function of each var polypeptide is not currently known there are several observations that constrain the possiblities. First, neither varl nor var3 is likely to be a component of cytochrome oxidase (cyt ox); neither protein is immunoprecipitated by a cyt ox antiserum[11]. Also three mit loci, each affectint cyt ox, map elsewhere in the genome[3]. As of the moment it is not clear whether there are any mit or other loci mapping between ery and olil. Thus varl has real promise of being an addition to the currently stable list of known mitochondrial translation products.

The situation is somewhat more clearcut for var3. This locus maps near and is probably allelic with oli1. We have previously shown that var3 is an aggregate which then dissociated by NaOH migrates as a low molecular protein[11]. This result is reminiscent of Tzagoloff's studies of a subunit of ATPase[14]. Thus it appears likely that var3 (and probably oli1) is the structural gene for a small subunit of ATPase.

It should be emphasized that these variants were detected in standard laboratory stocks without prior mutagenesis and selection. Even though the proteins are physically altered it is clear that they must either be functional or provide a largely gratuitous function. As reported previously, we have analyzed the mitochondrial translation products of over thirty strains and have detected six different electrophoretic forms of var1[11]. No new variants were observed in that sample. It is curious that this type of mutation should be apparently limited to the var1 locus; however, this dilemma may be resolved once we have established the molecular basis of the mobility alteration.

ACKNOWLEDGMENTS

The expert technical assistance of Mary Townsend and Elaine Kendrick is gratefully acknowledged. This work was supported by Grant GM 19090 from the U.S. Public Health Service and Grant NP-128E from the American Cancer Society to R.A.B. and Grant 19607 from the U.S. Public Health Service to P. S. Perlman. M.G.D.'s present address is Abteilung Biochimie, Biozentrum, CH-4056 Basel, Switzerland.

REFERENCES

1. Birky, C. W. Jr. (1975) in Genetics and Biogenesis of Mitochondria and Chloroplasts (Birky, C. W., Jr., Perlman, P. S., and Byers, T. J., eds.) Ohio State University Press, Columbus, Ohio 182-224.
2. Clavilier, L. (1976) Genetics 83, 227-243.
3. Slonimski, P. P. and Tzagoloff, A. (1976) Europ. J. Biochem. 61, 27-41.
4. Faye, G., Kujawa, C. and Fukuhara, H. (1974) J. Mol. Biol. 88,185-203.
5. Faye, G., Kujawa, C., Dujon, B., Bolotin-Fukuhara, M., Wolf, K., Fukuhara, H. and Slonimski, P. P. (1975) J. Mol. Biol. 99, 203-217.
6. Casey, J. W., Hsu, H. J., Getz, G. S., Rabinowitz, M. and Fukuhara, H. (1974) J. Mol. Biol. 88, 735-747.
7. Gouhier, M. and Mounolou, J. C. (1973) Molec. Gen. Genet. 122, 149-164.
8. Storm, E. M. and Marmur, J. (1975) Biochem. Biophys. Res. Commun. 64, 752-759.
9. Mahler, H. R., Bastos, R. N., Flury, U., Lin, C. C. and Phan, S. H. (1975) in Genetics and Biogenesis of Mitochondria and Chloroplasts (Birky, C. W., Jr., Perlman, P. S. and Byers, T. J., eds.) Ohio State University Press, Columbus, Ohio 66-116.

10. Douglas, M. C. and Butow, R. A. (1976) Proc. Nat. Acad. Sci. (USA) 73, 1083-1086.

11. Douglas, M. C., Kendrick, E., Boulikas, P., Perlman, P. and Butow, R. A. (1976) in The Genetic Function of Mitochondrial DNA (Kroon, A. and Saccone, C., eds.) North Holland Publishing Company, Amsterdam, in press.

12. Lancashire, W. E. and Griffiths, D. E. (1975) Europ. J. Biochem. 51, 403-413.

13. Molloy, P. L., Linnane, A. W. and Lukins, H. B. (1975) J. Bact. 122, 7-18.

14. Tzagoloff, A. and Meagher, P. (1972) J. Biol. Chem. 247, 594-603.

15. Coen, D., Deutsch, J., Netter, P., Petrochilo, E., and Slonimski, P. P. (1970) in Control of Organelle Development, 24th Symp. Soc. Exptl. Biol. Cambridge University Press, Cambridge 449-496.

Genetics and Biogenesis of Chloroplasts and Mitochondria, Th. Bücher et al. eds.

REGULATION OF CYTOCHROME OXIDASE FORMATION BY MUTATIONS IN A MITOCHONDRIAL GENE FOR CYTOCHROME b

P. Pajot, M.L. Wambier-Kluppel, Z. Kotylak and P.P. Slonimski
Centre de Génétique Moléculaire du C.N.R.S.
F-91190 Gif-sur-Yvette, France

INTRODUCTION

We have shown recently (1) that a unique region of yeast mitDNA controls simultaneously the biogenesis and/or assembly of cytochromes b and a. Eight mutations, designated by the term BOX for cytochrome B and OXidase deficiency and belonging to the general class of mit$^-$ mutations (cf 2), were localized in the vicinity of the OLI1 locus and were found to be very closely adjacent to the mitDNA region COB1 which controls cytochrome b alone. By an extensive restoration mapping with the use of rho$^-$ deletions and by a complete matrix of recombination analysis, the BOX region has been subdivided into four loci. Loci BOX1 and 2 were found to be moderately linked by recombination and inseparable by rho$^-$ deletions. Loci BOX3 and 4 were even less linked by recombination and inseparable by rho$^-$ deletions, while the separation between BOX1,2 and BOX3,4 could be achieved more frequently. In order to explain the joint control of cytochromes a and b we proposed a hypothesis according to which one of the structural genes, or one of the polypeptide chains, of cytochrome b would play a regulatory role for cytochrome oxidase biosynthesis and/or assembly (1).

In the present communication we report further studies which strengthen this point of view. We show that mutations at the loci BOX1,2 & 4 increase the sensitivity of the cytochrome-oxidase forming system to catabolite repression caused by the fermentation of glucose.

MATERIAL AND METHODS

The isolation, characterization of mutants and the genetic methods are described in (1). Two series of isonuclear (i.e. of identical nuclear genetic background) and isomitochondrial (i.e. of identical mitochondrial genetic background) strains, carrying the different BOX mutations, were constructed. The first series comprises the eight original haploid mutants, all derived by Mn^{++} mutagenesis (3) from the strain 777-3A which carries the nuclear mutation op1$^-$. In this series (called the op1$^-$ series) all the cells are rho$^+$, for the op1$^-$ mutation makes rho$^-$ cells lethals. The second one (called the OP1$^+$ series) was constructed as follows. From the strain IL126-1C, which carries the wild type, OP1$^+$, nuclear allele, a rho^0 clone has been isolated after a prolongated growth in 40 mg/ml of ethidium bromide. This clone, IL126-1C/52 was crossed to the eight original mutants of the first series and the diploid clones isolated. These diploid clones are isonuclear and $op1^-/OP1^+$, the latter allele being dominant on the former, and have an identical mitochondrial genome to the strains of the first series. Cultures of the $OP1^+$ series of strains contain rho^- cells but in a small proportion (~10%)

Yeast cells were grown aerobically at 28°C on media containing 1% yeast extract (Difco), 1% bactopeptone (Difco) and 20 mg/ml adenine to which,either glucose (2%), or galactose (2%) and glucose(0.1%)were added as carbon and energy sources.Cultures in the liquid media ,200 ml in 1 liter erlenmeyer flask,were shaken on a reciprocal shaker. Plates were prepared by adding 2.3% agar to the liquid medium. For the isolation of mitochondria, cells were broken in a Braun homogenizer and the particles sedimented at 25 000 x g for 15 min. The mitochondrial fraction was suspended at about 20 mg/ml in the isolation medium containing 0.6 M sorbitol, 50 mM Tris-HCl, 0.25 mM EDTA, pH 7.2 and kept frozen at - 70°C.

Absorption spectra were recorded at liquid nitrogen temperature according to Claisse et al. (4) with whole cells and according to Claisse and Pajot (5) with mitochondrial suspensions.

Cytochrome oxidase activity was measured at 28° in following the cyanide-sensitive oxidation of cytochrome c in a 1 cm optical cuvette containing 1 ml of a solution 50 mM phosphate buffer, 2 mM EDTA, 30 μM reduced horse heart cytochrome c and 10 - 20 μg mitochondrial protein. 5 mM KCN was added after 2 minutes.

RESULTS

Catabolite repression brought about by a rapid fermentation of sugars is one of the main regulatory mechanisms in the formation of cytochromes and other mitochondrial enzymes. In certain strains of *Saccharomyces cerevisiae* the threshold for inhibition of biosynthesis of respiratory enzymes can be as low as 6 mM glucose (6). As long as growth occurs in the presence of concentrations above the threshold, the biosynthesis is low and derepression takes place during the deceleration and the stationary phases of growth (7). Sugars which are fermented more slowly, like galactose, are less effective in repression than the rapidly fermented ones like glucose (6, 8).

In order to study the effect of catabolite repression on the synthesis of cytochromes, we have compared the cells grown under rigorously identical conditions except for the nature of the energy source, galactose vs glucose. Cultures were inoculated at 1.5×10^6 cells/ml density using glucose-grown precultures and harvested after 20, 40 and 72 hours of growth. The first sampling corresponds to the end of the exponential growth phase and to the beginning of the stationary phase. It occurs at the density of 2×10^8 cells/ml for the $op1^-$ series of strains grown on galactose. During the next 20 and 50 hours the density increases very slightly, equivalent to 0.2 cell generation only. All the strains of the $op1^-$ series have identical growth rates and growth yields, including the BOX^+ mitochondrial wild type (mean generation time is 2.1 hours on glucose and 2.4 hours on galactose).

This is not surprising because the nuclear mutation $op1^-$, which prevents the effective utilisation of respirable substrates acts in an epistatic manner on the mitochondrial mutations. In the $OP1^+$ series of strains, the mitochondrial wild type BOX^+ has a higher growth yield than the BOX^- mutants.

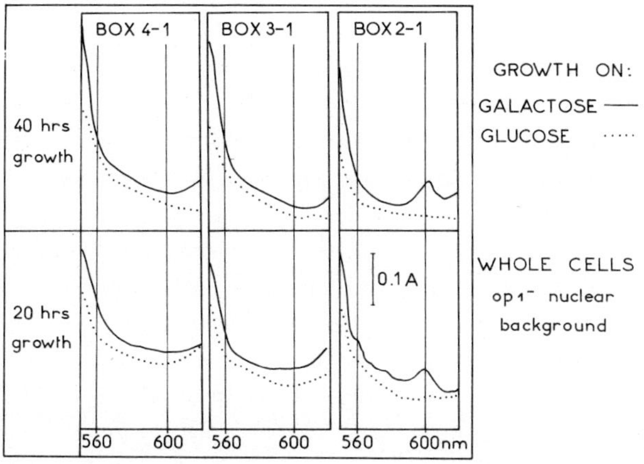

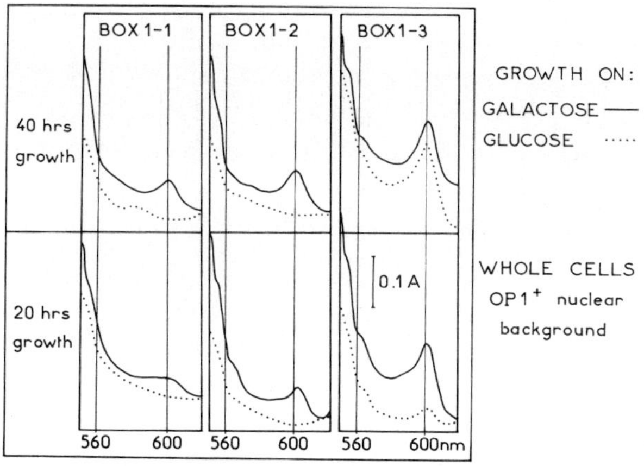

Fig. 1. Absolute absorption spectra of fully dithionite reduced cells, at liquid nitrogen temperature.

Figure 1 shows that derepression of the formation of cytochromes aa_3 occurs in the BOX mutants. Galactose grown cultures contain a higher cytochrome aa_3 peak than the corresponding glucose grown ones. This derepression is specific for the mutated allele. For instance, mutants BOX1-3 and 2-2 are the most easily derepressed:

Mutants BOX 1-1, 1-2 are derepressed but more slowly, while mutants BOX3-1 and 3-2, are not derepressed at all under the same conditions. Furthermore, the rate and extent of derepression depend on the nuclear genetic background. The derepression is greater in the OP1^{+} context, the mitochondrial genome being the same, than in the op1^{-} one. However, the specific dependance on the nature and location of the mutated allele is conserved within each isonuclear series of strains eg. the degree of derepression is always :

BOX 1-3 > BOX 1-2 > BOX 4-1 > BOX 3-1

Several points should be stressed :

1) Cytochrome b is never seen in the BOX mutants whatever are the derepression conditions or changes in the nuclear genome used to bring about the formation of cytochrome oxidase. It is reasonable to conclude therefore, that the BOX mutations affect a structural and not a regulatory mitochondrial gene for the cytochrome b formation. Cytochrome b is absent in the absorption spectra of intact cells (Fig.1), as well as in the spectra of isolated mitochondria (Fig. 2). The absence of the pigment is confirmed by the absence of, antimycin sensitive, NADH-cytochrome c reductase activity, as well as the absence of the succinate cytochrome c reductase activity.

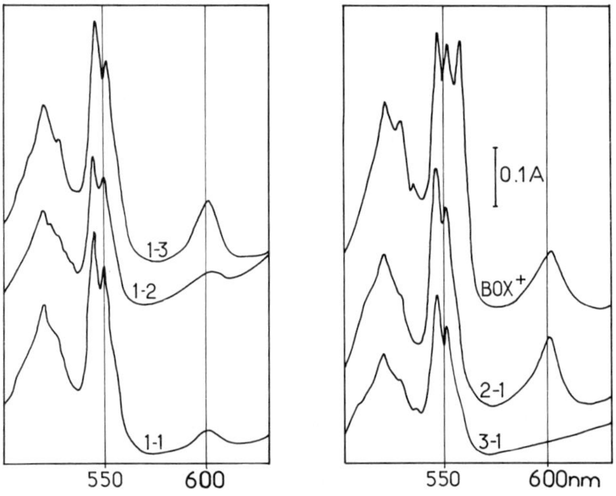

Fig.2. Absolute absorption spectra of fully dithionite reduced mitochondria, isolated from galactose-grown cells harvested in the stationary phase.

2) In all cases, the BOX mutants have a greater sensitivity to catabolite repression of the cytochrome oxidase forming system than the corresponding wild type mitochondrial genome. This is proven by the fact that the BOX^+ strains form already a considerable amount of cytochrome aa_3 under conditions where no or very little cytochrome aa_3 is detectable in the BOX mutants. After 20 hours of growth on glucose (and a fortiori on galactose) the BOX^+ strain has a greater amount of cytochrome aa_3 than the most easily derepressible mutant BOX1-3, which in turn has more cytochrome aa_3 then all the other BOX mutants. This is true for the $op1^-$ as well as for the $OP1^+$ series of strains.

3) Studies on the fine structure of the mitDNA region where the BOX mutations are located have shown that, within the locus BOX1, there exist three distinct, albeit closely linked hereroalleles BOX1-1, BOX1-2 and BOX1-3 (1). Table 1 shows that the heteroallele BOX1-2 is more closely linked to BOX1-3 than to BOX1-1. It is important to underline that this linkage is extremely tight for only 0.03 % wild type recombinants are found. Such a tight linkage must correspond to a physical distance of a few, or at most, a few dozen base pairs only. Nevertheless, the sensitivity to catabolite repression is quite different for the two heteroalleles : BOX1-3 shows a very low and BOX1-2 a medium sensitivity (see Fig. 1 and 2). Furthermore, BOX1-2 and 1-1 heteroalleles have similar sensitivity to catabolite repression,

TABLE 1

FREQUENCY OF WILD TYPE RECOMBINANTS IN CROSSES BETWEEN MUTANTS AT THE LOCUS BOX1

		a strain		
		BOX1-1	BOX1-2	BOX1-3
α strain	BOX1-1	0.00 %	0.52 %	0.33 %
	BOX1-2	0.48 %	0.00 %	0.02 %
	BOX1-3	0.28 %	0.04 %	0.00 %

Three mutants have been crossed in all possible pairwise combinations. In each cross the original α $op1^-$ mit^- strain was crossed to the a $OP1^+$ mit^- strain constructed through meiosis. After a synchronous mating in a full medium diploid cells were selected by prototrophy in a minimal medium, the two parental strains being deficient for complementary auxotrophies.The diploids were allowed to multiply for ca 15 cell generations in a medium containing high (10 %) concentration of glucose and under practically anaerobic conditions. After the completion of growth, the proportion of wild type i.e. ρ^+ mit^+ and mutant i.e. ρ^+ mit^- cells was determined in a sample of the population by the aliquot plating technique on glycerol vs glucose containing media. For further details see (1).

albeit they are more distant. This proves that the regulatory property, expressed in an increased sensitivity to catabolite repression, has a functional fine structure resulting both from the molecular nature and the localisation of the mutational change. It is not a "all or none" property. The fact (shown here for the first time for the mit_i^- x mit_j^- crosses) that the frequencies of wild type recombinants are practically identical in the reciprocal crosses (table 1) brings substance to the notion that we are dealing here with a parameter directly related to the physical distance.

4) Table 2 gives results on the activity of cytochrome oxidase compared to the height of the, fully reduced by dithionite, cytochrome aa_3 band in purified mitochondria. In all the cases but one, there is a parallel derepression of the activity and spectrum, i.e. the ratio of activity to spectrum is similar for the wild type and BOX mutations at the loci 1, 2 and 3.Some fluctuations due to the cumulation of errors, are found but they are smaller than the factor of two. A notable exception is the mutant BOX4-1. This mutant is so sensitive to the catabolite repression that it does not derepress its cytochrome aa_3 spectrum when grown on galactose in the $op1^-$ context (Fig. 1, Table 2). In the $OP1^+$ nuclear background the derepression is spectroscopically very small, although the activity is already considerable (Table 2). We have verified that this activity is KCN sensitive. More work is

TABLE 2

CYTOCHROME aa_3 AND CYTOCHROME C OXIDASE ACTIVITY IN MITOCHONDRIA OF BOX MUTANTS

Nuclear genome	Mitochondrial genome	cyt.c oxidase µmole cyt.c/min.mg	O.D. at 600 nm. 20 mg/ml suspension 2 mm cuvette 70°K
$op1^-$	BOX^+	0.35	0.070
"	BOX 1-1	0.053	0.0175
"	BOX 1-2	0.069	0.018
"	BOX 1-3	0.25	0.067
"	BOX 2-1	0.29	0.079
"	BOX 2-2	0.48	0.075
"	BOX 3-1	0.000	< 0.002
"	BOX 3-2	0.000	< 0.002
"	BOX 4-1	0.051	< 0.002
$OP1^+$	BOX 4-1	0.21	0.0073

Mitochondria purified from galactose grown cells harvested in the stationary phase.

obviously needed, but the mutant BOX4-1 offers an interesting possibility. It is plausible that during its derepression an unbalanced biosynthesis and/or assembly of various structural elements of the cytochrome oxidase complex takes place, in which the catalytically active components are out of pace in comparison to the spectrally detectable component. The comparison of the content of cytochrome a_3 vs cytochrome a will be of interest.

5) The last point concerns the reversibility of changes. We have verified that all cultures of BOX mutants, whether glucose or galactose grown, are devoid of BOX^+ revertants. This has been done by measuring the frequency of glycerol positive haploid cells in the $OP1^+$ series of strains and by measuring the frequency of glycerol positive diploid cells (issued from the cross by an $OP1^+$ rho^0 tester, cf.1) in the $op1^-$ series of strains. The frequency of BOX^+ revertants was found to be smaller than 10^{-4}.

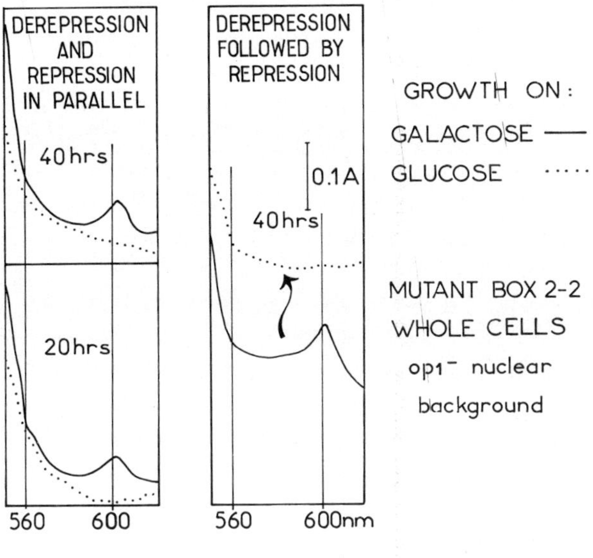

Fig. 3. Reversibility of the repression-derepression. (for details see text).

Fig. 3 shows that the changes in the cytochrome aa_3 content are reversible. A glucose grown preculture of the mutant BOX2-2 was inoculated into two parallel cultures, one on glucose , the other on galactose (left part of the figure).Within some 6 cell generations a cytochrome aa_3 spectrum appears under derepressing conditions. The right part of the figure shows a converse experiment. A derepressed preculture (lower spectrum) is inoculated into a glucose medium. The cytochrome aa_3 spectrum disappears within few generations (upper spectrum). It is important to stress that under identical conditions the mitochondrial wild type strain, BOX^+, develops a considerable amount of the cytochrome aa_3. This proves that the mutant has an increased sensitivity to catabolite repression and that the changes occur in both directions (repression → derepression, derepression → repression) within a small number of cell generations. This rules out hypotheses according to which the changes were due to selection and accumulation of mutants. The changes are due to a regulatory mechanism of the cytochrome aa_3 formation and this mechanism has been modified, in an allele specific way, by presumably structural mutations of a cytochrome b gene(s).

DISCUSSION AND CONCLUSIONS

Previous work (1) has shown that the most probable order of genetic loci on the mitDNA is :

OLI1 - COB1 - (BOX1, BOX2) - (BOX3, BOX4) - OLI2

the loci within brackets could not have been oriented until now. Present work shows that the mutations which occur at the loci BOX1 and 2 produce two simultaneous effects: they cause an absolute deficiency in cytochrome b and they increase the sensitivity to catabolite repression of cytochrome oxidase formation. This increase in sensitivity is allele specific, but of a moderate degree on the whole. The only mutation known presently at the locus BOX4 increases enormously the sensitivity to catabolite repression, while for the mutants occuring at the locus BOX3 the derepression has not been achieved at all until now. A more rigorous derepressing condition , like a sugar limited chemostat growth (cf. 9), should be tried in order to see whether the derepression is possible.

In view of theseresults, it is tempting to correlate the localisation of a mutation with its regulatory properties in respect to the catabolite repression sensitivity. According to this correlation mutants at the COB1 locus would not increase the sensitivity, mutants at the loci BOX1 and 2 would increase it moderately, mutants at the locus BOX4 would increase it to a greater degree and mutants at the locus BOX3 to an even much greater extent than all the previous ones. Obviously, more mutants have to be studied in this mitDNA region to confirm or infirm this idea. It is clear however, that mutations in this segment of mitDNA express themselves in modifications of the sensitivity to catabolite repression. It is the first time, to our knowledge, that a specific segment of mitDNA is identified as a target of the catabolite repression processes. In bacteria it is known (cf. 10) that specific proteins react with certain DNA sequences and regulate the catabolite repression phenomena. It will be of interest to search for such proteins in mitochondria. The mitDNA segment comprising the BOX loci, amplified selectively in a special class of rho^- mutants, may be a useful tool in this respect.

Acknowledgements. M.L. Wambier-Kluppel greatfully acknowledges the receipt of a fellowship from the Federal University of Parana (Brasil) during her leave of absence from that University. Z. Kotylak received a long term fellowship granted by the European Molecular Biology Organization during his leave of absence from the Institute of Microbiology, University of Wroclaw, Poland. This work was supported by DGRST grant N° 75.70748.

REFERENCES

1. Kotylak, Z., and Slonimski, P.P. (1976) International Conference on the Genetic Function on Mitochondrial DNA, Bari, North Holland, Elsevier, 143-154.
2. Slonimski, P.P., and Tzagoloff, A. (1976) Eur. J. Biochem., 61, 27-41.
3. Putrament, A., Baranowska, M. and Prazmo, W. (1973) Mol. Gen. Genetics, 126, 357-366.
4. Claisse, M., Pere-Aubert, G., Clavilier, L. and Slonimski, P.P. (1970) Eur. J. Biochem., 16, 430-438.
5. Claisse, M. and Pajot, P.(1974) Eur. J. Biochem., 49, 49-59.
6. Slonimski, P.P. (1956). Adaptation respiratoire : développement du système hémoprotéique induit par l'oxygène. Proc. of the Third International Congress of Biochemistry, Brussels, 1955. Cl. Liebecq Editor. Academic Press. New-York p. 242-252.
7. Ephrussi, B., Slonimski, P.P., Yotsuyanagi, Y. and Tavlitzki, J. (1956) C.R. Trav. Lab. Carlsberg Ser. Physiol. 26, 87-102.
8. Polakis, E.S., Bartley, W. and Meek, G.A. (1965) Biochem. J. 97, 298-302.
9. Clavilier, L., Pere-Aubert, G., Somlo, M. and Slonimski, P.P. (1976) Biochimie 58, 155-172.
10. Rickenberg, H.V. (1974) Ann. Rev. Microbiol., 28, 353-369.

Genetics and Biogenesis of Chloroplasts and Mitochondria, Th. Bücher et al. eds.

MITOCHONDRIAL GENES DETERMINING CYTOCHROME b (COMPLEX III) AND CYTOCHROME OXIDASE FUNCTION

G.S. Cobon, D.J. Groot Obbink, Ruth M. Hall, R. Maxwell, Mark Murphy,
J. Rytka and Anthony W. Linnane
Department of Biochemistry, Monash University,
Clayton, Victoria, 3168, Australia

INTRODUCTION

It is well established that three mitochondrial inner-membrane enzyme complexes, oligomycin-sensitive ATPase, coenzyme QH_2-cytochrome c reductase (complex III) and cytochrome c oxidase, contain a number of mitochondrially-synthesised protein subunits and considerable evidence is now accumulating which indicates that these proteins are indeed gene-products of mtDNA (for review see 1). It now remains to identify the structural genes for the mitochondrially-synthesised proteins and to determine the arrangement of these genes on the mitochondrial chromosome. This information must necessarily come from the demonstration of a gene-product relationship using mutants with a single mutation which specifically affects the structure of a single protein, and from the mapping of such mutations. We have already used this approach to identify the mitochondrial genes for proteins of the mtATPase complex. Detailed analysis of a number of oligomycin resistant mutants of *S. cerevisiae,* using physical and genetic mapping procedures, revealed three distinct mitochondrial oligomycin resistance loci[2,3] (see Fig. 1). In addition, evidence for a gene-product relationship between two of these loci and two of the membrane sector proteins of the oligomycin-sensitive ATPase, subunits 6 (20 Kdalton) and 9 (9.5 Kdalton), has been obtained[3].

In this paper we present details of a similar analysis of a number of mutants which have a specific defect in either complex III or the cytochrome oxidase complex. We have previously described a number of *mit*⁻ mutants with mitochondrial mutations which affect one or more of complex III, cytochrome c oxidase and mtATPase[3]. However, such mutants are only useful for the identification of gene-product relationships if the defects result from a single small lesion in the mtDNA. We have therefore selected for biochemical analysis *mit*⁻ mutants in which the mutation is located within a single physically defined segment of the mtDNA[4].

Four classes of genetically defined *mit*⁻ strains are described herein. Mutants of one class, *cyb*1, have lesions located within the segment 22-25 mtDNA units and are specifically defective in complex III function. One of these mutants has low NADH cytochrome c reductase activity and retains spectrally detectable cytochrome b, with an absorption maximum which differs from that observed in the parent.

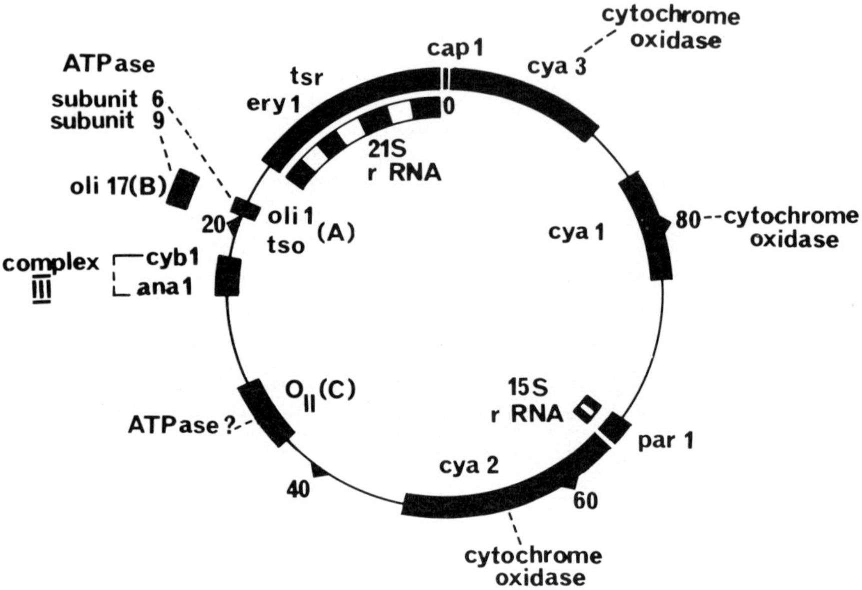

Fig. 1. Map of the mitochondrial chromosome. Bars represent the region within which each mutation has been mapped[4,9,14]. The markers are described in the text and elsewhere[9,14]; *tsr* is a mutation which results in temperature-sensitive mitochondrial protein synthesis.

This indicates that a protein of complex III is altered in this mutant, and thus we conclude that at least one structural gene for a complex III protein is located within the mtDNA segment 22-25 units. The *ana*1 mutation which confers resistance to antimycin A is also located in this 3 unit (2.25 Kbase) segment of the mtDNA[3]. The remaining mutants all have defects in cytochrome oxidase. The mutations are located within three distinct widely separated segments of the mtDNA, 76-84 units (*cya*1), 47-63 units (*cya*2) and 89-100 units (*cya*3). One of the mutants (*cya*1) retains a low level of spectrally detectable cytochrome $\underline{aa}_3$ and of cytochrome oxidase complex. The other two loci probably also represent structural genes for proteins of cytochrome oxidase. Thus the genes for the three proteins of cytochrome oxidase which are mitochondriall-coded[5,6,7] appear to be widely separated on the mtDNA.

CHARACTERISATION OF *MIT*⁻ STRAINS

The strains described in this communication are all *mit*⁻, as they are unable to grow on non-fermentable carbon sources but retain mitochondrial protein synthetic capacity[8]. All strains have mutations which are located within one of four different segments of the mitochondrial chromosome: 22-25 units (*cyb*1), 76-84 units (*cya*1), 47-63 units (*cya*2) and 89-100 units (*cya*3), as shown in Fig. 1. The procedures used for the isolation and mapping of mutants are described elsewhere[4,9].

These genetically defined *mit*$^-$ mutants have been screened for defects in oligomycin-sensitive ATPase, NADH cytochrome c reductase and cytochrome c oxidase activities. In preliminary studies using batch cultures with 2% glucose as carbon source we found that in some *mit*$^-$ strains the levels of cytochromes b and aa_3 differed markedly in cells harvested during log phase and stationary phase indicating that, compared to the parent strain J69, the cytochrome levels in these *mit*$^-$ strains are extremely sensitive to catabolite repression. This effect was further investigated by studying the levels of cytochromes in whole cells during growth in glucose-limited chemostat culture[10], a procedure which is routinely used in this laboratory to obtain derepressed cultures of respiratory deficient cells. Fig. 2 illustrates the effects of catabolite derepression on one *mit*$^-$ mutant 41-2-4. The innoculum was grown in batch culture (2% glucose) and cells were repressed; no cytochrome b or aa_3 was detectable (Fig.2a). A significant level of cytochrome aa_3 was detected in cells taken from the chemostat culture shortly prior to establishment of the steady state (Fig.2b), and in fully derepressed cells from the steady state culture high levels of cytochrome aa_3 were present but no cytochrome b was detectable (Fig.2c). Similar effects have been observed with some other *mit*$^-$ strains. Since catabolite repression can markedly affect the composition of the mitochondria in *mit*$^-$ strains, the interpretation of the results of biochemical analysis using mitochondria isolated from repressed or partially repressed *mit*$^-$ cells is at best highly questionable. We have therefore routinely used chemostat grown cells for biochemical studies. A further advantage of this procedure is that mitochondrial preparations from derepressed cells show more defined secondary structure and are less contaminated with non-mitochondrial membrane material than those from repressed or partially repressed cells.

1. Mutants with lesions at the *cyb*1 locus

Four *mit*$^-$ strains, 41-2-4, 17-62-1, 17-62-11 and 37-16-6, which have mitochondrial lesions at the *cyb*1 locus have been characterised. Low temperature (77°K) spectra of mitochondria isolated from derepressed cells of these strains and the parent strain J69 are shown in Fig.3. The enzyme activities of isolated mitochondria are shown in Table 1. Strains 41-2-4, 17-62-1 and 17-62-11 show levels of cytochromes c, c_1 and aa_3 comparable to those in the parent J69 but completely lack cytochrome b. These three mutants fall into two classes on the basis of their response to catabolite repression. Repressed cells of both 41-2-4 and 17-62-11 apparently lack spectral cytochrome aa_3 (see Fig. 2), while a significant level of cytochrome aa_3 is detectable in repressed cells of 17-62-1. These strains, which lack detectable spectral cytochrome b, show extremely low NADH cytochrome c reductase activities (<10% wild type) which is completely antimycin A insensitive.

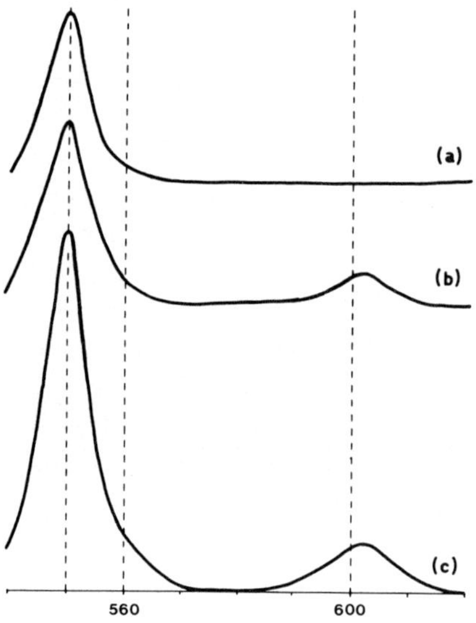

Fig.2. Effect of catabolite repression on spectral cytochromes in strain 41-2-4. Reduced whole cell spectra were determined at room temperature[13]. Stationary phase cells (a) grown in batch culture (2% glucose) were used to innoculate a glucose-limited chemostat culture[10] (dilution rate 0.1 hr^{-1}). Cells were harvested at (b) 24 h [early steady state] and (c) 48 h [steady state].

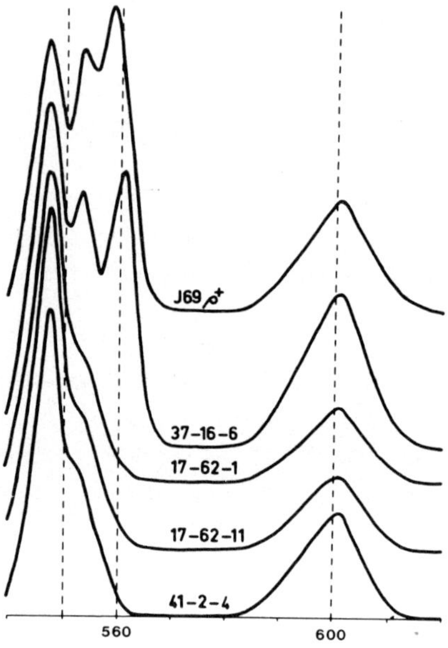

Fig.3. Low temperature (77°K) spectra of mitochondria from derepressed cells of *cyb* mutants. Difference spectra of mitochondria (5 mg protein/ml) were determined at 77°K.

TABLE I

ENZYME ACTIVITIES IN MITOCHONDRIA FROM *CYB* MUTANTS

Strain	ATPase (% oligomycin sensitive)		NADH cytochrome *c* reductase	Cytochrome *c* oxidase
J69 (ρ^+)	0.3 - 0.7	(80)	1 - 3	3 - 6
17-62-1 (*cyb*1)	0.12	(70)	0.08	1.8
17-62-11 (*cyb*1)	0.16	(65)	0.01	1.4
41-2-4 (*cyb*1)	0.17	(60)	0.03	1.8
37-16-6 (*cyb*1)	0.54	(75)	0.18	6.0

Mitochondria were isolated from derepressed cells grown in a glucose-limited chemostat. The enzyme assays used were: ATPase[15]; NADH cytochrome *c* reductase[16] and cytochrome *c* oxidase[17]. Specific activities were expressed as μmoles substrate/min/mg protein.

Strain 37-16-6 contains normal levels of all cytochromes (Fig.3), however, the absorption maximum for cytochrome b at low temperature (77°K) is at a significantly higher wavelength (λ max = 560.5 nm) than that observed in the parent strain J69 (λ max 558 nm). This mutant, 37-16-6, shows a level of NADH cytochrome c reductase activity which is <20% of wild type activity. Moreover, only 20% of the mutant activity is antimycin A sensitive. The defect in this mutant is therefore clearly associated with complex III function. We conclude that the lesion in 37-16-6 affects the mitochondrially coded structural gene for a protein associated with complex III, possibly the haem-associated protein of cytochrome b.

The mutants of 41-2-4, 17-62-11 and 17-62-1 are also clearly defective in complex III function, as they lack both spectral cytochrome b and antimycin A sensitive NADH cytochrome c reductase activity. We conclude that the mitochondrial lesions in each of these mutants also affect a structural protein of complex III. The activities of oligomycin-sensitive ATPase and cytochrome oxidase observed in these mutants are lower than in 37-16-6 which has levels comparable to the parent strain. This suggests that maximal activities of other mitochondrial functions depend on the presence of fully integrated complex III in the mitochondrial membrane.

We have been able to reconstitute antimycin A-sensitive NADH cytochrome c reductase activity and cyanide sensitive NADH oxidase activity in sub-mitochondrial particles isolated from the mutant 17-62-1. A partially purified cytochrome b extract from wild type sub-mitochondrial particles was added to cholate treated sub-mitochondrial particles from the mutant and dialysed overnight. Significant levels of activity for antimycin A sensitive NADH cytochrome c reductase (0.42 μmoles cyt c/min/mg protein) and for cyanide-sensitive NADH cytochrome oxidase (0.71 μmoles NADH/min/mg protein) were detectable in the reconstituted mixture. No activities were observed in either the cytochrome b extract or sub-mitochondrial particles of 17-62-1 before or after overnight dialysis. This result demonstrates that one or more proteins of wild type complex III may be integrated into the mitochondrial membrane from the mutant to form a complete, functional respiratory chain, and confirms the proposal that a structural protein of complex III is either non-functional or absent in this mutant.

Slonimski, Tzagoloff and coworkers have described *mit*$^-$ mutants with defects either in complex III alone (COB) or in both complex III and cytochrome oxidase (BOX)[8,18]. Both classes map within a broad region of the genome which is linked to o_I (a class A oligomycin resistance locus - see Fig. 1). (COB) probably corresponds to the *cyb* locus. Though we have observed an apparent BOX phenotype in mutants 41-2-4 and 17-62-11 using repressed cells, these mutants are clearly defective only in cytochrome b (complex III) when derepressed cells are used for analysis. Whether the mutants described as BOX by Slonimski and Tzagoloff are indeed defective in both cytochrome b and cytochrome oxidase, or whether they are similar to the mutants described here, which are extremely sensitive to catabolite repression, is not clear at this stage.

The exact number of mitochondrially coded and synthesised proteins which are associated with complex III is not known; estimates ranging from 1 to 3 have been reported[11,12]. As the mtDNA segment 22-25 units (2.25 Kbase) is sufficiently large to encode more than one protein, it is uncertain whether the same protein is affected in each of the four *cyb* mutants described here, and in the antimycin A resistant mutant which has a mutation, *ana*1, located also within this segment of mtDNA[3]. This question is further discussed in an accompanying paper[4].

2. Mutants with lesions at *cya* loci

Whole cell and mitochondrial spectra of 46-3-1 (*cya*1), 45-1-2, 45-2-1 and 10-15 (*cya*2) and 45-3-1 (*cya*3) are shown in Fig.4. Enzyme activities of mitochondria isolated from these strains are shown in Table II. In whole cell spectra of 46-3-1 a low level of cytochrome aa_3 is present and an absorption band at 580-590 nm is also observed. Spectra of mitochondria from this strain show a broad absorption band (570-610 nm) which we have interpreted as indicating the presence in the mitochondria of trace amounts of both cytochrome aa_3 and the 580-590 nm absorbing material, which spectrally resembles a type haemochromogen. The majority of 580-590 nm absorbing material is detected in the post-microsomal supernatant. All other mutants clearly lack cytochrome aa_3, though some 580-590 nm absorbing material is observed in spectra of whole cells but not mitochondria

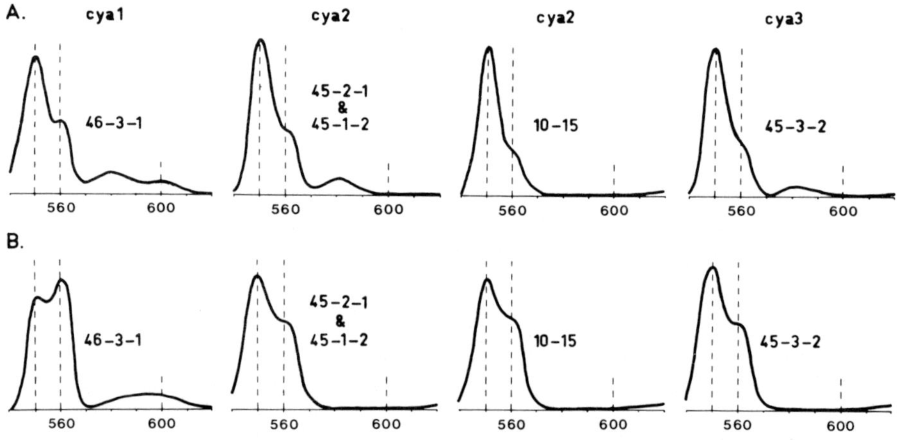

Fig. 4. Spectra of whole cells and isolated mitochondria of *cya* mutants. (A) Reduced cytochrome spectra of whole cells, and (B) Difference spectra of isolated mitochondria (5 mg protein/ml) were measured at room temperature[13].

TABLE II

ENZYME ACTIVITIES IN MITOCHONDRIA FROM *CYA* MUTANTS

Strain		ATPase (% oligomycin sensitive)	NADH cytochrome c reductase	Cytochrome c oxidase
J69	(ρ^+)	0.3 - 0.7 (80)	1 - 3	3 - 6
46-3-1	(*cya*1)	0.2 (65)	3.2	0.30
45-1-2	(*cya*2)	0.5 (60)	0.9	0.03
45-2-1	(*cya*2)	0.5 (65)	1.0	0.02
10-15-1B	(*cya*2)	0.2 (60)	0.8	0.03
45-3-2	(*cya*3)	0.35 (60)	1.2	0.02

Mitochondria were isolated from derepressed cells grown in a glucose-limited chemostat. The enzyme assays used were: ATPase[15]; NADH cytochrome c reductase[16] and cytochrome c oxidase[17]. Specific activities were expressed as μmoles substrate/min/mg protein.

of strains 45-1-2 and 45-3-1. All mutants are clearly defective in cytochrome oxidase activity; the activities of the other enzymes are similar to those normally observed in the parent strain J69. Since strain 46-3-1 contains low levels of cytochrome aa_3 and cytochrome oxidase activity, the mitochondrial lesion in 46-3-1 (*cya*1) clearly affects the structural gene for a protein of the cytochrome oxidase complex. Lesions at other loci (*cya*2 and *cya*3) may also affect structural genes for proteins of the cytochrome oxidase complex.

Tzagoloff and Slonimski[8] have reported a large number of *mit*⁻ mutants which affect cytochrome oxidase. On the basis of recombination analysis the mutations have been classified into three distinct groups designated *OXI*1, *OXI*2, and *OXI*3. Although the precise physical location of these groups on the mitochondrial genome has not been determined, the genetic studies indicate that *OXI*3 is located within the same region of the genome as the *cya*2 locus; *OXI*1 and *OXI*2 are located in a segment which includes the *cya*1 and *cya*3 loci.

It is well known that three proteins of the cytochrome oxidase complex are mitochondrially synthesised[5,6] and coded[7], and the three *cya* loci may well correspond to the structural genes for these three proteins. A more detailed study of the biochemistry of these mutants, aimed at establishing a gene-product relationship, is currently in progress.

REFERENCES

1. Hall, R.M. and Linnane, A.W. (1976) In: Cell Biology. A Comprehensive Treatise (Goldstein,L. and Prescott, D.B.eds.). Academic Press, New York, (in press).

2. Trembath, M.K., Molloy, P.L., Sriprakash, K.S., Cutting, G.J., Linnane, A.W. and Lukins, H.B. (1976) Mol. Gen. Genet. 145, 43-52.

3. Groot Obbink, D.J., Hall, R.M., Linnane, A.W., Lukins, H.B., Monk, B.C., Spithill, T.W. and Trembath, M.K. (1976) In: The Genetic Function of Mitochondrial DNA. (Saccone, C. and Kroon, A.M. eds.). North-Holland, Amsterdam pp 163-173.

4. Rytka, J., English, K.J., Hall, R.M., Linnane, A.W. and Lukins, H.B. (1976) In: Genetics and Biogenesis of Chloroplasts and Mitochondria (Bucher, T., Neupert, W., Sebald, W. and Werner, S. eds.). North-Holland, Amsterdam (this volume).

5. Rubin, M.S. and Tzagoloff, A. (1973) J. Biol. Chem. 248, 4275-4279.

6. Mason, T.M. and Schatz, G. (1973) J. Biol. Chem. 248, 1355-1360.

7. Padmanaban, G., Hendler, F., Patzer, J. and Rabinowitz, M. (1975) Proc. Nat. Acad. Sci. USA, 72, 4293-4297.

8. Slonimski, P.P. and Tzagoloff, A. (1976) Eur. J. Biochem. 61, 27-41.

9. Nagley, P., Sriprakash, K.S., Rytka, J., Choo, K.B., Trembath, M.K., Lukins, H.B. and Linnane, A.W. (1976) In: The Genetic Function of Mitochondrial DNA (Saccone, C. and Kroon, A.M. eds.). North-Holland, Amsterdam pp 231-242.

10. Marzuki, S., Cobon, G.S., Haslam, J.M. and Linnane, A.W. (1975) Arch. Biochem. Biophys. 169, 577-590.

11. Katan, M.B. and Groot, G.S.P. (1975) In: Electron Transfer Chains and Oxidative Phosphorylation (Quagliariello, E. et al., eds.). pp. 127-132. North-Holland, Amsterdam.

12. Marjanen, L.A. and Ryrie, I.J. (1976) Arch. Biochem. Biophys. 172, 679-684.

13. Clark-Walker, G.D. and Linnane, A.W. (1967) J. Cell Biol. 34, 1-4.

14. Sriprakash, K.S., Molloy, P.L., Nagley, P., Lukins, H.B. and Linnane, A.W. (1976) J. Mol. Biol. 104, 485-503.

15. Cobon, G.S. and Haslam, J.M. (1973) Biochem. Biophys. Res. Commun. 52, 320-326.

16. Hatefi, Y. and Rieske, J.S. (1967) Methods in Enzymology X (Eastbrook, R.W. and Pullman, M.E., eds.). pp 225-231, Academic Press, New York.

17. Wharton, D.C. and Tzagoloff, A. (1967) Methods in Enzymology X (Eastbrook, R.W. and Pullman, M.E. eds.). pp 245-250, Academic Press, New York.

18. Slonimski et al. and Tzagoloff et al. (1976) In: The Genetic Function of Mitochondrial DNA (Saccone, C. and Kroon, A.M. eds.). North-Holland, Amsterdam pp 143-154 and pp 155-161.

Genetics and Biogenesis of Chloroplasts and Mitochondria, Th. Bücher et al. eds.

ANTIMYCIN- AND FUNICULOSIN-RESISTANT MUTANTS IN SACCHAROMYCES CEREVISIAE: NEW MARKERS ON THE MITOCHONDRIAL DNA

B. Lang; G. Burger; W. Bandlow; F. Kaudewitz and R.J. Schweyen
Genetisches Institut der Universität München
Maria-Ward-Str. 1 a; D-8000 München 19; Germany

INTRODUCTION

In order to find new markers on the mitochondrial DNA in S.cer. we isolated mutants resistant to antimycin and funiculosin (1). (Both are inhibitors in the bc_1-segment of the respiratory chain.) In Schizosaccharomyces pombe, antimycin-resistant mutants with mitochondrial inheritance of the resistance marker were described (2,3) which are in-vitro cross-resistant to funiculosin. The mutants in S.cer. described in this communication differ from the mutants in S.pombe because they are in-vitro resistant to either funiculosin or antimycin.

MATERIALS AND METHODS

Biochemical tests and recombination analysis are described in reference (4), deletion mapping in (5). Funiculosin was kindly provided by Dr. Bollinger and ubiquinone-3 by Dr. Solms. Abbreviations used: HQNO = 2-heptyl-4-hydroxiquinoline-N-oxide; Q-3 = ubiquinone-3; BSA = bovine serum albumine.

RESULTS AND DISCUSSION

A mutant of S.cer. was selected (ana^r-101) which is resistant to antimycin (4). The mutation conferring antimycin-resistance is located on the mitochondrial genome. In recombination analysis, ana^r-101 is not linked to the loci CAP, ERY, OLI1, OLI2, PAR, COB, and OXI3 (TABLE I).

By rho^- deletion mappings the mutation site could be located between OLI1 and PAR1 and is slightly more distant from OLI1 than from mutation cob^{ts}-1282 (Fig. 1). (Deletion mapping was performed as described in (5) except that strain RM 400-9B tsp 25 C^R_{321} O^P_1 A^R_{101} P^R_{454} was used.)

TABLE 1

MULTIFACTORIAL CROSSES INVOLVING MUTATION ana^r-101

% recombinants							
ERY	OLI1	OLI2	PAR	ANA	COB	OXI3	
11	26	n.d.	20.9	25.5	25	22.2	CAP
	19.3	n.d.	18.4	19.4	n.d.	n.d.	ERY
		n.d.	23.1	23.3	11.6	22.4	OLI1
			18.6	22.6	n.d.	n.d.	OLI2
				22.9	n.d.	n.d.	PAR
					24.2	28.2	ANA
						n.d.	COB

The progeny of zygotes was analyzed by replicaplating colonies of essentially homoplasmic diploid cells on media containing 4 mg chloramphenicol/ml, 2 mg erythromycin/ml, 3 µg oligomycin/ml, 2 mg paromomycin/ml, or 0.01 µg antimycin/ml. Alleles used correspond to the following loci: RIB1 (C^R321), RIB3 (E^R221), OLI1 (O^R1), OLI2 (O^R144), PAR1 (P^R454), COB (cob^{ts}-1282) (6) and OXI3 (oxi-4131) (6).

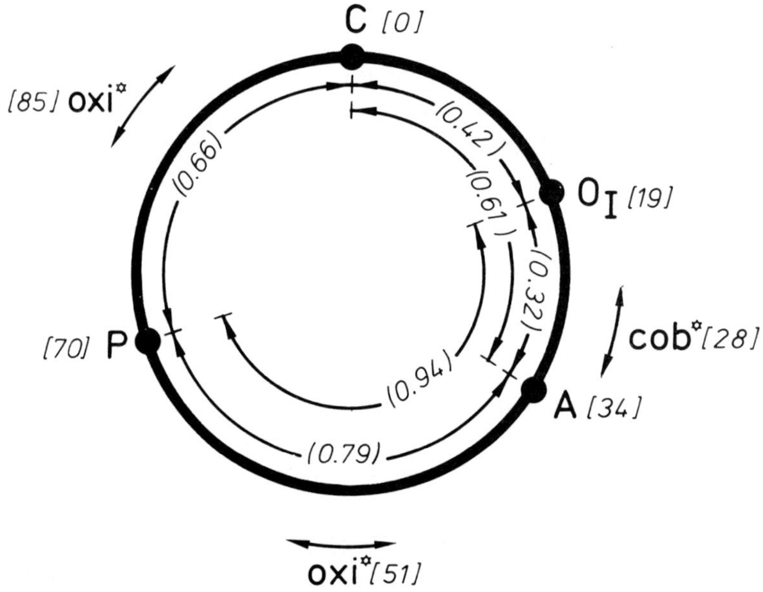

Fig. 1. Deletion mapping of ana^r-101.

Legend to Fig. 1:
The analysis is based on 90 rho⁻ clones
(n) : frequency of separation (e.g. $X^+Y^o + X^oY^+$)
n = 1 maximal frequency (= absence of linkage)
n = o closest linkage
[m] : relative distance of a marker from CAP (maximum = 100)
* Position was determined in a separate experiment (5), which showed identical relative positions for CAP, OLI1 and PAR.

In-vitro resistance of NADH oxidase can only be demonstrated by using competitors of antimycin: Q-3 or BSA. Q-3 increases the antimycin tolerance of the NADH oxidase activity in mutant but not in wild-type mitochondrial particles. Mitochondrial respiration inhibited by antimycin is restored by BSA in ana^r-101 only (Fig. 2, 3). These results indicate that resistance is due to a weaker binding of the drug.

The mutant ana^r-101 is in-vitro cross-resistant to HQNO (Fig. 4) but not to funiculosin (Fig. 5).

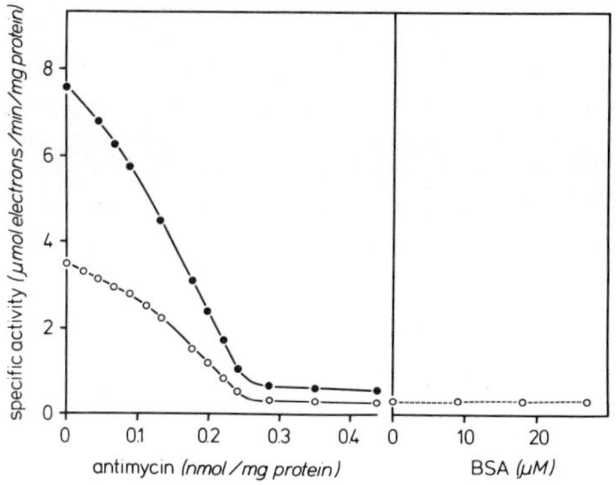

Fig. 2. Inhibition of the NADH oxidase by antimycin in the wild-type (o——o), effect of ubiquinone -3 on the inhibition of the NADH oxidase (●——●), and restoration of the inhibited (90%) activity by BSA (o---o).

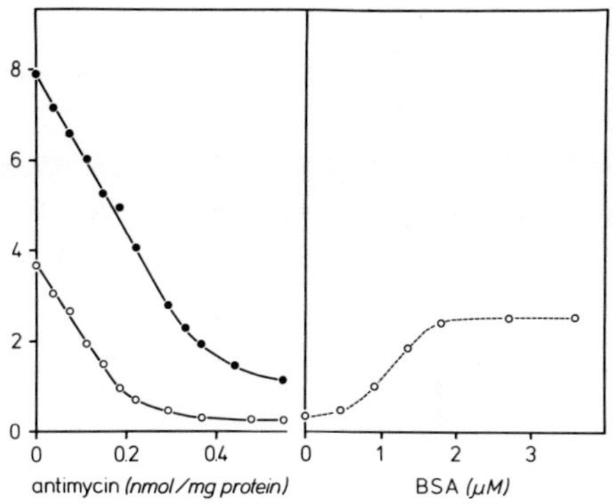

Fig. 3. Inhibition of the NADH oxidase by antimycin in ana^r-101 (o——o), effect of ubiquinone-3 on the inhibition of the NADH oxidase (●——●), and restoration of the inhibited (90%) activity by BSA (o---o).

We have compared ana^r-101 with two other mutants, which are in-vivo resistant to antimycin:

a) fun^r-202 was selected for growth on glycerol in the presence of 2 µg funiculosin per ml (the wild-type is almost completely inhibited by 0.5 µg/ml). The mutation fun^r-202 is mitochondrially inherited. It confers in-vitro resistance to funiculosin (Fig. 5) but not to antimycin. In vivo, it is slightly resistant to antimycin.

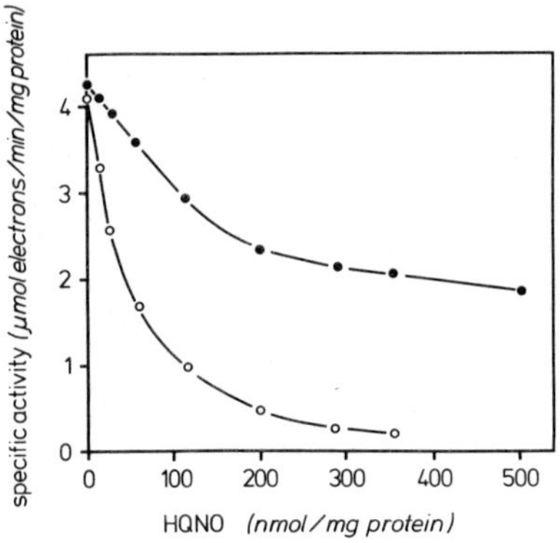

Fig. 4. Inhibition of the NADH oxidase by HQNO in the wild-type (o——o) and in ana^r-101 (●——●).

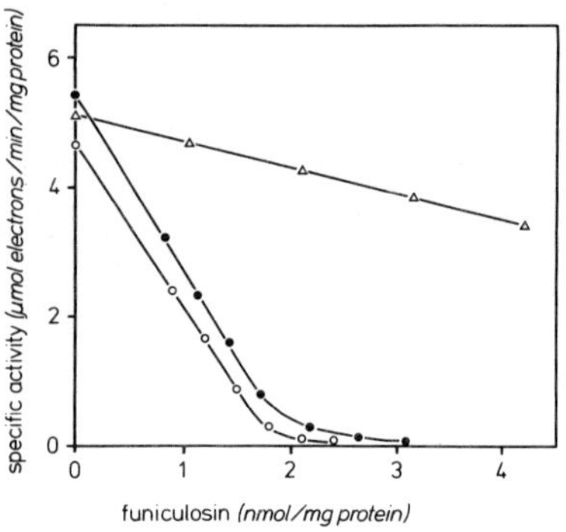

Fig. 5. Inhibition of the NADH oxidase by funiculosin in the wild-type (o——o), in ana^r-101 (●——●) and in fun^r-202 (△——△).

b) mik^r-1 was isolated and described as a mitochondrially inherited mikamycin-resistant mutant. Recently, it was reported to be resistant to antimycin (7).

The results of a comparison between the three mutants ana^r-101, fun^r-202, mik^r-1 and the antimycin-resistant mutant ana^r-8 in S.pombe are summarized in TABLE 2.

TABLE 2

RESISTANCE OF SEVERAL MUTANTS TO INHIBITORS IN THE bc_1-SEGMENT OF THE RESPIRATORY CHAIN COMPARED TO THE WILD-TYPE

		In vivo (growth)		in vitro (NADH oxidase)		
		antimycin	funiculosin	antimycin	funiculosin	HQNO
ana^r-101	(S.c.)	+++	——	+++	——	+++
fun^r-202	(S.c.)	++	+++	——	+++	——
mik^r-1	(S.c.)	++	++	——	+++	——
ana^r-8	(S.p.)	++++	+++	++++	++	——

(+)...(++++): degree of resistance. (——): sensitive

This communication provides evidence that mitochondrial mutations causing *in-vivo* resistance to inhibitors in the bc_1-segment are not necessarily related to *in-vitro* detectable functional alterations of the respiratory chain (e.g. the antimycin resistance of fun^r-202 and mik^r-1 *in-vivo* is not found *in-vitro*).

ACKNOWLEDGEMENTS

The authors are obliged to G. Dörr, E. Habild and B. Heinerich for excellent technical assistance. This work was supported by Deutsche Forschungsgemeinschaft.

REFERENCES

1. Moser, M.K. and Walter, P. (1975) FEBS Letters 50, 279-282.
2. Lang, B. et al. (1975) Molec. gen. Genet. 137, 353-363
3. Wolf, K. et al. (1976) Molec. gen. Genet. 144, 75-81
4. Burger, G. et al. (1976) submitted for publication
5. Schweyen, R.J. et al. (1976) In: The genetic information of mitochondrial DNA (C. Saccone and A.M. Kroon eds.) Elsevier, Amsterdam, p. 251-258
6. Schweyen, R.J., unpublished results
7. Nagley, P. et al. (1976) In: The genetic information of mitochondrial DNA (C. Saccone and A.M. Kroon eds.) Elsevier, Amsterdam, p. 231-242

Genetics and Biogenesis of Chloroplasts and Mitochondria, Th. Bücher et al. eds.

TWO MITOCHONDRIAL ANTIMYCIN A RESISTANCE LOCI IN SACCHAROMYCES CEREVISIAE

Elke Pratje and Georg Michaelis
Physiologisch-Chemisches Institut der Universität, Koellikerstr. 2, D-87 Würzburg, West Germany

Antimycin A is a highly specific inhibitor of the mitochondrial respiratory chain acting on the cytochrome bc_1 complex. In yeast it is known that antimycin A inhibits growth on non-fermentable substrates. We have utilized this property to isolate a large number of antimycin A resistant mutants of Saccharomyces cerevisiae on media containing glycerol or ethanol as sole carbon source. Some of these mutants were shown to result from mutations localized in the mitochondrial DNA (1).

In this communication five mitochondrial antimycin A resistant mutants have been tested for allelism. The mutations derived from independent mutagenic events fall into two loci (A_I and A_{II}) which are linked. Two mitochondrial mutations conferring resistance to funiculosin, another specific inhibitor of the cytochrome bc_1 complex, have been characterized and shown to map at the A_{II} locus.

Allelism studies of mitochondrial antimycin A resistant mutants

Resistance and sensitivity to antimycin A were determined by replicating colonies onto N1 medium (2% ethanol, 1% yeast extract, 1% bactopeptone, 2.3% agar, 0.05M Sörensen buffer pH 6.25) and N1 containing 0.1 µg/ml antimycin A. The mitochondrial antimycin A resistant (A^R) strains used in these studies were : G13, G1a and G25 from D8-22A (α his4 ade2) described by Michaelis (1); and A32 and A84 from D243-4A (a ade1 lys2). The five A^R mutants can grow in the presence of 0.1 µg antimycin A per ml, but growth is completely inhibited in N1 medium containing more than 0.5 µg/ml of the drug.

A standard test for allelism was made by crossing two A^R strains of opposite mating type. The results of the crosses performed are shown in Table 1. It can be seen that there are very few or no sensitive recombinants when mutants A32 and A84 are crossed either to G13 or to G1a. It seems that this four mutants constitute one locus, which we have called locus A_I. However, mutant G25 when crossed to mutants A32 and A84 produces an average of about 7.8% sensitive recombinants. This suggests that mutant G25 defines another locus which we have designated A_{II}.

TABLE 1

ALLELISM TESTS BETWEEN ANTIMYCIN A RESISTANT MUTANTS

Cross A^R	x	A^R	A^R	A^S	Total colonies
G13	x	A32	1534	1	1535
G13	x	A84	1297	0	1297
G1a	x	A32	1836	0	1836
G1a	x	A84	1493	0	1493
G25	x	A32	2226	214	2440
G25	x	A84	2044	147	2191

Matings were performed by prototrophic selection and the resulting diploid progeny analysed for the presence of A^R/A^S cell types by replica plating onto N1 medium without and with 0.1 μg antimycin A per ml.

The linkage or map distance between the two loci A_I and A_{II} can be estimated from the frequency of the two reciprocal recombinants if they are present in equal numbers. We have been unable to detect the expected double-resistant recombinants on N1 plates containing higher concentrations of antimycin A. However, the two reciprocal recombinants are produced in crosses between the antimycin A resistant mutants and mitochondrial funiculosin resistant mutants, as described later in this article. Based on these crosses one has to postulate that the two reciprocal recombinant classes $A_I^S A_{II}^S$ and $A_I^R A_{II}^R$ are of similar sensitive phenotype and occur with equal frequencies. This leads to a recombination frequency of 7.8% (4631 colonies scored) between loci A_I and A_{II}.

Mitochondrial mutations conferring resistance to funiculosin

It has been reported recently that the antibiotic funiculosin is another specific inhibitor of the cytochrome bc_1 complex of the mitochondrial respiratory chain (2). We have found that funiculosin inhibits growth of yeast on non-fermentable substrates. We therefore set out to obtain mutants resistant to the action of this drug. Twenty independent mutants were selected on N1 medium containing

TABLE 2

GENETIC CHARACTERIZATION OF TWO MUTANTS RESISTANT TO FUNICULOSIN

Mitotic segregation of FUN^R/FUN^S among diploids from the cross			
$\rho^+FUN_5^R$ x ρ^+FUN^S	28% FUN^R	72% FUN^S	540 colonies
$\rho^+FUN_5^R$ x ρ^0	0%	100%	413
$\rho^+FUN_9^R$ x ρ^+FUN^S	27%	73%	593
$\rho^+FUN_9^R$ x ρ^0	0%	100%	553
Meiotic segregation in tetrads			
from FUN_5^R diploids	4 : 0	FUN^R/FUN^S	(17 tetrads)
from FUN_9^R diploids	4 : 0	FUN^R/FUN^S	(8 tetrads)
from FUN^S diploids	0 : 4	FUN^R/FUN^S	(9 tetrads)
Elimination of FUN^R by ethidium bromide			
ρ_1^-(from FUN_5^R) x ρ^+FUN^S	0% FUN^R	100% FUN^S	839 colonies
ρ_2^-(from FUN_5^R) x ρ^+FUN^S	0%	100%	780
ρ_3^-(from FUN_9^R) x ρ^+FUN^S	0%	100%	726
ρ_4^-(from FUN_9^R) x ρ^+FUN^S	0%	100%	871

Diploids from the crosses indicated in the table were tested for mitotic segregation by replica plating onto N1 and N1 containing 5 µg funiculosin per ml. Strains: funiculosin resistant mutants Fun5 and Fun9 from D243-4A (a ade1 lys2); D8-22A (α his4 ade2 FUN^S); ρ^0 from D8-22A; the four petites (presumed to be ρ^0) were isolated after ethidium bromide treatment (25 µg ethidium bromide per ml; growth for 24 hrs).

5 µg funiculosin per ml. They were derived from strain D243-4A. The two funiculosin resistant mutants D243-4A/Fun5 and D243-4A/Fun9 were characterized as mitochondrial by the following criteria : existence of mitotic segregation for FUN^R versus FUN^S cell types amongst the diploid progeny of FUN^R x FUN^S crosses, the absence of mitotic segregation in crosses of the type ρ^+FUN^R x ρ^0, lack of meiotic segregation of the purified progeny from ρ^+FUN^R x ρ^+FUN^S crosses, and association of the FUN^R marker with the ρ factor. Extensive ethidium bromide treatment eliminates the FUN^R marker. Table 2 summarizes the results obtained for the two mutants Fun5

and Fun9.

Mitochondrial funiculosin resistance maps at the A_{II} locus

The two mitochondrial FUN^R mutants described express resistance to funiculosin and a low degree of resistance to antimycin A. The same is true for the mitochondrial A^R mutations conferring resistance to antimycin A and a weak cross-resistance to funiculosin. Nevertheless the phenotypes of FUN^R and A^R mutants are different enough to be distinguished on N1 medium containing 0.1 µg antimycin A or 5 µg funiculosin per ml.

In Table 3 the results of crosses involving two FUN^R and two A^R mutations are presented. Sensitive recombinants which can easily be recognized are produced when mutant Fun5 is crossed to mutant G1a

TABLE 3
ALLELISM TESTS BETWEEN ANTIMYCIN A RESISTANT MUTANTS AND FUNICULOSIN RESISTANT MUTANTS

Cross A^R x FUN^R	FUN^RA^S	FUN^SA^R	FUN^SA^S	FUN^RA^R	Total colonies
G1a x Fun5	339	979	65	52	1435
G25 x Fun5	464	1396	1	3	1864
G25 x Fun9	526	1565	0	4	2095

Random diploid progeny were obtained by the prototrophic mating technique, and analysed by replica plating onto N1, N1 with 5 µg funiculosin per ml, and N1 with 0.1 µg antimycin A per ml.

belonging to locus A_I. The reciprocal double-resistant recombinants occur with comparable frequency in this cross. The $FUN^RA_I^R$ recombinants do not only grow on the two antibiotic media indicated, but they represent the only colony type of the cross which tolerates up to 0.5 µg antimycin A per ml. No or very few FUN^SA^S recombinants are recovered when mutants Fun5 and Fun9 are crossed to mutant G25 of locus A_{II}. The few FUN^RA^R colonies observed in these crosses were sectored and did not grow in the presence of 0.5 µg antimycin A per ml. Some of them are probably not true recombinants but rather the result of a late segregation. It seems therefore that the two FUN^R mutations have to be grouped at the A_{II} locus. The recombination frequency between locus A_I and A_{II} is the sum of the two equal frequencies of reciprocal recombinants. This leads to a frequency of 8% between loci A_I and A_{II} as deduced from the FUN^R x A_I^R crosses. A similar value of 7.8% has been observed for the sensitive recombinants in the isonuclear A_I^R x A_{II}^R crosses. This value can best be explained if one assumes that the two reciprocal recombinants $A_I^R A_{II}^R$ and $A_I^S A_{II}^S$ occur with equal frequency and are both of the double-sensitive phenotype.

Concluding remarks

In this study we have demonstrated that mutations conferring resistance to antimycin A can arise at least at two mitochondrial loci which are linked. The first locus (A_I) has been characterized by four A^R mutations, the second locus (A_{II}) by one A^R mutation. Two funiculosin resistant mutants have been isolated and characterized as mitochondrial. The two FUN^R mutations map at the A_{II} locus. The recombination frequency between A_I and A_{II} is 7.9%. There are two possibilities to explain the functions of the two loci, either both code for one very large component or separately for two components of the cytochrome bc_1 complex. In case the two loci specify one large protein two specific regions have to exist in this molecule where alterations can selectively confer antibiotic resistance. On the other hand the recombination value found is high enough to postulate the existence of two distinct mitochondrial genes specifying two components of the cytochrome bc_1 complex.

Acknowledgements

The authors are indebted to Dr. P. Bollinger from Sandoz AG in Basel for his co-operation. Professor Dr. E. Wintersberger provided stimulating discussions. This investigation was supported by the Deutsche Forschungsgemeinschaft.

References

1. Michaelis, G.: Cytoplasmic inheritance of antimycin A resistance in Saccharomyces cerevisiae. Molec. gen. Genet. in press (1976).
2. Moser, U.K. and Walter, P.: Funiculosin, a new specific inhibitor of the respiratory chain in rat liver mitochondria. FEBS Lett. 50, 279-282 (1975).

Genetics and Biogenesis of Chloroplasts and Mitochondria, Th. Bücher et al. eds.

MITOCHONDRIAL INHERITANCE OF MUCIDIN RESISTANCE IN YEAST

Július Šubík
Food Research Institute, 89813 Bratislava
Czechoslovakia

INTRODUCTION

Mucidin is an antifungal antibiotic /1/ which inhibits the respiration as well as the growth of yeast on non-fermentable substrates /2/. This antibiotic like antimycin A having no effect on phosphorylation system in isolated mitochondria was found to inhibit the mitochondrial electron transport at the level of $\underline{bc}_1$ complex /3/ the biogenesis of which is under the control of both mitochondrial and nuclear genes /4/. Since isolation of antimycin A-resistant mutants of *S. cerevisiae* providing a new insight into ubiquinol-cytochrome *c* reductase biogenesis and function have not yet been reported we tried to find mutants which show resistance to mucidin. In order to get mutants useful also for mapping mtDNA we were particularly interested in those mutants which are likely to inherit the resistance extrachromosomally. In this paper the properties of *S. cerevisiae* mutants, resistant to mucidin as a result of mutation in the mitochondrial genome, are described.

RESULTS AND DISCUSSION

Mutants of *S. cerevisiae* resistant to mucidin were isolated on the basis of their ability to grow on non-fermentable media in the presence of mucidin. Mutant designated 3D /α adel lys2 ρ^+ M^R_{3D}/ arose spontaneously from parental strain D 225-5A /5/ whereas mutant 101-4B /α adel lys2 ρ^+ M^R_{101}/ was isolated through meiosis from spontaneously arisen mucidin-resistant mutant strain 101.

Some properties of these strains are summarized in Table 1. *In vivo* both mutants were specifically resistant to mucidin and did not show cross-resistance to other inhibitors of mitochondrial functions, e.g. antimycin A, HQNO /2 heptyl-4 hydroxyquinoline-N-oxide/, chloramphenicol, erythromycin and oligomycin. In both mutants resistance to mucidin was expressed also at the level of isolated mitochondria. According with results of *in vivo* study a significant change in the sensitivity of respiration to other inhibitors of ubiquinol-cytochrome *c* reductase such as antimycin A and HQNO was not observed. The specific resistance to mucidin in yeast appears to be caused by modifications of mucidin binding or

TABLE 1

RESISTANCE OF PARENTAL AND MUTANT STRAINS TO INHIBITORS

Cells were grown and respiratory activity of mitochondria determined as described previously /5/.

Strain	Genotype	Tolerance to mucidin *in vivo* /μg/ml/[+]	Inhibitor I_{50} values[++] /μg/mg protein/ Mucidin	Antimycin A	HQNO
D 225-5A	M^S	0.1	0.015	0.035	0.35
3D	M^R_{3D}	2.0	1.5	0.042	0.28
101-4B	M^R_{101}	1.0	0.45	0.053	0.26

\+ Defined as the maximum concentration allowing normal growth of the strain on YEPGlyc plates.

\++ Defined as the amount of inhibitor required to inhibit the respiratory activity of isolated mitochondria to 50 % of the maximal control activity.

inhibitory sites in the inner mitochondrial membrane. These sites are specific for mucidin and different from those of antimycin A and HQNO.

Genetic analysis by several independent tests /Table 2/ clearly demonstrated that mutations conferring resistance to mucidin in the strains 3D and 101-4B are cytoplasmically inherited:/a/ sensitivity to mucidin of cells growing on non-fermentable substrates was not restored when the mutants had been mated to a ρ^o tester strain which contained a normal complement of nuclear genes for mitochondrial functions, /b/ the genetic determinant of mucidin

TABLE 2

GENETIC CHARACTERISATION OF MUTANTS RESISTANT TO MUCIDIN

Strains were mated and the diploids from the crosses indicated in the table were analysed by replica plating onto glycerol and glycerol plus mucidin /0.5 μg/ml/ containing media. For tetrad analysis diploid cells were sporulated, asci digested with snail gut enzyme and dissected with a micromanipulator.

Strain	Mitotic segregation of M^R vs. M^S among diploids from the cross: ρ^+M^R with ρ^o from M^S	ρ^o from M^R with ρ^+M^S	ρ^+M^R with ρ^+M^S tester	Meiotic segregation in tetrads M^R vs. M^S
D 225-5A	NO /all M^S/	NO /all M^S/	NO /all M^S/	0:4
3D	NO /all M^R/	NO /all M^S/	YES	4:0 /3:0/
101-4B	NO /all M^R/	NO /all M^S/	YES	4:0 /3:0/

resistance was lost after extensive treatment with ethidium bromide to produce ρ^o cells, /c/ mitotic segregants were observed when mutants had been crossed to a ρ^+ strain, /d/ resistance phenotype of mutants segregated in a non-Mendelian fashion during meiosis.

Recombination analysis /Table 3/ have shown that mucidin-resistant mutants 3D and 101-4B are not allelic. A significant level of wild-type /M^S/ and on non-fermentable media not growing cytochrome b deficient /mit$^-$/ recombinants was found among progeny diploids in the crosses between mutants 101-4B and strain 3DC-1B /isolated as the a mating type after meiosis from mutant 3D/.

TABLE 3

ALLELISM TEST BETWEEN MUCIDIN-RESISTANT MUTANTS

Standard cross conditions. Quantitative analysis by replica plating of random diploid cells onto YEPGlyc, mucidin /0.5 μg/ml/ containing YEPGlyc and YEPGlu media.

Cross $\alpha\ M^R_{101} \times$ a M^R_{3D}	Diploids /%/ M^R	M^S	mit$^-$	Total colonies
101-4B x 3DC-1B	84.87	10.17	4.96	423

This indicates that mucidin-resistant mutations can arise at least at two genetic loci on mtDNA designated MUC1 and MUC2 corresponding to mutations M^R_{3D} and M^R_{101} respectively. These loci are apparently not closely linked as the frequency of recombination between both mucidin mutations was significantly higher than that observed between loci RIB1 and RIB3 /6, see also Table 4/. This indicates that at least two gene products of mtDNA are concerned in specification of resistance to mucidin. It seems very probable that these gene products take part in the formation of the mitochondrial ubiquinol-cytochrome c reductase of which deficiency in yeast can result from the mutation as nuclear /7,8/ as mitochondrial genome /9/.

In order to find out the relation of mucidin loci to ribosomal and ATPase loci on mtDNA the mucidin-resistant mutants were crossed with the strains carrying the corresponding mitochondrial mutations C^R, E^R and O^R /Table 4/. In the three factor crosses involving C^R, E^R and M^R mutations no significant polarity of recombination between C^R and E^R markers was observed when the strains 3D and 101-4B were crossed both with standard ω^+ and ω^- strains. This indicates some peculiarity of the ω locus, perhaps its presence as ω^n allele /10/, in the tested mucidin-resistant strains.

TABLE 4

RECOMBINATION BETWEEN RIB1, RIB3, OLI1 and MUC LOCI IN INDICATED CROSSES

Standard cross conditions. Quantitative analysis by replica plating of random diploid cells onto antibiotic containing YEPGlyc media according to the markers present in the cross. The following concentrations of antibiotics in media were used: chloramphenicol /4 mg/ml/, erythromycin /4mg/ml/, oligomycin /10 μg/ml/ and mucidin /0.5 μg/ml/.

Cross $\alpha \times$ a	Transmission /%/ C^{α}	E^{α}	M^{α}	O^{α}	Recombination between pairs of markers /%/ C-E	C-M	E-M	O-M	Total colonies
3D x IL8-8D $M^R_{3D} \times \omega^+ C^R_{321} E^R_{514}$	28.8	24.8	29.8		9.6	20.3	22.9		1139
3D x IL126-1C $M^R_{3D} \times \omega^- C^R_{321} E^R_{221}$	37.1	32.2	30.7		4.5	16.7	18.1		995
3D x 55R5-3C/1 $M^R_{3D} \times \omega^- O^R_1$			19.1	17.5				10.5	2654
101-4B x IL8-8D $M^R_{101} \times \omega^+ C^R_{321} E^R_{514}$	32.3	28.9	32.6		13.1	21.4	25.6		1172
101-4B x IL126-1C $M^R_{101} \times \omega^- C^R_{321} E^R_{221}$	36.7	34.6	35.5		6.6	23.0	24.1		1108
101-4B x 55R5-3C/1 $M^R_{101} \times \omega^- O^R_1$			20.4	19.8				7.8	1628

The high frequencies of recombination between the mucidin mutations and C^R, E^R mutations indicate that MUC1 and MUC2 loci map outside of the closely linked C-E region. On the other hand, the frequencies of recombination between the mucidin mutations and O^R_1 mutation were found to be significantly lower than that between the pairs of markers C-M and E-M indicating that MUC1 and MUC2 loci are linked closer to OLI1 locus. These results as well as the observed covariance of transmission of MUC1 and MUC2 loci with loci RIB1, RIB3 and OLI1 are thus consistent with the idea that mucidin-resistant mutations are located on the same mitochondrial DNA molecule as ribosomal and ATPase loci. Because of mutations leading to cytochrome b deficiency in yeast appear to have similar location as MUC loci, it is conceivable that at least one of the M^R mutations may be situated at the COB region on mtDNA, responsible for cytochrome b deficiency /11/.

In conclusion, mucidin resistance in *Saccharomyces cerevisiae* can be correlated with two loci in the mitochondrial genome, MUC1 and MUC2. The resistance phenomenon can also be correlated with changes in the mucidin binding or inhibitory sites at the level of mitochondria. At least two mitochondrial gene products, responsible for mucidin resistance/sensitivity in yeast, take part in the formation of the cytochrome bc_1 region of the mitochondrial respiratory chain.

ACKNOWLEDGEMENTS

I am indebted to Dr L. Kováč for stimulating discussions and constructive criticism. I also wish to thank Professor P.P. Slonimski for the yeast strains used in recombination studies. The devoted participation in the work reported in this paper of Dr V. Kováčová and Dr G. Takácsová is also gratefully appreciated.

REFERENCES

1. Musílek, V. et al. /1969/ Folia Microbiol. 14, 377-387.
2. Šubík, J. et al. /1974/ Biochim. Biophys. Acta 343, 363-370.
3. Šubík, J. et al. /1974/ Biochem. Biophys. Res. Commun. 57, 17-22.
4. Kroon, A.M. and Saccone, C. /1974/ The Biogenesis of Mitochondria, Academic Press, New York.
5. Šubík, J. /1975/ FEBS Letters 59, 273-276.
6. Netter, P. et al. /1974/ Genetics 78, 1063-1100.
7. Šubík, J. et al. /1972/ Biochim. Biophys. Acta 283, 146-154.
8. Tzagoloff, A. et al. /1975/ J. Biol. Chem. 250, 8228-8235.
9. Tzagoloff, A. et al. /1975/ J. Biol. Chem. 250, 8236-8242.
10. Dujon, B. et al. /1976/ Molec.Gen. Genet. 143, 131-165.
11. Slonimski, P.P. and Tzagoloff, A. /1976/ Eur. J. Biochem. 61, 27-41.

Genetics and Biogenesis of Chloroplasts and Mitochondria, Th. Bücher et al. eds.

BEHAVIOUR OF SACCHAROMYCES CEREVISIAE MUTANT RESISTANT TO JANUS GREEN

A. Kruszewska, B. Szczesniak
Institute of Biochemistry and Biophysics
Polish Academy of Sciences, Warsaw, Poland

Janus Green at 1-10 μg/ml concentration inhibits growth of wild type yeast strains in non-fermentable carbon source. Spontaneous Janus Green-resistant (J^R) mutants occur with a frequency of $\pm 1 \times 10^{-7}$. When cells from newly-isolated J^R colony are plated on N3J medium (complete glycerol medium with Janus Green) only a small fraction of cells can grow into colonies. But after $\geqslant$22 cell generations in N3J the J^R character becomes stable. The J^R mutant is cross-resistant to oligomycin, but not to chloramphenicol or cycloheximide.

J^R diploids from crosses $J^R x J^S$ were sporulated. The segregation of J^R character was $2J^R:2J^S$ in 2 tetrads, $3J^R:1J^S$ in 2 tetrads, and $4J^R:0J^S$ in 3 tetrads.

Transmission of J^R in crosses with J^S rho^+ and J^S rho^o:

Cross:	percent J^R transmission:
J^R rho^+ x J^S rho^+	0.40 ± 0.40
J^R rho^+ x J^S rho^o	0.14 ± 0.03

Transmission of J^R, C^S and E^S:

Cross:	Percent transmission of:		
	J^R	C^S	E^S
4.2 J^R rho^+ $^+$ C^S E^S			
x IL458-1C J^S rho^+ ω^+ C^R E^R	8.7	45.2	44.6
xIL8-8D J^S rho^+ ω^+ C^R E^R	0.2	22.5	24.6
x 23-9D J^S rho^+ ω^+ C^R E^R	0.1	61.1	58.2

Percent of four classes of mitochondrial genotypes in the progenies of cross 4.2 J^R rho^+ $^+$ C^S E^S x IL458-1C J^S rho^+ ω^+ C^R E^R

Sample of diploid cells	Genotypes				Total colonies
	C^R E^S	C^S E^R	C^R E^R	C^S E^S	
Random	5.5	6.0	49.3	39.2	383
J^R	7.6	5.2	45.3	42.5	327

Mode of segregation in tetrads, together with segregation in diploids suggest that the J^R character is extranuclear. The absence of coordinate transmission with known mitochondrial markers together with equal transmission in crosses with rho^+ and rho^O strains suggest that the J^R character is extramitochondrial.

Acknowledgement.
This work was supported by the Polish Academy of Sciences within Project 0.9.3.1., and in part by the PHS Grant No. 05-001-0

Genetics and Biogenesis of Chloroplasts and Mitochondria, Th. Bücher et al. eds.

MITOCHONDRIAL MUTATIONS CONFERRING HEAT OR COLD SENSITIVITY IN SACCHAROMYCES CEREVISIAE

William E. Lancashire
Department of Molecular Sciences
University of Warwick
Coventry CV4 7AL

In recent years the isolation of mitochondrially inherited mutations in Saccharomyces cerevisiae has been of great importance in the understanding of the genetic and biochemical functions associated with mitochondria. The first major type of specific mutations were those conferring resistance to inhibition of mitochondrial processes such as protein synthesis and oxidative phosphorylation. Genetic mapping studies with these markers involved attempts to demonstrate linkage relationships by techniques such as recombination transmission and polarity mapping [1,2,3]. These studies proved suitable for mapping the closely linked loci RIB1, RIB2 and RIB3, but other weakly linked loci such as PAR1 [4], OLI1 and OLI2 [5,6] could not be located on the genome by recombination studies. Attempts to isolate more drug resistant loci to possibly close some of the wide gaps in the genome yielded three further loci, namely OLI3 [6], OLI4 [7] and MIK1 [8]. However, recombination studies including these loci could not relatively locate them either on a linear or circular genome; therefore, four recombinational linkage groups were proposed [6,9]. Group 1 consisted of the ωRIB1, RIB2, RIB3 segment, Group 2 the PAR1 segment, Group 3 the OLI1, OLI3 segment and Group 4 the OLI2, OLI 4 segment. More recent mapping studies, including new loci relating to respiratory deficiencies and utilising petite deletion mapping [8,10], have been able to locate these four linkage groups on a circular genome. The evidence for the circular map has also been substantiated by elegant physical mapping techniques [11,12], which have also located many of the mitochondrial transfer RNA genes.

This study forms part of a programme aimed towards isolating a large series of mutants having deficiencies, both conditional (temperature sensitive) and unconditional, in their oxidative metabolism, in order to define more genes on mitochondrial DNA and correlate them with specific biochemical lesions. A systemmatic biochemical genetic study of temperature sensitive mutations could lead to a revealing of genes in mitochondrial DNA whose products and their functions at present may be unknown and unexpected.

The selected strains are shown to contain mutations of the mitochondrial genome and on the basis of recombination and petite deletion data their mapping relations to some of the known loci are determined.

MATERIALS AND METHODS

Strains

The temperature sensitive (either heat sensitive or cold sensitive) mutants were derived from *Saccharomyces cerevisiae* strains D22 a ad and D6 α arg met. Strain D41 α ur his was also used as a tester strain in some crosses. All three parental strains have the mitochondrial genotype $\rho^+ \omega^+ C^S E^S V^S O^S P^S$, the letters referring to the mitochondrial loci RIB1, RIB3, OLI1, OLI2 and PAR1 which determine sensitivity/ resistance to the inhibitors chloramphenicol, erythromycin, venturicidin, oligomycin and paromomycin, respectively. The venturicidin strains D22/60 a ad V^R_{60} and D22/61 a ad V^R_{61} carrying markers of loci OLI1 and OLI3 respectively were used in recombination mapping studies and have been described previously[6]. Constructed haploid strain DL160-1C a ad $\rho^+ \omega^+ V^R_{60} O^R_{144}$ carrying markers of loci OLI1 and OLI2 was also used. The stable, spontaneous ρ^- petite strains used in the deletion mapping experiments were isolated from haploid strains DL191-13A a ad DL191-22A a ad, DL191-12C a ad, DL191-6C a ad and DL191-18Dαhis, which are meiotic segregants of the constructed diploid DL191 which has the mitochondrial genotype $\rho^+ \omega^+ C^R_{321} E^R_{514} V^R_{60} O^R_{144} P^R_{454}$. The construction of these strains and the isolation of the variously deleted petites will be described in a future publication. The presence or absence of resistance mutations in the remaining DNA fragments of petites was determined using the 'marker rescue' procedure described previously[13], which essentially involves velveteen replica plating the petites onto a lawn of ρ^+ inhibitor sensitive tester strain on minimal medium and after diploid formation analysing for resistance in the diploids by further replica plating onto series of plates containing the drugs.

Media

The standard media used were as follows: N-1% yeast extract, 1% peptone, 0.05M sodium potassium phosphate, pH 6.25; NO- N+ 2% glucose; N1-N + 2% ethanol; N3 -N+ 3% glycerol, WO-0.67% yeast nitrogen base (Difco) 2% glucose. Solid media consisted of the above with the addition of 2.3% agar. Media containing antibiotics and drugs was prepared as described previously[13]. Percentage of petites in a population was estimated by plating on N2 medium, which is N3 plus 0.1% glucose.

Cytochrome Spectra : Mitochondria were prepared as described previously[13]. Sodium dithionite reduced, potassium ferricyanide oxidised difference spectra were recorded at room temperature and at -196° using a Pye Unicam SP 1800 spectrophotometer.

ATPase: Preparation of yeast mitochondria and the assay conditions for oligomycin sensitive ATPase were as previously described[14].

Growth Analysis: Growth on solid media and in liquid media using baffled conical side-arm flasks has already been described [13,14]. The temperatures, however, were strictly controlled at either 17° or 30°.

Genetic Procedures : The mating procedure, determination of proportion of each mitotic segregant in zygotic progeny, and the loss or retention of antibiotic resistance mutations concomitant with the spontaneous induced petite state were essentially as already described[13].

Isolation of Mutants: Parental strains D22 and D6 were mutagenised with N-methyl-N'-nitro-N-nitrosoguanidine or 2-aminopurine to give 5-10% survival and after growing on in glucose medium (NO), temperature sensitive mutants were isolated by two methods:a) plating on N3 medium containing 5 mg/ml erythromycin followed by screening of resistant papillae for temperative sensitivity, b) directly plating onto glucose plates (NO) at 17° and 30° and after colony formation identifying mutants by replicating onto N3 plates and incubating replicas at the two temperatures.

RESULTS

General Characteristics of Mutants: From the two isolation procedures described above many mutants unable to grow at 17° ($COLD^{S}$) were obtained and just one mutant (D22/26L) obtained unable to grow at 30° ($HEAT^{S}$). The general properties of a small selection of mutants is given in Table 1. The strains grow normally on solid glucose medium at both 17° on 30° whereas on ethanol solid medium growth only occurred at the permissive temperature for each strain. Growth in liquid medium follows the same pattern as for solid medium and Figure 1 demonstrates that the aerobic phase of the mutants (in this case D6/5) is conditional upon temperature, but not the fermentative phase. Plating experiments of samples from liquid culture show that the effect is not due to cell death or to an over-production of cytoclasmic petites. The temperature shift experiment shown in Figure 2 demonstrates this reversibility. A complete study of the respiration rates, oligomycin sensitive ATPase levels, cytochrome levels and other mitochondrial enzyme levels in strains grown at 17° and 30° and assayed at both temperatures will be presented in a forthcoming publication. The results are, however, summarised in Table 1. There are no absolute changes in any of the components in any of the strains although levels do vary somewhat and may be secondary effects, possibly of a stoichiometric effect, the cause of which may be revealed in analysis of mitochondrial proteins by polyacrylamide gradient slab gel electrophoresis.

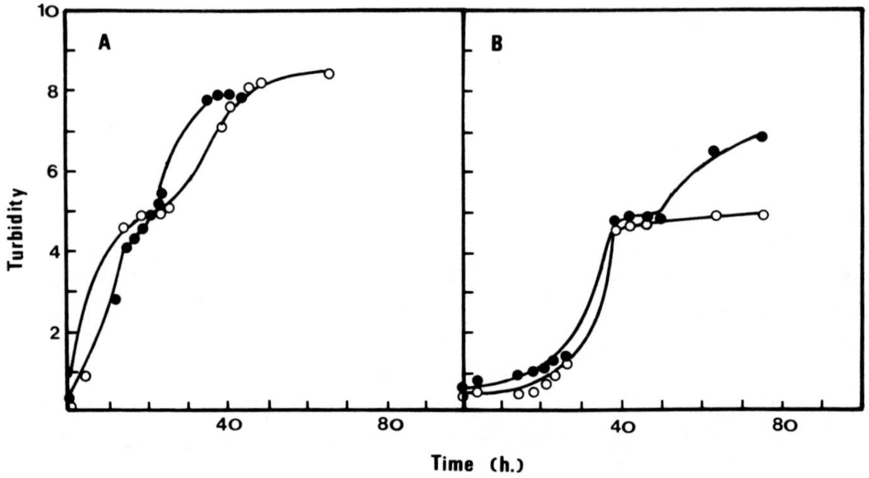

FIG.1 LIQUID GROWTH CHARACTERISTICS:Strain D6(●——●) and mutant strain D6/5 (○——○) were grown in 250 ml baffled conical flasks containing 50 ml of glucose complex medium (NO). Flasks were incubated at 17° (B) or 30° (A) shaking vigorously. Cell density was estimated turbidimetrically.

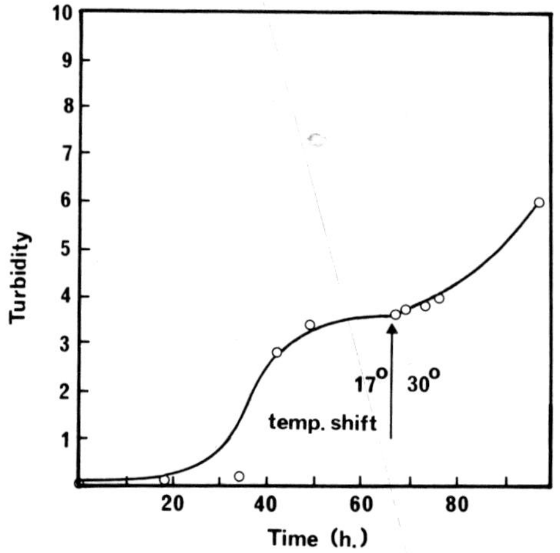

FIG 2. REVERSIBILITY OF PHENOTYPE Strain D6/5 was grown in glucose medium, as for Figure 1, at 17°C. After growth ceased at the end of fermentation, the flask was transferred to 30° and growth followed turbidimetrically.

Genetic Location of Mutations : Mutations in yeast are classified as cytoplasmic on the basis of three criteria as follows: (a) a mixed population of diploids is formed during post zygotic division in a cooss of a mutant and wild type strain, (b) meiotic segregation of 4:0 and 0:4 (mutant:wild type) is obtained in mutant and normal diploids respectively, and (c) petite strains can be produced in which the mutation or its allele is deleted. These tests were carried out on many of the temperature sensitive (TP^{S}) strains obtained, and both nuclear and cytoplasmic inherited mutations were observed. Table 2 summarises these

TABLE 1

GENERAL CHARACTERISTICS OF TEMPERATURE SENSITIVE MUTANTS

Growth on ethanol at 17° and 30° was estimated in both solid and liquid culture on complex medium (N1). Triphenyl tetrazolium chloride (0.1%) in 1.5% agar was overlayed onto plate cultures (NO medium) and the colour reaction recorded after incubation for 3 hours (at the same temperature as growth). Cytochromes were estimated as described in Methods. ATPase was assayed at 30°C. Temperatures in the table refer to growth temperature.

Strain	Growth on Ethanol		Tetrazolum Reaction		Cytochromes		ATPase	
	17°	30°	17°	30°	17°	30°	17°	30°
D22	+	+	+	+	+	+	+	+
D6	+	+	+	+	+	+	+	+
D22/96	-	+	+	+	low aa_3	low aa_3	low	low
D22/107	-	+	-	+	low aa_3	low aa_3	low	low
D6/5	-	+	+	+	low b, aa_3	low b, aa_3	low	low
D6/N32	-	+	+	+	+	+	N.E.	N.E.
D6/26L	+	-	+	+	low b, aa_3	low b, aa_3	N.E.	N.E.

N.E. Not estimated

TABLE 2

GENETIC CHARACTERISTICS OF MUTANTS

See Methods for experimental procedures. Mitotic segregation of temperature sensitive (TP^S) and temperature resistant (TP^R) strains was determined on solid ethanol media (N1) at 30° for heat sensitive mutations and 17° for cold sensitive mutations.

Mutant Strain (Phenotype)	Mitotic Segregation in TP^S x TP^R crosses	Meiotic Segregation TP^S : TP^R	Progency of Crosses $\rho^o TP^S \times \rho^+ TP^R$	Progency of Crosses $\rho^+ TP^S \times \rho^o TP^R$
D22/96 ($COLD^S$)	Yes	4 : 0	All $COLD^R$	All $COLD^S$
D22/107 ($COLD^S$)	Yes	4 : 0	All $COLD^R$	All $COLD^S$
D6/5 ($COLD^S$)	Yes	4 : 0	All $COLD^R$	All $COLD^S$
D6/N302 ($COLD^S$)	Yes	4 : 0	All $COLD^R$	All $COLD^S$
D6/26L ($HEAT^S$)	Yes	4 : 0	All $HEAT^R$	All $HEAT^S$

findings with the five mutants chosen for this presentation. Clear mitotic segregation of temperature sensitive and normal diploid segregants was observed in all strains at the restrictive temperature, i.e. 30° for strain D6/26L and 17° for other strains. The transmission values in several selected crosses of each strain are given in Table 3.

TABLE 3

MITOTIC SEGREGATION IN CROSSES OF TEMPERATURE SENSITIVE (TP^S) AND TEMPERATURE RESISTANT STRAINS

Mating and quantitative random diploid analysis were carried out as described in Methods. The diploid colonies were replicated onto complex ethanol medium (N1) and glucose medium (NO) and a plate of each type incubated at 17° and 30°. Mitotic segregation of temperature sensitive and temperature resistant phenotypes was scored on the ethanol medium (N1) at the restrictive temperature of the mutation in the cross, i.e. 30°C for crosses with D6/26L and 17° for all other crosses.

Cross	Percentage of Total Diploids		Total Colonies
TP^S x TP^R	TP^S	TP^R	
D22/96 x D6	69.2	30.8	390
D22/96 x D41	50.3	49.7	254
D22/96 x DL191-6C	48.5	51.5	276
D22/107 x D6	53.9	47.1	170
D22/107 x D41	46.5	53.5	286
D22/107 x DL191-6C	43.4	56.6	233
D6/5 x D22	43.3	56.7	254
D6/5 x D22	39.0	61.0	740
D6/5 x DL160-1C	57.1	42.9	142
D6/5 x DL191-13A	43.2	56.8	463
D6/N302 x D22	96.0	4.0	240
D6/26L x D22	58.3	41.7	108
D6/26L x DL191-13A	18.5	81.5	135

The behaviour of the mutations in two inverse types of petite cross indicate that the loci of temperature sensitivity are situated in an ethidium bromide sensitive genome, i.e. cytoplasmic.

Mapping of Mutations on the Mitochondrial DNA: To examine the relationships between the TP^S mutations and mitochondrial loci, many homosexual cross (ω^+ x ω^+) were performed[6] between the temperature sensitive mutants and suitable strains containing various combinations of mutations of loci RIB1, RIB3, OLI1, OLI2 and PAR1 (C^R, E^R, V^R, O^R and P^R respectively). The sum of frequencies of reciprocal recombinants (percentages) between the pairs of loci are given in Table 4, together with the total numbers of diploid colonies analysed. The complete data for the crosses will form part of a forthcoming paper. The only clear case of close linkage is with strain D22/96 where only 0.7% recombination occurs with mutations at locus RIB3. What is also evident is that for any particular temperature sensitive (TP^S) mutation no random re-assortment with any of the mitochondrial loci is observed, i.e. recombination values of 40-50% are never observed, in fact 20% appears to be the maximum value observed with any pair of markers. This has been used as a test for new mutations being located on mitochondrial DNA and not some other non-linked cytoplasmic genome[6,13]. If one were to place significance on particular-

TABLE 4

RECOMBINATION FREQUENCIES BETWEEN TEMPERATURE SENSITIVE MUTATIONS AND MITOCHONDRIAL DRUG RESISTANCE LOCI

Pairs of strains carrying temperature sensitive and drug resistance markers were mated and the diploid progeny scored for temperature sensitivity and resistance as described in Methods. In some cases the percentage recombination value (upper figure) is derived from the data of more than one experiment. The lower figure is the total number of diploid colonies analysed.

	RIB1	RIB3	OLI1	OLI2	PAR1
D22/96	N.E.	0.7 301	N.E.	N.E.	N.E.
D22/107	16.3 346	8.3 298	N.E.	N.E.	N.E.
D6/5	17.6 264	18.0 264	16.8 142	20.3 142	14.6 264
D6/N302	20.6 581	21.0 581	14.5 581	13.6 581	19.3 581
D6/26L	17.8 435	18.1 435	16.9 646	17.2 531	11.2 435

N.E. Not estimated.

ly low recombination values it could be said that there is evidence of linkage between the following pairs of loci: D22/107-RIB2, D6/5-PAR1, D6/N302-OLI2 and D6/26L-PAR1.

In an attempt to relatively locate the TP^S mutations and the resistance loci on the same genome use was made of a series of petites having various characterised deletions in mitochondrial DNA. The petites were produced originally as part of a programme for the examination of methods of petite production, amplification of the retained segments and biochemical characterisation of their coding potential and gene products, and full details will appear elsewhere. In brief, however, the petites utilised here were spontaneous isolates from haploid strains constructed to contain the five resistance mutations C^R_{321}, E^R_{514}, V^R_{60}, O^R_{144} and P^R_{454} corresponding to loci RIB1, RIB3, OLIi, OLI2 and PAR1 respectively. Each petite was subcloned and checked for stability of retained genetic fragments as described in Methods. Petites having marked instability s were not chosen for this particular study. Several petites whose remaining genome was characterised with respect to co-retention or co-deletion of markers were chosen and crossed by velveteen replica plating onto lawns of the temperature sensitive mutant strains. The resulting diploids were then replicated onto ethanol media and incubated at restrictive temperatures to see if complimentation had occurred, in which case the petite would have retained the TP^R allele of the mutation under test. The results obtained with a small sample of petites are shown in Table 5. If one assumed that the deletions in the petites

TABLE 5

RETENTION OR LOSS OF TEMPERATURE SENSITIVE GENES
IN DELETED PETITE STRAINS

The first two columns indicate the names of petites and the drug resistance loci retained in their genome fragments. The production of petites from haploid five marker grandes and the procedure for testing them for the presence or absence of temperature sensitive mutant alleles, is described in Methods. The horizontal lines represent a linear presentation of the genome - the dotted areas correspond to deleted fragments.

Petite	Marker Retained
J3	CEP
L25	CEP
J15	CEV
K14	CEVP
M21	CEV
J6	CEVP
M3	VO
J12	VO
K6	VOP
K29	OP
M13	OP
M9	P

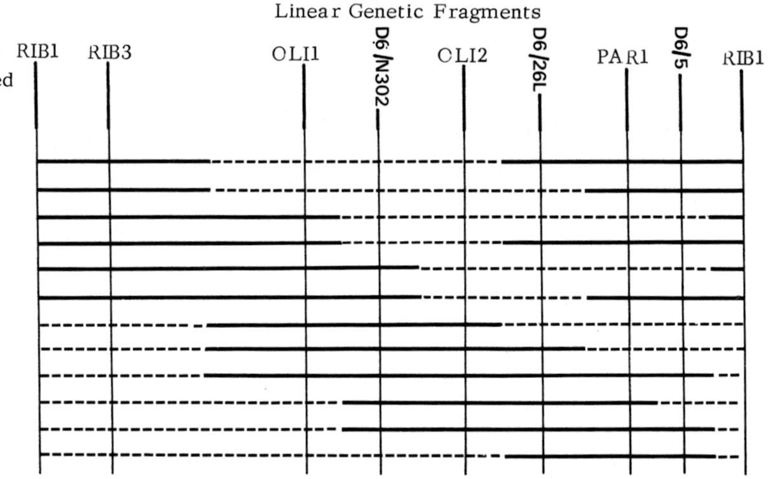

are the results of single events, then the deletions in the petites of the resistance markers are consistent with the order shown in the table from left to right-RIB1-RIB3-OLI1-OLI2-PAR1 and that the genome is circular. The positions of the three of the TP^S mutants are given in the table in positions compatible with the retention or loss of their alleles in the petites and are as follows: ---OLI1, –D6/302–OLI2---, ---OLI2 D6/26L -PAR1 ,and ---PAR1 - D6/5- RIB1--- .

DISCUSSION

A selection of mutants have been described that grow normally at both 17° and 30° , but growth on oxidisable substrates (glycerol and ethanol) completely ceases at a restrictive temperature, either 17° or 30° depending whether the mutation is cold sensitive or heat sensitive respectively. This indicates that the temperature sensitive component in these strains is associated with oxidative (mitochondrial) metabolism. The phenotypes are completely reversible and not due to permanent genetic changes such as conversion to cytoplasmic petite or to loss of respiratory activity as reported in the case of other temperature sensitive yeast mutants[15,16]. It is not improbable that the lesions may be in the ATP synthase complex but biochemical activity studies have not revealed

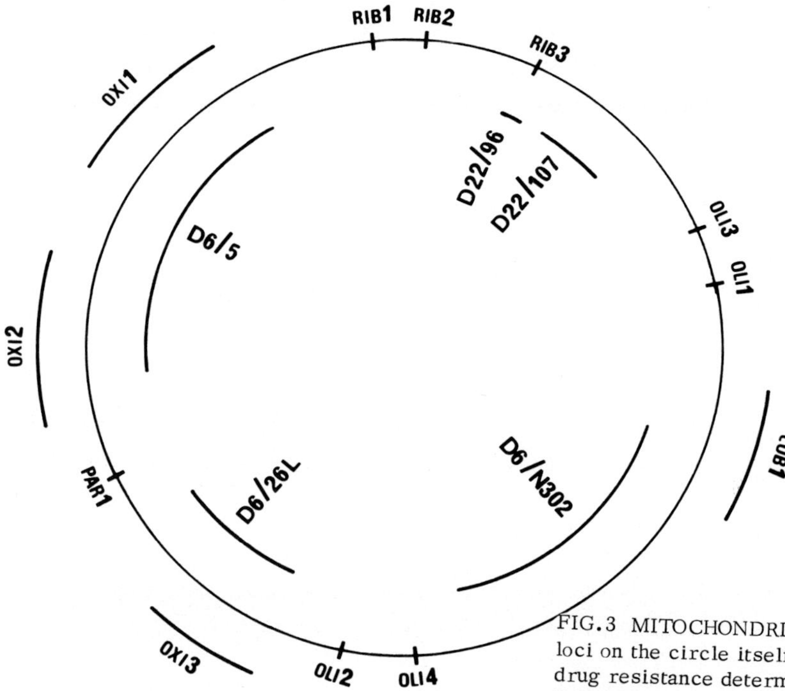

FIG.3 MITOCHONDRIAL GENOME: The loci on the circle itself are the mitochondrial drug resistance determinants [2,4,5,6,7]. Outside the circle are the mit⁻ regions as mapped by Tzagoloff and Slonimski[10]. The relative locations of the temperature sensitive mutation are on the inside of the circle.

any marked lesions, although polyacrylamide gradient studies of mitochondrial proteins in these strains is under study, which may reveal their nature. Although reduced levels of oligomycin sensitive ATPase are found in some strains the effects are not as extreme as those reported by Trembath <u>et al</u> in a cold sensitive mutant reported to be an assembly mutant.

Genetic tests reveal that the five mutants obey the criteria for cytoplasmic inheritance and recombination studies with mitochondrial loci RIB1, RIB3, OLI1, OLI2 and PAR1 indicate that the mutations are mitochondrial rather than episomal, unlike other recently described cytoplasmic mutants [6,13,19]. Studies with specific deleted cytoplasmic petites enabled the mutations to be located on the mitochondrial genome relative to the mitochondrial resistance loci. This is summarised to Figure 3, showing their relationship to markers located by other workers. The resistance loci are positioned on the circle itself, various <u>mit</u>⁻ regions [10] are positioned outside the circle and the temperature sensitive mutants described here are drawn on the inside. Clearly there are overlapping regions common to the OX1 regions and the temperature sensitive loci and detailed manipulations between these strains are needed to define their relative positions. These studies, together with analysis of protein differences in the mutants and of gene products of amplified petites containing these regions will contribute significantly to the understanding of the function of the mitochondrial genome and the incorporation of its gene products into the functional mitochondrial energy conserving membrane.

Acknowledgements

I am grateful to the Science Research Council for the award of a Postdoctoral Research Fellowship. I would also like to thank Dr.D.E.Griffiths for his continued interest in this work.

References

1. Thomas,D.Y. and Wilkie, D.(1968) Biochem.Biophys.Res.Comm.30 ,368-372

2. Howell, N.,Trembath, M.K., Linnane, A.W., and Lukins, H.B.(1973) Molec. Gen.Genet. 122 , 37-51.

3. Netter, P., Petrochilo, E., Slonimski, P.P., Bolotin-Fukuhara, M.,Coen,D. Deutsch, J. and Dujon, B. (1974) Genetics, 78 ,1003-1100.

4. Wolf, K., Dujon, B. and Slonimski,P.P.(1973) Molec.Gen.Genet.125 ,53-90.

5. Avner, P.R., Coen,D., Dujon ,B. and Slonimski, P.P.(1973) Molec.Gen.Genet. 125 , 9 -52.

6. Lancashire, W.E. and Griffiths, D.E.(1975) Eur.J.Biochem. 51 , 403-413.

7. Clavilier, L. unpublished results.

8. Trembath, M.K., Molloy, P.L., Sriprakash, K.S., Cutting, G.J., Linnane,A.W. and Lukins , H.B.(1976) Molec.Gen.Genet. 145 , 43-52.

9. Plischke, M.E., Borstel, R.C. von, Mortimer, R.K., Cohn, W.E. in Handbook of Biochemistry and Molecular Biology, 3rd ed.,(G.D.Fasman, ed.), Cleveland, Ohio, Chemical Rubber Co. press.

10. Slonimski, P.P. and Tzagoloff, A. (1976) Eur.J.Biochem. 61 , 27-41.

11. Sriprokash, K.S., Molloy, P.L., Nagley, P., Lukins, H.B. and Linnane,A.W. (1976) J.Molec.Biol. in press.

12. Fukuhara, H., Bolotin-Fukuhara, M. Hsu, H-J, and Rabinowitz, M. (1976) Molec.Gen.Genet. 145 , 7 - 17.

13. Lancashire, W.E. and Griffiths, D.E.(1975) Eur.J.Biochem. 51 , 377-392.

14. Griffiths, D.E., Houghton, R.L., Lancashire, W.E. and Meadows, P.A.(1975) Eur. J.Biochem. 51 , 393-402.

15. Hanwerker,A., Schweyen, R.J., Wolf, K. and Kaudewitz, F.(1973), J.Bacteriol. 113, 1307 -1310.

16. Butow, R.A., Ferguson, M.J. and Cederbaum, A.(1973) Biochemistry 12 ,158-164.

17. Trembath, M.K., Monk, B.C., Kellerman, G.M. and Linnane, A.W. (1975) Molec.Gen.Genet. 141 , 9-22

18. Griffiths, D.E., Lancashire, W.E. and Zanders, E.D.(1975) FEBS Letters, 53 , 126-130.

Genetics and Biogenesis of Chloroplasts and Mitochondria, Th. Bücher et al. eds.

HIGH SPONTANEOUS PETITE FREQUENCY STRAINS OF <u>SACCHAROMYCES CEREVISIAE</u> GENERATED IN COMPLEMENTATION TESTS.

G.D. Clark-Walker, K.M. Oakley
C.R. Mc Arthur and G.L. Gabor Miklos
Departments of Genetics and Population Biology
Research School of Biological Sciences
Australian National University,
Canberra, Australia.

During our studies on the restoration of respiratory competence in crosses between spontaneous petite mutants (ρ^-) of Saccharomyces cerevisiae (1) it was observed that some of the resulting diploids were highly unstable with regard to their production of petite mutants. Although most of the respiration competent (ρ^+) diploids arising in experiments had spontaneous petite frequencies of 1-2 %, similar to the parental strains, some colonies were observed to be highly sectored on the selective glycerol plates. These colonies can have spontaneous petite frequencies ranging from a few percent up to 80 % per generation. This "instability" of the mitochondrial genome has been shown to be a cytoplasmically inheritable property of some isolates. Subsequently we have performed complementation tests with petites derived form two "unstable" haploids isolated from a single ascus. Such a complementation experiment revealed a pattern of "hits" in the matrix which practically never overlapped. Additionally when petites from these unstable strains are crossed to those derived from a normal strain used previously, a pattern of hits is obtained which we interpret to mean that unstable ρ^+ strains have a mitochondrial genome which contains a redundancy. The possibility is discussed that a redundancy may account for the high frequency of petite formation in such strains.

MATERIALS AND METHODS

Growth conditions and methods used for performing complementation tests have been described previously (1) except for the following alterations. The glucose concentration in the mating medium has been increased to 3 % to promote mating. A replication device, consisting of milled steel rods imbedded in perspex, is issued for the transfer of mated cells to the testing plates. The presence of respiratory competent cells is assayed by replicating on to both glycerol and glucose plates. The glycerol medium allows only the respiratory competent cells to appear as colonies whereas on glucose plates respiratory competent colonies can be distinguished by their buff colour and also by their overgrowth of the paler respiratory deficient ones.

Yeast strains. Two haploid yeast strains RCI α lys 2^- and RC4 a ade 8-18 (Reconstructed from Complementation) are segregants of a single ascus from a respiratory competent diploid arising in a complementation experiment using spontaneous petites from strains T (a ade 8-18, lys 2) and F (α ade 1, arg 4). The spontaneous petite frequencies of RCI and RC4 are 86 and 84 % respectively.

RESULTS

Figure 1 shows the pattern of respiratory competent colonies arising as a result of crossing spontaneous petites from the unstable RCI and RC4 strains. The black diamonds correspond to ρ^+ colonies found on both glycerol and glucose plates whereas the clear diamonds represent growth of ρ^+ colonies on glucose only. When these two media are compared for their ability to reveal ρ^+ cells in the mating mixture, glucose produces the highest number of hits or positive events. The matrix has a number of distinctive features and differs markedly from our previously

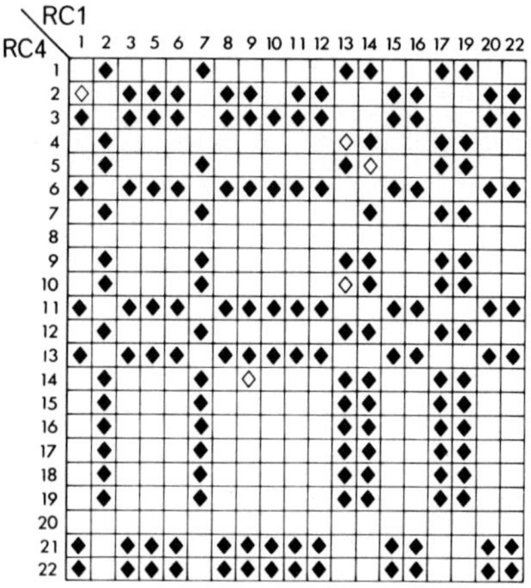

Figure 1

published experiment (1) which utilized petites from normal ρ^+ strains. Firstly, the amount of complementation (38 percent) greatly exceeds the 9 percent found previously. Secondly, the matrix is highly ordered in that there are apparently only two types of petites from both RCI and RC4 which show complementation. A third type of petite, exemplified by RC4/8 and RC4/20, shows no complementation.

In the petites showing complementation, which we dénote by types A and B, type A is illustrated by RCI/1 and RC4/1 and type B by RCI/2 and RC4/2. All other petites in the matrix fall into either category with three exceptions considered from either the RCI or RC4 sides of the matrix. Even though these petites each lack a single "expected" hit, we still regard petite RCI/10 as a type A and RCI/7 and /13 as type B. Furthermore with a single exception, type A petites never have any hits which overlap with those of type B. This pattern differs markedly from the overlapping patterns seen in complementation matrixes from normal strains. The single exception showing an overlap, at co-ordinates RCI/10: RC4/14, was recorded as a result of two ρ^+ colonies on a glucose plate. Finally, the type A petites, 13 for both RCI and RC4, occur in a higher frequency than type B petites, 6 for RCI and 7 for RC4. This situation may not have been expected if reciprocal excision products had an equal chance of being retained.

When an experiment is performed by crossing petites derived from both an unstable strain (RCI) and a normal strain (F) to the other unstable strain (RC4) two different patterns of complementation hits are obtained (Fig. 2).

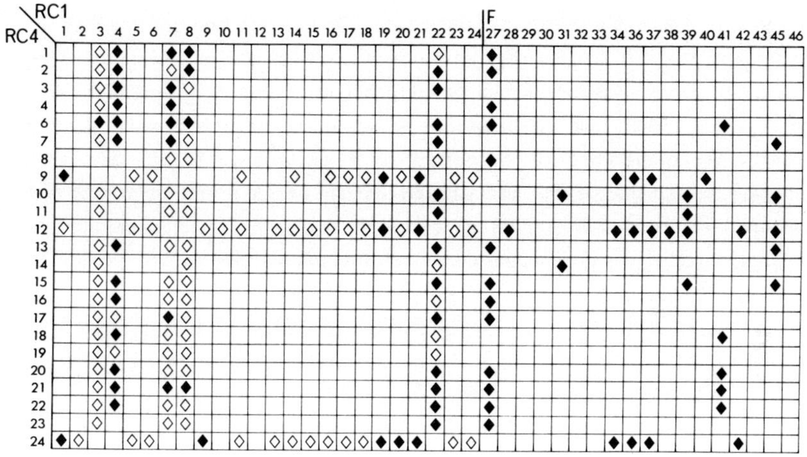

Figure 2

Once again in crosses between RCI and RC4 petites, a pattern of hits is found which does not overlap. However there are more "gaps" in this experiment than in the previous one, gaps being places where it might be expected that a complementation event would be found based on the assumption that there are only two types of petites in RCI and RC4. That this assumption may not be a valid one is discussed below. The non overlapping pattern again allows petites to be classified as type A (RCI/1 and RC4/1) and type B (RCI/3 and RC4/9) and here, with only 3 type B petites of RC4 and 5 of RCI, the asymetry in the occurrence of types A and B is even more marked than in the first experiment.

In the area of the matrix which contains the events obtained by crossing RC4 petites with those from F, a complex pattern is obtained. However petites F/34/36/37 have an identical pattern of hits to type A of RCI whereas F/27 resembles type B but with some gaps. There are also other petites in the sample which contain further gaps but which can still be placed in either category A or B without ambiguity and 6 petites which show no complementation. Nevertheless the most notable feature of the pattern with strain F petites is that illustrated by petites F/39 and 45 which contain hits overlapping between type A and type B petites of RC4. These petites can therefore be classified for the present as type AB although they differ between themselves in the pattern of hits.

DISCUSSION

The pattern of complementation events obtained with petites from unstable yeast strains differs from that previously obtained with petites from normal strains. The non overlapping pattern can be interpreted to indicate that the mitochondrial genome in the unstable strains is fragmenting in a non random manner to yield defective genomes which apparently contain reciprocal deletions. This apparent non random breakdown of mitochondrial DNA, coupled with the instability of the ρ^+ phenotype could be due to at least two possibilities. Firstly the respiratory competence of the parent could occur because of the presence of two separate molecules in the cell which together contain the information for respiratory competence. These molecules may then readily segregate into buds consequently producing a high frequency of petites which contain either one or the other defective genomes.

Secondly the mitochondrial genome in the unstable ρ^+strains may be present on one molecule but this DNA may for some reason contain sequences at which excisions invariably occur at a high rate. This alternative is presently favoured for reasons discussed later.

Regardless of the explanation the present data reveals another important fact concerning petites namely that a bias exists in the type that is present. For example in the above two experiments there is always an excess of one type of petite over the other in both RCI and RC4 strains. This unfortunate selectivity in the retention of defective genomes may equally extend to petites from normal strains such as T and F used previously. This uncertainty introduces an unknown variable into complementation experiments obtained with normal strains and renders impossible the interpretation of the patterns based on simulation tests involving numbers of specific excision sites and so on which in turn rests on the assumption that reciprocally deleted excision products are retained with an equal frequency.

In the compound experiment illustrated in figure 2, type A and type B petites of RC4 are identified in the first half of the matrix and are then used to test the identity of petites from F. The resulting pattern is quite complex in that there are no clearly identifiable type B petites of strain F although 3 type A petites are found. The fact that no clearly defined type B petites are present may be due to gaps in the matrix as a result of failure of the technique to identify all hits. This point is presently under investigation. However if the gaps are real then it indicates that RC4 petites of type A at least are heterogeneous within themselves even though such a heterogeneity is not apparent in the RCI and RC4 side of the matrix. A clue to analysing this problem may lie in considering the reason for the overlapping patterns of hits found with petites F/39 and 45 which allows them to be classified as type AB. If it is assumed that the original mitochondrial genome in strain F is unique then the overlap can only arise because of a redundancy in the mitochondrial genome of the unstable strains. One model which can account for this observation consists of a single circular mitochondrial molecule which contains an inverted region part of which at least is redundant. Excisions would most likely occur within the redundant region and possibly at a high frequency because of base sequence complementarity and, because of the inversion, two defective molecules would be produced. The presence of an inverted redundant region in such a genome may however produce variable excisions within such a region which could explain the apparent heterogeneity of type A petites in RC4. Thus the previously discussed model postulating a single molecule of DNA is favoured over the one involving two complementary but defective genomes co-existing in the one cell.

REFERENCES

1) Clark-Walker G.D. and Miklos G.L.G. (1975) Proc. Nat. Acad. Sci. USA., 72. 372-375.

Genetics and Biogenesis of Chloroplasts and Mitochondria, Th. Bücher et al. eds.
© 1976 Elsevier/North-Holland Biomedical Press, Amsterdam, The Netherlands.

EXTRACHROMOSOMAL INHERITANCE IN A PETITE - NEGATIVE YEAST - SCHIZOSACCHAROMYCES POMBE

K. Wolf, G. Seitz, G. Lückemann, B. Lang, G. Burger, W. Bandlow, and F. Kaudewitz

Genetisches Institut der Universität, Maria Ward - Str. 1 a, 8000 München 19, BRD

SUMMARY

Methods and results of research in extrachromosomal inheritance of drug resistance and respiratory deficiency in the fission yeast Schizosaccharomyces pombe are briefly described. Some prospects are presented how to proceed towards an understanding of the basic problem of petite- negativity in yeasts.

INTRODUCTION

The mitochondrial genetic system of wild- type yeast contributes in a modest, but indispensable way to the biogenesis of mitochondria. It is of considerable interest, why the eukaryotic cell requires the gene- consuming and vulnerable device of this separate genetic system for the construction of functional mitochondria. The vulnerable character of the mitochondrial genetic system is proofed in Saccharomyces cerevisiae by the observation that in many laboratory strains mitochondrial DNA without lethal effects is lost spontaneously at a high frequency and quantitatively after treatment with ethidium bromide. The occurrence of rho^- petites, specific for the so - called petite- positive yeasts, is lacking in the petite- negative group. However, the term petite- negative tolerates that chromosomal petites turn out to be viable in these yeasts. Thus the difference between petite- positive and petite- negative strains very likely resides in the mitochondrial DNA itself. It may differ structurally or by some indispensible functions coded by the mitochondrial DNA of petite- negative yeasts, the deletion of which would be lethal.

Mitochondrial DNA of all petite - negative yeasts so far analyzed is smaller than the 25 μm found in Saccharomyces cerevisiae. O'Connor et al. [1] reported that in Schizosaccharomyces pombe a fraction of large particles contains 6 μm circles and they tentatively concluded that these represent mitochondrial DNA.

The small difference in density of 0.006 g / cm^3 between nuclear and mitochondrial DNA [2] renders a pure preparation of mitochondrial DNA of Schizosaccharomyces pombe rather difficult so that data on mitochondrial DNA of Schizosaccharomyces pombe should be regarded with caution [3]. In Saccharomyces cerevisiae 50 % of the mitochondrial genome are AT - rich spacers, responsible for the high AT - content (83 %) of the whole genome [4]. In contrast, AT represents only about 67 % of the Schizosaccharomyces pombe mitochondrial genome [2]. If we accept 6 μm as size of Schizosaccharomyces pombe mitochondrial DNA, then this mitochondrial DNA would have hardly more information than animal mitochondrial DNA (5 μm), which is obviously the minimal size of all known mitochondrial DNAs. As a consequence mitochondrial DNA of Schizosaccharomyces pombe would be devoid of spacers, which in Saccharomyces cerevisiae are probably responsible for some regulatory functions concerning recombination and rho^- - induction. Therefore Schizosaccharomyces pombe may represent a model mitochondrial DNA for that of higher organisms which is more prone to give information about general features of mitochondrial DNA than the highly exceptional mitochondrial DNA of Saccharomyces cerevisiae.

We study this mitochondrial genome of Schizosaccharomyces pombe genetically and biochemically to learn about its coding capacity and recombination behaviour. Data obtained in this work may provide a better understanding of mechanisms and structures which lead to the difference between petite- negative and petite- positive yeasts.

METHODOLOGY

A prerequisite of mitochondrial genetics are appropriate genetic markers and a methodology for testing mitochondrial inheritance and mapping of genes. Extrachromosomal inheritance can be proved in Schizosaccharomyces pombe by similar methods as used in Saccharomyces cerevisiae. Diploids immediately sporulate in the haplontic Schizosaccharomyces pombe. In order to achieve stability and allow mitotic segregation of diploids, meiotic division therefore has to be blocked by use of an appropriate mutant (mei 1)[5]. Since mei 1 also prevents sporulation, tetrad analysis has to be replaced by chemical haploidization. Lack of segregation of markers during haploidization of Schizosaccharomyces pombe is the equivalent to lack of segregation of markers in ascospores of Saccharomyces cerevisiae. However, sporulation of Schizosaccharomyces pombe can also be prevented by keeping diploids in continuous logarithmic culture, thus allowing mitotic segregation - and tetrad- analysis of homoplasmic diploids. Stabilizing diploids by this method cannot be used as routine - procedure.

RESULTS AND DISCUSSION

Extrachromosomal mutations in Schizosaccharomyces pombe

On to 1974 in Schizosaccharomyces pombe a considerable number of mutants of the wild- type were obtained conferring resistance to chloramphenicol, erythromycin, neomycin, paromomycin, and oligomycin, and respiratory- deficiency. Then in our group we isolated mutants resistant to antimycin which were extrakaryotically inherited. Spontaneously and after mutagenic treatment from these antimycin - resistant strains mutants of all the above mentioned phenotypes were obtained. On to now all of them (more than 300) turned out to be extrakaryotic, as shown by mitotic segregation, mitotic haploidization, and tetrad analysis [6, 7, 8, 9, 10]. In analogy to Saccharomyces cerevisiae, these mutants are very likely mitochondrial. Some of the extrachromosomal respiratory- deficient mutants are candidates for mitochondrial point mutations, since they show mitotic segregation, are able to revert, and posess mitochondrial protein synthesis [11].

Effects of ethidium bromide on mitochondrial DNA of the petite- negative yeasts Kluyveromyces and Schizosaccharomyces pombe

Mahler and Bastos [12] claimed, that the first step in ethidium bromide - induced petite - mutagenesis is the covalent binding of the drug to mitochondrial DNA, a reaction which cannot be accomplished by petite - negative yeasts. In Kluyveromyces mitochondrial DNA after growth in ethidium bromide - containing medium is no longer detectable with analytical methods. It however reappears after removal of the drug [13]. In contrast, Bandlow and Kaudewitz [14] observed no diminution of the bulk of mitochondrial DNA of Schizosaccharomyces pombe during growth in ethidium bromide - containing medium. A common feature of both yeasts, however, is the appearance of a petite - like, reversible phenotype [15]. In the case of Kluyveromyces this can be attributed to the transient disappearance of mitochondrial DNA. In Schizosaccharomyces pombe it may be caused by a block in transcription or translation of mitochondrial DNA.

In contrast to the wild - type of Schizosaccharomyces pombe, the antimycin - resistant mutants produce up to 1 % extrachromosomal respiratory - deficient mutants spontaneously, and up to 20 - 30 % during growth in ethidium bromide - supplemented medium [8]. Strains carrying other resistance mutations (chloramphenicol - or erythromycin - resistance) do not throw off petites either spontaneously or ethidium bromide - induced [9]. At present it is unknown why mitochondrial DNA of the antimycin - resistant mutants is able to harbour stable mutational alterations. One might, however, speculate that as one possible mechanism a structural change in the inner mitochondrial membrane, caused by the mutation to antimycin - resistance, is responsible. It would lead to a distortion of the DNA - replication complex : membrane - polymerase - DNA and as mutator effect to errors in replication.

Segregation, transmission, and recombination of mitochondrial markers

A heteroplasmic zygote of *Saccharomyces cerevisiae* soon produces homoplasmic buds by a mechanism which is not yet understood. A sorting out of one or the other genome in a zygote or first daughter cells also operates in *Schizosaccharomyces pombe*, since in spite of regular fission, a considerable percentage of zygotes produces phenotypically homoplasmic daughter cells.

Transmission of a particular mitochondrial marker in *Schizosaccharomyces pombe* as in *Saccharomyces cerevisiae* can be influenced by ethidium bromide - treatment of haploids before mating or by glucose - repression [7].

The recombination rate of mitochondrial DNA molecules is dependent on the rate of mixing in the zygote, the efficiency of pairing, and the number of mating-rounds. In *Schizosaccharomyces pombe* stable mitotic and meiotic recombinants could be obtained for all markers tested, thus providing a basis for mapping [9]. It is, however, premature to start mapping until not all the parameters stated above are under experimental control.

Using the meiotic system, which has been applied for mapping chloroplast genes in *Chlamydomonas*, a number of two - factor - crosses in trans - configuration was performed with markers conferring resistance to antimycin, chloramphenicol, and erythromycin. As a common feature of all these crosses, double resistant recombinants were strongly biassed. However, reproducible and distinct values for double sensitive recombinants were found with the markers mentioned above. When experimental conditions for a recombination analysis will be worked out, mapping of the genome can be started using both the meiotic (as used in *Chlamydomonas*) and the mitotic system (as used in *Saccharomyces cerevisiae*).

ACKNOWLEDGEMENTS

We acknowledge the skilled technical assistence of Miss G. Dörr, Miss M. Schropp, and Miss R. Stets and appreciate valuable discussions with Dr. A. - M. Colson and the colleagues in the Genetische Institut. The work was supported by the Deutsche Forschungsgemeinschaft.

REFERENCES

1) O'Connor, R. M., Mc Arthur, C. R., Clark - Walker, G. D. (1975) Europ. J. Biochem. 53, 137

2) Bostock, C. J. (1969) Biochim. biophys. Acta 195, 579

3) Borst, P., Sanders, J. P. M., Heyting, C. (1976) in: Genetics, Biogenesis, and Bioenergetics of Mitochondria, Walter de Gruyter & Co., Berlin and New York, in press

4) Prunell, A., Bernardi, G. (1974) J. Mol. Biol. 86, 825

5) Egel, R. (1973) Molec. gen. Genet. 122, 339

6) Lang, B., Burger, G., Wolf, K., Bandlow, W., Kaudewitz, F. (1975) Molec. gen. Genet. 137, 353

7) Wolf, K., Burger, G., Lang, B., Kaudewitz, F. (1976) Molec. gen. Genet. 144, 67

8) Wolf, K., Lang, B., Burger, G., Kaudewitz, F. (1976) Molec. gen. Genet. 144, 75

9) Wolf, K., Seitz, G., Lang, B., Burger, G., Kaudewitz, F. (1976) in: Genetics, Biogenesis, and Bioenergetics of Mitochondria, Walter de Gruyter & Co., Berlin and New York, in press

10) Lang, B., Burger, G., Wolf, K., Bandlow, W., Kaudewitz, F. (1976) in: Genetics, Biogenesis, and Bioenergetics of Mitochondria, Walter de Gruyter & Co., in press

11) Bandlow, W. unpublished observations

12) Mahler, H. R., Bastos, R. N. (1974) Proc. Nat. Acad. Sci. USA 71, 2241

13) Luha, A. A., Sarcoe, L. E., Whittaker, P. A. (1971) Biochem. biophys. Res. Commun. 44, 396

14) Bandlow, W., Kaudewitz, F. (1974) Molec. gen. Genet. 131, 333

15) Schwab, R., Sebald, M., Kaudewitz, F. (1971) Molec. gen. Genet. 110, 361

Genetics and Biogenesis of Chloroplasts and Mitochondria, Th. Bücher et al. eds.

THE MITOCHONDRIAL GENOME OF YEAST : ORGANIZATION AND RECOMBINATION

Giorgio BERNARDI

Laboratoire de Génétique Moléculaire,
Institut de Biologie Moléculaire,
Paris, France.

Introduction. Previous investigations (1-6) led us to the recognition of two sequence elements in the mitochondrial genome of wild-type *Saccharomyces cerevisiae* cells : 1) AT-rich stretches (GC < 5 %), essentially formed by short dAdT:dAdT and dA:dT sequences ; these stretches are characterized by a remarkable compositional homogeneity, and are very largely responsible for the "anomalous" properties of yeast mitochondrial DNA ; 2) GC-rich stretches (average GC = 32 %), characterized by an extreme compositional heterogeneity ; the GC level in these stretches ranges from 22 % in the ribosomal cistrons (7) to 65 % in double-stranded fragments which can be prepared by micrococcal nuclease digestion and which represent 10 % of mitochondrial DNA (6).

These two sequence elements each represent 50 % of the mitochondrial genome of yeast, are interspersed with each other, and their average size is larger than $1.6.10^5$. We called the AT-rich stretches "spacers", and proposed that the GC-rich stretches correspond to genes with their regulatory elements (6). Both sequence elements were first postulated on the basis of the physical and chemical properties of yeast mitochondrial DNA (1,2), and then characterized by studying the mitochondrial DNA hydrolysates obtained by degradation with spleen acid DNase (3,4) and micrococcal nuclease (6) or by depurination (5).

An analysis of mitochondrial DNA hydrolysates obtained with two restriction enzymes, Hae III and Hpa II (referred to hencefrom as Hae and Hpa), led us (8,9) to the identification of two new sequence elements, which we call restriction site clusters and GC-rich clusters. Both sequence elements are present in the micrococcal nuclease fragments richest in GC, which belong to the GC-rich stretches. These new sequence elements appear to play a regulatory role.

Restriction site clusters. A clustering of Hae and Hpa sites among themselves and with each other can be demonstrated with two simple experiments : 1) Chromatography of either Hae or Hpa digests on hydroxyapatite revealed that close to 10^6 of DNA per genome unit of 50.10^6 were present in the digests as single-stranded oligonucleotides, which could only arise from the melting of DNA segments where Hae or Hpa sites were clustered. Double hydrolysates released

$1.3\text{-}1.5.10^6$ of DNA per genome unit ; the GC content of this material was 45-62 %. 2) The fragment pattern of Hae + Hpa double hydrolysates (8, 9) showed that almost all Hae sites are clustered with Hpa sites ; the number of such (Hae,Hpa) site clusters is therefore slightly lower than the number of Hae fragments, 71 in the case of S.carlsbergensis DNA. Since $1.3\text{-}1.5.10^6$ of DNA is present in 60-70 site clusters, each site cluster is formed, on the average, by 30 base pairs. It is unlikely that these melt unless they become shorter than about 10 base pairs; this implies that each site cluster contains, on the average, no less than 3-4 sites, namely 12-16 base pairs. Since Hae sites and Hpa sites (GGCC and CCGG, respectively) have a GC level of 100 %, and (Hae, Hpa) clusters a GC level of 45-62 %, then most base pairs alternating with the restriction sites in the clusters must be A:T base pairs.

GC-rich clusters. Degradation of the micrococcal nuclease fragments richest in GC with Hpa showed that, though these restriction sites were abundant in them, they were mostly formed by sequences resistant to Hpa (as well as to Hae and HhaI, an enzyme splitting the sequence GCGC) and having a GC content of 60 %. These GC-rich clusters have two important points in common with the (Hae, Hpa) clusters : 1) they are highly non-statistical in sequence since they do not contain any of the quadruplets GGCC, CCGG and GCGC, though their GC content is 60 % (close to that of the restriction site clusters) ; 2) they are not localized within the spacers (6), nor within the genes, which would otherwise contain such peculiar sequences ; the only possible localization remains that at the border of subsequent gene-spacer units, where GC-rich clusters and (Hae,Hpa) restriction site clusters may be contiguous to each other. The amount of DNA in (Hae,Hpa) clusters is about 3 %, that in GC-rich clusters about 5 %.

Significance of (Hae,Hpa) site clusters and GC-rich clusters. The restriction site clusters are characterized by 1) a high concentration in symmetrical nucleotide sequences, the Hae and Hpa restriction sites ; and 2) the alternation of short A:T and G:C sequences. The first feature suggests by itself a binding role for proteins having a dyad axis of symmetry ; such a symmetry relationship (10,11) has been found in all specific protein-DNA interactions (12). Both features found in the (Hae, Hpa) site clusters are also present in all prokaryotic promotor and operator sites sequenced so far (13). This suggests that the (Hae, Hpa) clusterscorrespond to regulatory sequences. Two additional points in favor of this view are 1) the localization and 2) the number of such sequences. In fact,this number is about twice as large as the number of yeast mitochondrial genes known so far (24 t-RNA genes (7,14), 2 r-RNA genes (15), and, very probably, 7 genes coding for polypeptide chains (16)). These genes, however, only account for half of the potentially coding DNA ; furthermore, 60-70 mitochondrial genes would mean an average gene size of about 350,000, which is a reasonable figure for a genome in which the small t-RNA genes represent over 30 % of the total.

Another possible role of the restriction site clusters is that they contain sites for the initiation of mitochondrial DNA replication. It is well known that the defective genomes of cytoplasmic "petite" mutants can arise by a deletion mechanism from any region of the wild-type mitochondrial genome, suggesting that multiple initiation sites for replication exist on the latter. Such sites are not likely to be located inside genes, nor inside spacers, and could well be in the clusters.

It is very probable that the GC-rich clusters also have the same biological role of the (Hae, Hpa) restriction site clusters; this is strongly suggested by both the sequence properties of the latter and by the contiguity of both types of clusters.

The genetic units of the mitochondrial genome. Several lines of evidence (9) suggest that genes and spacers are stoichiometric with each other and so are restriction site clusters and GC-rich clusters. If the stoichiometry and the topology of these four sequence elements are those just discussed, then the mitochondrial genome of yeast is organized in a number of genetic units, each one of which contains the four sequences (Figure). The relative order of (Hae,Hpa) site clusters and GC-rich cluster is unknown.

	GC rich cluster	site cluster	gene	spacer
MW	$\sim 3 \cdot 10^4$	$\sim 2 \cdot 10^4$	$\sim 3.5 \cdot 10^5$ (av.)	$\sim 4 \cdot 10^5$ (av.)
G + C	60%	45-62%	26%	< 5%

Evolution of the mitochondrial genome of yeast. The comparison of Hae and Hpa fragment patterns of mitochondrial DNAs from one S.carlsbergensis and three genetically unrelated S.cerevisiae wild-type strains revealed large differences (17-20). In particular, the fragment homology between S.carlsbergensis and S.cerevisiae DNAs is extremely small, almost all fragments having different electrophoretic mobilities. In apparent full contradiction with these results, a number of findings indicate a high level of sequence homology. For instance, (21) DNA-DNA hybridization experiments indicate 100 % homology, with no mismatch in the hybrids, for the DNAs from S.carlsbergensis and S.cerevisiae. Two, not mutually exclusive, possibilities can explain this apparent paradox : 1) Point mutations and/or base modifications at the restriction sites ; 2) Deletions and/or additions between restriction sites. The first possibility has been ruled out on several grounds (17-20) as a mechanism accounting for more than a minor part of the changes. Therefore, the differences in Hae or Hpa restriction patterns must be due to deletions and additions between restriction sites.

Recalling that practically all Hae fragments and most Hpa fragments correspond to the genetic units of the mitochondrial genome (8,9), the important question now is which one of the sequence elements forming the units is affected by deletions and additions. Obviously such deletions and additions cannot affect the genes themselves ; the two remaining possibilities are the AT-rich spacers or the site and GC-rich clusters. The first one is the most likely since 1) the AT-rich spacers represent 50 % of the genome, the clusters less than 10 % ; if the number of these sequences is the same, as it seems to be the case, the clusters are just too short to allow for the changes in length observed ; 2) if the clusters correspond to regulatory sequences, obviously they cannot be altered without functional losses ; in contrast the AT-rich spacers are not likely to play the same role nor to be under the same evolutionary constraints.

Several mechanisms may conceivably be responsible for the deletion-addition events occurring in the AT-rich spacers of the mitochondrial genome of yeast. The basic mechanism we favor is one involving unequal crossing-over events in the AT-rich spacers. Such a mechanism is likely to involve segment exchanges among genome units within the same cell. This process can potentially lead to extensive gene translocations, a phenomenon which is compatible with the Hae and Hpa patterns, since these enzymes unlink genetic units. Gene order is, however, very largely preserved, as shown by the great similarity of the band patterns produced from all DNAs by two restriction enzymes, Eco RI and Hind II + III, which only release ten fragments or so (8).

<ins>The cytoplasmic "petite" mutation</ins>. Several years ago, it was demonstrated (22,23) that the mitochondrial DNAs from two independent, acriflavine-induced petite mutants had a greatly altered base composition (GC = 4 %) compared to the DNAs from their parent wild-type cells (GC = 18 %). Subsequent investigations (1,2) showed that the DNAs from three spontaneous petite mutants 1) had GC contents in the 16 % - 13 % range ; 2) had multimodal melting profiles exhibiting the main low-melting component corresponding to dAT:dAT and dA:dT sequences, but lacking, to different degrees, high-melting components ; 3) renatured very rapidly.

On the basis of these data and of what was known on the structure of mitochondrial DNA from wild-type cells, it was proposed (24,4,6) that petite mutants had defective genomes in which mitochondrial genes had been deleted as a result of internal recombinations in the AT-rich stretches. The high frequency of the spontaneous petite mutation was considered to be the result of high recombination rates, the latter being due to the high level of sequence homology in the AT-rich stretches.

More recently, the organization of the mitochondrial genomes from several spontaneous petite mutants has been investigated again using the same approaches applied to the study of the mitochondrial genome from wild-type cells. Physical and chemical analyses and the use of enzymatic degradations by spleen acid DNase and by micrococcal nuclease has demonstrated that the basic organization of the petite genome is characterized, like that of the wild-type genome, by an interspersion of AT-rich stretches and GC-rich stretches. The most interesting results came, however, from the use of restriction enzyme degradations (25).

In one typical case, that of the spontaneous "petite" strain a_1 (18), the Hae and Hpa fragment patterns showed the following features : 1) practically all the bands had mobilities identical to those of the DNA from the parent strain A ; only 2 Hpa and 4 Hae bands had different mobilities ; 2) about one third of the bands (34 Hpa and 32 Hae bands) of the DNA from strain A were missing ; 3) a certain number of bands in the petite DNA had higher intensities than the corresponding bands in the wild-type DNA (8,25). In sharp contrast, 8 "new bands" out of 17, are seen in the Hind II + III pattern of DNA from "petite" a_1. Since the sum of the molecular weights of all Hind II + III bands (regardless of their multiplicity) is 70 millions, more than twice that of all Hae or Hpa bands, 30 millions, this result suggests that all or most of Hind II + III "new bands" contain tandem repeats of sequences present in the wild-type DNA ; such tandem repeats obviously lack Hind II + III sites, yet contain Hae and Hpa sites and lead, upon digestion with the latter enzymes, to the production of reinforced bands. These findings provide a direct evidence for the deletion and amplification phenomena characterizing the "petite" mutation.

As far as the mechanism of the "petite" mutation is concerned, the re- tite" genomes may conceivably occur in either of the two repetitive sequences of the wild-type genome, the AT-rich spacers or the GC-rich sequences, (restriction and GC-rich clusters). In the first case, the process would be essentially similar to that just described for the divergence of wild-type genomes, except that sequences essential for the respiratory function would have been deleted. The two processes, which are not mutually exclusive, have different consequences on the Hae and Hpa restriction patterns. In the case of crossing-overs in the AT-rich spacers, each event will cause the appearance in the "petite" pattern of a "new band", namely a band not present in the pattern of the parental wild-type strain; in the case of crossing-overs in the GC-rich sequences, no "new band" will be formed. It would be very interesting to assess the relative importance of the two processes by looking at the "new bands" present in the "petite" patterns. This is, however, made difficult by the heterogeneity of the mitochondrial genome of "petite" strains". The "petite" strain a_1 falls in this category, since clones issued from it show a different suppressiveness and a different Hae and Hpa restriction pattern.

Recombination in crosses. This was investigated (18,26) by studying the Hpa restriction patterns of cloned diploids issued from 1) zygotes obtained in mass mating experiments ; 2) individual buds from two cell lineages ; 3) individual zygotes. In all cases, crosses concerned the two S.cerevisiae wild-type strains A and B, previously investigated. The three major results of this work concern 1) the demonstration of physical recombination in crosses of wild-type yeast cells ; most bands from diploid clones are identical in mobility with those of either parent strain ; 2) the demonstration of "new bands" namely bands absent in either parent DNA ; these "new bands" appear to be due to a recombinational event in the spacers of two "homologous Hpa segments" of different length in the two parent DNAs; 3) the rate of the process ; recombination in zygotes, and possibly in vegetative cells as well, is an extremely fast process compared to point mutation. Results concerning the segregation of the mitochondrial genome into the buds will not be mentioned here.

Recombination in mitochondrial DNA fragments cloned in E.coli. Experiments (27) in which Eco RI fragments of mitochondrial DNA from S.carlsbergensis were cloned in E.coli revealed, in addition to plasmids which harbor some of the smallest Eco RI fragments, plasmids carrying mitochondrial DNA fragments showing an electrophoretic mobility different from those of the original Eco RI fragments. These cloned fragments are likely to represent deletion products originating from the larger Eco RI fragments which were not found in any of the plasmids. Since such deletion processes do not take place in the same rec A^- host-vector system for any of the Eco RI fragments obtained from λ DNA, it is likely that repetitive sequences once more play a role in the recombination process.

Conclusions. Two main conclusions can be drawn from the results briefly summarized here. The first one is that the mitochondrial genome of yeast has an eukaryotic and not a prokaryotic type of organization. In fact, this genome, like the nuclear genome of eukaryotes (28), is an interspersed system of unique and repetitive nucleotide sequences. In round figures, 60 to 70 gene sequences, averaging 530 base pairs in length, alternate with repetitive sequences, averaging 680 base pairs in length. These repetitive sequences are formed by the long AT-rich spacers (~ 600 base pairs) and by the short (Hae,Hpa) site clusters (~ 30 base pairs) and GC-rich clusters (~ 50 base pairs). The AT-rich spacers appear to play an important role in recombination phenomena ; the other two elements may correspond to promotor and operator sequences and may also contain initiation sites for DNA replication. Again as in the case of the nuclear genome of eukaryotes, the transcription of the mitochondrial genome appears to be monocistronic : it is known (29), that the two ribosomal RNA cistrons are very far apart from each other ; on the other hand no evidence has been found for transcription products originating from AT-rich spacers. The coordination of the expression of the mitochondrial genome does not, therefore, take place through polycistronic transcription, but rather through the simultaneous transcription at possibly identical promotor sites all over the genome ; similarly, the repression of this genome might occur through the interaction of repressors, possibly of nuclear origin, with a number of operator sites.

The second main conclusion is that in an interspersed system of unique and repetitive sequences, like the mitochondrial genome of yeast, recombination phenomena taking advantage of the sequence homology in the repetitive sequences, play a very important role. Genome divergence and evolution proceed essentially by such a mechanism, which is orders of magnitude faster than the point mutation mechanism. The production of the defective "petite" genomes seems to be just a very frequent accident in the general phenomenon of evolution ; these mutants, "lethal" in terms of mitochondrial function, can be observed simply because yeast can dispense with functional mitochondria.

An important implication of these results is that the mitochondrial genome of yeast really is a simple model of the nuclear genome of eukaryotes. A knowledge of its organization, expression and evolution may therefore be of great value in providing interesting clues for probing and understanding the much more complex nuclear genome. In this connection, the features of the evolution of the mitochondrial genome of yeast suggest that recombination processes taking place at a high rate at the interspersed repetitive sequences of the nuclear genome of eukaryotes may have played a very important role in evolution. A key feature of evolution by recombination at interspersed repetitive sequences is that it can generally affect the organization and the expression of the genome by causing translocations, inversions, duplications and deletions of genes, whereas it does not affect the nucleotide sequences of the latter ; only the much slower process of point mutation is operative on these. A major difference between prokaryotes and eukaryotes might precisely be that the latter developed an interspersed system of repetitive sequences allowing them to evolve much more rapidly than prokaryotes.

REFERENCES

1. Bernardi G., Faurès M., Piperno G., Slonimski P.P. (1970), J. Mol. Biol., 48. 23-42
2. Bernardi G., Timasheff S.N. (1970) J. Mol. Biol., 48. 43-52
3. Bernardi G., Piperno G., Fonty G. (1972) J. Mol. Biol., 65. 173-189
4. Piperno G., Fonty G., Bernardi G. (1972) J. Mol. Biol., 65. 191-205
5. Ehrlich S.D., Thiery J.P., Bernardi G. (1972) J. Mol. Biol., 65. 207-212
6. Prunell A., Bernardi G. (1974) J. Mol. Biol., 86. 825-841
7. Reijnders L., Borst P. (1972) Biochem. Biophys. Res. Comm., 47. 126-133
8. Bernardi G., Prunell A., Fonty G., Kopecka H., Strauss F. (1976) in The Genetic Function of Mitochondrial DNA (Saccone C., Kroon A.M.,eds.), North-Holland, Amsterdam

9. Prunell A., Bernardi G. (1976) J. Mol. Biol., submitted

10. Bernardi G. (1965) J. Mol. Biol., 13. 603-605

11. Bernardi G. (1968) in Advances in Enzymology (F.F. Nord,ed.), 31. 1-49, Wiley, New York

12. Sobell H.M. (1973) in Advances in Genetics (E. Caspari,ed.), 17. 411-490, Academic Press, New York

13. Dykes G., Bambara R., Marians K., Wu R. (1975) Nucleic Acids Res., 2. 327-345

14. Casey J., Cohen M., Rabinowitz M., Fukuhara H., Getz G.S. (1972) J. Mol. Biol., 63. 431-

15. Reijnders L., Kleisen C.M., Grivell L.A., Borst P. (1972) Biochim. Biophys. Acta, 272. 396-407

16. Schatz G., Mason T.L. (1974) Ann. Rev. Biochem., 43. 51-87

17. Bernardi G., Prunell A., Kopecka H. (1975) in Molecular Biology of Nucleocytoplasmic Relationships (S.Puiseux-Dao,ed.), Elsevier, Amsterdam, 85-90

18. Bernardi G. (1975) in Proceedings of the 10th FEBS Meeting, vol. 38 (G. Bernardi and F. Gros,eds.), North-Holland, 41-56

19. Bernardi G. (1976) in Euratom Symposium on "Mitochondrial Genetics Biogenesis and Bioenergetics" De Gruyter, Berlin (in press)

20. Prunell A., Kopecka H., Strauss F., Bernardi G. (1976) J. Mol. Biol., submitted

21. Groot S.P., Flavell R.A., Sanders J.P.M. (1975) Biochim. Biophys. Acta, 378. 186-194

22. Bernardi G., Carnevali F., Nicolaieff A., Piperno G., Tecce G. (1968) J. Mol. Biol., 37. 493-505

23. Mehrotra B.D., Mahler H.R. (1968) Arch. Biochem. Biophys., 128. 685-703

24. Bernardi G. (1969) Unpublished

25. Prunell A., Fonty G., Bernardi G. (1976) J. Mol. Biol., to be submitted

26. Fonty G., Kopecka H., Wilkie D., Bernardi G. (1976) J. Mol. Biol., to be submitted

27. Bernardi A., Bernardi G. (1976) In preparation

28. Davidson E.H., Galau G.A., Angerer R.C., Britten R.J. (1975) Chromosoma, 51. 253-259

29. Sanders J.P.M., Heyting G., Borst P. (1975) Biochem. Biophys. Res. Comm., 65. 699-707.

Genetics and Biogenesis of Chloroplasts and Mitochondria, Th. Bücher et al. eds.

THE VARIABILITY OF THE MITOCHONDRIAL GENOME OF Saccharomyces STRAINS

J.P.M.SANDERS, C.HEYTING AND P.BORST

Section for Medical Enzymology and Molecular Biology, Laboratory of Biochemistry, University of Amsterdam, Eerste Constantijn Huygens-straat 20, Amsterdam, The Netherlands

INTRODUCTION

mtDNAs of different organisms vary considerably in size and base sequence[1]. Even among the mtDNAs of various yeasts[2,3] or the sub-species of Tetrahymena pyriformis[4] large differences have been found. An interesting example of intra-species variation was recently uncovered by Bernardi and coworkers[5,6]. They found extensive differences between the mtDNAs from different strains of the yeasts Saccharomyces cerevisiae and the closely related Saccharomyces carlsbergensis, using the restriction endonucleases HpaII and HaeIII*. Less striking differences were observed in the fragment patterns of these DNAs obtained with endonucleases HindII+III, EcoR1 (refs 8,6,9) or BamHI (ref. 9). It is unlikely that these differences are due to gross heterogeneity among the available laboratory strains of Saccharomyces cerevisiae and carlsbergensis. These strains are probably all closely related and, where tested, the mtDNA of one strain has been shown to be fully functional in the nuclear background of other strains. Moreover, DNA.DNA hybridization experiments have failed to show any sequence difference between the mtDNA of S. carlsbergensis and that of a S. cerevisiae strain[3]. This suggests that the main differences between strains do not involve genes coding for structural RNAs or proteins but sequences between these genes.

A more detailed study of the nature of this sequence variation may provide information on regulatory sequences in mtDNA, on the molecular mechanism of sequence evolution in mtDNA and on the forces determining the size and gene order of mtDNA. We have started this study by analyzing three questions:

1. Are there differences in the gene order in the mtDNAs from different Saccharomyces strains?

2. Which parts of the mitochondrial genome are variable? Are the variable sections clustered or regularly spread over the DNA?

---

* Nomenclature of restriction enzymes according to Smith and Nathans[7]: Hind, Haemophilus influenzae Rd; Hpa, Haemophilus parainfluenzae; Hap, Haemophilus aphrophilus; Hae, Haemophilus aegyptius; EcoR, Escherichia coli RY13; BamH, Bacillus amyloliquefaciens H.

3. How are the variable sections located in relation to known mitochondrial genes?

To answer these questions we have constructed physical maps of mtDNAs from two S. cerevisiae strains[9] and compared these with the map of S. carlsbergensis mtDNA[8,9]. The results of this comparison are summarized in this paper.

RESULTS

Comparison of the HindII+III fragments of mtDNAs from various Saccharomyces strains

Among the HindII+III fragmentation patterns of 16 different yeast strains we have found 4 different classes. These are shown in Fig. 1. Some of the differences between the classes may be explained by the absence or presence of HindII+III recognition sites (see ref. 8). Also differences in total genome length are found that cannot be explained in this way. The mtDNA of group II is about 2000 base pairs longer than that of group I and the mtDNA of group III about 7500 base pairs.

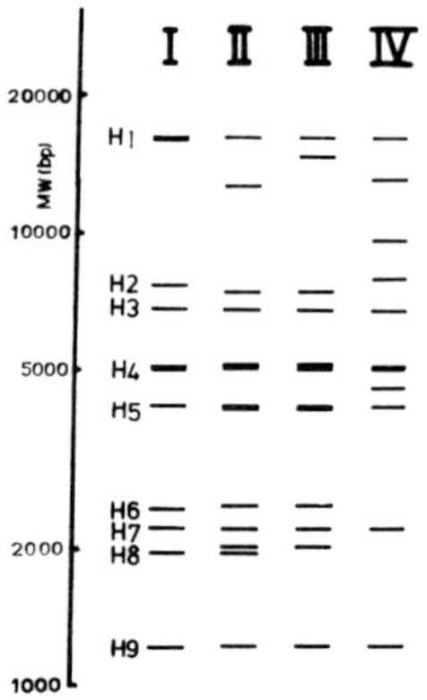

Fig. 1. Schematic representation of the agarose gel patterns obtained with HindII+III fragments of mtDNAs from 16 yeast strains. M9-94 and M9-101 (group II) have a fragment H2 which is approximately 100 base pairs larger than the fragment H2 of the other strains in that group. Group III can be divided into two subgroups, one consisting of strains that are ω^+ having a fragment H6 of 2450 base pairs and the other consisting of strains that are ω^- with a fragment H6 of 2400 base pairs (see ref. 8 for experimental details). I: S. carlsbergensis. II: JS1-3D; DP1-1B; DP1-1B/514; M9-94; M9-101; IL17. III: KL14-4A; 55R5-3C; IL46; IL46.250; M12; D85-4A; IL324-3A/2; IL324-3A/82. IV: H1.

Comparison of the physical maps of mtDNAs from three Saccharomyces strains

The size distribution of HindII+III fragments of the mtDNAs from groups I-III shows an extensive overall resemblance. This suggested that the main differences between these DNAs might be restricted to a few Hind fragments, simplifying the comparison of physical maps. We, therefore, constructed physical maps of the mtDNA of a representative of group II and group III and compared these with the map of *S. carlsbergensis* mtDNA. The result, shown in Fig. 2, confirms that the major differences are indeed restricted to three Hind fragments, clustered in one quarter of the map. In this region JS1-3D mtDNA (group II) is about 1000 base pairs longer than *S. carlsbergensis* mtDNA (group I), whereas KL14-4A (group III) has two additional insertions of 2600 and 3000 base pairs. Notwithstanding these alterations, the overall gene order is the same in these three mtDNAs, at least for the genes that have been located thusfar.

Using additional restriction enzymes[9], the "extra" sequences of KL14-4A mtDNA have been located on restriction enzyme fragments that are not much larger than the extra sequences themselves. For instance a fragment of 3600 base pairs of KL14-4A mtDNA contains 2600 base pairs of "extra" sequences (see also Fig. 4). This makes it likely

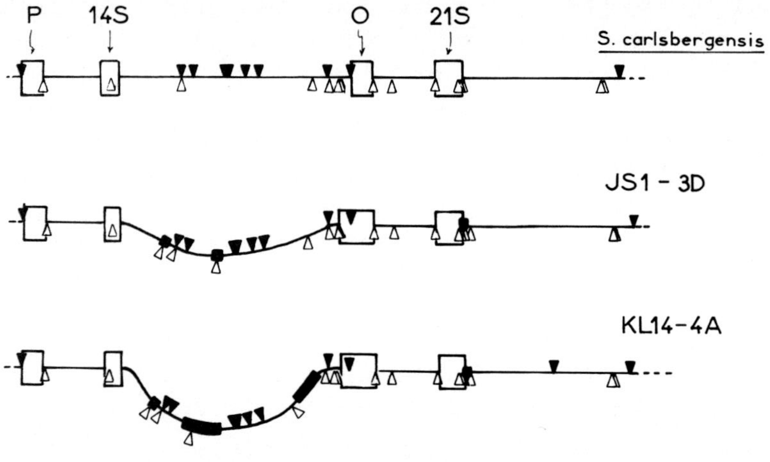

Fig. 2. Simplified representation of the physical maps of three *Saccharomyces* mtDNAs. The maps have been drawn aligning equivalent DNA sequences (open boxes) and HindII+III (Δ) and EcoR1 recognition sites (▼), considered equivalent. The closed blocks indicate DNA sequences present in JS1-3D and KL14-4A, that are absent in the same map position in *S. carlsbergensis* mtDNA. (From ref. 9.)

that these are large insertions in this region of the mtDNA of KL14-4A (or large deletions in S. carlsbergensis mtDNA) rather than many smaller ones.

Further analysis of the "extra" sequences

For a further analysis of the variable regions it is necessary to isolate the "extra" sequences. The large-scale isolation of DNA fragments from gels is difficult and in the case of yeast mtDNA the problems are accentuated because this mtDNA cannot be isolated intact. Hence, the fragments obtained with restriction endonucleases are always contaminated with pieces of larger fragments. We, therefore, tried to isolate the "extra" sequences on poly(rU)-Sephadex.

At high salt concentrations poly(rU) can form a triple helix with duplex poly(dA.dT). This property has been used by Flavell and Van den Berg[10] to specifically bind DNA containing (dA.dT) clusters of >10 base pairs to poly(rU)-Sephadex. The DNA is simply eluted from the column by lowering the salt concentration.

Yeast mtDNA contains many AT-rich stretches and Bernardi and co-workers[11] have proposed that these have a "spacer" function. As it seemed reasonable to expect that the variations in yeast mtDNA will mainly be in these "spacer" regions rather than in genes coding for proteins, rRNAs or tRNAs, we hoped to preferentially bind the "extra" DNA sequences to poly(rU)-Sephadex.

In initial experiments we indeed found that yeast mtDNA binds to poly(rU)-Sephadex (Flavell, R.A. and Sanders, J.P.M., unpublished). The application of this technique to restriction enzyme digests of mtDNA is shown in Fig. 3. The EcoR1 fragments R3-R6 and R7 of S. carlsbergensis (group I) bind to poly(rU)-Sephadex and, therefore, contain (dA.dT) stretches; the other fragments do not bind. This means that the number of long (dA.dT) stretches in S. carlsbergensis is limited or that these stretches are clustered. The former interpretation fits the results of Ehrlich et al.[13], who concluded that there are only very few long (dA.dT) stretches in yeast mtDNA.

The three EcoR1 fragments of S. carlsbergensis mtDNA that bind to poly(rU)-Sephadex are adjacent fragments lying in the same quadrant of the genome where we have located the large differences in size between the three wild-type physical maps (Fig. 2). The EcoR1 fragments of JS1-3D and KL14-4A which bind to poly(rU)-Sephadex also lie in this region of the map. Fragment R4 of KL14-4A, which contains two large insertions, binds to the column while the corresponding R4 fragments of S. carlsbergensis and JS1-3D do not. Therefore, the larger insertions in the KL14-4A mtDNA contain or are close to (dA.dT)

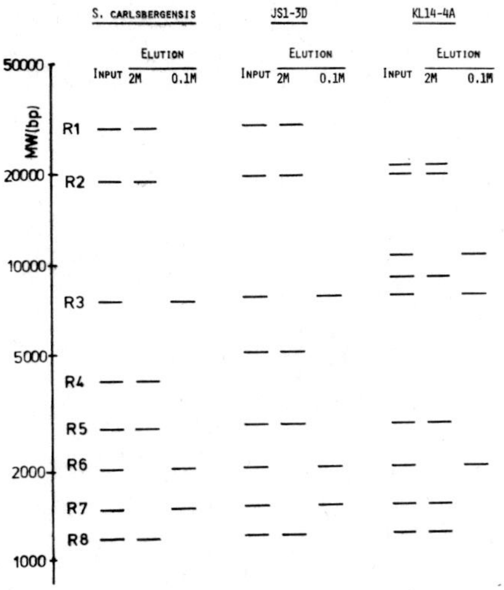

Fig. 3. Schematic representation of agarose gels of EcoR1 fragments of mtDNA fractionated on poly(rU)-Sephadex. 3 μg of S. carlsbergensis, JS1-3D, KL14-4A mtDNA digested with endonuclease EcoR1 was bound to poly(rU)-Sephadex columns as described in ref. 10. The column was washed with 2 M LiCl, 50 mM Tris-HCl (pH 7.5) and the bound fragments were eluted with 0.1 M LiCl, 10 mM Tris-HCl (pH 7.5). The fractions were dialysed, concentrated with iso-butanol/ether[12] and analysed on 0.75% agarose slab gels together with an unfractionated EcoR1 digest. Small amounts of the R1 and R2 fragments of all mtDNAs and R7 of KL14-4A bound to the column at 2 M LiCl; this is not shown in the figure.

clusters. The results of additional experiments on the distribution of these clusters in KL14-4A mtDNA are summarized in Fig. 4. After fragmentation of the DNA with endonucleases HindII+III, fragment H1b bound tightly to poly(rU)-Sephadex, as expected from the tight binding of EcoR1 fragments in this region. In addition, partial binding was observed for fragments H3, H4 and H5, indicative of the presence of shorter (dA.dT) clusters.

DISCUSSION

Our analysis of sequence variation in Saccharomyces mtDNA has led to three new findings:

1. mtDNAs from representatives of three groups of yeast strains have the same overall gene order, notwithstanding the substantial differences in the size distribution of their restriction fragments. This is fortunate, because it suggests that mapping results obtained with different strains can be directly compared and that there is one mitochondrial gene order characteristic for all Saccharomyces

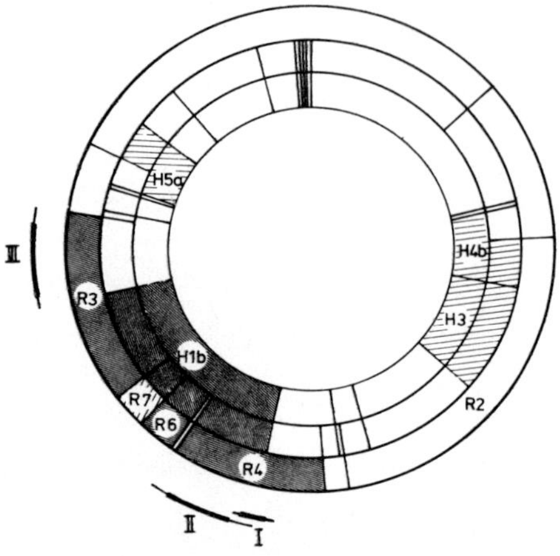

Fig. 4. Distribution of long (dA.dT) clusters in the mtDNA of S. cerevisiae strain KL14-4A. The inner ring, outer ring and middle ring show fragments obtained with endonucleases HindII+III, endonuclease EcoR1 and both enzyme preparations, respectively (see ref. 9). Fragments that bind to poly(rU)-Sephadex in 2 M LiCl are cross-hatched; weakly binding fragments are lightly cross-hatched. The "extra" sequences (see text) are indicated by the roman figures I-III. Data from Fig. 3 and unpublished experiments.

cerevisiae and carlsbergensis strains. This is in line with recent results with animal mtDNA which indicate that the gene order is strongly conserved[14]. It should be noted, however, that our analysis is based on a limited comparison of only three strains. Moreover, because of the presence of large insertions gene distances may differ in mtDNAs from different strains.

2. Large insertions occur and in the strains analysed these are clustered in one quadrant of the map.

3. Saccharomyces mtDNA contains (dA.dT) stretches long enough to bind the DNA to poly(rU) in 2 M LiCl. These stretches are clustered in the same quadrant of the map as the large insertions. This association may be fortuitous, but it is also conceivable that long (dA.dT) stretches could facilitate faulty recombination or that such stretches would be located primarily outside structural genes in regions tolerant to alteration. Whatever the nature of the association may be, it provides an additional handle to study the large insertions. For instance, insertion II takes up 75% of the DNA segment shared by fragment H1b and fragment R4. This segment quantitatively binds to poly(rU)-Sephadex and this should simplify the large-scale isolation of this insertion and the characterization of the DNA sequence that it contains.

The large insertions that we have found only occupy 10% of the genome. Therefore, they cannot account for all the differences in HapII and HaeIII fragments, that Bernardi and coworkers[6] observe in their comparison of mtDNAs from different yeast strains. Additional changes elsewhere in the genome must have occurred and, in fact, we also find small differences in fragments from other quadrants of the genome, when we compare the HindII+III x EcoR1 digestion products of mtDNAs from S. carlsbergensis, JS1-3D and KL14-4A (unpublished experiments). All our results are, therefore, compatible with Bernardi's proposal that most of the differences between mtDNAs from different strains can be attributed to multiple insertions and deletions.

To further analyse the nature of the variable regions in mtDNA and their relation to structural genes it will be necessary to make a fine-structure map that covers a structural gene with adjacent DNA segments and locate the alterations on this map. Such a map has been constructed for the region of the 21S rRNA gene[15] and we intend to study the variations in this region in the mtDNAs available.

ACKNOWLEDGEMENTS

We thank Mrs H.H.Menke for expert technical assistance and Dr. R.A.Flavell for introducing us to the poly(rU)-Sephadex technique. This work was supported in part by a grant from The Netherlands Foundation for Chemical Research (SON) with financial aid from The Netherlands Organization for the Advancement of Pure Research (ZWO).

REFERENCES

1 Borst, P. and Flavell, R.A. (1976) in Handbook of Biochemistry and Molecular Biology (Fasman, G.D., Ed.), Vol. 2, 3rd edn., CRC Press, Cleveland, Ohio, pp. 363-374.

2 Sanders, J.P.M., Weijers, P.J., Groot, G.S.P. and Borst, P. (1974) Biochim. Biophys.Acta 374, 136-144.

3 Groot, G.S.P., Flavell, R.A. and Sanders, J.P.M. (1975) Biochim.Biophys.Acta 378, 186-194.

4 Goldbach, R.W., Borst, P., Arnberg, A.C. and Van Bruggen, E.F.J. (1976) in The Genetic Function of Mitochondrial DNA (Saccone, C. and Kroon, A.M., Eds), North-Holland, Amsterdam, pp. 137-142.

5 Bernardi, G. (1975) in Proceedings of the 10th FEBS Meeting (Bernardi, G. and Gros, F., Eds), Vol. 38, North-Holland, Amsterdam, pp. 41-56.

6 Bernardi, G., Prunell, A., Fonty, G., Kopecka, H. and Strauss, F. (1976) - same as ref. 4 - pp. 185-198.

7 Smith, H.O. and Nathans, D. (1973) J.Mol.Biol. 81, 419-423.

8 Sanders, J.P.M., Borst, P. and Weijers, P.J. (1975) Mol.Gen.Gen. 143, 53-64.

9 Sanders, J.P.M., Heyting, C., DiFranco, A., Borst, P. and Slonimski, P.P. (1976) - same as ref. 4 - pp. 259-272.

10 Flavell, R.A. and Van den Berg, F.M. (1975) FEBS Letters 58, 90-93.

11 Prunell, A. and Bernardi, G. (1974) J.Mol.Biol. 86, 825-841.

12 Stafford, D.W. and Bieber, D. (1975) Biochim.Biophys.Acta 378, 18-21.

13 Ehrlich, S.D., Thiery, J.-P. and Bernardi, G. (1972) J.Mol.Biol. 65, 207-212.

14 Kroon, A.M. and Saccone, C. (1976) - same as ref. 4 - pp. 343-347.

15 Heyting, C. and Sanders, J.P.M. (1976) - same as ref. 4 - pp. 273-280.

Genetics and Biogenesis of Chloroplasts and Mitochondria, Th. Bücher et al. eds.

RESTRICTION ENDONUCLEASE MAPPING AND ANALYSIS OF GRANDE AND MUTANT YEAST MITOCHONDRIAL DNA

R. Morimoto*, A. Lewin*, S. Merten, and M. Rabinowitz
Departments of Medicine, Biochemistry and Biology
The University of Chicago and
The Franklin McLean Research Institute**
Chicago, Illinois 60637, U.S.A.

Petite mtDNA has been established to contain variable sized deletions of the grande mitochondrial genome[1,2], and deletion mapping with collections of petite mtDNAs representing different mtDNA segments has been widely used[3-6]. However, examination of petite mtDNAs by renaturation kinetics[7], hybridization[8,9], electron microscopic[1,10], restriction endonuclease[11,12], and genetic analysis[4,13] shows that some petites mtDNAs are not formed as the result of a single contiguous deletion having sequences colinear with grande mtDNA, but contain multiple internal deletions, amplifications or rearrangements of segments, and in some cases inverted repeats. Since these changes may result in serious errors in mapping, we have attempted to map the mtDNA of a genetically characterized strain by restriction endonuclease fragment analysis. The derived fragment map can be used to unequivocally localize genetic markers, and as a standard for the analysis of sequence arrangement in various petite strains.

We have analyzed the mtDNA from strain MH41-7B constructed by H. Fukuhara and M. Bollotin-Fukuhara[4], which contains five antibiotic resistance markers (C, E, O_I, O_{II}, P), and from which they have derived a large number of genetically characterized petite clones. Restriction endonuclease analysis of yeast mtDNA is somewhat difficult because intact 25μ molecules cannot be isolated, which results in the larger restriction fragments being present in less than molar ratios and in considerable background in gels separating the fragments. We have used four procedures to order fragments; 1) analysis of digests of closely related grande strains; 2) co-digestion with enzymes such as SalI or PstI which cleave at only one site on the mtDNA; 3) redigestion of isolated restriction fragments with another endonuclease; and 4) partial digestion analysis.

Digestion patterns of mtDNAs of several related grande strains, illustrated in Figure 1, were used to help order three of the larger EcoR1 fragments, and two of the HpaI fragments. The mtDNA of these strains contain one less restriction site

*Supported in part by research and training grants from the USPHS National Institutes of Health and the Louis Block Fund of the University of Chicago
**Operated by the University of Chicago for the U.S. Energy, Research and Development Administration.

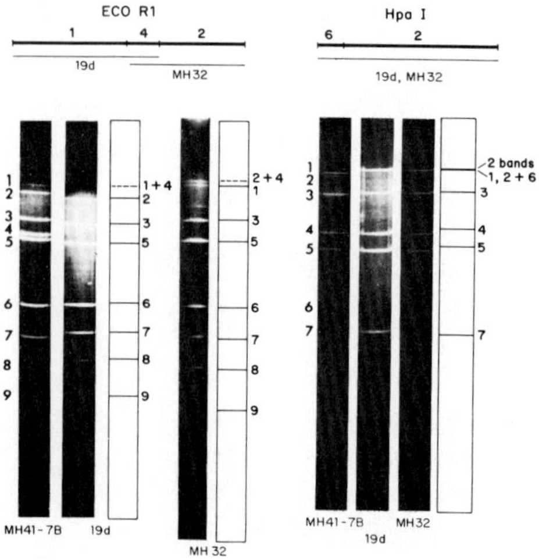

Fig. 1. Agarose (0.5%) gel electrophoresis of EcoR1 and HpaI digests of mtDNA from grande strains MH41-7B, MH32 and 19D. Fragments were further identified by co-migration of grande mtDNA digests; fragment size was estimated using λDNA digests as internal markers.

Fig. 2. Linearized EcoR1, HindIII, HpaI, BamHI fragment maps of MH41-7B mtDNA. The sites of cleavage by SalI and by PstI are indicated. Fragment order derived from analysis of grande strains and from SalI, PstI co-digestions are shown.

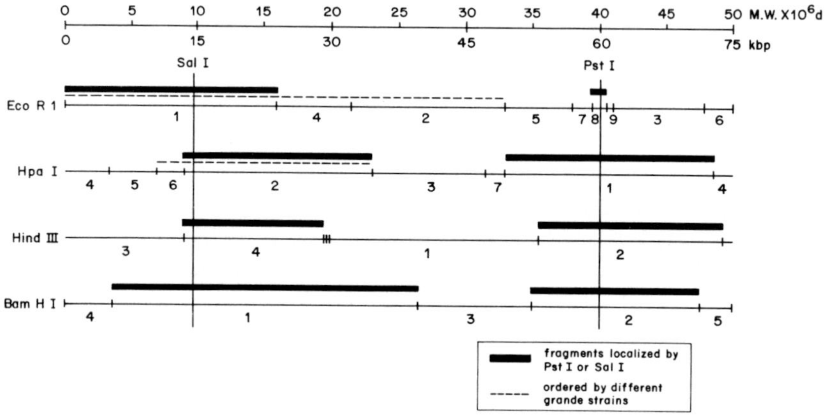

for EcoR1 or HpaI than MH41-7B, which leads to the presence of a new fragment composed of the sequences of two missing fragments. The EcoR1 digest of MH32 mtDNA differs from MH41 only in that two bands (2 and 4) of MH41 are missing. These are replaced by a single band with a molecular weight equal to the sum of the molecular weights of the two missing bands. We can conclude that bands 2 and 4 are contiguous. Similar relationships are observed in strain 19D, where bands 1 and 4 of MH41 are missing and are replaced by a large additive band, indicating that bands 1 and 4 are contiguous. Similarly HpaI fragments 2 and 6 are shown to be contiguous by this analysis (Fig. 1).

A second extremely useful approach first used by Sanders et al. for yeast mtDNA[14] is double enzyme digestion with an enzyme such as SalI or Pstl, which cleave the mtDNA of this strain at only one site together with another multisite endonuclease. Co-digestion with PstI or SalI provides information about the overlap of fragments in digests by different endonucleases, and the orientation of the affected fragments when SalI or PstI is added to double digests. Figure 2 shows our restriction fragment map for four endonucleases, EcoR1, HindIII, HpaI, and BamHI. The portion of the map indicated by interrupted lines can be derived from SalI and PstI co-digestion, and the heavy lines indicate the order of fragments derived from analysis of different grande strains.

The order of these fragments was confirmed and the order of the remaining fragments was established primarily by redigestion of isolated restriction fragments with the second restriction endonuclease. Partial digest data was used only to order the smaller EcoR1 fragments and to confirm the order of several other fragments.

Redigestion of EcoR1 fragments by HpaI, and of HpaI fragments by EcoR1 (Table 1) was used to complete the EcoR1 and HpaI map. For example, digestion of HpaI fragment 2 with EcoR1 results in fragments of molecular weight 6.7, 5.5, and 2.35 x 10^6, which are also derived from HpaI digests of EcoR1 fragments 1, 4 (undigested by HpaI) and 2, confirming the EcoR1 fragment order 1,4,2. Similar cross digestion data for all four endonucleases was also obtained for fragments generated by BamHI, EcoR1 and HpaI (results not shown) which allowed us to map all but the smaller of the EcoR1 fragments.

Partial digest analysis was used to order some of the smaller EcoR1 fragments as illustrated in Figure 3. The derived fragment order was the only one consistent with the data. Of note was the presence of a partial digest band of molecular weight 2.8 x 10^6 which could not be explained by the fragments available. Analysis of digests of larger amounts of mtDNA on 4% acrylamide gels

TABLE I

EcoR1 FRAGMENTS REDIGESTED BY HpaI (M.W. x 10^6D)

	Double Digest Fragments	1 (15.8)	2 (11.5)	3 (6.7)	4 (5.5)	5 (5.2)	6 (2.3)	7 (1.6)	8 (1.1)	9 (0.6)
1	8.9		8.9							
2	6.7	6.7								
3	6.7			6.7						
4	5.5				5.5					
5	5.1					5.1				
6	3.7	3.7								
7	3.6	3.6								
8	2.35		2.4							
9	2.1	2.1								
10	1.53		1.53							
11	1.5							1.5		
12	1.4						1.4			
13	1.12								1.12	
14	0.9						0.9			
15	0.6									0.6
Total M.W.	51.8	16.1	12.8	6.7	5.5	5.1	2.3	1.5	1.12	0.6

HpaI FRAGMENTS REDIGESTED BY EcoR1 (M.W. x 10^6D)

	Double Digest Fragments	1 (15.8)	2 (13.5)	3 (9.0)	4 (4.7)	5 (3.7)	6 (2.1)	7 (1.5)
1	8.9			8.9				
2	6.7	6.7						
3	6.7		6.7					
4	5.5		5.5					
5	5.1	5.1						
6	3.7				3.7			
7	3.6					3.6		
8	2.35		2.35					
9	2.1						2.1	
10	1.53							1.53
11	1.5	•						
12	1.4	•						
13	1.12				1.0			
14	0.9	•						
15	0.6	•						
Total M.W.	51.8	15.3	14.5	8.9	4.7	3.6	2.1	1.53

• EXPERIMENT NOT COMPLETED

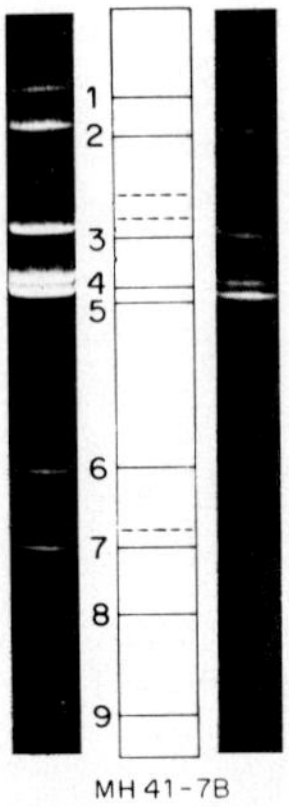

M.W. x 10^6 d of partial	DERIVED FRAGMENT ORDER
9.5	9 3 6
9.0	3 6
8.3	5 7 8
7.4	9 3
2.8	10 – 7 8 – 10
2.66	7 8
1.75	8 9

Fig. 3. Partial digest bands repeatedly observed (>10 times) in EcoR1 digests and the fragment order derived from these digests.

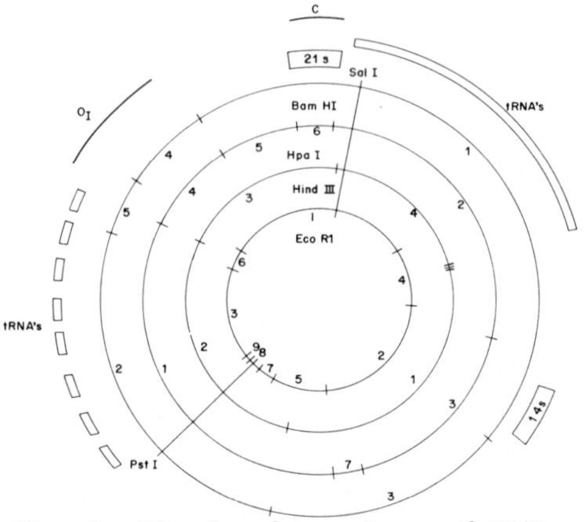

Fig. 4. Circular fragment map of MH41-7B mtDNA. The site of hybridization with ^{125}I-21S and 14S rRNA, ^{125}I-4S, and ^{3}H-cRNAs transcribed from petites C_5 and O_I-2 (see ref. 4).

however revealed the presence of a tenth EcoR1 fragment with molecular weight 0.2 x 10^6. Thus the unexplained partial digest band probably represents a combination of fragments 7, 8 and 10.

The EcoR1 fragment order shown in Figures 2 and 4 differs from our tentative map presented two months ago[15] only in the change in location of fragment 3. We also establish the order of HindIII fragments 3 and 4 on the basis of SalI digestion. Those fragments have similar molecular weights and can only be separated after prolonged electrophoresis. The EcoR1 map is identical with that presented by Sanders et al. for strain KL14-4D[14], except for the location of fragment 9.

We have also hybridized ^{125}I labeled 21S and 14S rRNA to single and double restriction digests using the Southern technique[16]. Results summarized on the circular fragment map (Fig. 4) confirm the observations of Faye et al.[13] and Sanders et al[17] that the two rRNA subunits are transcribed from widely separated sequences on the mtDNA. Hybridization of total ^{125}I 4sRNA, and of cRNA transcribed from low kinetic complexity petite clones that retain the C or O_I markers are also shown in Figure 4. The localization of antibiotic resistant markers is similar to that obtained from restriction analysis by Sanders et al.[14] and from deletion mapping[4], 4s RNA sequences are localized primarily between C and P markers, but near the O_I or O_{II} marker as well, similar to our previous localization of tRNA genes by deletion mapping[4].

Using the four enzyme fragment map as a basis, we examined the digestion patterns of a series of petite mtDNAs retaining various antibiotic resistance markers. By this means we could obtain information about the order and location of the genetic markers retained in these petites, and also about rearrangement of sequences present in some petite mtDNAs.

Analysis of the restriction digestion patterns of 12 petite mtDNAs with reference to the linearized restriction fragment map is shown in Figure 5. Eight of the 12 selected strains analyzed had sequences essentially colinear with those of grande mtDNA. These strains are therefore appropriate for deletion mapping and can be used to order and to determine the approximate location of antibody resistance markers. For example, the C and E markers shared by strains CEP-2 and CEO_I-2 must be localized to the region of overlap and are so indicated in Figure 5. The E marker is further localized rather precisely to overlapping sequences between strains EO_I and CEP-2. Similarly the O_I marker is localized at 4 to 10 million daltons to the left of the E and C markers. The O_{II} marker is narrowly localized on the map, since strain O_{II}-2 contains EcoR1 fragment 8. This fragment is cleaved by the enzyme PstI both in grande and in petite O_{II}-2. The P marker is similarly localized using data from strains CEO_I-2, $O_I(O_{II})P_2$ and CEP-2. The localization is similar to that obtained with restriction fragment

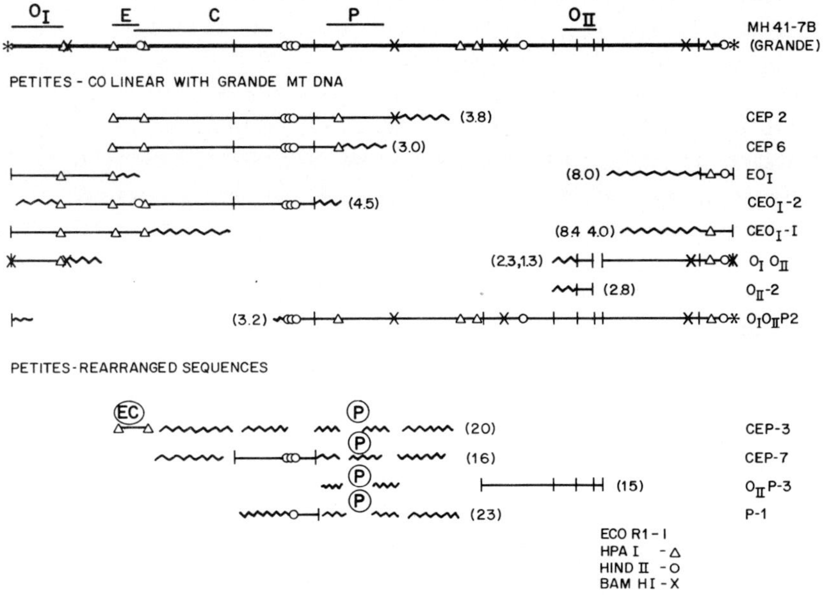

Fig. 5. Restriction fragments (EcoR1, HindIII, HpaI, BamHI) present in mtDNA of petite mutants retaining various combinations of antibiotic resistance markers, plotted in relation to the grande fragment map. [—— retained fragment, ～～ extra fragment which may be at either end of the molecule, ( ) molecular weight of extra fragment.] The antibiotic resistance markers are localized by deletion mapping data alone in this figure. Petites CEO_I-1 and O_IO_{II} have two new fragments which may be due to heterogeneity at the deleted region.

mapping (as in Figs. 4 and 5)+ Rearranged petite mtDNAs are seen on the lower portion of Figure 5. These are obviously not colinear with the grande fragment map and contain internal deletions as well as other arrangements. It is clear that analysis of petite mtDNAs by restriction endonuclease is essential to evaluate their suitability for use in deletion and recombination mapping.

In summary we have derived a physical map of mtDNA of a genetically characterized yeast strain and localized 21S and 14S rRNA, and total 4S genes as well as 4 antibiotic resistance markers by hybridization techniques. The map has been used to characterize petite mtDNAs. The localization of antibiotic resistance markers by deletion mapping using petite mtDNA's which have sequences colinear with grande agrees well with the restriction analysis.

REFERENCES

1. Locker, J. et al. (1974) J. Mol. Biol. 88, 489-502.
2. Michel, F. et al. (1974) J. Mol. Biol. 85, 411-431.
3. Slonimski, P.P. and Tzagoloff, A. (1976) Eur. J. Biochem. 61, 27-41.
4. Fukuhara, H. et al. (1976) Molec. Gen. Genet. 145, 7-17.
5. Faye, G. et al. (1974) J. Mol. Biol. 88, 185-203.
6. Sriprakash, K.S. et al. (1976) J. Mol. Biol. 104, 485-503.
7. Casey, J. et al. (1974) Biochemistry 13, 1059-1063.
8. Gordon, P. et al. (1974) Biochemistry 13, 1067-1075.
9. Lazowska, J. et al. (1974) J. Mol. Biol. 85, 393-410.
10. Locker, J. et al. (1974) Proc. Nat. Acad. Sci. US 71, 1366-1370.
11. Morimoto, R. et al. (1975) Proc. Nat. Acad. Sci. US 72, 3868-3872.
12. Bernardi, G. (1976 in press) The Genetic Function of Mitochondrial DNA, Eds. C. Saccone and A.M. Kroon, Elsevier Publishing, North-Holland, Amsterdam.
13. Faye, G. et al. (1975) J. Mol. Biol. 99, 203-217.
14. Sanders, J. et al. (1976 in press) The Genetic Function of Mitochondrial DNA, Eds. C. Saccone and A.M. Kroon, Elsevier Publishing, North-Holland, Amsterdam.
15. Rabinowitz, M. et al. The Genetic Function of Mitochondrial DNA, Eds. C. Saccone and A.M. Kroon, Elsevier Publishing, North-Holland, Amsterdam.
16. Southern, E.M. (1975) J. Mol. Biol. 98, 503-517.
17. Sanders, J. et al. (1975) Bioch. Biophys. Res. Comm. 65, 699-707.

+This data corrects our previously reported[15] preliminary assignment of antibiotic resistance markers to restriction fragments,

Genetics and Biogenesis of Chloroplasts and Mitochondria, Th. Bücher et al. eds.

THE CONTROL OF MITOCHONDRIAL DNA SYNTHESIS IN YEAST PETITE MUTANTS

P.BORST, C.HEYTING AND J.P.M.SANDERS

Section for Medical Enzymology and Molecular Biology, Laboratory of Biochemistry, University of Amsterdam, Eerste Constantijn Huygens-straat 20, Amsterdam, The Netherlands

INTRODUCTION

Cytoplasmic rho$^-$ petite mutants of yeast are deletion mutants and in some of these more than 99% of the mitochondrial genes are lost. Nevertheless, the total amount of mtDNA per cell in these mutants is the same as in the wild-type yeast from which they were derived. In this paper we present a speculative model for the control of mtDNA synthesis in which we try to account for this puzzling observation.

Replication intermediates

The mode of replication of yeast mtDNA has not been established. Putative intermediates in replication - branched Y-type molecules and circles with a duplex tail - were observed in electron micrographs of yeast mtDNA[1], but it is not possible to exclude that these molecules were intermediates in recombination, or aberrant side products of either replication, recombination or petite mutant formation. It is, therefore, necessary to turn to other systems than yeast to learn how mtDNAs replicate in general.

In animal mitochondria two modes of replication have been found: in mouse L-cells (see refs 2,3) replication starts at a fixed point of the DNA circle with the uni-directional synthesis of one strand (the H-strand). When this is about half completed synthesis of the other strand (the L-strand) starts, probably also at a unique position. Both strands grow in a 5'-3' overall direction, but it is not known whether synthesis is continuous or discontinuous by Okazaki-type fragments. In rat liver (see refs 4,5) the mode of synthesis is similar, but the synthesis of the L-strand may start earlier and close to the initiation point of H-strand synthesis. Hence, intermediates are also found that contain a fully duplex eye, varying in size from 4% to over 80% of the circle circumference[4]. These can only arise if the synthesis of the L-strand is discontinuous; from a more detailed analysis of these replicative intermediates Wolstenholme and coworkers[4] have concluded that the synthesis of the H-strand is discontinuous as well.

The mtDNA of Tetrahymena pyriformis (see ref. 6) is linear and replication starts at a unique point near the middle of the molecule and proceeds bi-directionally to the ends. In this case the replication intermediates have a duplex character, but the characteristic asymmetric single-stranded sections in the replication forks indicate that at least one strand is made discontinuously in each fork, as one would expect in a bi-directional mode of replication involving largely duplex intermediates.

Finally, replication of the linear DNA of Paramecium aurelia starts at one end and proceeds via duplex intermediates to the other end[7]. Again, one may expect that at least one of the strands is made in a discontinuous fashion.

Extrapolating from these results it seems likely that replication of yeast mtDNA will start at a unique position on the circle and that at least one of the strands is synthesized in a discontinuous mode.

Properties of mtDNA from petite mutant RD1A

In molecular terms the mtDNA from RD1A is the most extensively characterized mtDNA available and a summary of its properties may serve as an introduction to the properties of petite mtDNAs in general. RD1A was isolated by Hollenberg et al.[8] from wild-type yeast treated with high concentrations of ethidium bromide. More than 95% of its mtDNA consists of identical repeats of a 68-nucleotide sequence containing 2 GC and 66 consecutive AT pairs[9-11]. The complete nucleotide sequence of this repeat has recently been determined[11]. The repeats are arranged head to tail and from a melting analysis of the heteroduplex of RD1A mtDNA (or its complementary RNA) and wild-type mtDNA, the repeat unit appears to be a faithful copy of a segment of the wild-type mtDNA[11-13]. Pyrimidine tract analysis has shown the presence in RD1A mtDNA of two minor tracts accounting for less than 1% of the ^{32}P (refs 9,10). These tracts contain as yet unidentified unusual nucleotides[12] (not methyl-C). Whether they are present in (part of) the repeats or in inter-repeat "spacers" has not been determined.

The size and structure of RD1A mtDNA in situ is not known. When isolated, the DNA consists of large networks and we attribute this to branch migration of linear molecules during isolation[8]. In alkaline gradients the DNA sediments in a broad band, like wild-type mtDNA, with molecules up to 7000 nucleotides long, i.e. 100 times the repeat unit[8]. We think that the bulk of RD1A mtDNA is present as 25-μm circles in vivo, like wild-type mtDNA, and that these circles are broken during isolation. Small circles are rare in this DNA and they comprise less than 0.1% of the mtDNA mass[8].

The structure of petite mtDNA

The present views on the origin of rho$^-$ mtDNAs are schematically depicted in Fig. 1. We shall briefly consider the features of this scheme that bear on mtDNA replication:

1. Petite mtDNAs contain only segments of wild-type mtDNA. Evidence for this has come from hybridizing petite DNAs with excess wild-type mtDNA[13,15,16] and more recently from restriction enzyme analysis[17-19]. Only one exception to this rule has been reported[15] and we are not convinced (see ref. 13) that the inability of wild-type mtDNA to bind all mtDNA of this petite mutant in hybridization experiments, could not be due to a secondary amplification of part of

A b c d e f g h i j k l m n o p q r s t u A

↓ Fragmentation

A b c d e f g h i j k + l m n o p q + r + s t u

↓ Illegitimate recombination with replicator and compensatory amplification

A b c d e f g h i j k b c d e f g h i j k A

\+

A r r r r r r r r r r r r r r r r r r r A

\+

A b A b A b A b A b A b A b A b A b A b A

↓ etc.

↓ Cloning

A b c d e f g h i j k b c d e f g h i j k A

↓ Rearrangements, secondary deletions and compensatory amplification

A b c d f g g'f'd'c'b'b c d f g g'f'd'c'b'A

⇑↓ Excision of monomer repeats

b c d f g g'f'd'c'b'b

Fig. 1. Speculative model for the origin of petite mutants (modified from ref. 14). A = replication sequence; underlined terminal letters indicate circularity, *i.e.* in the top line A is joined to b and to u.

the petite sequence since the structure of this petite mtDNA is complex[20]. We think, therefore, that it is reasonable to conclude that petite mtDNAs in general contain only reshuffled wild-type sequences. It follows that petites should use the same origin for replication (the replicator) as wild-type mtDNA.

2. Petites contain amplified pieces of wild-type mtDNA[1,8,20,21] and in most petites this piece is a contiguous segment of the rho$^+$ genome with no or minor rearrangements[18,19].

3. Petites can be isolated containing sections of any part of the mtDNA map (see refs 16,18). If replication starts at a unique replicator site and if petites use the same replicator as wild-type mtDNA, it follows that an illegitimate recombination must take place to join the replicator with the amplified segment, unless the segment contains the replicator. The presence of additional (weak) replicator sequences in wild-type mtDNA that are activated when the true replicator is destroyed, does not affect the argument unless there are very many of these alternative replicators. This seems unlikely.

It is not known whether the replicator is always part of the amplified segment. Sanders et al.[13] have tried to test this by cross-hybridizing the mtDNAs of three petites with a very short repeat. No cross-hybridization was found, but this does not necessarily mean that the repeats of these DNAs do not share a common replicator, because recent model experiments have shown[11,12] that these AT-rich DNAs require a perfect sequence homology of >35 nucleotides to hybridize. Although the size of replicators is unknown, there is no reason why 30 nucleotides would not be enough.

4. Petites contain the same amount of mtDNA as the wild-type strain from which they were derived, whatever the fraction of the wild-type mtDNA that is deleted. This was noted in early work on petite mutants[22-24] and it has recently been confirmed by Hall et al. (ref. 25). They found only one exception to this important rule, petite mutant RD1A, which contained only 1/3rd of the level found in its parent. In our hands, however, the yield of mtDNA from this strain has been variable for two reasons: Under some conditions it throws off rho^0 petites and if these are not eliminated, mtDNA yields are low. In addition, the tendency of this DNA to form large networks leads to variable losses, that are difficult to control[8]. The low yield of extracted RD1A mtDNA may, therefore, not be due to a low concentration per cell.

The amount of mtDNA per cell in yeast is controlled by the nucleus and at least two nuclear genes are involved in this control[25]. The control is not tight, because temporarily mtDNA synthesis can continue when nuclear DNA synthesis is blocked[26].

5. The induction of petite mutants with intercalating dyes is accompanied by a decrease in the size of the DNA that can be extracted (refs 27,28) and probably by fragmentation of the DNA in situ.

6. Petite mtDNA is circular and these circles vary in size from 25 μm, the size of wild-type mtDNA[29,30], down to the size of the repeat unit(s). In all cases the bulk of the mtDNA is present as 25-μm circles. The evidence for this conclusion is thin and mainly indirect. Michels et al.[31] analysed the sedimentation of petite and wild-type mtDNAs in sucrose gradients under conditions minimizing damage. They found no difference and since the DNAs sedimented in a broad band around the position expected for linearized 25-μm circles, they concluded that the size and structure of the mtDNAs of these petite mutants was the same as that of the wild-type parent. The presence of 25-μm circles was not demonstrated by electron microscopy, however, in this work or in other labs. Usually, the bulk of the petite mtDNA is found to be linear with a size distribution similar to that of wild-type mtDNA. This is compatible with the idea that these linear molecules are derived from 25-μm circles, but they could obviously also come from 15-μm or 50-μm circles.

In addition to long DNA, all petite mutants contain small mtDNA circles (see refs 1,8,20,23,32). These circles form oligomeric series, the smallest ones having the size of the repeat unit(s). The fraction of the total mtDNA present as small circles varies and reported values for different mutants range from 0.1 to 28 weight percent.

Lazowska and Slonimski[32] have recently found for a set of petites that the frequency distribution of circles in oligomeric series follows a simple rule: Frequency of n-meric circles = 1/n frequency of monomeric circles for n = 2, 3 and 4. They conclude that on a mass basis each circle class represents the same fraction of total mtDNA and that the mitochondrial genome has the same probability to constitute one monomeric circle or to be a part of a n-meric circle. They interpret this to mean that in vivo all mtDNA is circular.

The extrapolation of the "frequency rule" of Lazowska and Slonimski[32] to higher values of n, leads to results that we find difficult to accept. For instance, for RD1A mtDNA with a repeat unit length equal to 0.1% of the wild-type mtDNA, it implies that 10% of this DNA should be present as circles with a contour length of 2.5 μm or smaller. This has not been found and although one can never be certain that a sizeable fraction of mtDNA is not preferentially lost during isolation, this seems rather unlikely in this case in view of the effort made to find circles (see ref. 8). Moreover, the frequency rule implies that each size class contains 0.1% of the total DNA. Since the haploid yeast cell only contains mtDNA for about hundred 25-μm circles, there should be less than one circle per cell for all classes >2.5 μm. Finally, we note that in our (unpublished) experience the amount of circular mtDNA in a large series of petites with repeat units around 10% of the wild-type genome is much lower than expected from the frequency rule. These petites were analysed on agarose gels, on which the heterogeneous linear DNA forms a broad smear, whereas the circles give sharp bands that can easily be seen even if 1% of the input DNA is present in such a band.

We, therefore, think that the frequency rule does not apply to all petites and that it can not be extrapolated to higher values of n.

7. Recombination of mtDNA can occur in petites in the absence of functional mtDNA. This has been proven by Michaelis et al.[33], who demonstrated mtDNA recombination in a cross between two cytoplasmic petite mutants. It is possible that the recombination machinery must be activated by mating, but we consider this unlikely. We think, therefore, that the monomer and oligomer circles in petite mtDNA continue to recombine in the absence of further mating.

8. Petite mtDNAs usually have a lower mole percent GC than the wild-type mtDNA and with highly repetitive petite mtDNAs this is invariably so (see refs 13,21,23,24). This could mean that AT-rich segments have a better chance of linking up to the replicator, either because the statistical chance of illegitimate recombination is better in a DNA segment with <5% GC than with 50% GC, or because the replicator itself is AT-rich. It is possible, however, that GC-rich petite mtDNAs exist but are systematically overlooked because the DNA bands in CsCl under the nuclear DNA peak.

Control of mtDNA; speculations

Mitochondrial protein synthesis is absent in cytoplasmic petite mutants and all proteins required for mtDNA synthesis are, therefore, made on cell-sap ribosomes. It seems likely that the genes for all these proteins are located in the nucleus and that the highly repetitive petite mtDNAs - like RD1A - do not contain functional genes in addition to a replicator sequence. Proof for this is difficult to obtain, because one gene of 700 base pairs in a genome of 70 000 cannot be detected in the absence of a specific probe for this gene. For the moment it seems reasonable, however, to accept that yeast mtDNA is a satellite replicon, that only contributes a replicator sequence to its replication.

How is the amount of mtDNA in the cell kept constant? The simplest mechanism is provided by a positive control, e.g. an initiator protein that acts stoicheiometrically and is inactivated in the initiation event[34]. Since mtDNA synthesis can proceed for at least one full round after cell-sap protein synthesis is blocked[26], there must be a delay between the synthesis of initiator and its action. In the steady state the number of mtDNA molecules per cell is between 1 and 2 times the number of initiator molecules made per cell cycle (disregarding turnover of mtDNA). According to this model the amount of mtDNA per cell is controlled by controlling the number of molecules that replicate per cell cycle.

How then is the amount of mtDNA in petite mutants kept constant and equal to the wild-type strain from which they are derived? We think that this can only be accomplished if the petite mutation does not affect the size of the replicating molecules. It is obvious, however, that molecules smaller than 25 μm are present in petite mitochondria, e.g. the small circles. Since recombination occurs in petites, there is no reason why such molecules could not pick up a replicator sequence and replicate. We conclude, therefore, that replication of mtDNA involves a sizing step, which discriminates against molecules smaller or larger than 25 μm.

If fragmentation of mtDNA is an obligatory step in petite induction by intercalating dyes, sizing should eliminate these fragments and lead only to rho^0 cells (i.e. cells without mtDNA) rather than a mixture of rho^0 and rho^- petites. There are three ways to reconcile the apparent replication of these fragments with our model. First, it is possible that small circles replicate with low efficiency and that size discrimination is relative rather than absolute. Second, it is possible that recombination will lead to reconstruction of faulty 25-μm circles and that subsequent repeated excision and reinsertion of segments leads to one type of repeat. Evidence for reconstruction of wild-type mtDNA from fragments during recovery from ethidium mutagenesis has recently been presented[35,36]. Finally, it is possible that sizing is stringent but that only molecules entering new replication units are sized. Molecules fragmented after entering would be able to replicate indefinitely and provide the monomer repeats required to build up by recombination the 25-μm circles required to fill new replication units.

How does the mitochondrion size its DNA? We see two possible mechanisms, packaging and measuring length. The packaging mechanism would be analogous to encapsidation of viral DNA. DNA that fits would be replicated. Measuring length could occur by multiple attachment of condensed mtDNA molecules to the mitochondrial inner membrane. Obviously, the two mechanisms are similar and hybrid mechanisms can be envisaged. If mtDNA remains condensed during replication, this could be a way of preventing recombination. In a system where the recombination frequency is high[37], it could be of advantage if replication and recombination are mutually exclusive to avoid replication of nicked strands.

We assume that every replicating circle contains a single replicator sequence, but the presence of multiple replicators would not necessarily affect the amount of mtDNA per cell in this model. We further assume that replication will be discontinuous and this will require additional recognition sites. Although these sites may not be very specific, this requirement may put a lower limit on the repeat sequence that can be present in petite mtDNA. It is possible that the unusual nucleotide found in RD1A mtDNA[12] is involved in this.

Predictions

The model makes a number of useful predictions:

1. *In situ* the bulk of mtDNA in all petites consists of 25-μm circles.

2. The small circles do not replicate and arise by internal recombination events.

3. Some petites cannot exist because none of the oligomers of the basic repeat comes near 25 μm. The apparent instability of petites containing more than 50% of the wild-type genome might be due to this.

4. Replication of mtDNA involves a condensed form of mtDNA.

5. The frequency rule of Lazowska and Slonimski[32] does not apply to all petites and cannot be extrapolated to higher values of n.

6. DNA segments close to the origin of replication have a better chance of ending up in a petite than the remainder of mtDNA.

ACKNOWLEDGEMENTS

Our work is in part supported by a grant from The Netherlands Foundation for Chemical Research (SON) with financial aid from The Netherlands Organization for the Advancement of Pure Research (ZWO).

REFERENCES

1 Locker, J., Rabinowitz, M. and Getz, G.S. (1974) J.Mol.Biol. 88, 489-502.
2 Kasamatsu, H., Grossman, L.I., Robberson, D.L., Watson, R. and Vinograd, J. (1974) Cold Spring Harbor Symp.Quant.Biol. 38, 281-288.
3 Berk, A.J. and Clayton, D.A. (1976) J.Mol.Biol. 100, 85-102.
4 Wolstenholme, D.R., Koike, K. and Cochran-Fouts, P. (1974) Cold Spring Harbor

Symp.Quant.Biol. 38, 267-280.
5 Borst, P. and Flavell, R.A. (1972) in Mitochondria: Biogenesis and Bioenergetics; Biomembranes: Molecular Arrangements and Transport Mechanisms (Van den Bergh, S.G., Borst, P., Van Deenen, L.L.M., Riemersma, J.C., Slater, E.C. and Tager, J.M., Eds), Vol. 28, North-Holland, Amsterdam, pp. 1-19.
6 Arnberg, A.C., Van Bruggen, E.F.J., Clegg, R.A., Upholt, W.B. and Borst, P. (1974) Biochim.Biophys.Acta 361, 266-276.
7 Goddard, J.M. and Cummings, D.J. (1975) J.Mol.Biol. 97, 593-609.
8 Hollenberg, C.P., Borst, P., Flavell, R.A., Van Kreijl, C.F., Van Bruggen, E.F.J. and Arnberg, A.C. (1972) Biochim.Biophys.Acta 277, 44-58.
9 Van Kreijl, C.F., Borst, P., Flavell, R.A. and Hollenberg, C.P. (1972) Biochim. Biophys.Acta 277, 61-70.
10 Mol, J.N.M., Borst, P., Grosveld, F.G. and Spencer, J.H. (1974) Biochim. Biophys.Acta 374, 115-128.
11 Van Kreijl, C.F. and Bos, J.L., unpublished results.
12 Mol, J.N.M. (1976) Ph.D.Thesis, The Analysis of (dA.dT) Clusters in DNAs, University of Amsterdam.
13 Sanders, J.P.M., Flavell, R.A., Borst, P. and Mol, J.N.M. (1973) Biochim. Biophys.Acta 312, 441-457.
14 Borst, P. (1974) in The Biogenesis of Mitochondria (Kroon, A.M. and Saccone, C., Eds), Academic Press, New York, pp. 147-156.
15 Gordon, P., Casey, J. and Rabinowitz, M. (1974) Biochemistry 13, 1067-1075.
16 Nagley, P., Sriprakash, K.S., Rytka, J., Choo, K.B., Trembath, M.K., Lukins, H.B. and Linnane, A.W. (1976) in The Genetic Function of Mitochondrial DNA (Saccone, C. and Kroon, A.M., Eds), North-Holland, Amsterdam, pp. 231-242.
17 Rabinowitz, M., Jakovcic, S., Martin, N., Hendler, F., Halbreich, A., Lewin, A. and Morimoto, R. (1976) - same as ref. 16 - pp. 219-230.
18 DiFranco, A., Sanders, J.P.M., Heyting, C., Borst, P. and Slonimski, P.P. (1976) - same as ref. 16 - pp. 291-304.
19 Heyting, C. and Sanders, J.P.M. (1976) - same as ref. 16 - pp. 273-280.
20 Locker, J. and Rabinowitz, M. (1976) - same as ref. 16 - pp. 313-324.
21 Faye, G., Fukuhara, H., Grandchamp, C., Lazowska, J., Michel, F., Casey, J., Getz, G.S., Locker, J., Rabinowitz, M., Bolotin-Fukuhara, M., Coen, D., Deutsch, J., Dujon, B., Netter, P. and Slonimski, P.P. (1973) Biochimie 55, 779-792.
22 Fukuhara, H. (1969) Europ.J.Biochem. 11, 135-139.
23 Hollenberg, C.P., Borst, P. and Van Bruggen, E.F.J. (1972) Biochim.Biophys.Acta 277, 35-43.
24 Nagley, P. and Linnane, A.W. (1972) J.Mol.Biol. 66, 181-193.
25 Hall, R.M., Nagley, P. and Linnane, A.W. (1976) Mol.Gen.Gen. 145, 169-175.
26 Grossman, L.I., Goldring, E.S. and Marmur, J. (1969) J.Mol.Biol. 46, 367-376.
27 Goldring, E.S., Grossman, L.I., Krupnick, D., Cryer, D.R. and Marmur, J. (1970) J.Mol.Biol. 52, 323-335.
28 Perlman, P.S. and Mahler, H.R. (1971) Nature New Biol. 231, 12-16.
29 Hollenberg, C.P., Borst, P. and Van Bruggen, E.F.J. (1970) Biochim.Biophys.Acta 209, 1-15.
30 Sanders, J.P.M., Borst, P. and Weijers, P.J. (1975) Mol.Gen.Gen. 143, 53-64.
31 Michels, C.A., Blamire, J., Goldfinger, B. and Marmur, J. (1974) J.Mol.Biol. 90, 431-449.
32 Lazowska, J. and Slonimski, P.P. (1976) Mol.Gen.Gen. 146, 61-78.
33 Michaelis, G., Petrochilo, E. and Slonimski, P.P. (1973) Mol.Gen.Gen. 123, 51-65.
34 Borst, P. (1975) in Molecular Biology of Nucleocytoplasmic Relationships (Puiseux-Dao, S., Ed.), Elsevier, Amsterdam, pp. 91-93.
35 Criddle, R.S., Wheelis, L., Trembath, M.K. and Linnane, A.W. (1976) Mol.Gen.Gen. 144, 263-272.
36 Perlman, P.S. (1975) in Genetics and Biogenesis of Mitochondria and Chloroplasts (Birky Jr., E.W., Perlman, P.S. and Byers, T.J., Eds), Ohio State University Press, Columbus, pp. 136.
37 Williamson, D.H. and Fennell, D.J. (1974) Mol.Gen.Gen. 131, 193-207.

Genetics and Biogenesis of Chloroplasts and Mitochondria, Th. Bücher et al. eds.
© 1976 Elsevier/North-Holland Biomedical Press, Amsterdam, The Netherlands.

A SEGMENT OF MITOCHONDRIAL DNA CARRYING OLIGOMYCIN RESISTANCE

Kazuhiko Wakabayashi
Faculty of Medicine, University of Tokyo,
Tokyo, Japan

INTRODUCTION

Mitochondrial oligomycin-resistance differs from erythromycin- and chloramphenicol-resistance on several points. The simultaneous loss of erythromycin- and chloramphenicol-resistance occurs frequently during the induction of petite, and its frequency is higher than the frequency of the loss of oligomycin resistance. Unequal recombinations occur between erythromycin- and chloramphenicol-resistance in crosses of mitochondria of different types, which do not, however, occur between erythromycin- and oligomycin-resistance. Thus the segment of mtDNA which carries oligomycin-resistance is studied in this report.

MATERIALS AND METHODS

Strains and culture media: Culture media and assays of antibiotic resistances in rho minus cells were described previously (1). A mitochondrial triple resistant haploid yeast, Saccharomyces cerevisiae 992 (a, ade,[oli],[ery],[cap]) was constructed from a cross of 909B and D2B (2) followed by a tetrad analysis.

Preparation of mitochondrial DNA: Protoplasts of cells were prepared as described previously (3). The protoplasts were homogenized, centrifuged at 15,000 x g for 30 minutes to remove the bulk of ribosomes. To the precipitates, suspended in SSC (0.15 M NaCl and 0.015 M sodium citrate) were added with Sarkosyl and $NaClO_4$ to make the final concentration at 3 % and 1 M respectively. After treatment with a mixture of chloroform and isoamylalcohol (24 : 1), mtDNA was precipitated from the water layer with cold ethanol. It was redissolved in 1 x SSC followed by the treatment with 30 μg of ribonuclease A, 0.3 μg of ribonuclease T_1, 150 μg of α amylase (per 1 g of cells respectively) at 30°C for 1.5 hour and by successive incubation with 600 μg of pronase (per 1 g of cells) for 1.5 hour. After treatment with a mixture of chloroform-isoamylalcohol in the same way, DNA was chromatographed with hydroxylapatite columns repeatedly essentially the same way as described by Bernardi et al.(4). Mitochondrial DNA was eluted with a phosphate buffer of continuous gradient at a concentration of about 0.27 M. In later experiments mtDNA was extracted from the mitochondrial fraction.

Melting curve: For precise measurement of the melting profile, a Shimazu double beam spectrophotometer UV 200 was used. A deuterium lamp for the 260 nm was stabilized using a custom-built stabilizer. Electrical noise in the out put of the spectrophotometer was cut using a custom-built low pass filter (below 0.3 Hz, -12 db per octave). Stability was thus kept within an absorbance range of 0.0005. The increase of the temperature for melting was controlled by a programmer Haake PG 11 at a rate of 0.3°C per minute. A thermister (Takara SSB) melted in a thin glass rod was immersed directly into DNA solution (1.2 x SSC) to be measured and was connected to a linearizer (Takara E311). An XY recorder was connected to both the linearizer and the spectrophotometer for automatic recordings of the melting profile. The absorbancy values were punched onto cards and processed by a HITAC 8800/8700 computer using 19th degree polynomials and quadruple precision.

ECoR I fragments: ECoR I endonuclease was obtained from Miles Laboratories Inc. About 1.5 µg of mtDNA was incubated with 0.8 µg of the enzyme in 10 mM Tris buffer pH 7.5 with 6.6 mM MgCl, 125 mM NaCl at 37°C for 1 hour. Additional 0.8 µg of the enzyme was added at 30 minutes to complete the cleavage. Electrophoresis was run with composite 0.6% polyacrylamide -0.7% agarose gel in 40 mM Tris acetate buffer pH 7.9 at 50 V, 1.6 mA per tube for 4 hours.

Electronmicrographs: Electronmicrographs were taken according to the method of Kleinschmidt et al.(5). Hitachi HS9 was used.

RESULTS

A spontaneous petite mutant, 992SPO, was isolated from a haploid of yeast, 992, which possesses mitochondrial triple resistance to oligomycin, erythromycin and chloramphenicol. The petite isolated retained oligomycin-resistance only and lost other resistances. After successive subclonings a stable oligomycin resistant petite was obtained. Mitochondrial DNA was isolated from protoplasts after repeated chromatography with hydroxylapatite. Thermal transition of mtDNA from the petite 992SPO and from the wild type strain 992 was studied. Changes of the absorbancies were calculated with a computer to obtain precise differential melting curves. It was found that mtDNA of the petite gave a reduced complexity. It showed one large peak between 71.5°C and 74.8°C with a shoulder and also another peak at 75.7°C, while it has very few components above 77°C. It differed significantly from the melting curve of mtDNA of 992, which gave several components above 77°C. The melting temperature of mtDNA of the petite was close to that of poly dAdT. The renaturation of mtDNA of the petite was complete and differed from that of wild type 992.

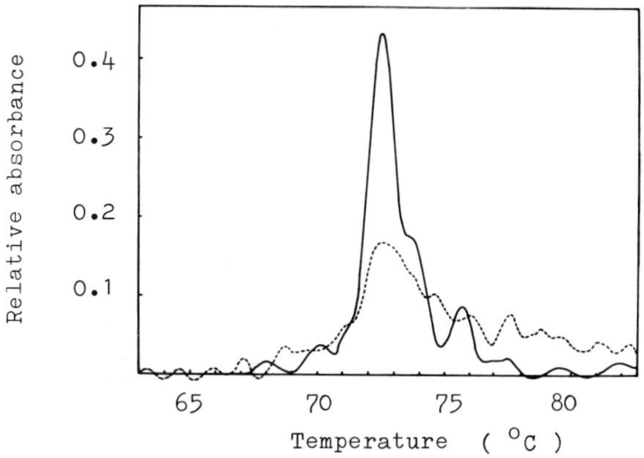

Fig. 1. Differential melting curves of mtDNA from the oligomycin resistant petite and the wild type strain. ——— 992SPO, - - - - - - 992.

It was observed that mtDNA of the petite gave a narrow band on gel electrophoresis, indicating that its size is homogeneous, whereas mtDNA of wild type 992 gave a broad band. When mtDNA of 992 was subjected to the cleavage by a restriction endonuclease ECoR I, at least five bands were observed, while mtDNA of 992SPO gave three bands (Plate 1). One of three fragments of the petite has no corresponding band in the fragments of mtDNA of wild type 992.

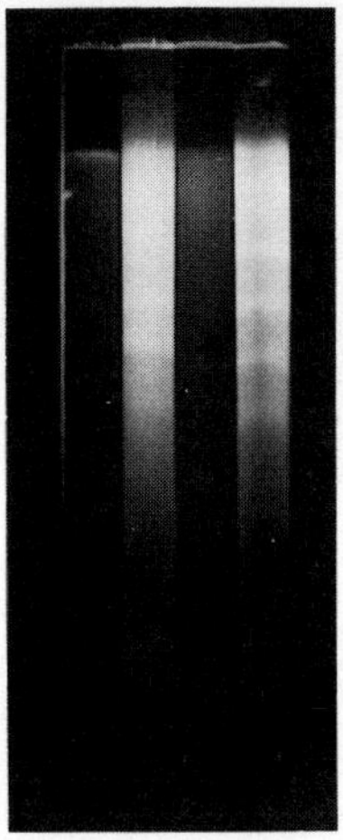

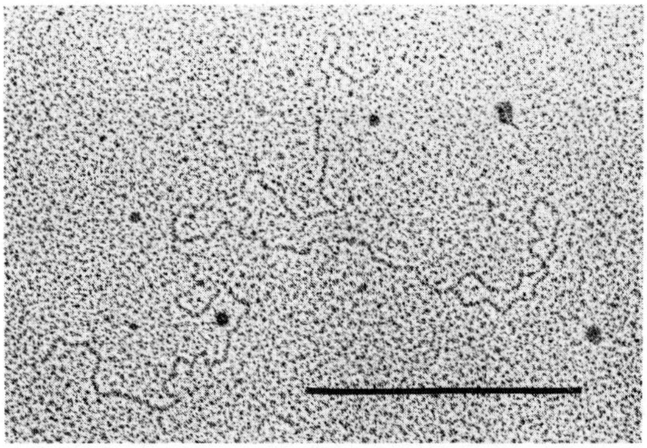

Plate 1. Gel electrophoresis of mtDNA stained by ethidium bromide and illuminated with u.v. light. 992SPO, 992SPO+ECoR I, 992, 992+ECoR I were shown from left to right.

Plate 2. Electronmicrograph of mtDNA of 992SPO. The bar indicates 1 μ.

The loss of the genetic markers, reduced heterogeneity in thermal transition as well as the loss of several ECoR I cleavage sites in mtDNA of the petite indicated the reduced size of the mitochondrial genome of the petite. The electrophoretic mobility of mtDNA of 992SPO, however, did not increase significantly. The nature of mtDNA of 992SPO with respect to renaturation and Tm, which is close to that of poly dAdT, supported the presence of the region having repeated sequences rich in AT. These sequences should be located in a segment close to the gene of oligomycin resistance. An erythromycin- and chloramphenicol-resistant petite gave melting profile of mtDNA close to that of wild type mtDNA. The amplification of the sequence may take place during the induction of petite. Such mechanism has been discussed (6,7). It is interesting that a lariat form of DNA was observed on mtDNA of 992SPO by electronmicroscope (Plate 2), which favours the rolling circle model.

ACKNOWLEDGEMENTS

The author thanks to Dr.M. Shioda in University of Tokyo for the electronmicrographic study and to Dr. S. Kamei in University of Tokyo for the isolation of a petite 992SPO.

SUMMARY

Mitochondrial DNA of a petite was studied, which possesses oligomycin-resistance and lost erythromycin- and chloramphenicol-resistance. It lost the heterogeneity in the melting profile and its Tm is close to that of synthetic poly dAdT. It also lost several ECoR I cleavage sites. The nature of the segment of mtDNA in the oligomycin resistant petite was discussed.

REFERENCES

1. Wakabayashi, K. (1974) J. Antibiotics 27, 373.
2. Wakabayashi, K. (1974) Proc. Japan Acad. 50, 396.
3. Wakabayashi, K., Takahashi,T., Kawai,K. and Hirosawa,K. (1972) J. Biochem. 72, 411.
4. Bernardi,G., Faures,M., Piperno,G. and Slonimski,P.P. (1970) J. Mol. Biol. 48, 23.
5. Kleinschmidt, A.K. and Zahn, R.K. (1959) Z. Naturfosch. B. 14, 770.
6. Sanders, J.M.P., Flavell, R.A., Borst, P. and Mol, J.N.M. (1973) Biochim. Biophy. Acta 312, 441.
7. Faye,G., Fukuhara,H., Grandchamp,C., Lazowska,J., Michel,F.,Casey,J., Getz,G.S., Locker,J., Rabinowitz,M., Bolotin-Hukuhara,M., Coen,D., Deutsch,J., Dujon,B., Netter,P. and Slonimski,P.P. (1973) Biochimie 55, 779.

Genetics and Biogenesis of Chloroplasts and Mitochondria, Th. Bücher et al. eds.
© 1976 Elsevier/North-Holland Biomedical Press, Amsterdam, The Netherlands.

GENE IDENTIFICATION BY COUPLED TRANSCRIPTION-TRANSLATION OF YEAST MITOCHONDRIAL DNA *in vitro*

A.F.M.MOORMAN AND L.A.GRIVELL
Section for Molecular Biology, Laboratory of Biochemistry, University of Amsterdam, Kruislaan 320, Amsterdam, The Netherlands

INTRODUCTION

Despite intensive genetic and biochemical analysis, much of the potential genetic information on yeast mtDNA remains unaccounted for. Genes for rRNAs, tRNAs and for the major products of mitochondrial protein synthesis occupy approximately 10-15% of the available information[1] and the question, therefore, arises whether the remaining sequences represent useful genetic information in terms of structural genes or regulatory elements, or whether they consist of nonsense accumulated in the course of evolution. This question has not been resolved by the recent construction of physical and genetic maps of this DNA[2-4]. Certainly, regions devoid of genetic markers and displaying a high degree of evolutionary variability exist and might represent "junk DNA"[5]. Further, loci simultaneously affecting both cytochrome _b_ and cytochrome _c_ oxidase formation (Box loci) have been found[6] and could be regulator sequences. However, elements of complexity have been added by the findings that at least seven genetically and physically separable loci are involved in the specification of the mitochondrial contribution to cytochrome _c_ oxidase and that some of these loci occupy segments of the genome considerably in excess of the coding requirements for polypeptides of molecular weights 42 000, 33 000 and 23 000 (refs 6,7).

Essential for our understanding of the organization of the mitocondrial genome is a better definition of the existing genetic loci in terms of structural genes and their associated regulatory elements. This could be achieved through isolation and characterization of new mutants. However, the finding[8] that with present selective techniques no new mt^--type mutations are isolable suggests that progress along these lines may be slow. An alternative approach to this problem consists of the identification and mapping of mitochondrial gene products by their synthesis _in vitro_ under the direction of mtDNA or defined segments thereof obtained either by restriction enzyme

---

Abbreviations: IgG, immunoglobulin G; SDS, sodium dodecyl sulphate; TCA, trichloroacetic acid.

cleavage, or from genetically-characterized petite mutants. In this article we describe this approach and its application to the characterization of the genes involved in the specification of cytochrome oxidase.

RESULTS AND DISCUSSION

Escherichia coli cell-free extracts transcribe and translate yeast mtDNA in vitro

An ideal system for the transcription-translation of mtDNA would be one derived from mitochondria. In our hands, however, mitochondrial extracts prepared in a variety of ways have not only proved to be completely inactive, even in the translation of poly(U), but are strongly inhibitory when added to active *E. coli* extracts, perhaps due to the presence of highly active nucleases and proteases (Moorman, A.F.M., unpublished).

Yeast mtDNA will act as template, however, in cell-free extracts derived from *E. coli*, provided that these are first depleted of endogenous DNA. This is most readily achieved by preincubation of the extracts with deoxyribonuclease immobilised on Sepharose, which can be subsequently removed by a simple centrifugation step[9]. Under optimal conditions, yeast mtDNA stimulates incorporation of (^{35}S)methionine into protein by a factor of 30-fold (Table I). Activity is linearly dependent on added DNA up to a concentration of 40 µg/ml (data not shown) and is completely dependent on transcription as shown by the strong inhibitory effect of rifampicin.

TABLE I

TRANSCRIPTION-TRANSLATION OF mtDNA *in vitro*

For details of protein synthesis and *E. coli* S-30 extract preparation see refs 9 and 10. mtDNA was prepared according to the procedure of Sanders *et al.* (11). Protein synthesis was for 15 min at 36°C. (^{35}S)Methionine (42 Ci/mmole) was 4.3 µM. Hot TCA-precipitable radioactivity was counted at 50% efficiency. Values given have been corrected for a zero time incorporation of 4873 cpm. Stimulation has been calculated as the increase of incorporation due to mtDNA compared with endogenous incorporation.

Additions	Incorporation (cpm/10 µl)	Stimulation due to mtDNA
None	5 213	
S. cerevisiae mtDNA (38 µg/ml)	164 274	30
S. cerevisiae mtDNA + rifampicin (2 x 10^{-5} M)	10 330	1

Yeast mtDNA directs the synthesis of material immunologically reactive as cytochrome c oxidase

Table II shows that a fraction of the product synthesized under the direction of yeast mtDNA is reactive with antibody raised against purified yeast cytochrome *c* oxidase. In a series of six independent experiments, 7.8 ± 0.8% of incorporated radioactivity could be precipitated in an indirect precipitation test by an anti-cytochrome *c* oxidase IgG fraction, whereas a control IgG preparation precipitated only 1.4 ± 0.6%. The interaction between *in vitro* product and anti-cytochrome *c* oxidase appears to be specific by three criteria:

1. As Table II shows, products synthesized under the direction of bacteriophage T_7 DNA show low precipitation with both control and anti-cytochrome *c* oxidase preparations (1.4 ± 0.8% and 1.5 ± 0.8%, respectively).

2. Addition of excess yeast, but not beef-heart cytochrome *c* oxidase, reduces radioactivity in the immunoprecipitate containing mitochondrial products by 80-90% (data not shown).

3. *In vitro* synthesized products react positively with antibodies directed against purified subunits I and II of cytochrome *c* oxidase (see below).

TABLE II

IMMUNOLOGICAL CHARACTERIZATION OF PRODUCTS SYNTHESIZED *in vitro*

The double precipitation was carried out as described in detail elsewhere (10). The input of *in vitro* product varied between 100 000-300 000 cpm per immunoprecipitation. Values given are the mean of six independent protein synthesis experiments and immunoprecipitations.

DNA added	% Precipitable by	
	Control IgG	Anti-cytochrome *c* oxidase IgG
None	1.7 ± 0.4	3.3 ± 0.5
mtDNA	1.4 ± 0.6	7.8 ± 0.8
Phage T_7 DNA	1.4 ± 0.8	1.5 ± 0.8

Table II also shows that material synthesized in the absence of added DNA consistently gives a low positive reaction with anti-cytochrome *c* oxidase. This is not serious in our system, since due to the high stimulation by mtDNA, endogenous products should represent only 4.5% of the total, but this could be a problem in systems where response to added templates is poorer (*cf*. ref. 12).

It is worth noting that the type of immunoprecipitation assay used has three main advantages over the more widely used direct precipitation test:

1. Addition of carrier antigen for formation of networks is unnecessary, so that artefacts due to aggregation (see below), or to competition of the added antigen with in vitro synthesized polypeptides can be avoided.

2. Aspecific precipitation, due to trapping of radioactivity in networks can be assessed, since both the composition and size of the precipitates is unaffected by the specificity of the primary antibody.

3. The synthesis of incomplete polypeptides can be detected.

Electrophoretic characterization of the in vitro product

Despite the success of the immunological approach to the characterization of the in vitro synthesized products, it is not yet clear whether the E. coli system is capable of the synthesis of a complete mitochondrial protein. Analysis of the products by SDS-acrylamide gel electrophoresis has been complicated by the fact that, in a conventional system such as the one described by Laemmli and Favre[13], specific aggregation takes place between newly-synthesized mitochondrial polypeptides and proteins present in E. coli extracts[14]. Less than 10% of the incorporated radioactivity is present in bands with M_r = 10-100 000 and mixing experiments show that the banding pattern can be influenced by the amount of E. coli proteins present in samples, or by addition of mitochondrial proteins as carrier or as internal markers.

Since pre-treatment of mitochondrial product with either urea or alkali was also without effect on the extent of aggregation (Moorman, A.F.M., unpublished), we turned to electrophoresis in SDS-urea in the hope of getting round this problem. Plate I compares the results obtained in this system with those in the conventional Laemmli system. A reduction in the amount of aggregation is indicated by the considerably simplified banding pattern (compare tracks b) and e)). Mixing experiments demonstrate, however, that protein:protein interactions are still present in that the distribution of radioactivity can be influenced by addition of E. coli or mitochondrial protein (track f)), or E. coli protein (data not shown). Although this aggregation phenomenon limits considerably the application of the coupled transcription-translation system to the identification of as yet uncharacterized mitochondrial gene products, it could conceivably be of use in the screening of different yeast strains for variants in the

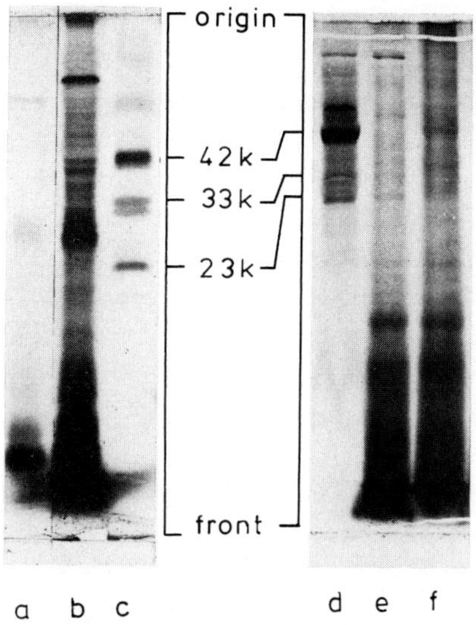

Plate I. Electrophoretic analysis of the (^{35}S)methionine labelled products synthesized in vitro by coupled transcription-translation of S. cerevisiae mtDNA. Sample preparation and electrophoresis was carried out according to the procedure of Laemmli and Favre (13) as described (10), except that sample mix, stacking gel and resolving gel of tracks d-f contained 8 M urea (Schwartz) extra. Tracks a-c represent a 15% gel and tracks d-f a 13.5% gel. a) Endogenous control; b) and d) in vitro products of S. cerevisiae mtDNA; c) and e) in vivo products of mitochondrial protein synthesis; submitochondrial particles were prepared from yeast cells pulse-labelled in the presence of cycloheximide (15); f) unlabelled submitochondrial particles mixed with labelled in vitro product of S. cerevisiae mtDNA.

products of mitochondrial protein synthesis, by providing a basis for phenotypic variation[16].

Genes for cytochrome c oxidase: The nature of the oxi loci

Of the many mt^- mutants isolated and characterized by Tzagoloff and coworkers[17], those affecting the formation of cytochrome c oxidase are probably the most interesting. So far, seven distinct loci have been found to be involved, of which three (oxi 1-3) are specific for cytochrome oxidase alone[6-8]. The exact relationship of these loci to the structural genes for the mitochondrially specified subunits of cytochrome c oxidase is not known and, indeed, is difficult to visualize in the light of the findings that:

1. The loci are widely spaced on the genome[7].
2. Each of the loci extends across segments of the genome far in excess of the coding requirements for three polypeptides of molecular weights 42 000, 33 000 and 23 000 (ref. 7).
3. Mutations at different loci result in apparently identical

phenotypes, namely the loss of the largest subunit of the cytochrome c oxidase complex[17].

In the hope of clarifying this situation, we have prepared mtDNA from mutants at the oxi loci and examined the products synthesized under their direction by immunoprecipitation with antisera directed against anti-cytochrome c oxidase holoenzyme, purified subunit II and against a mixture of subunits I and II (ref. 18). As controls on the precipitation, mtDNA from the wild-type strain (D273-10B) and from a mutant at the COB locus (M13-101) were used. The mutants used probably arise by small deletions of the wild-type genome. First, none of them reverts spontaneously; second, Rabinowitz et al.[19] have found, for mutant M5-16, alterations in the cleavage pattern produced by the restriction endonuclease Haemophilus haemolyticus I, consistent with a deletion of 4500 base pairs.

Table III shows that all mutant DNAs give products which react positively with all antisera. Both M13-101 (COB) and M9-94 (oxi-1) give precipitation values closely resembling those of the wild type. M11-15 (oxi-2) and M5-16 (oxi-3), on the other hand, give slightly reduced values with both anti-holoenzyme and anti-subunits I+II, but this reduction is probably not significant, in view of the scatter in the data. For M9-94 these results are consistent with data obtained in pulse-labelling experiments of intact cells in the presence of cycloheximide, since like all other mutants so far isolated

TABLE III

IMMUNOLOGICAL CHARACTERIZATION OF PRODUCTS SYNTHESIZED *in vitro* UNDER THE DIRECTION OF mtDNA FROM WILD-TYPE AND OXI$^-$ STRAINS

See the legend to Tables I and II for details of protein synthesis and immunoprecipitation, respectively. (^{35}S)Methionine (143 Ci/mmole) was 4.4 µM. Immunoprecipitates were dissolved for 10 min in 200 µl 0.5 M NaOH and then precipitated with 5% TCA on glass fibre discs. Values have been corrected for precipitation found with control sera (0.6-1.6%) and represent the mean of two independent experiments.

mtDNA	TCA-precipitable radioactivity per 5 µl	% Precipitable by		
		Anti-holoenzyme	Anti-subunits I+II	Anti-subunit II
M9-94 (oxi-1)	310 510	3.3	3.3	3.2
M11-15 (oxi-2)	272 290	2.9	2.8	3.3
M5-16 (oxi-3)	250 630	2.4	2.8	2.5
M13-101 (COB)	292 120	3.6	3.7	3.5
D273-10B (w.t.)	289 680	3.4	3.5	2.8

at this locus, this mutant displays a normal pattern of mitochondrial products[17]. This result is also in line with our finding[14] that mtDNA from the petite mutant F41, which retains the oxi-1 locus, as judged from its ability to restore wild-type function to M9-94 (Heyting, C., unpublished), fails to direct the synthesis of antigenic determinants of cytochrome oxidase. Nevertheless, the function of the oxi-1 locus and its relationship to the structural genes for cytochrome c oxidase remains enigmatic.

The positive reaction of products of M11-15 and M5-16 with all antisera is unexpected, however. Both mutants have been found to be deficient in the 42 000 molecular weight subunit of cytochrome c oxidase[17]. Possible explanations for this discrepancy are:

1. Both loci are regulatory genes concerned with the integration of subunit I into active holoenzyme. Absence of subunit I from intact cells results from rapid degradation of unintegrated polypeptide.

2. One of the loci is the structural gene for subunit I of cytochrome c oxidase. The mutant gene directs the synthesis of an aberrant or incomplete protein, which in vitro is immunologically reactive, but which in vivo fails to be integrated and hence is rapidly degraded.

Resolution of these two possibilities should, of course, be possible through further characterization of these and similar mutants. Additionally, localization of the genes for subunits of cytochrome c oxidase is now in progress, using both genetically-marked petite mtDNAs and restriction enzyme fragments of mtDNA to programme protein synthesis in vitro, followed by identification of products by means of subunit-specific antisera. This should provide independent confirmation of the nature of these loci and, when extended to the study of other mt^- mutations and mitochondrial gene products, will permit the construction of a fine-structure map of the mitochondrial genome. We trust that the results will be both interesting and informative.

ACKNOWLEDGEMENTS

We thank Professor P.Borst and Dr. G.S.P.Groot for helpful discussions and criticism; Dr. R.O.Poyton for generous gifts of antisera; Dr. A.Tzagoloff for providing the mutants used; Mrs C.M.Meiland for performing the experiments presented in Table II; Mrs N.Van Harten-Loosbroek for submitochondrial particles and Mrs G.Van Dam for technical assistance. This work was supported in part by a grant to P.Borst from The Netherlands Foundation for Chemical Research (SON) with financial aid from The Netherlands Organization for the Advancement of Pure Research (ZWO).

REFERENCES

1 Borst, P., Sanders, J.P.M. and Heyting, C. (1976) in EURATOM Symposium on Mitochondrial Genetics, Biogenesis and Bioenergetics, De Gruyter, Berlin, in press.

2 Sanders, J.P.M., Heyting, C., DiFranco, A., Borst, P. and Slonimski, P.P. (1976) in The Genetic Function of Mitochondrial DNA (Saccone, C. and Kroon, A.M., Eds), North-Holland, Amsterdam, pp. 259-272.

3 Nagley, P., Sriprakash, K.S., Rytka, J., Choo, K.B., Trembath, M.K., Lukins, H. B. and Linnane, A.W. (1976) - same as ref. 2 - pp. 231-242.

4 Schweyen, R.J., Weiss-Brummer, B., Backhaus, B. and Kaudewitz, F. (1976) - same as ref. 2 - pp. 251-258.

5 Sanders, J.P.M., Heyting, C. and Borst, P. (1976) This volume.

6 Kotylak, Z. and Slonimski, P.P. (1976) - same as ref. 2 - pp. 143-154.

7 Slonimski, P.P. and Tzagoloff, A. (1976) Europ.J.Biochem. 61, 27-41.

8 Tzagoloff, A., Foury, F. and Akai, A. (1976) - same as ref. 2 - pp. 155-161.

9 Moorman, A.F.M., Lamie, F. and Grivell, L.A. (1976) FEBS Letters, submitted.

10 Moorman, A.F.M., Grivell, L.A., Katan, M.B., Van Ommen, G.J.B. and Meiland, C. M. (1976) in EURATOM Symposium on Mitochondrial Genetics, Biogenesis and Bioenergetics, De Gruyter, Berlin, in press.

11 Sanders, J.P.M., Weijers, P.J., Groot, G.S.P. and Borst, P. (1974) Biochim. Biophys.Acta 374, 136-144.

12 Scragg, A.H. and Thomas, D.Y. (1975) Europ.J.Biochem. 56, 183-192.

13 Laemmli, N.K. and Favre, M. (1973) J.Mol.Biol. 80, 575-599.

14 Moorman, A.F.M. and Grivell, L.A. (1976) - same as ref. 2 - pp. 281-289.

15 Groot, G.S.P. and Van Harten-Loosbroek, N. (1976) in EURATOM Symposium on Mitochondrial Genetics, Biogenesis and Bioenergetics, De Gruyter, Berlin, in press.

16 Douglas, M.G., Kendrick, E., Boulikas, P., Perlman, P. and Butow, R.A. (1976) - same as ref. 2 - pp. 199-207.

17 Tzagoloff, A., Akai, A., Needleman, R.B. and Zulch, G. (1975) J.Biol.Chem. 250, 8236-8242.

18 Poyton, R.O. and Schatz, G. (1975) J.Biol.Chem. 250, 762-766.

19 Rabinowitz, M., Jakovcic, S., Martin, N., Hendler, F., Halbreich, A., Lewin, A. and Morimoto, R. (1976) - same as ref. 2 - pp. 219-230.

Genetics and Biogenesis of Chloroplasts and Mitochondria, Th. Bücher et al. eds.

Mitochondrial mutations that affect mitochondrial transfer ribonucleic acid in Saccharomyces cerevisiae.

G. Faye, M. Bolotin-Fukuhara[x] and H. Fukuhara
Fondation Curie Institut du Radium, Section de Biologie, Orsay, 91405, France.

[x]Laboratoire de Biologie Générale, Université Paris-Sud, Orsay, 91405, France.

1. Search for mitochondrial transfer RNA mutations

A large number of new mitochondrial mutations (mit$^-$ mutations) have been described recently by several laboratories (1, 2, 3, 4, 5). Some of these mutations have been shown to affect specifically the respiratory chain components of mitochondria. In this paper we will describe another type of mitochondrial mutations which appear to modify specifically the mitochondrial transfer RNA (tRNA) genes.

We have previously identified mitochondrial DNA-coded tRNAs for sixteen different amino-acids (5, 6). Most of the tRNA genes have been mapped between the two mitochondrial genetic markers, C and P (7) ; the positions of the tRNA genes relative to several mit$^-$ mutations have also been determined (5). To confirm and extend these studies, we have been looking for mitochondrial mutations which modify the mitochondrial tRNA genes. Since we have no direct way of screening tRNA mutants, we have adopted a somewnat unorthodox way of detecting tRNA mutations : first find mit$^-$ mutations that map within the mitochondrial DNA segment carrying most of the tRNA genes, then examine individually the mitochondrial tRNA species in these mutants. This approach looked tedious a priori, but, since the number of mutations found in the tRNA region is small, the analysis of tRNA is not the limiting step in practice.

To map mit$^-$ mutations, we have been using a set of eighty established rho$^-$ clones for which the extent of deletion in the mitochondrial genome is known in terms of five drug resistance markers (C, E, O_I, O_{II} and P). When a mit$^-$ mutant is crossed to rho$^-$ mutants which retained the DNA sequence allelic to the mit$^-$ mutation, respiratory competent recombinant cells are produced as detected by positive growth on glycerol medium (cf. 8) ; the mutation can be, therefore, located with respect to the drug resistance markers.

This procedure, although not devoid of uncertainties due to multiple deletions in rho$^-$ mutants, provides a rapid and convenient way of locating mit$^-$ mutations.

Recently we have mapped by this procedure about 200 mit$^-$ mutations (including temperature-sensitive and cold-sensitive mutants) with respect to drug resistance markers and tRNA genes (5). Many of these mutations have been localized on the already known mit$^-$ loci (oxi-1, oxi-2, oxi-3 etc, cf. 8). As shown in Table 1, the mit$^-$ mutants can be easily distinguished from each other by the crosses made with the standard set of rho$^-$ testers. The pattern of restoration of growth on glycerol is characteristic for each mit$^-$ loci. Several mutants, however, showed quite different and new patterns : they restored growth on glycerol in many of the crosses made with the rho$^-$ testers carrying C, CE or CEP markers. Since the region covering the P-C-E segment is known to carry most of the tRNA genes, this class of mutants could be candidates for a tRNA mutation. The growth restoration test was therefore extended in more detail using the rho$^-$ testers which have retained known tRNA genes.

2. Mapping of the mutations 170, M7-37 and ts832.

We have specifically studied three mit$^-$ mutations that showed very similar patterns of growth restoration in the test presented in table 1. The mutants (170, M7-37 and ts832) were crossed to a series of rho$^-$ testers for which the presence or absence of individual tRNA genes is already known (7). From the patterns of growth restoration, the positions of mit$^-$ mutations related to tRNA genes were deduced (for examples of the deletion mapping, see 7).

Figure 1 shows the result of such experiments. All of the three mutations were located in the region carrying phenylalanyl, alanyl, tyrosyl, aspartyl and arginyl tRNA genes (the order of these genes is not yet established). So far no other mitochondrial mutations are known in this region. We do not know how large this region is in the total mitochondrial genome, but an additional experiment indicated that the three mit$^-$ mutations are indeed very close to each other : when they were crossed with each other respiratory competent recombinants appeared with a very low but significant frequency (less than 0.02 %).

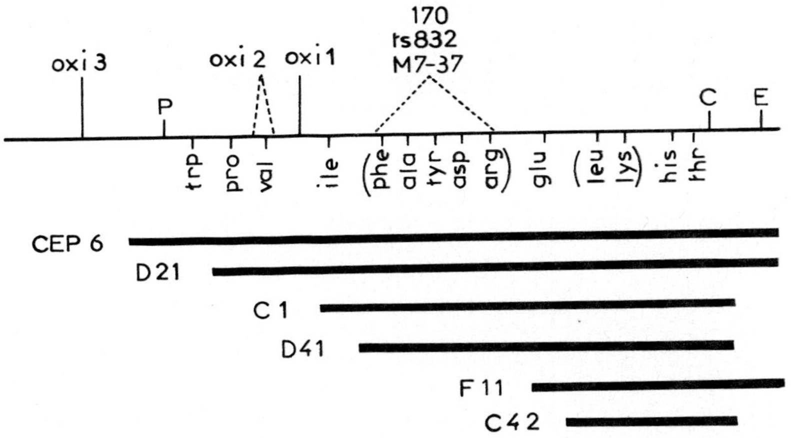

Figure 1. Positions of some mit$^-$ mutations with respect to mitochondrial tRNA genes.

The mitochondrial DNA sequences present in the rho$^-$ testers used for the mapping are represented by heavy lines. Distances between individual genes are arbitrary. Genes in parentheses have not been ordered. The gene trp may be to the left of P, and thr to the right of C.

3. <u>Mutation of mitochondrial aspartyl tRNA gene</u>

The three mit$^-$ mutations described above were analysed for their mitochondrial tRNAs. Highly purified mitochondria (cf. 9) were isolated from the mutants, and mitochondrial tRNA was extracted by phenol and LiCl. The preparations contained very little cytoplasmic tRNA species as checked by cytoplasmic acylation enzyme and RPC 5 chromatography of some ^{3}H aminoacylated tRNAs. The mitochondrial tRNAs were examined for their ability to accept individual ^{3}H-labelled amino-acids using a mitochondrial acylation enzyme isolated from purified wild type mitochondria. The extent of acylation was compared between the wild type mitochondrial tRNA and the mutant mitochondrial tRNA, under the conditions where the ^{3}H amino-acid uptake was proportional to tRNA input. Figure 2 illustrates some examples of the test.

The mitochondrial tRNA from the three mutants could be charged efficiently by all ^{3}H-aminoacids tested (phe, ala, tyr, thr, leu, val, pro, arg, ile, glu, his and lys) except by ^{3}H aspartic acid. The charge of the mutant 170 tRNA by this amino-acid was less than 0.5 % of the wild type tRNA for the same amounts of tRNA added. This low charge may even be due to a trace amount of cytoplasmic tRNA

Table 1

class	testers / strains	CEP1	CEP2	CEP3	CEP4	CEP5	CEP6	CEP7	CEP8	CEP9	CEP10	CEP11	CEP12
rho^-	C	R	R	R	R	R	R	R	R	R	R	R	R
	E	R	R	R	R	R	R	R	R	R	R	R	R
class markers	0_I	0	0	0	0	0	0	0	0	0	0	0	0
	0_{II}	0	0	0	0	0	0	0	0	0	S	0	S
	P^{II}	R	R	R	R	R	R	R	R	R	R	R	R
oxi-1	179	+	+	−	−	−	+	+	+	+	+	+	+
oxi-2	231	+	+	−	−	−	+	+	+	+	+	+	+
oxi-3	140	−	−	−	−	−	−	−	−	−	+	−	+
	134	−	−	−	−	−	−	−	−	−	+	−	+
	111	−	−	−	−	−	−	−	−	−	+	−	+
COB	226	−	−	−	−	−	−	−	−	−	−	−	−
	185	−	−	−	−	−	−	−	−	−	−	−	−
ATPase	C2-3	−	−	−	−	−	−	−	−	−	+	−	+
	b2-1	−	−	−	−	−	−	−	−	−	+	−	+
New	170	+	+	−	+	−	+	+	+	+	+	+	+
	ts832	+	+	−	+	−	+	+	+	+	+	+	+
	M7-37	+	+	−	+	−	+	+	+	+	+	+	+
	ts984	+	+	−	+	−	+	+	+	+	+	+	+
	cs909	+	+	+	−	+	+	+	+	+	+	+	+
	tcs902	+	+	+	+	+	+	+	+	+	+	+	+
	cs901	+	+	+	±	+	+	+	+	+	+	+	+
	cs990	−	−	−	−	−	−	−	−	−	−	−	−

class	testers / strains	CE01	CE02	CE03	CE04	CE05	CE06	CE07	CE08	CE09	CE010/2	CE011	CE012	CE013
rho^-	C	R	R	R	R	R	R	R	R	R	R	R	R	R
	E	R	R	R	R	R	R	R	R	R	R	R	R	R
class markers	0_I	R	R	R	R	R	R	R	R	R	R	R	R	R
	0_{II}	0	0	0	0	0	0	0	0	0	0	0	0	0
	P^{II}	0	0	0	0	0	0	0	0	0	0	0	0	0
oxi-1	179	−	−	+	−	+	−	−	−	−	−	−	−	−
oxi-2	231	−	−	+	−	−	−	−	−	−	−	−	−	−
oxi-3	140	−	−	−	−	−	−	−	−	−	−	−	−	−
	134	−	−	−	−	−	−	−	−	−	−	−	−	−
	111	−	−	−	−	−	−	−	−	−	−	−	−	−
COB	226	+	+	−	+	−	−	−	+	−	+	−	+	+
	185	+	+	−	+	−	−	−	−	−	+	−	−	−
ATPase	C2-3	−	−	−	−	−	−	−	−	−	−	−	−	−
	b2-1	−	−	−	−	−	−	−	−	−	−	−	−	−
New	170	−	+	+	−	+	+	−	+	+	+	+	+	+
	ts832	−	+	+	−	+	+	−	+	+	+	+	+	+
	M7-37	−	+	+	−	+	+	−	+	+	+	+	+	+
	ts984	−	−	+	−	+	−	−	+	+	−	+	+	−
	cs909	−	+	+	−	+	+	+	−	+	+	+	+	+
	tcs902	+	+	+	+	+	+	+	+	+	+	+	+	+
	cs901	+	+	+	+	+	+	+	±	+	+	+	+	+
	cs990	−	+	−	+	−	+	−	−	+	+	+	+	+

class	testers / strains	CE1	CE2	CE7	CE8	CE13	CE19	CE03/CE2	CE010/1	E01	P1	P2	P5	P9	P14	P22	P25	P27	Q1171
rho^-	C	R	R	R	R	R	R	R	R	0	0	0	0	0	0	0	0	0	0
	E	R	R	R	R	R	R	R	R	R	0	0	0	0	0	0	0	0	0
class markers	0_I	0	0	0	0	0	0	0	0	R	0	0	0	0	0	0	0	0	0
	0_{II}	0	0	0	0	0	0	0	0	0	0	0	0	0	0	0	0	0	0
	P^{II}	0	0	0	0	0	0	0	0	0	R	R	R	R	R	R	R	R	R
oxi-1	179	−	+	+	+	−	+	+	−	−	−	−	−	+	−	−	−	−	−
oxi-2	231	−	+	−	+	−	+	+	−	−	+	−	−	+	−	−	−	+	+
oxi-3	140	−	−	−	−	−	−	−	−	−	−	−	−	−	−	−	−	−	−
	134	−	−	−	−	−	−	−	−	−	−	−	−	−	−	−	−	−	−
	111	−	−	−	−	−	−	−	−	−	−	−	−	−	−	−	−	−	−
COB	226	−	−	−	−	+	−	−	−	−	−	−	−	−	−	−	−	−	−
	185	−	−	−	−	−	−	−	−	−	−	−	−	−	−	−	−	−	−
ATPase	C2-3	−	−	−	−	−	−	−	−	−	−	−	−	−	−	−	−	−	−
	b2-1	−	−	−	−	−	−	−	−	−	−	−	−	−	−	−	−	−	−
New	170	−	+	+	+	+	+	+	+	−	−	−	−	−	−	−	−	−	−
	ts832	−	+	+	+	+	+	+	+	−	−	−	−	−	−	−	−	−	−
	M7-37	−	+	+	+	+	+	−	+	−	−	−	−	−	−	−	−	−	−
	ts984	−	+	+	+	−	+	+	−	−	−	−	−	−	−	−	−	−	−
	cs909	+	+	+	+	+	+	+	+	−	−	−	−	−	−	−	−	−	−
	tcs902	+	+	+	+	+	+	+	+	−	−	−	−	−	−	−	−	−	−
	cs901	+	+	+	+	+	+	+	+	−	−	−	−	−	−	−	−	−	−
	cs990	−	−	−	−	+	−	−	−	+	−	−	−	−	−	−	−	−	−

class	testers / strains	P3	P12	$0_{II}P1$	$0_{II}P2$	Q3122
rho^-	C	0	0	0	0	0
	E	0	0	0	0	0
class markers	0_I	0	0	0	0	0
	0_{II}	S	S	S	S	S
	P^{II}	R	R	R	R	R
oxi-1	179	+	−	−	−	−
oxi-2	231	+	−	−	−	−
oxi-3	140	+	+	−	−	+
	134	+	+	−	+	+
	111	+	+	−	+	+
COB	226	−	−	−	−	−
	185	−	−	−	−	−
ATPase	C2-3	+	+	+	+	+
	b2-1	+	+	+	+	+
New	170	−	−	−	−	−
	ts832	−	−	−	−	−
	M7-37	−	−	−	−	−
	ts984	±	±	−	−	−
	cs909	−	−	−	−	−
	tcs902	−	−	−	−	−
	cs901	−	−	−	−	−
	cs990	−	−	−	−	−

Table 1
continued

class	testers strains	rho⁻ markers	O_IP1	O_IP2	$O_IO_{II}1$	$O_IO_{II}2$	$O_IO_{II}3$	$O_IO_{II}4$	O_I10	O_I13	O_I3	O_I5	O_I7	O_I8	O_I9	O_I11	O_I12	O_I14	$O_{II}1$	$O_{II}2$	$O_{II}3$	$O_{II}4$	$O_{II}5$	C 4	C 6	C 7
		C	0	0	0	0	0	0	0	0	0	0	0	0	0	0	0	0	0	0	0	0	0	R	R	R
		E	0	0	0	0	0	0	0	0	0	0	0	0	0	0	0	0	0	0	0	0	0	0	0	0
		O_I	R	R	R	R	R	R	R	R	R	R	R	R	R	R	R	R	0	0	0	0	0	0	0	0
		O_{II}	0	S	S	S	S	S	S	?	0	0	0	0	0	0	0	0	R	R	R	R	R	0	0	0
		P	R	R	0	0	0	0	0	0	0	0	0	0	0	0	0	0	0	0	0	0	0	0	0	0
oxi-1	179		-	-	-	-	-	-	-	-	-	-	-	-	-	-	-	-	-	-	-	-	-	-	-	-
oxi-2	231		+	+	+	+	-	+	-	-	-	-	-	-	-	-	-	-	-	-	-	-	-	-	-	-
oxi-3	140		-	+	+	+	+	-	-	-	-	-	-	-	-	+	-	-	-	-	-	-	-	-	-	-
	134		-	+	+	+	+	-	-	-	-	-	-	-	-	+	-	-	-	-	-	-	+	-	-	-
	111		-	+	+	+	+	-	+	-	-	-	-	-	-	+	-	-	-	-	-	-	+	-	-	-
COB	226		-	+	+	+	+	+	+	+	+	+	+	+	+	+	+	-	-	-	-	-	-	-	-	-
	185		-	+	+	+	+	+	+	+	+	+	+	+	+	+	+	-	-	-	-	-	+	-	-	-
ATPase	c2-3		-	+	+	+	+	+	+	+	-	-	-	-	-	-	-	-	+	+	+	+	+	-	-	-
	b2-1		-	+	+	+	+	+	+	-	-	-	-	-	-	-	-	-	+	+	+	+	+	-	-	-
New	170		-	-	-	-	-	-	-	-	-	-	-	-	-	-	-	-	-	-	-	-	-	-	-	-
	ts832		-	-	-	-	-	-	-	-	-	-	-	-	-	-	-	-	-	-	-	-	-	-	-	-
	M7-37		-	-	-	-	-	-	-	-	-	-	-	-	-	-	-	-	-	-	-	-	-	-	-	-
	ts984		-	-	-	-	-	-	-	-	-	-	-	-	-	-	-	-	-	-	-	-	-	-	-	-
	cs909		-	-	-	-	-	-	-	-	-	-	-	-	-	-	-	-	-	-	-	-	-	-	-	-
	tcs902		-	-	-	-	-	-	-	-	-	-	-	-	-	-	-	-	-	-	-	-	-	-	-	-
	cs901		-	-	-	-	-	-	-	-	-	-	-	-	-	-	-	-	-	-	-	-	-	+	+	+
	cs990		+	+	-	-	+	+	+	+	+	+	+	+	±	+	+	-	-	-	-	-	-	-	-	-

Table 1. Mit⁻ x rho⁻ crosses and restoration of growth on glycerol.
Most of the mit⁻ mutants were derived from a grande strain (FF 1210-6C, αura). Each mutant was crossed to the rho- testers listed on the top line. These testers have known genetic markers indicated on the second line. C, E, O_I and O_{II} and P designate drug resistance loci ; R (resistance), S (sensitivity) and O (deletion) are the allelic states of the marker mutations in the loci (see 7). The diploids issued from the crosses were replicated on complete glycerol agar plates and the growth was scored (+ and -). All crosses were repeated twice.
M7-37 was isolated by A. Tzagoloff from a different grande strain.

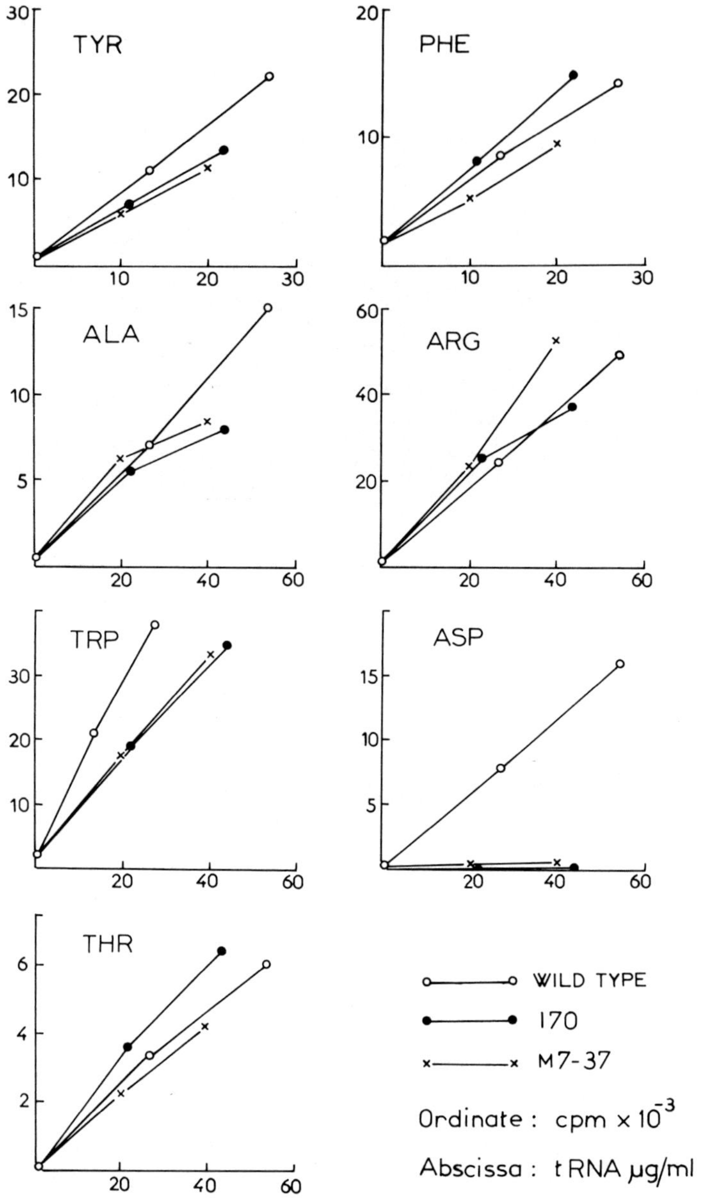

Figure 2. Acylation of mitochondrial tRNA with ^{3}H aminoacids.

The reaction conditions are essentially the same as those described earlier (6). The aminoacylated tRNA was precipitated with 5 % cold trichloracetic acid, washed and counted for radioactivity.

species or a non-specific charge. The charge of the mutant M7-37 mitochondrial tRNA was also very low (about 6 % of the wild type tRNA). The exact nature of this charge is under current study.

In the case of the temperature sensitive mutant ts832, the mitochondrial tRNA was prepared from a culture grown at permissive temperature (on galactose medium), and the acylation was tested at both permissive and non permissive (37°) temperatures. We found again that only the formation of aspartyl tRNA was very low (about 1 % of the wild type tRNA) even at the permissive temperature in the in vitro system. In all cases, the acylation of cytoplasmic aspartyl tRNA by cytoplasmic enzyme was found normal.

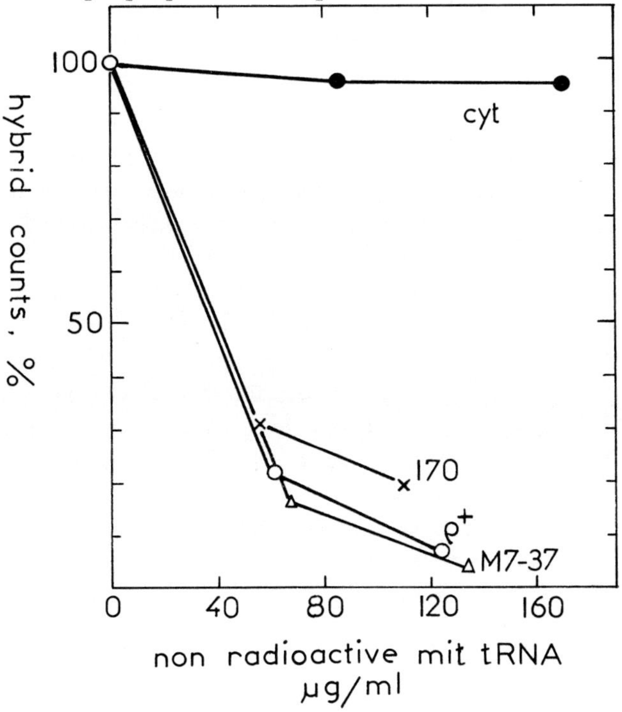

Figure 3. Competition hybridisation of aspartyl tRNA on mitochondrial DNA.

The conditions of hybridisation has been described earlier (6). ^{3}H-aspartyl tRNA (from the wild type mitochondria, 2182 cpm/μg), at constant and saturating concentration (16.5 μg/ml), was hybridized to the filter-bound mitochondrial DNA (wild type, 20 μg), in the presence of increasing concentrations of non radioactive tRNA from the mitochondria of wild type and of mit⁻ mutants. 100 % hybridized counts corresponds to actual counts of 772 cpm obtained in the absence of competitor. o—o wild type mitochondrial tRNA ; x—x 170 mitochondrial tRNA ; Δ—Δ M7-37 mitochondrial tRNA ; ●—● cytoplasmic tRNA.

The very low acylation of the mutant mitochondrial tRNA by aspartic acid may be due either to a modification of the aspartyl tRNA itself of to the absence of this tRNA species in the mutant mitochondria. We favor the first hypothesis because the tRNA fractions from both 170 and M7-37 mitochondria efficiently competed with the wild type aspartyl tRNA in a hybridisation reaction with mitochondrial DNA (Figure 3). Therefore, these mutations most likely represent a change in the DNA sequence of the aspartyl tRNA gene of mitochondria, producing a tRNA incapable of accepting aspartic acid. A few other possibilities, though less simple, remain also to be examined.

4. Discussion

The approach we have chosen to select mitochondrial tRNA mutants is based on two fundamental features of the yeast mitochondrial system : (i)most of the mitochondrial tRNA genes are located within one region which represents 25-35 % of the total mitochondrial genome ; (ii) mit^- mutations can be genetically localized, relatively easily, by means of a set of well characterized rho^- deletion mutants. Once a mit^- mutation is localized within the tRNA region, mitochondrial tRNA can be directly examined, since we can predict from the map position which tRNAs are possibly modified. The above examples show that the approach is a rewarding one. We believe that this is the first case of a tRNA mutation in a organelle system.

It is surprising that we have found three independent mutants all affected in the aspartyl tRNA gene. However, other mutations in the tRNA region do exist and a preliminary analysis has indicated that tRNA genes other than aspartyl tRNA can be modified.

The mutation of a mitochondrial tRNA would obviously block mitochondrial protein synthesis. In fact, the mutant M7-37 had been isolated by A. Tzagoloff and his collaborators as a protein synthesis negative mutant (A. Tzagoloff, personal communication, 1975), and it has a pleiotropic defect in the respiratory chain components. For the mutants 170 and ts832, however, we could not establish whether the mitochondrial protein synthesis is impaired, because these strains showed a variable response to antibiotics in the amino-acid incorporation tests in vivo.

Finally, the tRNA mutants described appear to be quite stable under normal conditions of growth. After some 30 cell divisions in a glucose complete medium, the production of rho^- cells did not exceed 20 %. Back mutation to the grande phenotype occured at a frequency a little less than 10^{-7}.

We appreciate the excellent technical assistance of Mrs. Denise Bornecque. We thank Dr. A. Tzagoloff who kindly gave us valuable strains and information.

References

1. Handwerker, A., Schweyen, R.J., Wolf, K. and Kaudewitz, F. (1973) J. Bacteriol. 113, 1307-1310.
2. Flury, U., Mahler, H.R. and Fedman, F. (1974) J. Biol. Chem. 249, 6130-6137.
3. Storm, E.M. and Marmur, J. (1975) Biochem. Biophys. Res. Commun. 64, 752-759.
4. Tzagoloff, A., Akai, A. and Needleman, R.B. (1975) Proc. Natl. Acad. Sci. U.S.A. 72, 2054-2057.
5. Bolotin-Fukuhara, M., Faye, G. and Fukuhara, H. (1976) in the Proceedings of the International Conference on the Genetic Function of Mitochondrial DNA, 25-29 May 1976, Bari. Edited by C. Saccone and A. Kroon, ASP Biological and Medical Press B.V., Amsterdam, The Netherlands.
6. Casey, J., Hsu, H.J., Rabinowitz, M., Getz, G.S. and Fukuhara, H. (1974) J. Mol. Biol. 88, 735-747.
7. Fukuhara, H., Bolotin-Fukuhara, M., Hsu, H.J. and Rabinowitz, M. (1976) Molec. Gen. Genet. 145, 7-17.
8. Slonimski, P. and Tzagoloff, A. (1976) Eur. J. Biochem. 61, 27-41.
9. Faye, G., Kujawa, C. and Fukuhara, H. (1974) J. Mol. Biol. 88, 185-203.

Genetics and Biogenesis of Chloroplasts and Mitochondria, Th. Bücher et al. eds.

STRUCTURE AND GENETICS OF THE 2 μm CIRCULAR DNA IN YEAST

Michel GUERINEAU[+], Claude Grandchamp[++], Piotr Slonimski[++]

[+] Laboratoire de Pharmacologie Moléculaire associé au CNRS n°147 Institut Gustave-Roussy 94800 Villejuif- France -

[++] Centre de Génétique Moléculaire du CNRS 91190 Gif-sur-Yvette - France -

It is well established now that Saccharomyces Cerevisiae contains in addition to the nuclear and mitochondrial DNAs a third, distinct molecular species composed of circular, covalently closed 2 μm long molecules. These molecules were first seen by Sinclair et al., in electron microscopy examination of total DNA extracts (1). They were isolated, purified and characterized as a unique molecular species by Hollenberg et al., (2) and Guerineau et al., (3). Subsequent studies (4, 5, 6) confirmed the existence of this new species of DNA but did not bring any suggestions as to its possible function.

The first clue comes from the study of a particular type of mutants displaying pleiotropic resistance to many drugs. Genetic studies disclosed very unusual features, characteristic both for cytoplasmic and nuclear inheritance(7). Furthermore, a one-to-one correlation between the presence of 2 μm circles and a specific genotype could be demonstrated. As a result of this study, we hypothesized that the 2 μm circular DNA may play an informational role in the biogenesis and/or function of cellular membranes(7). It may be therefore appropriate to discuss this DNA in a meeting devoted to the biogenesis of membrane rich organelles like mitochondria and chloroplasts.

CYTOPLASMIC AND NUCLEAR FEATURES OF INHERITANCE OF DRUG RESISTANCE.

When the original mutant clone DR19 is crossed by a standard mitochondrial cross technique (8) to a sensitive tester strain a clear cut mitotic segregation is observed amongst the diploid cells. Depending on the experiment and the tester strain used, 30 % to 70 % of diploid cells are simultaneously CAP^R, CYC^R and TET^R. The resistant cells are stable in vegetative multiplication and breed true to type in the absence of the drug. They are sensitive to oligomycin but in an unusual manner. After 2-3 days of incubation numerous papillae appear and later on, a confluent growth is resumed. This does not occur with a standard sensitive strain. When the original mutant DR19 is grown in a high concentration of ethidium bromide (40 μg/ml) for several cell generations and the same cross is repeated, a drastic diminution(or even complete elimination) of CAP^R, CYC^R, TET^R diploid cells is observed.

resistant to : cycloheximide, CYC ; chloramphenicol, CAP ; oligomycin, OLI ; venturicidine VEN ; triethyltin, TET.

On a superficial analysis the genetic determinants of drug resistance would be localised on the mit DNA because they display both the mitotic segregation and the $P^+ \rightarrow P^-$ elimination characteristics. A more complete study proves that this interpretation is false. Tetrad analysis of CAP^R, and TET^R diploids shows a 2 : 2 segregation of resistance vs sensitivity. Thus, a nuclear gene, heterozygous for its two alleles (designated henceforth $1^R/1^S$) is disclosed. All 1^R spores are also OLI^R and VEN^R. Surprisingly, tetrad analysis of CAP^S, CYC^S and TET^S diploid cells issued from the same cross also shows a meiotic segregation, two spores being triply resistant and two triply sensitive. Thus, a nuclear gene is disclosed again. This gene (designated henceforth $2^R/2^S$) does not confer OLI resistance but does lead to a weak VEN resistance. The two genes are not allelic and are probably on different chromosomes because both of them are centromere linked (ca 10 % linkage) and are not linked to each other.

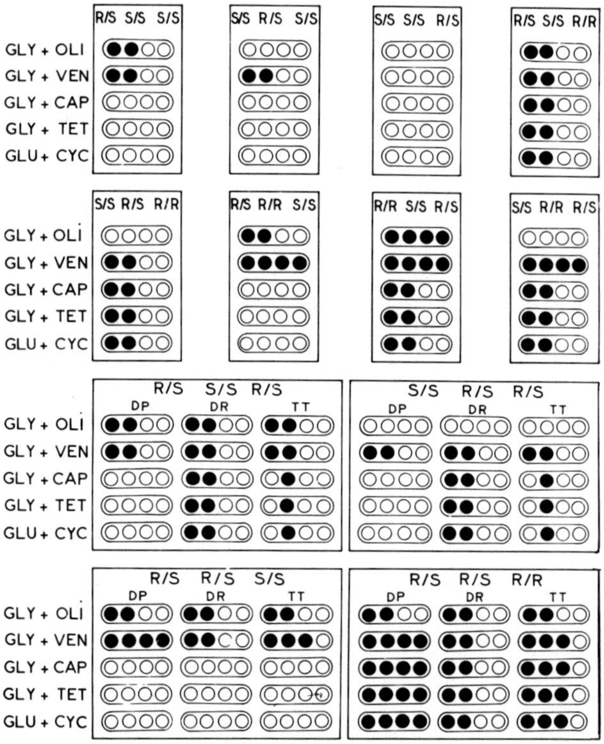

Figure 1

TYPES OF TETRADS OBTAINED FROM HYBRIDS HETEROZYGOUS FOR THE FIRST, SECOND AND THIRD RESISTANCE NUCLEAR GENE. (for details see Slonimski - in prep)

GLY = glycerol
GLU = glucose
● = resistant
O = sensitive
DP = diparental tetrad
DR = direcombinant tetrad
TT = tetratype tetrad

A third, centromere unlinked , nuclear gene is disclosed by further analysis. Numerous backcrosses, testcrosses and outcrosses have permitted us to establish the rules which govern the relations between the nuclear genotype and the resistance phenotype. They are schematically depicted in fig. 1.

We are thus faced with two sets of data : on the one hand the pattern of inheritance is cytoplasmic, on the other hand it is nuclear. The number of, experimentally reproducible, oddities and bizarreries being quite considerable, one is tempted to quitt. A successful search for a molecular correlation has enabled us to continue.

As shown in table I, 2 μm circles are absent from cells which have the 1^S 2^R nuclear genotype and are present in cells which have the 1^R 2^S genome. The first class of cells derives, from the latter one. Two types of hypotheses can be advanced in order to account for these findings : 1) Changes in the nuclear genes occur by an unknown mechanism and are responsible for the maintenance, or replication or extractibility of circular DNA, concomittantly with the drug resistance phenotype. 2) Circular DNA plays an informational role for cell membranes. It can mutate (by point mutations) and change the membrane permeability leading to drug resistance. It can behave like a cytoplasmic plasmid or like an episome (designated π, for permeability). When integrated in a chromosome it behaves like a nuclear gene.

Table I GENOTYPES, PHENOTYPES AND THE 2 μm CIRCULAR DNA.

NUCLEAR MENDELIAN GENOTYPE			EPISOME		RESISTANCE PHENOTYPE					2 μm CIRCLE
					OLI	VEN	CAP	TET	CYC	
1^R	2^S	3^S	II^R	T^S	+	+	-	-	-	+
1^S	2^R	3^S	II^-	T^S	-	+	-	-	-	-
1^R	2^S	3^R	II^R	T^R	+	+	+	+	+	+
1^S	2^R	3^R	II^-	T^R	-	+	+	+	+	+
1^S	2^S	3^R	II^S	T^R	-	-	-	-	-	+
1^S	2^S	3^S	II^S	T^S	-	-	-	-	-	+

It was, therefore, of interest to investigate the structure of the circular DNA. This study has shown that the molecule has precisely a structure which would facilitate the insertion-release faculty of an episome in the chromosome.

Structure of the π DNA circles.

We have shown by electron microscopy and by agarose gel electrophoresis of endo Eco RI digest that all molecules have two inverted repeats situated opposite each other on the perimeter of the circle (9).

1°)Characterization by electron microscopy of inverted repeats in π DNA.

After self-annealing three types of circular single stranded molecules are observed (1 type of monomers, 2 types of dimers). The monomers that fold back on themselves all looked like dumb-bells with a double stranded region between two single strand loops (Fig. 2A).This is the proof for the existence of two inverted repeat sequences in a singlestranded molecule. The dimers that fold back on themselves formed two different structures. (For details, see 9). The first form appeared as the result of attachement of two monomers. It had two apical single strand loops and two double strand stems separated by an internal single strand loop. The second form had also two apical single strand loops but they were separated by a continuous and long double stranded stem.

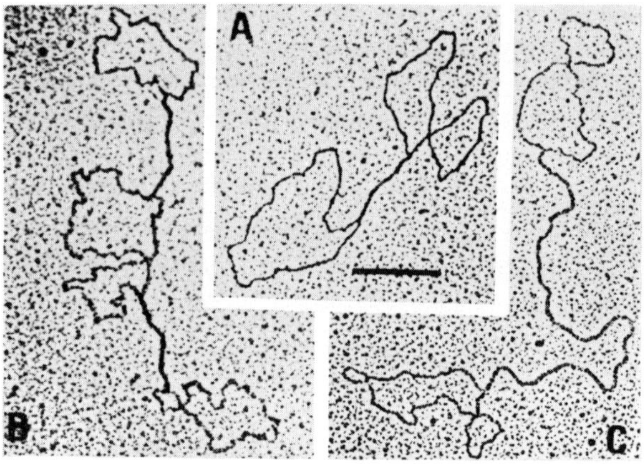

Figure 2

ELECTRON MICROGRAPHS OF SINGLE STRAND SELF ANNEALED π DNA

(bar = 0.2 μm)

A monomer

B dimer (DX type see fig. 4)

C dimer (DY type see fig. 4)

(for details see ref. 9).

Interpretation of our experimental results is presented in fig. 3 (for details, see 9). In a monomer, the inverted repeat sequences can be paired allowing an intramolecular reciprocal recombination. Such an event produces the inversion of the relative orientation of the non repeated segments bound by the inverted repeats and a new form of the monomer appears as depicted in the figure 3 (type A and B).

What are the different pathways which would allow the generation of the dimer from two monomers by intermolecular recombination ? We will only consider the reciprocal recombination between the inverted repeats. There are three possibilitiesei-ther AxA,AxB,or BxB.But in each case there are two ways of pairing. For example in the intermolecular recombination AxA, the inverted repeat A B C of one monomer bound by I and D could be paired either with the corresponding one of the other monomer (Fig. 4 left) or with the other one bound by g'and f' (Fig. 4 right) In the first case, I ABC D x I ABC D, the dimer (DX) possesses a head to tail arrangement of the two monomers. The self complementary regions are only the preexisting repeats from the parental monomers and the apical single strand loops of a self annealed dimer DX have the same size as monomers (Dimer DX Fig. 4) to be compared to the micrograph B Fig. 2. In the second case, I ABC D x g' ABC f', the arrangement of the noncomplementary sequences is different from the first case. These noncomplementary sequences from one monomer are inverted with respect to the other ones, creating a new inverted repeat sequence in addition of the preexisting ones (Fig.4 dimer D Y). After self annealing, a single strand circular dimer DY will appear as a dumb-bell with a very long double strand stem and two apical single strand loops of the same size as in monomers. Dimer DY Fif. 4 to be compared to the micrograph C fig. 2.

The other possibilities AxB BxB would give under the electron microscope the same two classes of selft annealed singlestrand dimers.

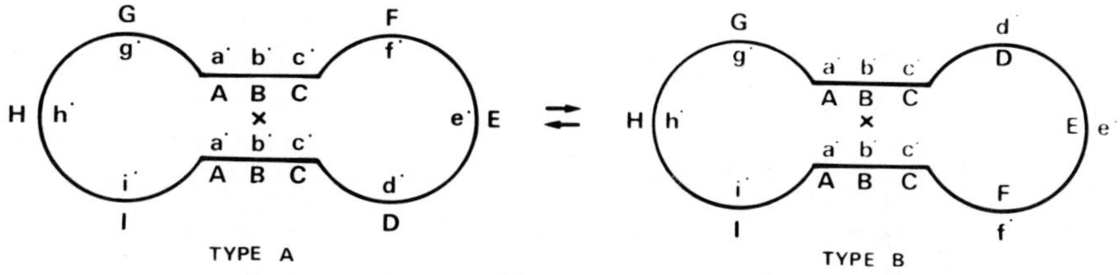

FIGURE 3 MODEL OF INTRAMOLECULAR RECOMBINATION WITHIN MONOMERIC CIRCLES.
The letters ABC and a' b' c' represent the inverted repeat sequence. The letters DEF and GHI represent the non repeated segments — double stranded DNA.

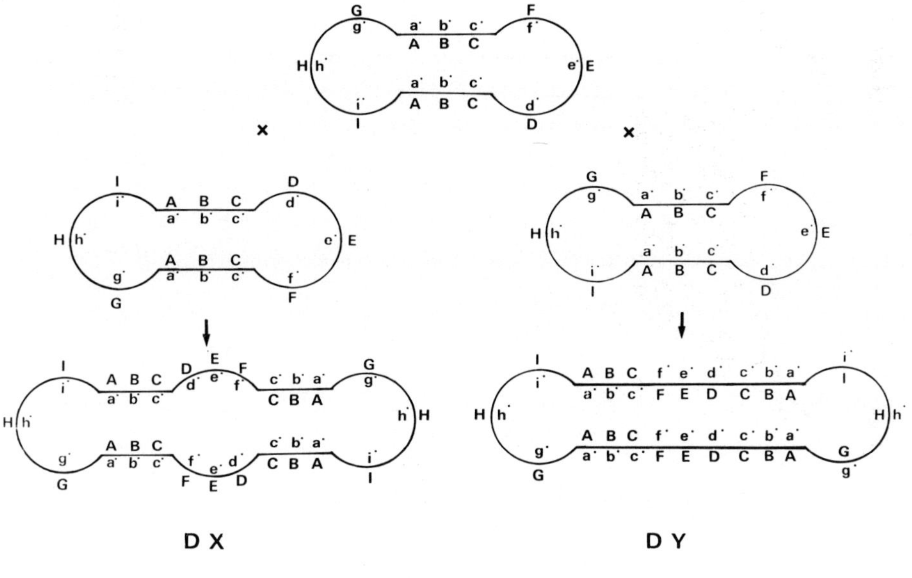

FIGURE 4 MODEL OF INTERMOLECULAR RECOMBINATION BETWEEN TWO MONOMERIC CIRCLES.
The two pathways of pairing and recombination between the inverted repeat sequences of two type A monomers are depicted — double stranded DNA.

2°) A MAP THE INVERTED REPEATS RELATIVE TO THE ENDO ECO RI SITES

To map the endo Eco RI sites with respect to the inverted repeats the circular DNA was partially digested to obtain a maximum number of molecules with only one nick. The denaturation self-renaturation experiments on the partial digest were then performedand the self annealed molecules were examined under the electron microscope. In order to minimize the size difference between the single and double strand DNA and for obtaining a precise map we combined two techniques (Fig. 5). Using these new conditions the linear densities of the single and double strand DNA were nearly the same. The length of the native double strand monomers was 2.13 µm (SEM 0,10 µm) and that of the self annealed 2.1 µm(SEM 0,10 µm). We photographed and measured the full length single strand linear molecules which self annealed. They could be classified unambiguously into two types showing that there are only two sites per molecule. A schematic representation of their structure is given in Fig. 6. One type (Fig. 5 A) has an apical single strand loop 0.92 µm (SEM 0.03 µm) long, one double stranded stem 0.20µ m (SEM 0.02 µ m)long two single strand tails unequal in length, the shorter was 0.12µ m (SEM 0.02µ m) and the longer 0.67 µm (SEM 0.02 µ m) (Fig. 6 RI 2). The whole length of the molecule was 2.11 µ m. The other type (Fig. 5 B) consisted of one apical single strand loop0.80(SEM 0.03 µ m) long, one double strand stem of 0.20 µ m and two single strand tails of 0.40 µ m (SEM 0.02 µ m) and 0.49 m (SEM 0.02 µ m) long. (Fig. 6 R I1). The whole length of the molecule was 2.09 µ m. One should note that if the intact loop has a length of 0.80 µ m the nicked loop adds up to 0.90 µ m. The contrary is also true i.e. when the intact loop has a length 0.92 µ m the nicked loop adds up to 0.80 µ m. The large loop (0.92µm) is always associated with the RI I. site and the small one (0.80 µ m) with the RI 2 site. Using these results the two possible maps of the inverted repeats relative to the endo Eco RI site can be constructed (Fig. 7). They correspond to the two monomeric forms A and B. As previously predicted (9) the site RI 1 is located near the axis of the inversion 23.8 % - 19.5 % = 2.1 % and the site RI 2 is located near the inverted repeats : 5.7 % (Fig. 7).

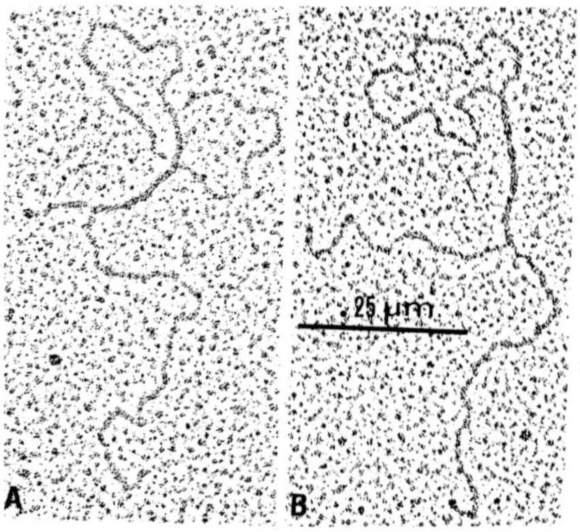

FIGURE 5

ELECTRON MICROGRAPHS OF SELF ANNEALED π DNA WITH ONE ENDO ECO RI NICK
(45 molecules measured)
A : RI_2 endo Eco RI nick
B : RI_1 endo Eco RI nick
The denaturation of DNA was carried out as in (10) and spreading, preparing, shadowing, and electron microscopy as in (11).

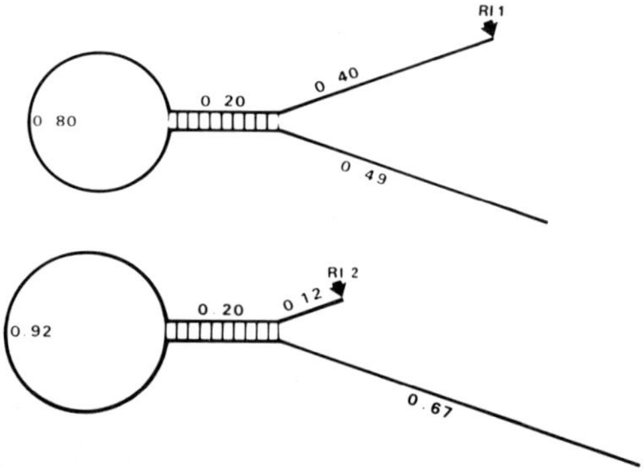

FIGURE 6

SCHEMATIC REPRESENTATION OF THE SELF ANNEALED π DNA WITH ONE ENDO ECO RI NICK.

What is the precision of the mapping of the inverted repeats relative to the endo Eco RI sites ? The self annealed linear molecules with only one endo Eco RI nick had the same length (2.10 μm) as compared to the single strand self annealed circles (2.10 μm). The full length linear molecules which fold back on themselves belong to two classes with no intermediates when one considers the single strand tails. It means that there are two endo Eco RI sites per monomer. The two monomeric forms A and B (Fig. 3) differ by the orientation of the non repeated segments and thus by the relative position of the two endo Eco RI sites. The two monomeric forms have been inserted and cloned on bacteriophage λ (12). A precise determination of the molecular weights of the endo Eco RI fragments has been done by agarose gel electrophoresis in each cloned form and was compared to the electron microscopic determinations. The type A is cleaved into two fragments which represents 61.5 and 38.5 % of the molecule to be compared to 61 and 39 % from the electron microscopic data (Fig. 7). The monomer type B is cleaved into two fragments which represents 64.1 % and 35.9 % of the molecule (12) to be compared to 65.3 and 34.7 from the electron microscopic data (Fig. 7).

FIGURE 7 - A MAP OF THE TWO ENDO Eco RI SITES RELATIVE TO THE INVERTED REPEATS. The two types of monomers are depicted (see fig. 3). Frangment sizes are indicated in % of the total molecule.

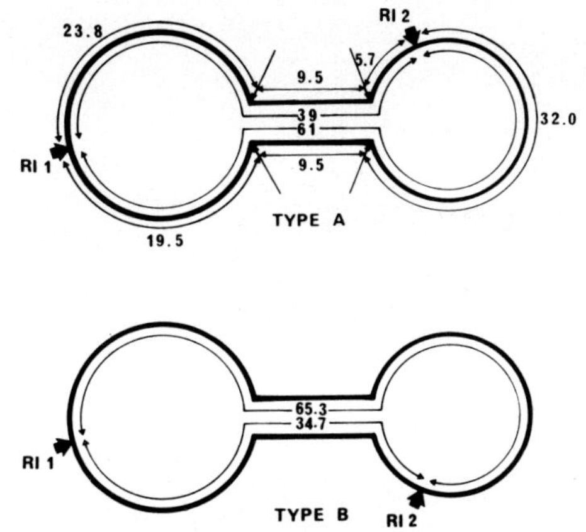

Inverted repeat sequences have been shown to promote the insertion and/or the translocation of defined genetic units at several specific sites either on the plasmid or on the bacterial chromosome(13-16).The existence of such inverted repeats in Π molecules may be responsible for their unusual genetic properties.

REFERENCES.

1. SINCLAIR J.H., STEVENS B.J., SANGHAVI P., RABINOWITZ M. (1967) Science, 156, 1234-1237.

2. HOLLENBERG C.P., BORST P., VAN BRUGGEN E.F.J. (1970) Biochim. Biophys. Acta, 209, 1-15.

3. GUERINEAU M., GRANDCHAMP C., PAOLETTI C., and SLONIMSKI P.P. (1971) Biochem. Biophys. Res. Commun., 42, 550-557.

4. STEVENS B.J., MOUSTACCHI E. (1971) Exp. Cell Res., 64, 259-266.

5. LETH-BAK A., CHRISTIANSEN C., CHRISTIANSEN G.(1972) Biochim., Biophys. Acta, 269, 527-530.

6. CLARK-WALKER G.D., (1973) Eur. J. Biochem. 32, 263-267.

7. GUERINEAU M. SLONIMSKI P.P., AVNER P. (1974) Biochem. Biophys. Res. Comm., 61, 462-469.

8. COEN D. DEUTSH J., NETTER P, PETROCHILLO E., SLONIMSKI P.P. (1970) Control of organelle development, Cambridge University Press, N° XXIV, 449-496.

9. GUERINEAU M., GRANDCHAMP C., SLONIMSKI P.P. (1976) Proc. Natl. Acad. Sci U.S.A., 73, in press.

10. LOCKER J., RABINOWITZ M., GETZ G.S., (1974) Proc. Natl. Acad. Sci U.S.A., 71, 1366-1370.

11. SCHNÖS M., INMAN R.B., (1970) J. Mol. Biol., 51, 61-73.

12. BEGGS J., GUERINEAU M., ATKINS J. (1976) Mol. Gen. Genet in press.

13. SHARP P.A., COHEN S.N., DAVIDSON N.(1973) J. Mol. Biol., 75, 235-255.

14. SAEDLER H., REIF H.J., HU S., DAVIDSON N. (1974) Mol. Gen. Genetics, 132, 265-289.

15. KOPECKO D.J., COHEN S.N., (1975) Proc. Natl. Acad. Sci. U.S.A. 72, 1373-1377.

16. HEFFRON F., RUBENS C., FALKOW S.(1975) Proc. Natl. Acad. Sci. U.S.A., 72, 3623-3627.

ELECTRON MICROSCOPICAL ANALYSIS OF NATIVE AND CLONED 2-μm DNA FROM SACCHAROMYCES CEREVISIAE

Cornelis P. Hollenberg and Hans-Dieter Royer
Max-Planck-Institut für Biologie, Abteilung Beermann
Spemannstraße 34, D-7400 Tübingen
Federal Republic of Germany

INTRODUCTION

Saccharomyces cerevisiae contains a class of closed circular DNA of 2-μm monomeric circumference and nuclear DNA density[1,2]. This circular DNA is located outside the mitochondria[2] and nucleus[3] and may be associated with a membrane fraction[4]. Longer molecules of around 3, 4 and 6 um have been reported[4,5]. The kinetic complexity is 4.7×10^6 daltons[6]. Cloning of 2-μm DNA in E. coli plasmid pCR1 directly showed the existence of two types of 2-μm monomers, which are differentiated by an inversion of about 40% of the molecule[7]. This inversion is locked in by the sequences of an inverted duplication.

In this communication we report the results of an electron microscopical analysis of yeast supercoiled DNA after gel fractionation. In addition to the normal monomer we found two types of molecules that consist of a tandem duplication of about one half of the normal monomer. We further show that the circular molecules of around 4 μm and 6 μm are dimers and trimers of the 2-μm monomer.

RESULTS AND DISCUSSION

Analysis of the monomer: Closed circular yeast DNA was purified via two sequential CsCl-ethidium bromide gradients and fractionated on an 0.7% agarose gel. At an input of 1.5 μg of DNA nine bands were visible (Fig.1), five of which were further analyzed. The heaviest bands, 9 and 5, contain the closed and open circular 2-um monomer, as we reported previously[7]. The length of 80 open circular monomers present in band 6 was measured with relaxed PM2 DNA as marker. The length distribution suggested three size classes: class I 1.76 μm ± 0.03 μm, class II 1.92 μm ± 0.03 μm and class III 2.03 μm ± 0.02 μm containing respectively 15, 77 and 8% of the measured molecules. Relaxed PM2 DNA had a mean size of 3.1 μm ± 0.05 μm.

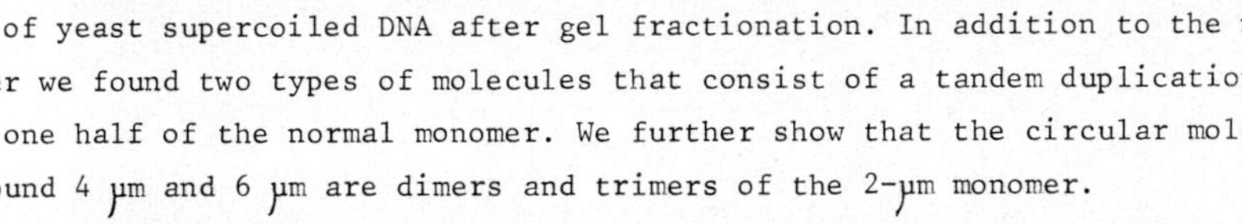

Fig. 1 Agarose gel of yeast supercoiled DNA. Electrophoresis was carried out in 0.7% agarose containing 0.5 μg ethidium bromide[8] per ml at 3 V/cm at 18°C. The gel was photographed on Polaroid type 55 P/N film with 366 nm light and a yellow and red filter. Strain: S. cerevisiae H1

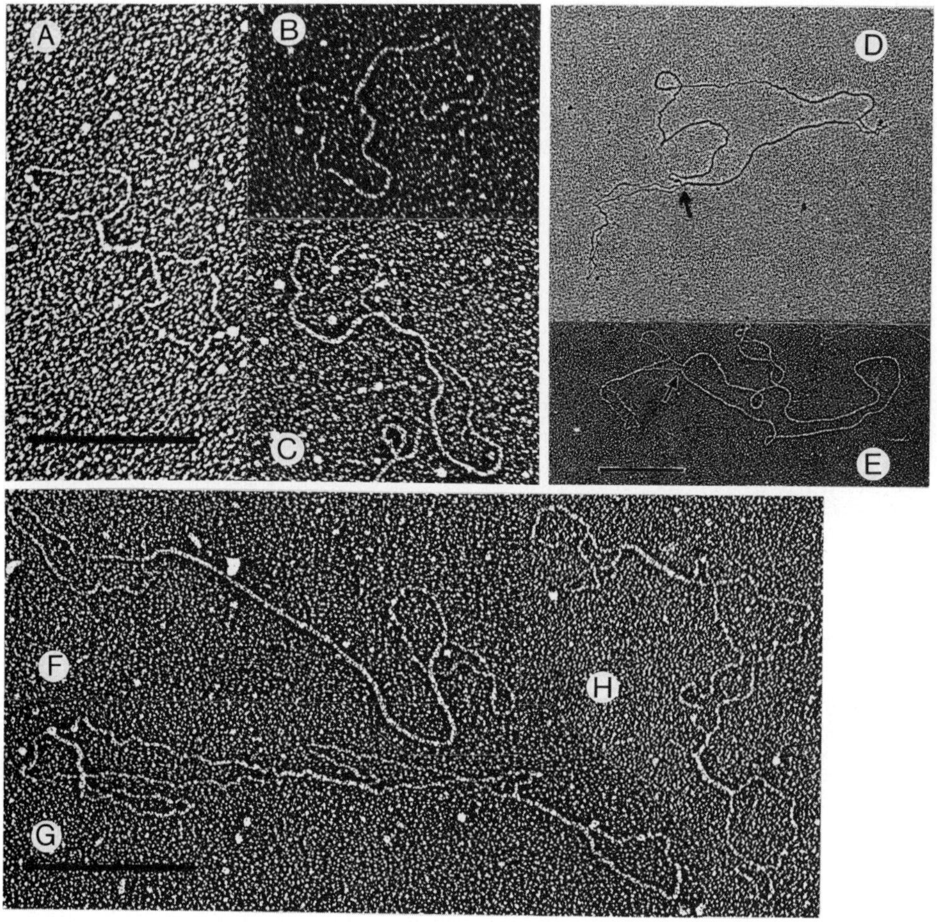

Fig. 2 Electron micrographs of self-annealed and heteroduplex molecules. A. 1.9 μm dumb-bell. B. Heteroduplex of normal monomer and HindIII fragment H1-H2. C. Double L-loop dumb-bell with both loops hybridized with fragment H1-H2. D. Heteroduplex of PTY75 and pCR1. E. Heteroduplex of PTY85 and pCR1. F. Dimer dumb-bell with long stem. G. Dimer dumb-bell with two stems. H. Trimer dumb-bell. Arrow in D and E points to connection site. Bar is 0.5 μm.

Upon denaturation and reannealing the 2-μm DNA circles occurred predominantly in a dumb-bell structure with two single stranded loops and a double stranded stem (Fig. 2A). This structure is formed by intramolecular renaturation of the sequences of a non-contiguous inverted duplication (id sequences) in the single stranded molecule (compare Fig.3). Contour length measurements on 40 dumb-bells with fully spread loops also indicated the presence of three size classes. Measurement of all loops present in these dumb-bells showed two discreet loop sizes: a small loop (S-loop) of 0.67 μm $\pm$ 0.06 μm and a large loop (L-loop) of 0.86 μm $\pm$ 0.05 μm. The length of the stem was 0.19 μm. Most dumb-bells had one S- and one L-loop, which reflects an asymmetric location of the two id sequences along the molecule. These dumb-bells all belonged to class II.

The remaining molecules, however, contained two loops of identical size. Class I dumb-bells had two S-loops and class III dumb-bells two L-loops. If each loop entity contains a fixed DNA sequence, the class I and III molecules each contain the same DNA sequence twice. Alternatively, insertion or deletion of a small DNA segment in the loop regions could result in the observed heterogeneity. In the latter case these circle types would be largely of unique DNA sequence. In a previous report[7] we had not yet distinguished the two different loop sizes, although the restriction fragment analysis indicated that the loop containing EcoRI site A could be larger. This loop also contains HindIII fragment H1-H2 (Fig.3) so that we could use heteroduplex mapping of this fragment on 2-um dumb-bells to show which loop contains site RIA. These heteroduplexes, further, could be used to identify the loops of double S- and double L-loop molecules. In all 30 heteroduplex dumb-bells analyzed, fragment H1-H2 was located in an L-loop (Fig.2B). In a number of double L-loop dumb-bells both loops had hybridized with fragment H1-H2 (Fig.2C), which demonstrates that the two large loops of at least part of the class III molecules have the same base sequence as the L-loop. No heteroduplex of fragment H1-H2 with class I molecules was detected, in agreement with the idea that these molecules are built up of two S-loops.

The formation of molecules with twice the same DNA segment in between the id sequences cannot easily be explained by conventional recombination among identical monomers. Legitimate intramolecular recombination within dimer molecules only gives rise to inversions. Reciprocal recombination between id sequences of two monomers with opposite polarity in the id sequences, however, could lead to double S- or double L-molecules[9].

<u>Mapping on cloned 2-μm DNA:</u> We have constructed a number of chimeric plasmids that consist of plasmid pCR1 and 2-μm DNA, joined at an EcoRI site[7]. Hetero-

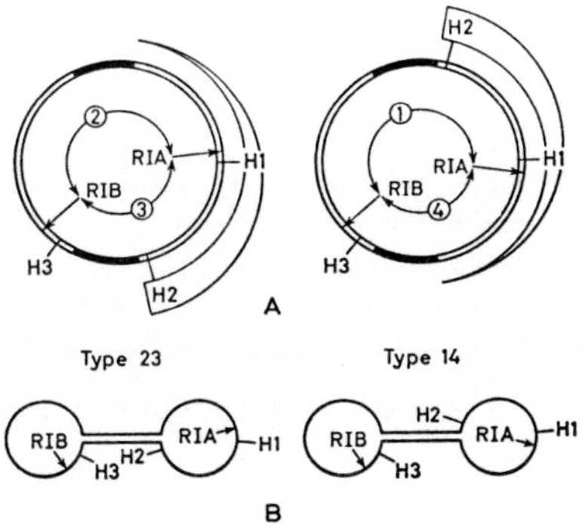

Fig. 3 A. Schematic outline of double stranded yeast monomer type 14 (consisting of EcoRI fragments 1 and 4) and type 23 (consisting of EcoRI fragments 2 and 3). The inversion that differentiates between types 14 and 23 is indicated with the half crescent. The id sequences are black. Numbers 1-4 refer to the EcoRI fragments; RIA and RIB refer to EcoRI sites; H1, H2 and H3 to HindIII sites. B. Schemes of reannealed single strands of the monomer.

duplexes with pCR1 can directly visualize the location of the joining EcoRI site on the dumb-bell. In PTY75 (Fig.2D) the two DNA molecules are linked at site RIA close to the middle of an L-loop and in PTY85 (Fig.2E) at site RIB in an S-loop at approximately 0.12 μm from the stem. In heteroduplexes of PTY85 and fragment H1-H2 a number of loops showed the hybridized fragment. In PTY75 all hybridized fragments were located at the loop-distal side of the id stem as expected for integration at site RIA in the L-loop. None of the molecules showed hybridization at both sides of the id stem indicating that 2-μm DNA segments with identical loop sequences did not arise in these integrated molecules.

Multimers: Band 6 represented supercoiled DNA of about 4 μm length[7]. Denaturation and reannealing of this DNA showed two types of dumb-bells: dumb-bells with a long stem (Fig.2F) and loop/stem/long-loop/stem/loop structures (Fig.2G) with contour lengths of 4.18 and 4.15 μm. The two-stem structures can be considered as dimers of identical 2-μm molecules, while the dumb-bell with long stem could arise from dimers that have undergone intramolecular recombination between id sequences, leading to the inversion of one loop segment. Dimer formation of a type 14 plus a type 23 monomer could also lead to the long stem structures. The same dumb-bell structures were found in material from band 3, which contained the open circular 4-μm DNA. Band 3 which sometimes looked heterogeneous and could be a double band, contained in addition dumb-bell structures of 6.9 and 8.8 μm. In the reannealed DNA preparation from band 4, which contained mainly supercoiled molecules of 6 μm lengths, again the remarkable alternating loop/stem structures were observed (Fig.2H). These results show that the longer molecules present in the supercoiled DNA fraction of yeast mainly represent multimers of the 1.9 μm monomer.

REFERENCES

1. Sinclair, J.H., Stevens, B.J., Sanghavi, P. and Rabinowitz, M. (1967) Science 156, 1234-1237.

2. Hollenberg, C.P., Borst, P. and van Bruggen, E.F.J. (1970) Biochim. Biophys. Acta 209, 1-15.

3. Clark-Walker, G.D. and Miklos, G.L.G. (1974) Eur. J. Biochem. 41, 359-365.

4. Guerineau, M., Grandchamp, C., Paoletti, J. and Slonimski, P. (1971) Biochem. Biophys. Res. Commun. 42, 550-557.

5. Clark-Walker, G.D. (1973) Eur. J. Biochem. 32, 263-267.

6. Leth Bak, A., Christiansen, C. and Christiansen G. (1972) Biochim. Biophys. Acta 269, 527-530.

7. Hollenberg, C.P., Degelmann, A., Kustermann-Kuhn, B. and Royer, H.D. (1976) Proc. Natl. Acad. Sci. USA, 73, 2072-2076.

8. Aay, C. and Borst, P. (1972) Biochim. Biophys. Acta 269, 192-200.

9. Royer, H.D. and Hollenberg, C.P., submitted for publication.

Genetics and Biogenesis of Chloroplasts and Mitochondria, Th. Bücher et al. eds.

THE STUDY OF THE GENETIC FUNCTION OF *PARAMECIUM* MITOCHONDRIAL DNA USING SPECIES HYBRIDS

A. Tait, J.K.C. Knowles, J.C. Hardy and H. Lipps

Institute of Animal Genetics
University of Edinburgh
Edinburgh, Scotland

INTRODUCTION

The study of the genetic function of mitochondrial DNA in *Paramecium* has been hindered by inability to obtain mitochondrial mutants affecting the respiratory chain and oxidative phosphorylation. Although mutants affecting the mitochondrial ribosomes[1,2] have been isolated and can be used to establish what the mitochrondrial DNA codes for, it is clear that in order to both map and determine the function of mitochondrial DNA further, either more mutants are required or an alternative approach must be sought. The development of the technique of microinjection[3] together with the injection of mitochondria from one species to another[4,5] has provided an alternative approach. In the case of the hybrid cells formed by the injection of species 1 mitochondria into species 7 cells, restriction enzyme studies have shown[6] that these hybrids only contain the mitochondrial DNA of species 1 and, of course, the nuclear DNA of species 7.

If variation between species 1 and species 7 of a particular protein or class of proteins can be detected, then the mitochondrial or nuclear control of the variation can be determined by examining the protein or class of proteins in the hybrid cells. If the hybrid cell mitochondria show the proteins characteristic of their nuclear parent (species 7) then the proteins are coded for by the nucleus, whereas if they are characteristic of their mitochondrial parent (species 1) then they are coded for by mitochondrial DNA. We have used this approach to study the genetic control of three types of protein, namely: (1) the soluble matrix enzymes (2) the chloroform methanol soluble membrane proteins and (3) the mitochondrial ribosomal proteins. The nuclear control of the soluble matrix enzymes, using fumarase as an example, has been demonstrated[4]; in this paper the results for the other two types of protein are presented.

RESULTS

The chloroform methanol soluble proteins were extracted from mitochondrial sub-particles, prepared by a modification of the technique of Tzagaloff and Akai[7]. This fraction contains many of the water insoluble hydrophobic proteins of mitochondrial membranes and we have shown by SDS polyacrylamide gel electrophoresis that it contains at least nine molecular weight classes varying in molecular weight from 66,000 to 9,000 daltons. Antisera were raised, in rabbits, against species 1 and species 7 proteins and the immunoglobulins were separated and labelled with ^{125}I. Species 1, species 7 and hybrid antigens (chloroform methanol soluble proteins) were incubated with the labelled immunoglobulins and the amount of immunoglobulins bound to the antigen measured. The results of these experiments are shown in Table 1.

TABLE 1

THE BINDING OF LABELLED IMMUNOGLOBULINS TO THE CHLOROFORM METHANOL SOLUBLE PROTEINS

Antigen	Antibody	Antibody bound CPM/unit antigen	Antibody bound % homologous binding [3]
Species 7		57,869	100
Hybrid	anti-7[1]	39,149	72
Species 1		31,290	61
Species 7		25,337	61
Hybrid	anti-1[2]	32,948	77
Species 1		36,075	100

1. Preabsorbed once with species 1 antigen
2. Preabsorbed once with species 7 antigen
3. Average of three experiments

It is clear that species 1 antigen binds more anti species 1 globulins than does species 7 antigen which, in turn, binds more anti species 7 globulins than species 1 antigen. Thus there is a clear immunological difference between the proteins of species 1 and species 7. When the antigen is prepared from hybrid cell mitochondria and incubated with either anti species 1 globulins or anti species 7 globulins (Table 1) it binds an amount of globulins which is intermediate between the homologous and heterologous binding. These results suggest that some of the antigens in the chloroform methanol soluble mitochondrial proteins are coded for by the nuclear genome and some by the mitochondrial genome.

A similar type of approach was taken to study the genetic control of the mitochondrial ribosomal using the same hybrid cells. Antisera were raised in rabbits to 55S mitochondrial ribosomal subunits (from species 1 and 7) prepared by puromycin high salt treatment followed by sucrose density gradient centrifugation. The interaction between the antisera produced and the mitochondrial ribosomal proteins (prepared by LiCl/urea treatment of the ribosomes) was studied by quantitative immuno-precipitation[8] of a fixed concentration of antigen (mitochondrial ribosomal proteins) by increasing amounts of antiserum. The amount of antigen-antibody complex formed was determined directly by protein estimation. Using this technique, the antigens of species 1 and species 7 cells were compared and as shown in Fig.1(A) a clear difference between them was observed. When mitochondrial ribosomal proteins prepared from hybrid cells are used as antigens their precipitation curve is intermediate between the homologous (species 7) and the heterologous (species 1) precipitation (Fig.1A). These results have also been confirmed using anti-species 1 sera where a similar result is obtained. Thus some of the antigens are coded for by the mitochondrial genome and some by the nuclear genome.

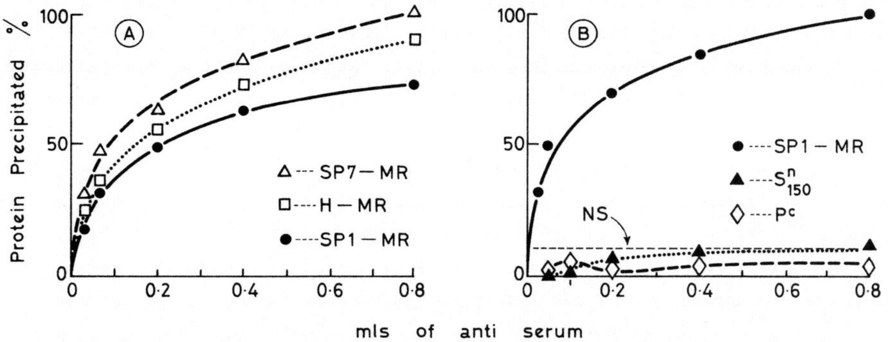

Fig. 1. Immunoprecipitation of mitochondrial ribosomal proteins and other mitochondrial proteins. The amount of precipitated protein (antigen-antibody complex) is expressed as a percentage of the maximum amount precipitated by the homologous reaction. (A) Precipitation of species 1, species 7 and hybrid (H) mitochondrial ribosomal proteins by anti-species 7 mitochondrial ribosomal serum. (B) Precipitation of mitochondrial ribosomal proteins and the two other fractions of the mitochondrial proteins (S^n_{150} and P^c) by anti-mitochondrial ribosomal serum.

A critical point in the interpretation of these data is the nature of the antigen used. Although the purity of our ribosome preparations is high (260/280 = 1.80), one can still not be certain that all the proteins attached to the ribosome are true ribosomal proteins[9]. In lieu of a functional test for each ribosomal protein we have tested whether the observed immunological reaction is due to ribosomal proteins by asking whether any other fractions of the mitochondria (S^n_{150} and P^c) isolated during the preparation of the ribosomes, contain antigens which react with the sera. These two other fractions contain the rest of the solubilised mitochondrial proteins[10] and were shown to be immunologically unreactive (Fig.1B). If the immunological reaction observed is due to non-ribosomal proteins; these proteins are selectively associated with the mitochondrial ribosomes and so are operationally mitochondrial ribosomal proteins.

DISCUSSION

The results present here suggest that some of the proteins of the mitochondrial membrane and of the mitochondrial ribosomes are coded for by *Paramecium* mitochondrial DNA. The chloroform methanol soluble proteins, used as antigens in this study, contain no RNA, but could be glyco- or lipo-proteins. Although the most likely conclusion is that the differences observed are due to differences in the primary structure of the proteins, it is possible that differences observed could be due to secondary modifications of the lipid or saccharide moieties. Further work is clearly required to establish which proteins are different and which of these are coded for by mitochondrial DNA as well as establishing that the differences observed are indeed due to alterations in the primary structure of the proteins concerned.

A considerable amount of research in other organisms has been directed towards examining whether any mitochondrial ribosomal proteins are coded for by mitochondrial DNA, or synthesised within the mitochondrion but these studies[9,11,12] do not conclusively eliminate the possibility that some mitochondrial ribosomal proteins are synthesised by the mitochondrion. Evidence suggesting a similar conclusion to ours has been presented for *Xenopus*[16], although, as the authors suggest, the presence of a heterozygous nuclear genome confuses the interpretation. Further work is currently being undertaken to identify the specific proteins responsible for the mitochondrially coded immunological difference observed.

It is clear from the data presented that the interspecies hybrid system developed in our laboratory is a valuable approach to the study of the genetic function of mitochondrial DNA and we hope, in the future, to extend our studies to other mitochondrial proteins.

ACKNOWLEDGEMENTS

We are indebted to Professor G.H. Beale for advice and discussion throughout this work and to the Science Research Council for financial support.

REFERENCES

1. Adoutte, A. (1974). The Biogenesis of Mitochondria. p.263, Academic Press, New York.
2. Spurlock, G., Tait, A. and Beale, G.H. (1975). FEBS Letters 56, 77.
3. Knowles, J.K.C. (1974). Exp. Cell Res. 88, 79.
4. Knowles, J.K.C. and Tait, A. (1972). Mol. Gen. Genet. 117, 53.
5. Beale, G.H. and Knowles, J.K.C. (1976). Mol. Gen. Genet. 143, 197.
6. Cummings, D.J., Goddard, J.M. and Maki, R.A. (1976). The Genetic Function of Mitochondrial DNA (in press), North Holland, Amsterdam.
7. Tzagoloff, A. and Akai, A. (1972). J. Biol. Chem. 247, 6517.
8. Stoffler, G. and Wittman, H.G. (1971). Proc. Natl Acad. Sci. U.S.A. 68, 2283.
9. Groot, G.S.P. (1974). The Biogenesis of Mitochondria. p.443. Academic Press New York.
10. Tait, A., Knowles, J.K.C. and Hardy, J.C. (1976). The Genetic Function of Mitochondrial DNA. (in press), North Holland, Amsterdam.
11. Lizardi, P.M. and Luck, D.J.L. (1972). J. Cell Biol. 54, 56.
12. Millis, A.J.T. and Suyama, Y. (1972). J. Biol. Chem. 247, 4063.
13. Leister, D.E. and Dawid, I.B. (1975). J. Mol. Biol. 96, 119.

Genetics and Biogenesis of Chloroplasts and Mitochondria, Th. Bücher et al. eds.

ORGANIZATION AND EXPRESSION OF THE MITOCHONDRIAL GENOME IN HELA CELLS

Giuseppe Attardi, Manfred Albring, François Amalric, Robert Gelfand, Jack Griffith,* Dennis Lynch, Christian Merkel, William Murphy and Deanna Ojala

Division of Biology, California Institute of Technology
Pasadena, California 91125, U.S.A., and
*Department of Biochemistry, Stanford University School of Medicine
Stanford, California 94305

In the last few years in our laboratory the organization and expression of the mitochondrial genome in HeLa cells have been the object of intensive investigations, utilizing a variety of molecular approaches. In particular, the study of mitochondrial RNA synthesis has provided a wealth of information concerning the mechanism of transcription of mitochondrial DNA (mit-DNA), the nature of the primary gene products, and some aspects of the regulation of expression of the mitochondrial genome.[1] Furthermore, electron microscope mapping of RNA-DNA hybrids and the use of restriction enzymes have made it possible to construct a partial genetic map of this genome. We want to report here on recent aspects of this work.

Evidence for complete symmetrical transcription *in vivo* of mitochondrial DNA in HeLa cells

Earlier work from this laboratory had shown that the heavy (H) strand of HeLa cell mit-DNA is completely transcribed *in vivo*.[2] In subsequent experiments, by analyzing the base sequence complementarity to separated mit-DNA strands of newly synthesized mitochondrial RNA and the ability of this RNA to form RNase-resistant RNA-RNA duplexes when self-annealed *in vitro*, it was found that the light (L) strand of mit-DNA is also transcribed over a considerable portion of its length;[3] however, the precise extent of this transcription could not be determined.

Recently, it has been possible to obtain direct evidence on this last point by carrying out saturation hybridization experiments between L strands and mitochondrial RNA enriched in L strand transcripts.[4] Both by analysis of the density in a Cs_2SO_4 gradient of the RNA-DNA hybrids formed with saturating amounts of RNA and by electron microscopy of these hybrids, it has been shown unambiguously that mitochondrial RNA contains sequences complementary to the entire or almost entire length of the L strand.

Figure 1 shows some examples of hybrids formed by annealing a preparation consisting of mostly intact, circular or linear, L strands with saturating amounts of L transcripts. The circular shape of the hybrids may reflect either the original structure of the DNA strands or a secondary circularization of intact linear strands by hybridized RNA. The length of the hybrids, as estimated by comparison

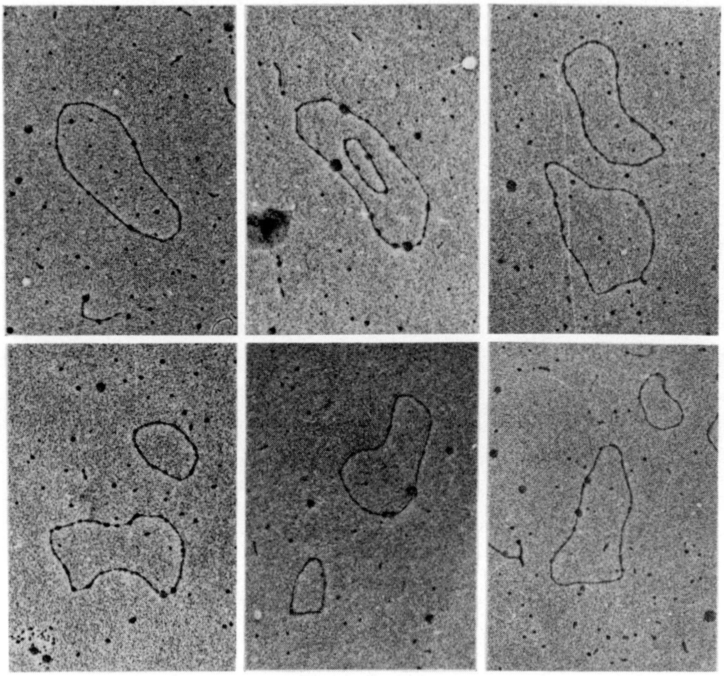

Fig. 1. Electron micrographs of hybrids between L strands of mit-DNA and saturating amounts of partially purified L strand transcripts, spread by the basic protein film technique. ØX174 RF DNA is included for size reference (taken from ref. 4).

with the internal size marker RF ØX174 (1.7 μ), is approximately 4.8 μ, as expected for fully base-paired RNA-DNA hybrids involving intact DNA strands.[4]

The significance of the symmetric transcription of mit-DNA is still elusive. After the initial discovery in HeLa mit-DNA, symmetric transcription has been shown to occur in SV40,[5] polyoma virus,[6] adenovirus[7] and herpes virus[8] DNAs. Pulse labeling experiments have shown that the rate of in vivo transcription of the H and L strands of HeLa mit-DNA is approximately equal, suggesting that symmetrical transcription is not a rare event in this system. The fundamental question which needs to be answered is whether and how the complete transcription of both the H and L strands of HeLa cell mit-DNA is related to the expression of the mitochondrial genes located on the two strands. If a direct relationship between the two phenomena does exist, one would like to know whether this mode of transcription plays a role in regulation, or whether, on the contrary, it represents a mechanism for increasing the amount of meaningful information stored in mit-DNA.

TABLE 1

Base sequence complementarity to heavy (H) and light (L) mitochondrial DNA strands of amino acyl-tRNAs from HeLa cell mitochondria

Amino acid specificity	Base sequence complementarity detected H strand	L strand	Neither strand	Distinctive mitochondrial synthetase activity†
Alanine		+		+
Arginine	+			+
Asparagine			+	
Aspartic acid	+			+
Cysteine		+		+
Glutamine			+	
Glutamic acid		+		+
Glycine	+			
Histidine			+	
Isoleucine	+			+
Leucine	+			+
Lysine	+			+
Methionine	+			+
Phenylalanine	+			+
Proline			+	
Serine	+	+		+
Threonine	+			
Tryptophan	+			+
Tyrosine		+		+
Valine	+			+

†Recognized by assaying the aminoacylating activity of the mitochondrial, as compared to the cytoplasmic, synthetase preparation versus mitochondrial and cytoplasmic tRNA.

Primary gene products of mitochondrial DNA

Many discrete RNA components coded for by HeLa cell mit-DNA have been identified in the last few years.[1] These include the two structural RNA components, 16S and 12S RNA, of mitochondria-specific 60S ribosomes, many 4S RNA species, representing mostly, if not exclusively, mitochondria-specific tRNA species, and several poly(A)-containing RNA components, which are presumably, at least in part, mRNA species or their precursors.

Recently, considerable attention has been devoted in our laboratory to the question of the number and origin of the tRNAs utilized for mitochondrial protein synthesis in HeLa cells. Earlier RNA-DNA hybridization experiments had indicated the presence in HeLa cell mit-DNA of only 11 or 12 sites for 4S RNA,[9,10] suggesting that the coding capacity for tRNA of HeLa cell mit-DNA may not be complete. To obtain further information on the above question, the amino acid specificity of the tRNA species coded for by HeLa cell mit-DNA has been analyzed by hybridizing amino acyl-tRNAs labeled in the amino acid with separated mit-DNA strands.[11] Seventeen tRNA species hybridizable with mit-DNA, which are specific for 16 amino acids, have been detected in HeLa cell mitochondria (Table 1); thus, it is likely that at least one gene for each of these tRNA species exists in HeLa cell mit-DNA. The data suggest the existence of two serine isoacceptors coded for by distinct mit-DNA genes. Formylation experiments have indicated that the methionyl-tRNA having sequence homology to mit-DNA consists all or in part of a formylatable species. For 14 of the 16 amino acids capable of charging tRNA species hybridizable to mit-DNA, amino acyl-tRNA synthetase activities distinct from the cytoplasmic ones have been identified in the organelles[11] (Table 1).

Four amino acids (asparagine, glutamine, histidine and proline) have consistently failed, in repeated tests carried out with different tRNA and amino acyl-tRNA synthetase preparations, to charge to any detectable extent mitochondrial tRNA species hybridizable with mit-DNA. These amino acids appear to be utilized for the synthesis, within the organelles, of discrete mitochondrial polypeptides.[12] Therefore, if the failure to charge with these amino acids tRNA species coded for by mit-DNA is not due to experimental difficulties, these results would imply that the tRNAs for these amino acids are nuclear DNA-coded and, therefore, must be imported from the cytoplasm. This possibility is open to direct experimental test.

Although the experimental limitations inherent in the above described type of approach would tend to make the number of tRNA genes indicated by this work a minimum one, it should be noted that recent results of experiments carried out with different approaches are in reasonable quantitative agreement with the hybridization data presented above. Thus, by electron microscope mapping with ferritin-labeled 4S RNA, seven 4S RNA genes have been mapped on the L strand and 12 on the H strand, relative to the ribosomal RNA (rRNA) gene positions[13] (see below).

Another area in which we are presently concentrating our efforts concerns the identification and characterization of discrete poly(A)-containing and non-poly(A)-containing RNA components coded for by mit-DNA, which may represent specific messenger RNAs or their precursors. A considerable difficulty in the fractionation of mitochondrial RNA from HeLa cells and in the study of its processing has been the great tendency of this RNA to aggregate, due in part to the presence of complementary sequences resulting from the symmetrical transcription of mit-DNA, in part, probably, to the poly(A) tails.[14] Recently, however, by

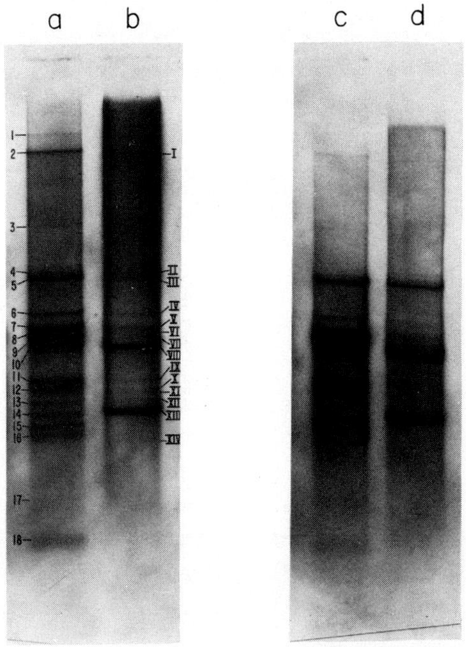

Fig. 2. Autoradiograms, after electrophoresis through agarose-CH_3HgOH slab gels, of oligo(dT)-cellulose bound (a,c) and non-bound (b,d) RNA from the structures of a mitochondrial Triton X-100 lysate which sediment in a sucrose gradient in the region of mitochondrial polysomes (c,d) and from the detergent insoluble structures (a,b) of HeLa cells labeled for 3 hr with ^{32}P-orthophosphate in the presence of 20 μg/ml camptothecin.

utilizing electrophoresis through agarose slab gels in the presence of methyl mercuric hydroxide as a denaturing agent,[15] in combination with autoradiography, we have been able to separate different species of ^{32}P-labeled mitochondrial RNA with a degree of resolution previously unsurpassed.

Figures 2a and 2b show the autoradiograms, after agarose-CH_3HgOH electrophoresis, of RNA extracted from the Triton X-100 insoluble mitochondrial structures of cells labeled for 3 hr with ^{32}P-orthophosphate in the presence of 20 μg/ml camptothecin (to block all high mol. wt. nuclear RNA synthesis); the RNA had been separated into an oligo(dT)-bound (Fig. 2a) and a non-oligo(dT)-bound fraction (Fig. 2b) by two consecutive passages through oligo(dT)-cellulose columns. At least 18 bands of varying intensity can be recognized in the autoradiogram of the oligo(dT)-bound RNA fraction and 14 bands in that of the non-bound RNA fraction. Most prominent among the latter are the two rRNA species, 16S (VIII) and 12S RNA (XIII); many of the other bands in this autoradiogram correspond in position to bands in Figure 2a, but there are some which do not show any obvious

TABLE 2

Hybridization with separated mitochondrial DNA strands of ^{32}P-labeled RNA components bound and non-bound to oligo(dT)-cellulose from HeLa cell mitochondria

Component	Triton X-100 insoluble structures		Structures sedimenting in polysome region	
	% cts/min hybridized		% cts/min hybridized	
	H strand	L strand	H strand	L strand
a) Oligo(dT)-cellulose bound				
5	64	26	77	8
6	55	26	67	8
7	67	21	82	6
9	64	16	75	10
10	74	12	81	7
11	57(75)	21(11)	77	11
12	61	20	78	14
13	61(81)	21(9)		
14	62(66)	20(<1)	77	15
15	62(76)	21(8)	79	11
16	61(73)	19(5)	78	11
17	50	31	69	15
18 (7S RNA)	34	62	38	59
b) Non-oligo(dT)-cellulose bound				
III	55	29	59	32
V	50	27	58	29
VIII (16S RNA)	91	3	95	<1
XIII (12S RNA)	95	<1	96	<1

The RNA components bound and non-bound to oligo(dT)-cellulose were separated on agarose-CH_3HgOH slab gels as shown in Fig. 2, and electroeluted from the gel. The RNA samples (150-200 cts/min) were hybridized in solution with an excess of DNA. The numbers in parentheses refer to hybridization data obtained with individual RNA components rerun on gels.

correspondence, like bands XI, XII, and XIV. It was previously shown that the oligo(dT)-bound material contains poly(A) stretches corresponding to about $4S_E$ (∿55 nucleotides).[16-19] Whether the RNA fraction not bound to oligo(dT) cellulose lacks poly(A) completely or contains shorter poly(A) chains is being investigated at present. Heterogeneous material can be seen as a fairly uniform background throughout the major portion of the gel of the oligo(dT)-bound RNA fraction, and, more abundant, in the upper third of the gel of the non-bound RNA fraction.

Figures 2c and 2d show the autoradiograms of the fraction of the RNA extracted

TABLE 3

Molecular weight of discrete RNA components bound and non-bound to oligo(dT)-cellulose, estimated from electrophoretic mobility

Oligo(dT)-cellulose bound RNA component	Molecular weight x 10^{-5}	Non-oligo(dT)-cellulose bound RNA component	Molecular weight x 10^{-5}
1	34	I	27
2	28	II	9.0
3	14	III	8.6
4	9.0	IV	6.7
5	8.6	V	6.2
6	6.7	VI	6.0
7	6.2	VII	5.8
8	6.0	VIII (16S RNA)	5.4**
9	5.8	IX	4.2
10	5.4	X	4.0
11	4.2	XI	3.9
12	4.0	XII	3.7
13	3.6	XIII (12S RNA)	3.5**
14	3.3	XIV	2.7
15	3.2		
16	2.9		
17	1.4		
18	0.93*		

Estimates of molecular weights were made on the basis of a standard curve constructed with 16S, 12S and 7S RNA.

*From ref. 20.

**From ref. 14.

from the Triton X-100 soluble mitochondrial structures sedimenting in the polysome region which was retained and, respectively, of that which was not retained on oligo(dT)-cellulose. In each of the two fractions one can observe in general the same components detectable in the corresponding fraction of the RNA extracted from the Triton X-100 insoluble structures, though with some differences in the relative intensity of the bands. This suggests that the Triton X-100 insoluble material may comprise RNA-containing structures similar or identical to those solubilized by the detergent, in particular, mitochondrial polysomes.

In control experiments carried out in the absence of camptothecin or other antibiotics, the ^{32}P-labeled mitochondrial RNA exhibited an electrophoretic pattern with all the discrete components shown in Fig. 2, indicating that these components are products of the normal mitochondrial RNA metabolism in HeLa cells.

The synthesis of the discrete components of the RNA fraction retained and of that not retained on oligo(dT)-cellulose from both the Triton X-100 soluble and the insoluble structures was found to be sensitive to 1 μg/ml ethidium bromide. This suggests that all of these components are coded for by mit-DNA. This conclusion was confirmed by RNA-DNA hybridization experiments carried out between the major discrete components, electroeluted from the gel, and separated strands of mit-DNA. As appears from Table 2, all discrete components tested, both of the RNA fraction retained and of that not retained on oligo(dT)-cellulose, were complementary in sequence to the H strand, with the exception of component #18, which hybridized predominantly to the L strand. The latter corresponds to the 7S RNA previously identified.[14]

The heterogeneous material detectable in the electrophoretic pattern of the RNA fraction retained on oligo(dT)-cellulose (Fig. 2a) hybridized to mit-DNA to an extent varying in different experiments between 20 and 50% of the RNA in the upper half of the gel and between 70 and 90% of the RNA in the lowest third: the nonhybridizable material presumably represents nuclear DNA-coded RNA which has escaped the camptothecin block. The heterogeneous RNA homologous to mit-DNA showed a predominant hybridization to the L strand in the portion of the gel corresponding to a mol. wt. range between 6×10^5 and 4.5×10^6, and, by contrast, a greater hybridization with the H strand than with the L strand in the region of lower mol. wts. A similar electrophoretic distribution of the heterogeneous material complementary to the L or the H strand was found for the RNA not retained on oligo(dT)-cellulose.

Table 3 shows the mol. wt. of the discrete electrophoretic components identified in the present work, as estimated by comparison with the electrophoretic mobility of known standards (16S, 12S, and 7S RNA).

Genetic and physical maps of HeLa cell mitochondrial DNA

Previous electron microscope mapping experiments with ferritin-labeled 4S RNA had indicated the presence of 9 sites for 4S RNA on the H strand of HeLa mit-DNA and 3 sites on the L strand.[10] In view of the above described hybridization results with amino acyl-tRNAs, which pointed to a larger number of tRNA genes in HeLa mit-DNA, the number and arrangement of 4S RNA genes on the H and L strands has been reinvestigated by using a new ferritin labeling technique (Fig. 3).[13] By this technique, seven binding sites for ferritin-4S RNA have been mapped on the L strand relative to the rRNA gene positions; furthermore, evidence suggesting the presence of three additional sites on the H strand has brought the total number of 4S RNA genes for this strand to 12. As shown in Fig. 3, the 4S RNA genes reveal a fairly uniform distribution on the two strands of HeLa mit-DNA. This contrasts with the arrangement of tRNA genes described for yeast mit-DNA: here, eleven out of twelve genes mapped so far appear to be located in about one-third of the genome.[22]

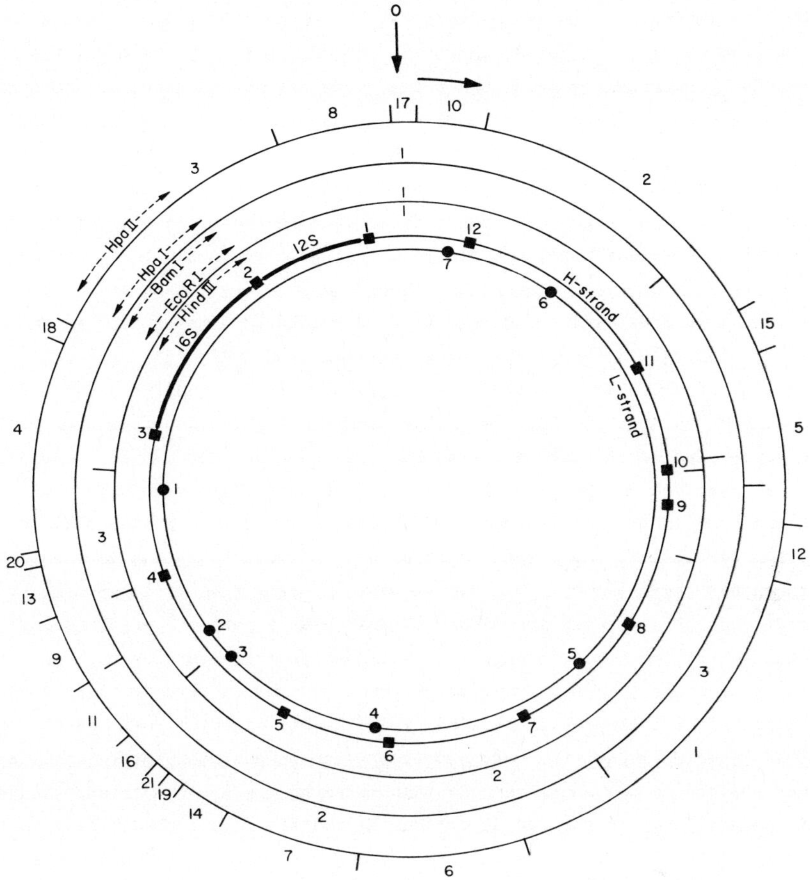

Fig. 3. Restriction maps of HeLa cell mit-DNA and map of the positions of the rRNA and 4S RNA genes. The restriction endonuclease cleavage maps determined in the present work with the enzymes Hpa II, Hpa I and Bam I and the previously available Eco RI and Hind III maps[21] have been aligned with the map of the positions of the complementary sequences for the 12S and 16S rRNA on the H strand and for 4S RNAs on the H and L strands (taken from ref. 13). 0: origin of mit-DNA replication.

Recently, a detailed physical map of HeLa cell mit-DNA has been constructed in our laboratory by ordering 21 fragments produced by the Hpa II restriction enzyme. This map has been aligned with the positions of rRNA and 4S RNA genes, with the origin of mit-DNA replication and with the previously available, less resolved maps obtained with the Eco RI and Hind III enzymes.[21]

The genetic map shown in Fig. 3 has interesting implications for the process of mit-DNA transcription. First, from the known direction of replication of the

L strand[21] (see Fig. 3), it can be inferred that transcription of the H strand proceeds from the 12S RNA gene towards the 16S RNA gene. Furthermore, the fairly uniform tRNA gene arrangement on the two strands of mit-DNA poses some restrictions as concerns the possible mechanism of transcription and processing of mitochondrial RNA and the interpretation of the multiple discrete RNA components resolved by electrophoresis through agarose-CH_3HgOH gels. Thus, if the complete transcription of the L strand is directly related to the expression of the tRNA genes located on this strand, the presence in the heterogeneous RNA labeled in a 3 hr ^{32}P pulse of molecules transcribed from L strand larger than any of the inter-4S RNA gene distances on the L strand (Fig. 2) would tend to exclude a transcription of the L-strand-coded 4S RNA genes starting from individual promoters. The data would, rather, suggest that transcription of the L strand produces a few large transcripts and, possibly, even one single RNA molecule of genome length.

The number and size of the discrete RNA species coded for by the H strand which were resolved in the present work are such that their sequences could not be accommodated in the portions of the H strand not occupied by the rRNA or tRNA genes. In particular, there appears to be no DNA stretch as long as the poly(A)-containing RNA component #5. These observations imply that at least some of the H-strand-coded tRNA species are also synthesized as a part of large transcripts, and that some of the discrete high mol. wt. RNA species separated on a gel are precursors to others. RNA-DNA hybridization experiments, already in progress, between restriction fragments produced by Hpa II and other enzymes and the individual RNA species identified here should provide in the near future conclusive evidence concerning the sequence relationship among the RNA electrophoretic components detected in the various mitochondrial fractions, and should lead to the assignment of their position on the genetic map. This analysis should also yield valuable information concerning the mechanism of transcription of mit-DNA in mammalian cells and the possible existence of spacers and regulatory sequences.

Membrane attachment of mitochondrial DNA at the site of the D-loop

In the course of E.M. investigations, by protein-free spreading techniques, of the possible association of HeLa cell mit-DNA with proteins, we have observed that almost all (>95%) of the mit-DNA molecules released by gentle lysis of mitochondria are associated with a single protein-containing structure varying in size between a small knob 100-200 Å in diameter (Fig. 4) and what appears to be a large membrane fragment up to 0.5-1 μm in maximum dimension.

Analysis of mit-DNA fragments produced by restriction endonucleases (Eco RI, Hind III, Hpa II) has provided unambiguous evidence that the protein structure is attached to the DNA in the region of the D-loop, and probably at or near the origin of mit-DNA replication (Figs. 3 and 4) (Albring, Griffith and Attardi, in preparation). The very high frequency of mit-DNA molecules with an attached "knob," as opposed to the normally low frequency of D-loops in HeLa cell mit-DNA,

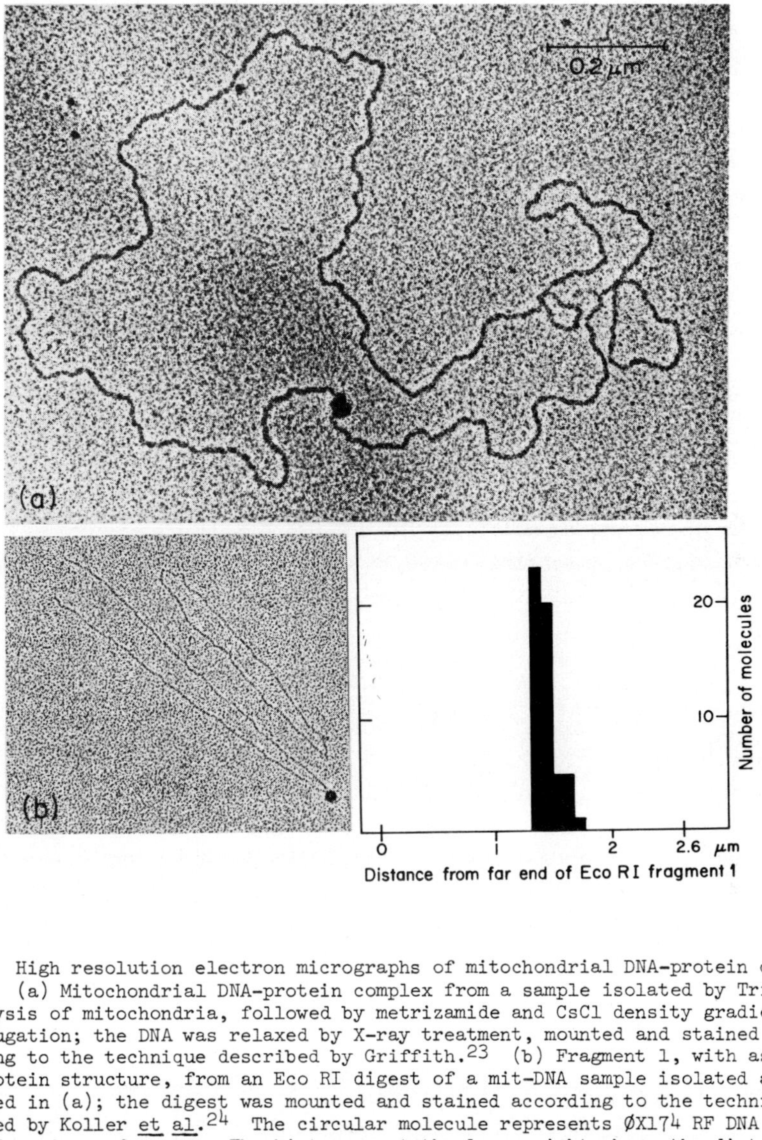

Fig. 4. High resolution electron micrographs of mitochondrial DNA-protein complexes. (a) Mitochondrial DNA-protein complex from a sample isolated by Triton X-100 lysis of mitochondria, followed by metrizamide and CsCl density gradient centrifugation; the DNA was relaxed by X-ray treatment, mounted and stained according to the technique described by Griffith.[23] (b) Fragment 1, with associated protein structure, from an Eco RI digest of a mit-DNA sample isolated as described in (a); the digest was mounted and stained according to the technique described by Koller et al.[24] The circular molecule represents ØX174 RF DNA included for size reference. The histogram at the lower right shows the distribution of the positions of the protein structure associated with Eco RI fragment 1: the lengths of the fragments were normalized to that of ØX174 RF DNA taken as 1.7 µm, and the position of the "knob" on each fragment determined on the basis of its relative distance from the farther end; 54 molecules were scored.

would tend to argue against the possibility that the presence of this "knob" in association with mit-DNA is artifactual, arising during mit-DNA isolation from sticking of protein complexes or membrane fragments to the single-stranded portion of the D-loop. Furthermore, the association with mit-DNA of the protein structure appears to withstand exposure to high salt (7 M CsCl) and 0.2% of the ionic detergent sodium lauryl sarcosinate. The most likely explanation of the above described observations is that HeLa cell mit-DNA is attached in vivo to the inner mitochondrial membrane at or near the origin of replication, and that a fragment of varying size of this membrane remains associated with the DNA during the isolation under the conditions used here.

The association of animal cell mit-DNA with the inner mitochondrial membrane had been surmised in the past on the basis of E.M. observations of osmotically lysed mitochondria or of ultrathin sections of these organelles;[25] however, the lack of suitable position markers in these early observations did not allow one to exclude non-specific adsorption phenomena.

An association with membrane, at the site of the origin of DNA replication, has been shown for the bacterial chromosome[26] and for the plasmid ColE1 DNA.[27] As in these systems, the membrane association of mit-DNA may have an important role in the regulation of mit-DNA replication and/or segregation of daughter molecules.

Acknowledgements

This work was supported by research grants from the U.S.P.H.S. (GM-11726) and NATO (5-2-0313-42).

References

1. Attardi, G., Costantino, P., England, J., Lynch, D., Murphy, W., Ojala, D., Posakony, J. and Storrie, B. (1975) In "Genetics and Biogenesis of Mitochondria and Chloroplasts (Birky, W., Perlman, P. and Byers, T., eds.), Ohio State University Press, Columbus, Ohio, p. 3-65.
2. Aloni, Y. and Attardi, G. (1971) J. Mol. Biol. 55, 251-270.
3. Aloni, Y. and Attardi, G. (1971) Proc. Nat. Acad. Sci. U.S.A. 68, 1957-1961.
4. Murphy, W., Attardi, B., Tu, C. and Attardi, G. (1975) J. Mol. Biol. 99, 809-814.
5. Aloni, Y. (1972) Proc. Nat. Acad. Sci. U.S.A. 69, 2404-2409.
6. Aloni, Y. and Locker, H. (1973) Virology 54, 495-505.
7. Lucas, J. J. and Ginsberg, H. S. (1972) Bioch. Bioph. Res. Comm. 49, 39-44.
8. Jacquemont, B. and Roizman, B. (1975) J. Virology 15, 707-713.
9. Aloni, Y. and Attardi, G. (1971) J. Mol. Biol. 55, 271-276.
10. Wu, M., Davidson, N., Attardi, G. and Aloni, Y. (1972) J. Mol. Biol. 71, 81-93.
11. Lynch, D. C. and Attardi, G. (1976) J. Mol. Biol. 102, 125-141.

12. Attardi, G., Amalric, F., Ching, E., Costantino, P., Gelfand, R. and Lynch, D. (1976) In "The Genetic Function of Mitochondrial DNA" (A. Kroon and C. Saccone, eds.) North-Holland, Amsterdam, in press.

13. Angerer, L., Davidson, N., Murphy, W., Lynch, D. C. and Attardi, G. (1976) Cell, in press.

14. Ojala, D. and Attardi, G. (1974) J. Mol. Biol. 88, 205-219.

15. Bailey, J. and Davidson, N. (1976) Anal. Biochem. 70, 75-85.

16. Perlman, S., Abelson, H. T. and Penman, S. (1973) Proc. Nat. Acad. Sci. U.S.A. 70, 350-353.

17. Attardi, G., Costantino, P., England, J., Lederman, M., Ojala, D. and Storrie, B. (1973) Acta Endocrinologica Suppl. No. 180, 263-293.

18. Ojala, D. and Attardi, G. (1974) J. Mol. Biol. 82, 151-176.

19. Hirsch, M. and Penman, S. (1973) J. Mol. Biol. 80, 379-391.

20. Robberson, D., Aloni, Y., Attardi, G. and Davidson, N. (1971) J. Mol. Biol. 60, 473-484.

21. Brown, W. M. and Vinograd, J. (1974) Proc. Nat. Acad. Sci. U.S.A. 71, 4617-4621.

22. Fukuhara, H., Bolotin-Fukuhara, H. M., Hsu, H. J. and Rabinowitz, M. (1976) Molec. Gen. Genetics, in press.

23. Griffith, J. D. (1973) In "Methods in Cell Biology (M. Prescott, ed.), vol. 7, 129-146. Academic Press Inc., New York.

24. Koller, T., Sogo, J. M. and Bujard, H. (1974) Biopolymers 13, 995-1009.

25. Nass, M.M.K. (1969) J. Mol. Biol. 42, 521-528.

26. Siegel, P. J. and Schaechter, M. (1973) Ann. Rev. Microbiol. 27, 261-282.

27. Sparks, R. B. and Helinski, D. R. (1976) Fed. Proc. 35, 1376.

Genetics and Biogenesis of Chloroplasts and Mitochondria, Th. Bücher et al. eds.
© 1976 Elsevier/North-Holland Biomedical Press, Amsterdam, The Netherlands.

FUNCTIONAL ORGANIZATION AND EVOLUTION OF ANIMAL MITOCHONDRIAL DNA

William B. Upholt* and Igor B. Dawid
Department of Embryology
Carnegie Institution of Washington
115 West University Parkway
Baltimore, Maryland 21210

Previous work with two species of *Xenopus* whose mtDNAs differ in 20-30% of their total bases has shown that these differences are not randomly distributed throughout the genomes. Certain regions, i.e. those coding for mt-rRNA and tRNAs, are nearly identical between these two species, whereas other parts of the genomes have diverged to the extent that no sequence homology could be detected (Dawid, 1972). More recent studies have used electron microscopy and restriction endonucleases to physically map specific genes and functional regions in these and other mtDNAs. The genes for the mt-ribosomal RNAs are located next to each other in Hela mtDNA (Wu et al., 1972), and in *Xenopus* and *Drosophila* mtDNA (Dawid et al., 1976). In addition, the position and polarity of the D-loop has been mapped with respect to the position of the rRNA cistrons in the two Xenopus species (Dawid et al., 1976). The 15 to 16 tRNA genes present on the H strand of *X. laevis* show a distribution similar to that found in Hela mtDNA (Wu et al., 1972; Dawid et al., 1976).

The above mentioned studies have primarily looked at those parts of the mitochondrial genome that are most highly conserved, the rRNA and tRNA genes. In order to study the part of the mitochondrial genome which is least conserved, we have chosen to compare the very closely related species where the majority of sequences are nearly identical and those segments of DNA which have changed most rapidly can be analyzed in some detail. After careful consideration of available evolutionary data, we chose sheep and goat as the species to be compared. Our primary analytical techniques have been the analysis by gel electrophoresis and electron microscopy of fragments generated by restriction endonucleases, and the analysis by electron microscopy of heteroduplexes formed from these fragments.

Polyacrylamide gel electrophoresis (Figure 1) of digests of mtDNAs from three individual sheep and two individual goats with the restriction endonuclease *Hae*III, shows that detectable differences exist between mtDNAs of various individuals of a single species, and that common fragments occur between sheep and goats. A detailed analysis of gels of these digests run under various electro-

---

*Present address: Department of Pediatrics and Biochemistry
Joseph P. Kennedy Jr. Mental Retardation Research Center
The University of Chicago
Chicago, Illinois 60637

Hae III 5' GG/CC 3'

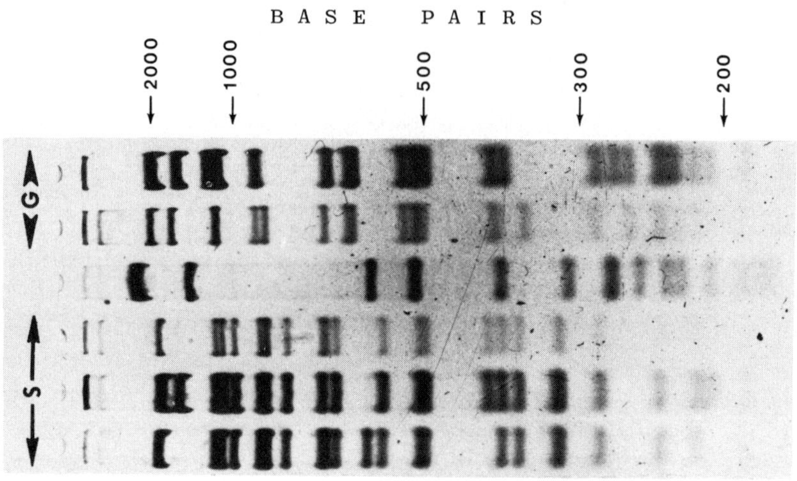

Figure 1. Polyacrylamide gels (4%) of HaeIII digests of sheep and goat mtDNAs. Three individual sheep and two goats were analyzed. The sixth lane contained reference DNA fragments. The largest fragments are about 2000 base pairs, and the smallest fragments shown are 150 bases. Smaller fragments occur but are not illustrated in this photograph. Migration is from left to right.

phoretic conditions show that sheep b and c each lack a single but different cleavage site when compared to sheep a, and each goat has probably lost two sites and gained two new ones when compared to the other goat. Sheep and goat mtDNAs have 8 to 11 HaeIII fragments in common, while the other 20 or so fragments differ. Differences in the fragment pattern of mtDNA between individuals of the same species were also observed by Potter et al., (1975).

A more comprehensive analysis of the cleavages by EcoRI and HindIII restriction endonucleases (which recognize more complex sites and thus make fewer cuts) has been done for the mtDNAs of these individuals and a map of the various cleavage sites is shown in Figure 2. The positions of the D-loops (Kasamatsu and Vinograd, 1974) on these fragments have also been mapped by electron microscopy. The relationships between the various cleavage sites have been determined by gel electrophoresis of double digests, and electron microscopic analysis of sheep-goat heteroduplexes formed from combinations of complete digests, partial digests and intact strands.

This map shows in finer detail results similar to those found with the HaeIII digests. First, the overall organization of sheep and goat mtDNAs is very similar and shows a high degree of conservation. Second, clear differences exist between sheep and goat, i.e. the existence of restriction sites unique to one or the other species and a difference in the position of one end of the D-loop

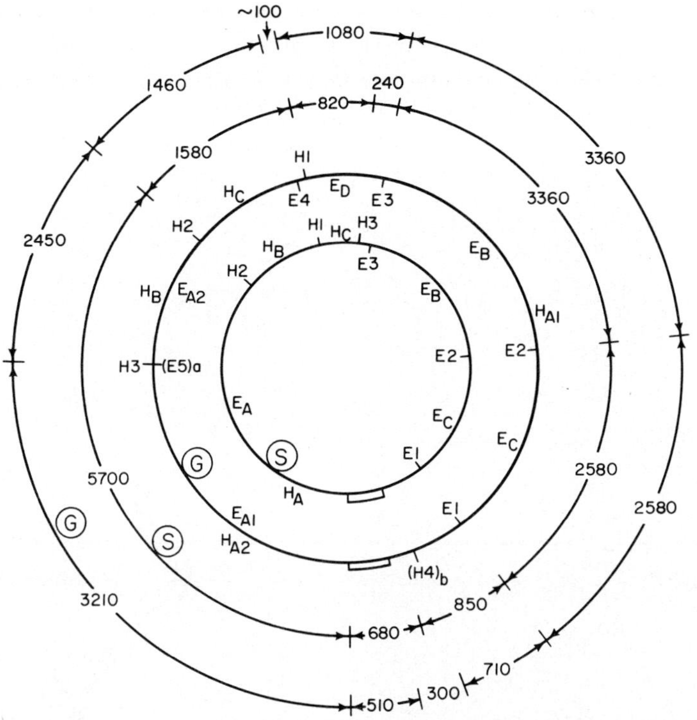

Figure 2. Maps of sheep and goat mtDNAs showing locations of D-loops and EcoRI and HindIII cleavage sites. The inner 2 circles show positions on sheep and goat mtDNAs, respectively. Cleavage sites are indicated by E or H (for EcoRI or HindIII) followed by Arabic numerals assigned counterclockwise. Sites indicated (E5)a and (H4)b are present only on goat a or goat b, respectively. Fragments are labeled by E or H followed by a letter subscript which are assigned according to decreasing size of the fragments. Fragments determined by cleavage sites present in only one goat are indicated as subfragments (H_{A1} and H_{A2}). The outer 2 circles show distances in base pairs between positions on the maps.

(see Figure 4). Third, differences between individual goats are also seen in EcoRI and HindIII sites, but the three sheep are identical in this respect.

Heteroduplexes between sheep and goat mtDNA formed at moderately stringent conditions show a single heterology loop at the position where the sheep D-loop is longer than the goat D-loop. This region of greatest evolutionary change between sheep and goat mtDNA was studied further by a threeway comparison of heteroduplexes formed by EcoRI restriction fragments from one sheep and two goats. The existence of a common EcoRI cleavage site very close to this region provides a convenient reference point for measuring position and sizes of heterology loops. These heteroduplexes were spread from solutions containing varying amounts of formamide to vary the criterion for duplex stability.

These studies allowed several conclusions. First, with one goat-sheep pair, two alternative heteroduplex forms were found at the nonconserved end of the D-loop. Figure 3a shows an example of the simpler of these two forms, with two symmetrical heterology loops at both ends of the D-loop region. The loop closer

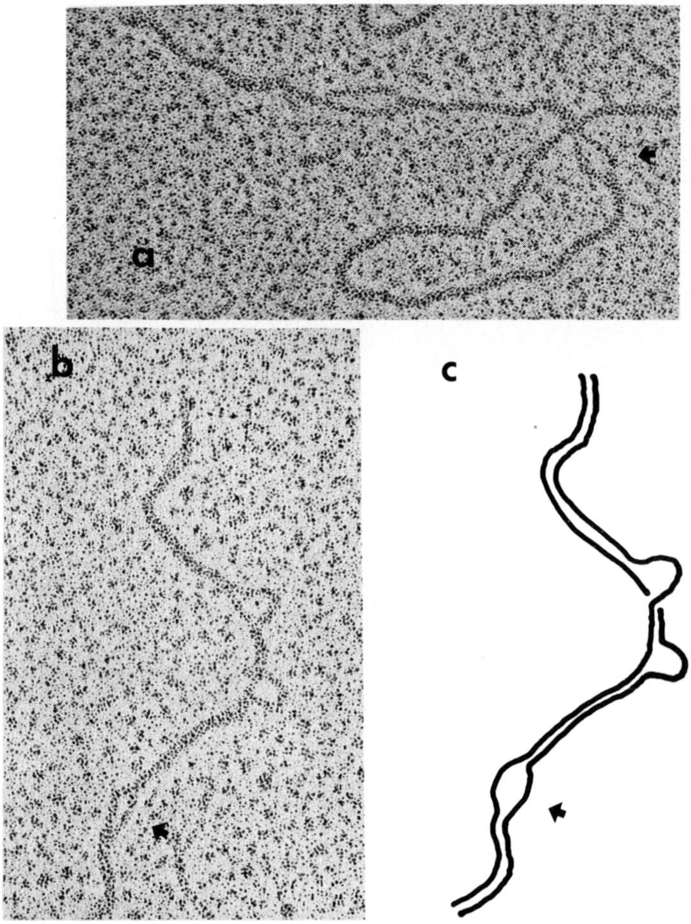

Figure 3. Electron micrographs of heteroduplex molecules formed from EcoRI fragments of sheep and goat mtDNA. Heteroduplexes were spread from 30% formamide onto water by a modification of the Kleinschmidt technique (Davis et al., 1971). Molecules terminate at the common EcoRI cleavage site E1 at the top of the figure. a) symmetrical loop; b) asymmetric loops; c) interpretative drawing of proposed strand continuity in asymmetric loops. An arrow points to the heterology loop beyond the conserved end of the D-loop, at 1500 to 1700 base pairs from the EcoRI site (see Figure 4).

to the EcoRI site also exists in an alternate asymmetric form in about 50% of the molecules (Figure 3b). In that form one branch of the loop remains unchanged in length, whereas the other branch is about half as long. A second asymmetric loop appears after a short duplex region. Figure 3c gives our interpretation of the strand continuity in the looped region. The presence of alternate forms of loops at the same position suggests that in one or both of these mtDNAs a duplicated segment occurs that can pair in either of two ways.

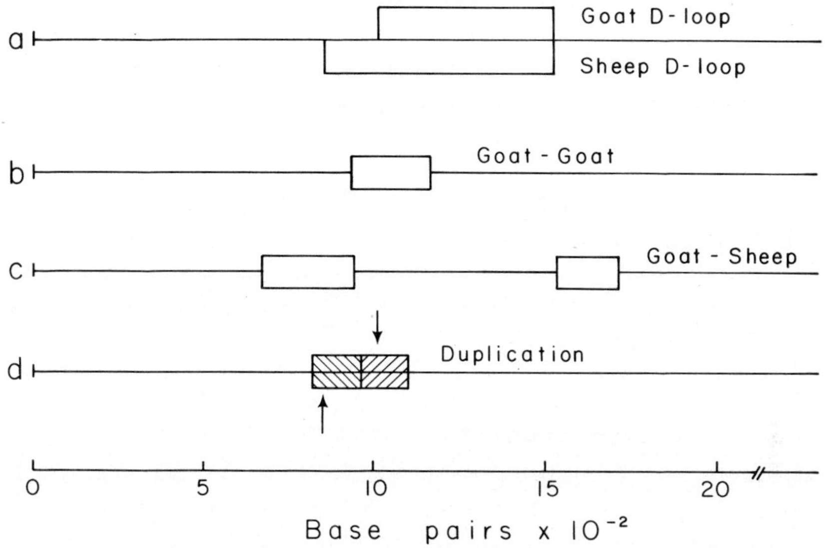

Figure 4. Evolutionary changes in the D-loop region of sheep and goat mtDNAs. The scale measures distance in base pairs from the common EcoRI cleavage site E1. a) position of goat and sheep D-loops. No differences were found between individuals of the same species; b) position of heterology loop in goat a-goat b heteroduplexes spread from 71% formamide; c) position and minimum sizes of heterology loops in sheep-goat heteroduplexes; d) position of duplicated segments forming duplex region between the asymmetric loops (see Figure 3b). Arrows indicate location of the ends of sheep and goat D-loops.

Figure 4 shows a summary of the data on the comparative structure of the region around the D-loop. Part a shows a diagram of the relative positions and sizes of the sheep and goat D-loops. Part b shows the location of the first heterology loop which occurs in goat-goat heteroduplexes. The estimated extent of mismatch in this region is 5-10%. Part c indicates the positions and sizes of the two regions of lowest homology between the sheep and goats. Although the loop at the conserved end of the D-loop is present in both sheep-goat heteroduplexes, the sequences in the two goats appear to have diverged from the sheep sequence to different extents since this loop becomes visible at different formamide concentrations in the two heteroduplexes. However, once this loop appears, its size remains relatively constant over a range of denaturation conditions. The loop at the nonconserved end, however, expands into the D-loop region to include the region of the goat-goat heterology loop as the stringency of the criterion for duplex formation is increased. Part d indicates the position of the two regions involved in the proposed duplication and indicates by arrows the ends of the

D-loops in relation to this duplication. The correlation between the positions of the duplicated sequences and the ends of the D-loops suggests that there may be some functional relationship between them.

In summary, we have found that the two least conserved portions of sheep and goat mtDNAs occur at the two ends of the D-loop. Goat and sheep mtDNAs have one conserved end of the D-loop and one end which is different. At the nonconserved end, there appears to be a duplication of a segment of 150 base in one or both of these species.

REFERENCES

Davis, R. W., Simon, M., and Davidson, N. (1971). Methods Enzymol. 21, 413-428.

Dawid, I. B. (1972). Develop. Biol. 29, 139-151.

Dawid, I. B., Klukas, C. K., Ohi, S., Ramirez, J. L., and Upholt, W. B. (1976). In: The Genetic Function of mitochondrial DNA, C. Saccone and A. M. Kroon, eds. Amsterdam: North Holland.

Kasamatsu, H. and Vinograd, J. (1974). Annu. Rev. Biochem. 43, 695-719.

Potter, S. S., Newbold, J. E., Hutchison, C. A., Edgell, M. H. (1975). Proc. Nat. Acad. Sci., U.S.A. 72, 4496-4500.

Wu, M., Davidson, N., Attardi, G., and Aloni, Y. (1972). J. Mol. Biol. 71, 81-93.

Genetics and Biogenesis of Chloroplasts and Mitochondria, Th. Bücher et al. eds.

PHYSICAL MAP AND REPLICATION OF RAT MITOCHONDRIAL DNA

K. Koike, M. Kobayashi, S. Tanaka and H. Mizusawa
Cancer Institute (Japanese Foundation for Cancer Research)
Kami-Ikebukuro, Toshima-ku, Tokyo 170, Japan

SUMMARY

The rat mtDNA was cleaved into unique segments by restriction endonucleases, EcoRI and Hha from E. coli and H. haemolyticus, respectively. The sizes of the segments were determined with data obtained from the electron microscope and the agarose gel electrophoresis. On the basis of these data and estimates of the sizes of the partially-digested products, the physical map of the mitochondrial genome was constructed. After mtDNA was pulse-labeled with [^{3}H]thymidine the distribution of label within the open-circular molecules was determined by analysis of segments produced by EcoRI and Hha, in order to deduce a temporal order of synthesis. Comparison of the temporal order with the physical order of the segments in the mtDNA molecule indicates that discontinuous replication of rat mtDNA begins at the fixed origin and proceeds unidirectionally around the circular genome, terminating at a region adjacent to the initiation origin.

INTRODUCTION

The genome of the rat mitochondria is a closed-circular molecule of about 10 million daltons that replicates discontinuously by means of a Cairns-type intermediate. We have been studying the process of mtDNA replication in vitro (1) and also analyzing the in vivo replicating structure with the products obtained by restriction endonucleases by means of electron micrographic and agarose gel electrophoretic techniques. As previously reported (2,3), the restriction endonuclease EcoRI creates six double-stranded breaks at specific sites in the molecule. By analyzing segments of in vivo Cairns-type intermediates, it was demonstrated that there is a fixed origin and the replication is unidirectional.

In this report, we describe recent results concerning the physical mapping of the six EcoRI segments in the rat mitochondrial genome and the terminus of replication by analyzing segments of the in vitro pulse-labeled mtDNA digested by EcoRI and Hha. We compare this temporal order with the physical sequence of the DNA segments and discuss a mode of replication of the mtDNA molecule.

MATERIALS AND METHODS

Isolation of rat mtDNA: Mitochondrial fraction was prepared from rat liver, and mtDNA was extracted by the SDS-phenol method (1). Closed-circular DNA was purified by equilibrium centrifugation in CsCl-EthBr (4).

Preparation of in vitro labeled mtDNA: As described previously (1), isolated mitochondria from newborn rat liver were labeled with [^{3}H]thymidine for 10 min.

After termination of the reaction, the labeled mtDNA was purified by the phenol extraction, followed by equilibrium centrifugation in Cs_2SO_4. The labeled mtDNA was, then, subjected to band velocity sedimentation in CsCl-EthBr (1), to separate closed- and open-circular DNA fractions.

Cleavage of mtDNA with restriction endonucleases: MtDNA was digested by restriction endonuclease EcoRI from E. coli or Hha from H. haemolyticus under the similar reaction conditions (5). For analysis of radioactive label in each DNA segments, [^{3}H]DNA was digested together with carrier mtDNA.

Agarose gel electrophoresis: The restriction product was subjected to electrophoresis in 1.4 % agarose gel containing 0.5 μg/ml EthBr (6). For quantitation of radioactive label in each segment, DNA bands were sliced under illumination with ultraviolet lamp, followed by counting.

Electron microscopy: Specimens were prepared by the formamide technique (7).

RESULTS AND DISCUSSION

Analysis of DNA segments generated by restriction endonuclease: The closed-circular mtDNA was completely cleaved by EcoRI, and the restriction products were examined by electron microscopy and agarose gel electrophoresis. Length distributions for the population of linear segments displayed six major components with average length of S1=2.01 μm, S2=1.28 μm, S3=0.95 μm, S4=0.60 μm, S5=0.23 μm, S6= 0.17 μm (3). The sum of these six segments is very close to the circular contour length, 5.26 μm. On the other hand, two and three EcoRI restriction sites have been reported in mouse and human mtDNA respectively (8,9).

Linear duplex segments generated by Hha digestion were also examined by agarose gel electrophoresis, which revealed four discrete bands. The DNA lengths of four segments were determined by their electrophoretic mobility relative to rat and mouse EcoRI segments as Hha1=2.85 μm, Hha2=1.41 μm, Hha3=0.63 μm and Hha4=0.33 μm.

Position of the initiation site and unidirectional replication: As described previously (2,3), restriction endonuclease cleavage provided a basis for examination of the initiation site of DNA replication. When the in vivo replicating molecules were extensively digested by EcoRI, the small replication eye was found exclusively on the segment S1 (2,3). The small eyes, randomly selected, were slightly heterogenuous in size and strandness. Electron micrographic analysis of those eyes indicated that 25 % of the linear replicative forms had two double-stranded branches, and the remaining forms had one each of single- and double-stranded branches. Alignment of these forms indicated that initiation of replication occurs on the largest segment at a position 0.47 μm away from the proximal restriction sites. Length measurements of these linear replicative forms also provided an array of molecules with increasing degrees of replication, indicating that discontinuous replication proceeds unidirectionally away from this proximal restriction site throughout the length of the largest EcoRI segment (3).

Physical mapping of the EcoRI restriction segments: In order to determine the order of the six DNA segments within the rat mtDNA, molecules containing the small eyes were cleaved by EcoRI under limited conditions and the resulting partially digested segments were examined by electron micrograph using the initiation site as a marker, since the initiation site is at a fixed position on the S1 segment. Based on the alignment of partially digested forms having small eyes, the EcoRI segments were constructed. Taking the proximal restriction site on the S1 segment as the zero point, four of the six cleavage sites could be accurately mapped in this manner and are located at 0.000, 0.382(S1), 0.625(S1-S2) and 0.806(S1-S2-S3) in order. Further data for this mapping were obtained by agarose gel electrophoretic analysis.

Molecules were digested by EcoRI under limited conditions and several intermediate species were obtained electrophoretically. The 0.42 μm intermediate contained two segments S5, S6 and the 0.85 μm intermediate consisted of the S4 and S5. It appears that the S5 segment lies between the S4 and S6 (3). Data about intermediate with size of 1.14 μm suggested a possible combination between two segments S3 and S6. As four Hha cleavage sites were found in the rat mtDNA, each EcoRI segment was further cleaved by Hha. Four EcoRI segments, S1, S3, S4 and S5 were digested into two pieces. On the basis of these observations, a combination of EcoRI segments S3-S4 was ruled out. In this way, cleavage map of the rat mtDNA was constructed as shown in Fig. 1.

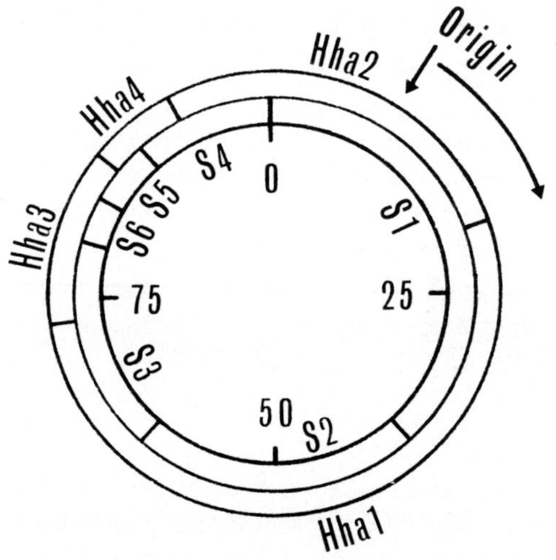

Fig. 1. Physical map of rat mtDNA.

Position of the termination site: As details were described previously (1), exposure of newborn rat mitochondria, which are actively synthesizing DNA, to a short pulse of [^{3}H]thymidine, resulted in incorporation of label at the growing point of replicating DNA. During a pulse time, open-circular molecules containing appropriate gaps will be labeled in those regions [i.e. a gap-filling synthesis after segregation of two daughter molecules (1,10)]. Then, there will be a gradient of labeling in such DNA, from which a temporal order of synthesis of different parts of the molecule can be deduced.

For this purpose, the [^{3}H]open-circular DNA of the pulse labeled daughter molecule was completely digested with EcoRI or Hha. Each segment was analyzed for its radioactivity after separation of the products by agarose gel electrophoresis. The specific activity of each segment is presented in Table I. As shown in the table, there is a consistent gradient of labeling. Segments with higher specific activity(S4 and S1 or Hha2 and Hha4) are from the terminal region of the DNA. Further digestion of the [^{3}H] labeled EcoRI segment S1 by Hha or other restriction enzyme is underway. It preliminary indicates that the specific activity was increased to maximal value in a small region of the segment S1, which is roughly located at a left side of the initiation origin (Fig. 1). Present data are consistent with an idea that replication of the rat mtDNA proceeds unidirectionally away from the origin, and terminates at the region adjacent to the initiation site, although the specific activity was not corrected for the slightly different thymidine contents of the segments. A comparison of the order of synthesis with the physical order of the segments may allow one to distinguish rat from mouse modes of replication. Rat mtDNA is similar to mouse mtDNA in having an unidirectional replication, but the degree of asymmetry is in contrast to that of mouse mtDNA, which is extremely asymmetric (10).

Table I Specific activity of each segment from pulse-labeled open-circular rat mtDNA

Segment		Specific activity cpm/μmDNA
EcoRI	S1	913
	S2	284
	S3	520
	S4	960
	S5	807
	S6	642
Hha	1	464
	2	1504
	3	963
	4	1448

Specific activity of pulse label is the counts/min divided by the μm DNA length of each segment, uncorrected for the thymidine content.

ACKNOWLEDGEMENT

We thank Dr. H. Sugano, Director of Cancer Institute, for his encouragement and generous support. Restriction endonuclease Hha was kindly provided by Dr. M. Takanami, Institute for Chemical Research, Kyoto University.

REFERENCES

1. Koike, K., Kobayashi, M. and Fujisawa, T. (1976) Biochim. Biophys. Acta, 425, 18-29.
2. Koike, K., Kobayashi, M. and Tanaka, S. (1975) in DNA Synthesis and Its Regulation, pp.578-585, Benjamin Inc., California.
3. Koike, K., Kobayashi, M., Tanaka, S. and Mizusawa, H. (1975) Proceedings of the 1975 Molecular Biology Meeting of Japan, pp.70-72, Kyoritsu Shuppan, Tokyo.
4. Radloff, R., Bauer, W. and Vinograd, J. (1967) Proc. Natl. Acad. Sci., 57, 1514-1521.
5. Yoshimori, R. N. (1971) PhD.Thesis, Univ. of Calif. at San Francisco Med. Center.
6. Sharp, P. A., Sugden, B. and Sambrook, J. (1973) Biochemistry, 12, 3055-3063.
7. Davis, R. W., Simon, M. and Davidson, N. (1971) in Methods in Enzymology, Vol. 21, pp.413-431, Academic Press, N.Y.
8. Brown, W. M. and Vinograd, J. (1974) Proc. Natl. Acad. Sci., 71, 4617-4621.
9. Robberson, D. L., Clayton, D. A. and Morrow, J. F. (1974) Proc. Natl. Acad. Sci., 71, 4447-4451.
10. Berk, A. J. and Clayton, D. A. (1975) in DNA Synthesis and Its Regulation, pp.560-577, Benjamin Inc., California.

Genetics and Biogenesis of Chloroplasts and Mitochondria, Th. Bücher et al. eds.

MEASUREMENT OF THE RELATIVE RATE OF MITOCHONDRIAL DNA SYNTHESIS UNDER EXPERIMENTALLY VARIED CONDITIONS

Daniel Bogenhagen and David A. Clayton
Laboratory of Experimental Oncology
Department of Pathology
Stanford University School of Medicine
Stanford, California 94305, U.S.A.

INTRODUCTION

Quantitative studies of mammalian mitochondrial DNA (mtDNA) replication have been complicated by several factors: (a) The quantity of mtDNA per cell is very small, so that direct physical quantitation is difficult; (b) There is a very large excess of nuclear DNA relative to mtDNA in almost all animal cells; (c) Simple radioactive labeling procedures using either radioactive thymidine or ^{32}P provide no assurance that the amount of radioactive label incorporated is exactly proportional to the amount of mtDNA synthesis. Clearly, it is especially difficult to assay the effect on the rate of mtDNA synthesis caused by various physiological manipulations or pharmacologic agents. Nevertheless, simple radioactive labeling experiments have been used to quantitate the effects of viral infection (1,2), drug treatment (3), and thermosensitivity (4) on the rate of mtDNA synthesis. The purpose of this paper is to present a general method for determination of the rate of mtDNA synthesis in which labeling artifacts are minimal.

The average suspension cultured mouse L-cell contains 1000 (5,6) mtDNA molecules of molecular weight 10^7 which replicate within the organelle by a semi-conservative mechanism. Each molecule requires only 1.0 hours of the 18 hour generation time for replication. Thus, to the extent that the replication time of a single molecule does not change greatly, measurement of the rate of mtDNA synthesis over a period of several hours is actually a measure of the number of mtDNA molecules which have replicated. This suggests that the overall rate of synthesis can be measured using bromodeoxyuridine (BrdUrd) density labeling in experiments patterned after the classical experiments of Meselson and Stahl (7). The methods described below employ ^3H-thymidine prelabeling of mtDNA followed by BrdUrd labeling. Replication of ^3H-labeled mtDNA molecules in the presence of BrdUrd produces molecules labeled in one strand with ^3H-thymine and in the other with bromouracil and which are of hybrid density in buoyant CsCl density gradients. The fraction of prelabel radioactivity shifted in buoyant density is directly proportional to the amount of mtDNA synthesized. Since density labeling occurs at high BrdUrd concentrations, the density shift is easily detected and the method is essentially independent of fluctuations in the rate of endogenous thymidine triphosphate production.

MATERIALS AND METHODS

All experiments were performed with mouse LA9 cells, unless otherwise noted. The thymidine kinase minus mutant L-cell line used in one experiment was the C2-1 subclone of the mouse LMTK$^-$ line described by Berk and Clayton (8). All experimental methods have been described previously (5,9,10).

RESULTS AND DISCUSSION

Two slightly different experimental protocols are presented in Figure 1. In method A, the technically simpler approach, mtDNA is isolated after progressively longer periods of BrdUrd labeling.

We have used this approach to determine the effect of methotrexate, in the presence of thymidine, on the rate of mtDNA synthesis in thymidine kinase minus mouse L-cells (9). In this experiment mtDNA was mass labeled before methotrexate treatment. After prelabeling was stopped, methotrexate and BrdUrd were added. mtDNA was isolated at 4, 8, 12 and 16 hours of BrdUrd labeling. Buoyant density gradient analysis showed that mtDNA synthesis can continue at approximately 60% of the theoretical control rate for at least 10 to 12 hours of virtually total inhibition of nuclear DNA synthesis. These results were in good agreement with independent ^{32}P labeling experiments. However, this approach has several disadvantages. First, after 8 to 10 hours of BrdUrd labeling, reentry of hybrid density molecules into replication becomes a complicating factor since replication produces a hybrid molecule indistinguishable from the parental molecule as well as a fully heavy but non-radioactive molecule which is not detectable. Second, prolonged exposure to BrdUrd is highly toxic to animal cells which contain a wild-type thymidine kinase activity.

These disadvantages are largely circumvented if the BrdUrd labeling time is minimized. Figure 1B describes an alternative experimental design which can sensitively detect a change in the rate of mtDNA synthesis following physiological changes such as those described above. This experimental design was originally used to study two related questions (10). First, in which periods of the mouse L-cell cycle is mtDNA synthesized? Second, are recently replicated mtDNA molecules randomly selected from the total population for a second round of replication? In order to achieve the greatest resolution within the cell cycle, only a 2-hour period of ^3H-thymidine labeling was used in order to identify molecules replicated in a short time interval. This label was then efficiently chased by addition of excess cold thymidine. At various chase times, aliquots of the culture were removed for four-hour periods of BrdUrd labeling. Thus, the 24-hour period following ^3H-thymidine labeling was essentially "scanned" with successive four-hour periods of BrdUrd labeling. By a theoretical analysis it was possible to show that random reselection of recently replicated molecules and

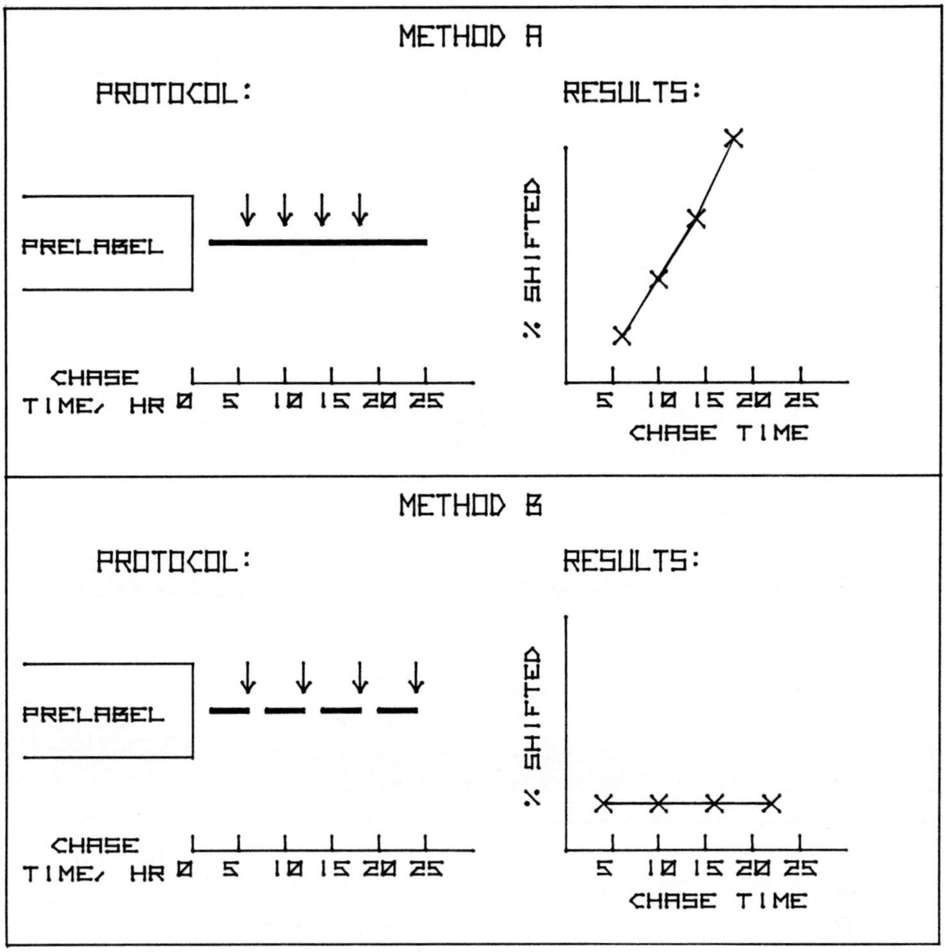

Figure 1. Methods for measurement of mtDNA synthesis using density labeling techniques.

A. Cells are prelabeled with ^{3}H-thymidine until addition of 20 μM cold thymidine to start the chase at time 0. BrdUrd labeling (80 μM) indicated by the heavy horizontal line is begun at +2 hours and equal culture volumes are removed for mtDNA isolation at the times indicated by the vertical arrows. Total mtDNA is isolated in propidium diiodide-CsCl gradients and analyzed in CsCl gradients for the fraction of prelabel radioactivity shifted in density by replication in the presence of BrdUrd. The plot of "fraction shifted vs. chase time" on the right illustrates the form of the results, with an arbitrary scale for the ordinate.

B. Prelabeling is identical to that for method A. At each of several chase times, aliquots are removed from the main culture vessel and incubated with BrdUrd for a standard time interval (as illustrated, 4 hours). Additional experimental procedures are as described for method A.

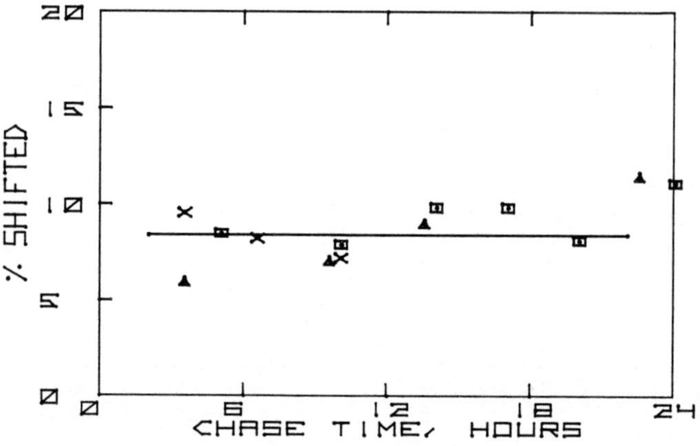

Figure 2. Density shift experiment following 2-hour ^{3}H-thymidine labeling.

The horizontal line indicates the theoretical expected value of 8.4% of mtDNA shifted in buoyant density by BrdUrd incorporation during 4 hours of density labeling. The scatter diagram shows the results of 3 separate experiments with a mean of 8.9 ± 1.7%.

TABLE 1

EFFECT OF HIGH DOSE ^{3}H-THYMIDINE ON THE RATE OF mtDNA SYNTHESIS

Experiment	BrdUrd labeling period, hours after chase	Percentage of mtDNA shifted in density, F(t)		
		exp.*	observed	% rate
A	2-6	8.4	7.5	89
	6-10	8.4	7.2	86
B**	1.5-7.5	11.7	7.84	67
	9-15	11.7	9.05	77
	16-22	11.7	7.72	66

* expected values are calculated as described (10) using the relation

$$F(t) = 1 - e^{-\beta \ln 2} \quad \text{where} \quad \beta = \frac{\text{labeling time - replication time per molecule}}{\text{cell doubling time}}$$

F(t) is multiplied by 0.77 to correct for a catenane frequency of 23% as described (10).

**The cell doubling time in this experiment was 21 hours.

continuous mtDNA synthesis during the cell cycle would result in a constant fraction of mtDNA shifted in density regardless of the length of time between tritium and BrdUrd labeling. Figure 2 shows that the experimental result, a relatively constant shift of 8.9 ± 1.7% of mtDNA, is in good agreement with the predicted value of 8.4%. It can also be shown that if mtDNA synthesis were restricted to a fraction of the cell cycle this plot of fraction shifted vs. chase time would show a distinct maximum at 16 to 18 hours of chase. No such variation was observed. This experiment thus shows that mtDNA molecules are randomly selected for replication throughout the cell cycle.

This experiment can serve as a control for further studies to test the effects of experimental variables such as hormones, drugs, metabolic toxins, radiation or nutritional factors. The experimental perturbation can be introduced after ^{3}H-thymidine labeling but before BrdUrd labeling. For most purposes, a long term ^{3}H-thymidine prelabeling is preferable in order to approximate mass labeling of mtDNA. In general, mass labeling is not necessary since molecules are randomly selected for replication. However, it is conceivable that some perturbations may alter this random character, so that recently replicated molecules may no longer be randomly selected for replication. If such an effect occurred, double labeling with short term ^{3}H-thymidine in addition to mass labeling with ^{14}C-thymidine could detect an asynchrony in the rate of rereplication of the recently replicated subpopulation of mtDNA molecules compared to the mass labeled population. Two examples of this method will be discussed below.

In the rereplication experiment described above the dose of ^{3}H-thymidine was minimized in order to prevent radiolytic damage to S-phase nuclei. Amounts of carrier free ^{3}H-thymidine greater than 0.2 mCi/liter produced a significant lag in the growth curve similar to that previously interpreted to represent mitotic arrest of cells which incorporated large amounts of ^{3}H-thymidine into nuclear DNA (11). This phenomenon raises the question of whether mtDNA replication continues normally when a large fraction of cells are arrested at mitosis. To address this question, we repeated the experiment of Figure 1 with a toxic concentration of ^{3}H-thymidine of 2 mCi/liter (46 Ci/mmol). As shown in Table 1, the relative rate of mtDNA synthesis was decreased in the two experiments to 87% and 70% of the expected rate for unperturbed mtDNA synthesis. However, only an insignificant fraction of the pulse-labeled nuclear DNA reentered replication, as would be predicted if mitotic arrest occurred. This continued synthesis of mtDNA is consistent with our previous observation that mtDNA synthesis continues independently of nuclear DNA synthesis during thymidylate starvation (9).

In an additional set of experiments we investigated the relative rate of mtDNA synthesis as cells enter the stationary stage of cell culture. Our initial

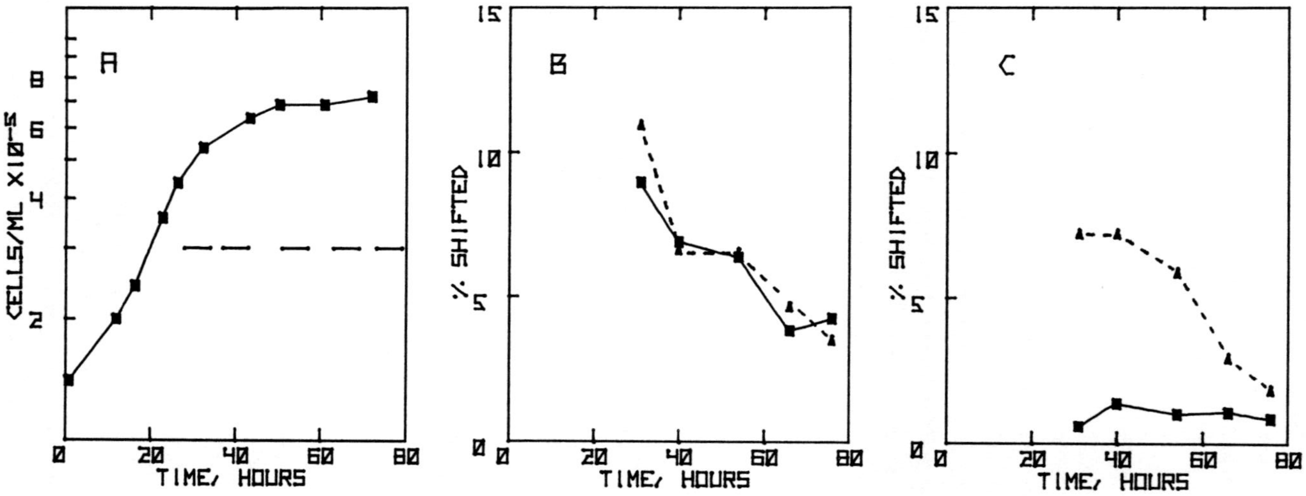

Figure 3. Rate of mtDNA replication in cells entering the stationary stage of suspension culture.

A. Cell growth curve and labeling protocol. A spinner culture of mouse L-cells prelabeled for four generations with 10 µCi/liter ^{14}C-thymidine was pulse-labeled with 50 µCi/liter ^{3}H-thymidine for 2 hours at a cell density of 6.5 x 10^5 per ml (arrow). The chase was initiated with the addition of 20 µM thymidine. One-fifth of the culture was removed at each of 5 time points for 6-hour periods of incubation with 80 µM BrdUrd (as in method B, Figure 1), indicated by horizontal bars.

B. Rate of mtDNA synthesis, expressed as the fraction of long-term ^{14}C(▲-▲) or pulse ^{3}H (■—■) prelabel radioactivity shifted in density, as a function of culture age.

C. Rate of nuclear DNA synthesis, expressed as for mtDNA in panel B, as a function of the culture age. Note that the ^{3}H-thymidine label was incorporated into nuclear DNA only in those cells in the S phase of the cell cycle when the culture was in late log stage. Nuclear DNA samples were sheared by repeated passage through a 25 gauge needle.

experiments, like those of Levine (2), used simple radioactive labeling methods to estimate the relative rates of both mtDNA and nuclear DNA synthesis. Our results with ^{3}H-thymidine and ^{32}P labeling of L-cells corroborated those of Levine for AGMK cells: The apparent rate of mtDNA synthesis, although reduced in comparison to the rate in growing cells, significantly exceeded the rate of nuclear DNA synthesis for approximately the first 2 days of stationary culture. However, these experiments, like those of Levine, could not be readily controlled for possible pool effects resulting from well-described changes in the specific activities of both thymidylate synthetase (12) and thymidine kinase (13) in stationary stage cells or for possible depletion of inorganic phosphate from the medium.

Density shift experiments were performed in an effort to corroborate the results of ^{3}H-thymidine labeling experiments. Mouse L-cells in suspension culture were mass-labeled with several additions of ^{14}C-thymidine during 4 generations of logarithmic growth and also pulse-prelabeled with ^{3}H-thymidine in late log stage. This double labeling protocol was used in order to determine whether recently replicated mtDNA molecules were selected for an additional round of replication in stationary cells. Density labeling followed the protocol of Figure 1B with 5 successive 6-hour BrdUrd incubations timed as indicated in Figure 3A. The results in panels B and C of Figure 3 show that the rate of replication for long term ^{14}C-thymidine labeled mtDNA and nuclear DNA progressively decreased during two days at the stationary cell density. However, nuclei labeled with ^{3}H-thymidine never engaged in a significant level of DNA synthesis. In contrast, mtDNA molecules labeled with ^{3}H-thymidine reentered replication at the same rate as the majority of the mtDNA population. The simplest interpretation of this result is that mtDNA synthesis continues in cells in the stationary stage of culture which no longer synthesize nuclear DNA. However, we note that the ^{14}C density shift data of Figure 3C reflect a significant overall rate of nuclear DNA synthesis for approximately the first 24 hours of the stationary stage. This may indicate a relatively high rate of cell turnover. This experiment may therefore be more clearly interpreted if repeated with a cell line which undergoes more stringent contact inhibition or cell density inhibition than does the mouse L-cell line.

SUMMARY

Density labeling techniques have been used to measure the rate of mtDNA synthesis in mouse L-cells. These experiments have demonstrated that mtDNA synthesis can proceed with little or no accompanying nuclear DNA synthesis under several conditions:

(a) in the G_1 and G_2 periods of the normal cell cycle;

(b) in thymidine kinase minus cells treated with methotrexate and thymidine to selectively diminish the thymidylate supply for nuclear DNA synthesis;

(c) in cells arrested at mitosis by incorporation of toxic quantities of ^{3}H-thymidine into nuclear DNA ("thymidine suicide");

(d) in cells in the stationary stage of suspension cell culture.

Furthermore, under all of these conditions, recently replicated mtDNA molecules are randomly selected from the total population for additional rounds of replication.

REFERENCES

1. Vesco, C. and Basilico, C. (1971). Nature 229, 336-337.
2. Levine, A.J. (1971). Proc. Natl. Acad. Sci. USA 69, 717-720.
3. Pica-Mattoccia, L. and Roberti, S. (1974). Biochem. Biophys. Res. Comm. 60, 882-893.
4. Burstin, S.J., Meiss, H.K. and Basilico, C. (1974). J. Cell Physiol. 84, 397-407.
5. Bogenhagen, D. and Clayton, D.A. (1974). J. Biol. Chem. 249, 7991-7995.
6. Nass, M.M.K. (1969). J. Mol. Biol. 42, 521-528.
7. Meselson, M. and Stahl, F.W. (1958). Proc. Natl. Acad. Sci. USA 44, 671-682.
8. Berk, A.J. and Clayton, D.A. (1974). J. Mol. Biol. 86, 801-824.
9. Bogenhagen, D. and Clayton, D.A. (1976). J. Biol. Chem. 251, 2938-2944.
10. Bogenhagen, D. and Clayton, D.A. Submitted for publication.
11. Whitmore, G.F. and Gulyas, S. (1966). Science 151, 691-694.
12. Conrad, A.H. (1971). J. Biol. Chem. 246, 1318-1323.
13. Littlefield, J.W. (1964). Biochim. Biophys. Acta 95, 14-22.

Genetics and Biogenesis of Chloroplasts and Mitochondria, Th. Bücher et al. eds.

USE OF ANTIBIOTIC INHIBITORS IN STUDIES OF REPLICATION AND REPAIR OF ANIMAL MITOCHONDRIAL DEOXYRIBONUCLEIC ACID

G.G.Gause,Jr.,V.S.Mikhailov,S.I.Tomarev,R.D.Zinovieva
Institute of Developmental Biology,Academy of Sciences of the USSR
Moscow,II7334,USSR

INTRODUCTION

In this paper we shall summarize in brief our findings concerning the following two main subjects: (I) action of antibiotics and other inhibitors of DNA template activity on the synthesis of mtDNA complementary strands in isolated rat liver mitochondria and (2) demonstration of the repair synthesis of mtDNA triggered in isolated mitochondria from the mature oocytes of Misgurnus fossilis by bruneomycin an antitumour antibiotic inducing single-strand breaks in DNA.

OLIVOMYCIN AND ANTHRAMYCIN SELECTIVELY INHIBIT SYNTHESIS OF mtDNA H-STRAND

MtDNA of animal cells replicates by a mechanism termed displacement replication. According to this model synthesis of the two daughter strands heavy (H-) and light (L-) is highly asynchronous.Elongation of the leading H-strand requires the opening of the parental duplex,while the elongation of the L-strand proceeds on the despiralyzed template converted to the single-stranded state during earlier phases of replication[I,2].

These differences in the template requirements for the synthesis of the H-strand and of the L-strand may be reflected in their differential sensitivity to ligands possessing different affinities to double-stranded (ds) and single-stranded DNA.

Previously Gause and Dolgilevich have shown that sibiromycin,antibiotic chemically related to anthramycin and binding specifically to dsDNA selectively inhibited synthesis of mtDNA H-strand in isolated rat liver mitochondria[3].In this paper we present our results on the action of several inhibitors of the DNA template function on synthesis of complementary strands of mtDNA.

Actinomycin D,rubomycin C,echinomycin 6270,olivomycin,anthramycin,tripaflavine and ethidium bromide were used.Anthramycin was a gift from Dr.Moses Tendler,Yeshiva University,USA.All other inhibitors were obtained from Prof.G.F.Gause,Institute of New Antibiotics,Moscow,USSR.

Isolation of mitochondria,their incubation with labeled precursors of DNA,isolation of mtDNA and its sedimentation analysis were performed as described previously[3,4].Strand separation of mtDNA was achieved by equilibrium ultracentrifugation in alkaline CsCl gradients.Hybridization of labeled mtDNA with separated strands was performed by a filter method in the presence of 50% formamide as described by others[5].

The action of studied inhibitors on ^{3}H-thymidine incorporation into mtDNA H- and L-strand separated by equilibrium centrifugation in alkaline CsCl gradients is illustrated by the results presented in Table I. The results are expressed as H/L ratio, defined as the ratio of ^{3}H-thymidine counts present in the H-strand to ^{3}H-thymidine counts present in the L-strand. This ratio is equal to I,36$\pm$0,I6 for the control mtDNA isolated from mitochondria incubated in the absence of inhibitors.

TABLE I

EFFECT OF SEVERAL INHIBITORS ON THE INCORPORATION OF ^{3}H-THYMIDINE INTO mtDNA HEAVY AND LIGHT STRANDS

Inhibitor	Concentration (μg/ml)	H/L ratio
None	---	I,36$\pm$ 0,I6
Actinomycin D	3,0	I,09
Rubomycin C	I,0	I,7
Tripaflavine	I,0	I,2
Ethidium bromide	0,5	I,7
Echinomycin 6270	3,0	0,5
Olivomycin	3,0	0,25
Anthramycin	3,0	0,I4

Mitochondria were preincubated with inhibitors for 5 min without radioactive thymidine, then tritiated thymidine was added and the incubation was continued for 50 min at 30°. Inhibitors were used at concentrations suppressing the total incorporation of radioactive thymidine into mtDNA by 60-70%. Composition of the incubation mixtures and details of ultracentrifugation are as in ref.3.

Among the tested inhibitors actinomycin D, rubomycin C, ethidium bromide and tripaflavine do not decrease the H/L ratio. Echinomycin 6270 inhibits H-strand synthesis somewhat stronger than the synthesis of the L-strand, but the selectivity of the H-strand inhibition is not very pronounced. Olivomycin and anthramycin display marked selectivity in the inhibition of the H-strand synthesis. In this respect two these antibiotics are similar to sibiromycin, the action of which on mtDNA strand synthesis was described by Gause and Dolgilevich somewhat earlier[3].

In further experiments the action of olivomycin was studied in greater detail.

Olivomycin in concentrations inhibiting mtDNA H-strand synthesis does not affect mitochondrial energy metabolism. The value of the respiratory control equal to 3,0 in the absence of olivomycin remained unchanged in its presence (3,0 μg/ml). This result is important because our preliminary experiments have shown that uncouplers of oxidative phosphorylation such as 2,4-dinitrophenol preferentially inhibit the incorporation of ^{3}H-thymidine into mtDNA H-strand as compared to the incorporation into the L-strand. The nature of this inhibitory effect is under study at present.

Since olivomycin does not uncouple mitochondrial oxidative phosphorylation we think that its preferential inhibitory action on the synthesis of mtDNA H-strand is due to the specific mode of its interaction with DNA.

Radioactive DNA with the density of the L-strand synthesized in mitochondria in the presence of olivomycin was characterized by hybridization with the separated H- and L- mtDNA complements.

Mitochondria were incubated with ^{3}H-thymidine without inhibitor and with olivomycin (0,3 and 0,6 µg/ml) for 60 min. MtDNA isolated from such mitochondria was centrifuged in equilibrium alkaline CsCl gradients. Radioactive DNA peaks resolved in these gradients were used in hybridization experiments. In the control two peaks with bouyant densities of the H- and L-strand were present. In mtDNA from olivomycin treated mitochondria only one radioactive peak with the density of the L-strand was present. The results of hybridization experiments are presented in Table 2.

TABLE 2

HYBRIDIZATION OF LABELED mtDNA PREPARATIONS WITH DNA IMMOBILIZED ON FILTERS

DNA on filter	Radioactive DNA added: H-control	L-control	Olivo-0,3	Olivo-0,6
	Percentage of input DNA radioactivity bound			
H-strand	5,I	35,0	40,5	33,0
L-strand	29,5	4,8	6,0	4,4
E.coli	0,8	--	0,I	I,0
Sea urchin sperm	0,3	0,6	0,6	I,I
No DNA	0,8	--	--	--
Radioactivity of input DNA (counts/min)	750	750	750	400

Hybridization was performed at 40° for I2 hours with shaking. After the incubation filters were rinsed with 2 SSC and thoroughly washed with 3mM Tris-HCl buffer, pH 9,4. Each filter contained either I µg of the specified mtDNA strand or 25 µg of the heterologous DNA.

Radioactive H- and L-strand from the control mitochondria effectively hybridize respectively with L- and H-strand immobilized on the filters. Hybridization in the L-L and H-H combinations is 6-7 times lower and is probably due to mutual contamination of the strands. Binding of the radioactive DNA to filters with heterologous DNA is low. This indicates that the specificity of hybridization under the used conditions is high.

Hybridization properties of labeled DNA synthesized in mitochondria in the presence of olivomycin are similar to those of the L-strand from the control mitochondria. On the basis of this result we conclude that labeled mtDNA synthesized in mitochondria during incubation with olivomycin is represented by the nucleotide sequences of the L-strand. Olivomycin thus represents specific inhibitor of the H-strand synthesis. Preliminary hybridization experiments with mtDNA synthesized in the presence of anthramycin (not shown) indicate that this antibiotic also specifically inhibits synthesis of the H-strand sequences in isolated mitochondria.

Olivomycin added to isolated mitochondria does not induce nicking or fragmentation of mtDNA. Sucrose gradient sedimentation profiles of the radioactive DNA isolated from ^{3}H-thymidine prelabeled mitochondria are essentially similar to profiles of mtDNA isolated from the control mitochondria incubated without olivomycin. The content of the radioactive superhelical 38 S component is equal to 74% for the control mtDNA and to 70% for mtDNA from mitochondria incubated with olivomycin. No radioactive DNA sedimenting slower than 27 S component was present. We conclude that the integrity of mtDNA replicative intermediates is not damaged when isolated mitochondria are incubated with olivomycin.

The model illustrating selective inhibition of mtDNA H-strand synthesis by olivomycin is presented in Fig. I.

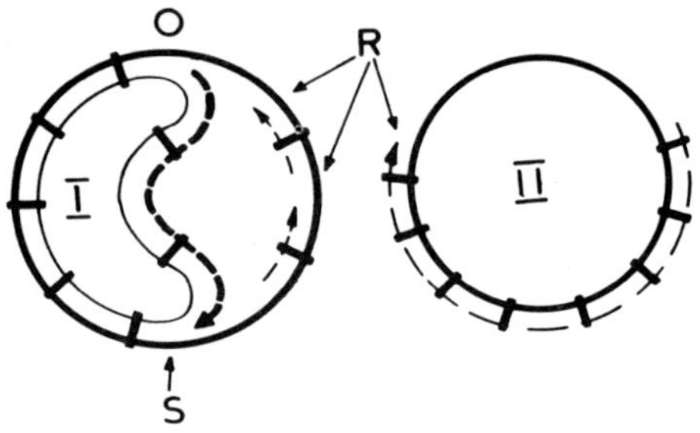

Fig. I. Model illustrating the selective inhibitory action of olivomycin on the synthesis of mtDNA H-strand in the branched (I) and gapped circular (II) replicative intermediates. Continuous thick and thin lines correspond to the H- and L-strands of the parental molecule. Dashed lines correspond to the newly synthesized strands. Black rectangles indicate olivomycin molecules. Note that the antibiotic is bound only to the double-stranded regions. S - point of the H-strand elongation sensitive to olivomycin inhibition. R - points of the L-strand elongation resistant to olivomycin. O - origin of replication.

According to this model the selective inhibition of the H-strand synthesis by olivomycin is due to its preferential interaction with dsDNA in the intermediates which replicate unidirectionally.The bound molecules of the antibiotic inhibit the movement of the replication fork along the template and therefore the elongation of the H-strand.Synthesis of the L-strand proceeding on the single-stranded template is more resistant to olivomycin since this antibiotic binds weaker to ssDNA.

This model explains the preferential inhibitory action of olivomycin on the H-strand synthesis and it is also applicable for the explanation of the selectivity in the inhibitory action of anthramycin.We do not know at present however the chemical nature underlying this selectivity.We do not know why olivomycin and anthramycin inhibit preferentially H-strand synthesis while actinomycin D,rubomycin tripaflavine and ethidium bromide do not possess this selectivity.

DEMONSTRATION OF mtDNA REPAIR SYNTHESIS TRIGGERED IN ISOLATED MITOCHONDRIA OF OOCYTES BY BRUNEOMYCIN

It is not known at present whether defects created in mtDNA by various treatments can be repaired.In 1973 we described some properties of the DNA-synthesizing system in mitochondria isolated from mature oocytes of the loach *Misgurnus fossilis*[6].Such mitochondria were almost completely inactive in DNA synthesis and incorporated very low amounts of radioactivity from ^{3}H-dTTP into mtDNA during incubation *in vitro*.However DNA synthesis in such mitochondria could be activated 30-50 times by bruneomycin an antibiotic inducing single-strand breaks in DNA.Bruneomycin itself does not interact with purified DNA but when it is added to cells or to mitochondria it is subjected to metabolic activation and as a result single-strand breaks are induced in DNA.

When the product of DNA synthesis triggered in isolated oocyte mitochondria was analyzed by gradient centrifugation 85% of the total DNA radioactivity was present in the fraction of open circular molecules,about 10-15% were in the fraction of the covalently closed circles.Linear molecules and fragments were absent.

Here we present evidence that the synthesis of mtDNA triggered in isolated mitochondria by bruneomycin is associated predominantly with mtDNA repair.

Mitochondria from embryos and oocytes of *M.fossilis* were incubated *in vitro* with 5-bromodeoxyuridine triphosphate (5-BrdUTP) and ^{3}H-dATP in the absence and in the presence of bruneomycin.Bouyant density distributon of mtDNA isolated from such mitochondria was analyzed by equilibrium centrifugation in neutral CsCl density gradients.

When mitochondria isolated from the late loach embryos (35 hours of development) were incubated with 5-BrdUTP in the absence of bruneomycin the mean bouyant density shift of the labeled mtDNA molecules relatively to the unlabeled one was equal to 16 mg/cm^3.We attribute this shift to replicative DNA synthesis since mtDNA is actively replicated in mitochondria from 35 hr embryos[7].

Bouaynt density analysis of mtDNA isolated from oocyte mitochondria incubated with bruneomycin (I μg/ml) in the presence of 5-BrdUTP is presented in Fig.2.

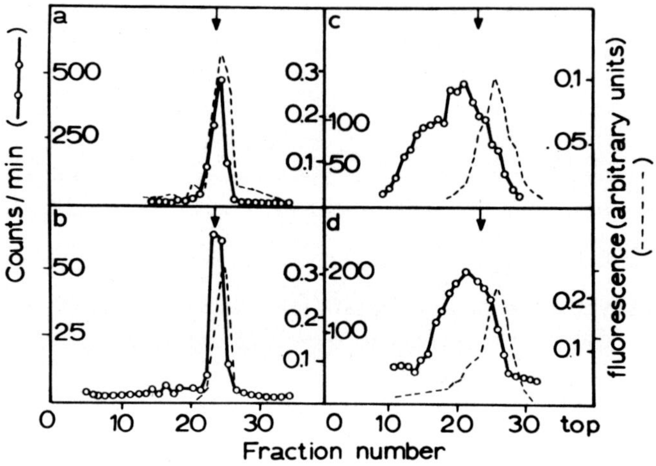

Fig.2.Centrifugation profiles in equilibrium CsCl density gradients of mtDNA from loach oocyte mitochondria incubated for 90 min with 50 μM 5-BrdUTP in the presence of bruneomycin (I μg/ml).(a) open circular fraction;(b) closed circular fraction; (c) as (a) but after sonication for IO min;(d) as (b) but after sonication for IO min. Continuous line - radioactivity;dashed line - fluorescence.Gradients were centrifuged for 60 hours at 40 000 rev/min in a Spinco Ti-50 rotor.Arrows show position of E.coli DNA labeled with radiocarbon.This DNA was used as an internal density marker.

Open circular and covalently closed mtDNA prepared by the fractionation of the total mtDNA in EthBr-CsCl gradients were analyzed separately.The bouyant shift of mtDNA from mitochondria incubated with bruneomycin and 5-BrdUTP is equal to 2-3 mg/cm^3 both for the open circular fraction (Fig.2 a) and for the covalently closed molecules (Fig.2 b).

The small magnitude of the shift is in agreement with the view that bruneomycin-triggered DNA synthesis is a repair process in which the density label is incorporated into relatively short DNA stretches.If this is the case sonication of DNA must give fragments with the higher relative content of the heavy analog.

When mtDNA synthesized in the presence of bruneomycin and 5-BrdUTP was fragmented by sonication to pieces of about 2 000 nucleotide pairs in length the density of the newly synthesized DNA increased markedly (Fig.2 c,d).Bouyant density of the unlabeled mtDNA remained virtually unchanged as can be seen from the fluorescence profiles.The density shift of radioactive DNA is equal to about I7 mg/cm^3 for the fragments obtained from the covalently closed fraction and to about 27 mg/cm^3 for the fragments derived from the open circular molecules.Thus the heavy label incorporated into mtDNA in the synthetic reaction triggered by bruneomycin is present in the relatively short DNA stretches. Calculations which will be presented elsewhere indicate that the average length of mtDNA fragment labeled in the bruneomycin-triggered reaction is equal to I 000 nucleotides for the covalently closed fraction and to 2 400 nucleotides for the open circular one. The meaning of this difference is not clear at present.

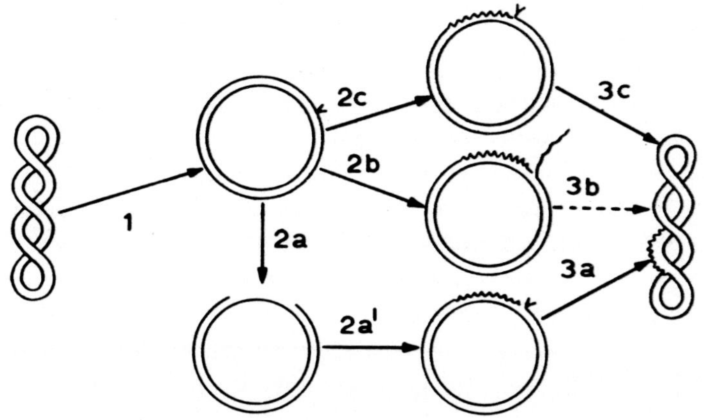

Fig. 3. Possible mechanisms of mtDNA synthesis induced by bruneomycin in loach oocyte mitochondria. Initial nick (I) is either converted to a gap (2 a) which is later filled (2 a^{I}) or directly serves for displacement synthesis (2 b) or for synthesis by nick translation (2 c). Completely double-stranded intermediates can be converted to the covalently closed state by DNA-ligase. Reaction (3 b) would require removal of the single-stranded tail. The significance of this reaction is questionable. Wavy lines show newly synthesized DNA stretches.

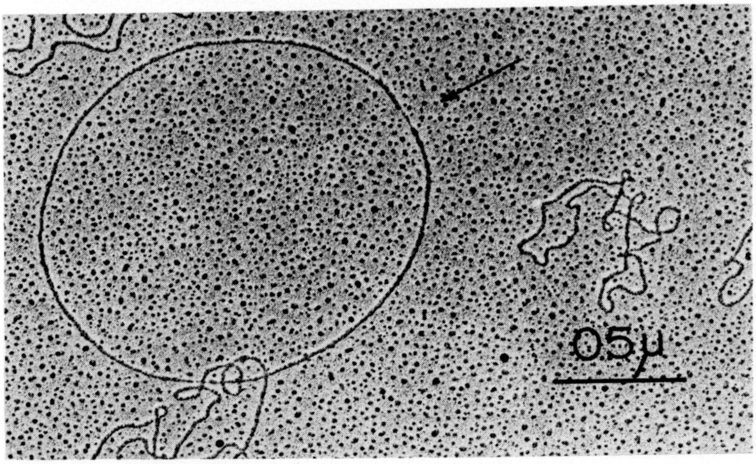

Fig. 4. Electron micrograph of mtDNA with single-stranded gap from mitochondria incubated with bruneomycin (I μg/ml). Gap is indicated by an arrow. Formamide spreading was used.

Equilibrium ultracentrifugation in alkaline CsCl gradients of the intact and sonicated 5-BU substituted mtDNA from mitochondria incubated with 5-BrdUTP has shown that the newly synthesized DNA stretches are covalently attached to the parental strands.

Possible pathways of the label incorporation into mtDNA in the bruneomycin induced synthetic reaction are presented in Fig.3.The relative importance of these pathways is under investigation.Gap filling synthesis and nick translation in this scheme correspond to DNA repair.

MtDNA from bruneomycin-treated mitochondria was studied using electron microscopy.The majority of the molecules was circular,several molecules however had single-stranded gaps.One of such molecules is presented in Fig.4.Since mitochondria from the mature loach oocytes do not contain mtDNA replicative intermediates[7] we think that the gapped molecules observed by us may represent intermediates in mtDNA repair proceeding via gap formation.

In addition to bruneomycin several other agents also activated mtDNA synthesis in mitochondria from loach oocytes.These were alkylating agents,alkyl nitrosoureas and X-irradiation.We think that DNA synthesis induced by these agents also represents repair of mtDNA.

REFERENCES

I; Robberson,D.L.,Kasamatsu,H.,Vinograd,J. (I972) Proc.Nat.Acad.Sci.US;69,737-74I.

2. Berk,A.J.,Clayton,D.A.(I974) J.Mol.Biol.86,80I-824.

3. Gause,Jr.,G.G.,Dolgilevich,S.M.(I975) Biochim.Biophys.Acta 383,9-I5.

4. Gause,Jr.,G.G.,Dolgilevich,S.M.,Fatkullina,L.G.,Mikhailov,V.S.(I973)Biochim. Biophys.Acta 3I2,I79-I9I.

5. Schmitt,H.,Beckmann,J.S.,Littauer,U.Z.(I974) Eur.J.Biochem.47,225-234.

6. Gause,Jr.,G.G.,Mikhailov,V.S.(I973) Biochim.Biophys.Acta 324,I89-I98.

7. Mikhailov,V.S.,Gause,Jr.,G.G.(I974) Develop.Biol.4I,57-7I.

Genetics and Biogenesis of Chloroplasts and Mitochondria, Th. Bücher et al. eds.

HORMONAL CONTROL OF MITOCHONDRIAL DNA REPLICATION IN MATURING OOCYTES

Monique BARAT, Catherine DUFRESNE, Hubert PINON,
Monique TOURTE and Jean-Claude MOUNOLOU
Laboratoire de Biologie Générale, Université Paris-Sud
Bâtiment 400, 91405 Orsay, France

INTRODUCTION

In rapidly dividing somatic cells mitochondrial multiplication is active and mitochondrial DNA (mtDNA) replicative intermediates are found in rather high proportions(see review by Kasamatsu and Vinograd, 1). On the contrary in resting cells replicative forms are absent and virtually all the mtDNA is stored as circular duplexes, this holds true also for sea urchin oocytes : replicative intermediates are abundant in the growing oocyte, none are present in the mature oocyte (2). A different situation is observed in Xenopus and in mouse unfertilized eggs : the percentage of mtDNA molecules with a D-loop levels up to 30 % in mouse (2) and 76 % in xenopus (3) as if the D-loop state could be a storage form.

We wondered whether this peculiar situation was the consequence of some hormonal control on mtDNA replication in xenopus oocytes. Before in vitro induction to maturation we observed a net mtDNA synthesis (2 % per 24 hours) corresponding to complete replications of the mitochondrial genome. When follicles are induced by human chorionic gonadotrophin (HCG) this synthesis in oocytes drops down ; E-strand is the main product made during this decrease and the proportion of mtDNA molecules with a D-loop rises (4). In this report we summarize these data obtained with a HCG treatment on follicles and compare them with some results obtained with progesterone induced oocytes.

MATERIALS AND METHODS

Experimental conditions have been fully described in ref 4 for in vitro induction by HCG, mtDNA purification and E-strand analysis, electron microscopy of mtDNA molecules. With progesterone the incubations conditions are the same as with HCG. Progesterone is added at the final concentration of 5μg/ml ; it is usually left for 6 hours in contact with the follicles without any damage to the oocytes (experiments 1 and 3 in table II) but an exposure of 1 hour to the steroid (experiment 2 in table II) is enough to get the same effect on maturation as judged by the germinal vesicle migration and breakdown 5 hours later. When a few oocytes have not matured after the 6 hours incubation period they are taken away one by one.

RESULTS

Accumulation of replicative intermediates upon maturation : All mtDNA molecules obtained from CsCl gradients are circular and 80 % of them are superhelical. Their average length is 5.3 microns. Replicative forms have been looked for in three kinds of oocytes : large ones, in vitro matured ones and oocytes treated by ethidium bromide, a good mtDNA synthesis inhibitor. Results of a typical experiment are presented on table I ; they show a clear accumulation of D-loop structures in HCG-matured oocytes and nothing of the kind for the other replication forms. Such an observation is clearly reproducible : in three independent experiments the percentages of mtDNA molecules with a D-loop were : 43 % out of 161 molecules examined (table 1), 44,3 % out of 400 molecules and 43,5 % out of 153 molecules.

Accumulation of replicative intermediates is also achieved by incubating non-stimulated oocytes with ethidium bromide. However some differences are clearly seen between the hormone and the drug treatments : the apparent length of the molecules is increased by about 10 % with ethidium (5), dimers are formed and there is a wider distribution of the D-loop structures with no real mode (table I).

TABLE I

ELECTRON MICROSCOPY ANALYSIS OF mtDNA FROM LARGE OOCYTES AND HCG TREATED OOCYTES

	Total molecules observed	average length µ	mode of D-loop extent (in % of mt DNA length)	% of replicative forms
large oocytes	152	5.3	3	18
in vitro HCG treated oocytes	161	5.4	2.5 to 3	43
EB* treated oocytes	109	5.8	no mode (+2 dimers)	53

Isolated large follicles are incubated in saline medium, alone, or with HCG (6 hours, 40 u/ml) or with ethidium bromide (20 hours, 25 µg/ml). Total incubation time is always 20 hours. After DNA extraction, residual EB* is not removed ; this explains the increase of the apparent average length (5) for the third batch.

*EB = ethidium bromide.

It is well established that HCG primarily acts on follicular cells which in turn induce the oocyte to mature. Maturation can be also achieved by a direct treatment of the oocyte with progesterone (6). We wondered whether these two hormonal treatments have the same effects on D-loop accumulation or not. The results of three different experiments performed on oocytes of three different females are given in table II. It appears that the accumulation of mtDNA with a D-loop depends on the female from which the oocytes are obtained. Moreover with one female (female 3 in table II) one can see that HCG was effective on the follicles and progesterone was not. On the whole it seems that the maturation events at the mitochondrial level can be induced by progesterone but with a poor efficiency.

TABLE II

ELECTRON MICROSCOPY ANALYSIS OF mtDNA MOLECULES FROM PROGESTERONE TREATED OOCYTES

Female	hormonal treatment of the oocytes	Total molecules observed	% of D-loop forms
1	Progesterone	35	34,2
2	Progesterone	142	38,7
3	- Progesterone	53	21
	- HCG	161	43

Conditions have been described in table I and material and methods. Progesterone concentration is 5 μg/ml.

Preferential E-strand synthesis upon maturation : The accumulation of D-loops after HCG treatment of the oocytes is confirmed by the analysis of the quality of the labelled molecules. When labelled mtDNA is layered on the top of a linear sucrose gradient (Fig. 1), the radioactivity is bound to entire mtDNA molecules. But if the DNA is previously heat-denatured and chilled a slower peak of radioactivity appears analogous to the 14S E-strand of Hallberg (3). Such a result has repetitively been obtained in four independent experiments out of five that have been performed (4). It appears that HCG treated oocytes mainly synthesize E-strands, when untreated ones replicate the whole mtDNA molecules.

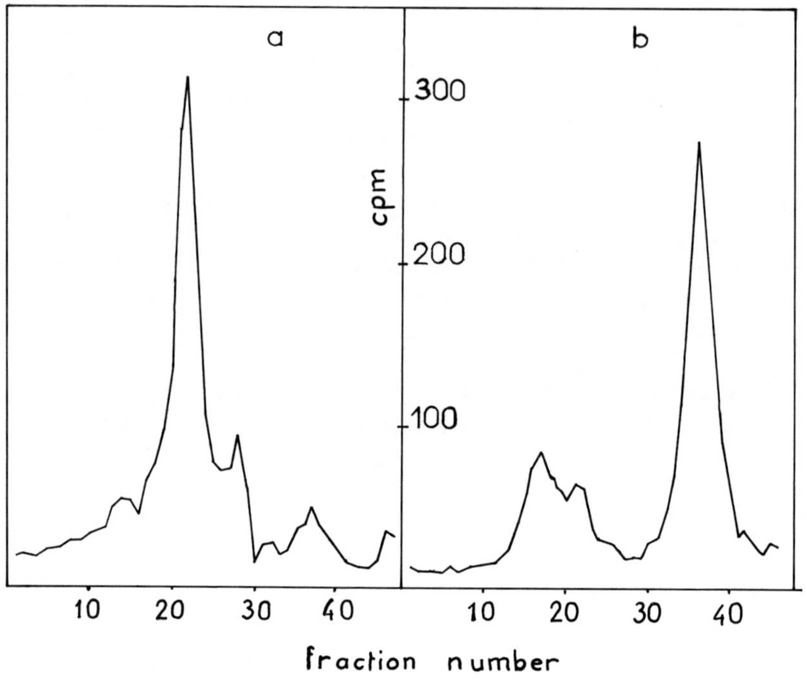

Fig. 1. Sucrose velocity pattern of labelled mtDNA in HCG treated oocytes. a) native DNA, b) heat-denatured DNA. Fractions are collected from the bottom (9). After heat-denaturation a large peack of radioactivity shifts from large molecules to isolated E-strands.

DISCUSSION

Chase et al, (7) observed that mtDNA replication is not carried on in young embryos and in fertilized eggs. Moreover molecules with a D-loop are accumulated in ovulated mature oocytes (3). Such a situation results from the expression of an hormonal control which switches the oocyte mitochondria to a state where DNA replication is blocked. This is supported by our results which show that HCG treatment induces an increase of the relative synthesis of E-strands and the accumulation of D-loops. The fact that ethidium bromide also increases the percentage of D-loops suggests that HCG acts by delaying E-strands elongation rather than by inducing a special E-strands synthesis. Preferential synthesis of E-strands after hormonal treatment corresponds to the accumulation of molecules with a D-loop and maybe also to the turnover of the E-strand forms as seen in mouse L cells (8).

A quantitative analysis of the tendency of the oocyte to control mtDNA replication upon a hormonal stimulation is difficult to achieve because of the variability of the animals. In fact as it is clearly seen in the ability to respond to a progesterone treatment all the females are not identical. Variations in the progesterone experiments can also be due to a differential efficiency of this steroid in influencing the nucleus and the mitochondria. If such is the case the choice of standard incubation conditions and periods can emphasize the discrepancies between oocytes of different females. Likewise as the conversion of duplex mtDNA molecules into D-loop intermediates upon maturation appears to be a progressive event, the length of the incubation period of the oocytes after HCG is probably too short to reach the state observed by Hallberg (3) in mature unfertilized eggs where 76 % of the mtDNA molecules have a D-loop.

At last nothing is known about the internal stimulation of each animal before any treatment and it is reasonable to imagine that the hormonal status is different in each case. This too can make it difficult to understand the meaning of any small variation of the percentage of mtDNA molecules with a D-loop in an oocyte and this is probably the reason for the difference between our observations on non-treated full grown oocytes (18 % of mtDNA with a D-loop) and Hallberg's data on whole ovaries (7 % of mtDNA with a D-loop) (3).

Although the biological meaning of the storage of mtDNA replicative forms at the end of oogenesis is not completely clear, the formation of a synchronized population of molecules could be a simple way to start again simultaneously the replication cycle of mtDNA in most cells of the young embryos.

We thank Prof. G. Brun in whose laboratory electron microscopy was carried on and C.N.R.S. (ATP 1848) and E.P.H.E. for their support.

REFERENCES

1. Kasamatsu, H. and Vinograd, J. (1974) Annual Review of Biochemistry, 43, 695-720.
2. Matsumoto, L., Kasamatsu, H., Piko, L. and Vinograd, J. (1974) J. Cell. Biol., 63, 146-159.
3. Hallberg, R.L. (1974) Dev. Biol., 38, 346-355.
4. Barat, M., Dufresne, C., Pinon, H., Tourte, M. and Mounolou, J.C. Dev. Biol. in press.
5. Butour, J.L. and Delain, E. (1975) J. Microscopie Biol. Cell., 24, 231-238.

6. Dennis Smith, L. (1975) in The Biochemistry of animal development, Weber, R. ed, Ac. Press, Vol. III, 1-46.
7. Chase, J.W. and Dawid, I.B. (1972) Dev. Biol., 27, 504-518.
8. Berk, A.J. and Clayton, D.A. (1976) J. Mol. Biol., 100, 85-102.
9. Kasamatsu, H., Robberson, D.L., Vinograd, J. (1971) Proc. Nat. Acad. Sci. USA, 68, 2252-2257.

TRANSCRIPTION AND TRANSLATION APPARATUS OF CHLOROPLASTS

Genetics and Biogenesis of Chloroplasts and Mitochondria, Th. Bücher et al. eds.

IN VITRO TRANSCRIPTION AND TRANSLATION OF CHLOROPLAST DNA OF *C. REINHARDI*.

S. J. Surzycki, J. A. Surzycki, and R. Lutz
Department of Plant Sciences, Indiana University
Bloomington, Indiana 47401

INTRODUCTION

A major mechanism for the regulation of gene expression is the selection of specific sites (promoters) at which transcription may be initiated. The initiation specificity of the most studied RNA polymerase, *E. coli* RNA polymerase, depends on the presence of sigma protein. A detailed study of the initiation of transcription has demonstrated that initiation is a three-step process: (1) RNA polymerase holoenzyme binds stably to a specific site on DNA (promoter) forming "I" complexes, (2) "I" complexes are spontaneously converted into "RS" complexes, perhaps by denaturation of stretches of DNA at the initiation sites, and (3) in the presence of ribonucleoside triphosphates, each "RS" complex catalyses the rapid formation of the first phosphodiester bond in a newly synthesised RNA chain[1]. *E. coli* RNA polymerase is able to recognize a limited number of initiation sites on DNAs not native to the enzyme and to initiate RNA synthesis at these sites with kinetics characteristic of initiation on native DNA promoters[2]. In this report we present results of experiments describing transcription of chloroplast DNA by *E. coli* RNA polymerase. Some preliminary results of translation *in vitro* of the mRNA obtained in this transcription are also presented.

RESULTS AND DISCUSSION

The transcription of chloroplast DNA by *E. coli* RNA polymerase is sigma-dependent. The synthesis of RNA was measured by incorporation of ^{3}H-UTP using core enzyme alone, and core enzyme in the presence of sigma protein. The stimulation of chloroplast RNA transcription by the presence of sigma protein was about ten-fold. A similar experiment with nuclear DNA of *C. reinhardi* showed only 1.7-fold stimulation. The stimulation of transcription occurred at the initiation stage, as shown by experiments restricting transcription to that occurring from rifampicin-resistant preinitiation complexes. These complexes can be formed only in the presence of sigma protein[3]. The level of nucleoside triphosphate incorporation from rifampicin-resistant pre-initiation complexes formed on phage DNAs exhibits a sigmoidal dependence on the temperature of pre-incubation[4,5]. The temperature dependent step is the conversion of "I" complexes to "RS" complexes[1]; differences in T^m values presumably reflect differences in the nucleotide sequences of the promoters. Initiation from rifampicin-resistant complexes on chloroplast DNA also exhibits a sigmoidal dependence on temperature of pre-incubation, with T^m equal to 25°. The "strength" of the promoters can be estimated by measuring the first order rate constant (k*) of initiation and half time (t1/2) of initiation as described by Mangel and Chamberlin[6]. The results of these experiments using native and chloroplast DNAs are presented in Table 1.

TABLE 1

QUALITATIVE AND QUANTITATIVE CHARACTERIZATION OF INITIATION ON CHLOROPLAST DNA BY E. COLI RNA POLYMERASE

Source of DNA	Sigma stimulation	T^m C°	k* sec^{-1}	t1/2 sec	Number of initiations per DNA Method		
					I	II	III
Chloroplast	10.5	25	0.75	0.9	29-59	29-59	61 (34 ATP/27 GTP)
T 7	10.3	25	0.5	1.4	3-6	3-6	3
Lamda b2	10.1	28	1.1	0.7	4.5-9.3	----	--

It is clear from these results that "promoters" on chloroplast DNA which are recognized by E. coli RNA polymerase are remarkably similar to those recognized on phage DNAs. Thus (a) formation of stable complexes with DNA is sigma dependent, (b) the frequency of initiation in the presence of rifampicin depends on the temperature of pre-incubation, with a transition temperature within the range of the transition temperatures observed for phage DNAs, and (c) the average half time for the formation of the first phosphodiester bond in the new RNA chains is similar to the half time determined under identical conditions using lambda and T7 DNAs.

The number of sites at which E. coli RNA polymerase can form stable, rifampicin-resistant pre-initiation complexes on chloroplast DNA was estimated (see TABLE 1) using three independent methods: (1) The RNA polymerase saturation method, where a constant amount of DNA is incubated with increasing amounts of enzyme[7] (the polymerase concentration at which no further increase in incorporation is observed in the presence of rifampicin is a measure of the number of polymerase molecules necessary to saturate the stable binding sites on the DNA), (2) an RNA polymerase binding method which takes advantage of the ability of RNA polymerase to promote the binding of the DNA to nitrocellulose filters[8], and (3) direct measurement of the number of initiating chains using γ-^{32}P labelled nucleoside triphosphates. In the last method, chains initiated with GTP and ATP can be measured separately.

RNA transcribed in vitro was further characterized by hybridization to chloroplast DNA. In saturation hybridization experiments, this RNA hybridized to 46.5% $\pm$ 0.2 of input DNA at an infinite concentration of RNA. This indicates that all DNA was transcribed by E. coli RNA polymerase and that transcription was asymetrical. Competition hybridization experiments with rRNA from chloroplast ribosomes showed that 0.3% of synthesized RNA can be competed off by rRNA, the amount expected if only one round of transcription is permitted and chloroplast DNA contains 3 ribosomal genes for each class of ribosomal RNA[9]. This indicates that transcription of chloroplast ribosomal genes by E. coli RNA polymerase using rapid starting pre-initiation complexes is correct.

The RNA obtained was also used in in in vitro translation systems using wheat embryo, Chlamydomonas, and E. coli as the source of S30 fraction. SDS gel electrophoresis of the products revealed about 50 bands ranging in molecular weight from 17,000 to 80,000 daltons. The products of translation were co-electrophoresed with labeled chloroplast membranes prepared as described by Chua[12]. Peptides of @ MW 20,000, 30,000, 50,000, and 66,000 daltons were found in both the membrane preparation and also in the products of *in vitro* translation using wheat embryo system and mRNA from *in vitro* transcriptions. *In vitro* translations using *E. coli* and *Chlamydomonas* systems showed fewer bands. All three systems synthesized polypeptides which have MW identical to membrane proteins of 20,000, 30,000, and 50,000 daltons.

It is therefore possible that *E. coli* RNA polymerase recognizes promoters on chloroplast DNA which are also recognized by the native enzyme. In view of this, it is interesting to mention that chloroplast DNA-dependent RNA polymerase from *C. reinhardi* resembles RNA polymerase of *E. coli* in its structure and stimulation of initiation by sigma-like protein [10, 11].

REFERENCES

1. Chamberlin, M. J. (1974) Ann. Rev. Biochem. 43: 721.

2. Surzycki, S. J., J. A. Surzycki, and G. N. Gussin (1976) Molec. Gen. Genet. 143: 167.

3. Bautz, E., F. Bautz (1970 Cold Spring Harbor Symp. Quant. Biol. 35: 227.

4. Zillig, W., K. Zechel, D. Rabussay, M. Schachner, V. S. Sethi, P. Palm, A. Heil, and W. Seifert (1970) Cold Spring Harbor Symp. Quant. Biol. 35: 47.

5. Bautz, E., F. A. Bautz, and E. Beck (1972) Mol. Gen. Genet. 118: 199.

6. Mangel, W. F., and M. J. Chamberlin (1974) J. Biol. Chem. 249: 2295.

7. Surzycki, S. J., J. A. Surzycki, W. DeLorbe, and G. W. Gussin (1976) Molec. Gen. Genet. 143: 177.

8. Hinkle, D. C., and M. J. Chamberlin (1972) J. Molec. Biol. 70: 187.

9. Surzycki, S. J. and J. D. Rochaix (1971) J. Molec. Biol. 62: 89.

10. Ratcliff, S. and S. J. Surzycki (1974) J. Cell Biol. 63: 281a (Abstract).

11. Surzycki, S. J. and D. L. Shellenbarger (1976) submitted for publication.

12. Chua, N-H., P. Bennoun (1975) Proc. Natl. Acad. Sci. USA 72: 2175.

Genetics and Biogenesis of Chloroplasts and Mitochondria, Th. Bücher et al. eds.

LOCALIZATION OF THE GENE CODING FOR THE LARGE SUBUNIT OF RIBULOSE BISPHOSPHATE CARBOXYLASE ON THE CHLOROPLAST GENOME OF CHLAMYDOMONAS REINHARDI

S. Howell, P. Heizmann and S. Gelvin
Department of Biology, University of California
San Diego, La Jolla, California 92093 (U.S.A.)

SUMMARY

Messenger RNA coding for the large subunit (LS) of ribulose bisphosphate carboxylase (RUBPCase) was isolated from *C. reinhardi* to use as a hybridization probe for identifying LS genes. LS mRNA has the properties that it sediments at about 14S, does not bind to oligo-dT cellulose and is translated *in vivo* on small chloroplast polyribosomes. Radioactive LS mRNA probe hybridizes to chloroplast DNA indicating that LS is encoded by the chloroplast genome. The probe hybridizes to the same $EcoR_1$ chloroplast DNA fragment (or at least to one with the same electrophoretic migration characteristics) as 16S and 23S rRNA. It is suggested the LS mRNA and chloroplast rRNA are synthesized at the same rate, if not coordinately, from the chloroplast genome.

INTRODUCTION

Because RUBPCase is such an abundant protein in the chloroplast, LS must be selectively synthesized in considerable excess over other potential polypeptide gene products coded for by the chloroplast genome. The chloroplast genome in *C. reinhardi* is sufficiently complex to code for 100-200 gene products (1) and, as shown in RNA-DNA hybridization experiments (2), virtually the entire chloroplast genome is transcribed. In an attempt to understand how selective synthesis of LS occurs in the chloroplast, we have isolated LS mRNA to serve as a hybridization probe for identifying and localizing the LS gene on the chloroplast genome.

RESULTS

We have used LS-specific antibodies (anti-LS) to detect LS mRNA activity in an *in vitro* translation system (derived from *E. coli*) and to identify polyribosomes engaged *in vivo* in LS synthesis. This has permitted us to determine the following properties of LS mRNA: a) LS mRNA sediments at about 14S, b) LS mRNA does not bind efficiently to oligo-dT cellulose and c) LS mRNA preparations which stimulate LS synthesis in an *E. coli* translation system do not do so in a eucaryotic translation system (derived from wheat germ).

We have used ^{125}I-labeled anti-LS to identify polyribosomes which are engaged *in vivo* in LS synthesis (Fig. 1). Labeled anti-LS binds predominantly to nascent polypeptide chains on small polyribosomes (average N = 2 or 3). This was unexpected because LS is a relatively large polypeptide (MW = 55,000) coded for by an accordingly large mRNA (14S) and so normally would be synthesized on large polyribosomes. Nevertheless, this observation was important for two reasons. First, it demonstrated that LS is probably synthesized rather inefficiently on a per polyribosome basis (that is under initiation rate-limiting conditions) but, second, it permitted a rapid way of isolating LS mRNA. LS mRNA would be an unusually large mRNA found in small polyribosome fractions. Thus, to produce the LS gene probe, we pulse-labeled cells with $^{32}P_i$, isolated RNA from small polyribosomes and separated the 14S LS mRNA from rRNA by electrophoresis on polyacrylamide gels.

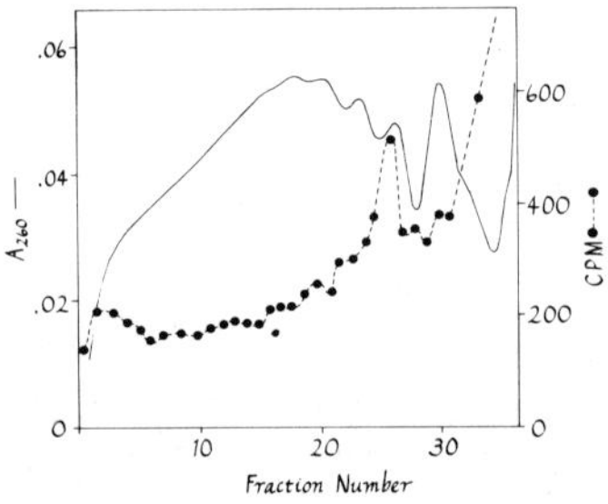

Fig. 1. Binding of ^{125}I-labeled anti-LS to whole cell polyribosomes.

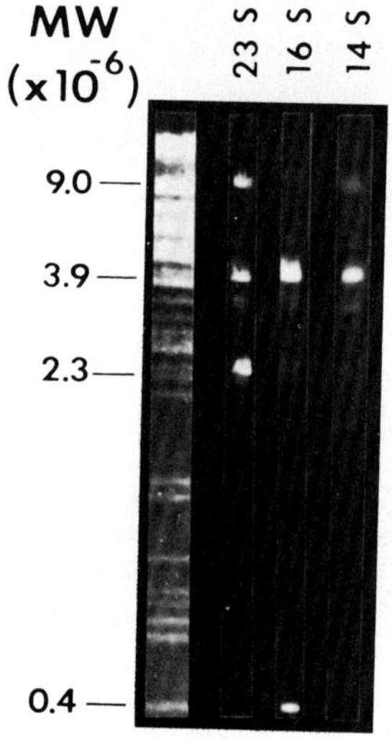

Fig. 2. Hybridization of ^{32}P-labeled rRNA and LSmRNA (14S) to $EcoR_1$ chloroplast DNA fragments.

The LS mRNA probe was used to determine if the LS gene is found on chloroplast DNA. The probe was hybridized to $EcoR_1$ chloroplast DNA fragments transferred to nitrocellulose filters according to Southern (3). The $EcoR_1$ digest pattern of C. reinhardi chloroplast DNA shows about 30 ethidium bromide stained bands (left channel, Fig. 2). The LS mRNA probe did, indeed, hybridize to chloroplast DNA and predominantly to one band with a molecular weight of 3.9×10^6 (Fig. 2). What was surprising was that 16S and 23S chloroplast rRNA also hybridized (in part) to the same DNA fragment (or to one with similar electrophoretic mobility). Preliminary competition experiments with unlabeled 16S and 23S chloroplast rRNA indicate that the 14S RNA is not just a rRNA fragment.

DISCUSSION

LS mRNA hybridizes to an $EcoR_1$ chloroplast DNA fragment that bears (portions of) the chloroplast rRNA cistrons. The DNA fragment is only 3.9×10^6 daltons and is present in about 100 copies per chloroplast (4). The apparent proximity of the LS to the rRNA genes on the chloroplast genome may be of some relevance to the question of how LS is selectively synthesized in the chloroplast. From the argument which follows, we would suggest that LS mRNA and rRNA are synthesized at the same high rate (if not coordinately) in the chloroplast.

A mature C. reinhardi chloroplast contains about 7×10^{-3} g or about 8×10^5 molecules of RUBPCase. Because each RUBPCase molecule contains 8 LS (as determined by SDS polyacrylamide gel electrophoresis analysis of pure RUBPCase), there are about 6×10^6 LS per chloroplast (or per cell). During 8 hrs of the cell cycle, a C. reinhardi cell makes nearly 3/4 of its LS (assuming LS is indefinitely stable) at an overall rate of about 9×10^3 LS/min. The time required to synthesize a single LS chain is about 2 min, since we have determined by other means that the approximate rate of chain elongation in the chloroplast is 5 AA/chain/sec (5). Thus, the number of LS chains growing at any instant would be 2×10^4 nascent chains/cell. Since we have shown here that LS is synthesized on small polyribosomes ($N \cong 3$), then the total number of polyribosomes engaged in LS synthesis at any one time is about 6×10^3 polyribosomes per chloroplast. To maintain 6×10^3 polyribosomes synthesizing LS at any one time requires a minimum steady-state level of the same number of LS mRNAs/chloroplast. That number is about 1/10 the steady-state number of chloroplast rRNA molecules (either 16S or 23S), and would be maintained if rRNA and LS mRNA were synthesized at the same rate but LS mRNA turned over with a half life of about 2 1/2 hrs. In preliminary experiments in which LS mRNA synthesis has been blocked by shifting light grown cells to the dark, we have found about a 2 1/2 hr decay half life.

Thus, the possible close relationship between LS and rRNA genes on the chloroplast genome may suggest a mechanism by which LS genes may be transcribed at the same high rate as rRNA cistrons.

This work has been supported by the National Science Foundation.

REFERENCES

1. Howell, S.H. and Walker, L.L. (1976). Biochim. Biophys. Acta 418, 249-256.
2. Howell, S.H. and Walker, L.L. in preparation.
3. Southern, E. (1975). J. Mol. Biol. 98, 503-517.
4. Heizmann, P. and Howell, S.H. in preparation.
5. Baumgartel, D.M. and Howell, S.H. in preparation.

Genetics and Biogenesis of Chloroplasts and Mitochondria, Th. Bücher et al. eds.

CHARACTERIZATION OF THE RNA COMPOUNDS SYNTHESIZED BY ISOLATED CHLOROPLASTS

Hans J. Bohnert, Albert J. Driesel and Reinhold G. Herrmann
Botanisches Institut der Universität Düsseldorf, FRG.

SUMMARY

Isolated unbroken chloroplasts from *Spinacia oleracea* incorporate radioactive ribonucleosides and ^{32}P-orthophosphate into high molecular weight RNA in a light dependent reaction. A transcript of M.Wt. of 2.7×10^6d is the precursor to chloroplast ribosomal RNA. This transcript contains the 23S,16S,5S and 4.5S RNA and approximately 30 % of transcribed spacer and is processed in several steps into the mature ribosomal RNAs. A processing scheme as originally established from acrylamide-gel patterns of RNA labelled in isolated chloroplasts has now been confirmed by saturation as well as competition-hybridization experiments.An arrangement of the rRNA sequences within the cistrons in spinach chloroplast DNA is proposed.

INTRODUCTION

In 1973 Hartley and Ellis[1] presented evidence that envelope covered plastids synthesize RNA species of a M.Wt of 2.7×10^6d in a light-dependent reaction. By using different conditions of organelle isolation and incubation we could demonstrate the appearance of several additional RNA species[2]. These were in part converted to RNA species resembling the ribosomal RNAs in molecular weight under dark-chase conditions. On the basis of these experiments we proposed a processing scheme for the maturation of the rRNAs. In this paper we present further evidence for the precursor-product relationship between these components by using the DNA/RNA hybridization approach.

MATERIALS AND METHODS

Spinach seedlings, 14-20 days old, were used throughout. The isolation of chloroplasts, the incubation assay, as well as the conditions for electrophoresis and handling of gels have already been described[1,2]. Hybridization with varying DNA to RNA ratios was carried out in 3 x SSC at 67°C for suitable times. The procedure of competition hybridization will be discribed elsewhere.

RESULTS AND DISCUSSION

RNA SYNTHESIS BY ISOLATED CHLOROPLASTS ; PRODUCT IDENTIFICATION

RNA of discrete size classes is synthesized by isolated chloroplasts in our standard sorbitol-KCl medium[2] (Fig.1). After 45 min. of labelling the most prominent peaks of radioactivity are found at positions of 2.7×10^6d and 1.2×10^6d of which the latter peak corresponds in M.Wt. to the immediate precursor (p23S) of 23S RNA. The immediate precursor to 16S rRNA and the ribosomal RNAs themselves are labelled to a much lesser extent. Three additional radioactive peaks are obvious. Two of them are located between the presumed transcript and the p 23 S peak. The third one, always coinciding with a peak in the optical density profile, was found at a M.Wt. of $0,45 \times 10^6$d. The label in the 2 and 0.45

megadalton regions appeared generally in the same ratio supporting the idea that both components are metabolically correlated. It should be mentioned that processing of the 2.7 x 10^6 d species is invariably incomplete. Evidence from double-labelling experiments

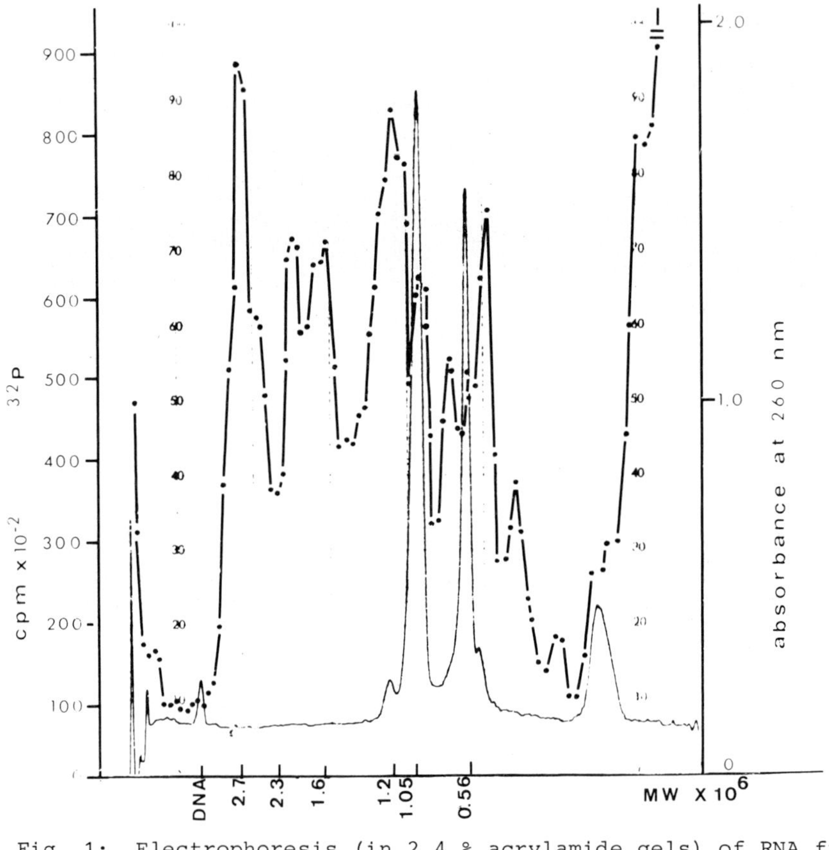

Fig. 1: Electrophoresis (in 2.4 % acrylamide gels) of RNA from chloroplasts incubated for 45 min. with $^{32}P_i$ (250 µCi/ml) $^{32}P_i$ was used instead of ^3H-nucleosides because it gives rise to higher specific activities (see also Bennett and Milewska[3]).

(Fig. 2) indicates that the processing rate diminishes during the incubation.

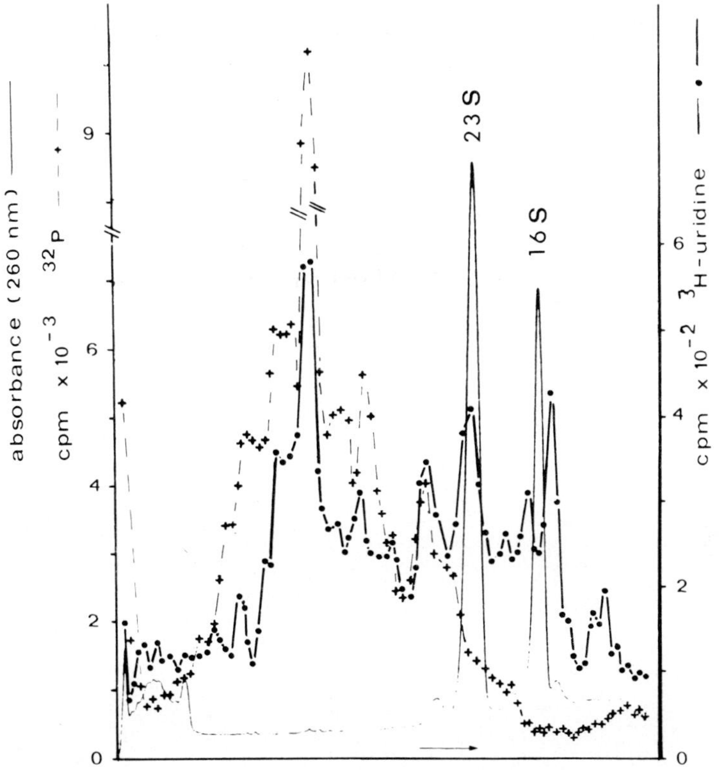

Fig. 2: RNA from chloroplasts separated by electrophoresis on 2.4 % acrylamide gels. The plastsids were incubated first with ^{3}H-uridine (---o---). After 30 min. they were washed in the presence of unlabelled uridine and then suspended in the incubation medium which contained all four nucleosides unlabelled and $^{32}P_i$. The incubation was continued for another 30 min. (---+---). Some tritium label reached the position of the 23 S and 16 S rRNAs, whereas only high M.Wt. RNA was labelled by ^{32}P. In control experiments where $^{32}P_i$ was added after a preincubation with unlabelled uridine also no processing could be observed after 2hours of total incubation.

To check that the 1.2 x 10^6d RNA species was indeed a fragment of the 2.7 x 10^6d transcript DNA/RNA hybridization experiments were undertaken.

HYBRIDIZATION EXPERIMENTS

The rationale of this approach is that the presumptive transcript of the ribosomal cistrons should be competed out by both the proposed processing intermediates and the mature rRNAs. It is likely that the 2.7×10^6d transcript contains non-conserved spacer sequences in addition to the mature rRNA species. These will not be competed out in the hybridization experiments to be described.

There were two prerequisites which had to be satisfied before we could begin the competition-hybridization studies. Firstly, we had to isolate all components and, secondly, the number of the 23 S and 16 S genes as well as the number of 2.7×10^6d transcripts per unit genome had to be measured.

The isolation of 23 S, 16 S, 5 S and 4.5 S RNAs is outlined in another paper of these Proceedings (Herrmann et al.). The presumed transcript and the intermediates were isolated from total chloroplast RNA preparations and purified by several cycles of sucrose-gradient centrifugations (Fig. 3).

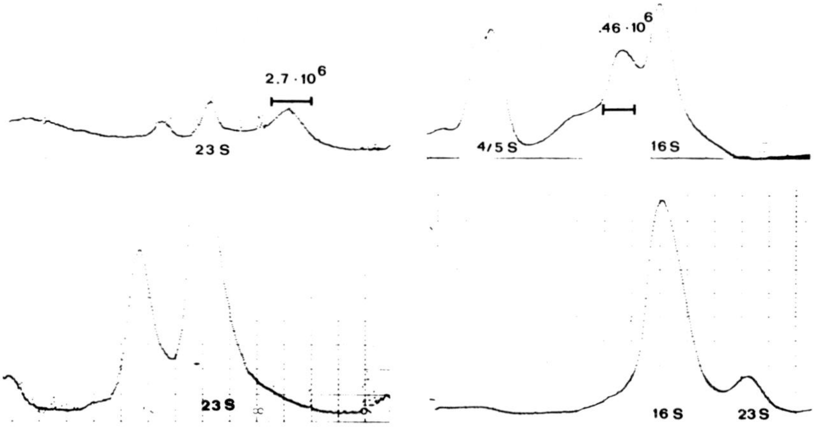

Fig. 3: Preparation of the 2.7 and the 0.45×10^6d RNA in linear sucrose gradients (13-40 %). The volume marked by a bar was pooled. 23 S and 16 S RNA separated in the same run are included for comparison. The 2.7 and 0.45×10^6d compounds labelled by isolated chloroplasts with ^{32}P could be obtained in sufficiently high specific activities to perform hybridization experiments. For some experiments these two RNA populations were also labelled with ^{125}I[4].

Two genes for each of the 23 S and 16 S r-RNAs were found by hybridization under saturation conditions[5], regardless of wether RNAs labelled in-vivo (^3H-uridine) or by iodination were used. In similar saturation-hybridization experiments iodinated 2.7×10^6d RNA was shown to be present twice per chloroplast DNA molecule (Fig. 4).

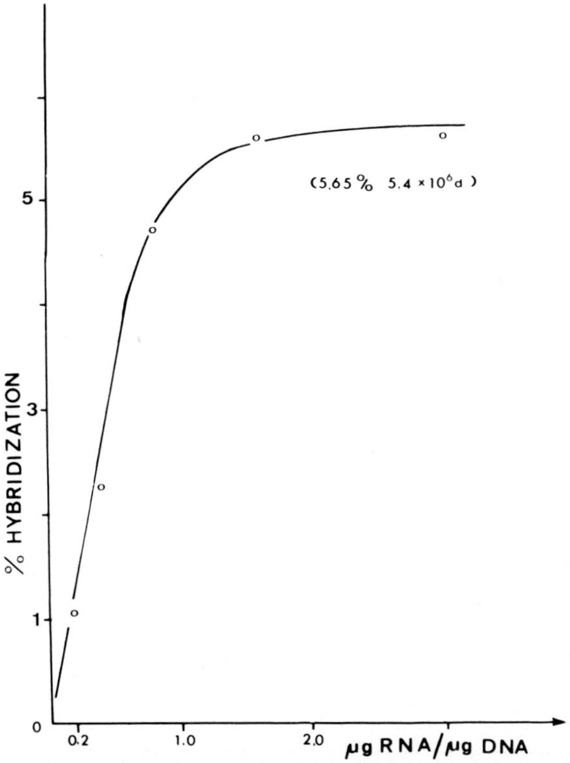

Fig. 4: Saturation-hybridization of ^{125}I labelled 2.7 x 10^6d RNA to spinach chloroplast DNA.

The competition hybridization experiments shown in Figs 5 and 6 revealed that the 23 S, 16 S and p 23 S RNA molecules are present in a 1 : 1 molar ratio each in the proposed transcript. However, when using ^{32}P-labelled 2.7 x 10^6d RNA synthesized by isolated chloroplasts the competition saturation level for 16 S rRNA was reached at very high RNA/DNA ratios only. A similar observation was recently reported by M. Hartley[6]. The most probable reason for this would be that the 16 S rRNA is at or near the 5'-end in the transcript and is not labelled or only partially in isolated chloroplasts, pointing to a very low or even zero rate of initiation of RNA chains together with slow elongation.

The size-heterogeneous RNA of the 2 megadalton region isolated from sucrose-gradients competed also with the 2.7 x 10^6d RNA.

In contrast to the ratios observed above, 1.5 to 2 copies of the 0.45 x 10^6d RNA population are present per proposed transcript both with ^{32}P- or ^{125}I-labelled transcript. From saturation-hybridization experiments (0.45 x 10^6d RNA iodinated) we have evidence that sequences for a molecule of thin size are present about four times on chloroplast DNA.

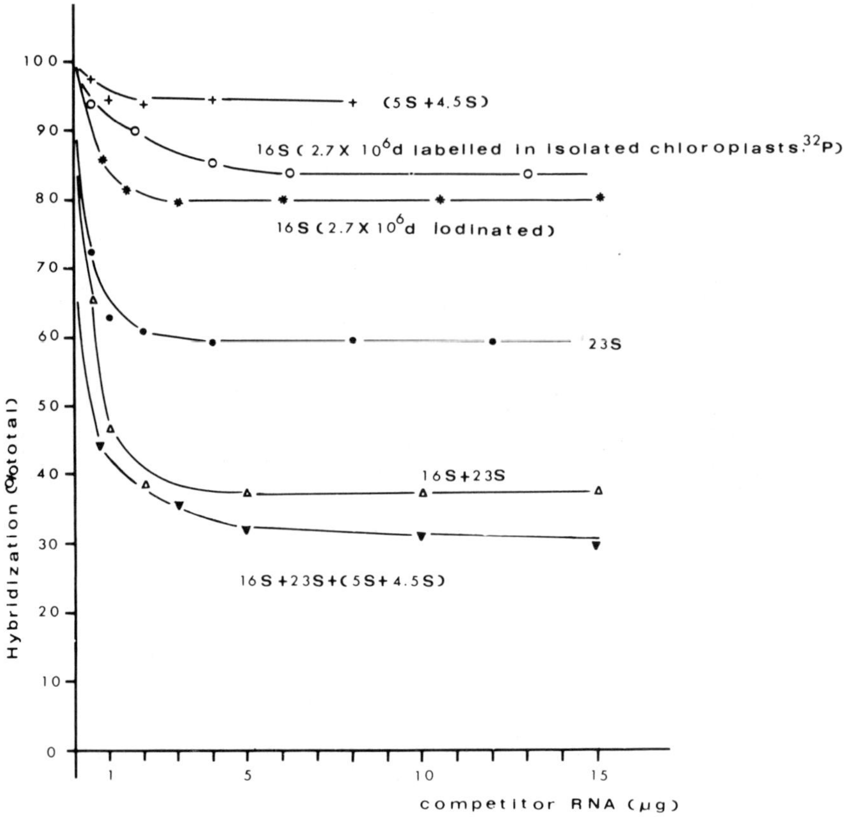

Fig. 5: Competition hybridization of labelled 2.7 x 10^6d RNA with rRNA and proposed processing intermediates from spinach chloroplasts.

We have further evidence from hybridization of this RNA fraction to SalI-restriction endonuclease fragments of spinach cp-DNA that some components are located on fragments which contain rRNA genes, while other components hybridize to other regions of the genome (in preparation).

In addition to the major rRNAs chloroplast ribosomes contain RNA of low molecular weight. We found a 5S and also a 4.5S RNA[7] which are both located in the large subunit of the ribosome; neither is a breakdown product of the 23S rRNA. The 4.5S RNA was independently found also by the group of T. Dyer in Cambridge and by P. Whitfeld and colleagues in Canberra. When hybridized, these molecules (not yet separated from one another) compete with the 2.7 x 10^6d RNA. Two copies of each of the molecules appear to be present in plastid DNA according to the saturation level in the absence or presence of unlabelled 23S and 16S rRNA, suggesting that both cistrons contain one copy of each of these low M.Wt. molecules.

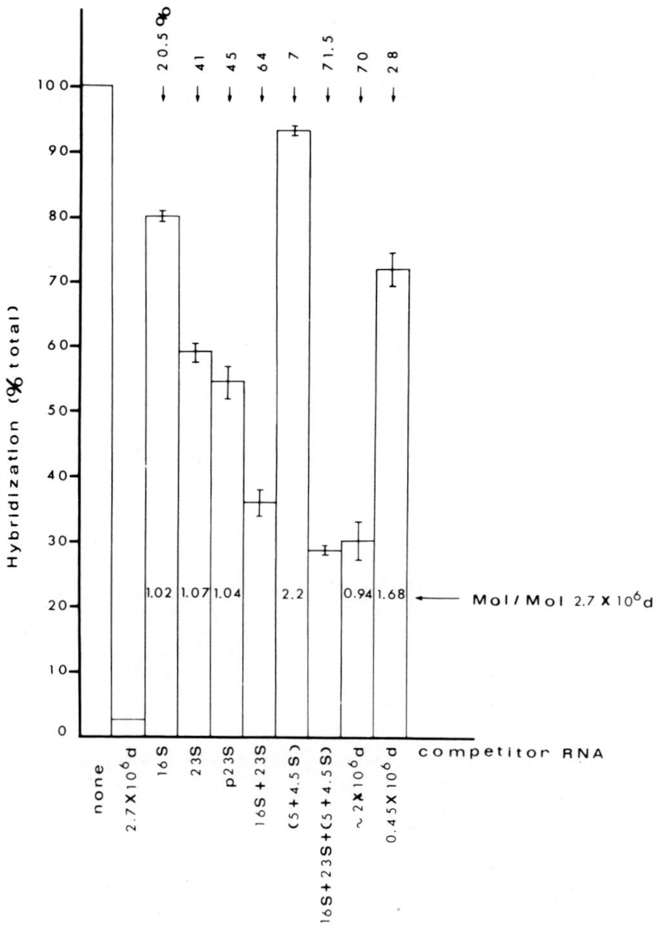

Fig. 6: Histogram of hybridization competition experiments.

Our hybridization data are summarized in Fig. 6. It is evident that the 2.7 x 10^6d RNA synthesized by isolated chloroplasts contains the sequence for the four RNAs found in the spinach chloroplast ribosome. A subpopulation of the molecules with M.Wt. of 0.45 x 10^6d appear to be metastable processing end-products. They do not compete with pure rRNA and are therefore most likely transcribed spacer. Finally, the processing intermediates proposed from labelling patterns are now verified by hybridization experiments. It is still unknown, however, whether all intermediates arise in the plant too. Additionally, it is not excluded that they may reflect differences between the two ribosomal cistrons. The current map of the topology of the ribosomal cistron of spinach chloroplast DNA is presented in Fig. 7. Its micromorphology using restriction endonuclease fragments of the chloroplast DNA (Herrmann et al., these Proceedings) should be established to solve the questions still remaining.

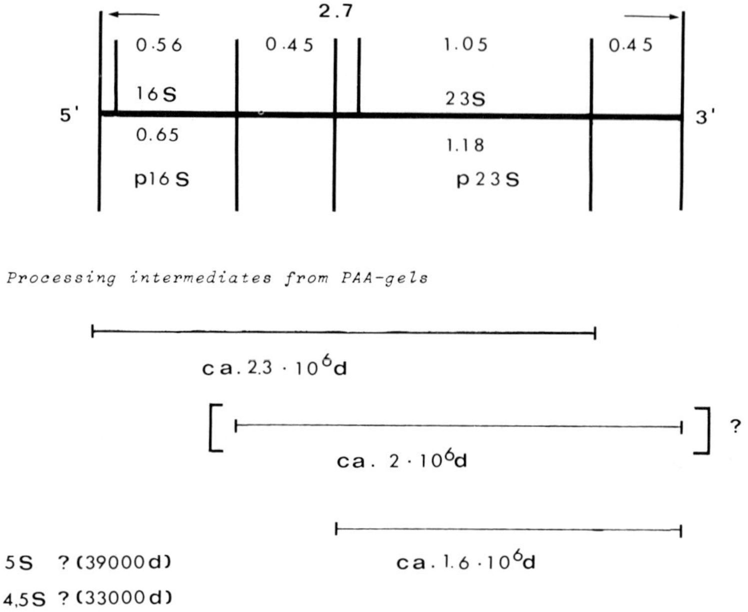

Fig. 7: Schematic presentation of the current map of the rDNA genes in spinach chloroplasts.

ACKNOWLEDGEMENTS

We gratefully acknowledge the assistance of our technicians and students. These investigations were supported by the Deutsche Forschungsgemeinschaft.

REFERENCES

1. Hartley, M.R. and Ellis, R.J. (1973) Biochem. J. 134, 249-262.
2. Bohnert, H.J., Schmitt, J.M. and Herrmann, R.G. (1974) Portugal. Acta Biol. 14, 71-90.
3. Bennett, J. and Milewska, Y. (1976) These Proceedings.
4. Sriprakash, K.S., Choo, K.B., Nagley, P. and Linnane, A.W. (1976) Biochem. Biophys. Res. Commun. 69, 85-91.
5. Bohnert, H.J., Driesel, A.J. and Herrmann, R.G. (1976) in Proc. FEBS-Course "Nucleic Acid and Protein Synthesis in Plants" (J.H. Weil and L. Bogorad, Eds.) Strassbourg, in press.
6. Hartley, M.R. (1976) at the same place.
7. Driesel, A.J. (1975) Diplomarbeit, Universität Köln

Genetics and Biogenesis of Chloroplasts and Mitochondria, Th. Bücher et al. eds.
© 1976 Elsevier/North-Holland Biomedical Press, Amsterdam, The Netherlands.

INCORPORATION OF ^{32}P-ORTHOPHOSPHATE INTO NUCLEOSIDE 5'-TRIPHOSPHATES AND RNA BY ISOLATED PEA CHLOROPLASTS

John Bennett and Yvonne Milewska
Department of Biological Sciences
University of Warwick
Coventry CV4 7AL
United Kingdom

SUMMARY

Isolated, enveloped pea (*Pisum sativum*) chloroplasts incorporated ^{32}P-orthophosphate into ATP, CTP, GTP and UTP in the light. Most of the incorporation was in to β- and γ-phosphates, but substantial incorporation into the α-phosphate position also occurred, as revealed by direct analysis of purified ^{32}P-ATP and by the labelling of *in vitro* RNA transcripts. Isolated pea chloroplasts elongated and terminated the 2.7 million dalton RNA species characteristic of chloroplasts but no evidence for *in vitro* initiation of this RNA species was obtained.

INTRODUCTION

The first direct and convincing evidence that chloroplasts synthesise discrete species of RNA was provided by Hartley and Ellis[1] who showed that isolated spinach chloroplasts incorporated ^{3}H-uridine into RNA with a molecular weight of about 2.7 million daltons. The incorporation occurred in plastids which had retained their envelope and their ability to photophosphorylate. Presumably, uridine penetrated the envelope of the isolated organelles, was phosphorylated successively to UMP, UDP and UTP, and was incorporated into growing RNA chains as a uridylyl group.

One implication of this pathway is that it should be possible to label *in vitro* transcripts with ^{32}P-orthophosphate. The crucial step would be the transfer of the γ-phosphate of ATP (formed from orthophosphate and ADP by photophosphorylation) to uridine or any other of the four nucleosides that constitute RNA, and requires that a nucleoside pool should exist within the chloroplast.

In this paper, we provide conclusive evidence for the *in vitro* labelling of the 2.7 million dalton transcript of pea chloroplasts after exposure to ^{32}P-orthophosphate. Labelling is dependent upon photophosphorylation but there is no need to supply exogenous nucleosides to the isolated organelles.

MATERIALS AND METHODS

Plants: Pea seedlings (*Pisum sativum* L. var. Feltham First) were grown under a 12 h light regime[2] and buds were harvested at a time of rapid chloroplast RNA synthesis (8-9 days after sowing).

Chloroplasts: Plastids were extracted from buds by homogenization in sucrose medium and incubated at 20°C in isotonic KCl medium[2]. For every 100-400μg chlorophyll, 20-40 μCi [5-^{3}H] uridine or 200-500 μCi of [^{32}P] orthophosphate (Amersham, Bucks) were used.

Nucleoside triphosphates: Nucleotides were extracted from incubated chloroplasts by adding to the incubation mixture an equal volume of ice-cold 10% trichloroacetic acid (TCA), removing the precipitate by centrifugation and neutralizing the extract with ether. The extract was lyophilized and the powder was dissolved in water prior to two dimensional thin layer chromatography[3] on sheets of polyethylene-imine-impregnated cellulose (Macherey-Nagel and Co., Düren). Labelled compounds were detected by autoradiography and NTPs were identified by co-chromatography with authentic samples.

RNA: Total nucleic acids were extracted from incubated plastids by the phenol procedure and ethanol precipitation[1]. For alkaline digestion of radioactive RNA, the ethanol precipitate was resuspended in 0.1M triethylammonium carbonate (TEC) (pH 8.5) and passed onto a DEAE-cellulose column (20 x 9 mm) equilibrated with the same buffer. Low molecular weight contaminants were eluted with 1.0M TEC (pH 8.5) and RNA was eluted with 2.0M TEC (pH 10.0). Radioactivity in column fractions was detected by Cerenkov counting[4]. Fractions containing RNA were lyophilized and dissolved in 0.3M KOH and incubated for 40 h at 37°C. Under identical conditions, there was no breakdown of γ-^{32}P-UTP (Amersham, Bucks). The alkaline digest was neutralized with a cation exchange resin and either subjected to high-voltage paper electrophoresis in 20 mM trisodium citrate (pH 3.4) or column chromatography on DEAE cellulose (150 x 9 mm) with elution by a linear gradient of TEC (pH 8.5, 0.1-0.5M). Digestion products were located by autoradiography and Cerenkov counting, respectively. Tritiated ATP (Amersham, Bucks) was used as a marker in the DEAE-cellulose column. The molecular weight of the labelled RNA was determined by resuspending a crude chloroplast RNA preparation in half strength electrophoresis buffer[1] containing 7M urea and 10% sucrose, heating to 70°C for 5 min, cooling rapidly and fractionating the RNA by electrophoresis through 2.4% polyacrylamide gels. Unlabelled RNA was detected by scanning at 260 nm, while ^{3}H-RNA was detected by scintillation counting and ^{32}P-RNA by Cerenkov counting.

RESULTS

Isolated pea chloroplasts incorporated ^{32}P-orthophosphate into numerous acid-soluble compounds, including the four ribonucleoside-5'-triphosphates (Fig. 1). In the light ATP was most heavily labelled triphosphate, followed in order by GTP, CTP and UTP. It is not known whether these marked differences in labelling reflect differences in NTP abundance or differences in labelling kinetics. In the dark, little labelling of triphosphates was detected

RNA prepared from ^{32}P-labelled plastids by phenol extraction and ethanol precipitation was heavily contaminated by low molecular weight phosphate esters, including NTPs. These were removed by column chromatography on DEAE-cellulose. RNA accounted for less than 5% of the radioactivity in the ethanol precipitate.

Alkaline hydrolysis of RNA purified on DEAE-cellulose yielded four labelled

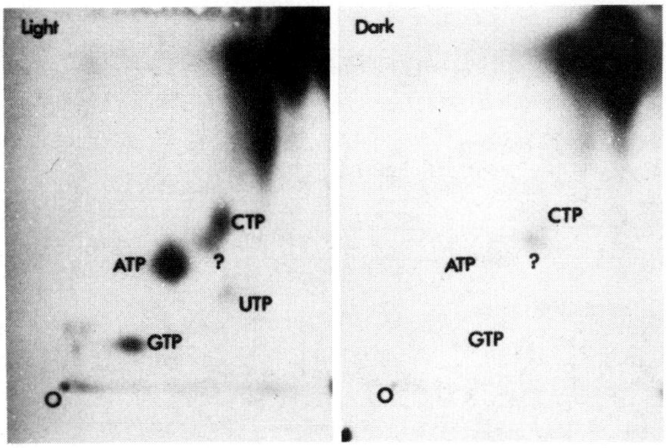

Fig. 1. Two-dimensional thin layer separations of acid-soluble, P-32-labelled compounds extracted from pea chloroplasts after incubation in the light or dark for 15 min with P-32-orthophosphate. The compound indicated by the question mark appeared in approximately equal amounts in both light and dark chromatograms and may be pyrophosphate contaminating the radioactive orthophosphate.

monophosphates which were separated by paper electrophoresis (Fig. 2. inset). This

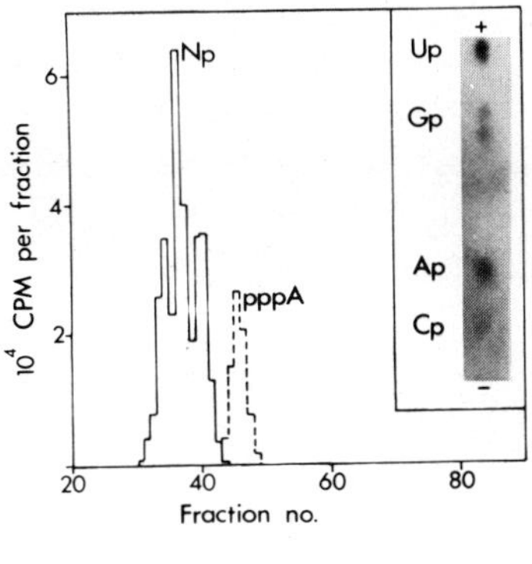

Fig. 2. Analysis of an alkaline digest of P-32-RNA prepared from pea chloroplasts incubated in the light for 15 min with P-32-orthophosphate. The histograms represent P-32-labelling in the digestion products (solid line) and H-3-labelling in marker ATP (broken line) separated by DEAE-cellulose column chromatography. The inset shows an autoradiogram of digestion products separated by paper electrophoresis.

confirms that the material was RNA and indicates that some at least of the label observed in the NTPs was in the α-phosphate. Direct analysis of ^{32}P-ATP, purified from chromatograms such as that shown in Fig. 1. indicated that labelling of the α-phosphate accounted for 1-5% of the label in that particular triphosphate. In principle, if the labelled RNA had been initiated _in vitro_ and had escaped processing by phosphates or nucleases, some label originating from the β and γ-phosphates of the initiating nucleotide would have been expected in the 5'-terminus (released as pppNp by alkali). However, no convincing evidence has been obtained for the presence in the alkaline digest of a tetraphosphate (Fig. 2.). When such a digest was fractionated on a DEAE-cellulose column by gradient elution, more than 99% of the label was recovered in monophosphates (eluted before ^{3}H-ATP marker) rather than in a 5'-terminal tetaphosphate (expected to be eluted after ATP).

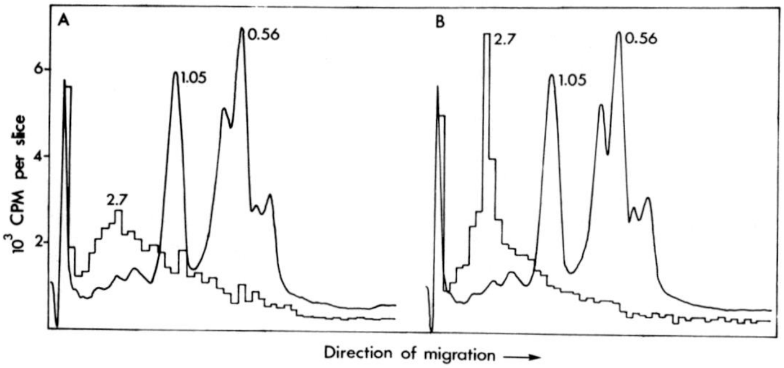

Fig. 3. Electrophoretic fractionation of RNA extracted from isolated pea chloroplasts that had been incubated in the light for 30 min with P-32-orthophosphate (A) or for 60 min with H-3-uridine (B). The numbers indicate the approximate molecular weights of the RNA species (in millions of daltons).

Electrophoresis of ^{32}P-RNA on 2.4% polyacrylamide gels revealed considerable labelling* of high molecular weight nucleic acid (Fig. 3A). Indeed, the pattern of incorporation into RNA was very similar to that observed with ^{3}H-uridine (Fig. 3B), with much of the isotope, in both cases, incorporated into the 2.7 million dalton transcript first reported in spinach chloroplasts by Hartley and Ellis[1].

DISCUSSION

Isolated, enveloped chloroplasts from spinach[1] and pea (see above) elongate and terminate the 2.7 million dalton transcript. However, it is not clear whether the chloroplasts initiate chains in vitro. This is an important point as far as chloroplast - cytoplasmic interactions are concerned. We have not been able to detect initiation in the present experiments. This might reflect an absence of initiation in isolated organelles, or it might be due to the processing of the 5'-terminus with the loss of characteristic 5'-pppNp grouping.

REFERENCES

1. Hartley, M.R. and Ellis, R.J. (1973) Biochem. J. 134, 249-262.
2. Bennett, J. (1976) Phytochemistry 15, 263-265.
3. Randerath, K. (1968) Thin Layer Chromatography, second edition, Academic Press, New York.
4. Mardh, S. (1975) Anal. Biochem. 63, 1-4.

*Not observed in dark-incubated or osmotically ruptured plastids.

Genetics and Biogenesis of Chloroplasts and Mitochondria, Th. Bücher et al. eds.

PHYLOGENETIC ORIGIN OF CHLOROPLAST 16S RIBOSOMAL RNA

D. E. Buetow*, M. S. Kissil* and L. Zablen**†
Departments of Physiology and Biophysics*,
Genetics and Development**, and Microbiology†,
University of Illinois
Urbana, Illinois 61801 U.S.A.

INTRODUCTION

Chloroplasts show striking structural and functional similarities with prokaryotes (e.g., 1), and an evolutionary relationship between these organelles and prokaryotic cells has often been postulated (e.g., 2). Phylogenetic relationships among prokaryotes have been derived by means of primary structural characterizations of 16S ribosomal RNAs (3). Thus, if the chloroplast arose from an ancestral prokaryote, the chloroplast 16S rRNA (the structural gene for which is located in the chloroplast) should carry the organelle's pedigree. Considered here will be a T_1-ribonuclease "fingerprint" analysis of the chloroplast 16S rRNA of *Euglena gracilis* (4, 5).

MATERIALS AND METHODS

Organism and growth conditions: *E. gracilis* strain Z was grown in the dark with limited phosphorus, supplemented with 10-25 mCi carrier-free $^{32}P_i$ (as H_3PO_4, New England Nuclear Co.), and incubated in the light as described (4, 5).

Isolation of chloroplast 16S rRNA: Chloroplasts were isolated and lysed as described (6). 16S rRNA was prepared as described (4, 5).

Sequence characterization of 16S rRNA: T_1-RNase digests of 16S rRNA were "fingerprinted" by the Sanger method as modified (7). Oligomer sequences were determined by secondary cleavage with endonucleases (7). Identified oligomers are given in refs. 4, 5.

RESULTS

Under greening conditions, *Euglena* take up 88% of the available isotope in 88 hr. About 220 x 10^6 cpm are recovered as chloroplast (ct) total RNA and 21% of this as ct 16S rRNA (45.4 x 10^6 cpm). The electrophoretic "fingerprint" of this latter RNA yields over 100 spots (Fig. 1). Only pentamer and larger fragments

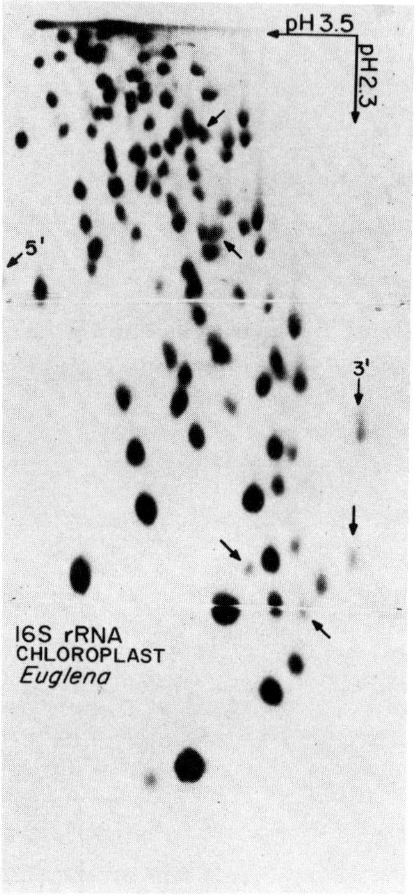

Fig. 1. Two dimensional electrophoretic fingerprint of chloroplast 16S rRNA. Post-transcriptionally modified oligomers and 3' and 5' termini denoted by arrows.

were assessed since these allow phylogenetic relationships to be established (e.g., 3). One hundred one individual pentamers and larger oligomers were found and 97 of these were characterized.

Of the 50 oligomers (hexamers and larger, and modified) identified in the *Euglena* ct 16S rRNA, only 2 are found also in yeast or mouse L-cell cytoplasmic 18S rRNA (Table 1). Of all 95 ct 16S rRNA oligomers listed, only 8 are also found in *P. cruentum* (a red alga) cytoplasmic 18S rRNA, but 39 are found in *P. cruentum* total 16S rRNA. This latter rRNA is from both chloroplasts and mitochondria, but the chloroplast 16S rRNA predominates (8). Further, of the 23 distinct nonmodified pentamers found in the *Euglena* chloroplast, 18 occur also in *Porphyridium* 16S rRNA, and 3 of these occur in no other organism so far studied (4, 5). Also, of the hexamer and larger oligomers, again the largest number is shared by the pair *Euglena* chloroplast - *Porphyridium* 16S rRNA.

The next highest degree of common oligomers occurs in the pair *Euglena* chloroplast - *Anacystis* (a blue-green alga) 16S rRNA followed by the pair *Euglena* chloroplast - bacilli 16S rRNA (Table 1). The chloroplast also has the typical "prokaryotic" capacity for base modification since it contains modified oligomers found so far only in prokaryotic 16S rRNAs (Table 1).

Some oligomers are designated as prokaryotic "universals" because they occur in 90% of the more than 30 prokaryotes, but in none of the eukaryotes, so far examined (9). The *Euglena* ct 16S rRNA contains at least 12 of these "universals" (4, 5), indicating that chloroplast rRNA is "prokaryotic". Also, the ct 16S rRNA contains the modified sequence $UCA\overset{*}{C}ACCA^{U}_{C}G$ (asterisk denotes modified base) found only in Bacillaceae and *Anacystis* (e.g., 4, 5) and lacks the modified sequence $\overset{*}{A}\overset{*}{A}CCUG$ characteristic of the Vibrio-Enteric group, *Rhodopseudomonas*, and many other gram-negative organisms (e.g., 4, 5).

DISCUSSION

Chloroplast 16S rRNA is prokaryotic because a significant number (4, 8) of sequences are shared by the chloroplast and at least certain prokaryotes, but not by the chloroplast and eukaryotes (Table 1). Also, the conserved sequences in the ct 16S rRNA occur in regions of high sequence conservation in prokaryotic 16S rRNAs (4, 5, 9), and the chloroplast modifies certain 16S rRNA sequences in typical prokaryotic fashion (Table 1).

The chloroplast is, at least, distantly related to the blue-green algae and perhaps also to the Bacillaceae because the ct 16S rRNA shares more sequences with these 2 groups than with other prokaryotes (Table 1) and it contains the modified sequence characteristic of these 2 groups.

ACKNOWLEDGEMENTS

Supported by NIH grant GM 19641 and NSF grant GB 41928.

TABLE 1

EUGLENA CHLOROPLAST OLIGOMERS COMMON TO OTHER rRNAs

Euglena ct 16S rRNA	A*	B*	C*	D*	E*	F*	G*
4 modified**	0	0	3	4	3	3	3
23 pentamers	-†	7	18	14	13	12	14
17 hexamers	2	1	6	5	3	5	3
13 heptamers	0	0	4	2	2	2	2
16 octamers	0	0	3	2	4	3	4
11 nonamers	-†	0	2	4	3	3	2
11 decamers and larger	-†	0	3	3	2	0	0
Total	2	8	39	34	30	28	28

* A, yeast and L cell 18S rRNA (4); B, *Porphyridium cruentum* 18S (8); C, *P. cruentum* 16S rRNA (8); D, *Anacystis nidulans* 16S; E, Bacillaceae 16S; F, *Rhodopseudomonas spheroides* 16S; G, Vibrio-Enteric bacteria 16S (see 4).

**Post-transcriptionally (see 4).

† Not available.

REFERENCES

1. Taylor, D. L. (1970) *Int. Rev. Cytol.* 27, 29-64.
2. Flavell, R. (1972) *Biochem. Genet.* 6, 275-291.
3. Zablen, L., Bonen, L., Meyer, R. and Woese, C. R. (1975) *J. Mol. Evol.* 4, 347-358.
4. Zablen, L., Kissil, M. S., Woese, C. R. and Buetow, D. E. (1975) *Proc. Nat. Acad. Sci., U.S.* 72, 2418-2422.
5. Buetow, D. E. (1976) *J. Protozoology* 23, 41-47.
6. Kissil, M. S. and Buetow, D. E. (1974) *J. Cell Biol.* 63, 169a.
7. Uchida, T., Bonen, L., Schaup, H. W., Lewis, B. J., Zablen, L. and Woese, C. R. (1974) *J. Mol. Evol.* 3, 63-77.
8. Bonen, L. and Doolittle, W. F. (1975) *Proc. Nat. Acad. Sci., U.S.* 72, 2310-2314.
9. Woese, C. R., Fox, G. E., Zablen, L., Uchida, T., Bonen, L., Pechman, K., Lewis, B. J. and Stahl, D. (1975) *Nature (London)* 254, 83-86.

Genetics and Biogenesis of Chloroplasts and Mitochondria, Th. Bücher et al. eds.

A SEQUENCE ANALYSIS OF LOW-MOLECULAR-WEIGHT rRNA FROM CHLOROPLASTS OF FLOWERING PLANTS

Tristan A. Dyer and Catherine M. Bowman
ARC Unit of Developmental Botany,
181A Huntingdon Road, Cambridge, CB3 0DY, U.K.

INTRODUCTION

At least five different types of low-molecular-weight RNA occur as structural components of the ribosomes of flowering plants. The ribosomes of the cytoplasm, chloroplasts and mitochondria each contain a distinctive species of 5S rRNA and, in addition, the cytoplasmic ribosomes contain a 5.8S rRNA and the chloroplast ribosomes a 4.5S rRNA. Some of the characteristics of these molecules are reviewed elsewhere[1]. Sequence analyses have been carried out in our laboratory on examples of all but the mitochondrial 5S rRNA and the results which relate particularly to the chloroplast RNAs are described here.

Sequence analyses can be expected to provide information about the structure[2] and function[3] of RNAs. Also, differences and similarities in the sequence of homologous molecules reflect the way in which they and the organisms of which they are part may have evolved from a common ancestor[4]. The comparison made in this paper between chloroplast and cytoplasmic 5S rRNA is of particular interest in this context.

MATERIALS AND METHODS

An axenic culture of duckweed (*Lemna minor*) was kindly provided by Dr. C.J. Leaver. The fronds were grown in conical flasks at 27°C on an inorganic medium[5] under a mercury vapour lamp (Philips 500W M.B.T.L. bulb). When the nucleic acids were to be labelled, the non-radioactive phosphate in the medium was replaced with Tris and between 10 and 30 mCi ^{32}P-orthophosphate added per experiment. The fronds were incubated for 6 days and about 12 hr. before harvesting, potassium phosphate (pH 7.0) was added to a concentration of 2.3 mM. The addition of this non-radioactive phosphate was to lower the specific radio-activity of contaminating material in the rRNA preparations. All operations to this stage were carried out under aseptic conditions.

The fronds (about 5 g. fresh wt.) were collected, chilled for at least 1 h, then macerated in a dual glass homogenizer in 50 ml. of a medium containing 50 mM KCl, 10 mM $MgCl_2$, 10 mM dithiothreitol and 50 mM Tris-HCl (pH 7.8 at room temperature). Details of the preparation of cytoplasmic monosomes and of chloroplast large ribosomal subunits[6] and fractionation of low-molecular-weight rRNA are given elsewhere[7].

Standard procedures were used for the enzymatic digestion of the low-molecular-weight rRNA, the two-dimensional fractionation of the products by electrophoresis and the analysis of products by further digestion[8].

RESULTS AND DISCUSSION

Two factors have been of great importance in this study of chloroplast low-molecular-weight rRNA. Firstly, the choice of flowering plant from which to extract the RNA, and secondly, the method of obtaining chloroplast and cytoplasmic 5S rRNA preparations that were not cross-contaminated.

Duckweed is a small floating aquatic plant. As grown for the experiments described here it absorbs up to 95% of the radioactive phosphate from the medium in six days. Most of this ^{32}P is accumulated in rapidly growing photosynthetic tissue and, as a result, the specific activity of the chloroplast 5S rRNA becomes sufficiently high for its complete sequence to be determined by standard procedures. The chloroplast nucleic acids of several other flowering plants were labelled but in none of these did the RNA become radioactive enough for our purpose. Another advantage of duckweed is that it can be maintained readily in axenic culture, so eliminating the chance of spurious results being obtained from the RNA of contaminating micro-organisms.

Concerning the method for preparing 5S rRNA, the usual techniques for fractionating low-molecular-weight rRNA, at best, only partially separated the

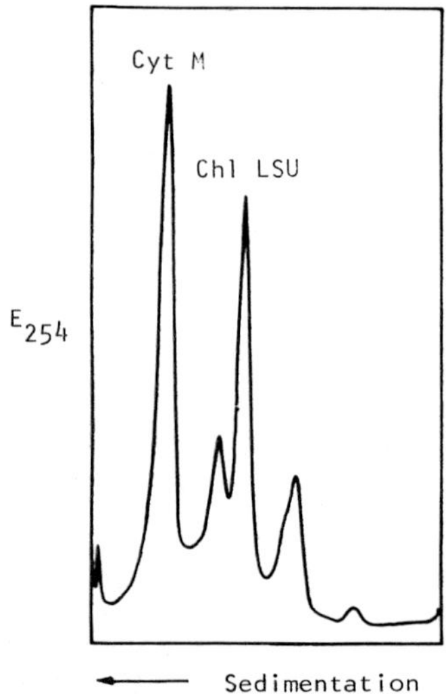

Fig. 1. Fractionation by sucrose density gradient centrifugation of a ribosome preparation from duckweed. Cyt M - cytoplasmic monosome; Chl LSU - chloroplast large ribosomal subunit.

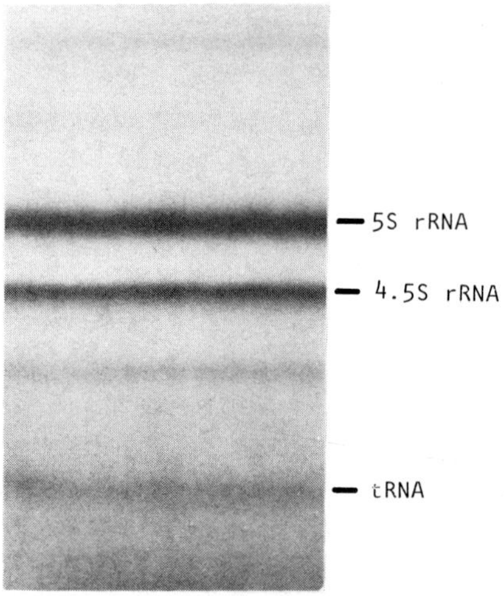

Fig. 2. Autoradiograph of chloroplast large ribosomal subunit RNA which had been fractionated by polyacrylamide gel electrophoresis. The top of the gel containing high-molecular-weight RNA was removed before autoradiography.

chloroplast and cytoplasmic 5S components. However, by separating cytoplasmic monosomes and chloroplast large ribosomal subunits and extracting the RNA from these, 5S rRNA preparations could be obtained with virtually no cross-contamination. The essential feature of the preparation of ribosomes was that they were first sedimented by centrifugation in a medium containing 10 mM Mg^{2+}, then resuspended and fractionated by sucrose gradient centrifugation in a medium containing 3 mM Mg^{2+}. Chloroplast monosomes were selectively dissociated when the magnesium concentration was lowered and their subunits could be completely separated from the cytoplasmic monosomes (Fig. 1). It is then relatively easy to purify the low-molecular-weight rRNA from the ribosomal particles by gel electrophoresis (Fig. 2). Although quite a high proportion of 5S rRNA is probably lost during its preparation, it does not appear to have been fragmented at all during this procedure.

The complete nucleotide sequence of rye cytoplasmic 5S rRNA has recently been determined[9]. Although the fingerprints of oligonucleotides produced by T_1 and pancreatic A ribonuclease digestion of the cytoplasmic 5S rRNA of duckweed were very similar to those of rye, the sequences of the duckweed oligonucleotides were determined independently with the exception of P18 and P28. These gave the same

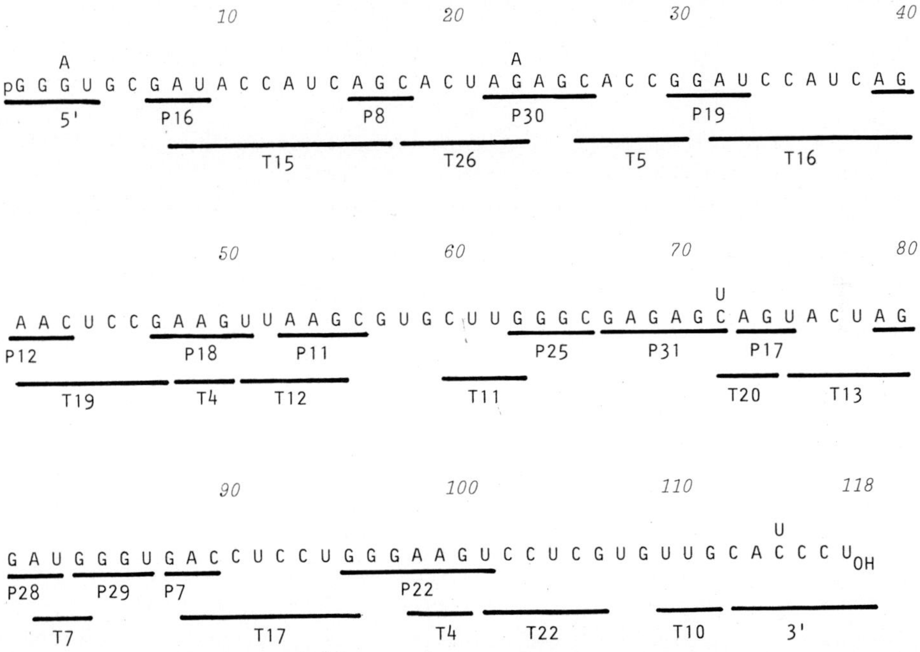

Fig. 3. Probable sequence of cytoplasmic 5S rRNA from duckweed. Rye nucleotides, where different, are shown above the duckweed sequence. The position in the entire molecule of the larger oligonucleotides produced by T_1 and pancreatic A digestion are given below the sequence.

products as P18 and P28 from rye after digestion with T_1 ribonuclease and therefore the rye and duckweed oligonucleotides were assumed to have the same sequence. The probable sequence of duckweed cytoplasmic 5S rRNA was deduced by aligning its oligonucleotides with homologous sequences in the rye 5S rRNA (Fig.3).

The sequence of duckweed cytoplasmic 5S rRNA is typical of that in flowering plants. It differs in only eight positions from the most dissimilar that we have studied (broad-bean) and in only four positions from rye[9]. Thus in higher plants, as in animals, 5S rRNA has been highly conserved suggesting that there are severe constraints on changes occurring in its sequence. The similarity between flowering plant cytoplasmic 5S rRNA and that of the unicellular green alga *Chlorella pyrenoidosa*[10] is much less pronounced than amongst the flowering plants themselves although there are several long sequences which the flowering plant and algal 5S rRNAs have in common.

The determination of the complete sequence of duckweed chloroplast 5S rRNA is still being carried out. Nearly all the sequences of oligonucleotides produced by T_1 and pancreatic A ribonuclease digestion of the molecule are known and now partial digests of the molecule are being analysed. From a comparison of the fingerprints of rye and of duckweed chloroplast 5S rRNAs it may be concluded that approximately the same number of nucleotides differ in the two as in their cytoplasmic counterparts.

TABLE 1

Oligonucleotides produced by enzymatic digestion of chloroplast 5S rRNA from duckweed

	T_1 ribonuclease		Pancreatic A ribonuclease
T1	Gp	P1	Up
T2	C-Gp	P2	Cp
T3	A-Gp	P3	A-Cp
T4	A-C-Gp	P5	A-A-A-Cp
T5	C-C-A-Gp	P6	G-Cp
T6	U-Gp	P7	A-Up
T7	U-A-Gp	P9]	2(G-A-Cp)
T8	C-U-C-Gp	P10]	+A-G-Cp
T9	U-C-C-U-Gp	P11	G-A-A-Cp
T10	U-C-C-U-A-Gp	P13	A-A-Up
T11	A-A-A-U-A-Gp	P14	G-Up
T12	A-A-C-U-U-Gp	P15	pUp
T13	A-U-A-C-U-Gp	P16	pCp
T14	pU-A-U(2U,C)Gp (5')	P18	G-A-Up
T15	pC(2U,C)Gp (5')	P19	A-G-G-Cp
T16	U-U-A-A-A-C-U-C-U-A-C-U-Gp	P20	G-G-Up
T17	A-A-C(A-U-C-C-A,C-C-A,C-C-A,A)	P23	A-G-A-G-G-A-A-Cp
	U-C-C-C-Gp	P24	G-A-A-A-Up
3'(a)	A-A-U_{OH}	P25	(2G, A-G)Up
3'(b)	A-A_{OH}	3'(a)	A-G-A-A-U_{OH}
		3'(b)	A-G-A-A_{OH}

Duckweed chloroplast 5S rRNA is obviously very different from the cytoplasmic 5S rRNA as they have very few oligonucleotides in common (Fig. 3 and Table 1). This has also been shown for the equivalent two 5S rRNAs from *Chlorella pyrenoidosa*[11]. Preliminary data from partial digests of the duckweed chloroplast 5S rRNA suggests that it resembles to a much greater extent the 5S rRNA of the blue-green alga *Anacystis nidulans*[12]. Both appear to have identical sequences of nucleotides towards the ends of the molecule. This does not necessarily indicate that the blue-green algae were the progenitors of chloroplasts but merely that the ancestors of the chloroplasts were very likely to have been prokaryotes. The chloroplast 5S rRNA has other prokaryotic characteristics. It contains the -pyr-G-A-A-C- sequence which is thought to interact with the -G-T-Ψ-C-pur- in tRNA[3] and several large oligonucleotides are produced when it is digested with T_1 ribonuclease. In contrast, the cytoplasmic 5S rRNA contains a -pur-G-A-A-C- sequence and none of the oligonucleotides produced by T_1 ribonuclease digestion are very large.

Although the chloroplast and cytoplasmic 5S rRNAs differ to a considerable degree, they do have several features in common which confirms that they are true homologues of one another. As well as being very similar in size, neither contain any modified nucleotides. In both, the bases in the terminal regions are complementary to one another, which suggests that these two segments of the molecule are paired. There is probably substantially more base pairing than this in 5S rRNA. For example, the cytoplasmic 5S rRNA could be folded as predicted by a recent universal model for the secondary structure of 5S rRNA[2], and our preliminary data also suggests that chloroplast 5S rRNA can be folded in the same way.

A feature of special interest in chloroplast 5S rRNA is that two different oligonucleotides are obtained from each end of the molecule when it is digested enzymatically. These are present in half-molar amounts. From this it would seem that there are at least two distinct cistrons on which this 5S rRNA is synthesised. This may be of particular significance, as hybridization studies have already indicated that there are two cistrons for high-molecular-weight rRNA per copy of chloroplast DNA[13].

Concerning the smaller of the two chloroplast low-molecular-weight rRNAs, it has several features which show that it is quite different from any of the other rRNAs. For instance it has only been detected in the ribosomes from chloroplasts of flowering plants and it is not of constant size. While in several flowering plants it contains about 103 nucleotides, the molecule which appears to be its homologue in some legumes only has about 80 nucleotides. A comparison of the fingerprints of the 4.5S rRNA from rye and from duckweed clearly indicates that the molecules are related to one another but that there is more of a difference between the 4.5S rRNAs than between the chloroplast 5S rRNAs or the cytoplasmic

5S rRNAs of these plants. As the 4.5S rRNA occurs in approximately equimolar amounts with the 5S rRNA (Fig. 2) it is probably a structural component of the ribosome. There are no obvious long complementary sequences in the oligonucleotides produced by T_1 and pancreatic A digestion of 4.5S rRNA (Table 2) to show that it is like the RNA of similar size in *E. coli*[14] in which there is a lot of base pairing. Whereas all 5S rRNAs studied so far have been phosphorylated at their 5' end, the duckweed 4.5S apparently does not have a 5' phosphate. This may be because the 4.5S and 5S rRNA molecules are processed differently during their synthesis.

At this stage one can only speculate as to the nature of the 4.5S rRNA. It could possibly be the chloroplast equivalent of cytoplasmic 5.8S rRNA but in contrast to that molecule it contains no methylated bases. A further difference between the two is that 5.8S rRNA is very uniform in size[15] and 4.5S rRNA is not. Alternatively, the 4.5S rRNA could be a fragment of the high-molecular-weight rRNA but as it is found in extracts made from plant tissues, using methods which rigorously suppress nuclease activity, fragmentation, if it occurs, must take place *in situ*. It appears more likely at present that 4.5S rRNA is a piece of transcribed spacer, perhaps with no particular role in the ribosome. This would be consistent with the observations that it is not the same size in all flowering plants and that, so far, it has only been found in a restricted range of organisms.

TABLE 2

Oligonucleotides produced by enzymic digestion of chloroplast 4.5S rRNA from duckweed

	T_1 ribonuclease		Pancreatic A ribonuclease
T1	Gp	P1	Up
T3	C-C-Gp	P2	Cp
T4	A-Gp	P3	A-Cp
T5	C-A-Gp	P4	A-A-Cp
T6	A-C-Gp	P6	G-Cp
T7	A-C-C-Gp	P7	A-Up
T8	A-A-Gp	P8	G-A-A-C_{OH} (3')
T9	C-A-A-Gp	P9	A-G-Cp
T10	U-Gp	P10	A-G-A-Cp
T11	C-U-Gp	P11	A-A-G-A-Cp
T12	A-U-Gp	P12	G-Up
T13	U-C-A-C-Gp	P13	G-G-Cp
T14	U-A-A-Gp	P14	G-A-G-Cp
T15	U-C-A-A-Gp	P15	G-A-G-A-C
T16	A-U-A-Gp	P16	G-G-A-Up
T17	U-A-U-Gp	P17	A-G-Up
T18	A-U-U-U-Gp	P18	A-A-G-Up
T19	A-C-A-U-C-C-U-A-A-C-A-Gp	P19	A-G-G-Up
T20	U-U-U-A-U-C-A-U-U-A-C-Gp	P20	A-A-G-G-Up
3'	A-A-C_{OH}	P21	(2G,A-A-G)Up
5'	not detected	P22	G-A-G-A-G-A-Up

Whether or not it has a function in the ribosome might be resolved by reconstituting ribosomes in its presence and absence.

It may be concluded from this study that the structural low-molecular-weight rRNAs of the chloroplasts are quite distinct from those of the cytoplasm. It is interesting to question whether this is merely the result of their having had separate origins or whether the particular features of the chloroplast RNAs make them especially well suited to participate in the synthesis of chloroplast proteins.

REFERENCES

1. Dyer, T.A., Bowman, C.M. and Payne, P.I. (1976) In *Nucleic Acids and Protein Synthesis in Plants*, Plenum Publishing Corp., New York, in press.
2. Fox, G.E. and Woese, C.R. (1975) J. Mol. Evol. 6, 61-76.
3. Richter, D., Erdmann, V.A. and Sprinzl, M. (1973) Nat. New Biol. 246, 132-135.
4. Bonen, L. and Doolittle, W.F. (1976) Nature 261, 669-673.
5. Huntner, S.H. (1953) *Growth and Differentiation in Plants* (Loomis, W.E., ed.) pp. 417-446, Iowa State College Press, Ames.
6. Dyer, T.A. and Koller, B. (1972) Proc. IInd Int. Congr. Photosynth. Res. pp. 2537-2544, W. Junk N.V., Den Haag.
7. Payne, P.I., Corry, M.J., and Dyer, T.A. (1973) Biochem. J. 135, 845-851.
8. Brownlee, G.G. (1972) *Determination of Sequences in RNA*, North Holland, Amsterdam.
9. Payne, P.I. and Dyer, T.A. (1976). Submitted to Eur. J. Biochem.
10. Jordan, B.R., Galling, G. and Jourdan, R. (1974) J. Mol. Biol. 87, 205-225.
11. Galling, G. and Jordan, B.R. (1972) Biochimie, 1257-1265.
12. Corry, M.J., Payne, P.I. and Dyer, T.A. (1974) FEBS Lett. 46, 63-66.
13. Thomas, J.R. and Tewari, K.K. (1974) Proc. Nat. Acad. Sci. USA 71,3147-3151.
14. Griffin, B.E. (1975) J. Biol. Chem. 250, 5426-5437.
15. Payne, P.I. and Dyer, T.A. (1972) Nat. New Biol. 235, 145-147.

Genetics and Biogenesis of Chloroplasts and Mitochondria, Th. Bücher et al. eds.

Chloroplast ribosomal proteins of *Euglena gracilis*. Immunological studies

Georges Freyssinet, François Morlé and Victor Nigon
Département de Biologie générale et appliquée
Laboratoire associé au CNRS
Université de Lyon I
69621 Villeurbanne
France

SUMMARY

Two antisera, one against cytoplasmic ribosomal proteins, the other against chloroplast ribosomal proteins, have been prepared and tested by double immunodiffusion and quantitative precipitation with the corresponding antigens. Results indicate that cross-reactions might exist between the two types of ribosomal proteins.

INTRODUCTION

Higher plants and green algae have three types of ribosomes, namely, cytoplasmic, chloroplast and mitochondrial ribosomes. If it is almost sure that the rRNA of the organelles are synthesized within these organelles, some of their ribosomal proteins appear to be coded by nuclear DNA and synthesized in the cytosol[1]. One can therefore ask whether some structural identities exist between proteins of different types of ribosomes, especially when these proteins are coded by nuclear DNA. The use of immunochemical techniques should allow an approach of this problem.

We present here preliminary results obtained with antiserum against cytoplasmic ribosomal proteins and antiserum against chloroplast ribosomal proteins of *Euglena*.

MATERIALS AND METHODS

Wild-type strain and a white mutant, W_3BUL, of *Euglena* have been used for these experiments. Wild-type cells were grown in the light on the NC medium deficient in vitamin B_{12}[2] and W_3BUL on the Hutner's medium[3].

Cytoplasmic ribosomes were extracted from the mutant according to the technique already described[4]. Chloroplast ribosomes were extracted from wild-type and purified on a 10 to 45 % sucrose gradient (Freyssinet, in preparation).

Ribosomal proteins used for antiserum preparation were extracted with acetic acid[4] and injected into rabbits according to Stöffler and Wittmann[5]. Injections consisting of 1 mg of ribosomal proteins in 2 ml of buffer[5] and 2 ml of complete Freund's adjuvant were done on the first, second, third, fifth and sixth weeks. Animals were bled on the seventh week and the immune serum collected. Prior to the first injection nonimmune serum was taken from each animal. Double immunodiffusion and quantitative precipitation were performed as described by Stöffler and Wittmann[5]. For these two tests ribosomal proteins have been extracted using the LiCl-urea technique[6].

Abbreviations : Ag Cy and Ag Ch = antigen from ribosomal proteins of cytoplasm and chloroplast respectively. Anti Cy and Anti Ch = antisera against cytoplasmic and chloroplast ribosomal proteins respectively.

RESULTS

1. *Antigens* : In order to limit cross contaminations the Ag Cy has been prepared from the *Euglena* white mutant W_3BUL. This mutant does not contain any chloroplast DNA[7] or chloroplast rRNA[8]. In the case of Ag Ch, we have checked the purity of ribosomal preparations used by analyzing their RNA content. We have then noticed, on gel electrophoresis, either the complete absence of cytoplasmic rRNA or the presence sometimes of a 20 S component and maybe a very small amount of 25 S. Fig. 1 A represents the preparation containing the maximum amount of these two components. The 20 S component corresponds either to the small cytoplasmic rRNA or to a degradation product of the chloroplast rRNA[9]. This should be the case for the major part of the 20 S analyzed in fig. 1 A as the amount of 25 S, the large cytoplasmic rRNA, is very small. As a control we represented on fig. 1 B the electrophoretogram of a mixture made of the RNA preparation analyzed in fig. 1 A and of 10 % of cytoplasmic rRNA extracted from ribosomes of W_3BUL. In this case the presence of 25 S cytoplasmic rRNA is obvious.

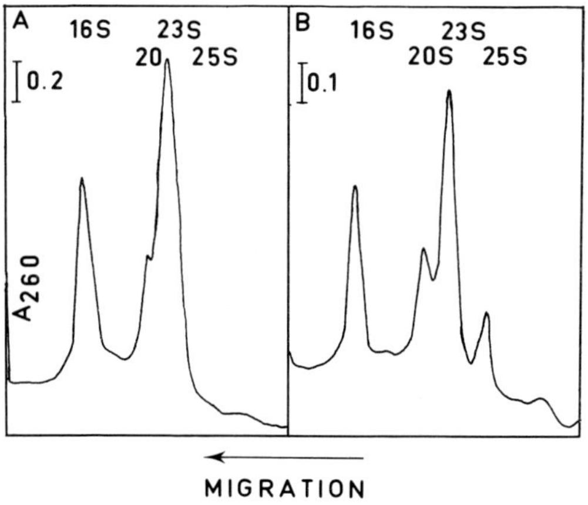

Fig. 1 : Electrophoretogram of rRNA extracted from *Euglena* ribosomes. A - rRNA from purified chloroplast ribosomes (2 O.D.). B - Mixture of rRNA from purified chloroplast ribosomes (1 O.D.) and rRNA from cytoplasmic ribosomes (0.1 O.D.). Ribosome preparations were treated with sodium dodecyl sulfate (2.5 %) during 2 min. and then layered onto a 2.8 % polyacrylamide gel prepared according to Loening[10]. Electrophoresis was run for 5 hrs. at 5 mA/gel. Gels were scanned at 260 nm with a Gilford Spectrophotometer.

2. *Immunological studies* : When antigens and antibodies were tested by double immunodiffusion we obtained in all cases, precipitation lines (fig. 2). The most important difference is found with Anti Ch ; there are two precipitation lines when tested with Ag Ch and only one with Ag Cy.

In order to determine more precisely the degree of cross reaction between cytoplasmic and chloroplast ribosomal proteins of *Euglena*, we have established the quantitative immunoprecipitation curves (fig. 3). For that, equal amounts of antigens were mixed with increasing quantities of antiserum in a constant volume of 1 ml. Reactions between Anti Cy-Ag Ch and Anti Ch-Ag Cy represent 30 percent of reactions between Anti Cy-Ag Cy and Anti Ch-Ag Ch respectively.

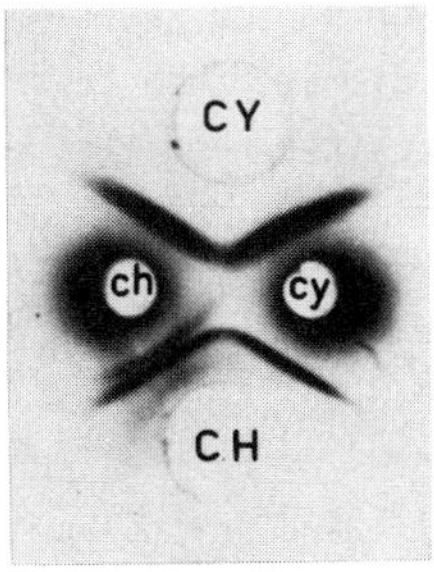

Fig. 2 : Ouchterlony double diffusion patterns of cytoplasmic and chloroplast ribosomal proteins. The center wells contain 25 µl of antiserum either to cytoplasmic ribosomal proteins (CY) or to chloroplast ribosomal proteins (CH). The peripheral wells contain 6 µl of either cytoplasmic ribosomal proteins (Cy) or chloroplast ribosomal proteins (Ch).
The agarose gel was prepared as described by Stöffler et al.[5]. Diffusion lasts 48 hrs. at 4°C. The distance from the edge of control well to that of the surrounding wells was 4 mm. The gel was stained with the Coomassie blue and then photographed with a Polaroid MP 4 camera.

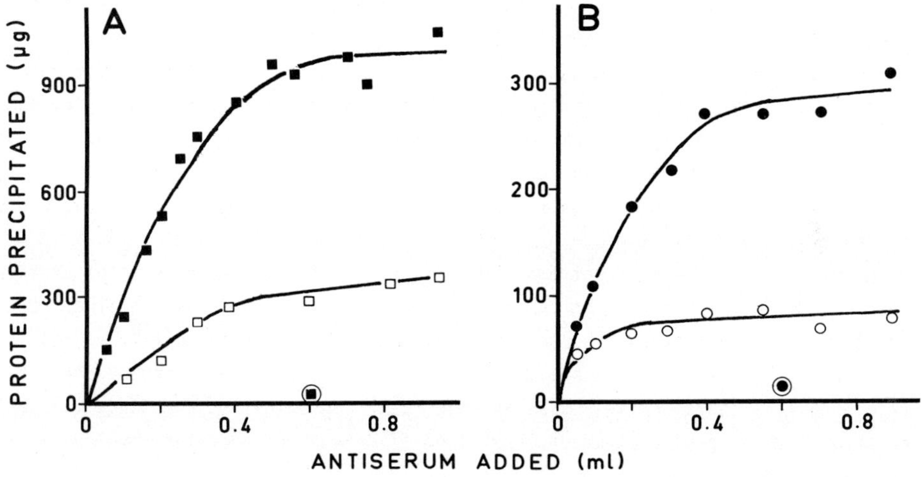

Fig. 3 : Quantitative immunoprecipitation of *Euglena* ribosomal proteins :
A - Antiserum directed against cytoplasmic ribosomal proteins with either cytoplasmic ribosomal proteins (■) or chloroplast ribosomal proteins (□).
B - Antiserum directed against chloroplast ribosomal proteins with either cytoplasmic ribosomal proteins (●) or chloroplast ribosomal proteins (○).
Ribosomal proteins (320 µg in A and 50 µg in B) were incubated, in 1 ml final volume, with the indicated amounts of antiserum. Pellets were washed twice with Tris-HCl 50 mM, pH 8.2 ; LiCl, 750 mM and dissolved in NaOH 0.1 N. Protein concentration was determined by biuret using bovine serum albumin as standard. As a control two nonimmune sera were tested against Ag Cy (▣) or Ag Ch (◉).

As our antigen preparations are formed of several proteins (70 in the case of Ag Cy[4]) we have tried to fractionate them on CM cellulose columns. Ag Cy was eluted with increasing concentrations of Na-acetate, fractions were recovered, precipitated (Morlè, Freyssinet and Nigon, in preparation) and tested against the two antisera by double immunodiffusion. Two conclusions can be done : a. Some fractions do not show precipitation lines against the Anti Cy, even when we used concentrations three times higher than that of the unfractionated Ag Cy ; b. One or two fractions give a weak precipitation line or even no precipitation line at all when tested against the Anti Ch while important precipitation lines were obtained when tested against the Anti Cy.

DISCUSSION

Results obtained indicate that there exists a partial immunological relationship between the two antigens. Similar results have been obtained with cytoplasmic and chloroplast ribosomes from higher plants[11] whereas mitochondrial and cytoplasmic ribosomes do not show cross reaction[12]. This relationship can be assigned either to cross contamination or to real immunochemical identities between some of the cytoplasmic and chloroplast ribosomal proteins.

The Ag Cy should not be contaminated as it is prepared from an aplastidic mutant. In the case of Ag Ch we cannot be sure that this antigen is free of cytoplasmic contamination which may come either from cytoplasmic ribosomes or from non ribosomal proteins bound to ribosomes during the extraction procedure. In particular, the ribosome sensitivity to high salt concentrations[4-13] has not allowed us to wash extensively the preparations.

Results obtained with fractions from CM cellulose columns have to be extented before a definitive conclusion can be reached. If they should be confirmed, cross reaction by contaminating ribosomal proteins would become difficult to explain.

Finally, since the cross reaction is so important, one third of the direct reaction, it appears unlikely that the eventual contaminating compounds will only be given by the purification and the identification of proteins involved in the reaction.

REFERENCES

1. Chua, N.H. and Luck, D.J.L. (1974). *In* Ribosomes, pp. 513-539. Eds. Nomura, Tissières and Lengyel. Cold Spring Laboratory.
2. Freyssinet, G. (1975). *Plant Sci. Letters* 5, 305-311.
3. Greenblatt, G.L. and Schiff, J.A. (1959). *J. Protozool.* 6, 23-28.
4. Freyssinet, G. and Schiff, J.A. (1974). *Plant Physiol.* 53, 543-544.
5. Stöffler, G. and Wittmann, H.G. (1971). *Proc.Nat. Acad. Sci. USA* 68, 2283-2287.
6. Kaltschmidt, E. and Wittmann, H.G. (1972). *Biochimie* 54, 167-175.
7. Edelman, M., Schiff, J.A. and Epstein, H.T. (1965). *J. Mol. Biol.* 11, 769-774.
8. Heizmann, P., Salvador, G. and Nigon, V. (1976). *Exptl. Cell Res.* 99, 253-260.
9. Zeldin, M.H. and Schiff, J.A. (1967). *Plant Physiol.* 42, 922-932.
10. Loening U.E.L. (1967). *Biochem. J.* 102, 251-257.
11. Gualerzi, C., Janda, H.G., Passow, H. and Stöffler, G. (1974). *J. Biol. Chem.* 249, 3347-3355.
12. Hallermayer, G. and Neupert, W. (1974). *FEBS Letters* 41, 264-268.
13. Delihas, N., Jupp, A. and Lyman, H. (1972). *Biochim.Biophys. Acta* 262, 344-351.

Genetics and Biogenesis of Chloroplasts and Mitochondria, Th. Bücher et al. eds.

A CHLOROPLAST MEMBRANE FRACTION ENRICHED IN CHLOROPLAST RIBOSOMES

Maurice M. Margulies and Jessie Weistrop
Radiation Biology Laboratory, Smithsonian Institution
12441 Parklawn Drive, Rockville, MD 20852
U. S. A.

INTRODUCTION

Chloroplast ribosomes can be associated with chloroplast thylakoid membranes (1-5). Fifty percent of chloroplast ribosomes of Chlamydomonas reinhardtii are associated with chloroplast membranes in cells treated with chloramphenicol (+CAP) to prevent runoff of chloroplast polyribosomes (6). Here we wish to report that treatment of these membranes with Triton dissolves most of the membrane proteins, leaving a membranous fraction which contains most of the ribosomes and a distinct set of membrane polypeptides. Our results confirm and extend previous observations that Triton-treatment of chloroplast membranes leaves a particulate component of residual membrane with attached ribosomes (5).

RESULTS

When +CAP membranes are treated with Triton and then are centrifuged on density gradients, most of the chlorophyll is found at the top of the gradient. Some ultraviolet absorbing material is found in the gradient, and a light green pellet is found at the bottom of the tube. The ultraviolet absorbance in the gradient does not show any polyribosome-like profile (Fig. 1). The pellet at the bottom of the tube (+CAP Triton pellet) contains 70% of the chloroplast ribosomal RNAs, 12% of the protein and 0.5% of the chlorophyll of the original +CAP membranes. The same amount of protein and chlorophyll are found in Triton pellet prepared from membranes of untreated cells (-CAP Triton pellet). Membranes from -CAP cells contain a low level of chloroplast ribosomes compared to membranes of +CAP cells. -CAP Triton pellet contains a correspondingly low level of chloroplast ribosomal RNAs compared to +CAP Triton pellet.

Unsectioned +CAP membranes show particles the size of ribosomes, usually in groups suggesting polysomal arrays. These polysomes are widely scattered on the membranes (Plate I A). Likewise unsectioned +CAP Triton pellet shows ribosome-size particles on membranes (Plate 1 B). However, the concentration of ribosomes on the +CAP Triton pellet is much higher than on +CAP membranes.

Treatment of +CAP Triton pellet with Nonidet and trypsin-chymotrypsin yields ribosomal subunits, ribosomes and polyribosomes in amounts comparable to those obtained when an amount of intact +CAP membranes, equal to that from which the Triton pellet is derived, are treated directly with Nonidet and trypsin chymotrypsin (Fig. 2). Thus, state and attachment of ribosomes are probably the same in +CAP membranes and +CAP Triton pellet.

To determine whether +CAP Triton pellet contains membrane polypeptides, the polypeptide composition of +CAP Triton pellet was compared with -CAP Triton pellet, chloroplast ribosomes, +CAP membranes and -CAP membranes. Only three of the polypeptides in +CAP Triton pellet appear at equal intensity in -CAP Triton

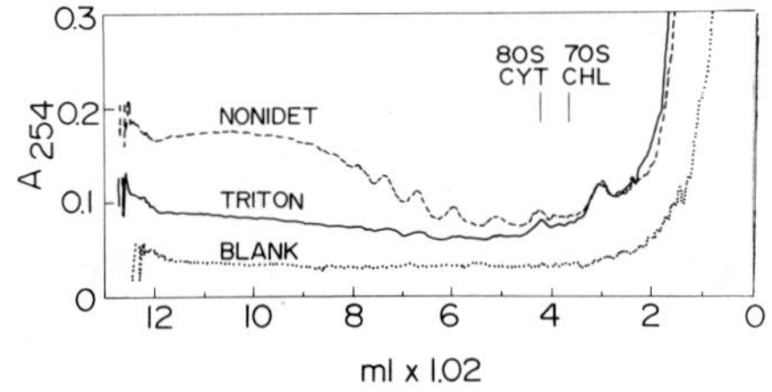

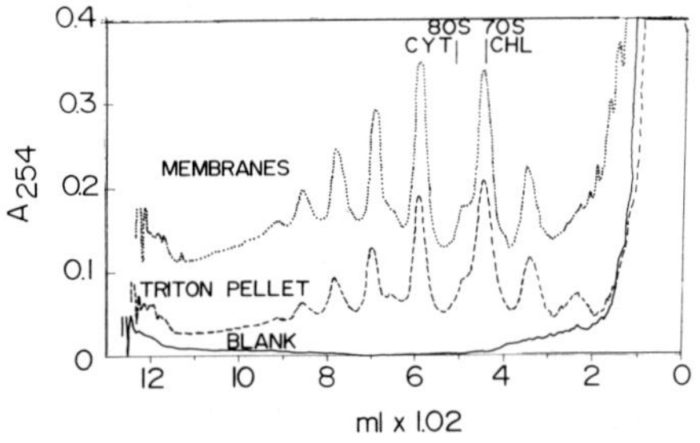

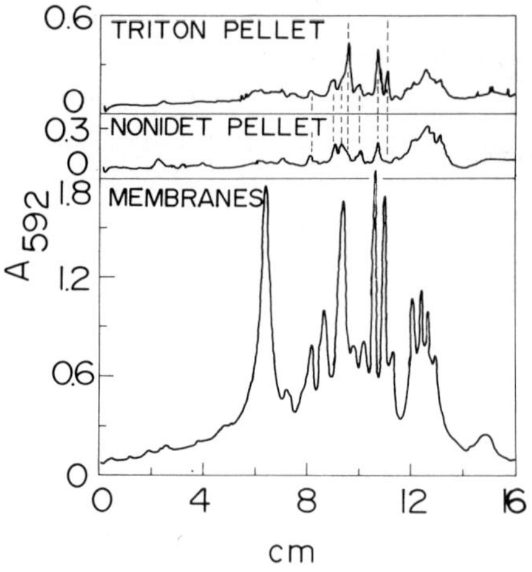

Fig. 1 (Top left). Release of polyribosomes from +CAP membranes by Nonidet P-40 but not Triton X-100. Membranes were treated with 3% (v/v) detergent and centrifuged 1.5 h at 40,000 rpm on sucrose density gradients (4).

Fig. 2 (Bottom left). Release of ribosomal particles from +CAP membranes, and from Triton pellet obtained from an equal amount of +CAP membranes. Samples were incubated 1 h at 0° with 3% (v/v) Nonidet and 6mg/ml each of trypsin and chymotrypsin and then centrifuged 2 h at 40,000 rpm.

Fig. 3 (Right). Polypeptide composition of +CAP membranes, +CAP Triton and +CAP Nonidet pellets derived from half the amount of +CAP membrane. Samples were treated with SDS and electrophoresed on acrylamide gradient gels (7) which were then stained with Coomassie Brilliant Blue and scanned.

PLATE I. A-C. Electron micrographs of unfixed samples stained with uranyl acetate. A. +CAP membranes. B. +CAP Triton pellet. C. Polyribosomes from +CAP membranes that were treated with Nonidet. D. Polypeptide composition of membranes, sub-membrane fractions and ribosomes. SDS acrylamide-gradient gels (7) stained with Coomassie Brilliant Blue: +MEM, +CAP membranes; -MEM, -CAP membranes; RIB, chloroplast ribosomes; +TRI, +CAP Triton pellet; -TRI, -CAP Triton pellet; -TRI RIB, a combination of -CAP Triton pellet and ribosomes about equal to the amount present in +TRI.

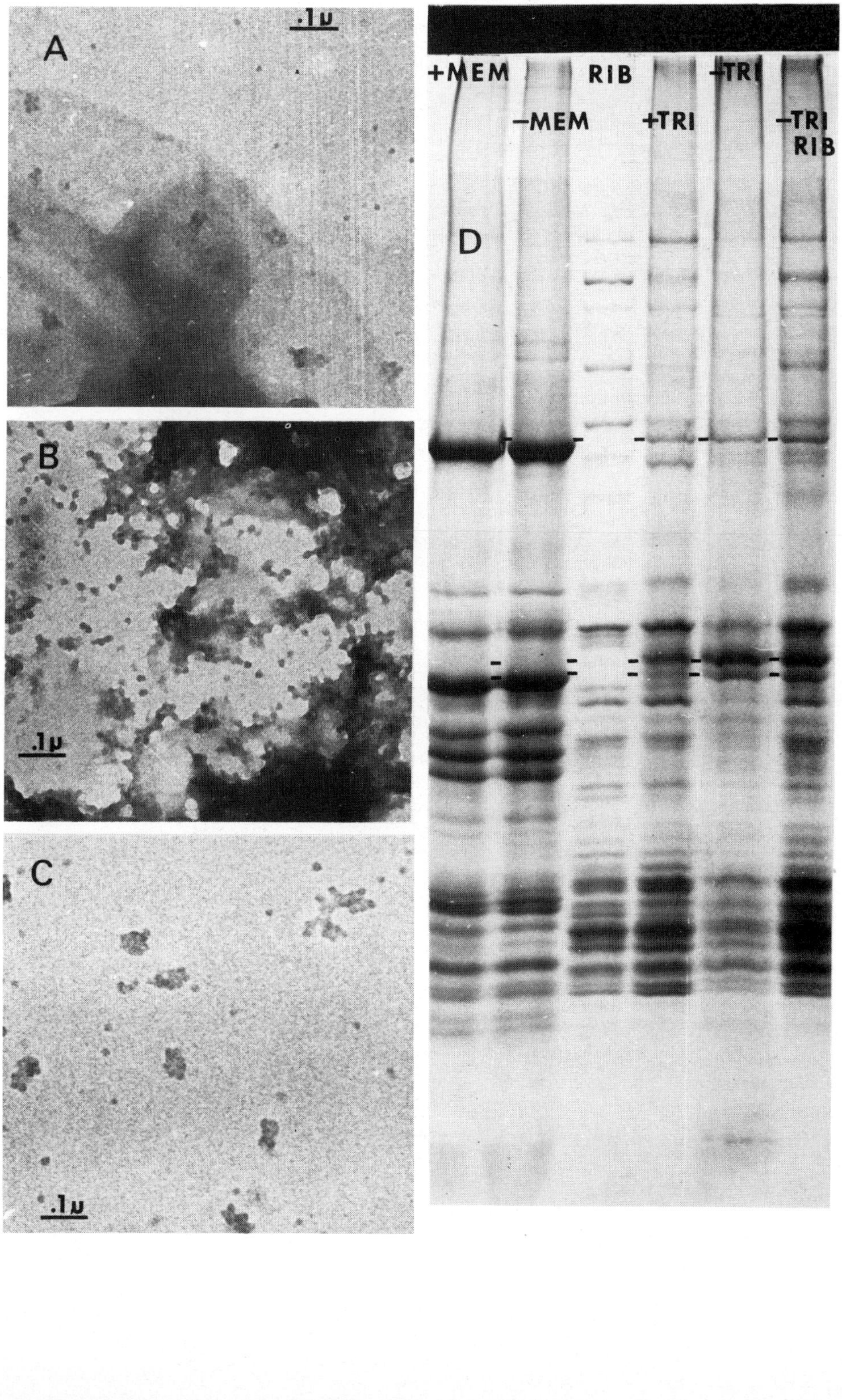
A
.1μ
B
.1μ
C
.1μ
D
+MEM
−MEM
RIB
+TRI
−TRI
−TRI
RIB

pellet and are absent from ribosomes (Plate I D). These are tentatively identified as membrane polypeptides associated with Triton pellet. They do not correspond to the major +CAP membrane polypeptides.

When +CAP membranes are treated with Nonidet and centrifuged, on density gradients, polyribosome-like absorbance profiles are obtained, in contrast to the result obtained with Triton (6) (Fig. 1). Polyribosome-like structures are found in the gradients (Plate 1 C). A pellet is obtained at the bottom of the density gradient tube (+CAP Nonidet pellet). The polypeptide composition of this Nonidet pellet differs qualitatively from that of +CAP Triton pellet (Fig. 3).

CONCLUSIONS

Nonidet, but not Triton, may be releasing ribosomes as large polyribosomes because the former removes some polypeptides more complely than the latter. The difference in membrane polypeptide composition of Triton pellet and intact membranes, and the high density of ribosomes on Triton membranes suggests that Triton dissolves portions of the membrane, leaving behind portions of distinct composition. The distinctness of these membrane regions is a function of the membranes themselves, since membrane residues of similar composition are obtained even with membranes of much lower ribosome content.

REFERENCES

1. Chen, J.L. and Wildman, S.G. (1970) Biochim. Biophys. Acta 209, 207-219.
2. Chua, N.-H., Blobel, G., Siekevitz, P. and Palade, G.E. (1973) Proc. Nat. Acad. Sci. U.S. 70, 1554-1558.
3. Falk, H. (1969) J. Cell Biol. 42, 582-587.
4. Margulies, M.M. and Michaels, A. (1974) J. Cell Biol. 60, 65-77.
5. Philippovich, I.I., Bezmertnya, I.N. and Oparin, A.I. (1973) Exp. Cell Res. 79, 159-168.
6. Margulies, M.M. and Michaels, A. (1975) Biochim. Biophys. Acta 402, 297-308.
7. Chua, N.-H., and Bennoun, P. (1975) Proc. Nat. Acad. Sci. U.S. 72, 2175-2179.

Genetics and Biogenesis of Chloroplasts and Mitochondria, Th. Bücher et al. eds.

THE tRNAs AND AMINOACYL-tRNA SYNTHETASES OF EUGLENA CHLOROPLASTS*

W. Edgar Barnett, Steven D. Schwartzbach,† and Lanny I. Hecker
Biology Division, Oak Ridge National Laboratory
Oak Ridge, Tennessee 37830

For a number of years our laboratory has been involved in studies relating to the molecular biology of cellular organelles. In this article we present a brief survey of observations concerning the tRNAs and aminoacyl-tRNA synthetases of Euglena gracilis chloroplasts.

Identification and regulation: The basic criteria[1, 2] we have used in the identification of chloroplastic tRNAs and synthetases are that they be the species which are found in purified chloroplast preparations and that they be essentially absent in cytoplasmic fractions from which chloroplasts have been centrifugally removed. In Figure 1, for example, hydroxyapatite chromatography has been used to identify the cytoplasmic (cyt) and chloroplastic (chl) species of valyl-tRNA synthetase on the basis of their presence or absence in the appropriate subcellular fraction. It is apparent from such data that the more rapidly eluting activity is the chl enzyme. Chromatography of this type also permits one to examine an interesting characteristic of chl synthetases, namely, their photoinduction.[1, 2] Simply stated, Euglena cells grown heterotrophically in the dark contain very low levels of the chl synthetases, and upon exposure to light rapidly begin to synthesize these enzymes. Photoinduction in exponentially growing cells results in up to a 100-fold increase in the chloroplastic synthetases.

The chl tRNAs, which are also chromatographically identifiable,[1, 3–6] are, as is the case with the synthetases, regulated by light[1, 3, 7] (Fig. 2). It is interesting to note that chl tRNA transcription begins almost immediately upon exposure to light (as might be expected in view of the role of this macromolecule in translation), and that synthesis rapidly ceases upon removal of light.[3]

Transcriptional and translational origin of chloroplastic synthetases: Given the observation that the chloroplasts of Euglena contain unique species of synthetases, we addressed the issue of the location of their translation, i.e., whether they are translated within the organelle and retained there, or whether they are translated upon cyt ribosomes and subsequently compartmentalized within the organelle. In these experiments[2] (in collaboration with Dr. J. A. Schiff),

*Research sponsored by the U.S. Energy Research and Development Administration under contract with the Union Carbide Corporation.
†Supported by the American Cancer Society, Postdoctoral Fellowship PF-1055.

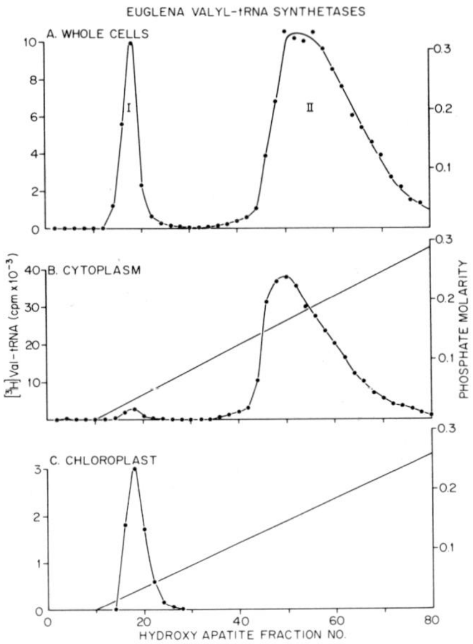

Fig. 1. Chromatography of Euglena valyl-tRNA synthetases.

we used "resting cells"[8] that contain proplastids which, upon exposure to light, are capable of developing into normal, functional chloroplasts in the absence of cell division. This latter consideration proved to be most important, since it has been our experience that any agents which affect the growth (or cell division) rate of Euglena dramatically reduce the levels of both organelle synthetases and tRNAs and obscure any "specific" effects of, for example, antibiotics.

The effects of cycloheximide[10] and streptomycin[8, 11, 12] (which inhibit translation of the cyt ribosomes of eukaryotes and the "prokaryotic-type" ribosomes of organelles, respectively) upon the photoinduction of the chl valyl-tRNA synthetase in resting cells may be seen in Figure 3. Precisely the same results were observed for the leucyl-, phenylalanyl-, and lysyl-tRNA synthetases, i.e., inhibition of translation by cyt ribosomes with cycloheximide completely blocks the appearance or synthesis of the synthetases, whereas inhibition of organelle translation with streptomycin has no effect on their photoinduction. It should be noted that under the conditions used for streptomycin exposure, the synthesis of other Euglena chloroplastic components such as chlorophyll, ribulose diphosphate carboxylase, and cytochrome 556 is inhibited.[2, 8, 11] Thus we conclude that the chloroplast-specific aminoacyl-tRNA synthetases are translated on cytoplasmic ribosomes and compartmentalized within the organelle.

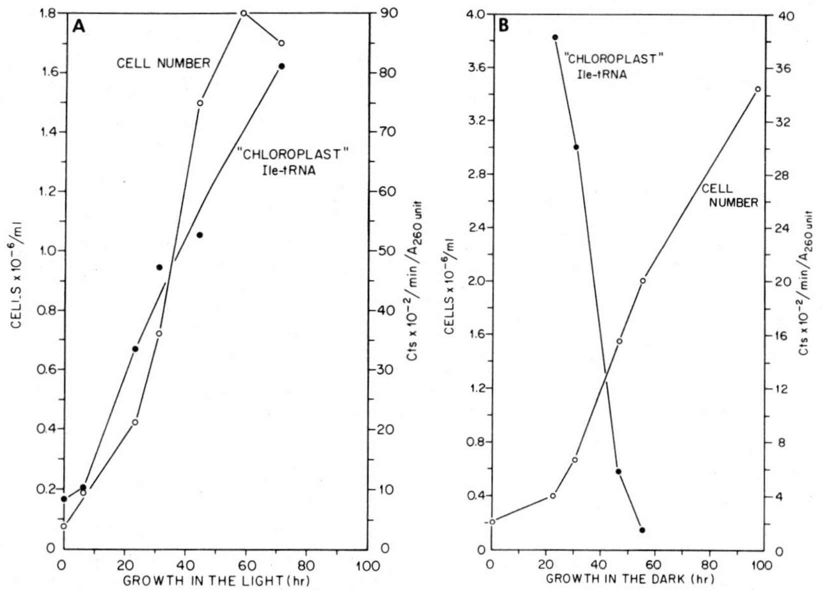

Fig. 2. Regulation of chl isoleucine tRNA by light. In (A) dark-grown cells were exposed to light at time zero; in (B) light-grown cells were removed from the light at time zero.[3]

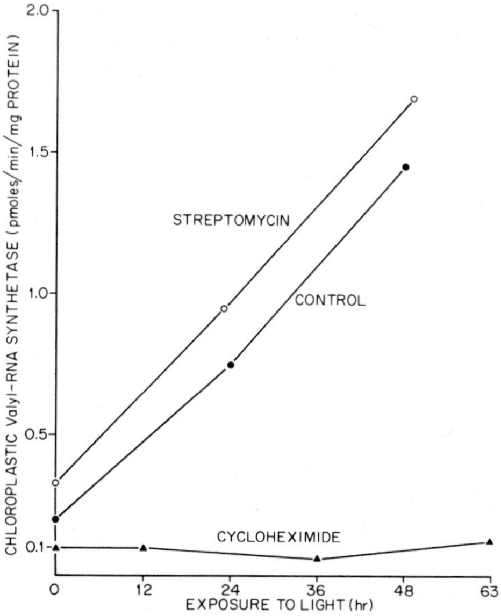

Fig. 3. Effects of streptomycin and cycloheximide on translation of chl valyl-tRNA synthetase.[2]

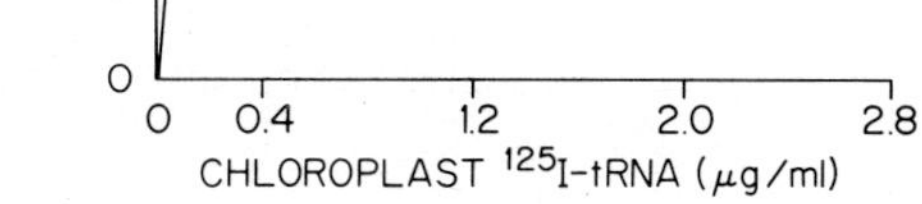

Fig. 4. Hybridization of total chloroplast tRNA to chloroplast DNA.[13]

Evidence that the structural genes for the chl synthetases reside within the nuclear genome comes from studies with the aplastidic mutant W_3BUL.[2] W_3BUL contains no detectable plastid structure or chl DNA, yet it contains "dark-grown" levels of the chl synthetases. In contrast, the chl tRNAs — transcripts of the chloroplast genome — are completely absent in W_3BUL cells. Thus the presence of chl synthetases in Euglena cells, which contain no chl DNA, permits us to conclude that their transcriptional origin is the nuclear genome.[1, 2]

Transcriptional origin of chloroplastic tRNAs: Euglena chl tRNAs hybridize with chl DNA.[13] Figure 4 shows a typical saturation plot, indicating that there are approximately 26 tRNA cistrons in the chloroplast genome. Similar experiments have also been performed with purified chl $tRNA^{Phe}$ and $tRNA^{Asp}$, the data indicating that for these individual species there is a single structural gene. This is, of course, consistent with the value of 26 for unfractionated tRNAs, and suggests either that there are 20 plus 6 additional isoaccepting tRNA species or that there is limited gene redundancy. The former appears to be the case since there are two isoaccepting tRNA species for at least methionine and valine. We have not characterized the tRNAs for all amino acids by chromatography.

We have also found that hybridization does not occur at significant levels between chl tRNA and nuclear DNA. However, although it appears unlikely from our results, it should be noted that our experiments do not rigorously rule out the possibility that there is a limited number of chl tRNA cistrons within the larger nuclear genome.

Hybridization of chl ^{125}I-tRNA in the presence of equimolar quantities of unlabeled chl tRNA results in a 42% decrease in the hybrid counts (Fig. 5), indicating that the hybrid formed is highly specific and occurs only between those regions of chl DNA whose base sequence is complementary to chl tRNA. From earlier chromatographic experiments[3, 7] with dark- and light-grown chl tRNAs, it was inferred that photoinduction of tRNAs involved induction of transcription. Competition experiments confirm this conclusion, i.e., the tRNA from dark-grown cells contains markedly less RNA capable of competing (in hybridization) with chl tRNA than does the tRNA from light-grown cells. We have also found (data not shown) that the high molecular weight RNA from dark-grown cells does not contain competing species. Thus it appears that photoinduction of tRNAs is at the level of transcription rather than maturation. With regard to the aplastidic mutant W_3BUL, competition experiments again verify the chromatographic data. W_3BUL contains no species of tRNA similar to chl tRNA.

Primary sequence studies of chloroplastic tRNAs: To better understand the evolutionary origin of chl tRNAs, we have, in collaboration with Drs. U. L. RajBhandary and Simon Chang and their co-workers, purified and sequenced the $tRNA^{Phe}$ of Euglena chloroplasts[14] (Fig. 6). Also in the process of being sequenced are the $tRNA^{Phe}$'s of the Euglena cytoplasm and of the

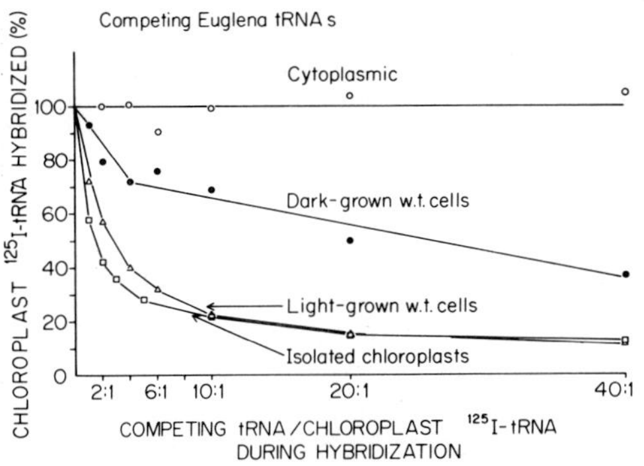

Fig. 5. Competition experiments. Saturating concentrations of chl ^{125}I-tRNA were hybridized with filters containing 1 μg of chl DNA in the presence of increasing amounts of unlabeled tRNAs.[13]

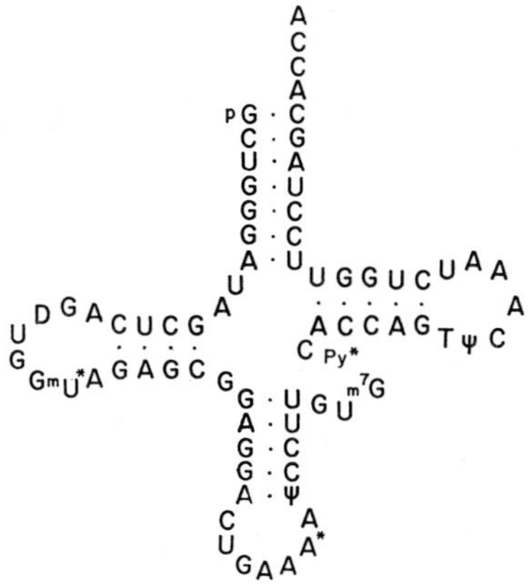

Fig. 6. Primary sequence of Euglena chl tRNA^{Phe}.[14]

blue-green algae Agmenellum quadriduplicatum. It is our intent to compare the degree of sequence homology between these tRNA^{Phe}'s and others of known sequence from other organisms in order to gain insight into their relatedness or evolutionary relationships. Ideally, one would like to be able to show that chl tRNA^{Phe} is extremely similar to one, and only one, of the tRNA^{Phe}'s from either the blue-green algae or the nuclear genome.

Interestingly, the sequences (exclusive of posttranscriptional base modification) of three mammalian tRNA^{Phe}'s are identical with each other,[15–17] as are those for tRNA^{Phe} from wheat germ[15] and pea.[18] Indeed, there is considerable homology between all tRNA^{Phe}'s, particularly in the single-stranded regions. At the present time we can conclude only that the minor, modified base content of chl tRNA^{Phe} resembles the pattern found in prokaryotic tRNA^{Phe}'s somewhat more closely than that found in eukaryotic tRNA^{Phe}'s;[5] we await the sequence completion of Euglena cytoplasmic and Agmenellum tRNA^{Phe}.

REFERENCES

1. Reger, B. J., Fairfield, S. A., Epler, J. L., and Barnett, W. E. (1970) Proc. Natl. Acad. Sci. USA 67, 1207-1213.

2. Hecker, L. I., Egan, J., Reynolds, R. J., Nix, C. E., Schiff, J. A., and Barnett, W. E. (1974) Proc. Natl. Acad. Sci. USA 71, 1910-1914.

3. Goins, D. J., Reynolds, R. J., Schiff, J. A., and Barnett, W. E. (1973) Proc. Natl. Acad. Sci. USA 70, 1749-1752.

4. Fairfield, S. A. and Barnett, W. E. (1971) Proc. Natl. Acad. Sci. USA 68, 2972-2976.

5. Hecker, L. I., Uziel, M., and Barnett, W. E. (1976) Nucl. Acids Res. 3, 371-380.

6. Parthier, B. and Krauspe, R. (1974) Biochem. Physiol. Pflanz. 165, 1-17.

7. Barnett, W. E., Pennington, C. J., and Fairfield, S. A. (1969) Proc. Natl. Acad. Sci. USA 63, 1261-1263.

8. Schiff, J. A. (1971) In: Autonomy and Biogenesis of Mitochondria and Chloroplasts (ed. by N. K. Boardman, A. W. Linnane, and R. M. Smillie), pp. 98-118. American Elsevier, New York.

9. Holowinsky, A. W. and Schiff, J. A. (1970) Plant Physiol. 45, 339-347.

10. Smillie, R. M., Bishop, D. G., Gibbons, G. C., Graham, D., Grieve, A. M., Raison, J. K., and Reger, B. J. (1971) In: Autonomy and Biogenesis of Mitochondria and Chloroplasts (ed. by N. K. Boardman, A. W. Linnane, and R. M. Smillie), pp. 422-433. American Elsevier, New York.

11. Bovarnick, J. G., Freedman, Z., and Schiff, J. A. (1970) Plant Physiol. 46, S21.

12. Schwartzbach, S., Freyssinet, G., and Schiff, J. A. (1973) Plant Physiol. 51, S27.

13. Schwartzbach, S. D., Hecker, L. I., and Barnett, W. E. (1976) Proc. Natl. Acad. Sci. USA 73, in press.

14. Chang, S. H., Hecker, L. I., Silberklang, S. L., Brum, C. K., RajBhandary, U. L., and Barnett, W. E. (1976) Manuscript in preparation.

15. Barrell, B. G. and Clark, B. F. C. (1974) Handbook of Nucleic Acid Sequences. Joynson-Bouvers, Ltd., Oxford.

16. Keith, G., Ebel, J. P., and Dirheimer, G. (1974) FEBS Lett. 48, 50-52.

17. Roe, B. A., Anandaraj, M. P. J. S., Chia, L. S. Y., Randerath, E., Gupta, R. C., and Randerath, K. (1975) Biochem. Biophys. Res. Commun. 66, 1097-1105.

18. Everett, G. A. and Madison, J. T. (1976) Biochemistry 15, 1016-1021.

Genetics and Biogenesis of Chloroplasts and Mitochondria, Th. Bücher et al. eds.

tRNAs AND AMINOACYL-tRNA SYNTHETASES IN PLANT ORGANELLES

J.H.Weil, G.Burkard, P.Guillemaut, G.Jeannin
R.Martin and A.Steinmetz
Institut de Biologie Moléculaire et Cellulaire
Université Louis Pasteur
15, rue Descartes, 67084 Strasbourg
France

INTRODUCTION

In a plant cell, protein synthesis takes place in three compartments, namely in the cytoplasm, in the chloroplasts and in the mitochondria. In many respects, protein synthesis in the organelles (mitochondria and chloroplasts) resembles protein synthesis in prokaryotic organisms (especially bacteria) rather than that taking place in the surrounding cytoplasm, as illustrated, for instance, by the small size of the organellar ribosomes, the formylation of organellar initiator methionyl-tRNA, or the sensitivity of organellar protein synthesis to certain antibiotics such as chloramphenicol (1). These observations are in agreement with the theory that the organelles have an endosymbiotic origin and have evolved from prokaryotic ancestors.

More arguments in favor of this theory have been obtained from comparative studies on plant cytoplasmic, chloroplast and mitochondrial tRNAs and aminoacyl-tRNA synthetases. In a number of cases it has been possible to separate (generally by hydroxyapatite or DEAE-cellulose chromatography) a cytoplasmic aminoacyl-tRNA synthetase from its chloroplastic or mitochondrial counterpart, or to separate (usually by reverse-phase chromatography) the cytoplasmic, chloroplast and mitochondrial isoacceptor tRNAs (2-17).

It has then become possible to perform homologous and heterologous aminoacylation reactions and therefore to compare the recognition of the tRNAs from the three plant cell compartments by homologous and heterologous (including prokaryotic) aminoacyl-tRNA synthetases. In general these cross-aminoacylation reactions have revealed a cross-recognition pattern between plant organellar (but not cytoplasmic) and prokaryotic tRNAs and aminoacyl-tRNA synthetases (14,15, 17-20), suggesting a similarity between organellar and prokaryotic tRNAs. More evidence in favor of such a similarity has recently come from structural studies on chloroplast tRNAs.

RESULTS

I) Base composition analogies between bacterial and chloroplast $tRNAs^{Phe}$.

Up to this year, little information was available on the structure of chloroplast tRNAs, but recently the first two base compositions were determined and published, that of Euglena chloroplast $tRNA^{Phe}$ (21) and that of Phaseolus chloroplast $tRNA_1^{Phe}$ (22). The latter is given in table I, where the nucleoside composition of E.coli and of wheat germ (cytoplasmic) $tRNAs^{Phe}$ have been listed for comparison.

TABLE I

NUCLEOSIDE COMPOSITION OF CHLOROPLASTIC $tRNA_1^{Phe}$ AS COMPARED TO THAT OF E.COLI AND OF WHEAT GERM (CYTOPLASMIC) $tRNAs^{Phe}$.

	E.coli $tRNA^{Phe}$ (23)	Bean chloroplastic $tRNA_1^{Phe}$	Wheat germ (cytoplasmic) $tRNA^{Phe}$ (24)
A	14	17	16
U	8	10 or 11	7
C	21	19	19
G	23	20	20
ms^2i^6A	1	1	-
m^1A	-	-	1
T	1	1	1
ψ	3	2	3
hU	2	2	3
acp^3U	1	1	-
s^4U	1	-	-
Cm	-	-	1
m^2G	-	-	1
m_2^2G	-	1	1
m7G	1	1	1
Gm	-	1 or 2	1
Y	-	-	1
	76	76(?)	76

The results are expressed as moles of nucleosides/mole tRNA. In the case of chloroplast $tRNA_1^{Phe}$ the calculations were made with the assumption that this tRNA is also 76 nucleotides long and the values were rounded off to the closest unit.

It can be seen that bean chloroplast $tRNA_1^{Phe}$ resembles E.coli $tRNA^{Phe}$ in that they have the same two hypermodified nucleosides, namely N^6-(Δ^2-isopentenyl)-2-methylthioadenosine (or ms^2i^6A) and 3-(3-amino-3-carboxypropyl) uridine (or acp^3U), and in that they both lack the nucleoside Y which is present in cytoplasmic $tRNAs^{Phe}$. A better comparison of chloroplast and bacterial $tRNAs^{Phe}$ will be possible when the nucleotide sequence of Euglena chloroplast $tRNA^{Phe}$, which has been determined by Barnett, Raj Bhandary and their co-workers, will be published.

II) Origin of the chloroplast specific tRNAs

The existence of chloroplast-specific tRNA species, different from their cytoplasmic counterparts, raises the problem of their origin : are these chloroplast tRNAs coded for by chloroplast or by nuclear genes ? This question can be answered by DNA-tRNA hybridization experiments (25-29) and we have studied the hybridization of bean cytoplasmic and chloroplastic ^{3}H-leucyl-tRNAs and ^{3}H-phenylalanyl-tRNAs to chloroplast DNA immobilized on nitrocellulose filters (0.45 μ) for 2 h. at 40° in the presence of 8 M urea (30) at pH 5 (in order to prevent deacylation).

Total tRNA from bean cytoplasm, chloroplasts or from E.coli was aminoacylated with high specific radioactivity ^{3}H-leucine or ^{3}H-phenylalanine (about 30 Ci/mmole) using homologous enzyme preparations. Hybridization to either chloroplast or E.coli DNA was then measured as a function of time. As shown on fig.1, chloroplast leucyl- and phenylalanyl-tRNAs hybridize to chloroplast DNA and, to a much lesser extent to E.coli (fig.1a,d). The reverse situation is true for E.coli leucyl- and phenylalanyl-tRNAs which hybridize preferentially to E.coli DNA (fig.1b,e). Cytoplasmic leucyl- and phenylalanyl-tRNAs do not hybridize to chloroplast or to E.coli DNA (fig.1c,f). It should be pointed out that, when chloroplast tRNAs are aminoacylated with leucine or phenylalanine using the homologous (chloroplast) aminoacyl-tRNA synthetase, only the chloroplast-specific isoacceptors are charged while the cytoplasmic tRNA species remain uncharged (14,17), so that the results of fig.1 suggest that chloroplast-specific $tRNAs^{Leu}$ and $tRNAs^{Phe}$, in contrast to their cytoplasmic counterparts, are coded for by chloroplast DNA.

These results were confirmed using the fractionated chloroplast-specific tRNA species, namely 3 isoacceptors in the case of $tRNAs^{Leu}$ (14) and 2 in the case of $tRNAs^{Phe}$ (22). Chloroplast specific $tRNA_1^{Leu}$

(fig.2a,b), $tRNA^{Leu}_2$ (fig.2a,c), $tRNA^{Leu}_3$ (fig.2b,c), $tRNA^{Phe}_1$ (fig.2d) and $tRNA^{Phe}_2$ (fig.2d) hybridize with chloroplast DNA.

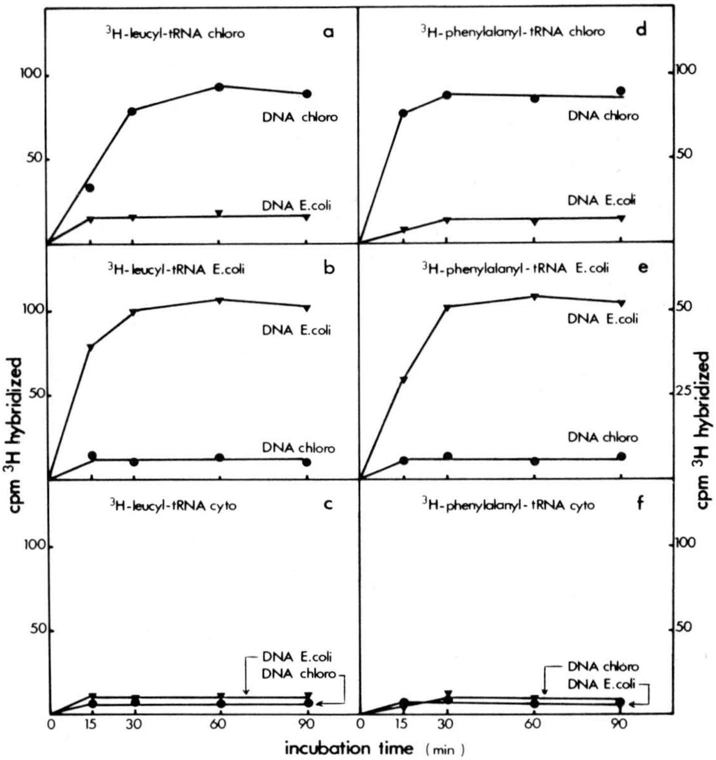

Fig. 1. Hybridization of {^{3}H} leucyl- or {^{3}H} phenylalanyl-tRNAs from bean chloroplasts or cytoplasm or from E.coli to chloroplast or E.coli DNA. 8 µg chloroplast or E.coli DNA was immobilized on each filter, and the input radioactivity was between 15,000 and 20,000 CPM.

In order to determine whether the chloroplast-specific isoacceptors are coded for by the same or by different gene(s), experiments were performed to check if the hybridizations of the isoacceptors are additive. As shown on fig.2a,b,c, all three combinations of two chloroplast-specific $tRNAs^{Leu}$ show no additivity, and the combination of the two chloroplast $tRNAs^{Phe}$ does not show any additivity either (fig.2d), whereas control experiments where one chloroplast $tRNA^{Leu}$ and one chloroplast $tRNA^{Phe}$ were mixed showed the expected additivity (fig.2e). These results suggest that the 3 chloroplast $tRNAs^{Leu}$ are coded by the same gene(s), and that this is also the case for the 2 chloroplast $tRNAs^{Phe}$.

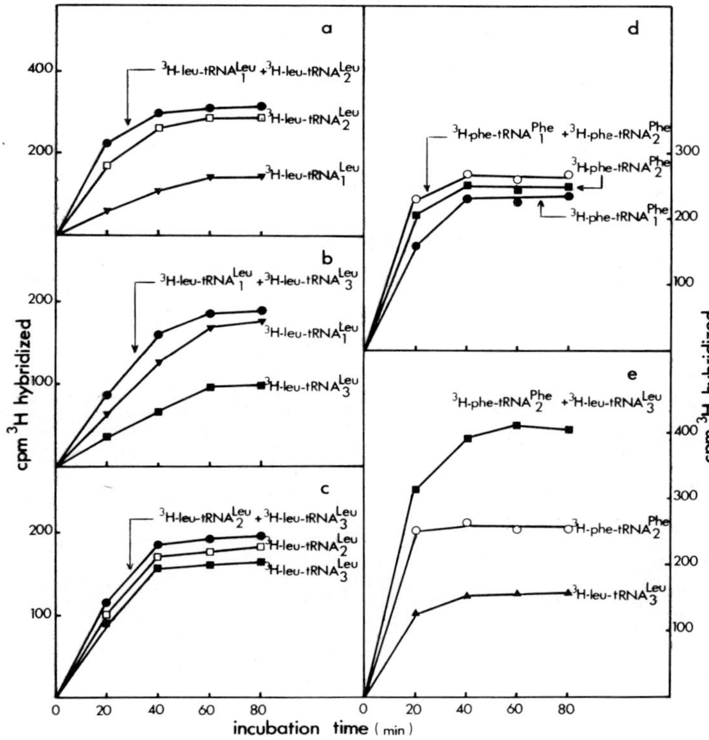

Fig. 2. Hybridization of chloroplast {^{3}H}leucyl- and {^{3}H}phenylalanyl-tRNA isoacceptors to chloroplast DNA : Comparison of the hybridization values obtained with a single isoacceptor and with a mixture of two isoacceptors (either specific for the same aminoacid, or for two different aminoacids). 8 µg chloroplast DNA was immobilized on each filter and the input radioactivity was between 15,000 and 20,000 in experiments with a single isoacceptor, and between 30,000 and 40,000 CPM in experiments where two isoacceptors were mixed.

This is confirmed by experiments where the competition between the chloroplast isoacceptors was determined in the hybridization with chloroplast DNA. In these experiments increasing amounts of non-radioactive uncharged tRNAs were added to each {^{3}H}leucyl or {^{3}H} phenylalanyl-tRNA isoacceptor in the hybridization mixture. The hybridization of each of the three chloroplast tRNAs^{Leu}, is decreased upon addition of the same isoacceptor or of any of the other two isoacceptors, but not upon addition of a tRNA^{Phe} isoacceptor (fig. 3a,b,c). Conversely competition occurs when uncharged tRNA$^{Phe}_{2}$ is added to {^{3}H}phenylalanyl-tRNA$^{Phe}_{1}$, but not upon addition of a tRNA^{Leu} (fig.3d).

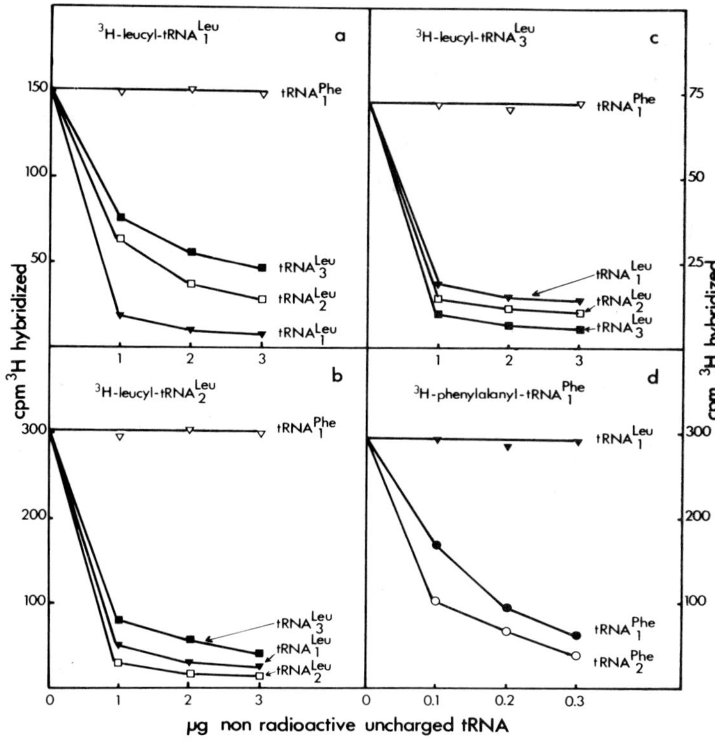

Fig. 3. Hybridization of chloroplast {^{3}H}leucyl- and {^{3}H}phenylalanyl-tRNA isoacceptors to chloroplast DNA : Competition observed upon addition of increasing amounts of non-radioactive uncharged isoacceptors (either specific for the same aminoacid, or specific for a different aminoacid). 8 µg chloroplast DNA was immobilized on each filter and the input radioactivity was between 15,000 and 20,000 CPM.

The fact that the hybridization of the three chloroplast tRNA^{Leu} isoacceptors or of the two chloroplast tRNA^{Phe} isoacceptors shows no additivity, and that the chloroplast isoacceptors compete with each other for the same sites on chloroplast DNA, suggests that these isoacceptors are coded for by the same gene(s) and differ only in the extent of post-transcriptional modification. It is known that reverse-phase chromatography is able to separate tRNAs which differ very little (even by one modified nucleotide only), so that the iso-acceptor chloroplast tRNAs which we have fractionated may indeed be transcribed from the same gene(s). In a recent study of the hybridization of Euglena chloroplast total tRNAs (labelled with ^{125}I) with Euglena chloroplast DNA, Barnett et al. were able to estimate that there are about 26 tRNA cistrons on Euglena chloroplast DNA (31).

Our hybridization experiments show that chloroplast leucyl- and phenylalanyl-tRNAs do not hybridize appreciably with E.coli DNA (fig.1a,d) and that E.coli leucyl- and phenylalanyl-tRNAs do not hybridize with chloroplast DNA (fig.1b,e). This is not in contradiction with the similarities observed between chloroplast and E.coli tRNAs when studying cross-aminoacylation reactions or base composition : some structural features could be identical, and allow cross-aminoacylation, but might not be enough to allow appreciable cross-hybridization. The lack of cross-hybridization does not rule out the possibility of an endosymbiotic origin of the chloroplasts, as the ancestor could have been a photosynthetic prokaryote rather than a bacterium, and as the tRNA structural genes may have been modified during evolution so as to no longer allow appreciable cross-hybridization.

DISCUSSION

The existence of organelle-specific tRNAs, coded by organellar DNA, and which can only be aminoacylated by the cognate organellar (or bacterial) aminoacyl-tRNA synthetase, is now well documented. But why are organelle-specific tRNAs necessary ? Could not cytoplasmic tRNAs species be imported into the organelles and allow organellar protein synthesis ? The presence of organelle-specific tRNAs is perhaps required to recognize certain codons which cannot be recognized by cytoplasmic tRNAs ; this hypothesis is presently being tested in our laboratory : we are studying the binding of the two cytoplasmic and of the three chloroplast leucyl-tRNAs (14) to the six leucine codewords in the presence of ribosomes. But it is also possible that chloroplast-specific tRNAs are required for efficient protein synthesis within the organelle, because these tRNAs (and not the cytoplasmic ones) specifically interact with some constituents of the chloroplast protein synthesis machinery ; to test this hypothesis we would need a cell-free protein synthesizing system from chloroplast which is tRNA-dependant, so as to allow a comparative study of the efficiency of chloroplast tRNAs and cytoplasmic tRNAs in the synthesis of a typical chloroplast protein made on chloroplast ribosomes.

But instead of asking the question why are there organelle-specific tRNAs and organelle-specific aminoacyl-tRNA synthetases, one could ask a more general question : why is there an organelle-specific protein synthesizing machinery ? And in view of the fact that some important organelle-specific proteins are coded by nuclear DNA and made on cytoplasmic ribosomes, why are some organellar proteins coded by organellar DNA and made on organellar ribosomes ?

Some of the polypeptides which are manufactured in the organelles are parts of organelle-specific enzymes (cytochrome oxidase or ATPase in the mitochondria, ribulose 1-5-diphosphate carboxylase in the chloroplasts) which must combine with other polypeptide chains, made in the cytoplasm, to form active enzymes. This of course raises the problem of the relationships between the nucleus, the cytoplasm and the organelles, a problem which will be discussed by other speakers during this conference.

REFERENCES

1. Boulter, D., Ellis, R.J. and Yarwood, A. Biol. Rev.47, 113-175
2. Aliyev, K. and Philippovich, I. (1968) Mol. Biol. 2, 364-373
3. Barnett, W.E., Pennington, C. and Fairfield, S. (1969) Proc. Natl. Acad. Sci. U.S. 63, 1261-1268
4. Reger, B., Fairfield, S., Epler, J. and Barnett, W.E. (1970) Proc. Natl. Acad. Sci. U.S. 67, 1207-1213
5. Burkard, G., Guillemaut, P. and Weil, J.H. (1970) Biochim. Biophys. Acta 224, 184-198
6. Kanabus, J. and Cherry, J. (1971) Proc. Natl. Acad. Sci. U.S. 68, 873-876
7. Merrick, W and Dure, L. (1972) J. Biol. Chem. 247, 7988-7999
8. Guderian, R., Pulliam, R. and Gordon, H. (1972) Biochim. Biophys. Acta 262, 50-65
9. Parthier, B., Krauspe, R. and Samtleben, S. (1972) Biochim. Biophys. Acta 277, 335-341
10. Kislev, U., Selsky, M., Norton, C. and Eisenstadt, J. (1972) Biochim. Biophys. Acta 287, 256-269
11. Hiatt, U. and Snyder, L. (1973) Biochim. Biophys. Acta 324, 57-68
12. Parthier, B. and Krauspe, R. (1974) Biochem. Physiol. Pflanz. 165 18-36

13. Wright, R., Kanabus, J. and Cherry, J. (1974) Plant Sci. Lett. 2, 347-355
14. Guillemaut, P., Steinmetz, A., Burkard, G. and Weil, J.H. (1975) Biochim. Biophys. Acta 378, 64-72
15. Guillemaut, P. and Weil, J.H. (1975) Biochim. Biophys. Acta 407, 240-248
16. Brantner, J. and Dure, L. (1975) Biochim. Biophys. Acta 414, 99-114
17. Jeannin, G., Burkard, G. and Weil, J.H. (1976) Biochim. Biophys. Acta 442, 24-31
18. Merrick, W. and Dure L. (1973) Biochemistry 12, 629-635
19. Beauchemin, N., Larue, B. and Cedergren, R. (1973) Arch. Biochem. Biophys. 156, 17-25
20. Parthier,B. and Krauspe, R. (1974) Biochem. Biophys. Pflanz. 165 1-17
21. Hecker, L., Uziel, M. and Barnett, W.E. (1976) Nucl. Ac. Res. 3, 371-380
22. Guillemaut, P., Martin, R. and Weil, J.H. (1976) FEBS Lett. 63, 273-277
23. Barrell, B. and Sanger, F. (1969) FEBS Lett.3, 275-278
24. Dudock,B., Katz, G., Taylor, E. and Holley, R. (1969) Proc. Natl. Acad. Sci. U.S. 62, 941-945
25. Williams, G. and Williams, A. (1970) Biochem. Biophys. Res. Comm. 39, 858-863
26. Guillemaut, P., Steinmetz, A., Burkard, G. and Weil, J.H. (1973) C.R.Soc.Biol. 167, 961-966
27. Tewari, K. and Wildman, S. (1970) Soc. Experim. Biol. Sympos. 24, Cambridge Univ. Press 147-179
28. Williams, G., Williams, A. and George, S. (1973) Proc. Natl. Acad. Sci. U.S. 70, 3498-3501
29. Schwartzbach, S., Hecker, L. and Barnett, W.E. (1975) Plant Physiol. 56, S 384
30. Kourilsky, P., Manteuil, S., Zamansky, M. and Gros, F. (1970) Biochem. Biophys. Res. Comm. 41, 1080-1087
31. Schwartzbach, S., Hecker, L. and Barnett, W.E. (1976) Proc. Natl. Sci. U.S. (in press)

TRANSCRIPTION AND TRANSLATION APPARATUS OF MITOCHONDRIA

Genetics and Biogenesis of Chloroplasts and Mitochondria, Th. Bücher et al. eds.

CHARACTERIZATION AND TRANSLATION OF YEAST MITOCHONDRIAL RNA

F. Hendler, A. Halbreich, S. Jakovcic
J. Patzer, S. Merten, and M. Rabinowitz
Departments of Medicine and Biochemistry and
The Franklin McLean Memorial Research Institute
The University of Chicago
Chicago, Illinois 60637 U.S.A.

For some years our laboratory has been engaged in the identification and characterization of transcripts of yeast mtDNA. In this paper we will summarize some of our recent work concerning the extent of transcription of yeast mtDNA, the translation of mitochondrial RNA in a cell free system and the visualization of mitochondrial ribosomal RNA genes by electron microscopy.

Although yeast mtDNA has a molecular weight of about 50 x 10^6 and thus potentially contains 5 times more genetic information than animal mtDNA[1], additional gene products have not been identified. Bernardi has suggested that much of yeast mtDNA is composed of high AT spacer having no genetic function.[2] HeLa cell mtDNA has been shown by Attardi and colleagues[3,4] to be completely and symmetrically transcribed. As an initial approach to evaluate the informational content and the nature of transcription of yeast mtDNA, we tried to determine the fraction of yeast mtDNA transcribed *in vivo* using mitochondrial RNA-DNA hybridization. The data are plotted as percent of DNA hybridized vs. R_0t, the product of the RNA concentration (R_0) and time of incubation (t), the latter on a logarithmic scale (Fig. 1). Approximately 35% of the DNA hybridized at high RNA inputs. Since it has not yet been possible to separate strands of grande yeast mtDNA, it is difficult to be certain whether these results indicate symmetric-

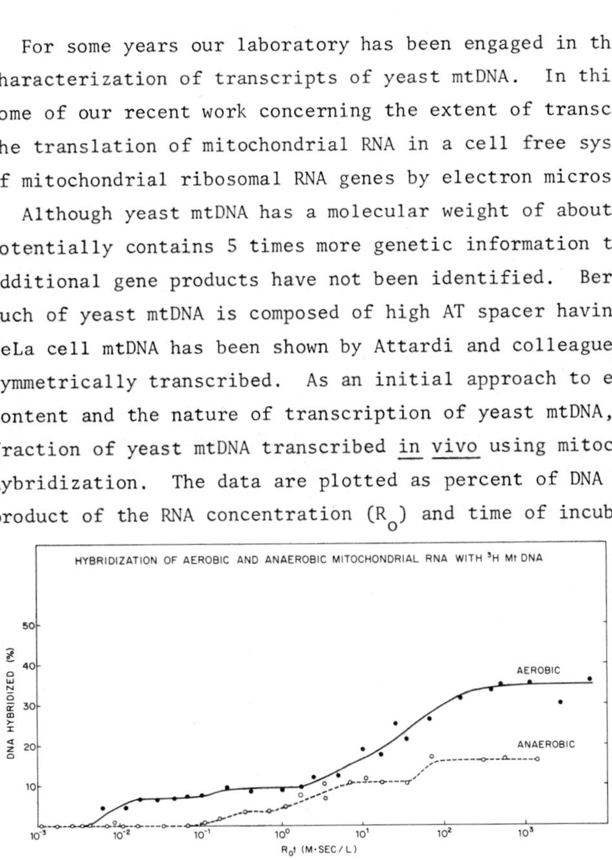

Fig. 1. Hybridization of mitochondrial RNA from aerobic and anaerobic grown cells with ^3H-mtDNA. The DNA was labeled to 4 x 10^6 cpm/μg with *E. coli* polymerase I.[10] Conditions for hybridization and analysis using *Neurospora* nuclease are as previously described.[5] Anaerobic RNA preparations were substantially contaminated with cytoplasmic rRNA by gel electrophoresis, but at least 20% to 50% of the RNA was judged to be mitochondrial. The curve presented for anaerobic RNA should thus be shifted 2-5 R_0t units to the left.

al transcription of about 35% of the DNA or asymmetrical transcription of 70%. However, low concentration symmetrical transcripts might be expected to hybridize with their RNA complements present in high concentration and thus be removed from the RNA-DNA hybridization system.

When the RNA was pre-annealed to a high R_0t prior to hybridization with mtDNA, hybridization values were slightly reduced from 35% to 29%[5]. When the RNA was similarly pre-annealed, but denatured prior to hybridization with DNA, the hybridization level increased to 46%[5]. If the pre-annealing of RNA effectively removed all presumptive low-concentration symmetrical transcripts then the remaining 30% hybridization would indicate that a minimum of 60% of the equivalent of one DNA strand is being transcribed. The possibility of symmetrical transcription must be evaluated as has been done in HeLa cells[3,4] by isolation of double-stranded RNA, and when strand separation of yeast mtDNA becomes technically feasible, by hybridization with separated DNA strands.

RNA isolated from anaerobically grown yeast reached an apparent hybridization plateau considerably lower (about 16%) than RNA from aerobically grown cells (Fig. 1). This difference in the percent mitochondrial DNA transcribed appears to be valid even if consideration is given to the substantially greater cytoplasmic contamination in anaerobic mitochondrial RNA preparations. The data indicate that some mitochondrial transcripts are either absent or are present in very low abundance in anaerobic cells.

We recently reported on the isolation of a mitochondrial RNA fraction from yeast containing short segments of poly(A) (20-30 nucleotides) using poly (U) sepharose columns[6,7]. The hybridization of mitochondrial RNA bound to poly (U) sepharose [poly(A)-containing] is compared with unbound RNA [poly(A) minus] in Figure 2. Similar hybridization plateaus were obtained. By gel electrophoresis mitochondrial rRNA was absent in the poly(A) RNA preparations. We may conclude

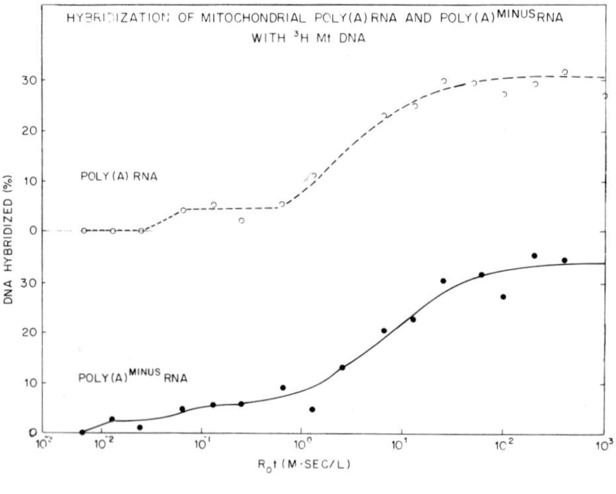

Fig. 2. Hybridization of poly(A) RNA and poly(A) minus RNA with mtDNA, as described in the legend for Figure 1.

that most of the other RNA sequences are shared by poly(A) RNA and poly(A) minus RNA fractions.

We have previously reported the translation of the mitochondrial poly(A) RNA in an E. coli ribosomal system[7,8]. About 3-5% of the labeled peptide was immunoprecipitated with antibody to cytochrome oxidase using the direct immunoprecipitation procedure, and the dissociated antibody-antigen complex showed three major peptides with molecular weights that corresponded to the three largest peptides of cytochrome oxidase on SDS-polyacrylamide gel electrophoresis. The possibility exists, however, that hydrophobic peptides synthesized in the cell-free system may non-specifically associate with the carrier cytochrome oxidase and result in non-specific immunoprecipitation in the direct reaction, particularly in view of the demonstration by Grivell et al. that such associations could persist even in SDS during electrophoresis[8]. We therefore have examined the product of the cell-free system by indirect immunoprecipitation, where carrier cytochorme oxidase is not added. The antigen-rabbit antibody complex is precipitated with goat antibody against rabbit serum.

Table I shows the immunoprecipitation by cytochrome oxidase antibody of peptides synthesized in the E. coli S30 ribosomal system upon the addition of mitochondrial poly(A) RNA and poly(A) minus RNA. Results obtained with the indirect and direct immunoprecipitation systems are compared. With both methods there are significant increments of labeled peptides immunoprecipitated when either poly(A) RNA or poly(A) minus RNA was used, although control values (MS2-RNA added, or immunoprecipitation with myosin antibody) were higher with the indirect procedure.

On SDS gel electrophoresis the dissociated antigen antibody complex from the indirect reaction contained peptides having the mobility of the two larger cytochrome oxidase peptides and perhaps the third as well (Fig. 3). Further analysis is necessary to verify the identity of these peptides. These data support the hypothesis that the three largest cytochrome oxidase peptides are products of the yeast mitochondrial genome. The data also indicate that mitochondrial cytochrome oxidase mRNA is present in both poly(A) RNA and non-poly(A)-containing mitochondrial RNA.

It should be noted that the translation of mitochondrial RNA fractions from both grande and petite yeast were stimulated considerably by the addition of Ca^{++} to the E. coli ribosomal system (Table II). Stimulation was maximum at 6mM Ca^{++}. Stimulation appeared to be relatively specific for mitochondrial RNA since there was inhibition of translation of a variety of other messenger RNAs (Table II). Since both total incorporation and immunoprecipitable radioactivity is increased by Ca^{++} (see Fig. 3), there appears to be a stimulation of specific translation.

TABLE I

ANTIBODY PRECIPITATION OF PEPTIDES SYNTHESIZED IN AN E. COLI S30 SYSTEM

1.5µg of MS2 RNA and 5µg of mitochondrial RNAs were translated in an *E. coli* cell free system[7]. The direct antibody precipitation was as previouslz described[7]. For the indirect precipitation the cell free system was diluted to a final concentration of 10mM $NaPO_4$ (pH 7.4), 15mM NaCl, 0.1% SDS, 1% Triton X-100, 1mM leucine and reacted with 5µg of monospecific cytochrome oxidase antibody at 4°C for 4 hours. 50µg of goat anti-rabbit antibody was then added and the reaction continued for 12 hours. The reaction mixture was diluted to 0.5ml of 20mM $NaPO_4$, 80mM NaCl, 0.1% SDS, 1% Triton X-100 and layered over a step gradient (50µl 0.5M sucrose, 100µl 1.0M sucrose) and centrifuged 15' at 10,000 G. This procedure was repeated three times.

	^{3}H-leucine Incorporated into Protein Precipitates			
	TCA	Antibody	Indirect	Direct
Additions	cpm		cpm (% input)	cpm (% input)
None	40,640	Cyt. ox	981	989
Mt poly(A) RNA	137,797	Cyt. ox.	4,928 (3.6)	5,052 (3.7)
		Myosin	2,469	
Mt poly(A) minus RNA	188,062	Cyt. ox.	4,705 (2.5)	8,340 (4.4)
		Myosin	2,314	
MS2 RNA	300,468	Cyt. ox.	3,376 (1.1)	979 (0.3)
		Myosin	2,415	

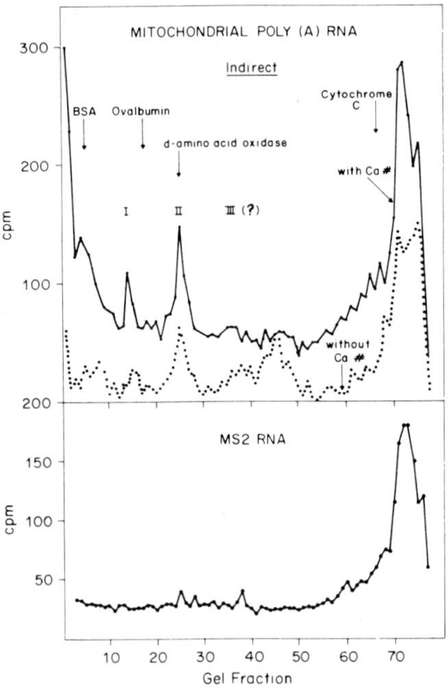

Fig. 3. SDS acrylamide gel electrophoresis of peptides synthesized in an *E. coli* S30 ribosomal system that are immunoprecipitated by monospecific rabbit cytochrome oxidase antibody, and anti-rabbit goat antibody. Synthesis directed by poly(A)-RNA in the presence of 6mM Ca^{++} ●——●, the absence of Ca^{++} ●---● and with MS2 RNA ●——●. The mobility expected for cytochrome oxidase peptides are indicated by I, II and III.

TABLE II

THE DIFFERENTIAL EFFECT OF CA^{++} IONS ON mRNA TRANSLATION IN AN *E. COLI* CELL FREE SYSTEM

		^{3}H-Leucine Incorporated into Protein Precipitates (cpm)					
		Grande	Petites			Viral	RNA
Additions	None	R_1	O_IP_2	CEP_2	CEO_I-1	MS2	T5
none	4,116	11,039	9,787	9,581	7,970	8,151	37,916
6mM Ca^{++}	4,127	19,100	16,318	18,039	13,920	6,149	24,162

5µg mitochondrial poly (A) RNA, 1µg MS2 RNA or 4µg T5 RNA were added to 50µl *E. coli* cell free system.

Our preliminary data show that petite mitochondrial RNAs are also translated in the *E. coli* S30 system, and are immunoprecipitated with cytochrome oxidase antibody using the indirect procedure (Table III). We are currently analyzing the peptides synthesized and examining the purity of the RNA preparations.

TABLE III

STIMULATION OF (^{3}H)-LEUCINE INCORPORATION IN AN *E. COLI* S30 SYSTEM BY PETITE mt-RNAs

Incorporation into Protein Precipitates (cpm)			
Additions	TCA	Cyt. Ox. Antibody	% input
none	28,730	651	2.2
T5	754,293	989	0.1
Mt-Poly(A) RNAs			
R_1 (grande)	393,835	3,336	0.9
CEO_I-1	131,560	1,278	1.0
O_IP_2	426,709	1,785	0.5
CEP_2	86,759	1,092	1.3

We have also attempted to map mitochondrial ribosomal RNA cistrons by electron microscopy. The displacement loop or R-loop technique described by Thomas, White and Davis (personal communication) was used to map 21S and 14S ribosomal genes in grande and petite yeast. Visualization of the R-loop with 21S rRNA in a linear molecule of petite F-11 is shown in Figure 4. 21S RNA has a repeat spacing of 4.5 microns similar to our previous analyses of monomer cycle contour lengths[9]. Mapping of 21 and 14S tRNA cistrons in grande mitochondrial DNA is now in progress.

In summary, we have shown that a substantial fraction of yeast mitochondrial DNA is transcribed, probably more than 60% of a single-strand equivalent, and that in anaerobic yeast the fraction transcribed appears to be decreased. Poly(A) RNA and poly(A) minus RNA appear to share many sequences and both contain messenger activity. Both have been translated in *E. coli* cell-free system and the products immunoprecipitated by cytochrome oxidase antibody by the indirect

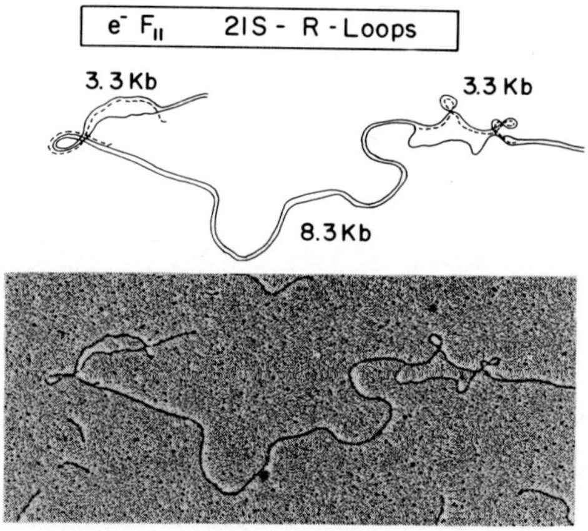

Fig. 4. Electron microscopic visualization of the mitochondrial 21S rRNA cistron was performed by the R-loop technique. Native petite F11 mtDNA was incubated with 21S rRNA at 39°C for 20 hours in 70% formamide. A spacing of 4.4 microns for the 21S rRNA is observed in the tandem-repeated linear molecule, identical with the contour length of the monomer circle[9].

procedure. Peptides migrating with the two largest cytochrome oxidase peptides were observed on SDS gel electrophoresis of the dissociated immunoprecipitate. Calcium stimulated the translation of mitochondrial RNAs, but not of several other RNAs. We have also visualized rRNA genes on petite mitochondrial DNA by the R-loop technique.

REFERENCES

1. Hollenberg, C.P. et al (1969) Biochim. Biophys. Acta 209: 1.
2. Bernardi, G. et al (1972) J. Mol. Biol. 65: 173.
3. Aloni, Y. and Attardi, G. (1971) Proc. Natl. Acad. Sci. (U.S.) 68: 1757.
4. Aloni, Y. and Attardi, G. (1971) J. Mol. Biol. 55: 251.
5. Rabinowitz, M. et al (1976 in press) The Genetic Function of Mitochondrial DNA. Eds. C. Saccone and A.M. Kroon, Elsevier Publishing, Amsterdam.
6. Hendler, F.J. et al (1975) Nature 258: 357.
7. Padmanaban, G. et al (1975) Proc. Natl. Acad. Sci. (U.S.) 72: 4293.
8. Grivell, L.A., et al (1976 in press) The Genetic Function of Mitochondrial DNA. Eds. C. Saccone and A.M. Kroon, Elsevier Publishing, Amsterdam.
9. Locker, J. et al (1974) J. Mol. Biol. 88: 489.
10. Jakovcic, S. et al (1975) Biochem. 14: 2043.

ACKNOWLEDGEMENT

This work was supported in part by grants from the National Heart and Lung Institute, United States Public Health Service, and a grant from the Louis Block Fund, University of Chicago. The Franklin McLean Memorial Research Institute is operated by the University of Chicago for the U.S. Energy Research Development Administration.

Genetics and Biogenesis of Chloroplasts and Mitochondria, Th. Bücher et al. eds.

THE MITOCHONDRIAL RNAs OF *NEUROSPORA CRASSA*: THEIR FUNCTION IN TRANSLATION AND THEIR RELATION TO THE MITOCHONDRIAL GENOME

A.M. Kroon, P. Terpstra, M. Holtrop, H. de Vries,
C. van den Bogert, J. de Jonge and E. Agsteribbe.
Laboratory of Physiological Chemistry, State University,
Bloemsingel 10, Groningen, The Netherlands.

INTRODUCTION

It is well recognized that the intrinsic mitochondrial nucleic acids play a key role in the formation of mitochondria within the cells of aerobic organisms. We are interested in the physicochemical and functional characteristics of the various components of the mitochondrial transcription-translation machinery. The aim of our studies is to unravel the contribution of the mitochondria to their own synthesis. It is hoped that a thorough knowledge of these processes will help to understand and to further explore the complicated and well-balanced interplay between the activities of various cell compartments in the formation of mitochondria. We have been concentrating on two experimental systems: the animal system of mainly rat-liver and the mould *Neurospora crassa*[1]. For *Neurospora* we have reported earlier on the characterization of the mitochondrial ribosomes as a class of 80S ribosomes[2], on the isolation of mitochondrial polysomes[3], and on the circularity of the mitochondrial DNA[4]. We have further shown that the mitochondrial DNA is heterogeneous in base composition by equilibrium density centrifugation and denaturation-renaturation techniques[5] and finally we have constructed the physical map of the mitochondrial DNA with the aid of a number of restriction endonucleases[6].

In this paper we will present our recent results with respect to the characterization of the mitochondrial rRNAs, tRNAs and mRNA and report on the localization of the genes for these RNAs on the mitochondrial DNA.

MITOCHONDRIAL RIBOSOMAL RNAs

It has been suggested by Lizardi and Luck[7,8] that the mitochondrial ribosomes of *Neurospora crassa* lack the 5S rRNA molecule that is present in bacterial as well as in cytoplasmic ribosomes from a variety of different organisms. So far 5SrRNA has neither been detected in mitochondria from other sources. Lizardi and Luck offer two possible explanations to cope with the difficulty that somehow a 5SrRNA should be retained in the ribosomes: they consider the possibility that the 5SrRNA function has been adopted either by a 4S-type RNA or by the large rRNA as the consequence of incomplete processing[7].

In our studies on the characterization of the mitochondrial ribosomes from *Neurospora crassa*, we have arrived at the conclusion that the native mitochondrial ribosome is an 80S particle, that can easily be transformed into a 73S particle if

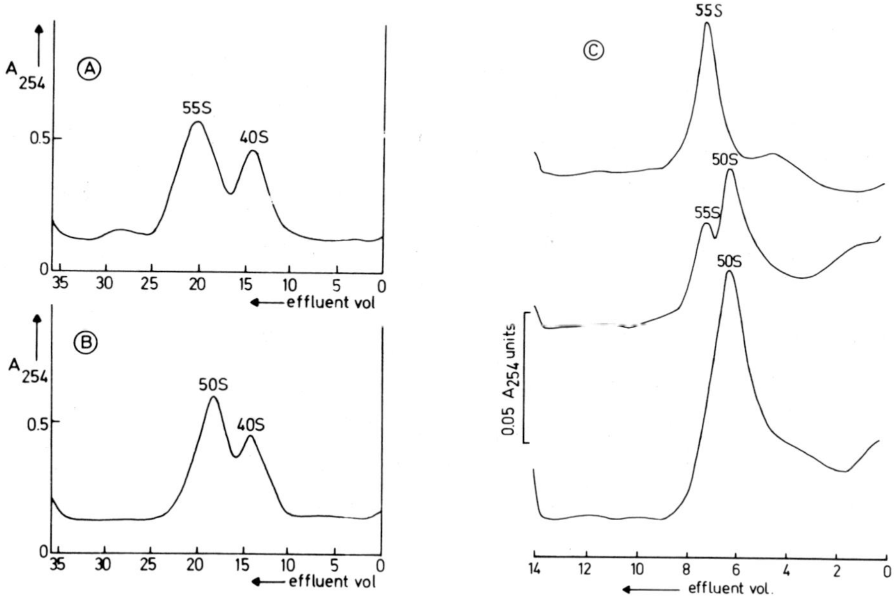

Fig. 1. A+B: Dissociation of *N. crassa* mitochondrial ribosomes, previously washed in $A_{500}M_{10}T_{10}$. A: 80S ribosomes; B: 73S ribosomes. Dissociation was in $A_{100}M_{1}T_{10}H$, centrifugation in isokinetic sucrose gradients (15-30 %) in the same buffer; SW 27 rotor, 15 h 23.000 rpm at 4 °C. C: Sedimentation profiles of 55S subunits before (upper trace) and after dialysis (16 h at 4 °C) against (middle trace) $A_{10}M_{10}T_{10}$ or (lower trace) $A_{100}T_{10}$; SW 27.1 rotor, 15-30 % isokinetic sucrose gradients in $A_{100}M_{10}T_{10}$. [S=sucrose; A=NH_4Cl; M=$MgCl_2$; T=Tris HCl, pH7.5; H=heparin].

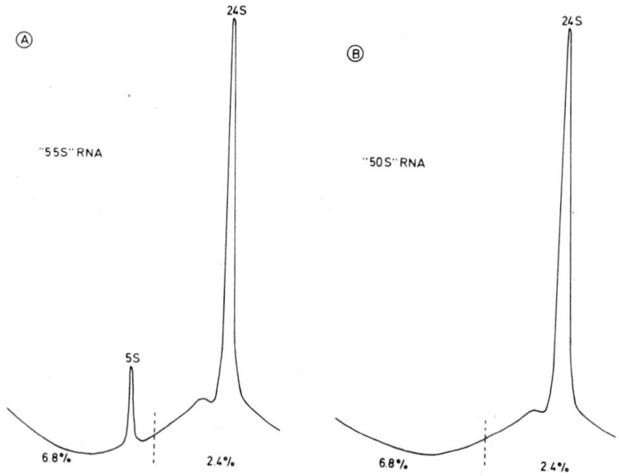

Fig. 2. Gel-electrophoresis analysis of RNA from 55S subunits (A) and 50S subunits. Discontinuous cylindrical gels of 2.4 % + 6.8 % polyacrylamide in 30 mM Tris-HCl, 0.1 mM EDTA, 2 mM Mg-acetate, pH 8.1 were used. Electrophoresis was at 4 mA/gel, until bromophenol blue was at 1 cm from the bottom. Gels were scanned at 260 nm in a Gilford spectrofotometer equipped with a linear transport system.

certain precautions for conserving the integrity of the ribosomes are neglected[2]. Both types of ribosomes contain the 24S and 17S rRNAs; the integrity of these RNAs is retained the best in the 80S particles. In preliminary experiments it was shown that one of the differences between the 80S and 73S ribosomes might be a 5S rRNA molecule, being present in the former and either absent or degraded in the latter ribosomes[9]. In 5 experiments the two large and the 5S rRNAs were present in about equimolar amounts; the 260 nm absorbancy ratio 24S:17S:5S was $25.2 \pm 0.74 : 11.0 \pm 0.33 : 1$. This is a similar ratio as found for cytoplasmic rRNAs and for bacterial rRNAs. Since the contamination of the mitochondrial ribosomes with cytoplasmic ribosomes is low and at worst of the order of 10%[10], this observation strengthened our conviction that the 5S RNA is an intrinsic component of the mitochondrial ribosomes.

Further support for this idea comes from the comparison and analysis of the large subunits of the 80S and 73S mitochondrial ribosomes. Washed 80S ribosomes dissociate into 55S and 40S subunits (Fig. 1A); 73S ribosomes into 50S and 40S subunits (Fig. 1B). Moreover, the 55S subunit can be transformed into a 50S particle (Fig. 1C). Neither the 50S subunit from 73S ribosomes, nor the 50S particle derived from the 55S subunit do contain a 5SRNA molecule, whereas the 55S subunit itself does. The analyses of the 55S- and 50S subunit-RNAs on discontinuous polyacrylamidegels are shown in Fig. 2. As outlined elsewhere[10,11] a further difference between the 55S and 50S subunits is found in the apparent loss of about 8 ribosomal proteins during the transition of 55S subunits into 50S particles.

Table I gives the results of a preliminary hybridization experiment. Although the radioactivity bound to the filters in the absence of competing 24S plus 17S rRNA was unexpectedly high the data are strongly suggestive for coding of the 5S rRNA on mitochondrial DNA. The residual hybridization of cytoplasmic 5SrRNA may be explained by a certain degree of homology between the two types of 5SRNA. It is well established that the basesequence of 5SRNAs is conserved quite stringently even in evolutionary remote organisms and organelles[12,13]. If we assume the presence of one 5SrRNA gene per mitochondrial genome, the maximum number of counts that could have bound to the filters under the experimental conditions used is 180. Further experiments to substantiate these observations are in progress.

TABLE I

FILTER HYBRIDIZATION OF 5S RNA FROM MITOCHONDRIAL RIBOSOMES WITH MITOCHONDRIAL DNA

0.2 μg (18,000 cpm) of electrophoretically purified [^{32}P]-5S RNA was incubated with filters containing 2 μg mtDNA. Hybridization was for 25 h at 65°C in 3xSSC, 0,2% SDS*. The values for blank or *E.coli*-DNA filters ranged from -4 to 10 cpm.

μg competitor rRNA (24S + 17S)	mitochondrial 5S RNA (cpm/filter)	cytoplasmic 5S RNA (cpm/filter)
0	430	288
5	106	31
10	100	22
20	88	27

* SSC=Standard Saline Citrate=0.15M NaCl-0.015M Na-citrate;SDS=Sodium Dodecyl Sulphate.

MITOCHONDRIAL TRANSFER RNAs

It is known for some time already, that mitochondria of *Neurospora crassa* as other mitochondria contain specific tRNAs[14]. Since there is no reason to believe that any of the aminoacids normally present in proteins, is excluded from incorporation by the mitochondrial proteinsynthetic system, the minimal number of tRNAs needed for complete read-out of the genetic code is 32. One would, therefore, expect to find at least 32 different tRNAs within mitochondria. If this number is reached, indeed, or perhaps even exceeded, an important question remains whether the genes for these tRNAs are present on the mitochondrial or on the nuclear genome. We have started experiments to obtain more precise information about the number of specific tRNAs present within the mitochondria, about their chemical characteristics and about their genetic localization. The data obtained so far show that the G + C content of the mitochondrial tRNAs from *N. crassa* (about 48%) is low as compared to the cytoplasmic tRNAs of the same organism (about 55 %), but high as compared to the mitochondrial tRNAs from the yeast *Saccharomyces cerevisiae* (33,4 %)[15]. This is remarkable in as much the former can be efficiently charged by yeast mitochondrial synthetases and not by *Neurospora* cytoplasmic enzymes. The recognition site for the mitochondrial aminoacyl-tRNA synthetases has apparently been conserved stronger than the overall base composition. In the preliminary experiments the ribo-thymidine content was less than one on a molar basis. This is of special interest in view of the function in other systems attributed to the TΨCG loop of tRNA in the interaction with the 5S ribosomal RNA[16]. Further analyses will be necessary to establish these data more firmly.

On two-dimensional polyacrylamide gels[17] the mitochondrial tRNA preparations could be resolved into 25 strong and 17 weak spots. It remains to be seen as to how far especially the weak spots are to be ascribed to cytoplasmic contamination. Anyway a slight "contamination" (about 10%) of the mitochondrial tRNAs with cytoplasmic species was revealed by RPC-5 chromatography. It cannot be decided yet whether this "contamination" has to be interpreted as an isolation artefact or as an indication for transport of tRNAs from the cytoplasm into the mitochondria.

MITOCHONDRIAL MESSENGER RNAs

Several attempts with various organisms have been made to isolate RNAs from mitochondria with messenger activity. Following the observations that eukaryotic messengers contain a long poly(A)stretch at their 3' end, it was investigated if poly(A)containing RNAs are present within mitochondria as well. For animal mitochondria this is indeed the case[18,19], for yeast the results are conflicting in the sense that poly(A) tracts of a length comparable to nuclear messengers have been reported[20] but that other authors can only detect short poly(A)stretches[21] or no 3' end A-enrichment at all[22]. In mitochondria from the fungus *Trichoderma*

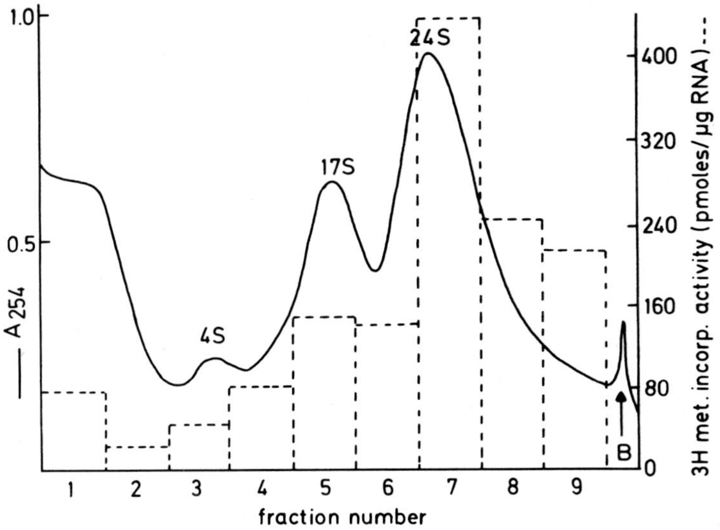

Fig. 3. Total RNA from S-AMT washed mitochondria was fractionated on a 5-30 % sucrose gradient in a SW41 rotor. Fractions were pooled and stimulation of 3H-methionine incorporation in a DNAse treated *E.coli* S30 was measured according to Moorman and Grivell[24]. 40 µg of RNA was added. Values were corrected for blanks without added RNA.

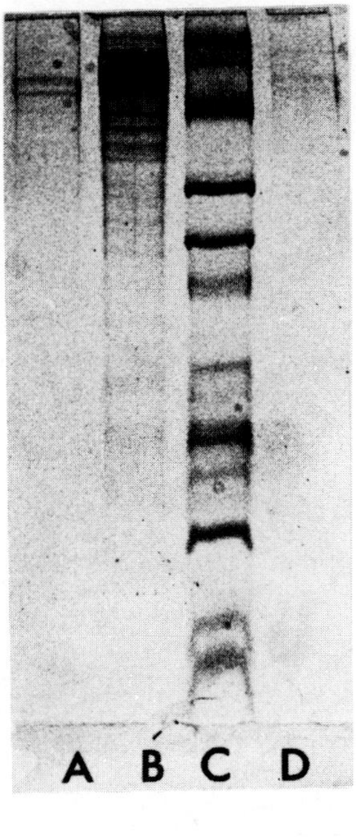

Fig. 4. Fractions 5 and 7 from the gradient of Fig. 3. were added to an *E.coli* S30 and incubated as described[3]. The reaction was terminated by adding antiserum against total *N.crassa* cytochrome *c* oxidase. After 15 min at room temperature immunoprecipitates were spun down in an Eppendorf centrifuge and washed twice with 0.15 M NaCl, 10 mM Tris.SO_4 pH 7.4, 0.5 % Triton X-100. Precipitates were dissolved in electrophoresis buffer containing 2.5 % SDS by heating at 100°C for 2 min. Electrophoresis was in a 0.7 mm slabgel containing 20 % acrylamide in 20 mM Tris pH 8.8 for one night at 15 V/cm. The gel was stained with coomassie blue. A: the immunoprecipitate of the incubation with the RNA of fraction 5; B: idem for fraction 7; C: peptide pattern of cytochrome *c* oxidase; D: immunoprecipitate after incubation of S-30 without added mtRNA.

viride, however, RNA containing short stretches of poly(A) has been found[23]. We have looked for the presence of poly(A)-containing RNA in mitochondria from *Neurospora crassa* with the aid of poly(U)-sepharose. If at all present, the poly(A) stretches are of the length of about 30 A residues at most. Moreover, neither total nor fractionated mitochondrial RNA hybridized to a significant extent to ^{3}H-poly(U). We have, therefore proceeded along the lines presented previously[3]. Isolation of run-off RNA from mitochondrial polysomes and the 12 - 5S fraction from gradients of total mitochondrial or polysomal RNA were tested for messenger-activity in an *E. coli* system and shown to stimulate aminoacid incorporation to a considerable extent[3,9]. Gradient fractions of total mitochondrial RNA also stimulated protein synthesis in an *E. coli* S30, the bulk of messenger activity sedimenting faster than 15S. This is shown in Fig. 3. It is worth noting that the purified mitochondrial rRNAs did not stimulate aminoacid incorporation under the same conditions. Moreover the fraction containing 20-25S RNA was shown to catalyze the synthesis of peptide material immunoprecipitable by antibodies against *Neurospora crassa* cytochrome c oxidase. This is illustrated in Fig. 4. Fraction 7 RNA gives a significant incorporation into immunoprecipitable polypeptides. There is, however, no clear correspondence of the peptides in these immunoprecipitates with the localization of the subunits of cytochrome c oxidase in slot C. This may be due either to aggregation of the newly synthesized products or to a lack of posttranslational modification. Similar observations have been made by others in the coupled system for transcription and translation of yeast mtDNA[24].

RNA GENES ON MITOCHONDRIAL DNA

We have recently constructed the restriction endonuclease map of mitochondrial DNA of *Neurospora crassa*[6]. The results of these studies are in good agreement with those of the combined denaturation/restriction enzyme mapping of Küntzel and coworkers[25]. The map of the restriction fragments obtained with the endonucleases Eco R I, Hind III and Bam H I is illustrated in Fig. 5. Preliminary experiments with the Eco R I fragments showed that both the 24S and 17S ribosomal RNAs hybridized exclusively with the Eco R I fragment 1 (see Fig. 6). Since this E1 fragment has a length of about 19.000 basepairs, this observation did not allow any conclusion about the minimal distance between the two large ribosomal RNA genes. We were interested to localize these genes more precisely on the one side because a 32S precursor RNA containing both rRNAs has been postulated[28], on the other side to establish either a difference or a similarity with the situation in yeast, where the two ribosomal RNA genes are far apart, at least in the species *Saccharomyces*[29]. By comparing the hybridization data obtained with separate stripfilters of Eco R I fragments, Hind III fragments and Bam H I fragments, we could localize the rRNA genes with a reasonable degree of accuracy as depicted in Fig. 7. The 24S rRNA hybridized with Bam H I, fragment 1 and with five of the Hind III fragments (H1, H7,

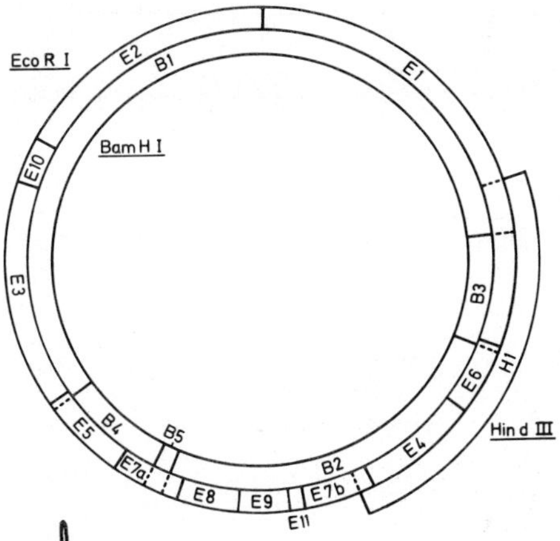

Fig. 5. Restriction fragment map of mitochondrial DNA from *Neurospora crassa*. The molecular weight as obtained from the sum of the restriction fragments is 41 x 10^6. The polarity of E8-E9-E11 and the sequence E3-E10-E2 have been established by the best fit between calculated and observed partial fragments and in the latter case also in combination with redigestion of partial fragments.

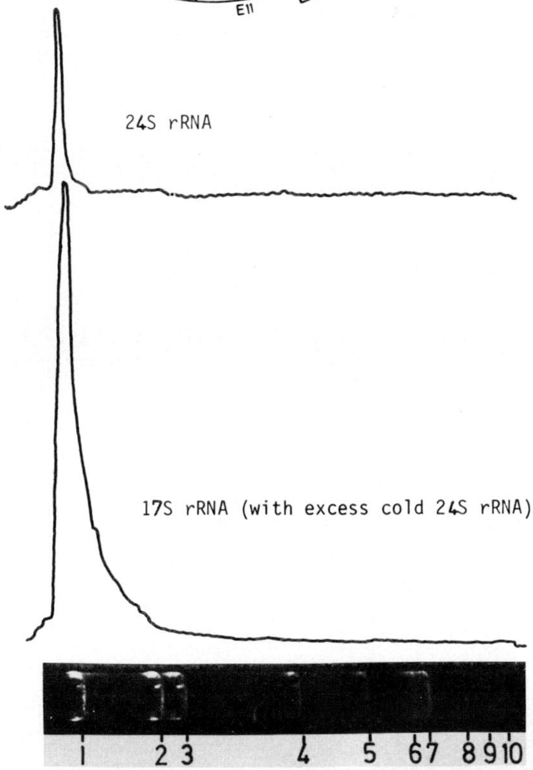

Fig. 6. Stripfilter hybridization of rRNA with Eco R I fragments. The ribosomal RNAs were extracted from purified 80-S ribosomes[2,10] separated on sucrose gradients (two cycles), iodinated *in vitro* [26], hybridized to a stripfilter[27] containing Eco R I fragments of 5 μg of mtDNA at a RNA/DNA ratio of about 1 in 3 x SSC, 0.2 % SDS at 65°C 18 h. After washing and RNAse-treatment the filters were subjected to autoradiography. At the bottom of this figure is a photograph of the ethidiumbromide stained gel from which the filters were made. The tracings represent optical scannings of the autoradiograms.

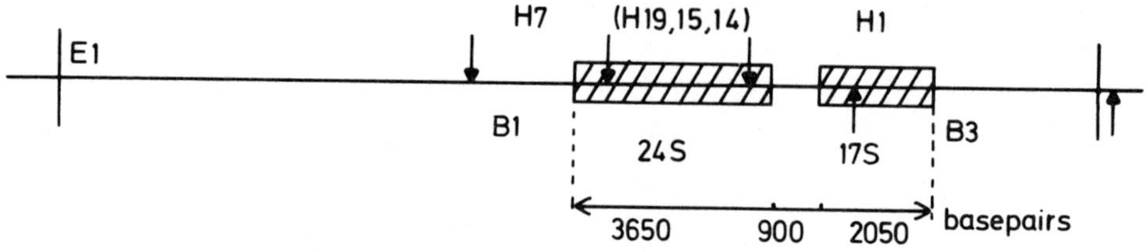

Fig. 7. Localization of the rRNA genes within Eco fragment E 1. The nomenclature of the restriction fragments is the same as in Fig. 5; H7, H14, H15, H19 have been added to the map. The gene length of the 24S and 17S rRNAs have been calculated from the published molecular weights of 1,28 x 10^6 and 0.72 x 10^6 respectively[30].

H14, H15 and H19). Although the map of the Hind fragments was not constructed, we are forced to conclude that the sequence H14-H15-H19 is in between H7 and H1, because the known length of H7 is incompatible with an internal position of the latter in the sequence of the five Hind fragments. The length of the 24S rRNA (about 3650 bases) is not enough to cover the complete H7 plus the smaller fragments H14, H15 and H19. As an additional control we have compared the amount of

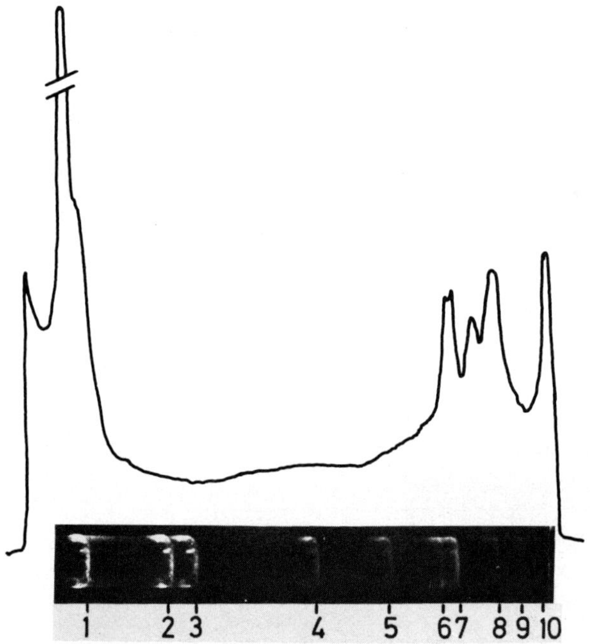

Fig. 8. Stripfilter hybridization of tRNA with Eco R I fragments. Total tRNA was extracted and purified[31,32] from extensively washed mitochondria, iodinated and hybridized to a stripfilter containing Eco fragments of mtDNA (see legend Fig. 6).

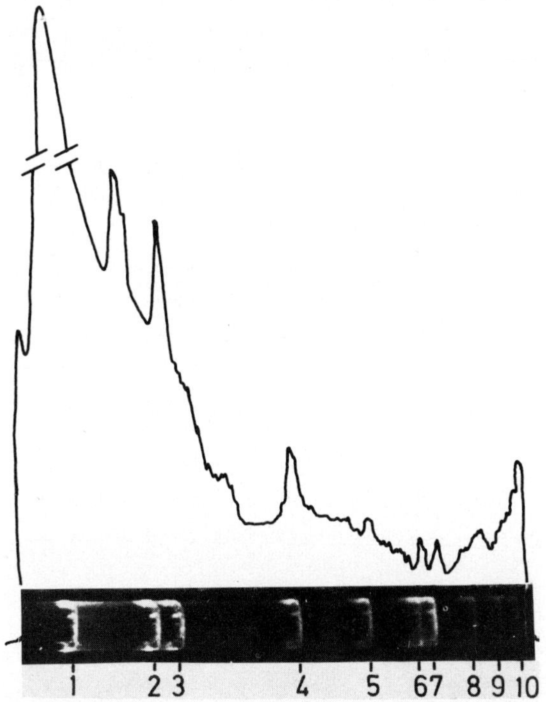

Fig. 9. Stripfilter hybridization of total mtRNA with Eco R I fragments. Total RNA was extracted from extensively washed mitochondria (see legend to Fig. 3), iodinated and hybridized to a stripfilter containing Eco R I fragments of mtDNA (see legend to Fig. 6).

radioactivity bound per unit fragment length. The results of these calculations lend support to a partial overlap of the 24S rRNA gene with the H1 and H7 fragments. Similar experiments and considerations have led to the localization of the 17S rRNA gene over the cleavage site of Bam H I between B1 and B3 as indicated in Fig. 7. Between the two genes there remains a gap of about 900 basepairs. This gap is small enough to allow that a precursor molecule with a sedimentation value of 32S and containing both the 24S and 17S rRNAs is formed as the primary transcript of the two rRNA genes[28]. There may be even some space left for posttranscriptional cleavage of short sequences at either end of the precursor.

With respect to the 5S rRNA, similar experiments did not yet lead to unambiguous results. Since small amounts of contaminating tRNAs or fragments of the other rRNAs can be picked up as hybrids with the techniques used, hybridization of 5S ^{125}I-rRNA had to be done in the presence of excess of unlabeled tRNA and rRNA. The supply of especially the tRNAs for this purpose has been prohibitive until now.

The results of the hybridization of total mitochondrial transfer RNA with Eco R I fragments is shown in Fig. 8. Most of the radioactivity coincides with the fragment E1, but there is also significant hybridization at the positions of E6, E7, E8 and E10. The percentage of total radioactivity hybridized to these smaller fragments was in the order of ten. If one assumes that the radioactivity is distributed equally over the different tRNAs and that the molar ratios for the different tRNAs is not extremely far from 1, it can be roughly estimated that there are about 3 tRNA genes at solitary positions on the mitochondrial genome, whereas 25 - 30 tRNA genes are more or less clustered on the E1 fragment. Of course more detailed studies are necessary to refine this conclusion further.

Finally we have looked for the hybridization properties of iodinated total mitochondrial RNA with the same stripfilter method. The results are shown in Fig. 9. The hybridization to E1 and to E6 - E10 can be conveniently explained by the presence of the rRNAs and tRNAs in the total mtRNA preparations, although it can not be excluded that also other RNA species are hybridizing in this regions of the genome. The hybridization to the Fragments E2, E3, E4 and E5 clearly show that these regions of the DNA are transcribed as well. The informational content of the RNAs hybridizing to these fragments remains to be elucidated, but it seems fair to speculate that at least the major part of it represents messenger-RNA species.

CONCLUSIONS

The results of the experiments presented above are summarized in Fig. 10. It is clear that the mtDNA of *Neurospora crassa* does not contain large regions which are devoid of genetic information, if the presence of transcripts is interpreted to mean expression of such information. Although the data for 5S rRNA can not be considered decisive as yet, the preliminary hybridization experiments indicate

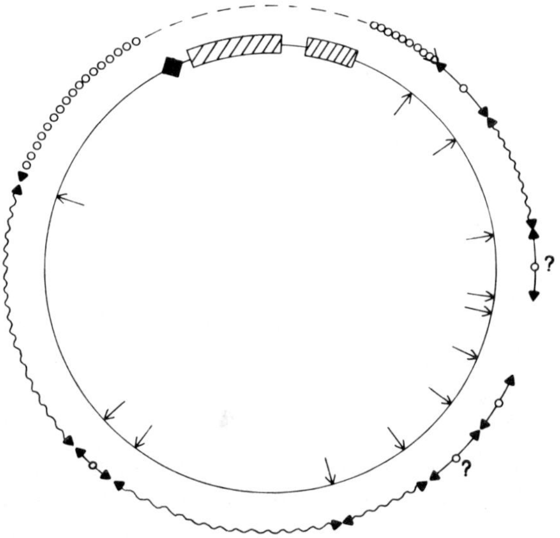

Fig. 10. Localization of genes on the restriction fragment map of the mtDNA of *Neurospora crassa*. *Inner circle:* Eco R I cleavage sites are indicated with arrows (*cf* Fig. 5). On this circle the hatched boxes represent the 24S and 17S rRNA genes (*cf* Fig. 7). The black diamond represents the tentative localization of the 80S ribosomal 5S rRNA. *Outer circle:* ←o–o→ represents the overall distribution of tRNA genes (*cf* Fig. 8) on the Eco R I fragments. ←∿∿∿→ represent the remaining Eco fragments that are labelled with total mtRNA (*cf* Fig. 9). The precision of gene localization on the outer circle is limited by the length of the corresponding Eco R I fragments.

binding to a Hind III fragment, H7. Although the H7 fraction contains 2 or perhaps even 3 different fragments, it is tempting to speculate that a 5S rRNA gene clusters with the 24S and 17S rRNA genes, a situation similar to that in other systems[33]. The exact number of tRNA genes remains to be estimated, but this may well be high enough to accomodate all tRNAs necessary for mitochondrial protein-synthesis. So the picture emerges that all RNA components of the mitochondrial translation machinery are encoded in mtDNA, whereas all protein components for this machinery are genetically localized in nuclear DNA[34]. This may indicate that the mitochondrial membranes are an impregnable barrier for RNAs, whereas they are not so for proteins. Finally it is worth mentioning that combination of the type of experiments described in Figs 3 & 4 with the stripfilter hybridization approach will allow us, at least in principle, also to localize the genes for the mitochondrial translation products on the mitochondrial genome in the near future.

ACKNOWLEDGEMENTS

The experiments on the characterization of the mitochondrial tRNAs were done in cooperation with the group of Prof. Dr. A. Stahl, University of Strassbourg. H. de V. was a recipient of a short term EMBO-fellowship. The authors wish to thank Prof. Dr. F.J. Loomeijer for his interest and encouragement. These studies were supported in part by a grant to A.M.K. from the Netherlands Foundation for Chemical Research (SON) with financial aid from the Netherlands Organisation for the Advancement of Pure Research (ZWO).

REFERENCES

1. Kroon, A.M., Agsteribbe, E. and De Vries, H. (1972), in L. Bosch (Ed), The Mechanism of Protein Synthesis and its Regulation, North Holland, Amsterdam, pp. 539-582.
2. Datema, R., Agsteribbe, E. and Kroon, A.M. (1974), Biochim. Biophys. Acta, 335, 386-395.
3. Agsteribbe, E., Datema, R. and Kroon, A.M. (1974), in A.M. Kroon and C. Saccone (Eds), The Biogenesis of Mitochondria, Academic Press, New York, pp 305-314.
4. Agsteribbe, E., Van Bruggen, E.F.J. and Kroon, A.M. (1972), Biochim. Biophys. Acta, 269, 299-303.
5. Terpstra, P., Holtrop, M. and Kroon, A.M. (1975), 10th FEBS Meeting, Paris, Abstracts, 143.
6. Terpstra, P., Holtrop, M. and Kroon, A.M. (1976), in C. Saccone and A.M. Kroon, (Eds), The Genetic Function of Mitochondrial DNA, North-Holland, Amsterdam, pp 111-118.
7. Lizardi, P.M. and Luck, D.J.L. (1971), Nature New Biol., 229, 140-142.
8. Lizardi, P.M. (1971), PhD Thesis, The Rockefeller University, New York.
9. Agsteribbe, E. (1974), PhD Thesis, University of Groningen.

10. Van den Bogert, C. and de Vries, H. (1976), Biochim. Biophys. Acta, in press.
11. De Vries, H. and Van den Bogert, C., see elsewhere in this volume.
12. Hori, H. (1976), Molec. gen. Genet., 145, 119-123.
13. Phillips, D.O. and Carr, N.G. (1975), FEBS Letters, 60, 94-97.
14. Barnett, W.E. and Brown, D.H. (1967), Proc. Nat. Acad. Sci. USA, 57, 452-458.
15. Martin, R., Schneller, J.M., Stahl, A.J.C. and Dirheimer, G. (1976), Biochem. Biophys. Res. Commun., 70, 997-1002.
16. Erdmann, V.A., Sprinzl, M. and Pongs, O. (1973), Biochem. Biophys. Res. Commun. 54, 942-948.
17. Fradin, A., Gruhl, H. and Feldmann, H. (1975), FEBS Letters, 50, 185-189.
18. Perlman, S., Abelson, H.T. and Penman, S. (1973), Proc. Nat. Acad. Sci. USA, 70, 350-353.
19. Ojala, D. and Attardi, G. (1974), J. Mol. Biol., 82, 151-176.
20. Cooper, C.S. and Avers, C.J. (1974), in: A.M. Kroon and C. Saccone (Eds), The Biogenesis of Mitochondria, Academic Press, New York, pp. 289-303.
21. Padmanaban, G., Hendler, F., Patzer, J., Ryan, R. and Rabinowitz, M. (1975), Proc. Nat. Acad. Sci., USA, 72, 4293-4297.
22. Groot, G.S.P., Flavell, R.A., Van Ommen, F.J.B. and Grivell, L.A. (1974), Nature, 252, 167-169.
23. Rosen, D. and Edelman, M. (1976), Eur. J. Biochem., 63, 525-532.
24. Moorman, A.F.M. and Grivell, L.A. (1976), in: C. Saccone and A.M. Kroon (Eds), The Genetic Function of Mitochondrial DNA, North Holland, Amsterdam, pp 281-289.
25. Bernard, U. and Küntzel, H. (1976), in: C. Saccone and A.M. Kroon (Eds), The Genetic Function of Mitochondrial DNA, North-Holland, Amsterdam, pp 105-109.
26. Prensky, W., Steffensen, D.M. and Hughes, W.L. (1973), Proc. Nat. Acad. Sci., USA, 70, 1860-1864.
27. Southern, E.M. (1975), J. Mol. Biol., 98, 503-517.
28. Kuriyama, Y. and Luck, D.J.L. (1973), J. Mol. Biol. 73, 425-437.
29. Sanders, J.P.M., Heyting, C. and Borst, P. (1975), Biochem. Biophys. Res. Commun., 65, 699-707.
30. Neupert, W., Sebald, W., Schwab, A.J., Pfaller, A. and Bücher, T. (1969), Eur. J. Biochem., 10, 585-588.
31. Accoceberry, B., Schneller, J.M. and Stahl, A.J.C. (1973), Biochimie, 55, 291-296.
32. Faye, G., Kujawa, C. and Fukuhara, H. (1974), J. Mol. Biol., 88, 185-203.
33. Yamamoto, M., Lindahl, L. and Nomura, M. (1976), Cell, 7, 179-190.
34. Kroon, A.M. and Saccone, C. , Eds (1974), The Biogenesis of Mitochondria, Academic Press, New York.

Genetics and Biogenesis of Chloroplasts and Mitochondria, Th. Bücher et al. eds.

Dual origin of mRNA associated proteins in Ehrlich ascites mitochondria

N. G. Avadhani, V. A. Aroskar,
F. S. Lewis, G. J. Hansel, and M. P. Wolf

Department of Animal Biology
School of Veterinary Medicine
University of Pennsylvania
Phila., Pa. 19174

INTRODUCTION:

Native mRNA components in diverse cell types ranging from bacteria to mammalian cells exist as specific ribonucleoprotein complexes (see 1-5). In general, mRNA-protein complexes (mRNP) are characterized by their low buoyant density in CsCl, resistance to high salt (such as 0.5M KCl) and to EDTA (2,5,6). The number of polypeptide species detected in mRNP range from 2 in the case of globin mRNA (3) to 10 in chick erythrocytes (7). It is believed that the mRNA associated proteins may play an important role in translational regulation by way of controlling the activity as well as the stability of mRNA molecules(2-6). With a view to determine the mechanism of translational regulation in mitochondria, we have extended these studies to Ehrlich ascites mitochondrial system. Our experiments show that intramitochondrial mRNAs exist as conventional mRNP and that majority of mRNA associated proteins (m-Proteins) are imported from the cytoplasm while at least two of the m-Proteins are translated within the organelle matrix.

MATERIALS AND METHODS

Preparation of pure mitochondria from Ehrlich ascites cells by digitonin technique (8) and isolation of mt polysomes by Mg^{+2} precipitation method (9) were as described before. Mitochondrial and cytoplasmic polysomes were dissociated either with 200 μg/ml puromycin or 10 μmoles/ml EDTA and used to isolate respective mRNP by oligo(dT) cellulose method (2,5). The oligo(dT) cellulose-bound mRNP were eluted with 15 ml of 10mM Tris-HCl (pH 7.5) and 0.25M NaCl containing 50% foramide (5), dialyzed overnight against 6 liters of 5mM Tris-HCl (pH 7.5), 60mM NH_4Cl and 3mM $MgCl_2$ and freeze-dried. The dry powder was stored at -70°C. For CsCl banding, mRNP was fixed with formaldehyde and analyzed on preformed gradients as described by Hamilton (10) and Irwin et al. (5). The protein and RNA contents of mRNP were calculated using the formula described by Hamilton (10).

For ^{125}I labeling of m-Proteins, the mRNP preparation was treated with 0.2N KOH for 4 hr at 37°C, dialyzed against 6 liters H_2O and freezed-dried. Labeling of m-Proteins with ^{125}I was carried out using the chloramine T method (11,12). Electrophoresis of proteins either on lab gels or on cylindrical gels was carried out according to Laemmli (13).

RESULTS AND DISCUSSION

In the initial experiments using mt polysomes pulse labeled with ^3H cytidine and adenosine it was observed that about 15-20% of the radioactivity tightly binds to the oligo(dT) column. RNA isolated from resultant mRNP resolved into 6-8 bands of 8Se to 16Se on polyacrylamide gels as previously observed for mt mRNA (8,9). Further, mt polysomes pulse labeled with ^3H leucine also showed similar oligo (dT) cellulose binding pattern. These results suggested that as in the cytoplasmic systems mt Poly(A)-containing mRNAs exist as RNA-Protein complexes.

The physical-chemical properties of mitochondrial mRNP somewhat differ from those of cytoplasmic mRNP (see Table I). The cytoplasmic mRNP has a buyant density of 1.42 to 1.45 gm/cm^3 while the mt mRNP bands with a density of 1.41 to 1.43 gm/cm^3. The cytoplasmic mRNP sediments as 15-66S particles while mitochondrial mRNP contains 10S to 35S particles as tested in exponential sucrose gradients. Based on the CsCl banding (10), the cyto mRNP contains 42% (W/W) RNA and 58% (W/W) Protein as against 34% (W/W) RNA and 66% (W/W) Protein for mitochondrial mRNP.

Table I

Physical and Chemical Properties of cytoplasmic and mitochondrial mRNP*

Origin of mRNP	Sedimentation value (S)	Density (gm/cm^3)	% Protein (W/W)
cytoplasmic	15S-66S	1.42-1.45	58
mitochondrial	10S-35S	1.41-1.43	66

*^3H-labeled cytoplasmic and mitochondrial mRNP were prepared from cells labeled with ^3H adenosine and cytidine. Other details were as described in the Materials and Methods.

In animal cell systems, it is now well known that mt ribosomes contain distinctly different proteins from cyto ribosomes (12,14), although almost all of them may be translated on the cytoplasmic ribosomes (see 15). To verify if a similar biogenetic relationship exists with respect to mitochondrial mRNP, m-Proteins derived from respective particles were labeled with ^{125}I and electrophoresed on slab gels containing 10% acrylamide. Radioautography presented in Fig. 1 shows that cyto mRNP contains 5 proteins having molecular weights of 84,000; 72,000; 46,000, 32,000 and 20,000. Out of these, two proteins having molecular weights of 84,000 and 20,000 appear to be the minor components. The mitochondrial mRNP on the other hand contains 6 proteins with the molecular weights of 64,000; 45,000; 40,500; 36,000; 24,000 and 17,500. As in the case of mt ribosomes, the mitochondrial mRNP contains 6 distinct proteins which are different from cytoplasmic m-Proteins.

The genetic origins of mt m-Proteins were determined by labeling mitochondria in vivo with ^{3}H Leu in presence of specific inhibitors. Electrophoretic pattern presented in Fig. 2 shows that at least two of the mt m-Proteins having the molecular weights 40,500 and 24,000 may be translated within mitochondria since they are labeled under conditions when 90% of cytoplasmic protein synthesis is inhibited by cycloheximide and emetine. The remaining m-Proteins may be of cytoplasmic origin since they are significantly labeled even when 90% of mitochondrial protein synthesis is inhibited using chloramphenicol.

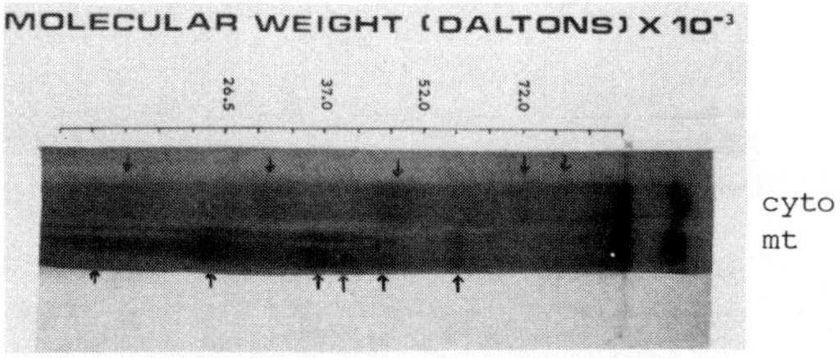

Fig. 1. Electrophoretic Patterns of m-Proteins. ^{125}I labeled m-Proteins were electrophoresed on slabs containing 10% polyacrylamide and exposed to Kodak no screen x-ray film for 3 hr.

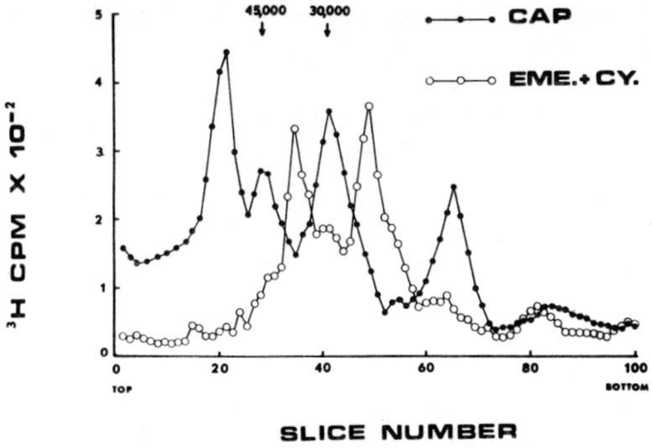

Fig. 2. Electrophoretic Profile of mt m-Proteins labeled in presence of chloramphenicol and cycloheximide/emetine. Cells were labeled with ^{3}H Leu for 3 hr in presence of inhibitors and mRNP prepared from total mt lysate as described in Methods. Proteins were electrophoresed on 10% cylindrical gels. Gel slices (1.5 mm) were swollen with NCS and counted as before (8).

In summary, we have presented evidence that intramitochondrial poly(A) containing RNA is associated with 6 polypeptides which differ from cytoplasmic mRNA associated proteins with respect to their size. Further, studies with inhibitors of mitochondrial and cytoplasmic protein synthesis indicate that at least two of the mitochondrial m-Proteins may be intramitochondrially translated while the remaining 4 may be imported from the cytoplasm. The dual origin of mito mRNA associated proteins may prove to be important in studying translational regulation and organelle biogenesis.

Acknowledgements: The authors are thankful to Professor R. J. Rutman for his continued interest and support. One of us (F. S. L.) is a recipient of USPHS Post-doctoral fellowship CA-1060. This research was supported in part by NSF grant BMS-75-04234.

REFERENCES

1. Spirin, A., (1969) Eur. J. Biochem. 10, 20.

2. Kumar, A., and Pederson, T. (1975) J. Mol. Biol. 96, 353.

3. Blobel, G. (1973) Proc. Natl. Acad. Sci. U. S. A. 70, 924.

4. Lindberg, U. and Sundquist, B. (1974) J. Mol. Biol. 86, 451.

5. Irwin, D., Kumar, A., and Malt, R. (1975) Cell 2, 157.

6. Bhorjee, J., and Pederson, T. (1973) Biochemistry 12, 2766.

7. Gander, E. S., Stewart, A. G., Morel, C. M., and Scherrer, K. (1973) Eur. J. Biochem. 38, 443.

8. Avadhani, N. G., Lewis, F. S., and Rutman, R. J. (1974) Biochemistry 13, 4638.

9. Lewis, F. S., Rutman, R. J., and Avadhani, N. G. (1976) Biochemistry, In press.

10. Hamilton, M. (1973) in "Methods in Enzymology", vol. 20, Moldave, K. and Grossman, L., Eds., p. 512, Academic Press, New York.

11. Greenwood, F. C., Hunter, W. M., and Glover, J. S. (1963) Biochem. J. 89, 114.

12. Leister, D. E., and Dawid, I. B. (1974) J. Biol. Chem. 249, 5108.

13. Laemmli, U. K. (1970) Nature, 227, 680.

14. O'Brien, T. W., Denslow, N. D., and Martin, G. R. (1974) in "The Biogenesis of Mitochondria" Kroon, A. M., and Saccone, C., Eds., p. 347, Academic Press, London.

15. Schatz, G., and Mason, T. (1974) Ann. Rev. Biochem. 43, 51.

Genetics and Biogenesis of Chloroplasts and Mitochondria, Th. Bücher et al. eds.

MITOCHONDRIAL TRANSCRIPTION IN RAT LIVER. STUDIES ON THE SYNTHESIS OF POLY(A)-CONTAINING RNA.

C. Saccone, P. Cantatore, G. Pepe, R. Gallerani,
C. De Giorgi and C. De Benedetto

Institute of Biological Chemistry, University of Bari, Bari, Italy.

INTRODUCTION

Studies over these last few years have clearly indicated that mitochondrial DNA possesses genetic information not only for ribosomal RNA species and a set of tRNAs but also for some messenger RNA species which could codify mitochondrial proteins having structural and/or regulatory function. This view is supported both by genetic and by biochemical evidences. The latter are essentially based on two kinds of experiments, namely 1) translation of mitochondrial RNA species 2) isolation of poly(A) containing RNA species from mitochondria (see ref. 1-4 for review). The presence of poly(A)-RNA hybridizing with mitochondrial DNA, firstly demonstrated by the groups of Attardi and Penman in HeLa cells (5,6) was confirmed in mitochondria of other animal cells (7,8). Also in some lower eukariotes their presence has been reported (9,10) although in the case of yeast there are controversial data probably due to the difficulty in detecting short polyadenylate sequences eventually attached to some messenger RNA species (10,11). With regard to the adenylation process of mitochondrial RNA few data are available at moment in the literature. In HeLa cells it has been demonstrated that the adenylation of RNA takes place within the mitochondrion and is a post-transcriptional process (5,6). On the other hand Avadhani et al. on the basis of the absence of poly(A) tracts in RNA synthesized by isolated Erlich ascites mitochondria and on the partial resistance of mitochondrial poly(A) to ethidium bromide, proposed that mitochondrial poly(A)-RNA may be, at least in part, of extramitochondrial origin (7). The presence of a mitochondrial poly(A) polymerase activity was firstly reported in rat liver by Jacob et al. in 1972 (12). Successively Rose et al. (13) succeeded in purifying this enzyme from Morris hepatoma and in the same paper suggested the purification of rat liver enzyme. However no data on the procedure of its purification and its properties were quoted. The presence of such enzymic activity was observed in our laboratory during the purification of mitochondrial DNA-dependent RNA polymerase and more recently we were able to isolate contemporaneously the two enzymes from rat liver mitochondria and to separate them by using two different chromatographic techniques clearly showing that the two activities are due to different protein molecules (14). The presence of such enzyme in the mitochondria of rat liver and other organisms (15) and our recent data on the in vitro synthesis of poly(A)-RNA (16) strongly indicate that mitochondrial messenger RNA species are polyadenylated inside the mitochondrion.

In this paper we will report the recent developments of our studies on the poly(A)-containing RNA species of rat liver mitochondria. Particularly the characteristics of the poly(A) synthesized in vitro by a mitochondrial poly(A) polymerase, partially purified in our laboratory, are here reported. Furthermore, in order to better investigate the physiological role of this enzyme in mitochondrial transcription, we have carried out the synthesis in vitro of RNA with isolated mitochondria and from the total mitochondrial RNA newly synthesized, we have isolated and characterized the fraction containing poly(A).

Synthesis of poly(A) by isolated poly(A)-polymerase.

Poly(A) polymerase was purified from rat liver mitochondria according to the technique reported in a previous paper (14) with the modifications indicated in the legend of Fig. 1. Using this procedure about a hundred times purification with respect to the solubilized extract was obtained, and the enzyme displayed the properties already reported, the most striking characteristics being its complete dependence on an endogenous factor of ribonucleotide nature already shown by Rose et al. (13), whose characterization is now in progress in our laboratory, and its inability to recognize any RNAs whatsoever as primer (14). Fig. 1 shows the time course of the reaction in our assay conditions. The acid insoluble material, already characterized as poly(A) by means of several criteria (14) was precipitated in diluted $CaCl_2$ solution according to the technique of Landes and Kitos (17) confirming its poly(A) nature. In order to characterize the size of the product synthesized in vitro we have used a 5-15% sucrose density gradient containing 0.5% sarkosyl. The results shown in Fig. 2, indicate that in these conditions a heterogenous population of molecules is synthesized. The major part of the product has a low molecular weight, but synthesis of poly(A) molecules having a sedimentation coefficient higher than 4S is also taking place.

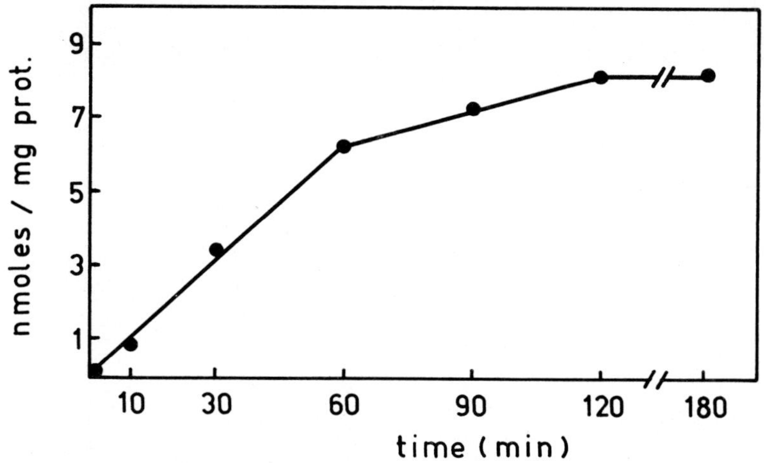

Fig. 1 - Kinetics of incorporation of ^{3}H-ATP into poly(A) by the poly(A) polymerase from rat liver mitochondria. The mitochondrial poly(A) polymerase was prepared as previously reported (14) with the following modification: the DNase treatment was omitted and the ammonium sulphate was added to a final concentration of 0.3 g/ml;after centrifugation at 80000 xg the pellet was chromatographed on phosphocellulose column. In order to measure the enzymatic activity, in a final volume of 110 μl of the incubation mixture (14), 2.4 μg of enzyme preparation were added. At each time 15 μl aliquot was assayed for acid-precipitable radioactivity.

Synthesis of poly(A)-containing RNA in isolated mitochondria

To study the synthesis of poly(A)-RNA *in vitro* we have incubated the isolated mitochondria in a medium suitable for RNA synthesis containing as labelled precursors either ^{3}H-ATP or ^{3}H-UTP and, in some cases, also the two nucleotides together. The radioactive profile of RNA newly synthesized, shown in Fig. 3, indicates that in these conditions the synthesis of two discrete species having the sedimentation coefficient of mitochondrial ribosomal RNAs, and of other species polydispersed in the region 12-4S, was taking place. In order to examine the possibility that the newly synthesized RNA contains poly(A) we have used two methods that are specific for poly(A)-RNA: the oligo dT-cellulose column and the millipore filters binding technique. As reported in Table I, when the RNA was passed through the oligo dT-cellulose column, while in terms of optical density in both cases 2-4% of the total RNA was linked to the column the radioactivity bound was 7% or 15% when ^{3}H-UTP or ^{3}H-ATP were respectively used as radioactive precursors. These data indicate that not only the chain of RNA but also the polyadenylate sequences are labelled *in vitro*. Treatment of labelled RNA with RNase confirm

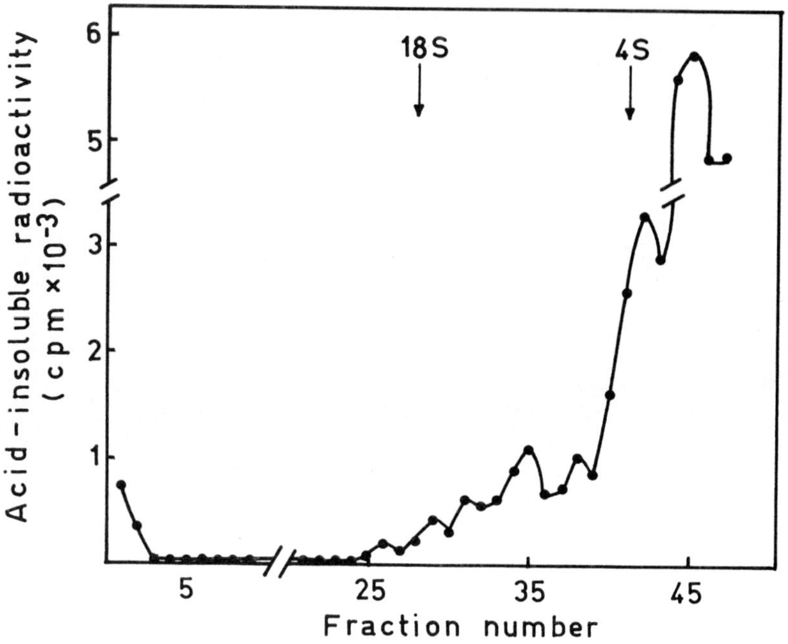

Fig. 2 - Sucrose gradient analysis of the product of the poly(A) polymerase. After 30' incubation the reaction was stopped by adding 0.5% sarkosyl and the mixture layered on 5-15% sucrose gradient. After centrifugation the gradient was fractionated and the acid insoluble radioactivity measured by adding to each fraction one volume of 10% TCA and 100 μg of bovine serum albumine.

this hypothesis. In fact, using RNA labelled with ^{3}H-ATP, RNase digestion reduced the binding by 30%, whereas in the case of ^{3}H-UTP labelled RNA, the binding disappeared completely. To characterize the poly(A)-RNA newly synthesized we used ^{3}H-UTP as radioactive precursor in order to label only the RNA chains avoiding the detection of "free" poly(A) labelled in vitro. For this purpose the poly(A)-RNA, after elution from oligo dT-cellulose column, was layered on a 15-30% sucrose density gradient as shown in Fig. 4. The radioactive profile appears to be polydispersed with some higher peaks which probably represent discrete species. The mitochondrial origin of poly(A)-RNA synthesized was determined by hybridization experiments. The total synthesized RNA, as well as the poly(A)-containing fraction specifically hybridize with mitochondrial DNA, whereas no counts were linked to nuclear or E. coli DNA. Furthermore, in order to localize these hybridization products on the physical map of mitochondrial DNA (18) we have performed in collaboration with A.Kroon

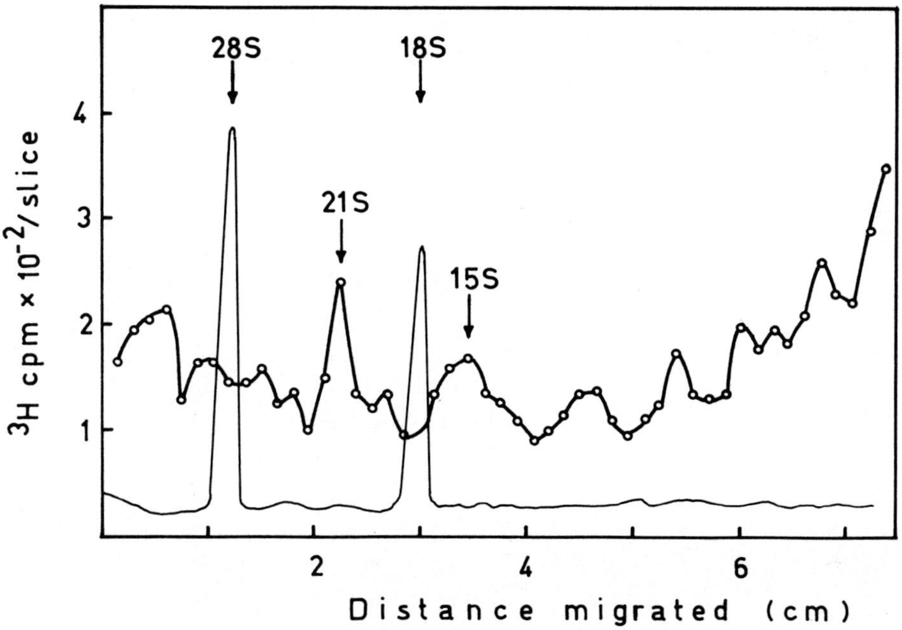

Fig. 3 - Polyacrylamide gel electrophoresis of <u>in vitro</u> synthesized RNA. The continuous line represent the scanning of a parallel gel containing RNA extracted by cytoplasmic ribosomes, used as marker.

TABLE I

BINDING OF IN VITRO SYNTHESIZED RNA TO MILLIPORE FILTERS AND OLIGO dT-CELLULOSE COLUMN.

The binding before and after the RNase treatment was carried out as described by some of us (16). The values reported in table indicate the percentage of cpm bound respect to input in the conditions used for the selective binding of poly(A)-RNA.

NUCLEOTIDE PRECURSOR	ADDITION	MILLIPORE FILTERS BINDING	OLIGO dT-CELLULOSE BINDING
^{3}H-ATP	none	12%	15%
	RNase	8%	10%
^{3}H-UTP	none	7%	7%
	RNase	1%	n.m.

and P. Terpstra of the University of Groningen, (manuscript in preparation), hybridization experiments with the isolated fragments of mitochondrial DNA obtained by digestion of intact circular DNA mole-

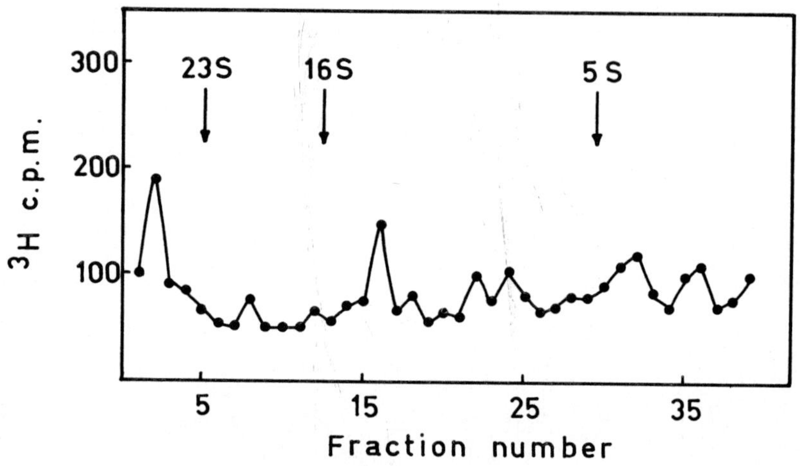

Fig. 4 - Sucrose gradient of in vitro synthesized ^{3}H poly(A)-RNA. The ^{3}H poly(A)-RNA synthesized in vitro, using ^{3}H-UTP as labelled precursor was layered on a 10-30% sucrose gradient (for experimental details see ref. 18 and the legend of Fig. 2).

cules with Eco RI or Hind III restriction enzymes. Digestion of mitochondrial DNA with these two enzymes gives origin to six fragments which can be separated by agarose gel electrophoresis. Hybridization of mitochondrial poly(A)-RNA synthesized in vitro was performed with the four larger Eco RI or Hind III fragments. The results of these experiments are reported in Fig. 5 and compared to the hybridization values obtained by annealing the poly(A)-containing mitochondrial RNA extracted from a mitochondrial preparation and labelled with ^{125}I with the same fragments (18). The results obtained with the two labelled RNAs agree fairly well showing that the poly (A)-RNA species hybridize at different degrees with almost all fragments. In the case of Eco RI fragments the lowest hybridization level was observed with the fragment D which contains the cistron for 16S rRNA (18) and which for its molecular weight, could not accommodate other genes. Hybridization with Hind III fragments on the other hand indicates very low complementarity between fragment A and poly(A)-labelled RNAs suggesting that this fragment, which contains the 12S rRNA cistron, might accommodate other genes like tRNA genes or regulatory regions.

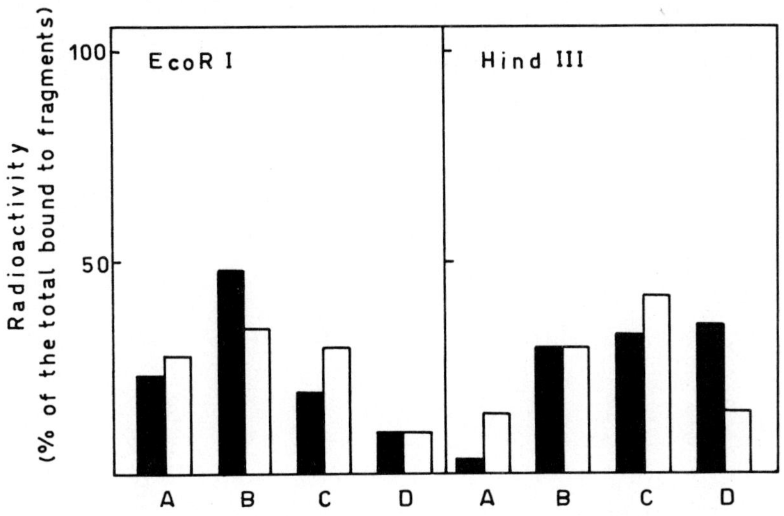

Fig. 5 - Hybridization of in vitro synthesized poly(A)-RNA and ^{125}I poly(A)-RNA with Eco RI and Hind III DNA fragments. Synthesis of poly(A)-RNA in vitro was carried out by using ^{3}H-ATP and ^{3}H-UTP as labelled precursors. DNA fragments were prepared by A.M. Kroon and P. Terpstra with the technique described by Southern (19). For the hybridization experiments of ^{125}I poly(A)-RNA and other experimental details see ref. 18. ■ Hybridization of ^{125}I poly(A)-RNA. □ Hybridization of in vitro synthesized ^{3}H poly(A)-RNA.

CONCLUSION

An enzyme able to synthesize polyadenylate molecules from ATP has been isolated and purified from mitochondria of rat liver. Its function should be the polyadenylation of mitochondrial messenger RNA species. However under our experimental conditions this enzyme is able to use as primer only an endogenous factor of oligonucleotide nature which can be extracted from mitochondria, whereas it recognizes neither poly(A) nor other RNAs of mitochondrial or extramitochondrial origin. The product synthesized in vitro by the partially purified enzyme, in these conditions, is a heterogeneous population of poly(A) molecules of different size thus suggesting that probably the enzyme in vitro has not been assayed in physiological conditions. A physiological role of this enzyme is however suggested by our experiments with isolated mitochondria which clearly indicate that poly(A)-containing RNA species are synthesized in vitro and furthermore that the polyadenylation is an intramitochondrially occurring process. Hybridization experiments with isolated fragments obtained by digestion of mitochondrial DNA with Eco RI or Hind III restriction endonucleases indicate that, in rat liver mitochondria, a population of poly(A)-RNA species whose genes are spread all over the mitochondrial genome is present.

REFERENCES

1. Saccone, C. and Quagliariello, E. (1975) International Review of Cytology 43, 125-166.
2. Kroon, A.M. and Saccone, C. (1974) The Biogenesis of Mitochondria, Academic Press, New York.
3. Avadhani, N.G., Lewis, F.S. and Rutman, R.J. (1975) Sub-Cellular Biochemistry 4, 93-145.
4. Saccone, C. and Kroon, A.M. (1976) The genetic function of mitochondrial DNA, North-Holland, Amsterdam.
5. Ojala, D. and Attardi, G. (1974) J. Mol. Biol. 82, 151-174.
6. Hirsch, M. and Penman, S. (1973) J. Mol. Biol. 80, 379-391.
7. Avadhani, N.G., Lewis, F.S. and Rutman, R.J. (1974) Biochemistry 13, 4368-4644.
8. Devlin, R. (1976) Developmental Biology 50, 443-456.
9. Rosen, D. and Edelman, M. (1976) Eur. J. Biochem. 63, 525-532.
10. Hendler, F.J., Padmanaban, G., Patzer, J., Ryan, R. and Rabinowitz, M. (1975) Nature 258, 357-359.
11. Groot, G.S.P., Flavell, R.A., Van Ommen, G.J.B. and Grivell, L.A. (1974) Nature 252, 167-169.
12. Jacob, S.T. and Schindler, D.G. (1972) Biochem. Biophys. Res. Commun. 48, 126-134.
13. Rose, K.M., Morris, H.P. and Jacob, T. (1975) Biochemistry 14, 1025-1032.
14. Gallerani, R., De Giorgi, C., De Benedetto, C. and Saccone, C. (1976) Biochem. J. in press.
15. De Giorgi, C., Gallerani, R. and Saccone, C. (1975) Boll. Soc. It. Biol. Sper. 51, 1342-1347.
16. Cantatore, P., De Giorgi, C. and Saccone, C. (1976) Biochem. Biophys. Res. Commun. 70, 431-437.
17. Landes, G.M. and Kitos, P.A. (1975) Biochem. Biophys. Res. Commun. 66, 1209-1217.
18. Saccone, C., Pepe, G., Cantatore, P., Terpstra, P. and Kroon, A.M. in The genetic function of mitochondrial DNA (Saccone, C. and Kroon, A.M. eds.) North-Holland, Amsterdam, in press.
19. Southern, E.M. (1975) J. Mol. Biol. 98, 503-517.

Genetics and Biogenesis of Chloroplasts and Mitochondria, Th. Bücher et al. eds.
© 1976 Elsevier/North-Holland Biomedical Press, Amsterdam, The Netherlands.

PROPERTIES AND PURIFICATION OF POLY(A) POLYMERASE FROM RAT LIVER MITOCHONDRIA

R. Gallerani, C. De Benedetto, C. De Giorgi and C. Saccone

Institute of Biological Chemistry, University of Bari, Bari, Italy.

Properties of mitochondrial poly(A)-polymerase.

Recently we reported the contemporaneous isolation of DNA dependent RNA polymerase and poly(A)-polymerase from rat liver mitochondria (1). To study the characteristics of mitochondrial poly(A)-polymerase we used the enzyme eluted from the phosphocellulose column prepared as reported elsewhere by Saccone et al. in this book. The modification of our original procedure (1) resulted in a larger yield of the enzyme and a purification of about a hundred times with respect to deoxycholate extract (2). The product of the reaction, sensitive to snake venom phosphodiesterase and insensitive to pancreatic ribonuclease, was characterized as poly(A) with the criteria described in ref. 1 and by Saccone et al. (this book). Table I shows the primer requirement of mitochondrial poly(A)-polymerase which is completely dependent on the addition of the endogenous factor already described by Jacob et al. (3) and unable to add poly(A) sequences to any kind of RNA or to poly(A). In Table II the substrate specificity of the enzyme either in presence or absence of other cold ribonucleoside triphosphates is reported. The inability of the enzyme to use ADP as substrate excludes the possibility of the mitochondrial poly(A)-polymerase activity being due to a coupling of ATPase and polynucleotide phosphorilase as found in another system (4).

TABLE I

SENSITIVITY TO DIFFERENT PRIMERS OF MITOCHONDRIAL POLY(A)-POLYMERASE
Mitochondrial and cytoplasmic RNAs were extracted from rat liver.

Addition	Concentrations (E_{260} Units/ml)	Incorporation (pmols/mg prot./30')
None	-	75
Endogenous Factor	1.9	3,800
Poly(A)	6.2	159
mit. RNA	5.4	116
cyt. RNA	8.2	115
Poly(A)-cyt.RNA	9.1	75

TABLE II

SUBSTRATE SPECIFICITY OF MITOCHONDRIAL POLY(A) POLYMERASE

When the three cold nucleotides were added together the concentration of each one was 0.1 mM

Labelled nucleotide	Cold nucleotide	nmols ^{3}H-substrate inc./mg prot./30'
ATP	-	10.4
ATP	GTP 0.1 mM	7.7
ATP	CTP 0.1 mM	9.5
ATP	UTP 0.1 mM	10.4
ATP	GTP+CTP+UTP	8.9
UTP	-	0
GTP	-	0
CTP	-	0
ADP	-	1.1

The K_m of the reaction studied in our conditions was determined by using the double reciprocal plot of Lineweaver and Burk and a value of about 90 µM could be calculated (Fig. 1).

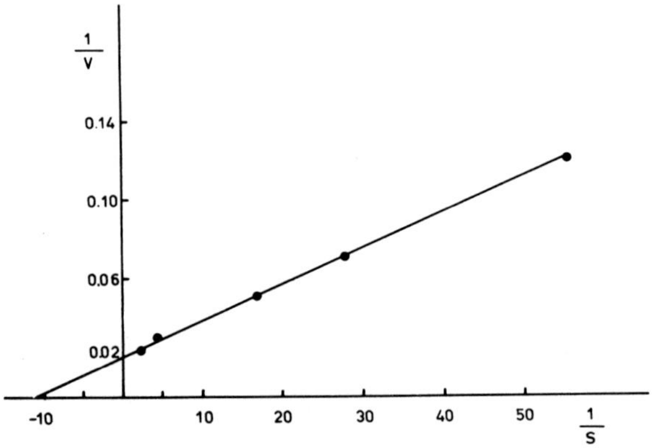

Fig. 1 - Double reciprocal plot of 1/V vs. 1/S for mitochondrial poly(A)-polymerase. Reaction mixtures contained 1.7 µg of protein purified through phosphocellulose column. V values are reported as pmols incorporated/30' incubation at 30°C; S values are expressed as ATP mM.

Purification of mitochondrial poly(A)-polymerase.

The enzyme was further purified after the phosphocellulose step by affinity chromatography. We used ATP covalently linked to an agarose gel through an hexane molecule (AHATP column). The elution pattern of this column is reported in Fig. 2. Poly(A)-polymerase was detected, in the wash fraction at very low specific activity, and in the two peaks shown in the Fig. 2. In the second peak the specific activity of poly(A)-polymerase was 27.9 nmols/mg protein. This procedure therefore increased the purification to about 350

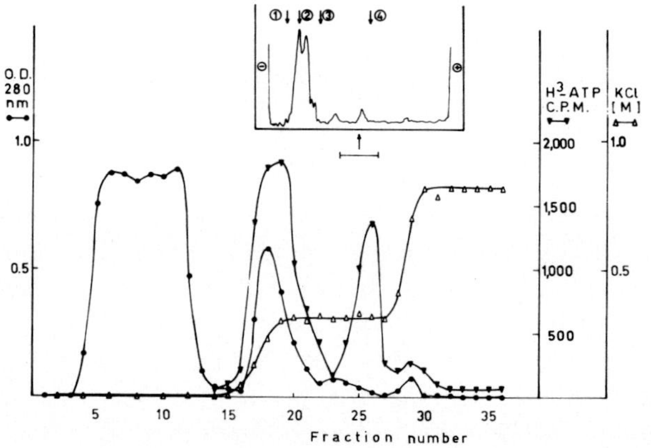

Fig. 2 - Elution pattern of agarose-hexane-ATP column. The active fractions of phosphocellulose column were pooled and the proteins precipitated by adding solid ammonium sulphate (0.3 gr/ml). The salt was removed and protein absorbed on AHATP column equilibrated with buffer B: Tris-HCl pH 7.4 25 mM, $MgCl_2$ 5 mM, DTT 0.1 mM glicerol 10% v/v. After absorbtion the column was washed with buffer B and eluted with the same buffer containing 0.3 M and 0.8 M KCl. Each fraction contained 0.4 ml, flow rate was 0.1 ml/min. Standard proteins used for SDS gel electrophoresis: 1) Bovine serum albumin, 2) Glutamate dehydrogenase, 3) Lactate dehydrogenase, 4) Ribonuclease.

times as shown in Table III. The fraction obtained from AHATP column when analyzed by SDS gel electrophoresis displayed two polypeptide chains having a molecular weight of 53000 and 50000 daltons (insert in fig. 2). The two polypeptides seem to be present in equimolar amount, but we were unable to establish whether they are both necessary for the enzyme activity.

TABLE III

PURIFICATION OF POLY(A)-POLYMERASE FROM RAT LIVER MITOCHONDRIA.

For technical details see ref. 1 and 2. The specific activity is expressed as pmols of ^{3}H-ATP incorporated in acid insoluble material /30' incubation/30°C/mg of protein. Protein concentration was measur ed according to Waddel (5).

Purification steps	Volume (ml)	Protein (mg)	Total Units	Specific Activity
DOC extract	15	154	12,936	84
Pellet after A.S. fract.(0.3 gr/ml)	10	85	108,800	1,280
Phosphocellulose P.11 peak	15.5	6.4	26,350	4,134
Agarose-hexane-ATP (0.3 M KCl 2nd peak)	1.2	0.14	3,953	27,946

The properties of poly(A)-polymerase purified from rat liver according to our procedure seem to be rather different from those reported by Rose et al. for the enzyme of Morris hepatoma (6). The major discrepancies concern the molecular weight of the enzyme which in Morris hepatoma is 60000 daltons and the ability of this enzyme to recognize poly(A) as primer. The physiological role of this enzyme is discussed in this book by Saccone et al.

REFERENCES

1. Gallerani, R., De Giorgi, C., De Benedetto, C. and Saccone, C. (1976) Biochemical J. in press.
2. De Benedetto, C., Gallerani, R., De Giorgi, C. and Saccone, C. (1976) Boll. Soc. It. Biol. Sper., in press.
3. Jacob, S.T., Rose, K.M. and Morris, H.P. (1974) Biochem. Biophys. Acta 361, 312-320.
4. Wuhderli, W., Hütter, R., Staehelin, M. and Wehrli, W. (1975) Eur. J. Biochem. 58, 87-94.
5. Waddel, W.J. (1956) J. Lab. Clin. Med. 48, 311-314.
6. Rose, K.M., Morris, H.P. and Jacob, T. (1975) Biochemistry 14, 1025-1032.

Genetics and Biogenesis of Chloroplasts and Mitochondria, Th. Bücher et al. eds.

THE POKY MUTANT OF NEUROSPORA CRASSA

Alan M. Lambowitz
Department of Biochemistry
Saint Louis University School
of Medicine
Saint Louis, Missouri 63104

The poky (mi-1) mutant of Neurospora crassa is one of the earliest examples of cytoplasmic inheritance in a eukaryote. The mutant was isolated by Mitchell and Mitchell in 1952 as a slow growing varient characterized by cyanide-resistant respiration and deficiencies in several cytochromes (1-3). These phenotypic changes are usually assumed to result from a mutation in mit DNA since, like mit DNA, they show maternal inheritance and can be transmitted to wild type by micro-injection of purified mitochondria (1,4,5).

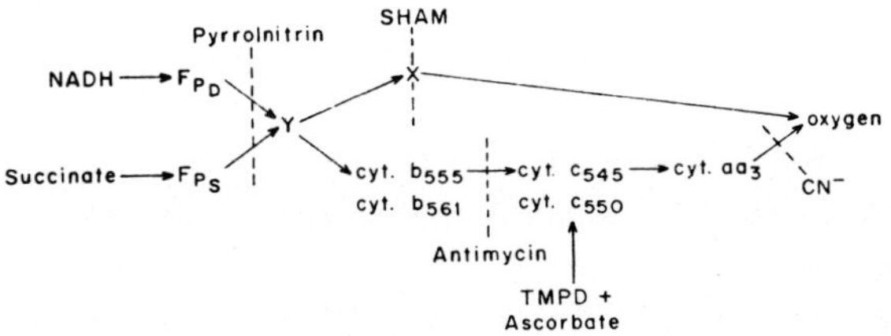

Fig. 1. Branched respiratory system of poky. X designates the cyanide-resistant alternate oxidase and Y designates the branch point. For experimental details see refs. 6-10.

Spectrophotometric studies by Lambowitz et al. (6-10; see also ref. 11) show that poky mitochondria respire via two terminal oxidase systems: one, the residual cytochrome system, deficient in cytochromes b-561, b-566, c_1 and aa_3; and the other, a cyanide-resistant, hydroxamic acid-sensitive alternate oxidase which carries 2/3 of the respiratory flux and which shunts electrons directly from the flavoprotein region to oxygen (Fig. 1). Growth in poky is dependent in part on the residual cytochrome system with oxidative phosphorylation at sites II and III and, in part on the alternate oxidase which does not itself carry out oxidative phosphorylation, but which accelerates substrate level phosphorylations in glycolysis and the Krebs cycle (8,9). (In principle, the alternate oxidase could also accelerate oxidative phosphorylation at site I, but under our growth conditions, site I is only marginally present in Neurospora (ref. 9; contrast ref. 11)). Supplementation by the alternate oxidase is presumably essential since the maximum activity of the poky cytochrome system is only 1/4 that of wild type (8). The alternate oxidase can be induced in wild type by blocking the cytochrome system (e.g., with antimycin) or by inhibiting mit protein synthesis (6). These results suggest that its presence in poky must be a secondary effect of the mutation.

The status of cytochrome *b* in *poky* has been of particular interest. Isolated cytochrome *b* from yeast and *Neurospora* has a single absorption maximum *ca*. 560 nm (25° C) and consists of one or two intramitochondrially synthesized polypeptides (12,13). In whole mitochondria, however, three *b*-type cytochromes can be distinguished by absorption maxima and midpoint potential (E_m)(ref. 10; Table 1).

Table 1. The *b*-cytochromes of *Neurospora* mitochondria. Numbers in parentheses indicate the proportion of total absorbance change at 559 *minus* 540 nm. See ref. 10 for experimental details.

component	α-peak 25°	E_m/wild type	E_m/poky
b-561	561 nm	78 mV (51%)	88 mV (33%)
b-566	566 nm	-73 mV (19%)	-50 mV (12%)
dithionite-reducible *b*	559 nm	-166 mV (30%)	-144 mV (55%)

Cytochromes *b*-561 and *b*-566 are present in a 1:1 ratio in both wild-type and *poky* mitochondria (10). However, the proportion of dithionite-reducible cytochrome *b* appears to be increased two-fold in *poky* (10). In view of its low E_m, it is conceivable that the dithionite-reducible *b* is a nascent or damaged species related to *b*-561 and/or *b*-566. In animal mitochondria, *b*-562 and b-566 are converted to low E_m forms which resemble dithionite-reducible *b* by treatments which disrupt mitochondrial membranes (14,15).

All of the studies on *poky*'s respiratory system have failed to reveal any striking functional alteration in the cytochrome components. As shown in Table 1, the E_m's of all 3 *b*-cytochromes agree within 10 to 20 mV between wild-type and *poky* mitochondria. Similarly, all of the *poky* cytochromes appear functionally normal as judged by spectral properties, inhibitor sensitivity and ability to carry out oxidative phosphorylation (6-11). (The absorption maximum of cytochrome aa_3 may be shifted in *poky*; ref. 7.) For this reason, attempts to elucidate the primary defect in *poky* have focused on mit ribosome assembly with the focus stemming primarily from 3 previous findings: (a) that isolated *poky* mitochondria have a diminished capacity for protein synthesis (16), (b) that chloramphenicol, a specific inhibitor of mit protein synthesis, produces *poky* phenocopies in wild-type *Neurospora* (17,6,18), and (c) that *poky* mitochondria are deficient in small ribosomal subunits so that the mass ratio of small to large subunits is 1:10 rather than 1:2 as in wild type (19,20). More recent insight comes from studies on mit rRNA processing. The 19 S and 25 S rRNAs in *Neurospora* are transcribed in tandem as a 32 S precursor molecule which is subsequently cleaved and trimmed to yield the mature species (21). 32 S RNA is synthesized in *poky* mitochondria, but the 19 S portion is rapidly degraded (22).

At the outset, abnormal processing of 32 S RNA could be accounted for by a defect in a mit rRNA or in a mit ribosomal protein. The first possibility was tested by fingerprint analysis, illustrated by the homochromatographic fingerprints in

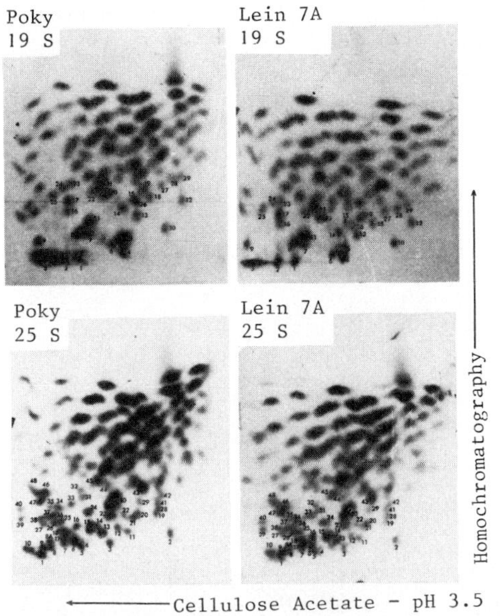

Fig. 2. Homochromatographic fingerprints of RNase T_1 digests of mit rRNAs from poky and wild-type Lein 7A. Lein 7A is the wild type from which poky was derived and the two strains presumably have closely isogenic mit DNA. See ref. 24 for details.

Fig. 2. Overall, 19 S and 25 S rRNAs from poky were compared with those from 3 different wild-type strains. To optimize resolution of different sets of sequences, the comparison employed 3 complementary fingerprinting systems--two-dimensional electrophoresis of T_1 plus phosphatase digests, homochromatography of pancreatic RNase digests, and homochromatography of RNase T_1 digests. In supplementary experiments, ^{32}P-labelled wild-type RNAs were cofingerprinted with ^{33}P-labelled poky to detect possible differences in the stoichiometries of oligonucleotides and, in other experiments, levels and patterns of methylated nucleotides were also compared. The result was that no alteration was found in either 19 S or 25 S RNA in poky. The data do not completely exclude certain types of mutations (mainly single base changes in large oligonucleotides) or mutations in unconserved regions of 32 S RNA. They do, however, strongly suggest that the primary defect in poky will not be found in the mit rRNAs (see refs. 23,24 for details).

In another light, the RNA studies showed that methylation of the *Neurospora* mit rRNAs is extremely sparse, apparently consisting of only 2 methylated ribose moieties on 25 S RNA and with 19 S RNA completely unmethylated (23,24). Low levels of methylation appear to be a general feature of mit rRNAs (25-27) to be contrasted with cytosolic rRNAs which always contain 1 to 2% methylated nucleotides (28). Methylation of cytosolic rRNAs has been implicated in RNA processing in eukaryotes (28) and in protein synthesis in prokaryotes (29). However, the results for *Neurospora* mitochondria suggest that methylation is not absolutely required for either ribosome processing or function since (a) there are too few methyl groups to direct processing of 32 S RNA and (b) all small subunit functions are carried out in the absence of methylation of 19 S RNA. Instead, the results are consistent with the idea that methylation of cytosolic rRNAs is required to facilitate correct RNA conformations which can be attained without methylation by the mit rRNAs.

As a tool for further analysis of poky and other mutants with defective mit ribosome assembly, procedures were developed for isolating mit ribosomal precursor

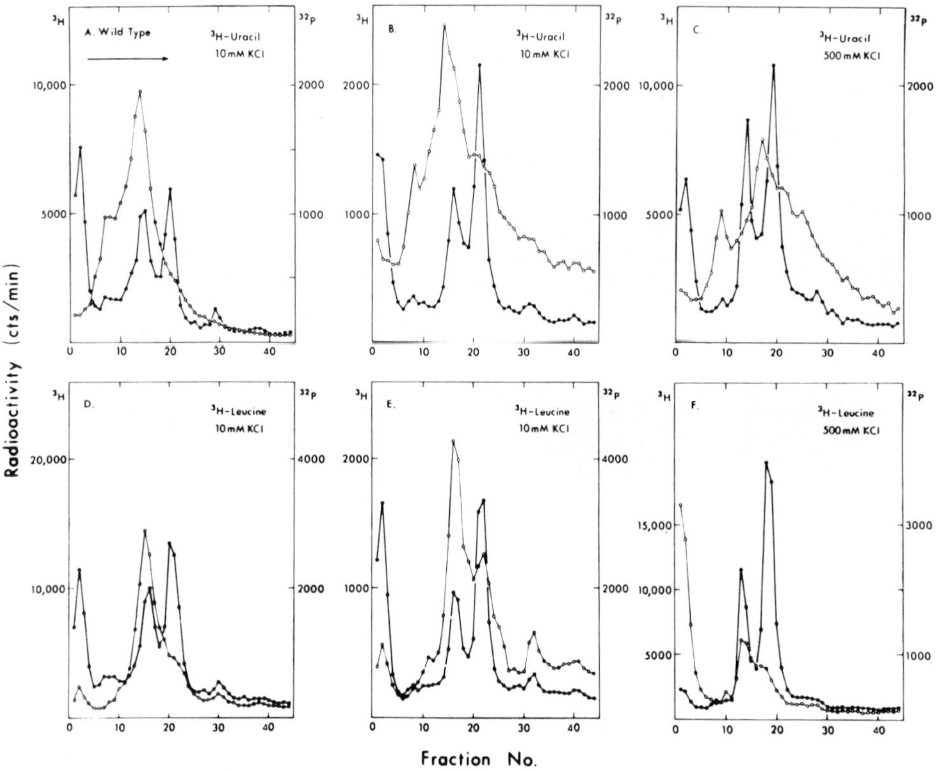

Fig. 3. Sucrose gradient analysis of mit ribosomal precursor particles. See ref. 30 for experimental details.

particles (30). At the outset, it seemed likely that the particles would be a minor species consisting of 32 S RNA and newly synthesized ribosomal proteins and that their identification would require specific labelling with short pulses of ^{3}H-uracil or ^{3}H-leucine followed by analysis on sucrose gradients (Fig. 3). The best results were obtained with gradients containing low salt buffer (10 mM KCl, 0.8 mM $MgCl_2$, 10 mM Tris·HCl, pH 7.5; Fig. 3A,B,D,E); the ^{32}P mass label (closed circles) traces the profiles of mature 30 S and 50 S subunits in all of the gradients. The pulse-label (open circles in Fig. 3A,D) show mainly the precursor particles which sediment slightly behind the small subunits. The profiles of newly synthesized RNA and protein (Fig. 3A and D, respectively) are roughly coincident, although the ^{3}H-uracil peak does show a shoulder at ca. 20 S which may be mRNA. The relatively slow sedimentation rate of the precursor particles suggests an extended conformation, a characteristic shared with ribosomal precursor particles from animal cell nucleoli (31). In other experiments, we confirmed that nearly all of the ^{3}H-uracil label in the precursor particle peak could be recovered as 32 S RNA (see ref. 30).

Sedimentation of the mit ribosomal precursor particles is extremely sensitive

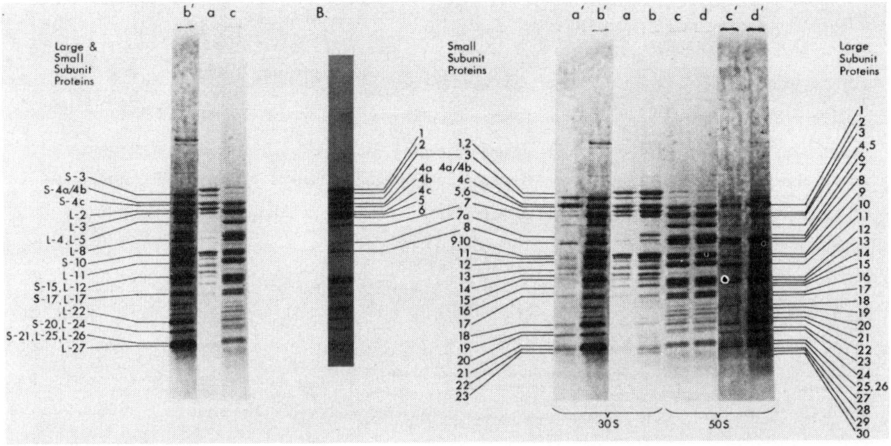

Fig. 4. Electrophoretic analysis of mit ribosomal and precursor particle proteins. Cells were mass-labelled with L-(4,5-^{3}H)lysine and pulse-labelled with L-(^{35}S)methionine. Slots a to d show proteins stained with Coomassie Blue and slots a' to d' show corresponding autoradiograms: slot a, 30 S fraction, "high salt" gradients; slot b, 30 S fraction, "low salt" gradients; slot c, 50 S fraction, "high salt" gradients; slot d, 50 S fraction, "low salt" gradients. Inset shows gel overrun to resolve S-4a, S-4b and S-4c. For experimental details see ref. 30.

to salt concentration in the gradient. At 50 mM KCl (5 to 10 mM $MgCl_2$), the particles sediment between the subunits (data not shown). A further increase in ionic strength results in destruction of the particles. At 500 mM KCl (25 mM $MgCl_2$), the ^{3}H-uracil label shows substantial aggregation (Fig. 3C) and the pulse-labelled protein is dissociated and appears in the load zone (Fig. 3F). The distinction, that proteins associated with mature ribosomal subunits do not dissociate in 500 mM KCl, was used in pulse-chase experiments. The salt-dissociable precursor particle protein was found to be "chased" into salt stable form in the time period required for processing of 32 S RNA (see ref. 30).

Fig. 4 shows mit ribosomal and precursor particle proteins separated on SDS-polyacrylamide gradient gels. The gel system resolves 23 small subunit proteins and 30 large subunit proteins (Fig. 4B, slots a and c, respectively) with the range of MWs from 10,000 to 60,000. Slot b' is an autoradiogram of precursor particle proteins pulse-labelled with ^{35}S-methionine. The proteins are seen to include both small and large subunit proteins as expected since 32 S RNA contains sequences for both 19 S and 25 S RNA. The comparison is most easily made in Fig. 4A where appropriate slots have been aligned. Specific activity measurements show that newly synthesized large subunit proteins appear initially in the precursor particle fraction (i.e., sedimenting *ca.* 30 S) before incorporation into mature subunits. Considered together, the results strongly suggest that we are dealing with *bona fide* precursor particles.

The ratios of pulse- to mass-label in proteins in mature subunits should give information about pool sizes and possible control points in mit ribosome assembly. By comparing specific activity (and with several different labelled amino acids), 9 small subunit proteins were identified which have exceptionally small free pools and which may thus play a key role in regulating the rate of mit ribosome assembly. 6 of the proteins (S-4c, S-17, S-18, S-19, S-21 and S-22) appear to be present in both precursor particles and mature small subunits whereas the remaining 3 proteins (S-3, S-4a and S-23) appear to be added later and were found only in mature small subunits. The high specific activities of S-3 and S-4 are qualitatively apparent from the autoradiogram slot a' of Fig. 4B.

To elucidate the role of the mitochondria in mit ribosome assembly, experiments were carried out to determine the site of synthesis of the mit ribosomal proteins. In the first experiment, wild-type cells were mass-labelled with ^{14}C-leucine, then treated with anisomycin (10 mM) to inhibit cytoplasmic protein synthesis, and pulse-labelled with ^{3}H-leucine. In this protocol, a protein synthesized inside the mitochondria would have a high ratio of $^{3}H/^{14}C$ radioactivity. As shown in Fig. 5, nearly all of the ^{3}H-leucine is incorporated into one band S-4, apparent MW 52,000. By overrunning the gels, S-4 is found to consist of 3 components (see inset Fig. 4) and all of the anisomycin-resistant incorporation is into one component, S-4a, already identified as one of the "special" proteins above. In complementary experiments, incorporation of ^{3}H-leucine into S-4a was specifically inhibited by chloramphenicol, thereby confirming intramitochondrial synthesis. Although the results do not establish a functional role for S-4a, several criteria do suggest that it is a "true" ribosomal protein: (a) it is specific for the small subunit and there is no evidence for association with the large subunit as might be expected for a "sticky" contaminant, (b) it remains associated with the small subunit after sedimentation through sucrose gradients containing 500 mM KCl and (c) with the gels extensively overrun to eliminate overlaps, S-4a has an apparent stoichiometry of ca. 1 molecule per ribosome (as judged by amount of mass-label radioactivity).

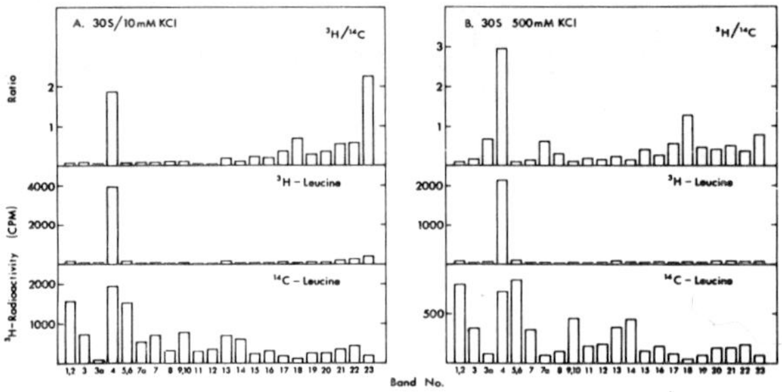

Fig. 5. Incorporation of ^{3}H-leucine into mit ribosomal proteins in the presence of anisomycin. Proteins were separated on gradient gels as in Fig 4. A and B show proteins from low and high salt sucrose gradients, respectively. None of the large subunit proteins showed significant incorporation. See ref. 30 for further details.

In the final experiments, the protein compositions of wild-type and poky precursor particles and mature ribosomal subunits were compared by SDS gel electrophoresis. The most striking result was that poky small subunits appear to be deficient in S-4a and in at least 2 other proteins, S-3 and S-8, with concentrations only ca. 50% those of wild type. By contrast, the stoichiometries of the large subunit proteins were consistently similar in wild type and poky (30).

The possibility that S-4a is a "true" ribosomal protein should be considered in light of previous studies. Lizardi and Luck (32) found a similar protein associated with *Neurospora* mit small ribosomal subunits, but not reproducibly. The consistency with which S-4a is now found is considered to reflect new methods of mit ribosome isolation which preserve the integrity of the mit rRNAs (24). In yeast, Davey et al. (33; see also ref. 34) reported that long term growth in the presence of chloramphenicol does not substantially decrease the rate of mit protein synthesis assayed *in vitro*. This finding has been considered evidence that all protein components required for mit protein synthesis are synthesized on cytosolic ribosomes. In fact, however, such experiments do not rule out intramitochondrial synthesis of a mit ribosomal protein for at least 2 reasons: (a) it is not clear how the rate of mit protein synthesis *in vitro* is related to that *in vivo* and (b) no data were presented on the rate of mit ribosome assembly during growth in the presence of the inhibitor. Recently, Groot (34) reported that a mitochondrially synthesized protein may be associated with the yeast mit small subunit, although the stoichiometry of the protein was not determined and it was not considered a ribosomal constituent. In general, intramitochondrial synthesis of a mit ribosomal protein would provide a means whereby the rate of mit ribosome assembly could be controlled by the rate of mit protein synthesis.

Our working hypothesis is that the primary defect in *poky* is in S-4a. Although we have no direct evidence for this hypothesis, a primary defect in S-4a could account for the phenotypic changes in *poky* with 3 assumptions: (a) that binding of mutant S-4a is rate-limiting in mit ribosome assembly in *poky*, (b) that failure to bind S-4a or binding of mutant S-4a leads to degradation of 19 S RNA and (c) that binding of mutant S-4a can affect binding of several other proteins, perhaps through a general effect on ribosome conformation. In terms of experimental precedents, the model is supported by the studies on reconstitution of *E. coli* ribosomes which show that omission of "key" assembly proteins results in particles having a drastically altered conformation and/or a deficiency of several other proteins (35). In addition, amino acid starvation experiments with HeLa cells suggest that the rate of ribosome assembly may be regulated by the rate of synthesis of some ribosomal proteins (36). Also, in HeLa cells, starvation for particular single amino acids can lead to abortive synthesis of one or the other ribosomal subunits, as is observed in *poky*, presumably by depleting pools for specific small or large subunit proteins (36). We are currently trying to obtain direct evidence that S-4a is the site of the primary defect in *poky*.

Acknowledgement--The author would like to acknowledge the support and collaboration of Drs. Carolyn W. Slayman, Walter D. Bonner, Jr. and David J.L. Luck in whose laboratories various portions of the work were carried out.

References

1...Mitchell, M.B. and Mitchell, H.K. (1952) Proc. Nat. Acad. Sci. U.S.A. 38, 442-449.

2...Mitchell, M.B., Mitchell, H.K. and Tissieres, A. (1953) Proc. Nat. Acad. Sci. U.S.A. 39, 606-613.

3...Tissieres, A., Mitchell, H.K. and Haskins, F.A. (1953) J. Biol. Chem. 205, 423-433.

4...Reich, E. and Luck, D.J.L. (1966) Proc. Nat. Acad. Sci. U.S.A. 55, 1600-1608.

5...Slott, E.F., Wilson, J.F. and Bates, W.K. (1975) Genetics 80 (suppl.), 76.

6...Lambowitz, A.M. and Slayman, C.W. (1971) J. Bacteriol. 108, 1087-1096.

7...Lambowitz, A.M., Slayman, C.W., Slayman, C.L. and Bonner, W.D., Jr. (1972) J. Biol. Chem. 247, 1536-1545.

8...Lambowitz, A.M., Smith, E.W. and Slayman, C.W. (1972) J. Biol. Chem. 247, 4850-4858.

9...Lambowitz, A.M., Smith, E.W. and Slayman, C.W. (1972) J. Biol. Chem. 247, 4859-4865.

10..Lambowitz, A.M. and Bonner, W.D., Jr. (1974) J. Biol. Chem. 249, 2886-2890.

11..von Jagow, G., Weiss, H., and Klingenberg, M. (1973) Eur. J. Biochem. 33, 140-157.

12..Weiss, H. and Ziganke, B. (1974) Eur. J. Biochem. 41, 63-71.

13..Katan, M.B., Pool, L. and Groot, G.S.P. (1976) Eur. J. Biochem. 65, 95-105.

14..Goldberger, R., Smith, A.L., Tisdale, H., and Bomstein, R. (1961) J. Biol. Chem. 236, 2788-2793.

15..Erecinska, M., Oshino, R., Oshino, N., and Chance, B. (1973) Arch. Biochem. Biophys. 157, 431-445.

16..Sebald, W., Bucher, T., Olbrich, B. and Kaudewitz, F. (1968) FEBS Lett. 1, 235-240.

17..Rifkin, M.R. (1969) Ph.D. Thesis, The Rockefeller University, New York, N.Y.

18..von Jagow, G. and Klingenberg, M. (1972) FEBS Lett. 24, 278-282.

19..Rifkin, M.R. and Luck, D.J.L. (1971) Proc. Nat. Acad. Sci. U.S.A. 68, 287-290.

20..Neupert, W., Massinger, P., and Pfaller, P. (1971) in Autonomy and Biogenesis of Mitochondria and Chloroplasts, pp. 328-338, North Holland Publishing Co., Amsterdam.

21..Kuriyama, Y. and Luck, D.J.L. (1973) J. Mol. Biol. 73, 425-437.

22..Kuriyama, Y. and Luck, D.J.L. (1974) J. Mol. Biol. 83, 253-266.

23..Lambowitz, A.M. and Luck, D.J.L. (1975) J. Mol. Biol. 96, 207-214.

24..Lambowitz, A.M. and Luck, D.J.L. (1976) J. Biol. Chem. 251, 3081-3095.

25..Vesco, C. and Penman, S. (1969) Proc. Nat. Acad. Sci. U.S.A. 62, 218-225.

26..Dubin, D.T. (1974) J. Mol. Biol. 84, 257-273.

27..Klootwijk, J., Klein, I. and Grivell, L.A. (1975) J. Mol. Biol. 97, 337-350.

28..Maden, B.E.H., Salim, M. and Robertson, J.S. (1974) in Ribosomes, pp. 829-839, Cold Spring Harbor Laboratory, New York.

29..Schlessinger, D. (1974) in Ribosomes, pp. 393-416, Cold Spring Harbor Laboratory, New York.

30..Lambowitz, A.M., Chua, N.-H. and Luck, D.J.L., J. Mol. Biol., in press.

31..Warner, J.R. (1974) in Ribosomes, pp. 461-488, Cold Spring Harbor Laboratory, New York.

32..Lizardi, P.M. and Luck, D.J.L. (1972) J. Cell Biol. 54, 56-74.

33..Davey, P.J., Yu, R. and Linnane, A.W. (1969) Biochem. Biophys. Res. Commun. 36, 30-34.

34..Groot, G.S.P. (1974) in Biogenesis of Mitochondria, pp. 443-452, Academic Press, London.

35..Nomura, M. (1973) Science 179, 864-873.

36..Maden, B.E.H. (1971) Prog. Biophys. Mol. Biol. 22, 127-177.

Genetics and Biogenesis of Chloroplasts and Mitochondria, Th. Bücher et al. eds.

THE PROTEINS OF *NEUROSPORA CRASSA* MITOCHONDRIAL AND CYTOPLASMIC RIBOSOMES

Hans de Vries and Coby van den Bogert
Lab. of Physiological Chemistry, Bloemsingel 10, Groningen, The Netherlands

INTRODUCTION

Mitochondrial ribosomes from *Neurospora crassa* are 80-S [1] when the medium for homogenization is SAMT (0.44 M sucrose, 100mM NH_4Cl, 10mM $MgCl_2$, 10mM Tris.HCl pH7.6), mitochondria are washed either 3 times in SAMT or 2 times in S-ET (0.44 M sucrose, 0.1mM EDTA, 10mM Tris.HCl pH7.6) and lysis is in AMTH (100mM NH_4Cl, 10mM $MgCl_2$, 10mM Tris.HCl pH7.6, 0.05% heparin). When the mycelium is homogenized and mitochondria are washed in S-ET, and lysis is in AMT, the ribosomes sediment at 73S [1,2]. Several data obtained in our laboratory [1,3] strongly support the conclusion that the 80-S ribosomes are not contaminating cytoplasmic ribosomes -which sediment at 77S- and that the 73-S particle is a modification product of 80-S. To compare further the two types of mitochondrial ribosomes with each other and with cytoplasmic ribosomes, we have analyzed the ribosomal proteins of their subunits by two-dimensional polyacrylamide gel electrophoresis.

MATERIALS AND METHODS

Isolation of the mitochondrial and cytoplasmic ribosomes was as described in ref. 4. Mitochondrial ribosomes were washed with $A_{500}MT$ (0.5 M NH_4Cl, 10mM $MgCl_2$, 10mM Tris.HCl pH7.6), and then dissociated in AM_1TH (100mM NH_4Cl, 1mM $MgCl_2$, 10mM Tris.HCl pH7.6, 0.05% heparin). Cytoplasmic ribosomes were not washed in $A_{500}MT$ -this treatment caused complete degradation of the large subunit [4]- but were dissociated at once in 100mM NH_4Cl, 0.1mM EDTA, 10mM Tris.HCl pH7.6, 0.05% heparin).

Subunits were isolated from 15-30% isokinetic sucrose gradients in the dissociation buffers by adding polyethylene glycol (10% w/v) to the appropriate fractions, after adjusting the $MgCl_2$ concentration to 25mM (mitochondrial) or 10mM (cytoplasmic). After 20h at 0^oC the subunits were spun down at 36000xg for 10 min. The extraction of ribosomal proteins and two-dimensional electrophoresis [5] were as described in ref. 4.

RESULTS

80-S and 73-S mitochondrial ribosomes, when washed previously with $A_{500}MT$, dissociate completely in 1mM $MgCl_2$; 80-S ribosomes yield a 55-S large subunit, whereas 73-S ribosomes give a 50-S particle. The small subunit sediments at 40S in both cases. Cytoplasmic 77-S ribosomes dissociate only partially at this $MgCl_2$ concentration and give 60-S and 40-S subunits.

The ribosomal protein patterns of the various subunits are shown in Figs. 1, 2 and 3. It is evident that the subunit proteins of 80-S and 73-S mitochondrial ribosomes are very similar to each other, but not identical; the cytoplasmic ribosomal protein patterns are completely different from the mitochondrial patterns. Hence, it is not

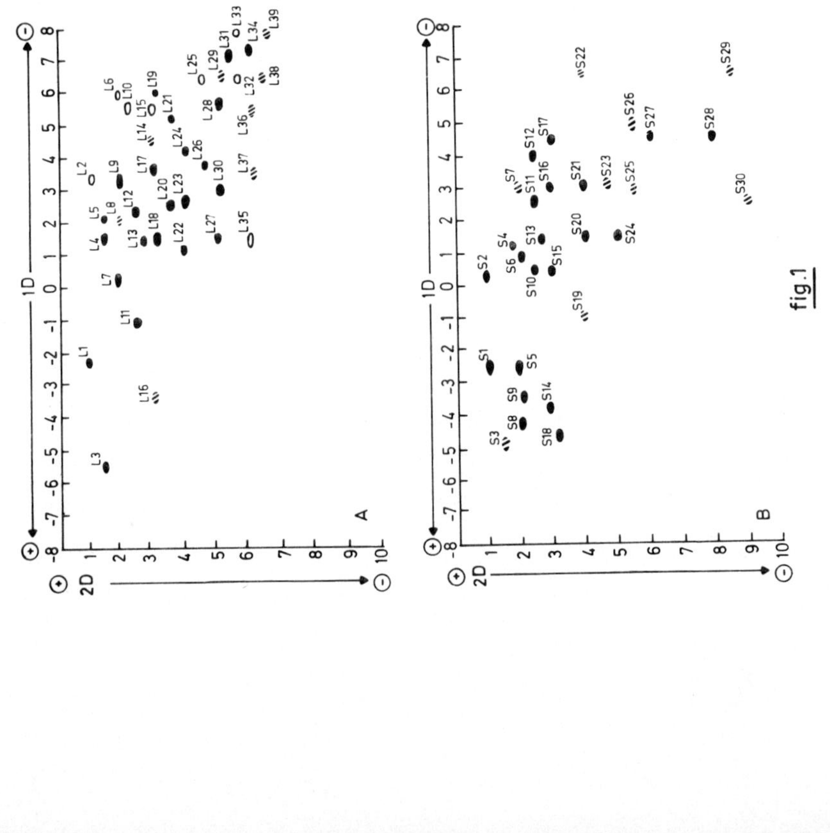

fig.1

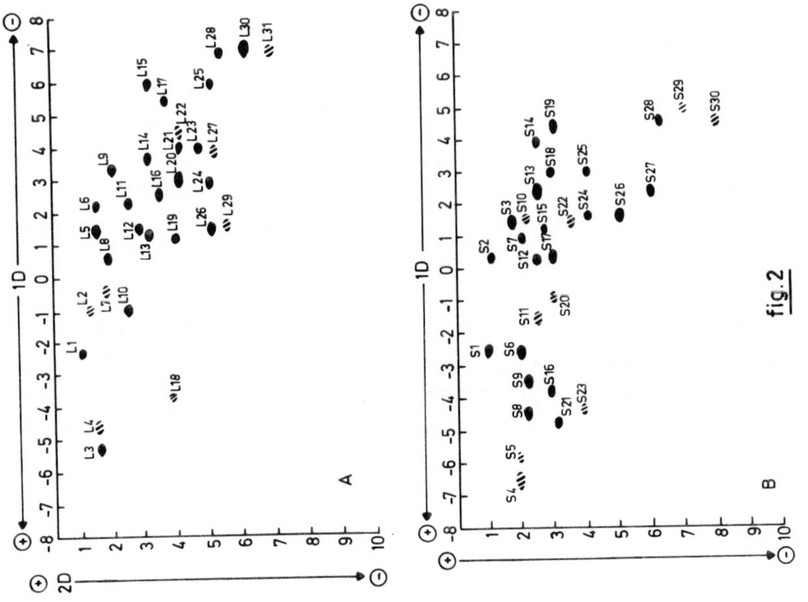

fig.2

Fig. 1. Electropherograms of the proteins from subunits of 80-S mitochondrial ribosomes. A. Large subunit. B. Small subunit. Closed spots are present in subunits of 80-S and of 73-S ribosomes; cross-hatched spots are present only in subunits of 80-S ribosomes; open spots are present only in subunits of A_{500}MT-washed 80-S ribosomes.

Fig. 2. Electropherograms of the proteins from A_{500}MT-washed 73-S mitochondrial ribosomes. A. Large subunit. B. Small subunit. Closed spots are present in subunits of 73-S and 80-S ribosomes; cross-hatched spots are present only in subunits of 73-S ribosomes.

Fig. 3. Electropherograms of the proteins from subunits of 77-S cytoplasmic ribosomes. A. Large subunit. B. Small subunit.

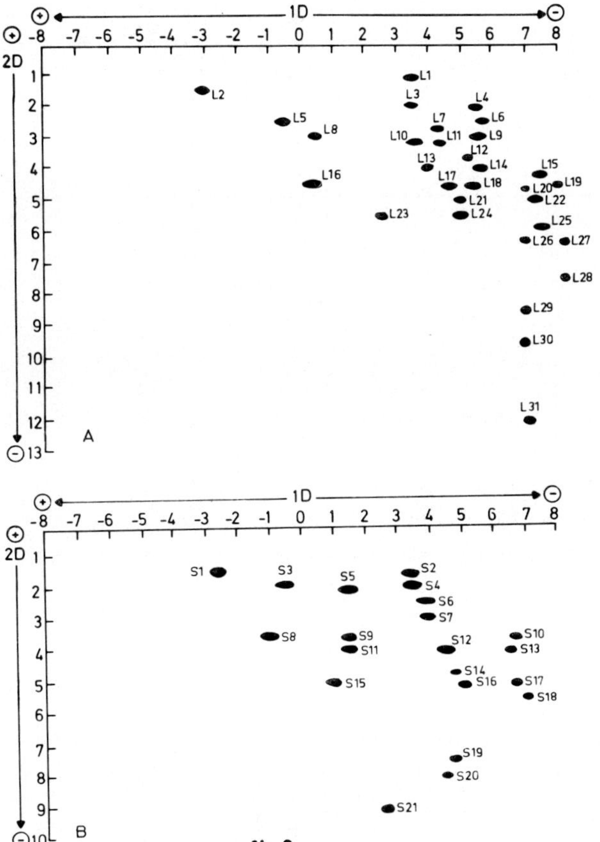

fig.3

possible that the mitochondrial ribosomes are contaminated for more than a few percent by cytoplasmic ribosomes.

If we compare the patterns of 80-S and 73-S ribosomes, it is clear that most of the spots are identical:for the small subunits 21 spots,for the large subunits 23 spots coincide.The non-coincidence of the remaining spots might be caused by some specific modification of the 80-S proteins.Moreover,it is apparent that the large subunit of 80-S contains 8 more proteins than the large subunit of 73-S.In this context it should be mentioned that unwashed 80-S ribosomes yield 50-S instead of 55-S large subunits.This 50-S subunit contains the same proteins as the 55-S subunit except that it lacks the 8 proteins drawn as open spots in Fig.1A.The similarities and differences between the 3 types of ribosomes described here are summarized in Table I

TABLE I

TOTAL NUMBER AND NUMBER OF COINCIDENT SPOTS IN ELECTROPHEROGRAMS

type of ribosome	total	coincident with		
		77-S cyto	73-S mito	80-S mito
80-S mito	69	13	44	-
73-S mito	61	8	-	
77-S cyto	52	-		

These results together indicate that the 55-S subunit of 80-S mitochondrial ribosomes can be converted to a 50-S particle,with the concomitant loss of 8 proteins. This transition is prevented by the high salt washing,the latter treatment possibly removes (a) proteolytic enzyme(s).A conversion of 55-S to 50-S particles indeed has been found *in vitro* [6].Furthermore,homogenization of the mycelium in S-ET (giving rise to 73-S ribosomes) also results in modification of the subunits:in the 40-S small subunit 9 proteins are modified,in the 50-S large subunit the 8 proteins mentioned above are again missing,and an additional 8 proteins are modified.It seems that the decrease in sedimentation coefficients (80S→73S,and 55S→50S) is either caused or accompanied by the splitting off of 8 proteins and,as is shown elsewhere in this volume [6],of a 5-S RNA molecule.The most probable explanation is that the low $MgCl_2$ concentration induced by homogenization in S-ET causes the release of a 5-S RNA protein complex from the 55-S large subunit.

REFERENCES

1.Datema,R.,Agsteribbe,E. and Kroon,A.M.(1974) Biochim.Biophys.Acta 335, 368-395

2.Lizardi,P.M. and Luck,D.J.L.(1971) Nat.New.Biol. 229, 140-142

3.Agsteribbe,E.,Datema,R. and Kroon,A.M.(1974) in the Biogenesis of Mitochondria (Kroon,A.M. and Saccone,C.,eds.),pp 305-314,Academic Press,New York

4.Van den Bogert,C. and De Vries,H.(1976) Biochim.Biophys.Acta, in press

5.Kaltschmidt,E. and Wittmann,H.(1970) Anal.Biochem.36, 401-412.

6.Kroon,A.M.,Terpstra,P.,Holtrop,M.,De Vries,H.,Van den Bogert,C.,De Jonge,J. and Agsteribbe,E., this volume.

Genetics and Biogenesis of Chloroplasts and Mitochondria, Th. Bücher et al. eds.

SIGNIFICANCE OF 80-S RIBOSOMES ASSOCIATED WITH NEUROSPORA CRASSA MITOCHONDRIA

R. Michel, G. Hallermayer, M.A. Harmey+, F. Miller and W. Neupert
Institut für Physiologische Chemie, Physikalische Biochemie und Zellbiologie
der Universität München, Germany

The mitochondrial ribosome of Neurospora crassa was first described as having a sedimentation coefficient of 73-S (subunits 50-S and 37-S) compared to 77-S for the cytoplasmic ribosome (subunits 60-S and 37-S) (1). The results of KÜNTZEL and NOLL (1) were accepted for some years. Recently however, DATEMA et al. (2) have described the mitochondrial ribosome as having an S-value of 80. They further stated that the 73-S mitoribosome of Neurospora is an artefact of the isolation procedure. It has been suggested (2) that washing mitochondria with EDTA-containing media results in the loss of essential components giving 73-S ribosomes, whereas isolation of mitochondria in the presence of magnesium and heparin yields intact ribosomes with an S-value of 80. Here we attempt to evaluate the significance of 73-S and 80-S ribosomes occuring in mitochondria isolated by different procedures.

1. Ultracentrifugation of ribosomes from mitochondria isolated in the presence and absence of magnesium ions

Mitochondria were isolated according to four different procedures as outlined in the table. Ribosomes were prepared from each preparation in an identical manner. Ribosomes from $SE_{1.0}T$ mitochondria show a prominent 73-S monomer with 50-S and 37-S subunits and a small amount of dimers (103-S) and trimers (fig. 1). In the case of SAMT ribosomes obtained from mitochondria isolated as described by DATEMA et al. (2), the gradient in addition to the 73-S component obtained in

TABLE

Designition of mitochondrial preparations	$SE_{1.0}T$	SAMT	SME	$SE_{0.1}T$
Isolation medium	A	B	B	C
Wash medium	2 x A	3 x B	2 x A	2 x C

A: 0.44 M sucrose, 1 mM EDTA, 10 mM Tris-HCl, pH 7.6; B: 0.44 M sucrose, 100 mM NH_4Cl, 10 mM $MgCl_2$, 10 mM Tris-HCl, pH 7.6; C: 0.44 M sucrose, 0,1 mM EDTA, 10 mM Tris-HCl, pH 7.6.

$SE_{1.0}T$ preparations, shows a prominent peak at 79-S with a smaller peak at 109-S (fig. 1). These additional peaks are coincident with mono- and dimers of cytoplasmic ribosomes. In the case of SME mitochondria (isolated in SAMT but twice washed

+M.A.H. (present address: Department of Botany, University College, Dublin (Irish Republic) wishes to thank CIBA GEIGY for a visiting fellowship.

with $SE_{1.0}T$), a profile qualitatively similar to that obtained from SAMT mitochondria is obtained. The amount of subunits is however greater and the content of 73-S, dimers and trimers is lower (not shown).

These data are in agreement with those of DATEMA et al. (2). We however, interpret them in a different manner. We conclude that the 79-S and the 109-S components can be attributed to contaminating cytoplasmic ribosomes.

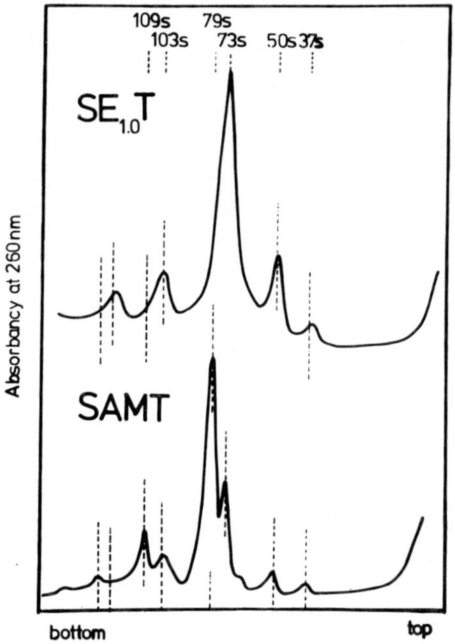

Fig. 1

Density gradient centrifugation of ribosomes isolated from mitochondria prepared in $SE_{1.0}T$ and SAMT buffer.

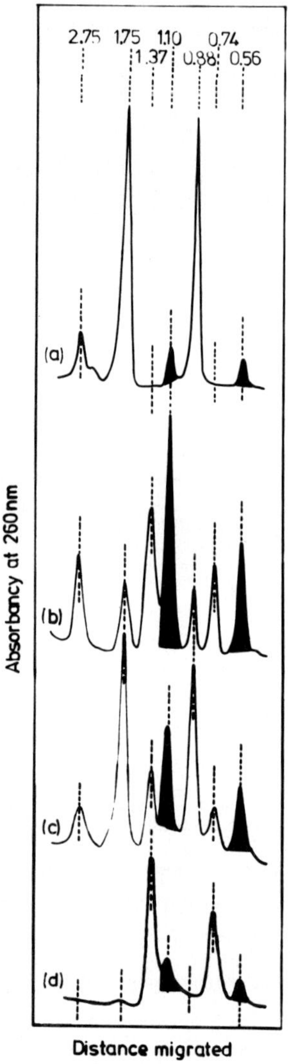

Fig. 2

Polyacrylamide gel electrophoresis of high molecular weight RNA extracted from mitochondria, which were isolated in different buffers: (a) $SE_{1.0}T$, (b) SAMT, (c) SME, (d) total cytoplasmic RNA. Apparent molecular weights were determined by comparison with E. coli ribosomal RNA co-electrophoresed with the RNA exracts. The gels were run at 25°C. Apparent molecular weights in daltons by 10^{-6} are shown at the top of the figure.

2. Ribosomal RNA from mitochondria isolated under different conditions

RNA was extracted from the different mitochondrial preparations and subjected to polyacrylamide gel electrophoresis. The $SE_{1.0}T$ preparation (fig. 2a) shows two ribosomal components with apparent molecular weights of 1.75 and 0.88 x 10^6 daltons and a small DNA peak. These components are clearly distinguishable from the cytoplasmic RNAs (apparent molecular weights 1.37 and 0.74 x 10^6daltons; cf. fig. 2d). SAMT and SME mitochondrial preparations (fig. 2b and 2c) show profiles, which are clearly composites of $SE_{1.0}T$ profile (fig. 2a) and that of the cytoplasmic RNA (fig. 2d). The SAMT profile (fig. 2b) also shows an extremely high DNA content. The individual RNA species are clearly distinguishable by their mobilities relative to E. coli ribosomal RNA (apparent molecular weight 1.1 and 0.56 x 10^6 daltons). The data in fig. 2 firmly support the conclusion that ribosomes from mitochondria isolated in the presence of magnesium consist of a mixture of cytoplasmic and mitochondrial ribosomes.

The presence of cytoplasmic ribosomes in SAMT preparations can be further shown by electrophoretic analysis of the low molecular weight RNAs. $SE_{1.0}T$ mitochondria show only a 4-S component and heating and cooling fails to show any further low molecular weight components. RNA from cytoplasmic ribosomes on the other hand shows only a 5-S RNA and on heating and cooling shows an additional peak of 5.8-S RNA. SAMT low molecular weight RNA shows the presence of 5-S and 4-S peak and on heat treatment shows in addition a 5.8-S peak. The latter peak is characteristic of cytoplasmic ribosomes and its presence in SAMT-RNA we attribute to the presence of cytoplasmic ribosomal RNA (3).

3. Electron microscopic examination of mitochondrial preparations

In view on the heterogeneity of the ribosomes obtained from SAMT mitochondria we compared the SAMT and $SE_{1.0}T$ mitochondrial preparations by electron microscopy.

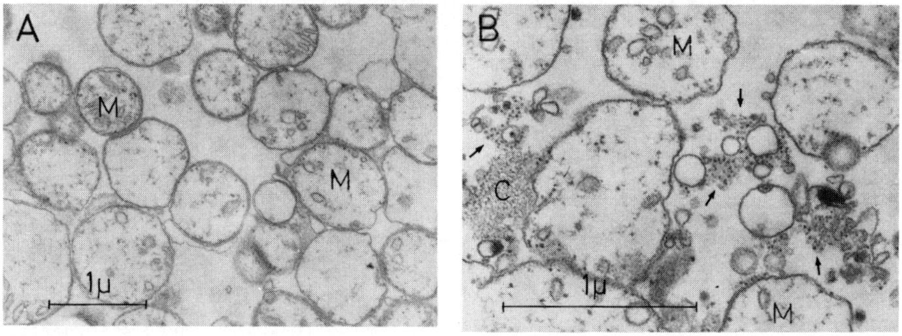

Fig. 3

Electron microscopy of mitochondrial preparations isolated in $SE_{1.0}T$ (A) and SAMT (B). Both fractions were treated with antibodies specific for cytoplasmic ribosomes.
M: mitochondria, C: tentatively identified chromatin; Arrows indicate agglutination of cytoplasmic ribosomes.

$SE_{1.0}T$ mitochondria can be seen to consist of a homogeneous population of organelles with well preserved inner and outer membranes. In contrast, the SAMT preparations are clearly heterogeneous containing much mitochondrial material, with patches of material resembling chromatin, and membrane attached ribosomes. The treatment of SAMT ribosomes with antibodies specific for cytoplasmic ribosomes led to the aggregation of much rough membrane vesicles and to the formation of large clumps of aggregated particles, which have the appearance of cytoplasmic ribosomes (fig. 3B). $SE_{1.0}T$ mitochondria treated with the same antibody show no evidence of either chromatin, rough vesicles or ribosome-like aggregates (fig. 3A). We therefore attribute the ribosomal heterogeneity of SAMT preparations (cf. fig. 1 and fig. 2b) to the presence of contaminating cytoplasmic ribosomes.

4. Effect of heparin on ribosomes from different mitochondrial preparations

DATEMA et al. (2) have stated that the isolation of the mitochondrial 80-S ribosomes requires the presence of magnesium ions and heparin during the lysis of the mitochondria to protect the ribosomes from degradation to the 73-S condition. The effect of heparin in the isolation media for ribosomes is shown in fig. 4

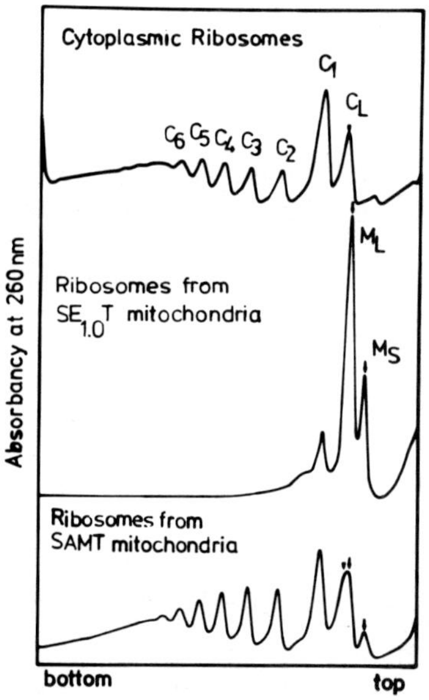

Fig. 4

Density gradient centrifugation of ribosomes isolated in the presence of heparin (0.5 mg/ml isolation medium)

Cytoplasmic ribosomes are effectively protected and only a small amount of 60-S subunit appears in heparin preparations. The effect of heparin on the $SE_{1.0}T$ preparations is quite striking. The ribosomes are almost completely dissociated and mainly subunits are obtained. In SAMT preparations the inclusion of heparin causes the disappearance of the 73-S ribosome with concomitant appearance of 50-S and 37-S subunits leaving the 80-S ribosome intact so that the gradients show only one monomer-type, viz. the cytoplasmic.

5. Immunoprecipitation of ribosomes from different mitochondrial preparations

We have previously described the preparation of specific antibodies against cytoplasmic and mitochondrial ribosomes (4). These antibodies were used to demonstrate in SAMT ribosomal preparations the presence of cytoplasmic and mitochondrial ribosomes. SAMT ribosomes were prepared from cells homogeneously labeled with ^{3}H-leucine. The ribosomes were subjected to density gradient centrifugation. The gradient was divided into 30 fractions and total radioactivity was measured (fig. 5A). Each fraction was then divided into two equal portions. One portion was treated with antibodies against cytoplasmic ribosomes and cold carrier cytoplasmic ribosomes, the other with antibodies against mitochondrial ribosomes and cold carrier 73-S ribosomes. Radioactivities in the precipitates and in the supernatants (fig. 5B and 5C) were dtermined. The antibody against the cytoribosomes completely precipitates the 79-S monomer and the 109-S dimer and leaves the 73-S monomer, the 103-S dimer and the 50-S subunit in the supernatant. Conversely, the antibody against mitochondrial 73-S ribosomes precipitates the 73-S monomer, the 103-S dimer and the 50-S subunit, leaving the 79-S particle in the supernatant.

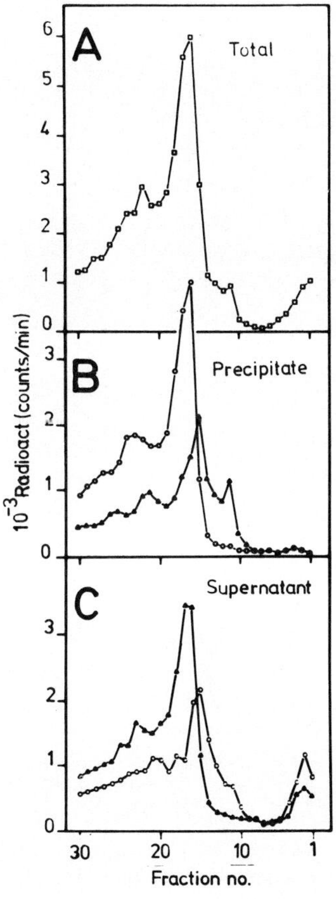

Fig. 5

Immunoprecipitation of mitochondrial and cytoplasmic ribosomes from density gradient fractions of ribosomes isolated from mitochondria prepared in SAMT buffer.

—o——o— : Radioactivity precipitated with antiserum against cytoplasmic ribosomes;

—▲——▲— : Radioactivity precipitated with antiserum against mitochondrial ribosomes.

6. Functional properties of ribosomes of mitochondria prepared in the presence and absence of magnesium ions

We have shown in an earlier publication that mitochondria prepared in $SE_{1.0}T$ under appropriate conditions are able to form subunits 1, 2 and 3 of cytochrome oxidase (5). This shows clearly that the ribosomes in $SE_{1.0}T$ mitochondria are capable of carrying out the synthesis of defined mitochondrial proteins. Since only one type of monomer is found in the $SE_{1.0}T$ mitochondria, viz. the 73-S, we conclude that this is the functional mitochondrial monomer. The isolated 73-S ribosome is able to react with puromycin and to release the nascent chains. In the absence of a crude preparation of G-factor, which may also contain a dissociation factor, after the release the ribosomes remain as 73-S monomer, in its presence they fall apart into 50-S and 37-S subunits.

Conclusions

The data presented clearly show that mitochondrial preparations obtained in the presence of magnesium contain both cytoplasmic and mitochondrial ribosomes. In our view, as first suggested by KÜNTZEL and NOLL (1), the 73-S ribosome is the real functional mitochondrial ribosome.

References

1. KÜNTZEL, H. & NOLL, H. (1967) Nature 215, 1340-1345
2. DATEMA, R. et al. (1974) Biochem. Biophys. Acta 335, 386-395
3. LEAVER, C.J. & HARMEY, M.A. (1976) Biochem. J. 157, 275-278
4. HALLERMAYER, G. & NEUPERT, W. (1974) FEBS Letters 41, 264-268
5. RUECKER, A.v. et al. (1974) FEBS Letters 47, 290-294

Genetics and Biogenesis of Chloroplasts and Mitochondria, Th. Bücher et al. eds.

COMPARATIVE STUDIES OF RIBOSOMES FROM MITOCHONDRIA, CHLOROPLASTS AND CYTOPLASM. MORPHOLOGY AND ELECTROPHORETIC BEHAVIOR

Barbara J. Stevens, Jean-Jacques Curgy,
Gérard Ledoigt⊕ and Jean André
Laboratoire de Biologie Cellulaire 4
Université de Paris-Sud, 91405 Orsay
and
⊕Laboratoire de Cytophysiologie de la
Photosynthèse, 91190 Gif-sur-Yvette (France)

INTRODUCTION

Efforts to characterize ribosomes from mitochondria (mitoribosomes) and chloroplasts (chlororibosomes) have provided data which point to the distinct character of each type of organelle ribosome and demonstrate basic differences from prokaryote and eukaryote cytoplasmic ribosomes (cytoribosomes)[1,2,3]. From the sedimentation coefficients of the monosomes and their subunits and from their rRNA components, mitoribosomes can be roughly grouped in three categories : 1) the 55-60S class from metazoan animals[4,5]; 2) the 70-74S class from fungi and *Euglena*[6,7,8]; and 3) the 80S class from ciliate protozoans[9,10,11]. Chlororibosomes fall into a separate class of 68-70S particles but show many differences from 70S prokaryote ribosomes[12].

In addition to the standard techniques for studying isolated ribosomes, such as sucrose gradient centrifugation, gel electrophoresis of RNA and protein, and analysis of protein synthetic activity, some valuable information has been obtained by two other techniques : polyacrylamide gel electrophoresis and electron microscopy of negatively stained particles. Prokaryote ribosomes can be clearly distinguished from eukaryote cytoribosomes by a faster electrophoretic migration rate relative to their S value[13], and by a significant size difference and several morphological details[14]. So far, analysis of mitoribosomes by electrophoresis has revealed in two cases, rat liver[15] and *Tetrahymena*[11], a lower charge/mass ratio and a slower mobility than those of cytoribosomes. Morphological studies of mito- and cytoribosomes from *Candida*[7] and *Locusta*[16] have shown some differences in form but little difference in size. However, observations on the 80S mitoribosome from *Tetrahymena*[11] have revealed a striking morphology and a clear size difference in comparison to cytoribosomes.

In this paper we have extended these studies to the mitoribosomes of a fish, *Tinca tinca*, and of two other ciliates, *Paramecium* and *Didinium*. The former ciliate is closely related to *Tetrahymena* but the latter is phylogenetically distant. Finally, we have examined the chlororibosomes from *Euglena* and from spinach. We have compared these results to the earlier data and we present a general survey of the electrophoretic and morphologic properties of the various ribosomes.

MITORIBOSOMES

Metazoan animals. The majority of data on the 55-60S mitoribosome is based on mammals, two amphibians and a few invertebrates[2,4,5,15,16]. We have examined the ribosomes from a lower vertebrate, the cyprinoid fish *Tinca tinca*. Mitochondria were isolated from ovaries, purified on a sucrose gradient and lysed for two hours with Triton X100. When analyzed by sucrose gradient centrifugation, cyto- and mitoribosomes showed sedimentation coefficients of 80S and 62S respectively. In spite of the apparent smaller size of mitoribosomes, they migrate more slowly in gel electrophoresis than the cytoribosomes, to a position intermediate between the cytoplasmic 80S monosomes and their 112S dimers[17] (fig. 1a).

Negatively stained mitoribosomes display the basic morphologic features of typical ribosomes, *i.e.*, slightly elongate particles, with a predominant frontal view in which the small subunit lies on top, apposed to the flattened base of the pyramidal large subunit. The junction of the subunits is marked by an electron dense spot, most often on the left side of the profile. Less frequent lateral views show the small subunit positioned asymmetrically on the large subunit and separated by a thin opaque line. When compared with their corresponding cytoribosomes, however, certain characteristic differences are revealed (Plate I). The mitoribosome is slightly smaller (Table 1) and has a more irregular contour with a forklike shape at the pointed end of the large subunit. The dense spot may be present on either side or in the center. The partition described in typical small subunits[14] is not apparent here and lateral projections are rare.

Plate I. Gallery of negatively stained cytoribosomes (a) and mitoribosomes (b) from the fish *Tinca tinca*. Top row : principal views (frontal) with dense spot at left ; middle row : rare dorsal views with dense line between subunits ; bottom row : enantiomorphic lateral views. Arrows show forked end on large subunit. (c) Field of negatively stained mitoribosomes from *Euglena gracilis*. Note regular contours and oblong shape.

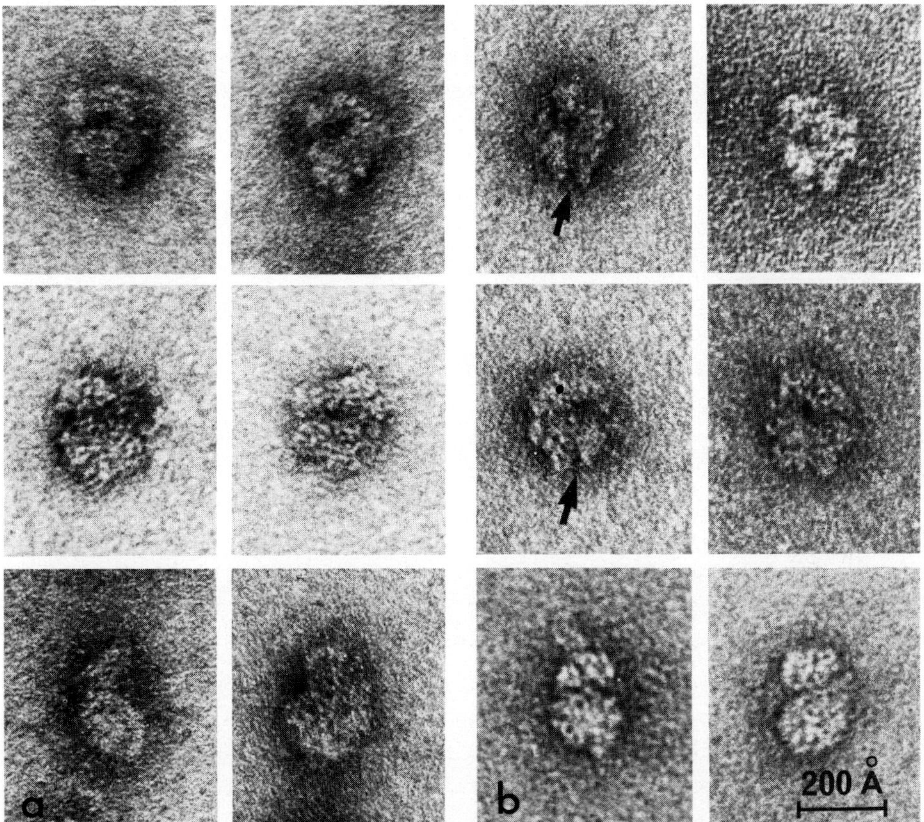
a
b
200 Å
c
500 Å

Preliminary observations of negatively stained rat liver mitoribosomes show greatly similar profiles with the same features : a forked end on the large subunit and an irregular contour. Measurements indicate smaller dimensions than those of corresponding cytoribosomes (Table 1). Images of mitoribosomes in the published study on *Locusta*[16] also show the features noted above, in particular the forked end.

Fungi and *Euglena*. The mitoribosomes in this group are close in their sedimentation coefficient to cytoribosomes but are distinct in a number of physico-chemical properties[6,7,8]. Our earlier study on *Candida*[7] revealed no significant difference in size and only minor differences in form between the two monosomes, but mitoribosomal subunits displayed shapes recalling those of bacterial particles. Preliminary observations of negatively stained mito- and cytoribosomes of *Euglena* (Plates I & III) reveal no particularities in form, although the dimensions of mitoribosomes are slightly smaller (Table 1). In general, however, mitoribosomes of both organisms can be recognized by their more oblong form, as opposed to the typical curvilinear triangular form of cytoribosomes.

Ciliate protozoans. The unusual sedimentation properties found for mitoribosomes from *Tetrahymena*[9] and *Paramecium*[10], *i.e.*, 80S particles similar to cytoribosomes and two 55S subunits, have been accompanied in *Tetrahymena* by data showing a slow electrophoretic migration, a peculiar morphology and a large size of the mitoribosome[11]. When ribosomes from *Paramecium* are analyzed by gel electrophoresis, the mitoribosome band migrates slowly, in a position close to the 160S band of cytoribosomal polysomes (fig. 1a). The morphology of these mitoribosomes is closely similar to those from *Tetrahymena* : clearly elongate particles bearing a prominant lobe on one subunit and a median cleft in the other subunit ; an electron-opaque spot or dense line delimits two nearly equal size subunits (Plate II) (similar to results of Knowles and Tait, pers. comm.). Mitoribosome particle length in both species is close to 370 Å, about 100 Å more than cytoribosomes.

Plate II. Gallery of negatively stained mito- and cytoribosomes from *Tetrahymena* (column a), *Paramecium* (column b) and *Didinium* (column c). Rows 1 & 2 : principal views of mitoribosomes showing cleft (arrow) and lobe (double arrow) ; row 3 : views of mitoribosomes showing dense line separating 2 equal size subunits ; row 4 : principal (frontal) views of cytoribosomes.

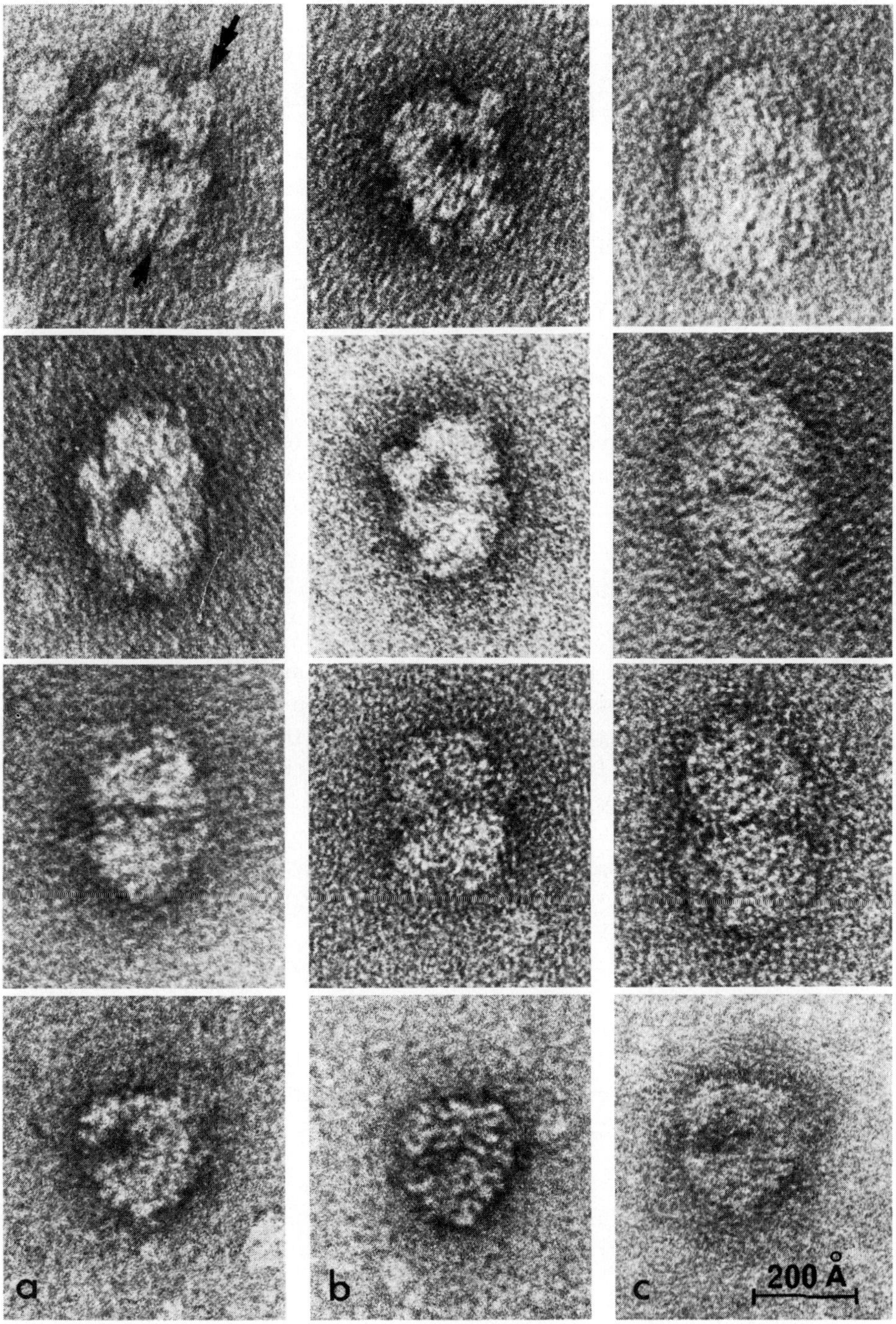
a
b
c
200 Å

The mitochondria from another ciliate, *Didinium nasutum*, contain numerous ribosomes and provide a source for comparison[18]. When a ribosomal pellet from isolated mitochondria is analyzed by gel electrophoresis, the mitoribosomes migrate slowly in a band which remains close to the origin in a position between the trimer and tetramer cytoribosomal polysome bands (fig. 1a). Extraction of RNA from the mitoribosomal pellet produces two major classes of 14-15S and 23-24S molecules[19]. Negatively stained mitoribosomes have a regular, elongate form with a spot or dense line delimiting two equal size subunits (Plate II). The lobe and cleft observed in the other two species are not visible and *Didinium* mitoribosomes are slightly larger, up to 400 Å long (Table 1).

Mitoribosomes of these three species are large, migrate slowly and have an elongate form composed of two equal size subunits. *Didinium* mitoribosomes are distinguished, however, by their large rRNA molecule (compared to 21S in *Tetrahymena* and *Paramecium*) and some morphological details.

CHLORORIBOSOMES

Ribosomes have been extracted from isolated chloroplasts of *Euglena*, purified on a gradient (68S) and compared to cytoribosomes (88S) and *E. coli* ribosomes (70S). Both chloro- and *E. coli* ribosomes migrate rapidly, with the former slightly behind. The slow mobility of 88S cytoribosomes differentiates them from the other two types and also from typical 80S cytoribosomes (fig. 1b). Ribosomes extracted from spinach chloroplasts likewise migrate rapidly, close to the prokaryote band. The spinach chlororibosome fractions were analyzed in SDS gels ; they yielded main peaks at 16S and 23S and little cytoribosomal contamination.

Both types of negatively stained chlororibosomes appear as small, rounded profiles, similar in size to *E. coli* ribosomes. One dimension is slightly longer than the other. A dense spot is sometimes observed but there is little indication of subunits (Plate III). Profiles of the corresponding 88S and 80S cytoribosomes display morphology and dimensions typical of this class.

Plate III. Fields of negatively stained chloro- (a & c) and cyto-ribosomes (b & d) from *Euglena* (a & b) and *Spinacia* (c & d).

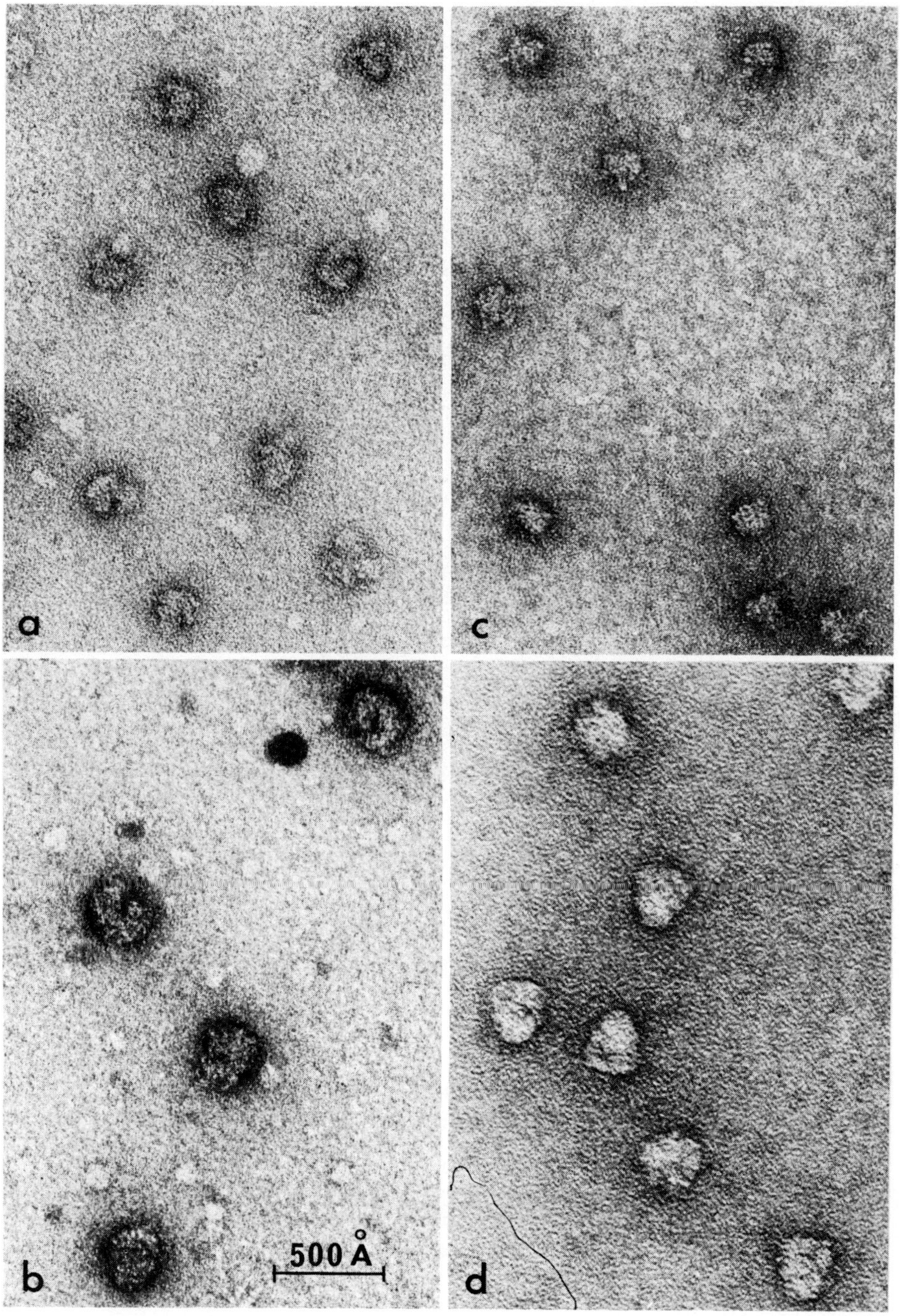
a
c
b
500 Å
d

TABLE 1

DIMENSIONS OF NEGATIVELY STAINED RIBOSOMES⊕ (Å)

Organism	Cytoribosomes	Mitoribosomes	Chlororibosomes
Eukaryotes	*80-88S*		
Tinca	263±26 × 224±18	*55-60S* 237±17 × 195±18	
Rattus	312±17 × 257±22	259±23 × 201±17	
Candida	262±16 × 222±17	*70-74S* 266±18 × 214±18	
Euglena	274±14 × 218±12	251±13 × 202±12	*68-70S* 207±13 × 169±18
Spinacia	234±10 × 195±12		198±12 × 164±14
Tetrahymena	274±14 × 227±13	*80S* 369±21 × 260±15	
Paramecium	287±14 × 224±17	373±19 × 250±15	
Didinium	267±10 × 226±13	400±12 × 247±13	
Prokaryotes	*70S*		
E. coli	213±10 × 178±13		

⊕Mean values and standard deviations (N 50) of the two major axes. Carbon films, except for *Spinacia* where ribosomes appear smaller on carbon-formvar films.

GENERAL COMPARISONS AND CONCLUSIONS

Electrophoretic behavior. Analysis by polyacrylamide gel electrophoresis can be used to distinguish four major categories of ribosomes : eukaryote cyto-, mito- and chlororibosomes, and prokaryote ribosomes. When the electrophoretic mobilities of polysomes, monosomes and subunits of prokaryote ribosomes are plotted as a function of their sedimentation coefficients, a logarithmic relation is found. Eukaryote cytoribosomal particles, with a slower migration rate, demonstrate a separate logarithmic relation[13] (fig. 1b).In comparison to these two types, the mobilities of mito- and chlororibosomes place them apart. The mitoribosomes examined, from three ciliates and one vertebrate, all manifest a very slow electrophoretic mobility. These particles migrate more slowly than their cytoplasmic counterparts. On the contrary, the 68-70S chlororibosomes migrate more rapidly than cytoribosomes, but slightly behind the 70S prokaryote ribosomes. While the electrophoretic mobility of ribosomes depends on their size and their spatial configuration, it also appears to be a function of their protein content. The slightly slower chlororibosomes have a higher (∿46 %) protein content than prokaryote ribosomes (∿35 %)[12]. Likewise, the high (up to 75 %) protein content of mitoribosomes can be directly related to their slow mobility[13].

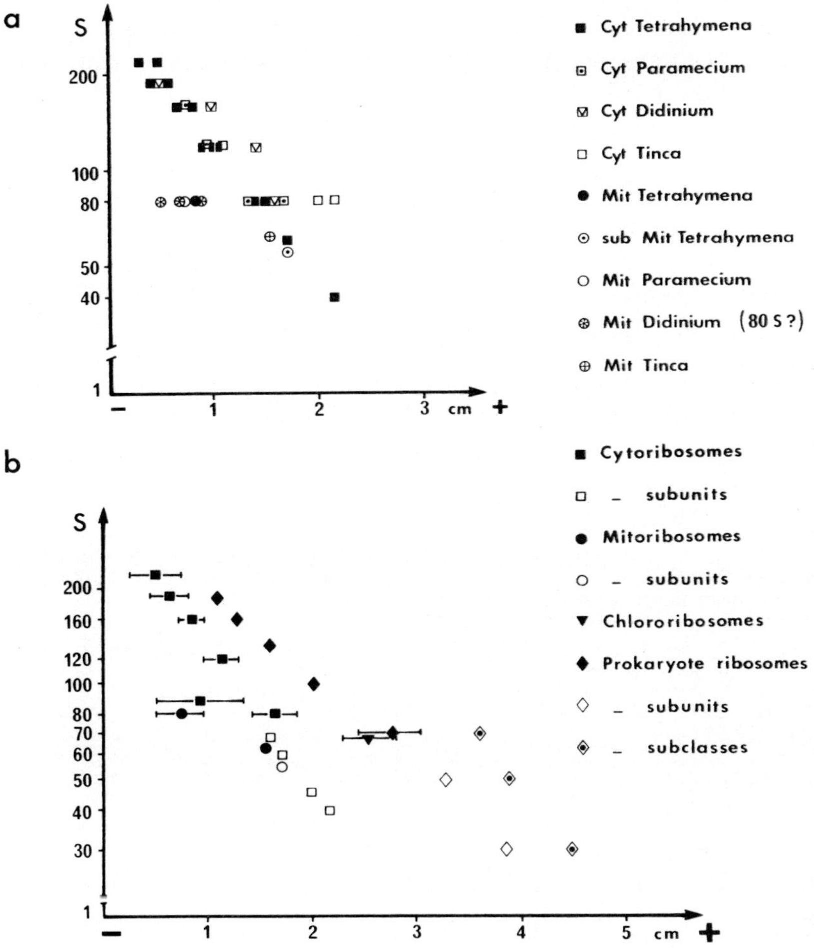

Fig. 1. Electrophoretic mobilities of ribosomes.
a) Mito- and cytoribosomes. b) All classes of ribosomes.
Abscissa : migration in cm. ; ordinate : S-values on logarithmic scale.
Isolated symbols : individual experimental points ; symbols with lines: mean values of several experiments and 95 % confidence limits.
Conditions as in ref. 13.

Morphology. The observations reported here indicate that the major categories of ribosomes can be identified on the basis of specific morphological characteristics. Furthermore, the 3 mitoribosome classes each present a particular form and size, a fact which strengthens their classification into 3 groups. The morphology of 70-74S mitoribosomes is closely similar to that of cytoribosomes, with little size difference but more oblong profiles. The 55-60S class is recognized by the characteristic forked end of the large subunit. Finally, the striking similarities in morphology and size of the three ciliate mitoribosomes examined emphasize the uniqueness of this group and raise discussion of its evolutionary significance. The chlororibosome is similar in size to prokaryote ribosomes, but the difficulties of defining its structural features may be related to a greater hydration or looser packing of its components.

It is satisfying to note that fine morphological details of the smallest cell organelle, the ribosome, depict evolutionary trends parallel to the established phylogenetic lines.

Acknowledgments. We are grateful to Cécile Couanon and Annie Charrier for faithful and skillful help in preparation of the manuscript. Work was supported by the C.N.R.S. (E.R.A. 174).

REFERENCES

1. Borst, P. and Grivell, L.A. (1971) *FEBS Lett. 13*, 73-88.
2. Dawid, I.B. (1972) *Mitochondria : Biogenesis and Bioenergetics ; Biomembranes : Molecular Arrangements and Transport Mechanisms.* (Van den Bergh, S.G., *et al.*, eds.), Vol. 28, pp. 35-51, North-Holland/American Elsevier, Amsterdam.
3. Smillie, R.M. and Scott, N.S. (1969) *Progress in Molecular and Subcellular Biology.* (Agranoff, B.W., *et al.*, eds.), Vol. 1, pp. 136-202, Springer-Verlag, Berlin.
4. O'Brien, T.W. (1971) *J. Biol. Chem. 246*, 3409-3417.
5. Swanson, R.F. and Dawid, I.B. (1970) *Proc. Natl. Acad. Sci. U.S. 66*, 117-124.
6. Küntzel, H. (1971) *Current Topics Microbiol. Immunol. 54*, 94-118.
7. Vignais, P.V., Stevens, B.J., Huet, J. and André, J. (1972) *J. Cell Biol. 54*, 468-492.
8. Avadhani, N.G. and Buetow, D.E. (1972) *Biochem. J. 128*, 353-365.
9. Chi, J.C.H. and Suyama, Y. (1970) *J. Mol. Biol. 53*, 531-556.
10. Tait, A. (1972) *FEBS Lett. 24*, 117-120.
11. Curgy, J.J., Ledoigt, G., Stevens, B.J. and André, J. (1974) *J. Cell Biol. 60*, 628-640.
12. Yurina, N.P. and Odintsova, M.S. (1974) *Plant Science Lett. 3*, 229-234.
13. Ledoigt, G., Curgy, J.J., Stevens, B.J. and André, J. (1975) *Biochim. Biophys. Acta 407*, 222-239.
14. Lake, J.A., Sabatini, D.D. and Nonomura, Y. (1974) *Ribosomes.* (Nomura, M., Tissières, A. and Lengyel, P., eds.), pp. 543-557, Cold Spring Harbor Laboratory, New York.
15. de Vries, H. and van der Koogh-Schuuring, R. (1973) *Biochem. Biophys. Res. Comm. 54*, 308-314.
16. Kleinow, W., Neupert, W. and Miller, F. (1974) *J. Cell Biol. 62*, 860-875.
17. Curgy, J.J., Clérot, J.C. and Stevens, B.J. (1976) *J. Microscopie Biol. Cell. 26*, 9a.
18. Stelly, N. and André, J. (1972) *J. Microscopie 14*, 93a-94a.
19. Curgy, J.J., Stelly, N. and Stevens, B.J. (1976) *J. Microscopie Biol. Cell. 26*, 9a.

Genetics and Biogenesis of Chloroplasts and Mitochondria, Th. Bücher et al. eds.

PROTEIN COMPOSITION OF THE BOVINE MITOCHONDRIAL RIBOSOME

Thomas W. O'Brien, David E. Matthews and Nancy D. Denslow
Department of Biochemistry and Molecular Biology
College of Medicine, University of Florida
Gainesville, Florida 32610
U.S.A.

INTRODUCTION

Several studies over the past few years have shown that the 55-60S ribosomes of animal mitochondria have unusual composition and physical-chemical properties (reviewed in reference 1). The first inference drawn from the low sedimentation rate of 55S ribosomes was that these mitochondrial ribosomes were smaller than other kinds of ribosomes ("mini-ribosomes") and this inference was reinforced by the observation of unusually small rRNA molecules in these particles. However, 55S ribosomes are actually slightly larger than 70S bacterial (E. coli) ribosomes, both on the basis of particle molecular weight[2-4] and physical dimensions[5,6]. It follows that a ribosome somewhat larger than bacterial ribosomes, but containing only about half as much rRNA, should have a correspondingly higher protein content.

Buoyant density measurements have been used widely to estimate the protein content of ribosomes, but the applicability of this method has come under question[7]. We have re-examined this relationship and have found that buoyant density measurements do indeed reflect the protein content of mammalian mitochondrial ribosomes and subribosomal particles.

Our interest in analyzing the protein component of these ribosomes arose from observations that they appear to contain a very large number (range, 70 to 107) of different proteins, as resolved by 2-dimensional electrophoresis[3,8-10]. These 55S ribosomes thus appear to contain more individual proteins than E. coli ribosomes (about 54)[11], or even eukaryotic cytoplasmic ribosomes (about 70)[12].

The presence of so many proteins in subribosomal particles derived from 55S ribosomes immediately raises some questions: 1) Are all of the putative r-proteins resolved by 2-dimensional electrophoresis bona fide ribosomal proteins? 2) Are all of the proteins in each subunit unique, or are some identical to proteins also detected in the other subunit?

In the case of bacterial ribosomes, such questions have been approached using a combination of ribosome reconstitution and immunochemical methods, both of which require the availability of relatively large amounts of individual r-proteins and, for this reason, are not presently feasible for mitochondrial ribosomes. We have, however, sought provisional answers to the above questions by several less direct methods, including an examination of the structural and functional properties of mitochondrial subribosomal particles after exposure to buffers of high ionic strength. We have chosen the bovine mitochondrial ribosome for this study, to develop a model for mammalian mitochondrial ribosomes in general, and to form the basis for future studies of the structure and function of these unusual ribosomes.

RESULTS AND DISCUSSION

Protein content of the bovine mitochondrial ribosome: Estimates for the protein content of 55S ribosomes have ranged from 65 to 75% depending on the particular analytical method used. We have used buoyant density measurements in CsCl to assess the protein content of mitochondrial subribosomal particles after washes in high salt buffers. This method has the advantage of reproducibility and sensitivity, requiring only small samples of particles for the measurement. A calibration curve relating the buoyant density of these ribosomes to their protein content was constructed according to the method of Hamilton[13], using the empirical value of 1.28 g/cm^3 determined for the density in CsCl of proteins extracted from bovine mitochondrial ribosomes. In this manner, we have determined that the buoyant densities of 55S bovine ribosomes (1.42 g/cm^3) and subribosomal particles (1.43 g/cm^3) correspond to protein contents of 69% and 67%, respectively[4]. Despite the apparent failure of this method when used with other ribosomes[7], our confidence in applying it to these protein-rich ribosomes is justified by the relationships in Figure 1.

In Figure 1a are plotted two different curves relating the molecular weight (MW) of mitochondrial ribosomes to their RNA content. One curve (dashed) was generated using values for the MW of the rRNA in mammalian mitochondria[3,14], while the other curve (solid) was generated from the sedimentation properties of the particle[4]. While the RNA content of the particle must be known in order to determine the particle MW, using either of the individual relationships alone, the two different curves relating to a given particle have a unique solution (intersection point), independent of any measurement of the particle RNA (or protein) content. As may be seen in Figure 1a, the "unique solution" MW of 2.82 x 10^6 daltons predicted for 55S bovine ribosomes agrees closely with the value of 2.83 x 10^6 daltons, obtained[4] using an RNA content of 31% determined as above from the particle buoyant density. Similarly, values for the RNA content of mitochondrial subribosomal particles, predicted by the intersection points in Figure 1b agree well with the values determined from their buoyant density measurements. Thus, the use of buoyant density measurements to calculate RNA (or protein) contents would seem justified, at least in the case of these protein-rich mitochondrial ribosomes.

Protein content of salt-washed subribosomal particles: To be suitable for analysis by 2-dimensional electrophoresis, ribosomes should be essentially free of adsorbed, non-ribosomal proteins. It has been amply demonstrated that exposure of ribosomes to media of high salt concentration may result in the removal of some proteins from the particles[15,16]. Appreciable loss of r-protein is readily detected by changes in the particle buoyant density. For comparative purposes in

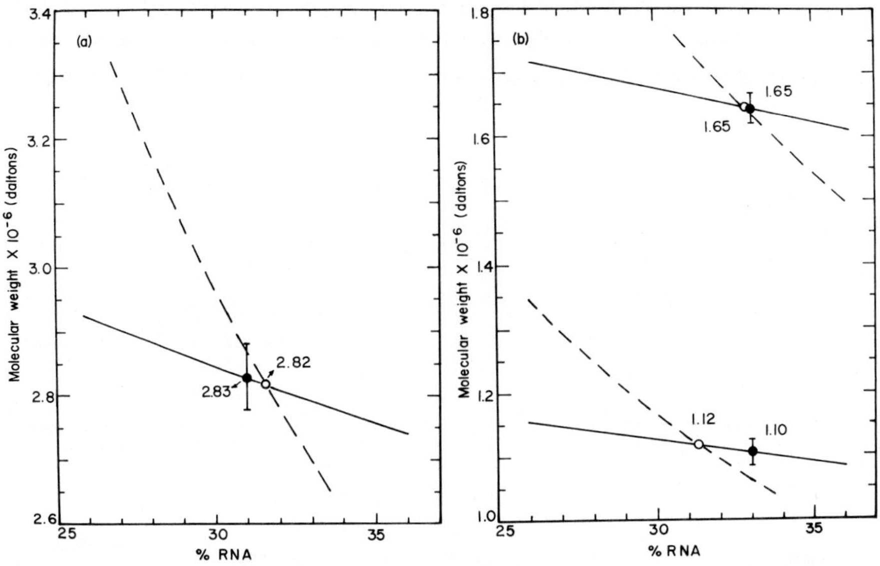

Figure 1. Relationships between the molecular weight of a ribosome and its RNA content. (a) 55S monoribosome; (b) large subunit (upper curves), and small subunit (lower curves). Dashed lines correspond to the equation

$$MW_{particle} = \frac{100}{\%RNA} \times MW\ RNA,$$

using values of 0.54×10^6 and 0.35×10^6 daltons for the molecular weights of the rRNA in mammalian mitoribosomal large and small subunits, respectively[14]. Solid lines correspond to the equation

$$MW_{particle} = \frac{(2RT/\omega^2)\ \frac{d\ \ln c}{d\ r^2}}{1 - \bar{v}\rho};$$

substituting the values obtained by high-speed equilibrium centrifugation[4], this equation becomes

$$MW = \frac{0.91}{1 - \bar{v}\rho} \quad \text{for the monoribosome,}$$

$$MW = \frac{0.535}{1 - \bar{v}\rho} \quad \text{for the large subunit, and}$$

$$MW = \frac{0.360}{1 - \bar{v}\rho} \quad \text{for the small subunit}^4.$$

$\bar{v}$ is calculated from %RNA by the equation

$$\bar{v} = \frac{\%RNA}{100}(\bar{v}_{RNA}) + \frac{100-\%RNA}{100}(\bar{v}_{protein}),$$

using the values $\bar{v}_{RNA} = 0.53$ ml/g and $\bar{v}_{protein} = 0.74$ ml/g.

this study, 80S extramitochondrial ribosomes obtained from lysates of partially purified mitochondria[17] were analyzed in parallel with mitochondrial ribosomes.

Bovine mitochondrial 55S and microsomal 80S ribosomes prepared in buffers containing 20mM Mg^{2+} and 0.1M KCl have buoyant densities of 1.42 g/cm^3 and 1.58 g/cm^3, respectively (Figure 2). We have studied the effects of high salt washes on the subribosomal particles derived from these 55S and 80S ribosomes. In general, for increasing ionic strength, at reduced Mg^{2+} levels, the buoyant density of the subribosomal particles increases, presumably due to the removal of loosely bound "non-ribosomal" components at intermediate $KCl/MgCl_2$ values (10 to 100), and putative ribosomal proteins at higher values.

Treatment of the particles in media of $KCl/MgCl_2$ values above 200 removes appreciable amounts of r-protein from both cytoplasmic subribosomal particles, causes marked destabilization of the small mitochondrial subribosomal particle, and produces a "core" particle of density 1.47 g/cm^3 from the large mitochondrial subribosomal particle (Figure 2).

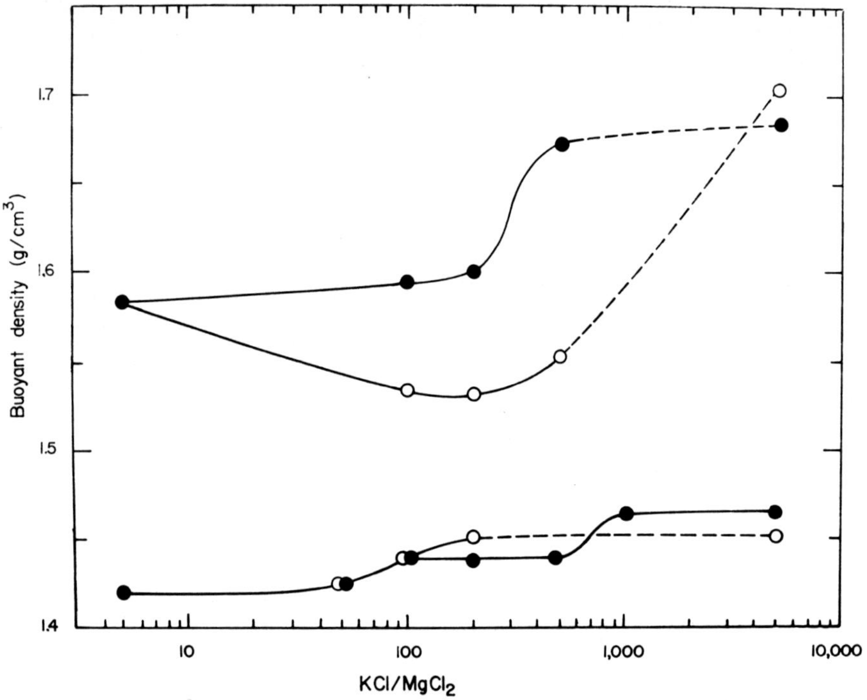

Figure 2. Effect of salt-washing conditions on the buoyant density of ribosomal particles. Mitochondrial (lower curves) and cytoplasmic (upper curves) ribosomes prepared in high Mg^{2+} buffer (20mM $MgCl_2$ and 0.1M KCl)[4] were dissociated and analyzed by sucrose density gradient centrifugation in various high salt buffers. ○ small subunits; ●, large subunits. Buffer composition ranged from 0.1 to 1M in KCl, and 20mM to 0.1mM in Mg^{2+}, with 5mM β-mercaptoethanol and 20mM triethanolamine, pH 7.5. Fractions containing the ribosome subunits were pooled and dialyzed in buffer D (5mM $MgCl_2$, 50mM KCl and 20mM triethanolamine, pH 7.5) before fixation with 5% formaldehyde. The fixed subunits were centrifuged to equilibrium in CsCl gradients[19] containing 0.3% formaldehyde in Buffer D.

Function of salt-washed subribosomal particles: To establish washing conditions adequate for the removal of non-ribosomal proteins, but not so stringent as to strip proteins required for ribosome function, we monitored the peptidyl transferace activity of subribosomal particles washed under the various conditions shown in Figure 2. Ribosomes washed in buffers having $KCl/MgCl_2$ values less than 100 retain peptidyl transferase activity, while ribosomes washed in buffers with $KCl/MgCl_2$ ratios of 200 or above are devoid of activity (Table 1). We conclude, therefore, that the first inflection point in the mitochondrial ribosome curve in Figure 2 corresponds to the removal of proteins required for ribosome function.

TABLE 1

PEPTIDYL TRANSFERASE ACTIVITY OF SALT-WASHED SUBRIBOSOMAL PARTICLES

Bovine mitoribosomal large subunits were prepared by sucrose density gradient centrifugation in buffers of the ionic compositions shown, and were assayed for peptidyl transferase activity by the modified fragment reaction[17].

Buffer condition ($KCl/MgCl_2$)	Peptidyl transferase activity (% of control)
5	100
50	73
200	3

Electrophoretic analysis of the proteins of bovine mitochondrial ribosomes: The most rigorous salt-washing condition ($KCl/MgCl_2$ = 50) under which mitochondrial ribosomes retain significant peptidyl transferase activity was selected to prepare ribosome subunits for analysis by 2-dimensional electrophoresis (Figures 3a and 3b). From multiple analyses we obtain a reproducible set of 53 proteins from the large subunit, and 41 from the small. This total of 94 proteins is considerably greater than the number of r-proteins found in other kinds of ribosomes, about 54 in *E. coli*[11], or about 70 in the case of eukaryotic cytoplasmic ribosomes[12]. Mitochondrial ribosomes of other animal species have also been found to contain exceptionally large numbers of proteins: 84 mitoribosomal proteins are reported for the toad, *Xenopus*[3], and 70[10] to 107[9] for the rat.

We have determined the MW of each of these 94 proteins from their mobility in the 2nd electrophoretic dimension (in SDS). The average MW of the large subunit proteins is 21,000 daltons, and 23,000 daltons for the small subunit. Values for the total protein content calculated for the large and small subunit, using these MW figures, are in reasonable agreement with those obtained independently by the physical-chemical means described above[18].

Are all the proteins in the large subunit distinct from those in the small subunit?

Some of the large-subunit proteins are found to have electrophoretic mobilities in both dimensions similar to those of some of the small-subunit proteins, raising the possibility that these proteins may be identical. This situation could arise

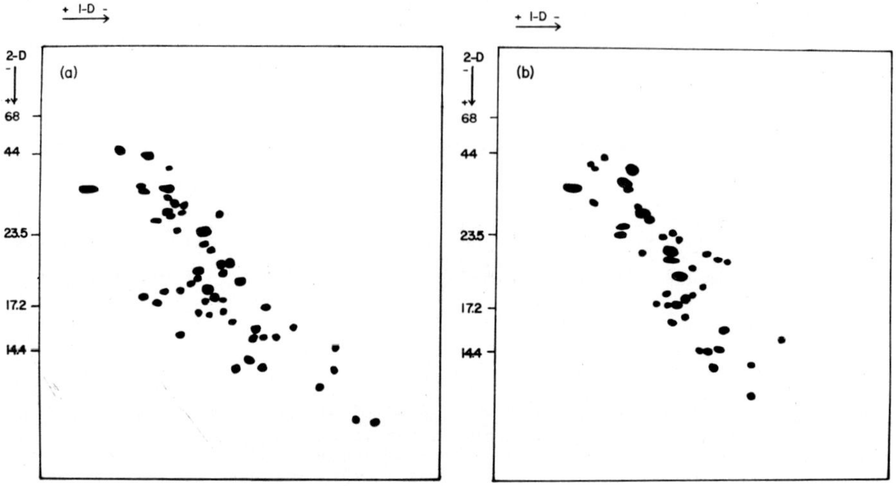

Figure 3. Two-dimensional electrophoretic patterns of bovine mitochondrial ribosomes. (a) Large subunit, (b) small subunit. Mitoribosomal subunits were prepared by sucrose density gradient centrifugation in a buffer condition of KCl/ $MgCl_2$ = 50. Ribosomal proteins were extracted with urea/lithium chloride and analyzed by two-dimensional polyacrylamide gel electrophoresis[18]. The first electrophoretic dimension (left to right) was in urea at pH 4.5, and the second (top to bottom) was in SDS. The second-dimension positions and molecular weights ($x10^3$ daltons) of several marker proteins are shown at the left of each diagram.

for example if some proteins, located perhaps at the subunit interface, partition to each of the two subunits when the ribosome is dissociated, instead of segregating exclusively to one subunit or the other. The most obvious ways to test this possibility are chemical analysis of the isolated proteins or the use of antibodies against the ribosomal proteins. We have used antibodies produced against the large

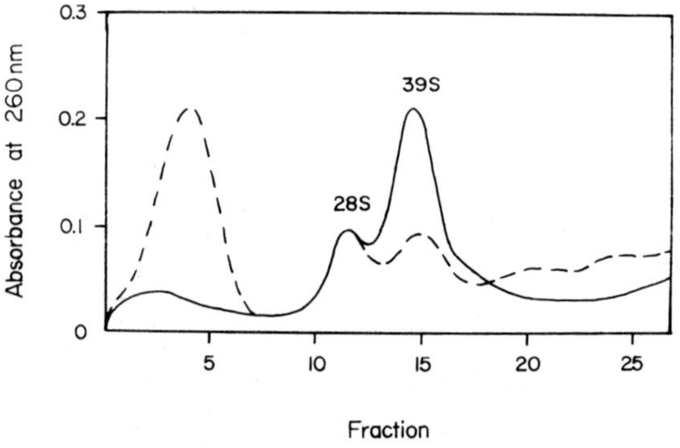

Figure 4. Sucrose gradient analysis of the interaction of anti-39S subunit antibodies with mitochondrial 28S and 39S subunits. The particles were incubated with (dashed curve) or without (solid curve) γ-globulin in a buffer containing 5mM $MgCl_2$, 200mM NH_4Cl, 6mM β-mercaptoethanol and 50mM Tris, pH 7.8, before analysis in sucrose density gradients of the same buffer composition.

subribosomal particles in preliminary experiments to test this possibility. As seen in Figure 4, these anti-large-subunit antibodies can interact with the large subunits to form aggregates of the particles. The failure of these antibodies to aggregate the small subunits suggests that the two subunits contain distinct sets of proteins. Confirmation of this conclusion, however, rests on the identification of the individual proteins recognized by these antibodies.

SUMMARY

To better define the protein composition of the protein-rich ribosomes of mammalian mitochondria, we have correlated the function of salt-washed bovine mitoribosomes with their protein composition. At present it appears that particles with a protein content of 67%, made up of about 90 individual proteins, represent the minimum functionally active form of the bovine mitochondrial ribosome. This unusual composition, so different from that of eukaryotic cytoplasmic ribosomes and especially from that of bacterial ribosomes, suggests that great differences will be found between the detailed structural and functional roles of the individual components of mammalian mitoribosomes and those of other kinds of ribosomes. The arrangement or "packaging" of the components must be fundamentally different in mammalian mitoribosomes, which have a protein:RNA ratio of 2:1, from that which is found in _E. coli_ ribosomes (protein:RNA = 0.6:1). For example, whereas all of the proteins in the 30S subunit from _E. coli_ are reported to be accessible on the surface of the particle, it seems likely that many of the mitoribosomal proteins will be found to be completely buried in the interior of the subunits. Such internal proteins would seem unlikely to be directly involved in any of the specific functional activities of the ribosome, probably serving a primarily structural role. Studies on the topography and functions of the individual proteins in this structurally unusual ribosome are in progress.

ACKNOWLEDGEMENTS

We would like to thank Mark Critoph, Warren Clark, and Mark Moffitt for their technical assistance. This work was supported by United States Public Health Service Grant GM 15438. DEM was supported by a National Science Foundation Graduate Fellowship.

REFERENCES

1. O'Brien, T.W. and Matthews, D.E. (1976) In: _Handbook of Genetics_, Vol. 5, Robert C. King (ed), Plenum Publishing Corp. New York.
2. Sacchi, A., Cerbone, F., Cammarano, P., and Ferrini, U. (1973) Biochim. Biophys. Acta _308_, 390-403.
3. Leister, D.E. and Dawid, I.B. (1974) J. Biol. Chem. _249_, 5108-5118.
4. Hamilton, M.G. and O'Brien, T.W. (1974) Biochemistry _13_, 5400-5403.

5. DeVries, H. and VanderKoogh-Schurring, R. (1973) Biochem. Biophys. Res. Commun. 54, 308-314.

6. DeVries, H. and Kroon, A.M. (1974) In: The Biogenesis of Mitochondria, Kroon, A.M. and Saccone, C. (eds), Academic Press, New York, pp 357-365.

7. McConkey, E.H. (1974) Proc. Natl. Acad. Sci. 71, 1379-1383.

8. O'Brien, T.W., Denslow, N.D., and Martin, G.R. (1974) In: Biogenesis of Mitochondria, Kroon, A.M. and Saccone, C. (eds), Academic Press, New York, pp 347-356.

9. Czempiel, W., Klose, J., and Bass, R. (1976) FEBS Letters 62, 259-262.

10. VandenBogert and DeVries, Biochim. Biophys. Acta, in press.

11. Wittmann, H. G. (1974) In: Ribosomes, Nomura, M., Tissieres, A., and Lengyel, P. (eds), Cold Spring Harbor Laboratory, pp. 93-114.

12. Wool, I.G. and Stöffler, G. (1974) In: Ribosomes, Nomura, M., Tissieres, A., and Lengyel, P., (eds), Cold Spring Harbor Laboratory, pp. 417-460.

13. Hamilton, M.G. (1971) Methods Enzymol. 20C, 512-521.

14. Robberson, D., Aloni, Y., and Attardi, G., (1971) J. Mol. Biol. 60, 473-484.

15. Warner, J.R. and Pene, M.G. (1966) Biochim. Biophys. Acta 129, 359-368.

16. Delaunay, J., Creusot, F., and Schapira, G. (1974) Biochimie, 56, 181-189.

17. Denslow, N.D. and O'Brien, T.W. (1974), Biochem. Biophys. Res. Commun. 57, 9-16.

18. Matthews, D.E. and O'Brien, T.W. (in preparation).

19. Brunk, C.F. and Leick, V. (1969) Biochim. Biophys. Acta 179, 136-144.

Genetics and Biogenesis of Chloroplasts and Mitochondria, Th. Bücher et al. eds.

TRANSFER RNAs OF YEAST MITOCHONDRIA*

Nancy C. Martin† and Murray Rabinowitz
Departments of Medicine and Biochemistry
The Franklin McLean Memorial Research Institute‡
The University of Chicago
Chicago, Illinois 60637, U.S.A.

Mitochondrial DNA has been shown to code for transfer RNAs in rat liver, HeLa cells, Xenopus, Tetrahymena, and yeast.[1] Estimates of the number of tRNA cistrons in mtDNA have ranged from 12 to 20 or 25 based on saturation-hybridization experiments using bulk labeled 4S RNA. The lower estimates raise the possibility that mtDNA of some species may not code for tRNAs for all of the 20 common amino acids and would make unlikely the presence of a substantial number of mtDNA coded heterogenic isoaccepting tRNAs, which have been found in all other protein synthetic systems to date. For some time, our laboratory has been engaged in characterizing the individual tRNA transcripts of yeast mtDNA to determine the number of amino acids accepted by mitochondrially coded tRNAs and to identify isoaccepting tRNAs involved in mitochondrial protein synthesis. We have also mapped many of the tRNA genes by deletion mapping in collaboration with H. Fukuhara.

Our laboratory has previously identified 14 aminoacyl tRNAs that are specified by yeast mtDNA using ^{3}H-aminoacyl tRNA-DNA hybridization.[2-7] Similar techniques have been used to demonstrate 4 tRNA genes in rat liver mtDNA[8,9] and, more recently, 16 tRNA genes in HeLa cells.[10] We now report that mitochondrially coded tRNAs in yeast accept 4 additional amino acids (Fig. 1), bringing the total number detected by this method to 18. We thus far have been unable to show hybridization of asparagine and glutamine tRNAs. While the inability to demonstrate asparagine tRNAs is probably due to instability or inactivity of the appropriate aminoacyl tRNA ligase, our inability to demonstrate glutamine (gln) tRNA appears to be due to the presence of a rather unusual pathway

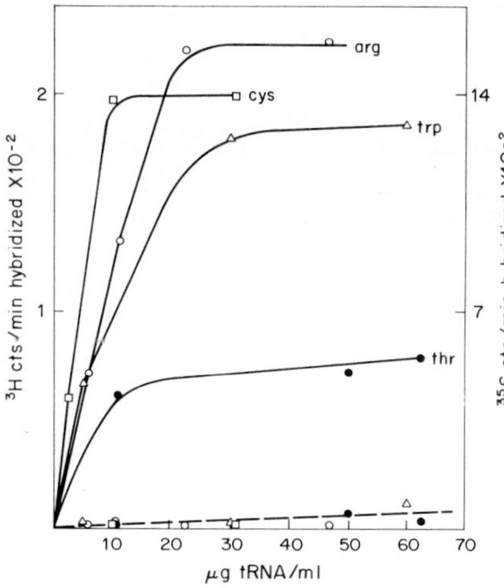

Fig. 1. Hybridization of mitochondrial tRNAs with mtDNA and nuclear DNA by methods previously described.[3]

*Supported in part by grants from the National Heart and Lung Institute, USPHS, and the Louis Block Fund of The University of Chicago.

†Postdoctoral Fellow of the American Cancer Society.

‡Operated by The University of Chicago for the U. S. Energy Research and Development Administration under Contract No. E(11-1)-69.

for the synthesis of gln $tRNA^{gln}$ in yeast mitochondria.

We have previously reported the presence of mitochondrial (mt) tRNAs acylated with glutamic acid (glu) and showed that they were transcribed from different sequences in widely separated regions of the genome.[11] In examining the codon responses of cytoplasmic (cyt) and mt tRNAs acylated with glu in yeast, we found that one of the mt glu tRNAs (previously designated $tRNA^{gluII}$) responded to oligonucleotides containing the CAA gln anticodon, while the other (previously designated $tRNA^{gluI}$) responded to oligonucleotides containing the GAA and GAG glu anticodons (Table 1). Cyt glu tRNAs responded to glu anticodons only. The tRNA responding to gln codons was directly aminoacylated with glu since the addition of unlabeled glu decreased incorporation of ^{3}H-glu into tRNA whereas unlabeled gln had no effect (Table 2). In contrast, unlabeled gln reduces the formation of cyt

TABLE 1

Codon Recognition of Yeast Glu tRNAs

Oligo-nucleotides	Aminoacyl tRNAs (cpm bound) cyt glu tRNA	mt glu $tRNA_I$	mt glu $tRNA_{II}$
none	63	38	32
U	96	28	40
GA	610	352	40
CA	109	38	227

Methods for binding of yeast glu tRNAs in response to oligonucleotides were essentially as described by Nirenberg and Leder.[13]

TABLE 2

Effect of Cold Amino Acids on the Acylation of Radioactive Amino Acids

tRNA	Cold amino acid	cts/min	% control
mt with 0.032 μg 3H glu	-	4792	100
	0.064 μg glu	3191	67
	0.124 μg glu	1695	35
	0.064 μg gln	4992	103
	2.92 μg gln	4848	101
cyt with 0.62 μg 14C gln	-	2432	100
	1.4 μg gln	1762	72
	2.9 μg gln	874	36
	5.8 μg gln	690	28

Cold glu or gln were added to acylation reactions.[7]

gln $tRNA^{gln}$ as expected (Table 2). The formation of gln $tRNA^{gln}$ in yeast cytoplasm proceeds as in most organisms where separate enzymes catalyze the acceptance of glu and gln by the corresponding species of tRNA. Yeast mt gln $tRNA^{gln}$, however, apparently is synthesized by a pathway in which a missence aminoacyl tRNA intermediate (glu $tRNA^{gln}$) is formed and then converted to gln $tRNA^{gln}$ by a transamidase. Such a reaction has been previously shown to occur in three bacterial species.[12] Studies designed to demonstrate the presence of an amido-transferase in yeast are currently in progress.

If all the sense codons in the universal code are to be recognized by mt tRNAs, then more than one tRNA for some amino acids is necessary. We have therefore sought to detect isoaccepting mt tRNA species by RPC-5 chromatographic analysis of tRNAs isolated from mitochondria charged with ^{3}H-amino acids and, in most cases, co-chromatographed with cyt ^{14}C-aminoacyl tRNAs isolated from a mtDNA-less

(ρ^o) petite. The results of RPC-5 chromatography of tRNAs acylated with 18 different amino acids are shown in Figs. 2 and 3. Table 3 summarizes these results and includes information on the transcriptional origins of some of the separated peaks. Five mt amino acyl tRNAs (ala, his, ileu, lys, and pro) showed only one peak by RPC-5 chromatography. In the case of ala, his, and ileu, the peaks migrated differently than the cyt isoaccepting tRNAs for that amino acid. The mt lys tRNA co-migrated with a cyt lys tRNA but its mitochondrial origin was ascertained by hybridization to mtDNA. Thus, co-migration with a cyt tRNA does not necessarily indicate a cytoplasmic origin.

TABLE 3

Yeast Mitochondrial tRNAs

Hybridization to mtDNA	
positive	19
not detected	asn
Isoaccepting tRNAs (RPC-5 chromatography)	
multiple peaks hybridized to mtDNA	*cys (2), *met (2), tyr (4), val (2), phe (2)
multiple peaks hybridization not yet tested	arg (2), thr (2), trp (3), gly (2), leu (2)
single species hybridized to mtDNA	ala, his, ile, lys, pro, asp, ser, glu, gln

*Different sequences.

Two other aminoacyl tRNAs (asp and ser) (Fig. 2 F, G) showed only one peak of unique mobility that hybridized with mtDNA but contained one or more tRNAs co-migrating with cyt species that did not hybridize with mtDNA. To the category containing only one isoaccepting peak can be added $tRNA^{glu}$ and $tRNA^{gln}$ as discussed above (Fig. 2H).

Four tRNAs for 5 amino acids contained two or more peaks that hybridized with mtDNA (cys, met, phe, tyr, val) (Fig. 3 A-E). The mobility of the mt tRNA peaks differed from cyt tRNAs, except for one val tRNA and one tyr tRNA. Schneller et al[14] have obtained similar results for val tRNAs although they also detected a small additional peak which hybridized with mtDNA. The hybridization of four tyr tRNA species to mtDNA is shown in Fig. 4. One peak (II) that co-migrates with cytoplasmic $tRNA^{tyr}$, as well as the three that do not, hybridize with mtDNA. Schneller et al detected only one tyr tRNA that hybridized with mtDNA.[14]

Trp and arg tRNA (Fig. 3 F, G) showed two peaks on RPC-5 chromatography which differed in mobility from cyt tRNA, but hybridization to establish their genetic origin has not yet been done. The tRNAs for 2 tRNAs (leu, thr) (Fig. 3 H-I) have multiple peaks, only one of which is clearly different from its cytoplasmic counterpart; however, hybridization studies are still incomplete. Finally, gly tRNA contained 2 peaks, both co-migrating with cyt tRNA (Fig. 3 J). It is also not clear whether all the cyt tRNAs observed in our preparations represent simple contamination or whether in some cases they reflect import of cytoplasmic species into the mitochondria.

The above data clearly establish the existance of isoaccepting tRNAs for five amino acids that hybridize with mtDNA. Mitochondrially specified isoacceptors are

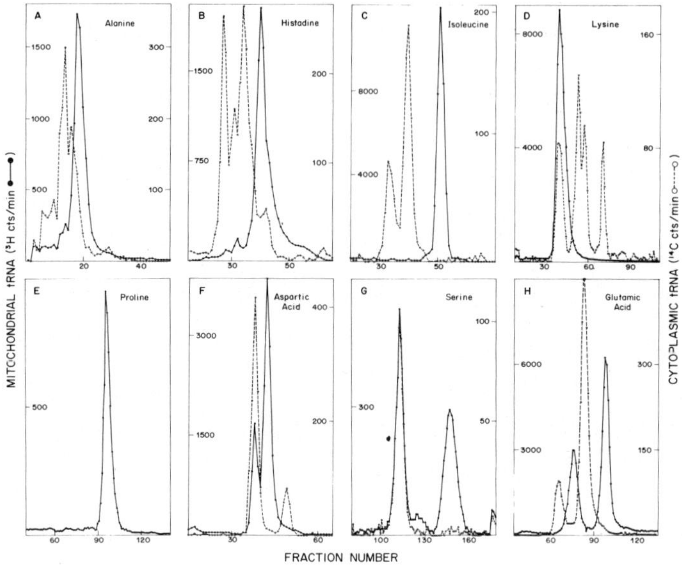

Fig. 2. RPC-5 chromatographs of ^{14}C cyt (o--o) and ^{3}H mt (o—o) tRNAs containing only one aminoacyl tRNA which hybridized to mtDNA; tRNAs were prepared and chromatographed as previously described.

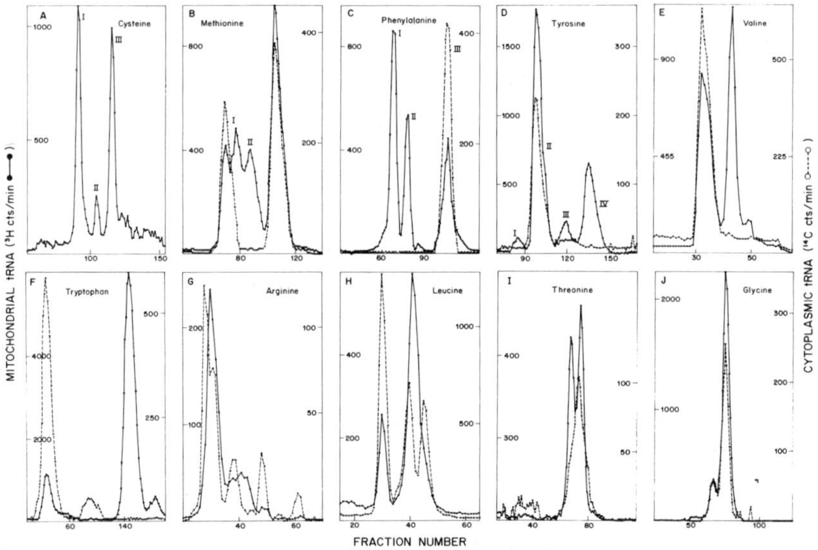

Fig. 3. RPC-5 chromatographs of ^{14}C cyt (o--o) and ^{3}H mt (o—o) tRNA containing more than one aminoacyl tRNA which hybridizes to mtDNA (A-E) or that remain to be tested by hybridization (F-J).

probably also present in some of the tRNAs containing multiple peaks that are still to be hybridized. Possibly additional species will be detected if growth conditions of the yeast are varied.

To obtain a valid estimate of the total number of tRNA genes in yeast mtDNA, it is necessary to determine whether the multiple tRNA species that hybridize with mtDNA are transcribed from separate mtDNA sequences, or whether they result from post-transcriptional modification. We have initiated experiments with several of the tRNAs to answer this question. For example, we recently presented data showing that the hybridization of the two mitochondrially specified cys tRNAs is additive, suggesting that they have different sequences.[15]

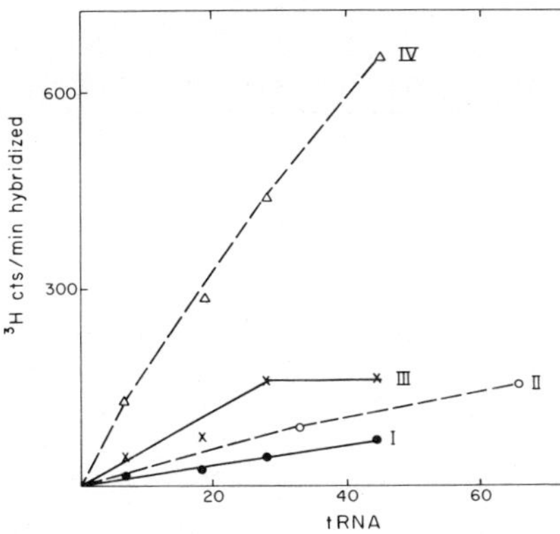

Fig. 4. Hybridization of tyr peaks to mtDNA. Blank filters fixed 1-10% of ^{3}H cts/min hybridized and these counts were subtracted from the experimental values. Hybridization of tyr II to mtDNA was competed by cold mt tRNA but not by cold cyt tRNA.

We have now also established that the two mitochondrially specified met tRNAs (designated I and II on the basis of elution properties on RPC-5 columns) also show additive hybridization (data not shown). We are currently trying to determine if one of these corresponds to an fmet initiator tRNA.[4] Furthermore, deletion mapping with a series of petite mtDNAs clearly shows that they are transcribed from different mtDNA sequences (Table 4). Met $tRNA_{II}$ hybridizes to several petite mtDNAs retaining the P marker (i.e., CEP-2, O_IP_2, and P_I), whereas met $tRNA_I$ hybridizes only with CEP-2 and CEP-4 mtDNAs. On the basis of our previous characterization of these petite mtDNAs,[16,17] we can conclude that met $tRNA_{II}$ maps near the val tRNA gene and met $tRNA_I$ lies further from P between val and isol tRNA genes. Since deletion mapping is a definitive way to demonstrate whether tRNA species are transcribed from separate mt genes, similar mapping studies are in progress for other isoacceptor tRNAs.

We have also used deletion mapping with petite mtDNAs, in collaboration with H. Fukuhara, to map the other newly identified mt tRNAs with respect to antibiotic resistance markers and previously mapped tRNA genes. All of the new tRNAs map in the CEP region of the genome with trp tRNA mapping close to the P marker, thr and cys tRNAs close to the C marker, and arg in the unordered set of tRNA genes located between the C and P markers. Fig. 5 shows our current deletion map of tRNA genes in yeast mtDNA.

In conclusion, we have shown that yeast mtDNA transcribes tRNAs for 19 and probably for 20 amino acids, and for at least 2 isoaccepting tRNA species. The two met tRNA isoacceptors are shown to map at different sites and thus be

TABLE 4

Hybridization of met $tRNA_I$ and met $tRNA_{II}$ to Petite mtDNA Relative to Grande mtDNA

Strain	mt loci					met $tRNA_I$	met $tRNA_{II}$
Grande IL8	C	E	O_I	O_{II}	P	1	1
CEP-2	+	+	-	-	+	4.5	5.5
CEP-3	+	+	-	-	+	0	0
CEP-4	+	+	-	-	+	2.5	0
R53	+	+	-	-	-	0	0
D41	+	-	-	-	-	0	N.T.
C42	+	-	-	-	-	0	0
D61	+	-	-	-	-	0	0
P_1	-	-	-	-	+	0	1.4
O_IP_2	-	-	+	+	+	0	1.0
O_IO_{II}-1	-	-	+	+	-	N.T.	0
EO_I	-	+	+	-	-	0	0

Petite strains and their use in deletion mapping have been described.[16,17]

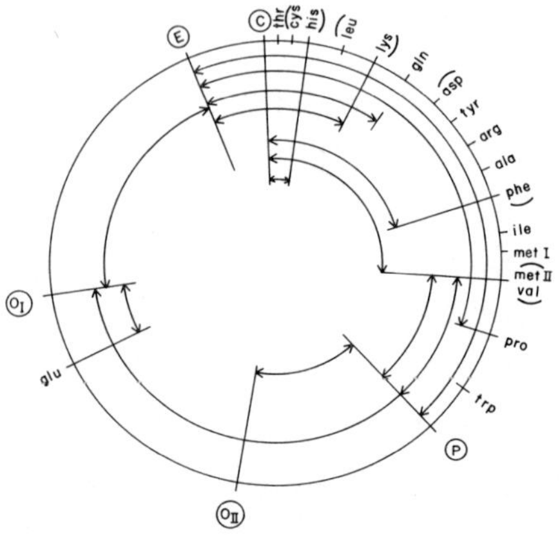

Fig. 5. Deletion map of yeast mt tRNA genes.

transcribed from different mtDNA genes. Several other isoaccepting species noted on RPC-5 chromatography are still to be mapped. We have also presented evidence for the presence of a mitochondrial $tRNA^{gln}$, which is probably synthesized by transamidation of a missense glu $tRNA^{gln}$ intermediate, a pathway present in some bacteria but not in the cytoplasm of yeast or other eukaryotic cells. Finally, we have mapped several additional mt tRNA genes.

REFERENCES

1. Suyama, Y. (1976) Cell Biology. Eds. P. L. Altmann and D. D. Katz, p. 228.
2. Casey, J. et al (1969) J. Cell Biol. 43, 18A.
3. Casey, J. et al (1972) J. Mol. Biol. 63, 431.
4. Halbreich, A. and Rabinowitz, M. (1971) Proc. Nat. Acad. Sci. USA 68, 294.
5. Cohen, M. et al (1972) J. Mol. Biol. 63, 441.
6. Cohen, M. and Rabinowitz, M. (1972) Biochim. Biophys. Acta 281, 192.
7. Casey, J. et al (1974) J. Mol. Biol. 88, 717.
8. Nass, M.M.K. and Buck, C.A. (1969) Proc. Nat. Acad. Sci USA 62, 306.
9. Nass, M.M.K. and Buck, C.A. (1970) J. Mol. Biol. 54, 187.
10. Lynch, P.C. and Attardi, G. (1976) J. Mol. Biol. 102, 125.
11. Martin, N. et al (1976) J. Mol. Biol., in press.
12. Wilcox. M.E. and Nirenberg, M.W. (1968) Proc. Nat. Acad. Sci. USA 61, 229.
13. Nirenberg, M.W. and Leder, P. (1964) Science 145, 1399.
14. Schniller, J. M. et al (1975) Biochimie, 57, 1051.
15. Rabinowitz, M. et al (1976) The Genetic Function of Mitochondrial DNA. Eds. C. Saccone and A. M. Kroon (in press).
16. Casey, J. et al (1974) J. Mol. Biol. 88, 717.
17. Fukuhara, H. et al (1976) Molec. Gen. Genet. 145, 7.

Genetics and Biogenesis of Chloroplasts and Mitochondria, Th. Bücher et al. eds.
© 1976 Elsevier/North-Holland Biomedical Press, Amsterdam, The Netherlands.

ISOACCEPTOR tRNA SPECIES IN YEAST MITOCHONDRIA - METHIONINE AND FORMYLMETHIONINE SPECIFIC tRNAs CODED BY MITOCHONDRIAL DNA

R. MARTIN, J.M. SCHNELLER, A.J.C. STAHL and G. DIRHEIMER
Institut de Biologie Moléculaire et Cellulaire du C.N.R.S.
15, rue Descartes, 67084 STRASBOURG CEDEX, FRANCE
and
Faculté de Pharmacie, 3, rue de l'Argonne, STRASBOURG, FRANCE

INTRODUCTION

It has been shown that yeast mit. DNA can hybridize with at least fourteen mit. tRNAs (1). We have reported the existence of mit. isoacceptor tRNAs having chromatographic behavior different from that of the corresponding cytoplasmic isoacceptors for Arg, Val, Tyr, Leu and Phe (2-6). Mit. $tRNA^{His}$ (7) and mit. $tRNA^{Glu}$ (8) have been separated form their cytoplasmic counterparts.

In this report we describe the separation by preparative chromatography of mitochondrial tRNA species corresponding to 11 aminoacids. We also describe experiments showing that mit. tRNA contains two isoaccepting species for Met, $tRNA^{Met}_{M}$ and $tRNA^{Met}_{F}$, both of which are coded by mit. DNA.

MATERIALS AND METHODS

Preparation of mit.tRNA from Saccharomyces cerevisiae ρ^+ IL8-8C and of cyt. tRNA from ρ° IL8-8C/H71, a petite lacking mit. DNA, were as described (6).

Aminoacylation of tRNAs and chromatography on RPC5 were as previously described (3). Chromatography on Sepharose 4B using a reverse salt gradient was performed as in (9), with minor modifications.

Formylation experiments were carried out as in (10) with 1-10-^{14}C formyltetrahydrofolate prepared according to SAMUEL et al. (11).

Hybridization procedures were reported in (5).

RESULTS AND DISCUSSION

1. Chromatography of bulk mit. tRNA on RPC5

RPC5 chromatography of uncharged mit tRNA (fig. 1) shows one peak for Ile, Gly, Arg and Tyr tRNAs, the two last sometimes showing an

---

Abbreviations : mit. : mitochondrial ; cyt : cytoplasmic.

additional minor peak. Two isoaccepting peaks were found for Met, Phe, Leu, Lys and Ser tRNAs and three isoaccepting peaks for Thr and Val tRNAs.

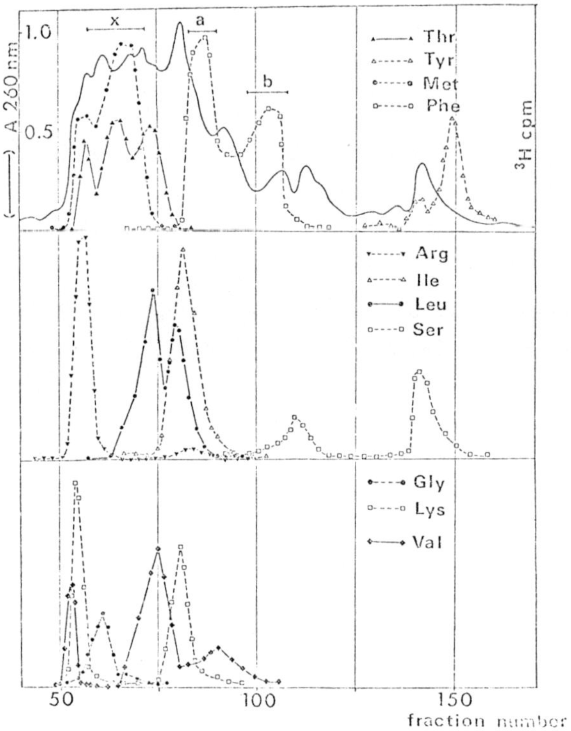

Figure 1 : RPC5 chromatography at pH 4.5 of mit. total tRNA.

These results are in good aggreement with those obtained by mit tRNA mapping on two dimensional gel electrophoresis (12). However, for $tRNA^{Leu}$ and $tRNA^{Phe}$ only one peak was found when RPC5 chromatography was done using charged tRNAs (5,6). Preliminary sequencing of purified $tRNA^{Phe}_{a}$ and $tRNA^{Phe}_{b}$ (see fig. 1) suggest that these two isoacceptors have the same primary structure and therefore may be two different conformations of the same mit. $tRNA^{Phe}$.

Hybridization of the different isolated isoacceptors will show whether they are products of different genes in the mit. DNA. We present now results obtained with mit. $tRNA^{Met}$.

2. <u>Evidence that mit. DNA codes for $tRNA^{Met}_{F}$ and $tRNA^{Met}_{M}$</u>

a) Cochromatography on RPC5 of ρ° ^{14}C Met tRNA and mit. ^{3}H Met tRNA.

Mit. $tRNA^{Met}$ elutes distinctly from ρ° $tRNA^{Met}$ (fig. 2) and does not contain any cyt. $tRNA^{Met}$.

Fraction X (see figure 1) was rechromatographed on a Sepharose 4B column using a reverse salt gradient, which improved the resolution of the two $tRNA^{Met}$ isoacceptors (figure 3).

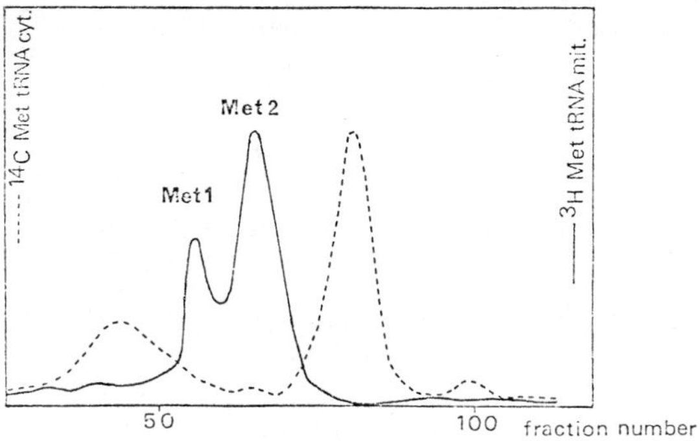

Figure 2

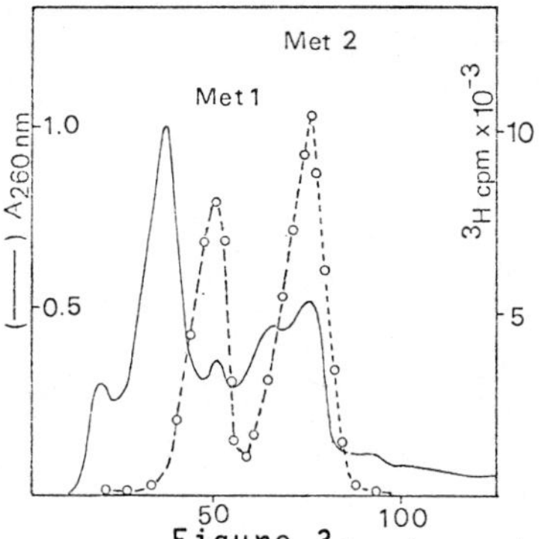

Figure 3

b) Characterization of mit. $tRNA^{Met}_{F}$

The two $tRNA^{Met}$ species were incubated with different enzymatic extracts and with 1-10-^{14}C formyltetrahydrofolate in an aminoacylation medium as described in (3). Results are shown in Table I.

TABLE I

AMINOACYLATION AND FORMYLATION OF MIT. $tRNA^{Met}_{1}$ AND $tRNA^{Met}_{2}$

	E.coli MRE600 Enzyme Aminoacylation (nmoles/mg)	E.coli MRE600 Enzyme Formylation (nmoles/mg)	Mit. Enzyme Aminoacylation (nmoles/mg)	Mit. Enzyme Formylation (nmoles/mg)
mit. $tRNA^{Met}_{1}$	11	0	10	0
mit. $tRNA^{Met}_{2}$	13	0	15	2.5
cyt. tRNA	1.4	0	1	0
pure E.coli $tRNA^{Met}_{F}$	33	18	20	3.2

Despite the low transformylase activity in mit. enzymatic extract, E.coli $tRNA^{Met}_{F}$, mit. $tRNA^{Met}_{2}$, but not mit. $tRNA^{Met}_{1}$, could be formylated. E.coli could formylate E.coli $tRNA^{Met}_{F}$ but showed no transformylase activity on mit. tRNA.

c) Hybridization of mit. $tRNA^{Met}$ with mit. DNA and nuclear DNA.

Figure 4 shows that $tRNA^{Met}_{1}$ could be hybridized to saturation with mit. DNA, whereas the plateau was not reached with $tRNA^{Met}_{2}$ which is not homogenous. However a hybridization plateau could be reached with $tRNA^{Met}_{2}$ in the presence of an excess of $tRNA^{Met}_{1}$. Under these conditions hybridization of the two $tRNA^{Met}$ showed additivity.

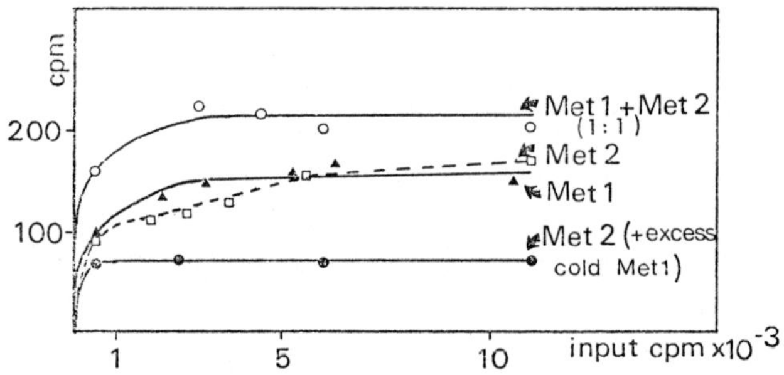

Figure 4 : Hybridization of mit.$tRNAs^{Met}$ or cyt. $tRNA^{Met}$ with 10μg mit. DNA. Absence of hybridization with nuclear DNA (20μg) and all $tRNAs^{Met}$ and with mit. DNA and cyt. tRNA (not shown).

It can be concluded that mit. $tRNA^{Met}_{1}$ is mit. $tRNA^{Met}_{M}$ and that mit. $tRNA^{Met}_{2}$ is mit. $tRNA^{Met}_{F}$. The two tRNAs are products of two distinct mit. DNA genes.

REFERENCES

1. CASEY, J.W., HSU, H.J., GETZ, G.S., RABINOWITZ, M. and FUKUHARA, H. (1974) J. Mol. Biol., 88, 735-747.
2. ACCOCEBERRY, B. and STAHL, A.J.C. (1971) Biochem. Biophys. Res. Commun., 42, 1235-1243.
3. ACCOCEBERRY, B., SCHNELLER, J.M. and STAHL, A.J.C. (1973), Biochimie, 55, 291-296.
4. SCHNELLER, J.M., ACCOCEBERRY, B. and STAHL, A.J.C. (1975) FEBS Lett., 53, 44-48.
5. SCHNELLER, J.M., STAHL, A. and FUKUHARA, H. (1975) Biochimie, 57, 1051-1057.
6. SCHNELLER, J.M., MARTIN, R., STAHL, A.J.C. and DIRHEIMER, G. (1975), Biochem. Biophys. Res. Commun., 64, 1046-1053.
7. BOGUSLAWSKI, G., VODKIN, M.H., FINKELSTEIN, D.B. and FINK, G.R. (1974) Biochemistry, 13, 4659-4667.
8. MARTIN, N., RABINOWITZ, M. and FUKUHARA, H. (1976) J. Mol. Biol., 101, 285-296.
9. HOLMES, W.M., HURD, R.E., REID, B.R., RIMERMAN, R.A. and HATFIELD, G.W. (1975) Proc. Nat. Acad. Sci. USA, 72, 1068-1071.
10. SAMUEL, C.E., McILROY, P.J. and RABINOWITZ, J.C. (1973), Biochemistry, 12, 3609-3615.
11. SAMUEL, C.E., D'ARI, L. and RABINOWITZ, J.C. (1970), J. Biol. Chem., 245, 5115.
12. MARTIN, R., SCHNELLER, J.M. (1976) Z. Physiol. Chem., 357, 298.

ISOACCEPTING $tRNA_{ser}$ IN MITOCHONDRIA FROM SACCHAROMYCES CEREVISIAE: MITOCHONDRIALLY CODED AND CYTOPLASMIC SPECIES

G.Baldacci, C.Falcone, L.Frontali, G.Macino and C.Palleschi
Centro Acidi Nucleici. Istituto di Fisiologia Generale
University of Rome (Italy)

In the last few years a fair amount of evidence has been collected concerning the number and identity of tRNAs coded for by mitDNA[1-4]. In HeLa and in yeast cells[3,5] the number of identified aminoacids which can be charged to mitochondrially coded tRNAs is 16, while, despite many efforts, no mitochondrial tRNAs have been found able to accept the remaining 4 aminoacids. Isoaccepting species coded for by mitDNA have been found in HeLa cells, in Tetrahymena and in yeast[3-8]. Two very important questions are therefore still requiring an answer, namely whether mitochondrial genome contains the information for the complete set of tRNAs to be used in mitochondrial protein synthesis and whether some cytoplasmic species are imported. In this communication we present some results which might be of interest for these questions.

In fig.1 reversed-phase chromatography of seryl-tRNAs is reported. Mitochondria were prepared from S.cerevisiae strain 42 as previously described[9]. The same profile is obtained when mitochondrial tRNA is acylated with mitochondrial or cytoplasmic ligase. On the contrary E.coli ligase acylates only peak V.

Two species are present in mitochondrial tRNA which are absent in the isogenic ρ_0 mutant and should therefore be mitochondrially coded. To assess this point, fractions corresponding to each peak were pooled and hybridized with mitDNA; results reported in table 1 show that only peaks IV and V hybridize with mitDNA. These two species might be the product of different mitochondrial genes or result from post-transcriptional modification of the same tRNA.

The first hypothesis is supported by the fact that species IV and V hybridize with mitDNA fragments having different base composition, obtained by sonication and fractionation on a CsCl gradient as previously described[9] (fig.2). Moreover the melting profiles of hybrids are slightly different (fig.3) and only species V is acylated by the ligase from E.coli (fig.1b).

Fig.4 shows that physiological conditions exist in which species IV is not detectable: no species IV is actually observed in the elution profile when tRNA is prepared from cells grown for 36 hours on 2% glucose, while the elution profile

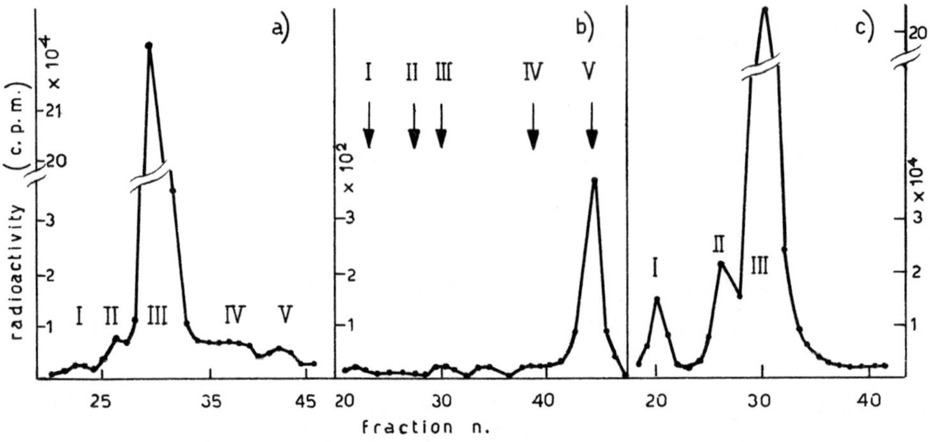

Fig.1. Elution profiles from RPC-5 columns of seryl-tRNA. Rapid reversed-phase chromatography of ^{3}H seryl-tRNA was carried out according to the method of Pearson et al.[10] . Columns, loaded with 100-200 μg ^{3}H seryl-tRNA, were developed with a linear NaCl gradient from 0.43 M to 0.65 M.

a) seryl-tRNA prepared from mitochondria of strain 42 cells grown for 15 hours on YEP medium containing 8% glucose and acylated with mitochondrial or cytoplasmic ligase as previously reported[9] (specific activity 500 cpm/μg). Acylated tRNA was fixed on a small DEAE column and eluted with 2 ml of 1.5 M NaCl, diluted to 0.43 M NaCl and loaded into a RPC-5 column.

b) The same seryl-tRNA preparation acylated with the E.coli ligase.

c) Total seryl-tRNA prepared from a ρ_0 mutant, acylated with mitochondrial or cytoplasmic ligase (specific activity 3500 cpm/μg).

TABLE 1

HYBRIDIZATION OF ISOLATED SERYL-tRNA SPECIES WITH mitDNA

The fractions corresponding to each of the five peaks of seryl-tRNA eluted from a RPC-5 column (loaded with 280 μg of mitochondrial tRNA) were pooled, seryl-tRNA fixed on a small DEAE column and eluted with 100 μl of 8xSSC pH 5.2 to which 100 μl H_2O and 200 μl formamide were added. Filters were incubated in this mixture for 5 hours at 33°C.

seryl-tRNA	input cpm	cpm retained on filters containing	
		30 μg mitDNA	no DNA
I	5,000	7	-
II	10,000	12	-
III	90,000	60	29
IV	12,000	175	4
V	17,000	130	3

of fig.1a has reproducibly been obtained with mitochondrial preparations from cells grown for 15 hours (up to $3\cdot10^7$ cells/ml) on high glucose.

A final important point is raised by the presence of cytoplasmic species in mitochondrial preparations. The elution profile of seryl-tRNA from mitochondria and

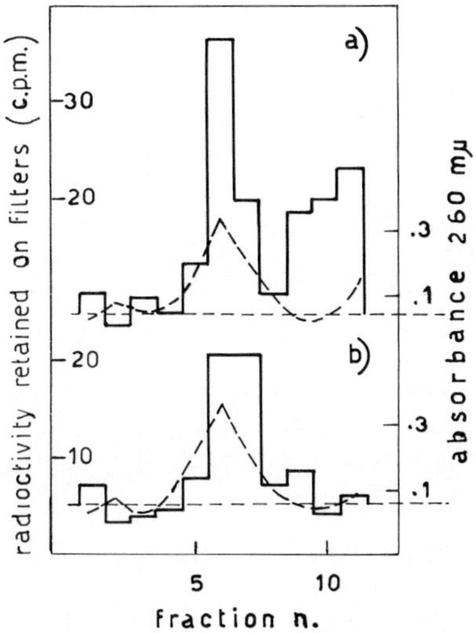

Fig.2. Annealing of seryl-tRNA on mitDNA fragments of different buoyant density. --- A_{260} of fractions of the CsCl gradient. a) seryl-tRNA preparation containing species IV and V. b) seryl-tRNA preparation containing only species V.

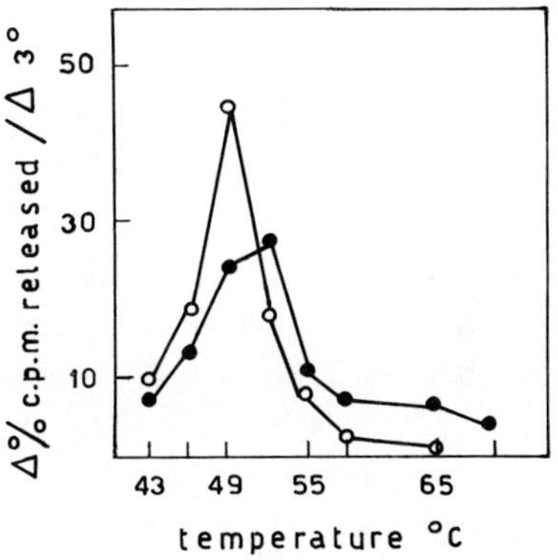

Fig.3. Differential melting profiles of hybrids between seryl-$tRNA_{IV}$ (●), seryl-$tRNA_V$ (o) and mitDNA. Melting procedure was performed as described by Martin et al.[7]

cytoplasm of the same cells (fig.4) shows that mitochondria contain a high proportion of species III, while species I and II are almost absent. Therefore we do not believe that the presence of peak III in this mitochondrial preparation can be explained only by cytoplasmic contamination. This contamination can be estimated around 15-20% as measured by the addition, during the preparation of mitochondria, of the cytoplasm of cells grown on labelled adenine. Treatment of mitochondria with RNAase or preparation of mitochondria from protoplasts[8] results in an increase in the proportion of mitochondrial species but does not alter the overall pattern.

These results are consistent with the presence in the organelle of mitochondrial and cytoplasmic isoacceptors as reported by Chiu et al.[6] and show that the presence of functional species IV is dependent from the physiological conditions of

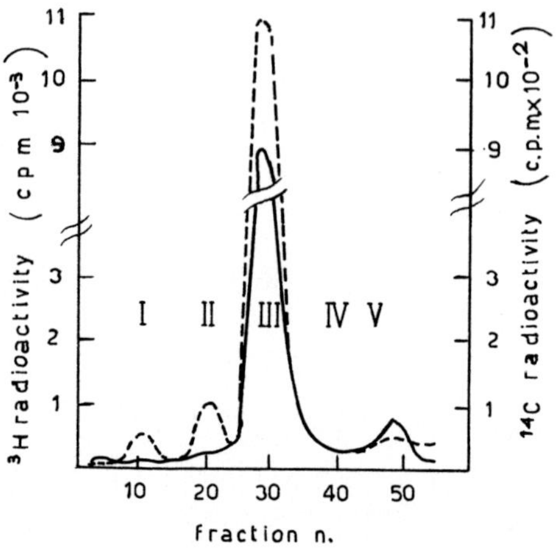

Fig.4. Elution profile from RPC-5 column of mitochondrial ^{14}C seryl-tRNA (——) and cytoplasmic ^{3}H seryl-tRNA (---) from the same cells grown for 36 hours on 2% glucose.

the cells, thus suggesting the possibility of a supply of tRNA to the mitochondrial protein synthesizing machinary, varying in connection with the functional conditions of the organelle.

REFERENCES

1. Dawid,I.B. (1972) J.Mol.Biol. 63, 201-216.
2. Casey,J., Hsu,H.J., Rabinowitz,M., Getz,G.S., Fukuhara,H. (1974) J.Mol.Biol. 88, 717-733.
3. Lynch,D.C. and Attardi,G. (1976) J.Mol.Biol. 102, 125-141.
4. Chiu,N., Chiu,O.S.A. and Suyama,Y. (1974) J.Mol.Biol. 82, 441-457.
5. Rabinowitz,M. (1976) In The Genetic Function of Mitochondrial DNA (Kroon,A.M., and Saccone,C. eds.) Elsevier-North Holland, Amsterdam.
6. Chiu,N., Chiu,O.S.A. and Suyama,Y. (1975) J.Mol.Biol.99, 37-50.
7. Martin,N.,Rabinowitz,M. and Fukuhara,H. (1976) J.Mol.Biol. 101, 285-296.
8. Schneller,J.M., Stahl,A., and Fukuhara,H. (1975) Biochimie 57, 1051-1057.
9. Baldacci,G., Carnevali,F., Frontali,L., Leoni,L., Macino,G. and Palleschi,C. (1975) Nucleic Acid Research 2, 1777-1786.
10. Pearson,R.L., Weiss,J.F. and Kelmers,A.D. (1971) Biochim.Biophys.Acta 228, 770-774.

Genetics and Biogenesis of Chloroplasts and Mitochondria, Th. Bücher et al. eds.

IMPORTED tRNA: ITS SYNTHETASE AS A PROBABLE TRANSPORT PROTEIN

Yoshitaka Suyama and Jin Hamada
Department of Biology
University of Pennsylvania
Philadelphia, Pennsylvania 19174
U.S.A.

Tetrahymena mitochondria contain functional tRNA species which are not coded for by the mt DNA (1,2). These tRNA species, by inference, must be transcribed from the nuclear DNA and transported in the mitochondria. We have coined the term "imported tRNA" for such tRNA, while those tRNA species which are found in the mitochondria, but coded for by the mtDNA, are called "native tRNA" (2).

From the combined chromatographic (RPC-5) and DNA/RNA hybridization analyses of individually aminoacylated mitochondrial tRNAs, we have to date tentatively classified mitochondrial tRNA species corresponding to 14 amino acids as in Table 1.

TABLE 1. NATIVE AND IMPORTED tRNA

	Amino acid (No. of isoacceptors)	
Native tRNA	Leu (3) ,	Phe (2)
	Trp (1) ,	Tyr (1)
Imported tRNA	Arg (1) ,	Lys (3), Val (2),
	Ala (2) ,	Asp (2), Gly (2 or 3),
	Gln (2) ,	His (2), Ile (2),
	Ser (4 ~ 5)	

Tetrahymena mitochondria must contain 30 or more isoacceptors for all 20 amino acids. Different isoacceptors known for native tRNA, three for Leu and two for Phe, are transcribed from the mtDNA and they are heterogeneric, *i.e.*, they are transcribed from different genes (2,3). Together with Trp and Tyr tRNAs identified here as native tRNAs, only 7 tRNA species are transcribed from the mtDNA, and all other tRNAs in the mitochondria are imported tRNAs. This is in contrast to the situation in other organisms where all tRNA species in mitochondria are so far hybridized only to mtDNA (4-9). An exception was noted, however, in *Saccharomyces* the presence in mitochondria of tRNA of nuclear-DNA origin was documented (10).

The discovery of the dual genetic origin of tRNA in mitochondria signifies further complex manifestations of the nuclear-organelle interaction; immediately apparent is the probable involvement of tRNA synthetases. In this paper we present our recent studies on the synthetase enzymes which acylate the native and imported tRNA. In order to interpret our results on the tRNA-synthetase relationship, we advance the hypothesis that the synthetases play a direct role in tRNA transport into the mitochondria.

RESULTS AND DISCUSSION

It has been shown (3) that mitochondrial native tRNA shows an isoacceptor pattern on RPC-5 which is clearly distinguishable from that of the corresponding cytoplasmic tRNAs. On the other hand, a chromatographic pattern for isoacceptors of imported tRNA is either identical to that of the cytoplasmic tRNA or different only in the relative abundance of different isoacceptors (Fig. 1).

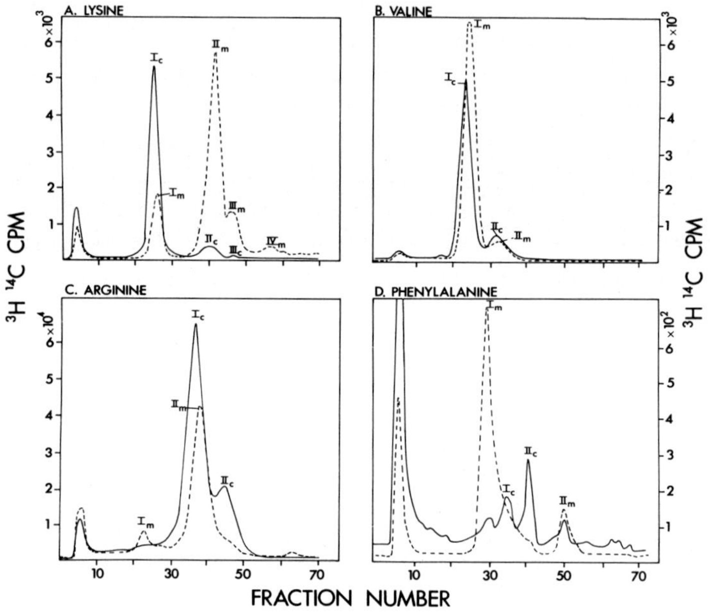

Fig. 1. Comparative RPC-5 profiles of isoacceptors of mitochondrial and whole-cell tRNAs aminoacylated with the homologous enzymes. Solid line, cytoplasmic tRNA; broken line, mitochondrial tRNA.

The differences in isoacceptor patterns could be ascribed either to intrinsic differences in tRNAs or to different charging enzymes between the mitochondria and the cytoplasm. The charging efficiencies for the two different tRNAs, Tetrahymena whole-cell (Twc) and *E. coli*, can be effectively used to differentiate the functional properties of mitochondrial (mt) and cytoplasmic (ct) tRNA synthetases (RS) (Table 2). For many amino acids, both the ct and mt RS's exhibit a preference for Twc tRNA rather than for *E. coli* tRNA, as charging preference is expressed as a ratio of charging rates for *E. coli* tRNA over Twc tRNA. The ct and mt RS's show a similar degree of preference for Ala, Arg, Asn, Asp, Gln, Ile, Lys, Met, Ser, Thr, and Val. On the other hand, some significant differences in charging were found for Cys, Glu, Gly, His, Leu, Phe, Pro, Trp and Tyr.

TABLE 2. COMPARISONS OF AMINOACYLATION EFFICIENCIES OF WHOLE-CELL OR E. COLI tRNA BY CYTOPLASMIC AND MITOCHONDRIAL SYNTHETASES.

Amino acid	Comparative charging efficiency (E. coli/wc)+		Maximum charging§	
	Cyto.	Mito.	Cyto.	Mito.
Ala	0.25	0.20	1,700	1,800
Arg	0.06	0.02	10,000	14,000
Asn	0.67	0.54	800	260
Asp	0.18	0.27	4,300	2,600
Cys	0.94	1.72	11,300	3,800
Gln	0.08	0.10	5,000	2,400
Glu	0.02	1.15	13,000	850
Gly	0.06	0.11	5,140	7,200
His	3.25	0.16	10,000	1,400
Ile	0.96	0.62	17,000	24,000
Leu	1.35	6.24	3,900	10,200
Lys	0.03	0.02	22,200	20,000
Met	1.57	1.53	6,270	4,200
Phe	0.04	9.49	14,000	1,600
Pro	0.08	1.16	2,400	2,800
Ser	0.15	0.17	11,000	14,500
Thr	1.38	1.02	110	340
Trp	0.74	1.31	1,600	2,500
Tyr	0.04	2.61	8,800	3,000
Val	0.01	0.03	11,000	8,800

+ Ratio of charging rates of E. coli and Tetrahymena whole-cell tRNAs with either cytoplasmic or mitochondrial synthetase preparations.

§ Maximum charging of whole-cell tRNA attainable for each amino acid with either cytoplasmic or mitochondrial tRNA synthetase preparations. The value is cpm of ^{3}H-amino acid (^{35}S for cys and met; ^{14}C for thr) incorporated into an arbitrary unit (2.0 μg) of whole-cell tRNA.

Among the latter class, Leu, Phe, Trp and Tyr RS's are of special interest, since the mt RS for each of them shows a greater charging preference for E. coli tRNA than the corresponding ct RS. For these amino acids, Tetrahymena mitochondria contain "native tRNAs" distinct from their cytoplasmic (non-mitochondrial) counterparts (Table 1). The preference of an RS for E. coli tRNA is not limited to (mt) RS, however; the ct His RS prefers E. coli tRNA more than the mt RS. Furthermore, the ct RS's for Ile, Leu, Met, Thr and Trp, all show very good charging abilities for E. coli tRNA as well as for cognate Twc tRNA.

The results above also demonstrated that mt RS's do not always prefer E. coli tRNA. In fact, mt RS often charges Twc or cytoplasmic tRNA as well as does the corresponding ct RS. This point is illustrated also in Table 2. For a number of amino acids the same levels of maximum charging of Twc tRNA could be attained when charged either with mt RS or with ct RS.

One important conclusion can be drawn from these observations concerning the functional properties of the mt and ct RS's. An mt RS that charges a "native tRNA" differs from the corresponding enzyme in the cytoplasm that charges cytoplasmic tRNA. This conclusion is amply supported by our previous data on Leu-RS in Tetrahymena (11,12). The mt and ct Leu-RS's differ immunologically as well as physicochemically. Differences in chromatographic behavior and charging efficiency have also been shown to exist between mitochondrial and cytoplasmic tRNA synthetases in a variety of organisms (4,13-15).

The conclusion reached above is not necessarily applicable to all other synthetases. Cursory examinations of Neurospora tRNA synthetases (3) reveal that charging efficiencies of mt and ct RS's for many amino acids are not clearly distinguishable. In Phaseolus, Guillemaut et al (14) demonstrated that the mt and ct Leu-RS's exhibited identical aminoacylation efficiency. In Tetrahymena a similar situation can be seen, particularly for those synthetases involving imported tRNA. As seen in Table 3, the ct and mt RS's for Val cannot be distinguished even on the basis of several kinetic properties examined. Heat inactivation rates and the linearity of the aminoacylation reaction also hold the same for both Val RS's. The mt and ct enzymes for Arg, Ile and Lys also are strikingly similar (Table 3). Although their properties show some indications of dissimilarity in these parameters, this level of experimental sophistication renders it impossible to differentiate between them. The only exception to this is found for Ser synthetases; the mt and ct RS's show significant differences in Km's for ATP and tRNA. Differences can also be noted in Mg^{+2} ion concentration and reaction temperature for optimal reaction.

TABLE 3. KINETIC PROPERTIES OF CYTOPLASMIC AND MITOCHONDRIAL SYNTHETASES FOR IMPORTED tRNAs.

	Km's						Reaction Optima			
	amino acid (10^{-6})		ATP (10^{-5})		$tRNA^{+}$ (10^{-6})		Mg. Conc.(mM)		Temperature (°C)	
RS	ct	mt	ct	mt	ct	mt	ct	mt	ct	mt
Leu*	7.2	6.5	110	170	-	-	10	2.5	35	20
Val	1.0	1.0	2.3	2.3	-	-	4.4	4.4	27.5	27.5
Ile	4.6	3.4	42	32	6.8	1.0	1.5	1.0	35	35
Ser	3.5	4.4	0.2	1.1	0.6	2.5	4.0	20	20	27.5
Arg	0.6	0.4	21	5.5	5.6	2.2	1.5	2.5	35	35
Lys	3.3	1.7	22	18	2.6	1.2	7.0	4.0	25	27.5

* Native tRNA; data from Chiu and Suyama (12).
\+ Twc tRNA was used. MW of tRNA is assumed to be 25,000.

The kinetic parameters, Km was determined from a Lineweaver-Burk plot using rates of aminoacylation reaction with varying concentrations of each substrate under saturating concentrations of other substrates and optimal Mg concentration at the optimal reaction temperature. The linear portion of aminoacylation was taken for the rate of reaction. Linearity of the reaction varies with the different enzyme, but no difference in the linearity of reaction was observed between the mt and ct synthetases.

Sucrose gradient centrifugation analyses of these synthetase preparations lend support to our previous notion that the mt RS's for native tRNAs are different from the corresponding ct RS's as revealed in sedimentation behavior (Table 4) Unimodal activity peak generally is observed as the basis to suggest that they are homogeneous enzymes. On the other hand, the ct and mt RS's for Arg, and Val show no significant differences in sedimentation. A difference in sedimentation may exist in these synthetases for Ile, Lys and Ser.

TABLE 4. SEDIMENTATION COEFFICIENTS (S)$^{+}$ OF SELECTED TETRAHYMENA tRNA SYNTHETASES

Amino acid	Cyto.	Mito.
Arg	4.0	4.0
Gly	6.0	5.7
Ile*	6.5	5.3
Leu	6.0	5.0
Lys*	6.0	5.3
Phe	8.0	3.7
Ser*	6.5	5.3
Tyr	11.5	4.7
Val	5.5	5.0

$^{+}$ Sedimentation values were calculated from activity peak positions in sucrose gradient fractions according to Chiu and Suyama (12). Leucine synthetases, ct RS (6S) and mt RS (5S) were used as standards for sedimentation.

* denotes amino acids for the tRNA synthetases for native tRNA.

The mt Val RS that charges imported $tRNA^{Val}$ in mitochondria, is by all probability identical to the ct Val RS. This is further strengthened by the fact that both mitochondrial and cytoplasmic tRNAs show an identical isoacceptor pattern on RPC-5, whether acylated with the homologous or with the heterologous enzyme. The other synthetases in mitochondria that charge imported tRNA, show a very close resemblance to the cytoplasmic synthetases in terms of several kinetic parameters applied here, but they may not be identical. In contrast for native tRNAs in mitochondria, there invariably exist the corresponding synthetases which are functionally and structurally distinct from the cytoplasmic synthetases, as exemplified in the case of Leu RS (11,12) and also as judged by very different sedimentation behaviors for Phe and Tyr mt and ct RS's. Furthermore, the native tRNAs are often preferentially charged by the mitochondria-specific enzymes, and their RPC-5 patterns differ significantly from those of the cytoplasmic tRNA.

These data, as a whole, suggest that there exists a mechanism which regulates the transport and distribution of tRNA molecules in accord with the compartmentation of their own synthetase enzymes. From all available evidence, the synthetases in mitochondria are most probably synthesized on the cytoribosomes and subsequently transported into the organelle. Since the synthetases show a relatively strong affinity (>10^7~10^9 M^{-1}) toward cognate tRNAs, they are likely to exist *in vivo* in the enzyme-tRNA complex form (16). It then follows that as the enzymes migrate into the organelle, tRNA can be carried in with them. Whether the acylation takes place at the site of the complex formation is not important, but the release of tRNA may be readily affected by different millieu in the mitochondria or by the probable conformation change that occurs at the site of the enzyme transport.

The hypothesis, that the synthetases serve as transport proteins, can circumvent the general conceptual difficulty in transporting such polar molecules as tRNA through the membranes. Furthermore, the specificity of tRNA to be transported as well as the selective transport of certain isoacceptors can be envisaged without invoking an unknown entity to serve these purposes.

The tRNA synthetases are relatively large and complex proteins; in all cases studied to date their molecular weights in functional forms are greater than 100,000. They are often dimeric and tetrameric, consisting of subunits (see also 16). In animal cells, the synthetases are known to form a highly agglomerated complex form with themselves and also with other components such as tRNA and ribosomes (16,17). Should such complex formation be interpreted as a mechanism to immobilize these enzymes for their optimal functioning, it should also render each enzyme molecule unavailable for binding to a possible transport site(s) on the mitochondrial membrane. Whether or not the enzymes have greater affinity for the complex formation or for the mitochondria determines their fate regarding the choice of compartmentation between mitochondria and cytoplasm. If the enzyme is relatively free from complex formation it could enter the mitochondrion, as would the Val RS's with tRNA bound to them. On the other hand, the ct Tyr or Phe tRNA synthetases might prefer the complex formation as their sedimentation values indicate, thereby preventing them from transport. Then, there must be different duplicate enzyme molecules which are needed to be transported into the mitochondria.

A peculiar property of *E. coli* Met RS may be of interest (18). Although the native form of this enzyme is known to be a tetramer of 43,000 M.W. subunit, trypsin treatment converts it to a dimer form consisting of a subunit M.W. of 32,000. Interestingly, the converted form of this enzyme exhibits no appreciable change in its acylating function or the Michaelis constants toward all the substrates. This example suffices in illustrating its applicability in reinforcing the present hypothesis. Some alteration of the cytoplasmic enzymes, such as is shown in this case, could account for the sufficient change from the complex form, that the original form preferred, to the mitochondrial form for charging "imported tRNAs." This could also explain the great similarity in property between the mitochondrial and cytoplasmic synthetases for these tRNAs, but with slightly reduced sedimentation values for the mitochondrial synthetases.

The hypothesis formulated here allows a possibility that mitochondrial tRNA and enzymes are not as pure as they are assumed to be. As it was noted earlier (10) in yeast, tRNAs of cytoplasmic origin were found associated with mitochondria. Buck and Nass (19) also observed that some cytoplasmic tRNA persists in their mitochondrial tRNA preparations as revealed in MAK column chromatography. In another example, Gross et al (20) reported that a Neurospora Leu RS mutant can grow despite the absence of the mitochondria-specific Leu RS. According to the present hypothesis, the cytoplasmic tRNA and its synthetase would be spilled into the mitochondria so that the mutant's mitochondria still function in protein synthesis. In the same vein, this seepage effect may be responsible for the presence of a full complement of tRNA molecules in the mitochondria of any organism where the general deficiency of tRNA genes in the mt DNA is apparent (5-8).

The hypothesis that tRNA synthetases serve as transport proteins of tRNA, undoubtedly requires rigid experimental support and scrutiny. As enumerated above, it can explain a number of observations accumulated in a variety of organisms. Thus although crucial evidence and support are lacking, we believe that tRNA and synthetase imports are a universally occurring phenomenon, albeit its differing magnitude of manifestation is dependent on the organisms. It is our hope that the present hypothesis will stimulate further elucidation of and account for the mechanism of tRNA transport, in addition to the evolution of tRNA and its synthetase, as well as their structural genes.

REFERENCES

1. Suyama, Y. (1967) Biochemistry 6, 2829-2839.
2. Chiu, N., Chiu, A., and Suyama, Y. (1975) J. Mol. Biol. 99, 37-50.
3. Chiu, N., Chiu, A., and Suyama, Y. (1974) J. Mol. Biol. 82, 441-457.
4. Nass, M.M., and Buck, C. (1970) J. Mol. Biol. 54, 187-198.
5. Dawid, I. (1972) J. Mol. Biol. 63, 201-216.
6. Aloni, Y., and Attardi, G. (1971) J. Mol. Biol. 55, 271-276.
7. Lynch, D.C., and Attardi, G. (1976) J. Mol. Biol. 102, 125-141.
8. Reijnders, L., and Borst, P. (1972) Biochem. Biophys. Res. Commun. 47, 126-133.
9. Rabinowitz, M., Casey, J., Gordon, P., Locker, J., and Hsu, H-J, and Getz, G.S. (1974) In the Biogenesis of Mitochondria. (A.M. Kroon, and C. Saccone, ed.). p 89, Acad. Press.
10. Casey, J.W., Hsu, H-J., Getz, G.S., Rabinowitz, M., and Fukuhara, H. (1974) J. Mol. Biol. 88, 735-747.
11. Chiu, A.O.S. and Suyama, Y. (1973). Biochim. Biophys. Acta. 299, 557-563.
12. Chiu, A.O.S., and Suyama, Y. (1975) Arch. Biochem. Biophys. 171, 43-54.
13. Barnett, W.E., Brown, D.H., and Epler, J.L. (1967) Proc. Natl. Acad. Sci. U.S. 57, 1775-1781.

14. Guillemaut, P., Steinmetz, A., Burkard, G. and Weil, J.H. (1975) Biochim. Biophys. Acta, 378, 64-72.

15. Kislev, N., Selsky, M.I., Norton, C., and Eisenstadt, J.M. (1972) Biochim. Biophys. Acta 287, 256-269.

16. Kisselev, L.L. and Favorova, O.O. (1974) In Adv. in Enzymol. (A. Meister, ed.). pp 141-238.

17. Vennegoor, C. and Bloemendal, H. (1970) In Methods in Enzymol. 29, E. 584-600.

18. Cassio, D. and Waller, J-P. (1971) Eur. J. Biochem. 20, 283-300.

19. Buck, C.A., and Nass, M.M.K. (1969) J. Mol. Biol. 41, 67-82.

20. Gross, S.R., McCoy, M.T., and Gilmore, E.B. (1968) Proc. Natl. Acad. Sci. U.S. 61, 253-260.

Genetics and Biogenesis of Chloroplasts and Mitochondria, Th. Bücher et al. eds.

CHARACTERIZATION OF rRNA AND tRNA FROM MITOCHONDRIA OF LOCUSTA MIGRATORIA

H. Feldmann
Institut für Physiologische Chemie, Physikalische Biochemie und Zellbiologie der Universität, Goethestraße 33, 8000 München 2
Germany

W. Kleinow
Zoologisches Institut der Universität, Weyertal 119, 5000 Köln 41
Germany

RNA from whole mitochondria of locust thoracic muscle was separated into three main fractions by gel electrophoresis [1, 2] or by gradient centrifugation (Fig. 1).

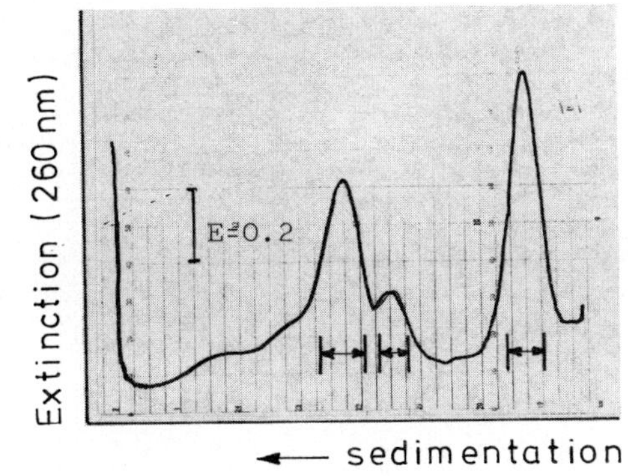

Fig. 1. Separation of the RNAs from locust mitochondria by gradient centrifugation. Mitochondria were purified and RNA was extracted as described [1, 3]. The gradients were prepared according to [6] from 5% and 20% RNase free sucrose in 0.1 M NH_4Cl, 0.03 M Tris-HCl, pH 7.6. 6 A_{260}-units were loaded onto the gradient and centrifuged for 17 h at 205 000 g. The indicated fractions were pooled for the estimation of the base composition and for the two-dimensional electrophoresis.

The two high molecular weight RNA species (520 000 and 28C 000 daltons) represent the rRNAs of the mitochondrial 60 S ribosomes [2]. The third fraction (35 - 50% from whole mitochondria) was shown to be tRNA. The purity of the mitochondrial preparation [2, 3] as well as the characteristics of the RNAs exclude a contamination of cytoplasmic compounds. The amount of the isolated tRNA was not changed upon preincubation of the mitochondria at 4° C up to 90 hours (Table 1).

TABLE 1

YIELD AND PROPORTION OF rRNAs AND tRNA OBTAINED FROM LOCUST MITOCHONDRIA AFTER DIFFERENT PREINCUBATION PERIODS.

Mitochondria were prepared and the RNA was extracted as described [3, 2]. In the control experiments the RNA was extracted immediately, in the other experiments the mitochondria were treated under the conditions indicated. After electrophoresis, the gels were stained with toluidine blue and the percentages of the different RNA species were calculated from densitograms. The total RNA content of the mitochondria was determined after hydrolysis in 0.2 N perchloric acid (10 min, 95° C) using spectrophotometry at 260 nm.

Conditions	% RNA extracted	large rRNA % in gel electrophoresis	large rRNA % RNA extracted	small rRNA % in gel electrophoresis	small rRNA % RNA extracted	tRNA % in gel electrophoresis	tRNA % RNA extracted
Control*)	23.2 ±2.0	27.6 ±4.3	7.1 ±1.6	13.7 ±2.3	3.5 ±0.9	47.2 ±3.8	11.8 ±1.0
37° C/ 18 h	23.6	13.0	3.0	12.0	2.8	62.0	14.4
37° C/ 27 h	14.0	7.0	1.0	25.0	3.5	69.0	9.7
4° C/ 27 h	20.5	35.0	7.1	17.0	3.5	48.0	9.8
4° C/ 91 h	21.5	32.5	7.0	20.0	4.3	47.0	10.1

*) average values from six experiments with standard deviations.

The apparent relatively high amount of isolated tRNA as compared to the amount of isolated rRNA seems to be due to an incomplete recovery of the rRNAs by the extraction procedure. The actual level of tRNA within the mitochondria amounts to only 10 or 15% of total mitochondrial RNA.

TABLE 2

NUCLEOTIDE COMPOSITION OF LOCUST MITOCHONDRIAL rRNAs and tRNA

1 to 3 A_{260}-units of RNA was incubated in 10 µl of 0.05 M Na-acetate, 0.01 M $MgCl_2$, pH 3.5, with 0.05 units of RNase T_2 (E.C. 3.1.4.23; Sankyo, Tokyo) in a sealed capillary tube for 4 hrs at 37° C. The sample was then charged onto a cellulose thin-layer plate (Merck, Darmstadt) as a 3 cm wide trace and chromatographed in the solvent system isopropanol-conc. HCl-H_2O (114:29:24, by volume) [7] together with reference markers. The material containing the nucleotides was scraped off the plates, extracted with 500 µl H_2O, and the UV-spectra were measured in a ZEISS PMQ II spectrophotometer. Moles percent were calculated from the corresponding extinction coefficients [8]. The spots for Gp, Ap, and Cp in this solvent system contain also modified nucleotides. For the detection of minor nucleotides, the material from the bands mentioned above was rechromatographed in the solvent system described in ref. 9 together with reference markers.

Moles %	Mitochondria large rRNA	Mitochondria small rRNA	Mitochondria tRNA	Cytosol tRNA
Gp	18.5	17.4	16.6	34.6
Ap	29.4	31.6	33.2	22.6
Cp	16.0	8.2	14.1	20.8
Up	36.1	42.8	30.6	17.2
Ψp	-	-	4.5	3.8
Tp	-	-	1.0	1.0
(Gp+Cp)	34.5	25.6	30.7	55.4

As shown in Table 2, the (G+C) content of the locust mitochondrial rRNAs was found to be extremely low and the base compositions of the large and the small rRNA were found to differ. A similar low (G+C) content was found for the mitochondrial tRNA, whereas the (G+C) content of the cytosolic tRNA was in the normal range. The mitochondrial tRNA was resolved by two-dimensional gel electrophoresis (Fig. 2) into about 25 major and some minor spots. Some of the spots were alloted to specific tRNA species.

We conclude that locust mitochondria contain a full complement of rRNAs and tRNAs. We suggest that these RNAs are specified by the mitochondrial genome. Similar findings as described here have been

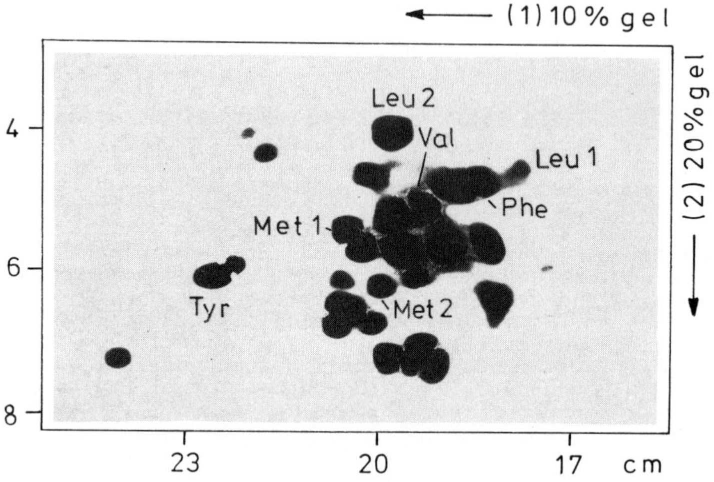

Fig. 2. Two-dimensional gel electrophoresis of locust mitochondrial tRNA, technical details see [10].

reported for mitochondrial rRNAs and for some of the tRNAs in other organisms which contain 60 S mitochondrial ribosomes, e.g.[4, 5]. Since the coding capacity even of a small mitochondrial genome would be sufficient for the two ribosomal RNAs and about 20 to 30 tRNA species, we feel it unlikely that some tRNAs should be specified by the mitochondrial genome, while others must be "imported".

REFERENCES

1. Kleinow, W. and Neupert, W. (1970) Z. Physiol. Chem. 351, 1205
2. Kleinow, W. (1974) in: A.M. Kroon and C. Saccone (eds.) "The Biogenesis of Mitochondria", Academic press: London
3. Kleinow, W. and Neupert, W. (1971) FEBS Letters 15, 359
4. Dawid, I.B. (1972) J. Mol. Biol. 63, 201
5. Wu, M. et al. (1972) J. Mol. Biol. 71, 81
6. Noll, H. (1967) Nature 215, 360
7. Wyatt, G.R. (1951), Biochem. J. 48, 584
8. Zachau, H.G. et al. (1966) Z. Physiol. Chem. 347, 212
9. Markham, R. and Smith, J.D. (1951) Biochem. J. 49, 401
10. Fradin, A. et al. (1975) FEBS Letters 50, 185

Genetics and Biogenesis of Chloroplasts and Mitochondria, Th. Bücher et al. eds.

IMMUNOLOGICAL STUDY OF YEAST MITOCHONDRIAL PHENYLALANYL-tRNA SYNTHETASE

J.M. SCHNELLER, C. SCHNELLER and A.J.C. STAHL
Laboratoire de Biochimie, U.E.R. de Sciences Pharmaceutiques,
Université Louis Pasteur - 3, rue de l'Argonne
67083 STRASBOURG, FRANCE

INTRODUCTION

We had previously shown (1) that yeast mitochondria contain aminoacyl-tRNA synthetases, specific for mitochondrial tRNA, and distinct from the cytoplasmic synthetases. Mitochondrial synthetases like the cytoplasmic enzymes are nuclear coded and are detected in the cytoplasm (1). The aim of this report is to look for an immunological difference between the two kinds of enzymes.

MATERIALS AND METHODS

Yeast mit tRNA was prepared as in reference 2. Yeast cyt tRNA was purchased from Boehringer (Mannheim). Yeast mitochondrial and total aminoacyl-tRNA synthetases extracts were obtained as earlier described (1, 3, 4). Cyt Phe-RS was purified to homogeneity as described in references : 5 and 6.

Cyt- and mit Phe-RS were chromatographed on hydroxylapatite (1).

Antibodies against pure cyt Phe-RS and against mit aminoacyl-tRNA synthetases were raised in rabbits, in four weeks, by injecting respectively four times 1.5 mg pure cyt Phe-RS and 5.4 mg, then three times 18 mg of mit enzymatic extract. The first injection was in a foot pad with complete Freund's adjuvant, the second in the muscle, the third and the fourth were intraveinous.

To assay the inhibition of the antibodies on the enzymatic activities, the measurements of the enzymatic activity were carried out as in (1), after preincubation of the enzymes with the antisera during 30 minutes at 37° C. Controls were performed in the same way by replacing the antisera by an equal volume of serum from a non-immunized rabbit.

Abbreviations : mit = mitochondrial, cyt = cytoplasmic, Phe-RS or Leu-RS = phenylalanyl - or leucyl-tRNA synthetases. anti cyt- or mit Phe-RS = antiserum directed toward pure Phe-RS or toward the enzymatic extract containing the mit Phe-RS.

RESULTS

1) Separation of cyt- and mit Phe-RS by hydroxylapatite chromatography of total yeast enzymes

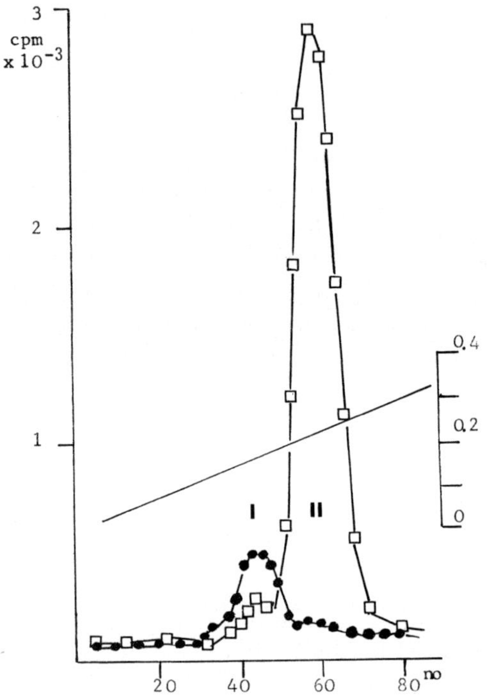

Figure 1 : Phe-RS activity in ρ^+ yeast total enzyme. The scale on the right represents the molarities of the K phosphate gradient in moles/l. The abscissa represents the number of the fractions. The activity was revealed with ●-●-● mit tRNA, or -□-□- cyt tRNA.

Two peaks are revealed. Peak II is the cyt Phe-RS. Peak I charges mit tRNA and is the unique one present in mitochondria (see figure 2a). Peak I represents therefore mit Phe-RS. As this peak was present in a $\rho°$ mutant lacking mit DNA and in ρ^+ yeast grown in presence of 10 % glucose, we concluded that mit Phe-RS is nuclear coded and not repressed by glucose (1).

2) Absence of cross contamination in pure cyt Phe-RS and in mit Phe-RS (i.e. enzymatic extract from highly purified mitochondria).

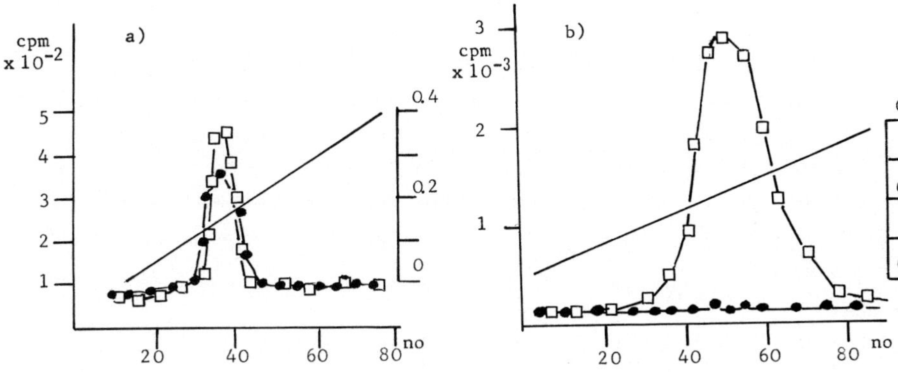

Figure 2 : Phe-RS activity in hydroxylapatite chromatographies of a) yeast mit enzyme, b) pure cyt Phe-RS. For legend see figure 1). Note that mit tRNA was incubated four times longer than cyt tRNA with the same aliquot of each fraction in b).

Mit enzyme contains only the mit Phe-RS (figure 2a). Pure cyt Phe-RS does not charge mit tRNA and is free of mit Phe-RS (figure 2b). Thus we could prepare antibodies against each synthetase.

3) <u>Inhibition of the Phe-RS activity by anti Phe-RS</u>

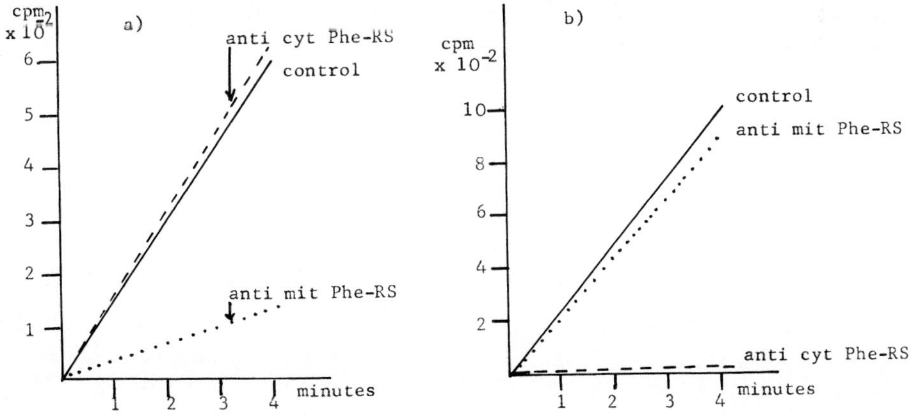

Figure 3 : a) Activity of mit Phe-RS tested on mit tRNA, b) activity of cyt Phe-RS tested on cyt tRNA.

Inhibition of enzymatic activities occurs only in homologous systems, i.e. anti mit Phe-RS with mit Phe-RS produces 77 % inhibition and anti cyt Phe-RS with cyt Phe-RS produces 100 % inhibition.

DISCUSSION

Mit- and cyt Phe-RS from Saccharomyces cerevisiae differ in 1) substrate specificity, mit $tRNA^{Phe}$ being charged only by mit Phe-RS, 2) chromatographic behaviour on hydroxylapatite, 3) inhibition by specific antibodies directed against each of the synthetases. Similar results have been obtained in the case of mit- and cyt Leu-RS from Tetrahymena pyriformis (7, 8). Our results lead to the conclusion that mit- and cyt Phe-RS are two distinct proteins which derive from two different nuclear genes. Work is actually in progress to gain similar knowledge for other synthetases. Preliminary experiments indicate that the same conclusion holds true for mit Leu-RS.

REFERENCES

1. SCHNELLER J.M., SCHNELLER C., MARTIN R. and STAHL A.J.C., (1976), Nucleic Acids Res., 3, 1151 - 1165.
2. SCHNELLER J.M., MARTIN R., STAHL A.J.C. and DIRHEIMER G. (1975), Biochem. Biophys. Res. Commun., 64, 1046 - 1053
3. ACCOCEBERRY B., SCHNELLER J.M. and STAHL A.J.C., (1973), Biochimie, 55, 291-296
4. SCHNELLER J.M., ACCOCEBERRY B. and STAHL A.J.C. (1975), Febs Lett., 53, 44 - 48
5. FASIOLO F., BEFORT N., BOULANGER Y. and EBEL J.P. (1970), Biochim. Biophys. Acta, 217, 305-318
6. FASIOLO F. and EBEL J.P. (1974), Eur. J. Biochem., 49, 257-263
7. CHIU A.O.S. and SUYAMA Y. (1973), Biochim. Biophys. Acta, 229, 557-563
8. CHIU A.O.S. and SUYAMA Y. (1975), Arch. Biochem. Biophys., 171, 43-54.

Genetics and Biogenesis of Chloroplasts and Mitochondria, Th. Bücher et al. eds.
© 1976 Elsevier/North-Holland Biomedical Press, Amsterdam, The Netherlands.

MITOCHONDRIAL PROTEIN SYNTHESIS IN HIGHER PLANTS

C. J. Leaver
Department of Botany
University of Edinburgh
Edinburgh EH9 3JH, Scotland

INTRODUCTION

It has been established that fungal and animal mitochondria contain distinct genetic and protein-synthesising systems, which are responsible for the synthesis of probably less than 10% of the total mitochondrial protein.

The 8-10 hydrophobic polypeptides synthesised on the fungal mitochondrial ribosomes, have been shown to form part of three enzyme complexes of the inner membrane, cytochrome c oxidase, oligomycin-sensitive ATPase and cytochrome b[1]. More recently it has been demonstrated that a similar range of polypeptides is synthesised in Hela cell mitochondria[2]. There is mounting evidence that these proteins, which form part of the inner mitochondrial membrane, are necessary for the correct assembly of the majority of mitochondrial proteins which are specified by nuclear genes and synthesised on cytoplasmic ribosomes.

Although the oxidative metabolism of higher plant mitochondria has been clearly defined, the characterisation and function of plant mitochondrial genetic and protein synthesising systems has, with few exceptions, been neglected. The lag in reliable information on the biogenesis of plant mitochondria has in part been due to technical difficulties in extracting suitable quantities of intact and uncontaminated mitochondria from plant tissue. It has been shown, however, that a range of higher plant mitochondrial DNAs have a remarkably constant buoyant density at around 1.706 - 1.707 g/cm^3 and a contour length of at least 30 μm[3,4] exceeding that of animal mitochondrial DNA by a factor of at least six and somewhat bigger than that reported for ascomycetes.

This raises the question as to why plant mitochondria apparently contain so much more genetic material than animal mitochondria, to perform the same functions, namely those of respiration and generation of energy.

With these questions in mind we have characterised the nucleic acids and protein synthesising system of higher plant mitochondria and examined whether or not plant mitochondria are capable of synthesing additional proteins to those synthesised by fungi and animal mitochondria.

RESULTS AND DISCUSSION

We have previously shown that a range of higher plant mitochondria contain ribosomes which sediment at 77-78S, and are quite distinct from the cytoplasmic 80S and chloroplast 70S ribosomes. These ribosomes can be dissociated into large and small subunits which contain discrete rRNA species which sediment at 24S and 18.5S and have estimated molecular weights of $1.12 - 1.26 \times 10^6$ and $0.69 - 0.78 \times 10^6$, depending on the plant species[3]. We have also demonstrated the presence of mitochondria-specific 4S RNA and a 5S (35,000 mol. wt.) rRNA component[5] which has not been reported in animal and ascomycete mitochondrial ribosomes[6], but has been shown to be present in all other types of procaryotic and eucaryotic ribosomes studied.

In terms of total cellular protein synthesis the mitochondrial contribution in higher plants is almost certainly less than 1%, which poses problems in studying the proteins specifically synthesised by these organelles. The most common approaches to overcoming this problem are either to use isolated organelles capable of *in vitro* protein synthesis or to analyse proteins labelled *in vivo* in the presence of selective inhibitors of cytoplasmic protein synthesis, such as cycloheximide. We have used both these procedures to study the proteins synthesised by mitochondria in mung bean (*Phaseolus aureus*) hypocotyl and artichoke (*Helianthus tuberosus*) tubers.

Mitochondria isolated from both mung bean and artichoke under sterile conditions were routinely checked for purity and intactness by electron microscopy, oxygen electrode polarography, spectrophotometry and bacterial counts. Optimal conditions for radioactive amino acid incorporation into protein by the isolated mitochondria were established[7] and the incorporation was shown to be completely dependent upon the intactness of the mitochondrion and coupled oxidative phosphorylation. The system exhibits characteristic substrate and co-factor requirements, and sensitivity towards inhibitors of respiration and protein synthesis.

The products of *in vitro* protein synthesis by isolated mitochondria from artichoke and mung bean, have been shown, by one dimensional sodium dodecyl sulphate polyacrylamide gel electrophoresis, to consist of at least eight recognisable polypeptides ranging in molecular weight from 12,000 to 55,000 (Fig. 1A). The polypeptides were shown to be components of the mitochondrial membrane system.

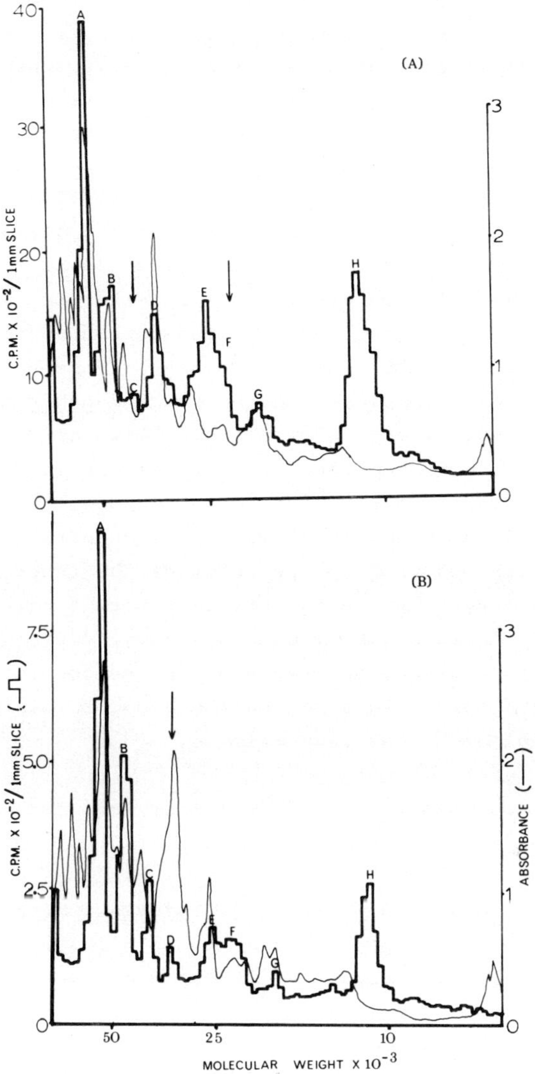

Fig. 1. SDS-polyacrylamide gel electrophoresis of the polypeptides synthesised (A) in vitro and (B) in vivo by mitochondria from artichoke tubers.

(A) Isolated mitochondria were labelled with (^{35}S)-methionine for 1 h in 200 mM-mannitol, 12.5 mM-KCl, 15 mM-$MgCl_2$, 5 mM-KH_2PO_4, 12.5 mM-Tris-HCl pH 7.2, 1 mM-ADP, 20 mM-malate, and 0.025 mM-amino acid mix minus methionine. (B) Artichoke tuber tissue was incubated in 20 µg/ml cycloheximide for 45 min, followed by the addition of (^{35}S)-methionine for 6 h. Mitochondria were isolated by standard procedures. Mitochondria solubilised with SDS were fractionated by SDS-PAG-electrophoresis on 15% gels for 6 h at 3 mA/gel.

The proteins synthesised *in vitro* by artichoke mitochondria were compared with those synthesised *in vivo* in the presence of cyclohex-imide at concentrations sufficient to completely inhibit cytoplasmic protein synthesis. Comparison of the labelled products of *in vivo* and *in vitro* synthesis analysed by gel electrophoresis (cf Fig. 1A and 1B) revealed that the radioactive profiles were strikingly similar, despite differences in the relative proportions of incorporation into three of the polypeptide components. Although it is not yet possible to make a functional identification of the discrete protein components detected here, it is interesting to notice that both the number and the size range of these components agree reasonably well with the number and size range of the polypeptide subunits that have been identified in the fungi *Saccharomyces* and *Neurospora*, and in animal cells, as products of mitochondrial protein synthesis. Our results together with others are suggestive, but by no means conclusive, evidence that there has been a conservation, during the evolutionary modification of mitochondria, of their ability to synthesise several key proteins of the inner mitochondrial membrane. This suggestion is supported by recent reports that the increased size of mitochondrial DNA in ascomycete fungi and higher plants is likely to be due to regulatory sequences rather than protein-coding sequences[8]. This raises the question as to why plants need more control sequences than animals.

REFERENCES

1. Schatz, G. and Mason, T. L. (1974) Ann. Rev. Biochem. 43, 51-87.
2. Constantino, P. and Attardi, G. (1975) J. Mol. Biol. 96, 291-306.
3. Leaver, C. J. and Harmey, M. A. (1973) Plant Mitochondrial Nucleic Acids in Biochem. Soc. Symp. 38, 175-193.
4. Kolodner, R. and Tewari, K. K. (1972) Proc. Nat. Acad. Sci. 69, 1830-1834.
5. Leaver, C. J. and Harmey, M. A. (1976) Biochem. J. 157, 275-277.
6. Borst, P. (1972) Ann.Rev. Biochem. 41, 333-376.
7. Leaver, C. J. (1975) Biogenisis of Plant Mitochondria in The Chemistry and Biochemistry of Plant Proteins (Harborne, J. B. and Van Sumere, C. F., eds.) 137-166. Academic Press.
8. Ellis, J. (1975) Nature 256, 617-618.

GENERAL ASPECTS OF MITOCHONDRIAL BIOGENESIS

Genetics and Biogenesis of Chloroplasts and Mitochondria, Th. Bücher et al. eds.

MITOCHONDRIAL PHOSPHOLIPID SYNTHESIS AND THE PHOSPHOLIPID EXCHANGE PROTEINS

K.W.A. Wirtz, R.H. Lumb, H.H. Kamp, G.M. Helmkamp, H. van den Bosch
L.L.M. van Deenen
Laboratory of Biochemistry, University of Utrecht,
University Center "De Uithof", Padualaan 8, Utrecht,
The Netherlands

INTRODUCTION

The principal phospholipids of the inner mitochondrial membrane from rat liver are phosphatidylcholine (PC), phosphatidylethanolamine (PE), phosphatidylinositol (PI) and cardiolipin (CL). A large bulk of evidence, however, indicates that in the rat liver PC, PE and PI are synthesized predominantly, if not exclusively, on the endoplasmic reticulum[1]. In analogy with mitochondrial proteins synthesized on the cytoribosomes we are faced with the problem of how these phospholipids become available for assembly during mitochondrial biogenesis. In this paper we present data on the role of phospholipid exchange proteins in the transfer of PC, PI and PE to the mitochondria. In addition, the biosynthesis of CL by the mitochondrion will be discussed.

PHOSPHOLIPID EXCHANGE PROTEINS

The cytosol of all eukaryotic cells investigated to date, contain phospholipid exchange proteins. It is these proteins that catalyze the transfer of phospholipids between membranes[2,3]. Incubation of, for example, a pH 5.1 supernatant from rat liver with ^{32}P-labeled microsomes and unlabeled mitochondria results in a transfer of [^{32}P] PC, [^{32}P] PI and [^{32}P] PE to the mitochondria (Fig. 1). If the pH 5.1 supernatant from rat brain is incubated with the same membranes, PC and PI are transferred at rates different from liver while transfer of PE is not observed. These data indicate that the phospholipid exchange proteins involved in the transfer of PC, PI and PE are present at different levels in the two tissues (see below).

Table 1 summarizes some characteristics of the phospholipid exchange proteins purified from rat liver[4]. In measuring the transfer of phospholipids from rat liver microsomes to PC vesicles it was found that the exchange protein with an isoelectric point of 8.4 transferred specifically PC; the exchange proteins with isoelectric points of 5.1 and 5.3 transferred PI and to a lesser extent PC. A protein that transfers PE has not been isolated yet. With an antiserum raised against the PC-exchange protein it was shown that 60% of the total PC-transferring activity of the rat liver pH 5.1 supernatant results from this protein. Similar studies have

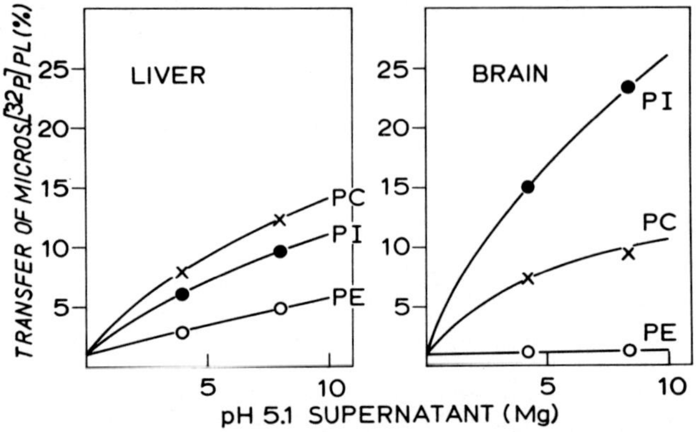

Fig. 1. The effect of the pH 5.1 supernatant protein from rat liver and brain on the exchange of PC, PI and PE between ^{32}P-labeled microsomes and unlabeled mitochondria. Exchange is expressed as the percentage of each microsomal [^{32}P] phospholipid pool transferred to the mitochondria. For experimental details, see similar experiments described in ref. 12.

TABLE 1

SOME PROPERTIES OF PHOSPHOLIPID EXCHANGE PROTEINS FROM RAT LIVER

Molecular weight	Isoelectric point	Fraction microsomal pool transferred (%)***		
		PI	PC	PE
16.000*	8.4	0.01	6.95	0.0
25.000-30.000**	5.1	6.93	0.76	0.0
25.000-30.000**	5.3	7.91	0.99	0.0

* Based upon sodium dodecylsulfate polyacrylamide gel electrophoresis and amino acid composition.

** Calculated from elution volume on Sephadex G-75

*** Rat liver microsomes (1.25 mg of protein) and phospholipid vesicles (1 μmole of phosphorus) were incubated for 30 min. at 37° with 0.21 μg protein (IEP 8.4), 3.76 μg protein (IEP 5.1) and 2.68 μg protein (IEP 5.1). For further experimental details see ref. 4.

been performed on the supernatants from bovine heart, liver and brain with antisera raised against the PC-exchange protein from bovine liver and the PI-exchange protein from bovine brain[5]. Whereas the antisera against these two proteins are not cross-reactive, the PI- and PC-exchange proteins are antigenically identical in heart, liver and brain, yet present at different levels.

The PC-exchange protein from bovine liver has a molecular weight of 22.000, an isoelectric point of 5.8, and contains one molecule of non-covalently bound PC upon purification[6]. Incubation of the protein with ^{14}C PC vesicles results in an incorporation of $[^{14}C]$ PC into the protein by an exchange mechanism (Fig. 2). Upon interaction of the $[^{14}C]$ PC-exchange protein with detergent micelles of sodium deoxycholate or Triton X-100 $[^{14}C]$ PC was recovered from the micelles. Then the delipidated exchange protein still has the capacity to function as a carrier of PC between membrane interfaces. This implies that the presence of PC on the protein is not a prerequisite for the protein to interact with a membrane. Moreover it shows that the delipidated protein can extract a PC-molecule from the membrane. These are conditions that have to be met for any phospholipid exchange protein to catalyze the net transfer of phospholipid from a donor to an acceptor membrane.

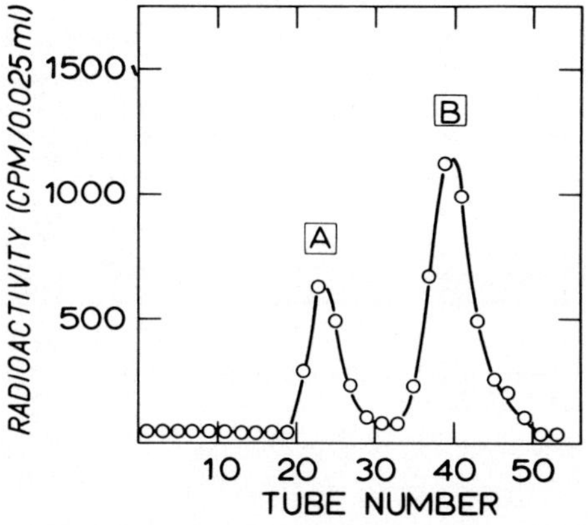

Fig. 2. Separation of $[^{14}C]$ PC-exchange protein (B) from $[^{14}C]$ PC vesicles (A) on Sephadex G-75. For experimental details, see ref. 6.

NET TRANSFER OF PHOSPHOLIPIDS

In general, if both donor and acceptor membrane contain PC, the phospholipid exchange proteins will catalyze an exchange of PC between these membranes. Net transfer of PC has been observed, however, with acceptor membranes prepared from PE, CL and hydrophobic protein[7]. The phospholipid exchange protein from bovine heart transported PC from PC-vesicles to these PC-deficient membranes. A concomitant protein-mediated transfer of PE and CL was not observed. This shows that the exchange protein can release PC into a membrane interface without binding a phospholipid molecule in return. It is feasible that this release of PC is governed by a competition between the exchange protein and the hydrophobic mitochondrial protein present in the acceptor membrane.

The PI-exchange proteins from bovine brain were found to catalyze the transfer of PI fromrat liver microsomes to PC vesicles deficient in PI[5]. Many processes that are controlled by the interaction of a stimulus with a receptor on the plasma membrane, are accompanied by an increased breakdown of PI in these membranes[8]. So it is observed that *in vitro* acetylcholine stimulates the turnover of PI in synaptosomes from rat brain. Synaptosomal plasma membranes, however, cannot synthesize PI and may depend on PI-exchange proteins to maintain the "steady state" of PI in this membrane. It is, therefore, of particular interest to find that synaptosomes are enriched in the PI-exchange proteins[9]. As compared to the 105.000 x g supernatant from total rat brain, 1 mg of soluble protein released from synaptosomes transfers 7-fold the amount of PI (Table 2).

TABLE 2

PI- AND PC-TRANSFER ACTIVITIES IN SYNAPTOSOMES AND CYTOSOL FROM RAT BRAIN

Protein fraction	Fraction microsomal pool transferred (%)*,**	
	PI	PC
Synaptosomal	28.2 ± 4.2	7.6 ± 1.5
105.000 x g supernatant	3.7 ± 0.4	2.0 ± 0.1

* Transfer of PI and PC were determined as described in ref. 5. Activity is expressed as the percentage of microsomal phospholipid transferred per 30 min. per mg of protein at 37°.

** Numbers ± s.d. are based on three experiments.

Mitochondria from rat liver lack the ability to synthesize phosphatidylserine (PS), yet contain in the inner membrane PS-decarboxylase[10]. Recently it was reported that the pH 5.1 supernatant from rat liver stimulates the transfer of [^{14}C] PS from microsomes or sonicated dispersions of microsomal lipids containing [^{14}C] PS to mitochondria[11]. The label in the mitochondria was recovered as [^{14}C] PE suggesting that [^{14}C] PS is transferred to the outer membrane followed by a transfer to the inner membrane to be decarboxylated. Since the mitochondria lack PS we are dealing here with a net transfer of PS. Thus, *in situ*, mitochondrial PE may in part originate from microsomal PS.

In the studies on the net transfer of PI and PS reported above, it has not been well established whether this transfer is compensated by a transfer of another lipid in the opposite direction. These studies do show, however, that phospholipid exchange proteins may play a role in the regulation of the phospholipid composition of membranes.

TRANSFER OF PC AND PE TO MITOCHONDRIAL INNER MEMBRANE

Administration of labeled phospholipid precursors to rats demonstrates that *in vivo* PC and PE synthesized on the endoplasmic reticulum, are very rapidly incorporated into the mitochondrial inner membrane[12]. This as well as the above experiment on the transfer of [^{14}C] PS raise the question as to how the inner membrane is accessible for phospholipids formed extra-mitochondrially.

Incubation of ^{32}P-labelled microsomes and unlabeled mitochondria with the pH 5.1 supernatant from rat liver results in an exchange of, for example, PC as indicated by the decrease of specific activity in the microsomes and the increase in the mitochondria (Fig. 3A). Subfractionation of the mitochondria showed that the outer membrane became more rapidly labeled than the inner membrane. In the reverse experiments where ^{32}P-labeled mitochondria were incubated with unlabeled microsomes, the drop in specific activity of mitochondrial PC was more pronounced for the outer than the inner membrane (Fig. 3B). These results suggest that *in vitro* the phospholipid exchange protein transfers PC between the microsomes and the outer membrane while inside the mitochondrion PC can move between outer and inner membrane. Direct proof for an intramitochondrial lipid movement is presented in Fig. 4. One hour after injection of [^{32}P] phosphate the phospholipids from the outer membrane have a 3 to 4-fold higher specific activity than the inner membrane. Incubation of these mitochondria has no effect on the specific activity of total PC and

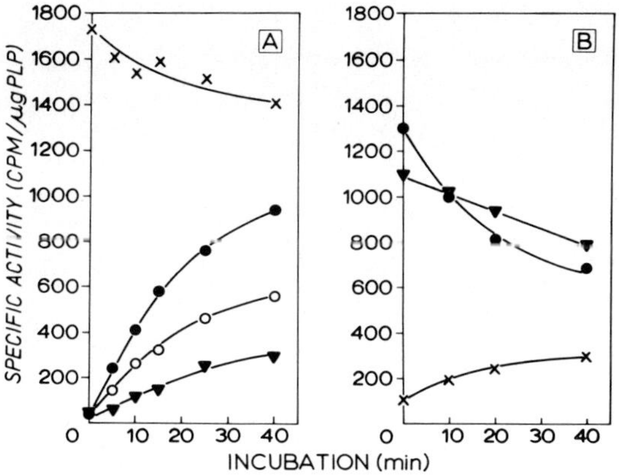

Fig. 3. Exchange of PC between microsomes and mitochondria from rat liver. Mitochondria (120 mg of protein), microsomes (40 mg of protein) and pH 5.1 supernatant (85 mg of protein) were incubated at 37° in a total volume of 30 ml 0.25 M sucrose - 1 mM EDTA - Tris (pH 7.8); ^{32}P-labeled microsomes in experiment A and ^{32}P-labeled mitochondria in experiment B. For further experimental details, see ref. 12.
X-X, microsomes; O-O, total mitochondria; ●-●, outer mitochondrial membrane; ▽-▽, inner mitochondrial membrane.

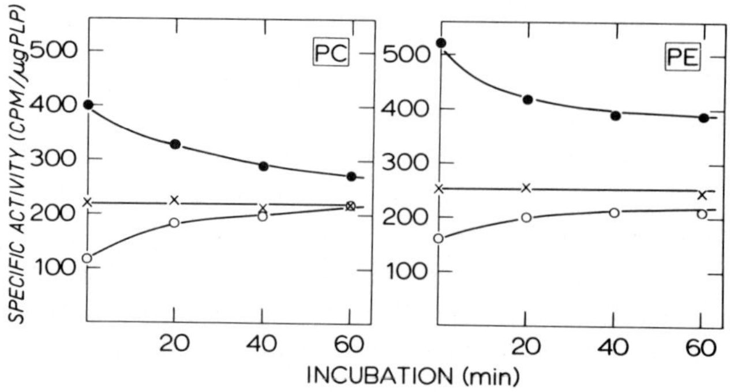

Fig. 4. Exchange of PC and PE inside the mitochondrion between outer and inner membrane. Mitochondria (120 mg of protein) were incubated at 37° in a total volume of 2.5 ml sucrose-EDTA-Tris. For further experimental details, see ref. 12.
X-X, total mitochondrion; ●-●, outer membrane; O-O, inner membrane.

PE. However, the specific activities of PC and PE increased in the inner membrane and decreased in the outer membrane. Since phospholipid exchange proteins were not detected in the intermembrane space, it is assumed that this intramitochondrial phospholipid movement reflects a direct contact between the outer and inner membrane. It is through these contacts that mitochondrial proteins synthesized on cytoribosomes may also be introduced into the inner membrane. Some evidence has been provided that in *Saccharomyces cerevisiae* cytoribosomes interact with the outer membrane at those sites where there exist a contact with the inner membrane[13]. Electron micrographs have provided additional evidence that such sites of contact exist[14].

BIOSYNTHESIS OF CARDIOLIPIN

In the rat liver cardiolipin is found exclusively in the mitochondria. This phospholipid is formed in the inner membrane according to the following reaction scheme:

I phosphatidic acid (PA) + CTP → CPD-diglyceride
II CDP-diglyceride + glycerol-3-phosphate → phosphatidylglycerol (PG)
III PG + CDP-diglyceride → cardiolipin

Whereas reactions I and II occur also in the microsomes, reaction III is limited to the mitochondria[1]. In view of the speculations on the bacterial origin of mitochondria it is worth noting that cardiolipin is synthesized in bacterial systems via the reaction:

2 phosphatidylglycerol → cardiolipin + glycerol

In the above sequence of reactions the subcellular origin of PA is not clear. It has been well established that microsomes unlike the inner membrane can synthesize PA[1]. It is still a point of controversy whether the mitochondrial outer membrane has a similar activity. PA can probably be transferred to the outer membrane as PA transfer actitivity has been observed in the cytosol from guinea pig brain[15] and with the PI-exchange proteins from bovine brain[5]. Transport of PA inside the mitochondrion to the side of cardiolipin biosynthesis in the inner membrane may follow a similar route as has been described above for PC and PE.

A tentative scheme is presented in Fig. 5 showing how the mitochondrial inner membrane may depend on the endoplasmic reticulum for its phospholipid constituents. In this concept the phospholipid exchange proteins control the transfer of phospholipids to the outer membrane whereas permanent and/or transient contacts between outer and inner membrane facilitate the supply of phospholipids to the inner membrane. Except for cardiolipin there is also a movement of phospholipids from the inner membrane to the endoplasmic reticulum (not indicated in Fig.). It is presumed that the continuous exchange of phospholipids between membrane structures in the intact cell reflects a mechanism by which the cell maintains a "steady state" of phospholipids in the various subcellular membranes.

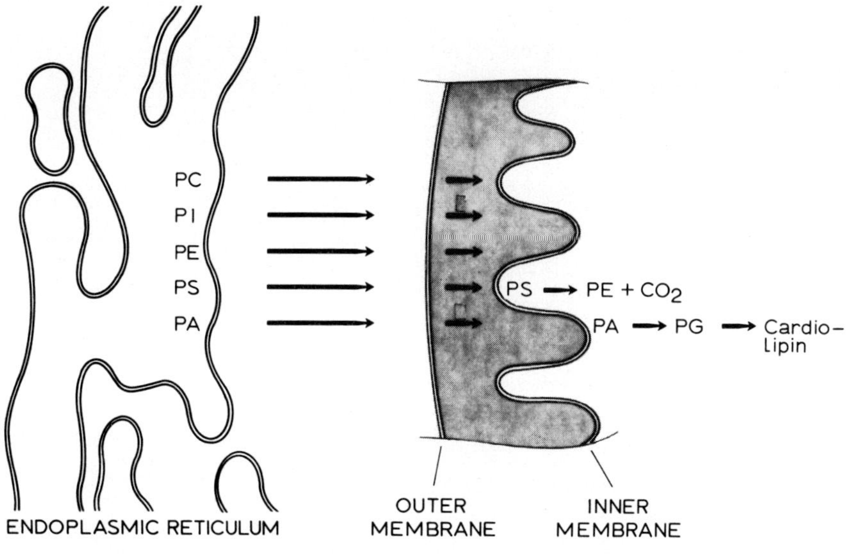

Fig. 5. Relationship between phospholipid biosynthesis in the endoplasmic reticulum and the mitochondrion of rat liver. A tentative scheme.

REFERENCES

1. van den Bosch, H. (1974) Ann. Rev. Biochem., 43, 243.
2. Wirtz, K.W.A. and Zilversmit, D.B. (1969) Biochim. Biophys. Acta, 193, 105.
3. McMurray, W.C. and Dawson, R.M.C. (1969) Biochem. J., 112, 91.
4. Lumb, R.H., Kloosterman, A.D., Wirtz, K.W.A. and van Deenen, L.L.M. (1976) submitted for publication.
5. Helmkamp, G.M., Nelemans, S.A. and Wirtz, K.W.A. (1976) Biochim. Biophys. Acta, 424, 168.
6. Kamp, H.H., Wirtz, K.W.A. and van Deenen, L.L.M. (1975) Biochim. Biophys. Acta, 398, 401.
7. Kagawa, Y., Johnson, L.W. and Racker, E. (1973) Biochem. Biophys. Res. Commun., 50, 245.
8. Michell, R.H. (1975) Biochim. Biophys. Acta, 415, 81.
9. Wirtz, K.W.A., Jolles, J., Westerman, J. and Neys, F. (1976) Nature, 260, 354.
10. van Golde, L.M.G., Raben, J., Batenburg, J.J., Fleischer, B., Zambrano, F. and Fleischer, S. (1974) Biochim. Biophys. Acta, 360, 179.
11. Butler, M.M. and Thompson, W. (1975) Biochim. Biophys. Acta, 388, 52.
12. Blok, M.C., Wirtz, K.W.A. and Scherphof, G.L. (1971) Biochim. Biophys. Acta, 233, 61.
13. Kellens, R.E., Allison, V.F. and Butow, R.A. (1975) J. Cell. Biol., 65, 1.
14. Hackenbrock, C.R. and Miller, K.J. (1975) J. Cell. Biol., 65, 615.
15. Miller, E.K. and Dawson, R.M.C. (1972) Biochem. J., 126, 832.

Genetics and Biogenesis of Chloroplasts and Mitochondria, Th. Bücher et al. eds.

INCORPORATION OF MITOCHONDRIAL MEMBRANE PROTEINS INTO LIPOSOMES

Gera D. Eytan*

Section of Biochemistry
Molecular and Cell Biology
Cornell University, Ithaca, New York 14853

ABSTRACT

Cytochrome oxidase and the oligomycin sensitive ATPase were incorporated into liposomes by incubating the isolated proteins with preformed liposomes. The liposomes contained phosphatidylethanolamine, phosphatidylcholine and an acidic phospholipid (cardiolipin, phosphatidylinositol or phosphatidylserine). The incorporation which occured without added detergent or sonication, resulted in unidirectional orientation of the proteins.

The hydrophobic proteins of the ATPase incubated in the presence of a mixture of liposomes which were prepared with and without cytochrome oxidase were preferentially incorporated into the cytochrome oxidase containing liposomes. On the other hand, cytochrome oxidase incubated in the presence of a mixture of liposomes which were prepared with and without the hydrophobic proteins was preferentially incorporated into the liposomes that did not contain the hydrophobic proteins.

INTRODUCTION

Biological membranes are formed by the insertion of polypeptides into preexisting membranes. Research into the mechanism of the actual incorporation of intrinsic proteins was limited by the absence of an appropriate in vitro system where the composition of the membrane and the inserted proteins can be varied at will. In the past few years procedures have been developed for reconstitution of biologically active vesicles[1,2]. For example, a Ca^{++} pump was reconstituted with the ATPase from sarcoplasmic reticulum[3] and an Na^{+} pump with an Na^{+}-K^{+} ATPase[4].

These reconstitution procedures involved either dissolving the liposomes with detergents[1,5] or disrupting them by sonication[2]. The present work will describe a procedure for the incorporation of proteins into preformed liposomes containing acidic lipids. The phospholipid composition of the liposomes suitable for incorporation resembled that of biological membranes. The incorporation took place without added detergents or sonication.

Intriguing features of the insertion of membrane proteins in vivo are the resulting unidirectional orientation of the proteins[6] and the selectivity of the process. Proteins are incorporated into the appropriate membrane in presence of other membranes[7]. As will be described in the present paper, proteins incorporated into liposomes were oriented unidirectionally. Presence of other proteins in the vesicles either decreased or increased the affinity of the proteins for the vesicles.

---

* Present address, Department of Biology, Technion-Israel Institute of Technology, Haifa, Israel.

The Incorporation Procedure

Unilamellar liposomes were prepared by drying down a solution of phospholipids (25 μmoles)in a test tube under a stream of nitrogen. 1 ml of 50 mM KPi (pH 7.0) was added and the suspension was sonicated to clarity in a bath sonicator(Laboratory Supplies Co., Hicksville, N.Y.). The liposomes were incubated for 30 minutes at room temperature with a mitochondrial membrane protein (1 mg) The activity of the incorporated proteins was assayed as described in the legends to the figures (see also 8, 9).

Incorporation of Cytochrome Oxidase

Isolated cytochrome oxidase was incorporated rapidly into preformed liposomes. Upon mixing the enzyme with liposomes, its activity decreased within few minutes (fig. 1). The full activity could be detected only in presence of either

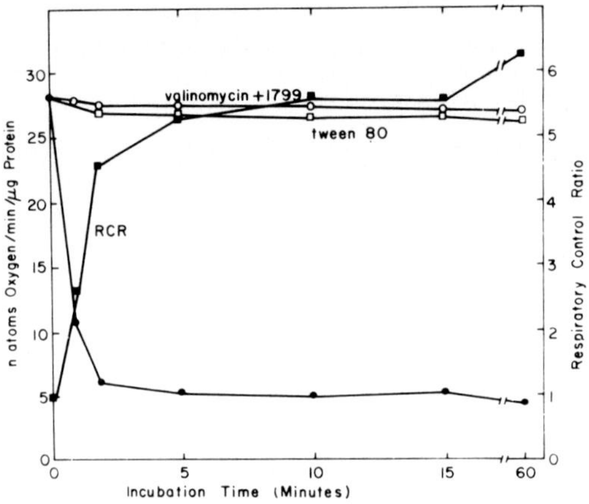

Fig. 1. Time course of cytochrome oxidase incorporation. Liposomes were prepared with phosphatidylethanolamine : phosphatidylcholine : phosphatidylserine (6:2:3). Purified cytochrome oxidase was added and the mixture was incubated at room temperature. Samples were withdrawn and assayed in the absence or presence of either Tween 80 (3%) or valinomycin (0.5 μg/ml) plus 1799 (20 μM). The respiratory control ratio (RCR) was calculated as the ratio of the enzyme activity in presence of uncouplers to that assayed in their absence.

detergents or uncouplers. The degree of incorporation was expressed as respiratory control ratio (RCR) which was defined as the ratio of the full activity to that assayed in absence of uncouplers or detergents. RCR of 5 implies that at least 80 percent of the cytochrome oxidase molecules were incorporated into the vesicles. Since the full activity could be assayed in presence of uncouplers (valinomycin and 1799) and the liposomes are impermeable to both cytochrome c and ascorbate, the enzyme must have been incorporated unidirectionally with its cytochrome c binding site exposed. This orientation resembles that observed in mitochondria[10]. Unidirectional orientation of the incorporated protein was also observed upon incorporation of the oligomycin sensitive ATPase of the inner mitochondrial membrane. In that case, all the ATPase was incorporated with its hydrolytic site exposed (unpublished results). This orientation resembles that observed in submitochondrial particles[10].

Phospholipid Requirements for Incorporation of Cytochrome Oxidase

The optimal phospholipids composition for incorporation contained acidic lipids. Cardiolipin was the most effective, its optimal concentration was 10-20 mole percent which is also its concentration in the mitochondrial inner membrane[11]. The acidic lipids, phosphatidylserine and phosphatidylinositol promoted incorporation at 30 mole percent. In the presence of cardiolipin, the only other phospholipid required for incorporation was phosphatidylcholine whereas in the presence of phosphatidylserine or phosphatidylinositol, an excess of phosphatidylethanolamine over phosphatidylcholine was required for optimal incorporation. Crude phospholipid extract from mitochondria was effective too, as well as phosphatidylethanolamine : phosphatidylcholine : cardiolipin at the ratios of 2:2:1. This composition resembles the reported composition of the inner mitochondrial membrane [11].

Are Detergents Required for Incorporation of Cytochrome Oxidase

The incorporation of cytochrome oxidase does not require the addition of detergents. Moreover, addition of cholate or lysolecithin, even at low concentrations, inhibited the incorporation (unpublished results). However, it should be mentioned that under different conditions both lysolecithin[9] and cholate[1] promote incorporation. The inhibition could not have been due to interference with the assay since these concentrations had no effect on the assay of already reconstituted vesicles. Another possibility that was considered was that the detergent bound to the purified enzyme plays a role in the incorporation procedure. Recently, Yu et al.[12] described a procedure for preparation of cytochrome oxidase using cholate as the only detergent. We repeated this procedure using radioactive cholate and found that the isolated enzyme, resuspended in detergent-free medium, still contained equal weight of cholate. Repeated precipitations of the enzyme with ammonium sulfate reduced the residual cholate to around 4 percent of the enzyme weight. Further washes or even extractions with ether or hexane did not remove this residual cholate. The cholate depleted enzyme was insoluble but could still be incorporated into liposomes containing acidic phospholipids (fig. 2).

Sequential Incorporation of Proteins into Liposomes

An interesting aspect of membrane biogenesis is the mechanism by which a protein "recognizes" the membrane of its destination. We approached this problem by exploring the affinity of a mitochondrial membrane protein to liposomes containing other proteins in comparison to its affinity to protein-free liposomes. As an assay for the presence of the hydrophobic moiety of the oligomycin sensitive ATPase in the same vesicle with cytochrome oxidase, we used the fact that hydrophobic protein serves as a proton channel[8,13]. Incorporation of hydrophobic protein into preformed cytochrome oxidase vesicle resulted in loss of respiratory control of the cytochrome oxidase activity. Half maximal uncoupling occurred at 0.25 mg/ml. The determining factor in the effectiveness of the hydrophobic

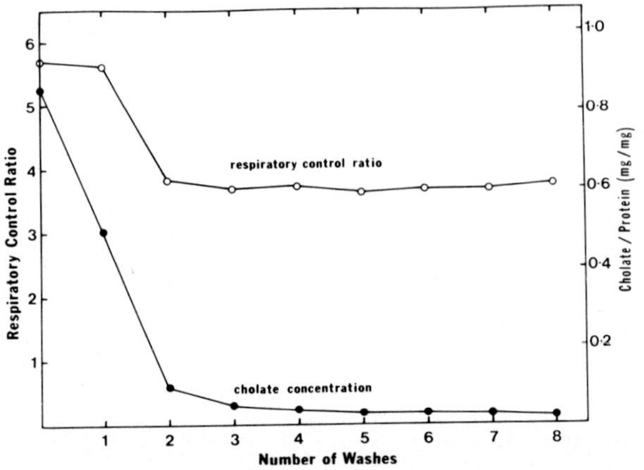

Fig. 2. Removal of cholate from cytochrome oxidase does not impair its subsequent incorporation into liposomes. Cytochrome oxidase (10 mg) was prepared according to Yu et al.[12] except that C^{14} cholate (40 µci/mmole) was used. The purified enzyme was resuspended in 1 ml 50 mM KPi (pH 7.0) - 0.4 MKCl and then precipitated twice at 35% ammonium sulfate saturation by centrifugation at 100,000 g for 20 min. After these washes the resuspended enzyme (10 mg/ml) was insoluble and further washes were carried out by pelleting the enzyme without ammonium sulfate. After every wash the pellet was resuspended in 1 ml of the above buffer and samples were withdrawn for radioactive determinations and incorporation into liposomes.

protein was not its absolute concentration but the ratio of this fraction to cytochrome oxidase vesicles. Thus, in the presence of increasing amounts of cytochrome oxidase vesicles, a given concentration of hydrophobic protein became less effective for uncoupling. The relative affinity of hydrophobic proteins to cytochrome oxidase vesicles was assessed by incorporating fixed amounts of hydrophobic protein into a mixture of fixed amount of cytochrome oxidase liposomes and increasing amounts of protein free liposomes. If the hydrophobic protein is randomly incorporated into both types of liposomes, the presence of excess of protein-free liposomes should decrease the uncoupling (i.e. raise the RCR) of the cytochrome oxidase vesicles. As shown in fig. 3 this was not the case. All of the added hydrophobic protein was incorporated only into the cytochrome oxidase vesicles. Essentially the same result was obtained when the cytochrome oxidase vesicles were prepared from an enzyme that had been almost totally depleted of lipids and detergents.

We then examined the affinity of cytochrome oxidase for hydrophobic protein liposomes. Cytochrome oxidase was incorporated into a mixture of fixed amounts of liposomes containing hydrophobic proteins and varying amounts of protein free liposomes. In the case of random distribution of the cytochrome oxidase the respiratory control should have increased with increasing amounts of protein free liposomes as increasing proportions of the enzyme would have been incorporated in

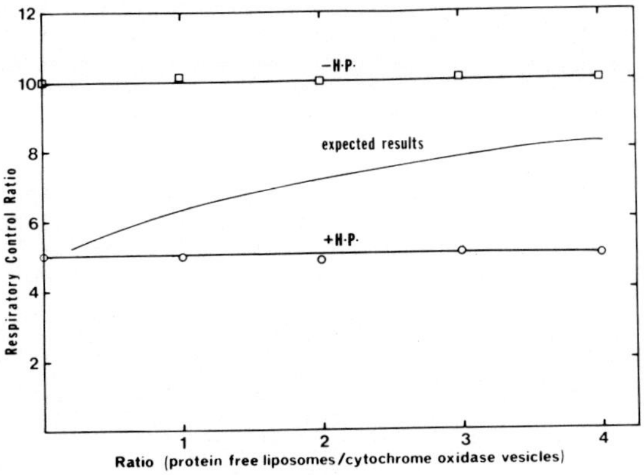

Fig. 3. Uncoupling of cytochrome oxidase vesicles by hydrophobic proteins in presence of excess protein free liposomes. Liposomes were prepared as described in the legend to fig. 1. Cytochrome oxidase (1 mg/ml) was incorporated into part of the liposomes. The rest of the liposomes were incubated under the same conditions with no protein. Cytochrome oxidase vesicles (2.5 mM lipids) and various amounts of protein-free liposomes were incubated with or without hydrophobic proteins (0.25 mg/ml) at room temperature for 30 minutes. Expected results for random distribution of the hydrophobic proteins among the protein free liposomes and the cytochrome oxidase vesicles were calculated assuming that in this case the uncoupling should be equal to that observed in the presence of cytochrome oxidase vesicles at a concentration equal to the combined concentrations of the liposomes and cytochrome oxidase vesicles.

the protein free liposomes. As shown in fig. 4, the respiratory control was even higher than expected for random distribution. The cytochrome oxidase exhibited high affinity towards the protein-free liposomes. This is in comparison to the

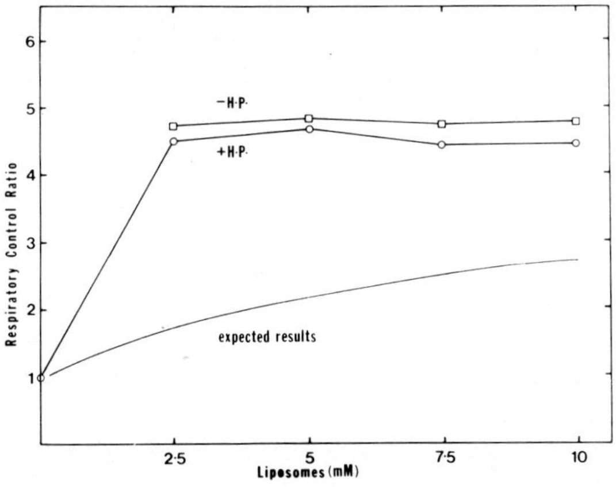

Fig. 4. Incorporation of cytochrome oxidase into hydrophobic proteins vesicles in the presence of excess liposomes. Hydrophobic protein vesicles were prepared by sonication to clarity of phofphatidylethanolamine : phosphatidylcholine : phospha-

tidylserine (3:1:2, 25 mM) in 50 mM KPi (pH 7.0) 1% cholate, adding hydrophobic proteins (2 mg/ml, final concentration) and dializing overnight at 4°C against 200 volumes 50 mM KPi (pH 7.0). Pure liposomes were prepared according to a similar procedure without the hydrophobic proteins. Various amounts of liposomes and cytochrome oxidase (0.1 mg/ml) were incubated with or without hydrophobic protein vesicles. After 30 min at room temperature samples were assayed for respiratory control of cytochrome oxidase activity. Expected results for random distribution of the cytochrome oxidase among the liposomes and hydrophobic protein vesicles were calculated assuming that cytochrome oxidase bound to hydrophobic protein vesicles is completely uncoupled while the respiratory control ratio of cytochrome oxidase incorporated into the liposomes is equal to that observed upon incorporation of the enzyme into liposomes with no hydrophobic protein vesicles present.

high affinity exhibited by hydrophobic protein towards cytochrome oxidase vesicles. The affinity of hydrophobic protein towards cytochrome oxidase liposomes could be abolished by minor modifications of the cytochrome oxidase, e.g. binding of cytochrome _c_ or partial reduction by ascorbate under anaerobic conditions (unpublished results).

CONCLUSIONS

The incorporation procedure described in this paper has advantages as a model for membrane biogenesis. It should be useful in answering a few questions of membrane biochemistry. What is the mechanism responsible for the assymetric incorporation of proteins into membranes? What is the role of acidic lipids in incorporation of cytochrome oxidase? Which parts of the proteins are responsible for the affinity and repulsion phenomena observed during the sequential incorporation of proteins into liposomes? How do slight modifications in the proteins change their affinity properties.

Such research should be useful in understanding the regulation of mitochondrial biogenesis by oxygen[14], glucose[15], heme[16] or mutations affecting single polypeptides[17]. However, one should be cautious in relating the results from a model system to membrane biogenesis _in vivo_. Indeed, it has been shown that mitochondrial membranes are not formed by incorporation of complete protein complexes such as cytochrome oxidase but probably by insertion of polypeptides[7] or even nascent chains[18]. However, the multitude of parallel phenomena in the incorporation into liposomes model and in membrane biogenesis _in vivo_ renders the model system attractive for future research.

ACKNOWLEDGEMENT

This investigation was supported by Grant No. PCM 73-01025A02 from the National Science Foundation. I am indebted to Dr. E. Racker for support and advice throughout this study.

REFERENCES

1. Kagawa, Y. and Racker, E. (1971) J. Biol. Chem. 246, 5477.
2. Racker, E. (1973) Biochem. Biophys. Res. Commun. 55, 224.
3. Racker, E. (1972) J. Biol. Chem. 247, 8198.
4. Goldin, S.M. and Tong, S.W. (1974) J. Biol. Chem. 249, 5907.
5. Racker, E., Chien, T.F. and Kandrach, A. (1975) FEBS Letters 57, 14.
6. Singer, S.J. (1974) Ann. Rev. Biochem. 43, 805.
7. Schatz, G. and Mason, T.L. (1974) Ann. Rev. Biochem. 43, 51.
8. Racker, E. (1972) J. Membr. Biol. 10, 221.
9. Eytan, G., Matheson, M.J. and Racker, E. (1975) FEBS Letters 57, 121.
10. Racker, E. (1970) in Essays in Biochemistry (P.N. Campbell and F. Dickens,eds.) Academic Press, vol. 6, pp. 1-22.
11. Fleischer, S., Klouwen, H. and Brierly, G. (1961) J. Biol. Chem. 236, 2936.
12. Yu, C-A., Yu, L. and King, T.E. (1975) J. Biol. Chem. 250, 1383.
13. Shchipakin, V., Chuchlova, E. and Evtodienko, Y. (1976) Biochem. Biophys. Res. Commun. 69, 123.
14. Slonimski, P.P. (1953) La formation des enzymes respiratoires chez la leuvere, Masson, Paris.
15. Mahler, H.R. (1973) CRC Crit. Rev. Biochem. 1, 381.
16. Gollub, E.G., Trocha, P., Lin, P.K. and Sprinson, D.B. (1974) Biochem. Biophys. Res. Commun. 56, 471.
17. Ebner, E., Mennucci, L. and Schatz, G. (1973) J. Biol. Chem. 248, 8043.
18. Kellems, R.E. and Butow, R.A. (1972) J. Biol. Chem. 247, 8043.

Genetics and Biogenesis of Chloroplasts and Mitochondria, Th. Bücher et al. eds.

THE ROLE OF MITOCHONDRIA-BOUND 80S RIBOSOMES IN MITOCHONDRIAL BIOGENESIS

W. F. Bennett, A. Gutierrez-Hartmann
and R. A. Butow
Department of Biochemistry, University of Texas
Health Science Center at Dallas, Dallas, Texas

Functional organelles such as mitochondria and chloroplasts are phenotypic hybrids resulting from the coordinate expression of two distinct gene pools. While it is now well established that the nuclear gene pool bears the major responsibility for determining organelle proteins (1), the mechanism by which the cytoplasmically synthesized polypeptides of organelles reach their final destination remains to be established. In the case of mitochondria, a number of distinct compartments must be served: the outer mitochondrial membrane, the intermembrane space, the inner mitochondrial membrane, and the matrix. Moreover, proteins appear specifically oriented with respect to the membrane and thus represent an additional level of organization which must be achieved by the biosynthetic apparatus. These considerations thus raise the possiblity that a variety of processes may operate to insure that cytoplasmically synthesized mitochondrial proteins reach their final destination within the organelle. Our approach to this problem has focused on the "secretory" model of higher eukaryotes (2), which states that proteins to be secreted from the cell are vectorially translated on ribosomes bound to the endoplasmic reticulum. A major feature of this model is that it provides a mechanism for the direct transfer of proteins across membranes (in the case of secretory cells, into the cisternae of the endoplasmic reticulum which is topologically equivalent to the exterior of the cell).

In our studies of mitochondrial biogenesis in yeast we have found cytoplasmic 80S polysomes associated with the outer mitochondrial membrane (3). Moreover, the binding appears specific for regions of the outer membrane in intimate contact with the inner mitochondrial membrane (4). A variety of biochemical studies have shown that these mitochondria-bound 80S polysomes have characteristics similar, if not identical, to ribosomes bound to the endoplasmic reticulum of higher eukaryotes, including their capacity to carry out vectorial translation (5) as well as their association with mitochondrial membranes through the nascent polypeptide chains in a salt stable linkage (5,6). Taken together, our results raise the possibility that mitochondria-bound 80S ribosomes in yeast might serve to facilitate the transfer of some mitochondrial polypeptides into the organelle. Some support for this view was obtained by immunological data suggesting that nascent F_1 ATPase, a cytoplasmically synthesized protein was enriched in mitochondria-bound versus free cytoplasmic polysomes (7); however, the extent of enrichment did not permit the unambiguous conclusion that the translation of F_1 ATPase was completely segregated within the yeast cell. Some alternative approaches to the characterization of mitochondria-bound and free cytoplasmic polysomes are presented in this paper which compares the *in vitro* translation products of cytoplasmic polysomes and poly A-containing mRNA extracted from mitochondria and the post-mitochondrial supernatant fractions. Data are also presented characterizing the optimum conditions for translation of yeast mRNA and polysomes.

In Vitro Translation of Polysomes and mRNA: A direct test for a functional difference between membrane bound and free cytoplasmic polysomes is the direct comparison of the products obtained from completion of nascent chains in pre-existing polysomes or *de novo* synthesis of polypeptide species directed by purified mRNA obtained from the bound and free fractions. Towards this end, we have examined some characteristics of translation of yeast mRNA *in vitro* by wheat germ extracts

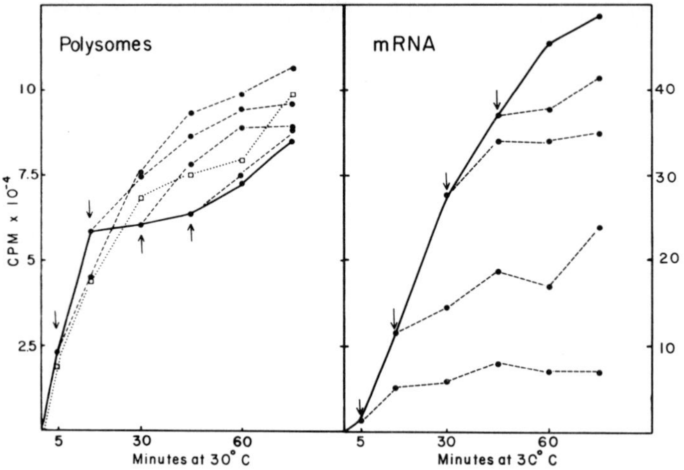

Fig. 1: Effect of aurintricarboxylic acid (ATA) on *in vitro* translation. Free polysomes were prepared from strain SB-1 spheroplasts by sedimenting the post-mitochondrial supernatant fraction through a discontinuous gradient of 0.5 M - 2.0 M sucrose prepared in 10 mM Tris-Cl pH 7.5, 1 mM MgCl, 5 mM NH_4Cl and 1 mM β-mercaptoethanol. Free polysomes were collected as a pellet. Messenger RNA was prepared by digesting the post-mitochondrial supernatant fraction with Proteinase K in 0.1 M NaCl and 0.5% SDS at 37° for 45 minutes. The RNA was precipitated with ethanol and poly A-containing mRNA obtained by chromatography on either poly U sepharose or oligo dT cellulose. Cell free translation in a wheat germ extract (19 A_{260}/ml) was carried out in a 500 μl reaction volume essentially as described by Marcu and Dudock (8). The final concentration of mRNA and polysomes was 36 μg/ml and 10 A_{260}/ml, respectively. [^{35}S]methionine was added to 20 μC/ml (420 Ci/mMole). At the times indicated by the arrows, 100 μl aliquots were removed and transferred to tubes containing ATA (at a final concentration of 0.1 mM). Kinetics of incorporation, expressed as total cpm were followed by removing 10 μl aliquots from the tubes at the times indicated onto Whatman 3 mm filter paper discs and precipitating with 10% TCA. After heating for 2 min at 90° in 10% TCA, the filters were washed, dried, and counted. —●—●—, no ATA. ---●---●---, plus ATA.

and the elongation and completion of nascent polypeptide in both the wheat germ system and the yeast pH 5 fraction (8,9).

The kinetics of [^{35}S]methionine incorporation into hot trichloroacetic acid precipitable counts supported by free cytoplasmic polysomes and poly A-containing mRNA extracted from the post-mitochondrial supernatant fraction, both incubated in a wheat germ extract, is compared in Fig. 1. The results show that active incorporation of label by polysomes and mRNA takes place throughout the 75 min assay period. The translation of mRNA involves continued reinitiation throughout the time course of incorporation as evidenced by the fact that aurintricarboxylic acid (ATA), a specific inhibitor of initiation of protein synthesis, effectively inhibits mRNA-supported incorporation regardless of when it is added during the experiment. On the other hand, ATA does not inhibit incorporation supported by polysomes but, in fact, appears to stimulate to some extent. While we have no explanation for this particular result, the data indicate that incorporation by polysomes represents primarily chain elongation and not reinitiation. Similar results have been obtained when polysomes are assayed in a yeast pH 5 fraction.

Recently, it has been shown that the translation of some mRNA species in cell free systems is dependent upon the presence of polyamines (10,11). To investigate whether the translation of yeast mRNA shows a similar dependence on polyamines, we have examined the effects of both spermine and spermidine on [^{35}S]methionine incorporation directed by mRNA in the wheat germ translation system. The results presented in Fig. 2 show that both the rate and extent of incorporation are markedly stimulated by polyamines (the concentration of spermine has been optimized at 0.05 mM whereas spermidine stimulation is a linear function of spermidine at all concentrations tested (up to 0.4 mM)).

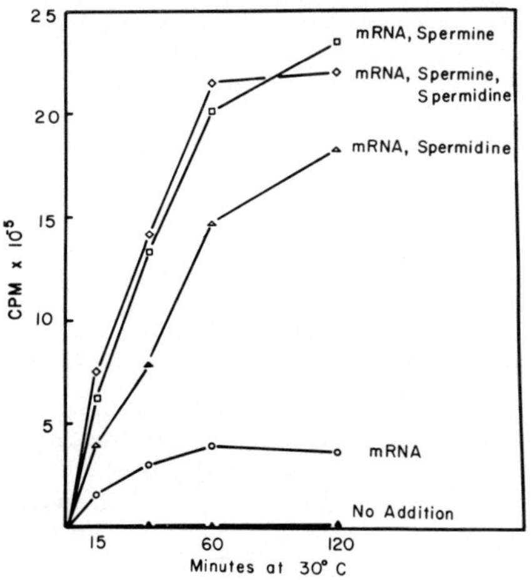

Fig. 2: Effect of polyamines on *in vitro* translation of mRNA. Incorporation of [^{35}S]methionine supported by mRNA (37 μg/ml) isolated from the post-mitochondrial supernatant fraction was carried out in the wheat germ extract described in the Legend to Fig. 1 in a total reaction volume of 100 μl. Spermine and spermidine concentrations were 0.05 mM and 0.25 mM, respectively.

SDS-Polyacrylamide Gel Analysis of In Vitro Translation Products: Cytoplasmic poly A - containing mRNA was extracted from whole yeast spheroplasts, isolated mitochondria and the post-mitochondrial supernatant fraction by the Proteinase K procedure described in the legend to Fig. 1. Following translation in the wheat germ system in the presence and absence of 0.05 mM spermine, aliquots were removed and compared by analysis in the presence of SDS on 12-18% exponential gradient polyacrylamide slab gels. In the absence of spermine, the profile of translation products programmed by mRNA's obtained from all fractions is limited to material migrating less than 25,000 daltons (Fig. 3B, D, and F). On the other hand, in the presence of 0.05 mM spermine, the amount of label appearing in the 25,000 to 100,000 dalton range is dramatically increased (Fig. 3C, E, and G). These results are similar to the effects of polyamines reported by Thang, et al., (11) on the in vitro synthesis of large molecular weight polypeptides encoded by alfalfa mosiac virus RNA.

A direct comparison of the translation products programmed by mRNA extracted from total spheroplasts, mitochondria and the post mitochondrial supernatant fraction reveals no obvious differences in the profile (Fig. 3C, E, and G). While to a first approximation these results would argue against the notion that mitochondria-bound and free cytoplasmic ribosomes are functionally in-

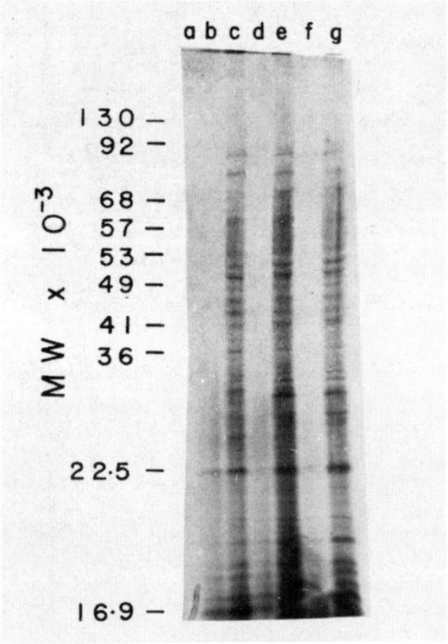

Fig. 3: Analysis of mRNA translation products by 12-18% SDS-gradient polyacrylamide slab gel electrophoresis. For the isolation of total yeast mRNA, spheroplasts were lysed in 0.1 M NaCl, 1 mM EDTA, 30 mM Tris-Cl pH 7.5, 0.5% SDS followed by digestion with Proteinase K for 45 minutes at 37°. The RNA was precipitated with ethanol and then washed with 3 M NaOAc, pH 6, followed by re-precipitation with ethanol. Poly A-containing total mRNA was isolated by chromatography on poly U sepharose or oligo dT cellulose. The mitochondria-bound and free cytoplasmic mRNA was isolated as described in Fig. 1. The various messages were translated in the wheat germ system for 60 minutes at 30° either in the absence or presence of 0.05 mM spermine as described in Fig. 1. At the end of the incubation period, an aliquot was removed and precipitated with 5% TCA. The precipitate was neutralized with 1 M Tris base and then dissolved in sample buffer (2% SDS, 0.002 M EDTA, 0.15 M 2-mercaptoethanol, 0.05 M Tris Cl, pH 6.8 and 10% (v/v) glycerol) and heated for 2 min at 90°. Electrophoresis was performed on gradient slab gels as described by Douglas and Butow (12). a) no added mRNA; b) total mRNA; c) total mRNA plus 0.05 mM spermine; d) mitochondria-bound mRNA; e) mitochondria-bound mRNA plus 0.05 mM spermine; f) mRNA from the post-mitochondrial supernatant fraction; g) mRNA from the post-mitochondrial supernatant fraction plus 0.05 mM spermine.

volved in translational segregation in yeast, a different picture emerges when products obtained from polysome chain completion are compared (Fig. 4). In this experiment, free cytoplasmic polysomes and bound-cytoplasmic polysomes extracted from mitochondria were isolated and nascent chains completed in vitro in a yeast pH 5 translation system in the presence of [^{35}S]methionine. The results presented in Fig. 4 show pronounced differences in the profile of polypeptide species between bound and free, particularly in the 40,000 to 70,000 molecular weight region of the gel. We note more than ten bands absent in the free polysome profile which are present in the bound. Work is currently in progress to determine if these "extra" polypeptide species represent specific mitochondrial products destined for import and integration into the organelle.

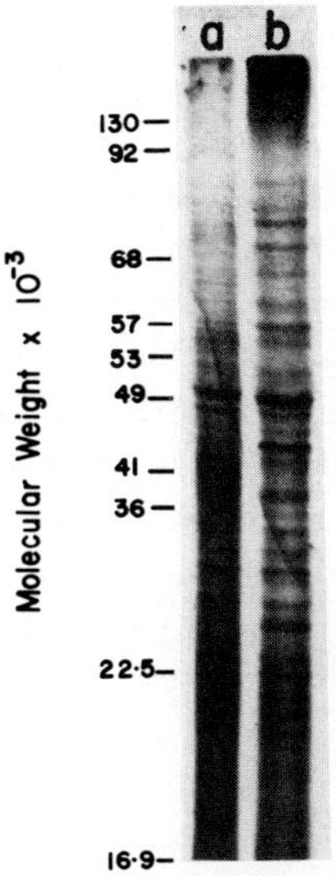

Fig. 4: In vitro elongation of free and mitochondria-bound polysomes. Free polysomes were prepared as in Fig. 1. Mitochondria were prepared by isopycnic centrifugation through a 30%-70% sucrose gradient containing 200 μg/ml cycloheximide. The purified mitochondria were then extracted with 2% Triton. The mitochondria-bound polysomes were collected by sedimenting the Triton extract through a discontinuous sucrose gradient as described in Fig. 1. Cell free elongation was carried out in the yeast pH 5 system in a final volume of 100 μl as described in Fig. 1. Samples for gels were prepared as in Fig. 3. a) Polysomes from the post-mitochondrial supernatant fraction; b) Mitochondria-bound polysomes.

DISCUSSION: In yeast, the study of the functional role of cytoplasmic ribosomes attached to the outer mitochondrial membrane is complicated by the fact that this polysome pool represents only a few percent of the total cellular polysomes (3). As a result, contamination by free cytoplasmic polysomes creates particular difficulties when attempting to assess the degree of translational segregation carried out by the mitochondria-bound polysome fraction. In the experiments reported here, we have tried to evaluate translational segregation by comparing directly the products synthesized in vitro from purified mRNA and by completion of nascent polypeptide chains in free and mitochondria-bound polysomes. While we are able to detect clear differences in the products in the chain completion system, no such differences were observed when the products of translation of purified mRNA's were compared. It should be noted that in these experiments mRNA was isolated directly from mitochondria by Proteinase K digestion and not from purified mitochondria-bound polysomes. (Although a direct comparison of the translation of polysomal mRNA's would be important, we have not yet succeeded in obtaining mRNA from purified mitochondria-bound polysomes which support translation in the wheat germ system). One possibility to account for the apparent discrepancy in the comparison between mRNA translation and the chain elongation system is the suggestion by Pogo and his collaborators (13,14) that membranes in yeast may be functional in the transport of newly synthesized mRNA to the cytoplasm. Consequently, if such putative membrane-bound nascent mRNA species were present in significant amounts in isolated mitochondrial fractions, then the translation products directed by mRNA obtained from Proteinase K digests of isolated mitochondria would represent not only products from polysomal mRNA but products from mRNA species which were not engaged in translation (i.e. polysomal) complexes. That non-polysomal mRNA may be present in significant amounts is suggested by preliminary results which show that the ratio of poly A-containing RNA to total RNA is greater when RNA is extracted directly from mitochondria or the post-mitochondrial supernatant fraction than from isolated polysomes (A. Gutierrez-Hartmann, unpublished results).

Acknowledgments: This work was supported by Grant GM 19090 from the U. S. Public Health Service and Grant NP-128E from the American Cancer Society to RAB.

References:

1. Schatz, G. and Mason, T. (1974), Ann. Rev. Biochem. 43: 51-87.
2. Siekevitz, P. and Palade, G. E. (1966) J. Biophys. Biochem. Cytol. 7:619-630.
3. Kellems, R. E. and Butow, R. A., (1972) J. Biol. Chem. 247:8043-8050.
4. Kellems, R. E., Allison, V. F., and Butow, R. A. (1975) J. Cell. Bio. 65:1-14.
5. Kellems, R. E., Allison, V. F., and Butow, R. A., (1974) J. Biol. Chem. 249:3297-3303.
6. Kellems, R. E., and Butow, R. A. (1974) J. Biol. Chem. 249:3303-3310.
7. Butow, R. A., Bennett, W. F., Finkelstein, D. B., and Kellems. R. E. (1975) in Membrane Biogenesis (Tzagoloff, A., ed.) p. 195. Plenum Publishing Corp. New York, N.Y.
8. Marcu, K., and Dudock, B. (1974) Nuc. Acids Res. 1:1385-1397.
9. Staehlin, T. & Falvey, A. K. (1971) Methods Enzymol. 20:433-440.
10. Thang, M. N., Thang, D. C., DeMaeyer, E., and Montagnier, L. (1975) Proc. Nat. Acad. Sci. 72:3975-3977.
11. Thang, M. N., Dondon, L., and Mohier, E. (1976) FEBS Letters 61:85-90.
12. Douglas, M. G. and Butow, R. A. (1976) Proc. Nat. Acad. Sci. 73: 1083-1086.
13. Shiokawa, K., and Pogo, A. O., (1974) Proc. Nat. Acad. Sci. 71:2658-2662.
14. Branes, L. and Pogo, A. O., (1975) Eur. J. Biochem. 54:317-328.

Genetics and Biogenesis of Chloroplasts and Mitochondria, Th. Bücher et al. eds.

STUDIES ON THE SYNTHESIS OF MITOCHONDRIAL PROTEINS IN THE CYTOPLASM AND ON THEIR TRANSPORT INTO THE MITOCHONDRION

G. Hallermayer and W. Neupert

Institut für Physiologische Chemie, Physikalische Biochemie und Zellbiologie
der Universität München
Germany

INTRODUCTION

It is well established that the majority of the mitochondrial proteins are synthesized outside the mitochondrion on cytoplasmic ribosomes. However, the mechanisms by which these proteins are transported and integrated into their functional locations remain unknown (for review see (1,2)). This gap in knowledge stems mainly from the experimental difficulties experienced in trying to detect and identify the trace amounts of proteins on their path to the mitochondrion. Furthermore, serious problems arise from artefacts produced by isolation of cellular components, viz. leakage of proteins from cell organelles, unspecific adsorption and proteolytic degradation.

In this report we present experimental data obtained from Neurospora cells by (a) kinetic dual labelling studies and (b) immunological location of mitochondrial proteins in subcellular fractions.

RESULTS

1. Pulse- and Pulse-Chase-labelling of the proteins of Neurospora cell fractions

Neurospora cells were grown in the presence of ^{35}S-sulphate to obtain homogeneous labelling of total cellular protein. After 16 hrs growth (exponential phase) the cells were cooled to 9 ^{o}C and maintained at this temperature for 2 hrs. A pulse of ^{3}H-leucine was given. Aliquots of the cultures were removed at the times indicated in Fig. 1. Cells were harvested and fractionated by homogenisation and differential centrifugation, and the ^{3}H/^{35}S-ratio determined in the various cell fractions. In addition to the pulse labelling a portion of the culture was given a chase of unlabelled leucine after 90 sec exposure to ^{3}H-leucine and the ^{3}H/^{35}S-ratio in the cell fractions was determined at the time intervals shown (Fig. 1). All fractions were characterized by electron microscopy and enzymatic analysis.

In the cell homogenate the ^{3}H/^{35}S-ratio increases linearly reaching a constant level after about 360-720 sec (when all of the added leucine is incorporated into cellular proteins). Leucine uptake is complete in about 60 sec under our experimental conditions. The effectiveness of the leucine chase can be seen by the immediate check, it imposes on the increase of the ^{3}H/^{35}S-ratio (Fig. 1). The cytoribosomes show a faster increase in the ^{3}H/^{35}S-ratio than the total homogenate;

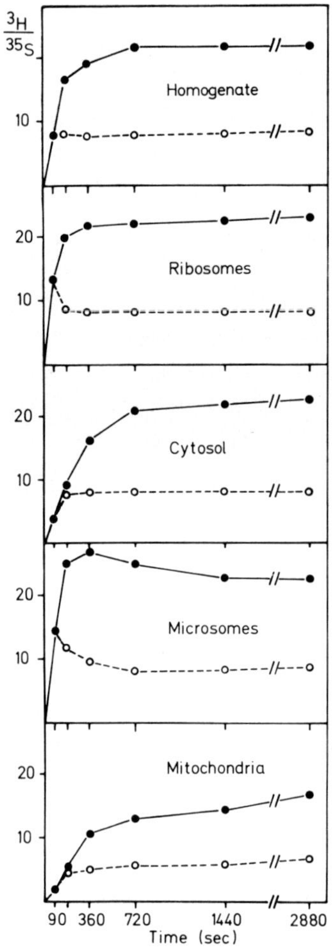

Fig. 1. Kinetics of the incorporation of ^3H-leucine into TCA-insoluble protein of total homogenate and subfractions of Neurospora crassa.
Cells were a) pulse labelled with ^3H-leucine -●——●-;
b) pulse labelled and chased with leucine after 90 sec -o----o-.
Specific ^3H-radioactivities of mitochondrial fractions are expressed by the $^3H/^{35}S$-ratio.

the addition of the chase causes a rapid decrease indicating a rapid replacement of radioactively labelled nascent polypeptide chains by unlabelled ones.

Labelling of the cytosolic proteins lags behind that of the ribosomes indicating that the ^3H-leucine has to pass through the pool of nascent chains on ribosomes, before it appears in the cytosol. During the first 90 sec after the chase, the $^3H/^{35}S$-ratio in the cytosol increases considerably and then stays constant, reflecting the rapid release of nascent chains from the ribosomes.

The microsomal fraction shows a unique labelling pattern, the $^3H/^{35}S$-ratio rises very rapidly to a maximum, before incorporation into whole cells is complete. This maximum is followed by a slow decline to a final level equal to that of the whole homogenate. A similar decrease is also seen after the chase. It should be noted that the decrease is slower than that due to the release of nascent chains from the ribosomes. This type of pattern would suggest that proteins are either synthesized or imported by the microsomal fraction and also exported.

The mitochondrial labelling is characterized by a pronounced lag with respect to both the total homogenate and the cytosolic proteins. The mitochondria have the slowest rate of increase in $^3H/^{35}S$-ratio. After 360 sec when the ^3H-incorporation of whole cells is almost completed, the ratio of the mitochondria is only about 1/2 of the final level of all other cell fractions. A plateau is not reached within 2880 sec. A similar slow increase is observed, when after 90 sec pulse labelling of proteins is stopped by a chase. Such a labelling pattern is consistent with the existence in the cell of an extramitochondrial pool of mitochondrial proteins. The time needed for synthesis appears to be small compared to the average time the mitochondrial proteins remain in the extramitochondrial pool.

2. Kinetics of labelling of individual protein fractions of the mitochondria

Immunoprecipitation was used to follow the labelling of individual mitochondrial proteins. The proteins selected were cytochrome c (3), the carboxyatractyloside-binding protein (CAT-protein) (4), mitochondrial ribosomal proteins and whole matrix protein. Cells were exposed to ^{3}H-leucine as in Fig. 1. The kinetics of appearance of these ^{3}H-labelled proteins in the mitochondria are shown in Fig. 2 and Fig. 3. They differ considerably from each other, indicating that the pools suggested from Fig. 1 are of different size, depending on the protein in question. Furthermore, the increase of the $^{3}H/^{35}S$-ratio after the chase in all mitochondrial proteins studied substantiates the conclusion, that extramitochondrial pools do exist.

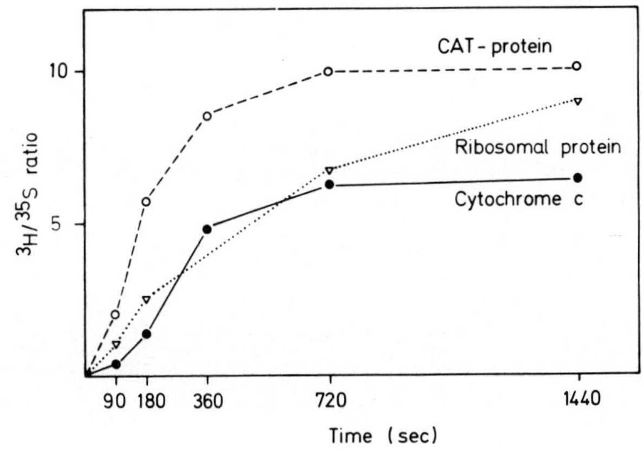

Fig. 2. Kinetics of the appearance of individual ^{3}H-labelled proteins in the mitochondria from cells which were exposed to ^{3}H-leucine at zero time.
CAT-binding protein, mitochondrial ribosomal protein and cytochrome c were immunoprecipitated from mitochondria at times indicated.

3. Assessment of extramitochondrial pools of mitochondrial proteins

The extramitochondrial pool of matrix protein was studied by monitoring the kinetics of appearance of ^{3}H-leucine labelled matrix in the various cellular fractions. Matrix protein was immunoprecipitated from cytosolic, microsomal and mitochondrial fractions. The label appeared quite rapidly in the cytosol and after about 360 sec decreased slowly. The mitochondria on the other hand showed the characteristic delayed labelling, reflecting a continuous accumulation (Fig.3). The radioactivities shown in Fig. 3 are related for each fraction to 0.1 g of Neurospora cells (wet weight). It thus appears that there is a quantitative relationship between the disappearance of label from the cytosol and the concomitant increase in the mitochondria. Furthermore, the sum of the radioactivities in cytosol and mitochondria follows the kinetics of total cell homogenate minus cytoplasmic ribosomes. In accordance with this finding the microsomal fraction shows

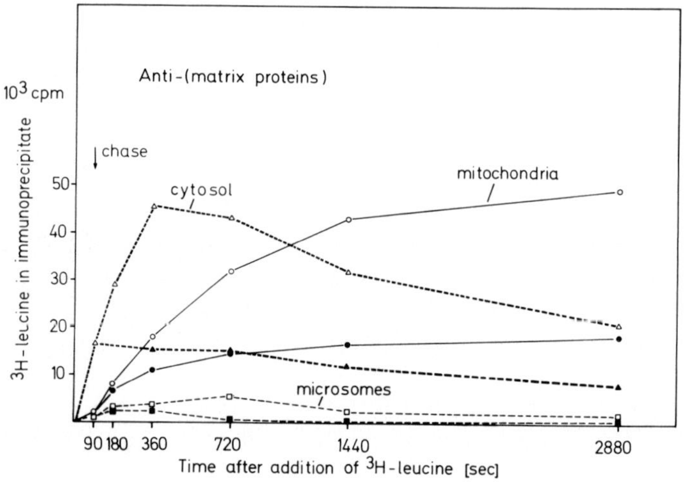

Fig. 3. The changes of ^{3}H-leucine radioactivity in mitochondrial matrix protein in subcellular fractions.
(a) pulse label (open symbols); (b) after 90 sec pulse of ^{3}H-leucine a chase of unlabelled leucine was given (solid symbols).

a relatively low labelling.

These data would suggest that there exists a cytosolic pool of mitochondrial matrix protein and that proteins from this pool are transported into the mitochondrion. However, it must be considered that the cytosol may not be the real location of this pool, since fractionation may result in artefactual redistribution of proteins.

4. Inhibition of translocation by carbonyl-cyanide-m-chlorophenylhydrazone (CCCP)

When a culture was pulse-labelled as described in Fig. 1 and CCCP was added together with the chase, it was found that the increase in the ^{3}H/^{35}S-ratio in the mitochondria was greatly reduced. This suggests a block in the transport of proteins into the mitochondrion. To substantiate this conclusion, the CAT-binding protein was precipitated from mitochondria of cells labelled in the following ways: a) pulse for 180 sec; b) pulse 180 sec and chase 600 sec; c) pulse 180 sec and chase plus CCCP for 600 sec. The ^{3}H/^{35}S-ratios measured in the immunoprecipitates were 7.0, 10.1 and 6.7, respectively. Electrophoresis of the immunoprecipitates reveals a single peak with an apparent molecular weight of 32,000, corresponding to the CAT-binding protein (Fig. 4). The higher ratio in the peak of the chased sample compared to the control suggests import of newly formed CAT-protein during the chase period.

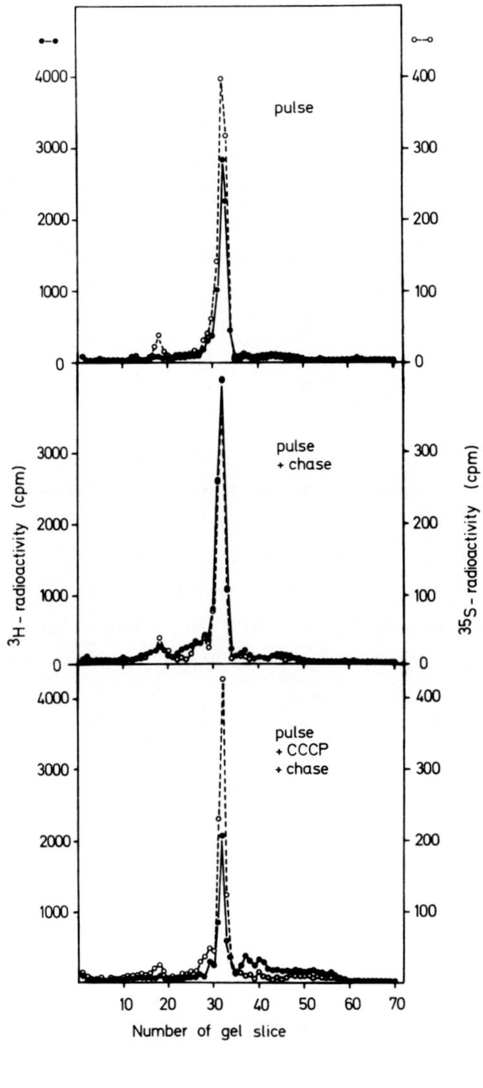

Fig. 4. SDS gel electrophoresis of immunoprecipitated CAT-binding protein from mitochondria of cells homogeneously labelled with ^{35}S.

a) 180 sec pulse of ^{3}H-leucine
b) 180 sec pulse followed by chase for 10 min
c) 180 sec pulse followed by simultaneous addition of CCCP and chase

In the CCCP treated sample ^{3}H-labelled low molecular weight material is also observed. Furthermore, the ^{3}H/^{35}S-ratio in the peak is relatively low in comparison to the ratio in the immunoprecipitate. The appearance of these low molecular weight components coupled with the low ^{3}H/^{35}S-ratio of the 32,000 MW peak suggests that inhibition of transport by CCCP is accompanied by degradation of newly synthesized protein. We are tempted to suggest that the newly synthesized protein is particularly labile when transport is inhibited, resulting in degradation to lower MW products. We have discussed a similar phenomenon concerning mitochondrially synthesized protein components on the basis of quite different experimental approaches (5).

DISCUSSION

The data presented show that extramitochondrial pools of newly synthesized mitochondrial proteins do exist. They further indicate that the sizes of these pools are different for different mitochondrial proteins. Newly formed mitochondrial matrix proteins were found to be localized predominantly in the cytosol fraction. This finding must be interpreted with caution, since redistribution of proteins may cause serious artefacts. Preferential susceptibility to proteolysis of proteins which are on the path to their final location in the mitochondrion is also a possible source of error. The latter is especially suggested by results obtained with the CAT-binding protein.

Recently a mechanism for the transport of proteins into the mitochondrion was suggested (6). According to this proposal, mitochondrial proteins are inserted into the mitochondrion directly by ribosomes attached to the outer mitochondrial membrane at sites where outer and inner membranes are fused. Such a mechanism would have the following characteristics: a) the kinetics of the labelling of mitochondrial proteins should parallel that of cytosolic proteins; b) the labelling of different mitochondrial proteins should also show parallel kinetic patterns (with the exception of proteins which are modified after transport with respect to their antigenic properties, e.g. cytochrome c); c) the $^{3}H/^{35}S$-ratio of mitochondrial proteins detected outside the mitochondria should not exceed that of the intramitochondrial proteins.

The experimental system here described has none of the above characteristics and we find it difficult to reconcile our findings with the direct insertion hypothesis as a general mechanism.

In an accompanying paper we describe the transport process in a cell free system, in which translocation of *in vitro* synthesized proteins can be followed and in which transport is blocked by CCCP as *in vivo*, but protein synthesis is not inhibited (7).

ACKNOWLEDGEMENTS

We thank Professor M. Klingenberg for his help and guidance in preparing the CAT-binding protein, and also for unpublished information on its site of synthesis. We would also like to acknowledge invaluable assistance of Heide Rothe and Dietlinde Rose.

This work was supported by the Deutsche Forschungsgemeinschaft, Schwerpunkt "Biochemie der Morphogenese".

REFERENCES

(1) Schatz, G. and Mason, T.L. (1974) Annu. Rev. Biochem. 43, 51-87.
(2) Gonzales-Cadavid, N.F. (1974) Sub-Cell. Biochem. 3, 275-309.
(3) Heller, J. and Smith, E.L. (1966) J. Biol. Chem. 241, 3158-3164.
(4) Klingenberg, M. et al. (1975) in "Electron Transfer Chains and Oxidative Phosphorylation", ed. E. Quagliariello et al., North-Holland Publishing Company, Amsterdam.
(5) Rücker, A.v., Neupert, W. and Werner, S. (1976) in "Genetics, Biogenesis and Bioenergetics of Mitochondria", ed. W. Bandlow et al., De Gruyter, Berlin, in press.
(6) Kellems, R.E., Allison, V.F. and Butow, R.A. (1975) J. Cell Biol. 65, 1-14.
(7) Harmey, M.A., Hallermayer, G. and Neupert, W., these Proceedings.

Genetics and Biogenesis of Chloroplasts and Mitochondria, Th. Bücher et al. eds.
© 1976 Elsevier/North-Holland Biomedical Press, Amsterdam, The Netherlands.

IN VITRO SYNTHESIS AND TRANSPORT INTO MITOCHONDRIA OF CYTOPLASMICALLY TRANSLATED PROTEINS

M.A. Harmey*, G. Hallermayer and W. Neupert
Institut für Physiologische Chemie, Physikalische Biochemie und Zellbiologie
der Universität München
Germany

INTRODUCTION

In a preceding paper the kinetics of synthesis and transport of cytoplasmically synthesized proteins into the mitochondrion in intact cells of Neurospora crassa have been described (1). These in vivo studies have shown the existence of extramitochondrial pools of proteins which are transported into the mitochondria. These studies however give only limited information on the transfer mechanism, as it is experimentally difficult to separate the processes of synthesis and transport in vivo. We describe here an in vitro system which allows discrimination between the synthesis of mitochondrial proteins on cytoplasmic ribosomes and their subsequent transport into the mitochondria.

RESULTS

1. In vitro synthesis of mitochondrial proteins

Neurospora cells were grown in the presence of ^{35}S-sulphate, after 14 hrs growth the cells were given a chase of unlabelled sulphate for one hr. The cells were then washed and filtered dry. They were ground with sand in incubation mixture as described (2), supplemented with an ATP regenerating system and the homogenate centrifuged twice at 3000xg for 5 min. The cell free supernatant thus obtained was used for in vitro protein synthesis.

The time course of incorporation of ^{3}H-leucine into total homogenate is shown in Fig. 1A. Incorporation starts without any lag and reaches a maximum level within 10-20 min. The detectable ^{3}H-radioactivity in protein is approximately 5×10^{6} counts x min^{-1} per ml of incubation mixture, corresponding to 0.5 g of cells (wet weight). These levels of incorporation enable us to detect the small amounts of mitochondrial proteins synthesized, while the double labelling of the proteins makes it possible to discriminate between newly synthesized and preexisting proteins.

The effect of carbonyl-cyanide-m-chlorophenylhydrazone (CCCP) on ^{3}H-leucine incorporation is also shown. The effect of CCCP is especially interesting in view of its inhibitory action on protein transport in vivo (1). The initial rate is unaffected by CCCP at levels which inhibit transport (cf. Fig. 4). The final

*M.A.H. (present address: Department of Botany, University College, Dublin (Irish Republic) wishes to thank CIBA GEIGY for a visiting fellowship.

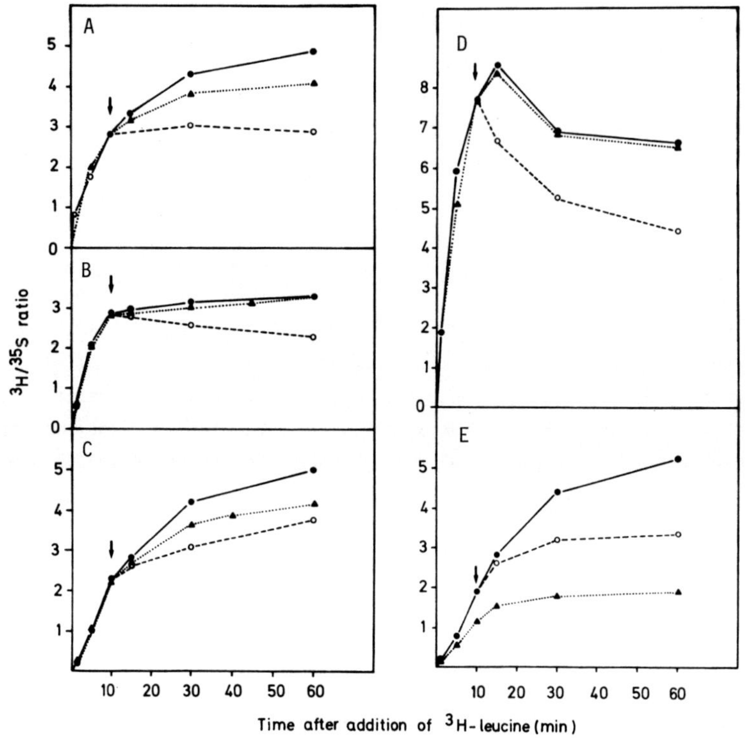

Fig. 1. Incorporation of ^3H-leucine into an ^{35}S-prelabelled cell free homogenate. Samples were withdrawn at the times indicated and fractionated. Labelling was as follows: a) continuous exposure to ^{3}leucine —●——●—; b) 10 min ^3H-leucine + chase --o----o--; c) continuous ^3H-leucine + CCCP (25 μM) ···▲·······▲··
A, homogenate; B, cytoribosomes; C, cytosol; D, microsomes; E, mitochondria.

incorporation level is depressed by 10-15%, however. Cycloheximide inhibits incorporation by approx. 95%, the remainder of the incorporation represents mitochondrial translation.

2. Kinetics of labelling of subcellular fractions

A chase of cold leucine given at 10 min immediately halts any further incorporation of ^3H-leucine into total homogenate. The *in vitro* system was sampled at time intervals for separation into subcellular fractions as shown in Fig. 1, B-E.

The $^3H/^{35}S$-ratio in the microsomal fraction rises rapidly reaching a maximum at 10 min. This maximum is followed by a gradual decline. Cytoribosomes also show a rapid increase in the $^3H/^{35}S$-ratio which reaches a plateau within 10-15 min. The addition of a chase causes a decrease in the ratio. The increase in the

$^{3}H/^{35}S$-ratio of the cytosol is relatively slow and is not checked by the addition of the chase.

The mitochondrial labelling pattern shows a slow increase in the $^{3}H/^{35}S$-ratio which is sustained over the experimental period. When a chase is given a parallel slow increase is maintained, showing no levelling off.

These results suggest that in the in vitro system synthesis of mitochondrial proteins and their subsequent transport into the mitochondria takes place.

The effect of CCCP on the progress of the $^{3}H/^{35}S$-ratios in the cell fractions is included in Fig. 1B-E. The addition of CCCP has no significant effect on the cell fractions except the mitochondria, where it reduces the increase in the $^{3}H/^{35}S$-ratio. The inhibition of mitochondrial labelling by CCCP is not complete. This can be attributed to mitochondrial translation which was found to continue in the presence of CCCP at the concentration level used (unpublished data). The in vitro system does not depend on endogeneous phosphorylation since an ATP regenerating system was added. Therefore it is reasonable to assume that CCCP inhibits transport of proteins into isolated mitochondria.

3. Transport of matrix proteins into the mitochondria

In vitro synthesis of defined mitochondrial proteins was shown by precipitation of mitochondrial matrix proteins with a specific antibody.We present in Fig. 2 electropherograms of the immunoprecipitates. We compared the profiles obtained from in vitro incorporation with those obtained from mitochondria isolated from pulse labelled whole cells (Fig. 3). There is good agreement between the two profiles indicating that the in vitro and in vivo synthesized proteins are the same. Similar results were obtained for the CAT-binding protein (3) and for cytochrome c (4) (see below).

To demonstrate transfer of matrix protein the following experiment was performed. A cell free system was prepared from ^{35}S prelabelled cells. Two parallel samples were set up. One was pulsed for 10 min with ^{3}H-leucine. The other was pulsed for 10 min and then chased for 30 min with cold leucine. Mitochondria and cytosol were prepared from both samples and matrix proteins immunoprecipitated. The $^{3}H/^{35}S$-ratios of the immunoprecipitates are presented in Table 1.

TABLE 1

$^{3}H/^{35}S$-ratios in immunoprecipitates of matrix proteins obtained from cytosol and mitochondria of a cell free system exposed to a pulse of 3H-leucine and to a pulse plus chase

	10 min pulse	10 min pulse plus 30 min chase
Cytosol	1.55	1.34
Mitochondria	.08	.72

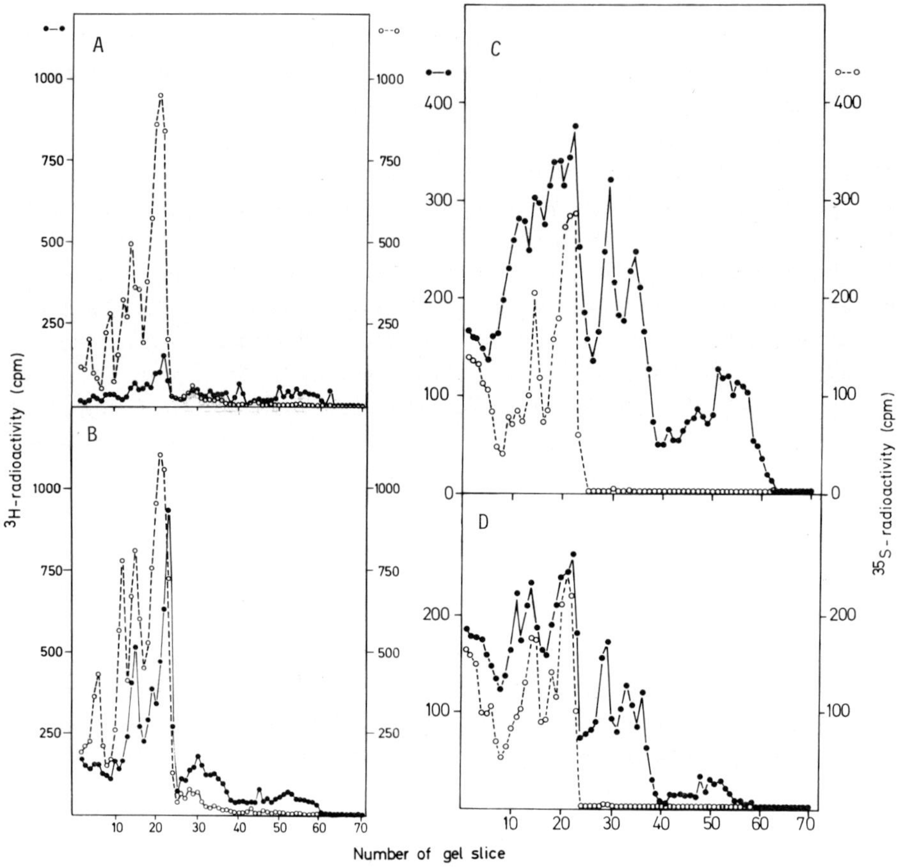

Fig. 2. SDS gel electrophoresis of immunoprecipitated matrix proteins from mitochondria (A,B) and cytosol (C,D). Fractions were isolated from the cell free homogenate after 10 min exposure to 3H-leucine (A,C) and following a 30 min chase (B,D).

In the mitochondria the ratio increased ninefold during the chase indicating that import of matrix proteins into the mitochondria took place during the chase. simultaneously, the ratio in the cytosol decreased, suggesting export of newly synthesized matrix proteins. It is significant that the $^3H/^{35}S$-ratio in the total cytosol increased appreciably during the chase whereas the immunoprecipitated matrix from the same fraction decreased (cf. Fig. 1).

SDS gel electrophoretic analysis of the immunoprecipitates is shown in Fig. 2. The ^{35}S-profiles confirm the occurence of matrix proteins in the cytosol fraction,

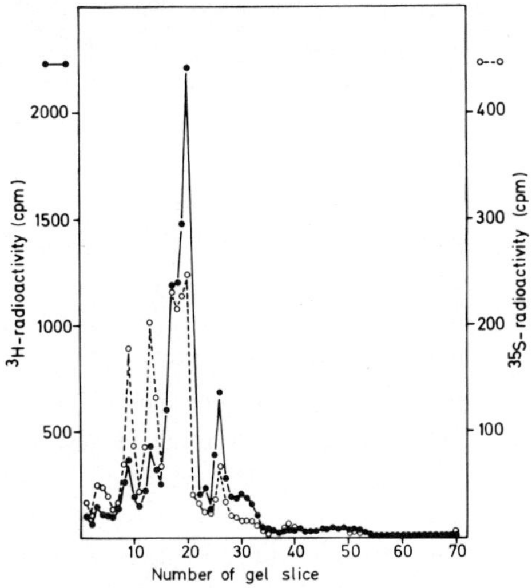

Fig. 3. SDS gel electrophoresis of matrix proteins from mitochondria isolated from pulse labelled cells. Cells were grown on 35S-sulphate and pulsed with 3H-leucine for 720 sec at 9^{0}C.

obviously due to leakage during cell fractionation. The protein peaks are similar in all four samples. The ^{3}H-profiles clearly show that matrix proteins formed during the pulse period are accumulated in the mitochondria during the chase period (Fig. 2A,B). The ^{3}H gel profiles from the cytosol indicate that also in this fraction newly formed matrix proteins are found (Fig. 2C,D).

4. Transport of cytochrome c and inhibition by CCCP

To demonstrate transport of one specific protein into the mitochondria the following experiment was carried out. Homogenates were incubated with ^{3}H-leucine in the presence and absence of CCCP for 40 min. Mitochondria were then isolated and cytochrome c precipitated with a specific antibody. The immunoprecipitates were analysed by SDS gel electrophoresis (Fig. 4). The distribution of ^{35}S in the gels shows that there is specific precipitation of cytochrome c (mol. weight 12,300 (4)). The control sample shows a prominent ^{3}H-peak demonstrating synthesis and import of newly synthesized cytochrome c (or the apo-molecule) in the *in vitro* system. In the mitochondria from the CCCP treated samples the appearance of the ^{3}H-peak is strongly inhibited. CCCP therefore appears to prevent the import of newly synthesized cytochrome c (or the apo-molecule) into the mitochondria.

CONCLUSIONS

The results presented here demonstrate the synthesis of defined mitochondrial proteins on cytoplasmic ribosomes in the *in vitro* system. They further show that in this cell free system the transport system is functional. The labelling

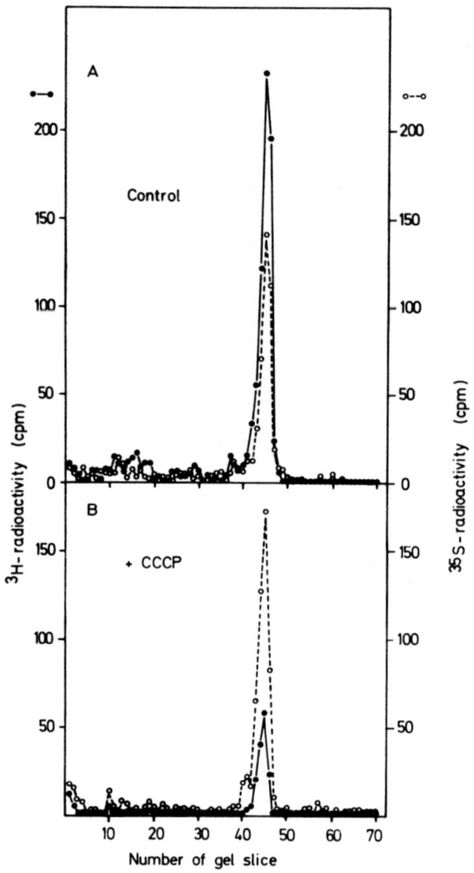

Fig. 4. SDS gel electrophoresis of immunoprecipitated cytochrome c of isolated mitochondria from cell free homogenate, exposed to 3H-leucine for 40 min in the absence and presence of CCCP (25 μM).

kinetics and the immunoprecipitation data demonstrate the existence of extramitochondrial pools of mitochondrial proteins, which are synthesized rapidly and exported relatively slowly into the mitochondria. The response of the system to CCCP makes it possible to inhibit transport while allowing synthesis of mitochondrial proteins to proceed unimpeded.

The need for a system to study synthesis and transport of cytoplasmically synthesized mitochondrial proteins has been expressed in a recent review (5). The system described here may partly meet this need. At present we are employing this system in an attempt to elucidate the molecular basis of the transport process.

ACKNOWLEDGEMENTS

This work was supported by the Deutsche Forschungsgemeinschaft, Schwerpunkt "Biochemie der Morphogenese". We wish to thank Dietlinde Rose and Heide Rothe for skillful technical assistance.

REFERENCES

(1) Hallermayer, G. and Neupert, W., these Proceedings
(2) Rücker, A.v., Werner, S. and Neupert, W. (1974) FEBS Lett. 47, 290-294
(3) Klingenberg, M. et al. (1975) in "Electron Transfer Chains and Oxidative Phosphorylation", ed. E. Quagliariello et al., North-Holland Publishing Company, Amsterdam
(4) Heller, J. and Smith, E.L. (1966) J. Biol. Chem. 241, 3158-3164
(5) Schatz, G. and Mason, T.L. (1974) Annu. Rev. Biochem. 43, 51-87

Genetics and Biogenesis of Chloroplasts and Mitochondria, Th. Bücher et al. eds.
© 1976 Elsevier/North-Holland Biomedical Press, Amsterdam, The Netherlands.

SPECIFIC LABELLING OF MITOCHONDRIALLY SYNTHESIZED PROTEINS IN YEAST CELLS IN THE ABSENCE OF ANTIBIOTICS

Wolfhard Bandlow
Genetisches Institut der Universität München
Maria Ward Straße 1A, D8 München 19
Germany

INTRODUCTION

Specific labelling of the products of the mitochondrial protein synthesizing machinery *in vivo* in the presence of cycloheximide or *in vitro* is a widespread tool in the study of the biogenesis of mitochondria. However, the techniques suffer some limitations: The incubation conditions are non-physiological, as i) the regulatory interrelationships of the nucleo-cytoplasmic system and the mitochondrial system are disturbed and ii) the mitochondria are disconnected from the supply of cytoplasmic translation products which may be a prerequisite for the integration into the membrane of at least some of the mitochondrially-made proteins (for review see 1). Although some methodological improvements have been achieved recently (2-4), it is desirable to escape the uncertainties and ambiguities of results obtained with such "uncoupled" systems.

In this communication I wish to compare two approaches to the labelling and identification of mitochondrial translation products *in vivo* in the absence of antibiotics. One which still employs an impaired system takes advantage of a nuclear mutation, 187 ts$^-$, which causes a temperature-sensitive, reversible inhibition of polypeptide chain initiation on cytoplasmic ribosomes. The other approach takes advantage of the fact that as in prokaryotes, and in contrast to cytoplasmic peptide chain initiation in eukaryotes (8,9), translation in mitochondria is initiated by amino-acylated N-fMet-tRNA (10). As a deformylase is lacking in mitochondria, N-terminal fMet is retained in newly synthesized polypeptides even after their integration into the mitochondrial inner membrane (11-13).

RESULTS AND DISCUSSION

In Figs. 1A and 1B SDS polyacrylamide gel electrophoretic patterns of mitochondrial proteins are compared after labelling yeast cells with amino acids in the presence of cycloheximide (solid lines in 1A and 1B) and with either formate in the presence of 5 mM unlabelled serine (but absence of cycloheximide, 1A) or N-formyl methionine (1B,

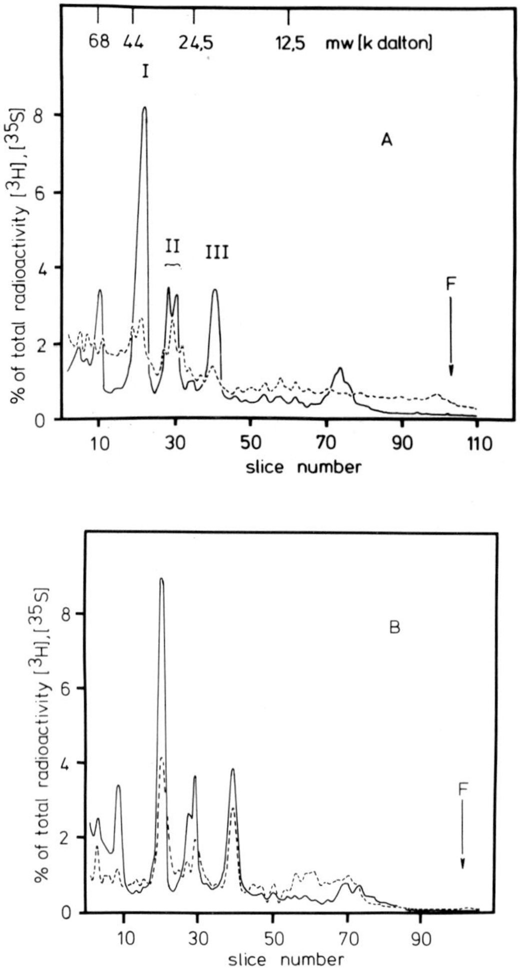

Fig. 1. Labelling pattern of mitochondrially synthesized proteins. Cells of *Saccharomyces cerevisiae* strain A 364 A, *a*, ade_1, ura_1, tyr_5, trp_5, cys_7, lys_2, gal_1 were grown in complete medium (YPG), harvested and transferred to synthetic medium as described (14) supplemented according to the auxotrophic markers and preincubated at 30°C under aeration for 3h. Labelling period 30min (20min for formate). Isolation of mitochondria, SDS polyacrylamide gel electrophoresis, gel fractionation and counting procedures were as described (14). 1A: 0.2mCi [^{35}S]-methionine *plus* cycloheximide (0.2 mg/ml, 3min prior to the label), 43.3oo total dpm (——); 0.8mCi [^{3}H]-formate *plus* 5 mM unlabelled serine, 56.ooo total dpm (---). 1B: each 0.1mCi [^{3}H]-leucine, isoleucine, valine, 0.05mCi phenylalanine *plus* cycloheximide (as above), 64.2oo total dpm (——); 0.4mCi [^{35}S]-N-formyl methionine, 5.2oo total dpm (---). The scale on top of Fig. 1A gives positions of marker proteins: cytochrome *c*, chymotrypsinogen, carbanhydrase, ovalbumin and bovine serum albumin. The apparent molecular weights of bands were calculated to be 42.ooo (I), 33.ooo + 30.ooo (II) and 22.5oo (III) dalton. F = dye front.

broken lines in 1A and 1B,respectively). It can be seen that essentially the same pattern of incorporation of radioactivity is obtained in all four cases, except that a low molecular weight fraction is absent (or small) after labelling with formate or N-fMet (broken lines in Figs. 1A and 1B). The pattern obtained after labelling with formate is unchanged, when the cold serine is omitted (labelling period 20 min) or when isolated mitochondria are labelled with N-fMet in the presence of post-polysomal cytoplasmic supernatant from a yeast cell homogenate (15, not shown).

The amount of label incorporated into distinct mitochondrial protein bands in the presence of formate or N-fMet is only once or twice the background level of incorporation. It cannot be increased by higher concentrations of radioactivity, pre-incubation with chloramphenicol (2) or post-incubation under chase conditions. The low specific radioactivity of the proteins may be mainly due to the fact that maximally one N-terminal residue per polypeptide chain can be labelled and that N-fMet is scarcely transported into yeast cells.

The labelling patterns shown in Figs. 1A and 1B cannot be simulated by either *in vivo* or *in vitro* incorporation of radioactive choline (A. Klein, unpublished observation). This excludes the possibility that the radioactivity retained in the protein bands after labelling with formate originates from contaminating lipids which were labelled after exchange of methyl groups and subsequently bound to mitochondrial proteins.

Thus formate (or N-fMet) may in fact help to identify mitochondrial translation products although the specific radioactivities of the labelled polypeptides are low.

Fig. 2 shows the distribution in SDS polyacrylamide gel electrophoretic patterns of radioactivity incorporated into mitochondria after labelling cells of the conditional nuclear mutant 187 ts^- (which is defective in cytoplasmic polypeptide chain initiation at 35°C, 6). Two aliquots of a cell culture were labelled with radioactive amino acids either at 23°C in the presence of cycloheximide or at 35°C in the absence of the drug. It is evident from this figure that at 35°C all the bands labelled at 23°C in the presence of cycloheximide are present (although with a lower proportion of the band with 42.000 dalton app. mw which is assumed to comprise band I of cytochrome oxidase). But in addition some bands with molecular weights between 27.000 and 70.000 dalton are found. Identical results are obtained, when a culture is first labelled at 23°C with $[^{14}C]$ -amino acids in the presence of cycloheximide and after removal of the drug again labelled at 35°C

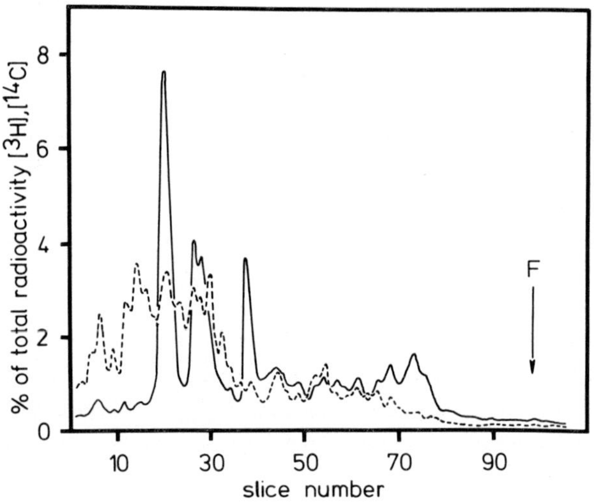

Fig. 2. Labelling patterns of mitochondrial proteins of mutant 187 ts$^-$ at permissive (23°C) and non-permissive (35°C) temperature. Conditions were as detailed under Fig. 1. (——), 23°C, 0.2 mg/ml cycloheximide 3 min prior to the label, each 40 µCi [^{14}C]-Leu, Ile, Val, 20 µCi Phe, 40.2oo total dpm. (---), 35°C, each 0.1 mCi [^{3}H]-Leu, Ile, Val, 0.05 mCi Phe, 221.7oo total dpm. (Preincubation period at 35°C, 20 min).

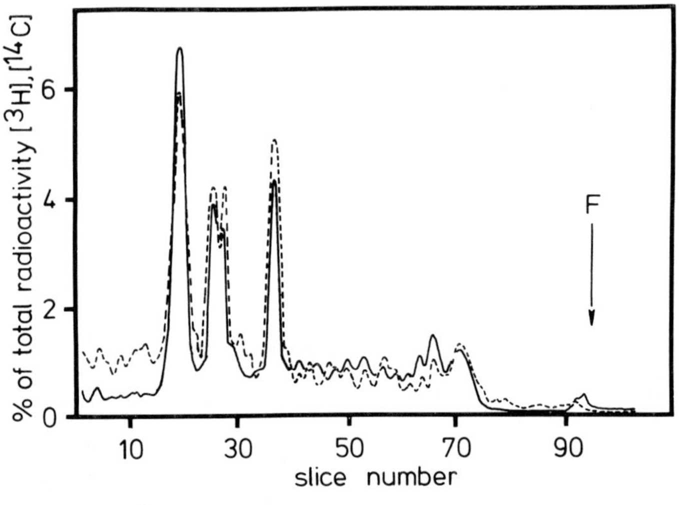

Fig.3. Labelling patterns of mutant 187 ts$^-$ in the presence of cycloheximide at permissive and non-permissive temperature. (——), 23°C, [^{14}C]-Leu, Ile, Val, Phe, 19.6oo total dpm; (---), 35°C, [^{3}H]-Leu, Ile, Val, Phe, 37.5oo total dpm. Other conditions as under Figs. 1 and 2.

with[³H]-amino acids (not shown). These obvious differences in labelling patterns could not be observed when i) wild type and mutant cells were both labelled at 23°C in the presence of cycloheximide or ii) at 35°C in the presence of cycloheximide (not shown) or when iii) the mutant was labelled at 23°C and 35°C both in the presence of cycloheximide, as shown in Fig. 3. Therefore the differences appear to be not a secondary consequence of the mutation 187 ts⁻ nor are they effected by the temperature shift in general, since no such proteins were found in the presence of cycloheximide at 23°C or at 35°C in the wild type or in the mutant.

In order to check wether these additional bands represent mitochondrial translation products or cytoplasmic proteins the synthesis of which is inhibited by cycloheximide but not inhibited at non-permissive temperature in the mutant, 187 ts⁻ cells were labelled at 35°C in the presence of chloramphenicol (solid line in Fig. 4). 187 ts⁻ cells were also labelled with formate at 35°C (in the presence of 5 mM unlabelled serine, broken line in Fig. 4).

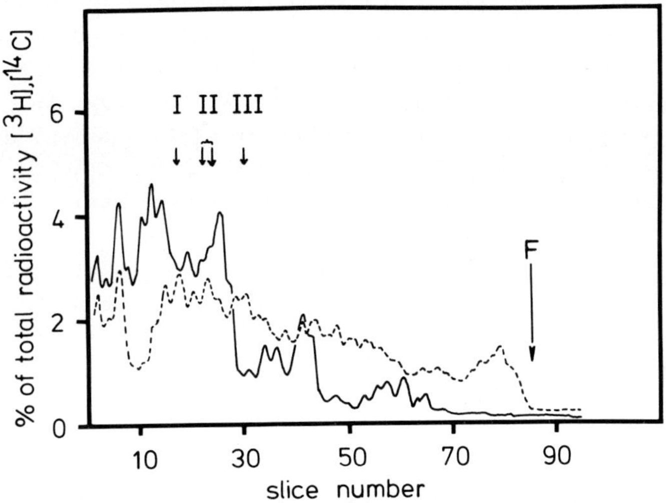

Fig. 4. Non-mitochondrial origin of pre-bands. Strain 187 ts⁻. (——), 35°C, (pre-incubation time at 35°C 20 min), chloramphenicol (4 mg/ml, pre-incubation time 20 min), [¹⁴C]-Leu, Ile, Val, Phe, 260.5oo total dpm; (---), 35°C (pre-incubation time 20 min), [³H]-formate *plus* 5 mM unlabelled serine, 3 h post-incubation in complete medium (5 mM formate) at 23°C, 44.ooo total dpm. Other conditions as under Figs. 1 and 2.

Fig. 4 shows that the polypeptide bands in question are not sensitive to chloramphenicol and are not labelled by formate at 35°C in the mutant. Moreover, these proteins persist during post-incubation under chase conditions after removing the chloramphenicol and, thus, are not precursors of mitochondrial proteins (not shown).

In the conditional mutant 187 ts^- a few mitochondrial protein bands are labelled at 35°C. Cytoplasmic ribosomal contribution to total cell protein synthesis is assumed to be shut off under these conditions. Three alternative explanations may be considered for the translational origin of these polypeptides: i) they are tranlation products of mitochondrial ribosomes the synthesis of which is non-specifically inhibited by cycloheximide, whereas the synthesis of the other mitochondrially governed proteins is not influenced by this drug. ii) these proteins owe their existence to a leak in the thermo-sensitive block of cytoplasmic ribosomes with a preference for certain messengers or iii) there exist two types of cytoplasmic ribosomes, both types being sensitive to cycloheximide, but only one type being temperature-sensitive in the mutant.

Alternative i) is unlikely since neither are the additional proteins significantly labelled by formate or N-fMet nor is the incorporation of radioactivity into these species sensitive to chloramphenicol. Although alternative ii) appears unlikely, too, it cannot be experimentally discriminated between possibilities ii) and iii) at present. Both interpretations are compatible with the observation that in the mutant at 35°C the temperature-induced block is less effective than the cycloheximide block (96% and 99.5% inhibition of cytoplasmic protein synthesis, respectively; 7 and unpublished data).

No evidence has been found so far for a physical or compositional heterogeneity of cytoplasmic ribosomes (R.A. Butow, personal communication), although evidence accumulates that cytoribosomes are compartmentalized and those bound to the outer surface of the outer mitochondrial membrane preferentially translate messengers bearing information for mitochondrial proteins in a vectorial manner (16-19). Evidence for a genetic heterogeneity of cytoplasmic ribosomes is as yet lacking.

In order to clarify this question it is desirable to examine more conditional mutants, temperature-sensitive in cytoplasmic ribosomal function. (Mutant 171 ts^- of Hartwell (5) appears to be iso-allelic with 187 ts^- and shows identical labelling patterns of mitochondrial protein). For identifying mitochondrial translation products both mutants are not particularily useful because i) mitochondrial translation and assembly are disconnected from the cytoplasmic translational system and ii) additional proteins are labelled the translational origin of which appears to be on cytoplasmic ribosomes.

ACKNOWLEDGEMENTS

The skillful technical help of Miss W. Hilscher is gratefully acknowledged. Strains A 364 A and 187 ts$^-$ were provided by L.H. Hartwell. I am indebted to D.Y. Thomas and R.J. Schweyen for revision and discussion of the manuscript and to the Deutsche Forschungsgemeinschaft for financial support.

REFERENCES

1. Schatz, G. and Mason, T.L. (1974) Annu. Rev. Biochem. 43, 51-87
2. Tzagoloff, A. (1971) J. Biol. Chem. 246, 3050-3056
3. Poyton, R.O. and Groot, G.S.P. (1975) Proc. Natl. Acad. Sci. USA. 72, 172-176
4. v. Rücker, A., Neupert, W. and Werner, S. (1976) In "Genetics, Biogenesis and Bioenergetics of Mitochondria" (Bandlow, W., Schweyen, R.J., Thomas, D.Y., Wolf, K. and Kaudewitz, F., eds) De Gruyter, Berlin, in the press
5. Hartwell, L.H. and McLaughlin, C.S. (1968) J. Bacteriol. 96, 1664-1671
6. Hartwell, L.H. and McLaughlin, C.S. (1969) Proc. Natl. Acad. Sci. USA. 62, 468-474
7. Schweyen, R.J. and Kaudewitz, F. (1971) Biochem. Biophys. Res. Commun. 44, 1351-1355
8. Lucas-Lenard, J. and Lipman, F. (1971) Annu. Rev. Biochem. 40, 409-448
9. Nombela, C., Nombela, N.A., Ochoa, S., Safer, B., Anderson, W.F. and Merrick, W.C. (1976) Proc. Natl. Acad. Sci. USA. 73, 298-301
10. Bianchetti, R., Lucchini, G. and Sartirana, M.L. (1971) Biochem. Biophys. Res. Commun. 42, 97-102
11. Mahler, H.R., Dawidowicz, K. and Feldman, F. (1972) J. Biol. Chem. 247, 7439-7442
12. Feldman, F. and Mahler, H.R. (1974) J. Biol. Chem. 249, 3702-3709
13. Mahler, H.R., Feldman, F., Phan, S.H., Hamill, P. and Dawidowicz, K. (1974) In "Biogenesis of Mitochondria" (Kroon, A.M. and Saccone, C., eds) pp. 423-441, Academic Press, New York
14. Bandlow, W. and Schweyen, R.J. (1975) Biochem. Biophys. Res. Commun. 67, 1078-1085
15. Bandlow, W., Kotzias, K. and Klein, A. (1976) in preparation
16. Kellems, R.E., Allison, V.F. and Butow, R.A. (1974) J. Biol. Chem. 249, 3297-3303

17. Gaitskhoki, V.S., Kisselev, O.J., Klimov, N.A., Monakhov, N.K., Mukha, G.V., Schwarzman, A.L. and Neifakh, S.A. (1974) FEBS Letters 43, 151-154
18. Butow, R.A., this volume
19. Michel, R., this volume

Genetics and Biogenesis of Chloroplasts and Mitochondria, Th. Bücher et al. eds.

INTEGRATION AND DISINTEGRATION OF PROTEINS SYNTHESIZED IN MITOCHONDRIA

H.-D. Hofmann, E. Hundt and B. Kadenbach
Biochemie, FB Chemie, Philipps-Universität
Lahnberge, D-3550 Marburg, Germany

INTRODUCTION

In earlier publications [1-5] it was suggested that the primary products of mitochondrial protein synthesis are small proteins with molecular weights of about 10 000 which are later integrated into functional proteins of higher molecular weights. This integration process can be blocked by inhibitors of energy synthesis [5] and of protein synthesis [4,5] as well. To understand the linkage within the final products of mitochondrial protein synthesis, attempts were made to disintegrate the high molecular weight proteins into the primary low molecular weight products by chemical treatment. In addition to experiments on in vitro labeled mitochondrial proteins, isolated subunits of cytochrome oxidase, which are synthesized in mitochondria, were studied.

MATERIALS AND METHODS

Isolation of rat liver mitochondria, incubation with labeled leucine and SDS-gelelectrophoresis was performed as described [5,8]. Cytochrome oxidase was isolated from rat liver mitochondria by use of Triton X-114 as described previously [9]. Ultrogel AcA 44 and AcA 54 were purchased from LKB (Gräfelfing).

RESULTS AND DISCUSSION

The change of molecular weight distribution of in vitro labeled mitochondrial proteins during a chase period of 60 minutes is shown in fig. 1. Labeled proteins with an apparent molecular weight of about 10 000 are converted to proteins of 20 - 40 000 molecular weight during the chase period. It was found that the radioactivity in the very high molecular weight range was due to heating of the dissolved membranes in the presence of mercaptoethanol prior to SDS-gelelectrophoresis. This aggregation was only found with mitochondrially synthesized proteins, whereas the pattern of coomassie-blue stained membrane proteins remained virtually unchanged. The tendency of mitochondrial translation products to aggregate in the presence of dodecylsulfate has been described [2,6,7].

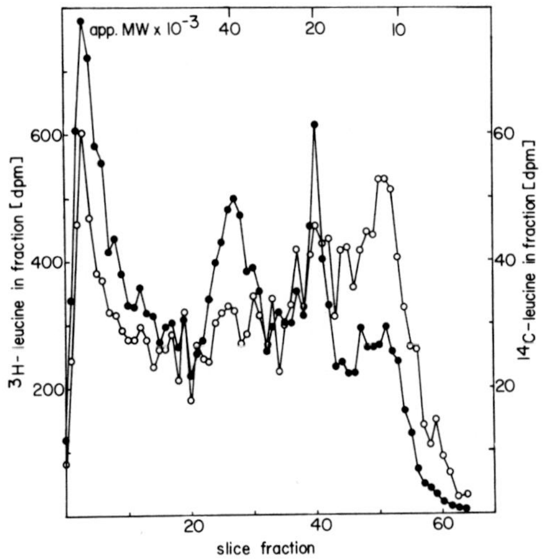

Fig. 1: Molecular weight patterns of in vitro labeled mitochondrial membrane proteins, analyzed by SDS-gelelectrophoresis. Labeling conditions: o--o: 10 min. incubation with ^{14}C-leucine. ●--●: incubation with ^{3}H-leucine, followed by 60 min. chase in the presence of 3 mM unlabeled leucine.

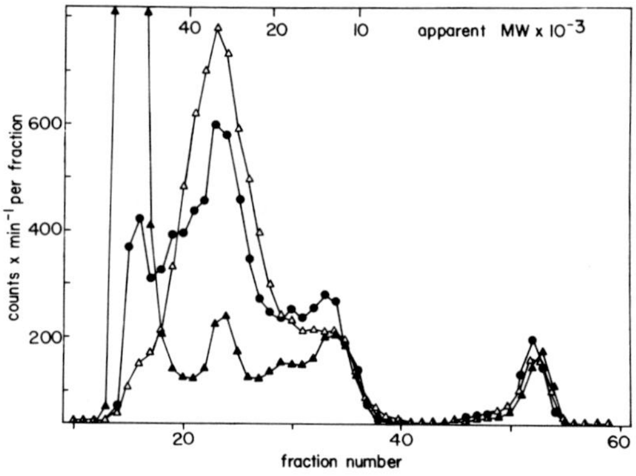

Fig. 2: Aggregation and partial disintegration of proteins synthesized in isolated rat liver mitochondria by chemical treatment. Rat liver mitochondria were labeled with ^{14}C-leucine for 40 min. in vitro. Aliquots of isolated membranes (about 0,5 mg protein) were dissolved in 0,1 ml buffer (50 mM Na-phosphate, pH 7,8, 8 M urea, 5 % SDS) and after treatment chromatographed on Ultrogel AcA 44-columns (0,5 x 150 cm). The elution buffer contained 0,1 M Na-phosphate, pH 7,8, 1 % SDS. △--△: addition of 5 % mercaptoethanol and 1 h at room temperature. ▲--▲: addition of 5 % mercaptoethanol and 10 min. at 100°C. o--o: addition of ICH_2COONa (100 mM) and 6 days at room temperature, then addition of 5 % mercaptoethanol and 10 min. at 100°C.

In fig. 2 the aggregation of labeled mitochondrial proteins due to heating in the presence of mercaptoethanol is shown. No aggregation of labeled proteins was observed, if the membranes, dissolved in dodecylsulfate-urea, are heated for 10 min. at 100°C in the absence of mercaptoethanol (not shown). In order to change the hydrophobicity of the proteins, carboxymethylation was per-

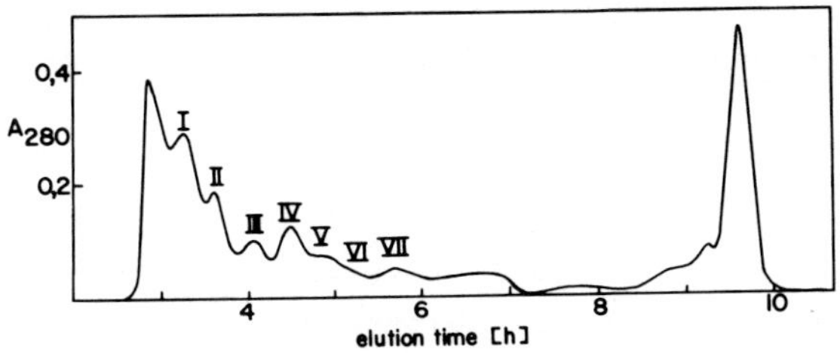

Fig. 3: Chromatography of cytochrome oxidase of rat liver on Ultrogel AcA 54. The column (0,8 x 150 cm) was eluted with 50 mM Tris-Cl, pH 7,4, 0,1 mM EDTA, 0,5 % SDS. Cytochrome oxidase (2,5 mg protein) was dissolved at room temperature in elution buffer containing 6 % SDS.

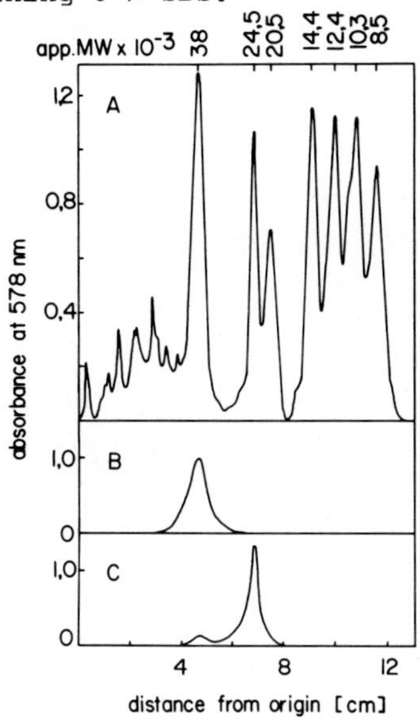

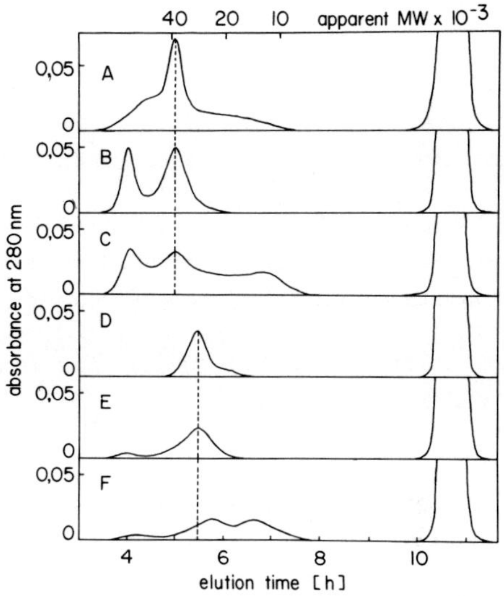

Fig. 4 (left): SDS-Gelelectrophoresis of rat liver cytochrome oxidase and isolated subunits. Subunit I (B) and II (C) were purified from cytochrome oxidase by rechromatography of the separated bands as shown in fig. 3.

Fig. 5 (right): Disintegration of purified subunits of cytochrome oxidase upon chemical treatment. Samples of 0,03 mg subunit I (A-C) and 0,01 mg subunit II (D-F) were dissolved in 0,1 ml buffer (0,5 M Tris-Cl, pH 8,0, 8 M urea, 2 mM EDTA 5 % SDS). A and D: 15 h at 50°C, then addition of dithiothreitol (80 mM) and 24 h at 50°C. B and E: addition of ICH_2COONa (80 mM) and 21 h at 50°C. C and F: addition of ICH_2COONa (80 mM) and 15 h at 50°C, then addition of dithiothreitol (160 mM) and 24 h at 50°C. The incubated samples were chromatographed on Ultrogel AcA 44-columns (0,5 x 150 cm). The elution buffer contained 50 mM Tris-Cl, pH 7,4, 0,1 mM EDTA, 0,5 % SDS and 0,02 % NaN_3.

formed with iodoacetate. If the membrane proteins were treated for 6 days at room temperature with iodoacetate in the presence of urea and SDS, the tendency to aggregate, was drastically reduced and a slight increase of radioactivity was observed in the molecular weight range of 10 000 (fig. 2).

In order to study the aggregation and disintegration of mitochondrial translation products on defined proteins, subunits I and II of rat liver cytochrome oxidase were isolated as shown in fig. 3. The purity of the isolated fractions was analyzed by SDS-gel-electrophoresis (fig. 4). Heating of cytochrome oxidase for 5 minutes at 100°C in the presence of 5 % mercaptoethanol resulted in complete aggregation of subunits I and III, but not of subunit II and of the cytoplasmically synthesized subunits. Treatment of subunits I and II with dithiothreitol or with iodoacetate resulted in partial aggregation of band I but not of band II (fig. 5). Treatment with dithiothreitol after treatment with iodoacetate resulted in partial disintegration of both subunits. A protein peak appeared in the molecular weight range of about 10 000.

The data of this investigation are most easily interpreted by the assumption that the mitochondrial translation products of higher molecular weight are composed of smaller polypeptides.

This work was supported by the Deutsche Forschungsgemeinschaft, SFB 103.

REFERENCES

1. Tzagoloff, A., and Akai, A. (1972) J. Biol.Chem. 247, 6517-6523
2. Michel, R. and Neupert, W. (1973) Eur.J.Biochem. 36, 53-67
3. Hawley, E.S. and Greenawalt, J.W. (1975) Eur.J.Biochem.54,585-601
4. Kuzela, S., Krempasky, V., Kolarov, J., and Ujhazy (1975) Eur. J. Biochem. 58, 483-491
5. Hofmann, H.-D. and Kadenbach, B. (1976) FEBS-Lett. 66, 27-30.
6. Küntzel, H. and Blossey, H.-C. (1974) Eur. J. Biochem.47,165-171
7. Küntzel, H., Pieniazek, N.J., Pieniazek, D., and Leister, D.E. (1975) Eur. J. Biochem. 54, 567-575
8. Kadenbach, B. and Hadvary, P. (1973) Eur. J. Biochem.32, 343-349
9. Kadenbach, B.: Autonomy and Biogenesis of Mitochondria and Chloroplasts (North-Holland, 1971) pp. 360-371

Genetics and Biogenesis of Chloroplasts and Mitochondria, Th. Bücher et al. eds.

SYNTHESIS OF MITOCHONDRIAL DNA, -PROTEINS AND -PHOSPHOLIPIDS IN THE YOUNG SEA URCHIN EMBRYO SPHAERECHINUS GRANULARIS.

Horst Bresch
Medizinische Hochschule Hannover, Abteilung für experimentelle Pathologie
Karl-Wiechert-Allee 9, 3 Hannover, West-Germany

INTRODUCTION

Large amounts of mitochondria are present in the sea urchin egg; several questions regarding these particles need to be clarified. 1. Are they only passively distributed to daughter cells after fertilization without a significant synthesis of structural components, or do the components replicate and, if so, at what stages? Severalfold multiplication of mitochondria has been reported as occurring from the mesenchyme to gastrula stages (1,2). 2. Is this multiplication connected with altered synthetic rates of structural components?

MATERIALS AND METHODS

Animals were taken from the Bay of Naples and gametes obtained and processed as described (3).

Mitochondria were isolated according to (3). Further details and characterization will be described elsewhere.

mtDNA synthesis: Isolation was performed as described (3). To dispose of nuclear DNA and other material the closed circular mtDNA (cmtDNA) was purified in four additional CsCl-ethidiumbromide (EB) gradients, until radioactivity in the upper band was lower and in a constant ratio to the lower band. EB was removed and concentrations of cmtDNA determined in CsCl gradients run in the analytical ultracentrifuge. Radioactivities were measured as described in (3). For electronmicroscopy total mtDNA (upper and lower band) were processed according to (4). Mitochondria were treated as described in (5), membrane fractions lysed in 1% SDS, 1% mercaptoethanol in 30mM H_2SO_4, 50mM Tris pH 6.1 and electrophoresis was run in a discontinuous SDS-polyacrylamide-gel system (6) with 3% upper and 12% lower gel. For measuring specific activities of bands, coomassie blue stained gels were photographed on Agfachrome 50L film and plates densitometrically evaluated. For measuring radioactivities, 0.8mm gel slices were burnt in a Packard oxidizer.

mtproteins: For determination of specific activities in membrane fractions, when embryos were incubated in high leucine concentrations, mitochondria were sonicated in 10mM phosphate buffer pH 7.0 and unbroken particles sedimented at 10000xg. The membrane fraction was obtained at 150000xg, 1hr; matrix was sedimented at 150000xg, 24 hrs. The sediments were lysed in 1M NaOH; protein determined according to Lowry and radioactivities measured in Unisolve scintillator.

Phospholipids were isolated from purified mitochondria essentially according to (7). Separation and quantitative determination was conducted as described (8).

Radioactivities were measured in aliquots of kieselgel eluates.

RESULTS

Results from thymidine incorporation are given in Table 1. Specific activities of cmtDNA were small compared to those of nuclear DNA and practically constant in all investigated stages. Calculations of replication rates based on these values were below 0.01%/h. In the second part of the investigation, mtDNA was examined by electronmicroscopy*. Unfertilized eggs, mesenchyme blastula- and gastrula stages were investigated. From one preparation at least 500 molecules were analysed. In all samples no D-loops could be observed. Less than two percent of the molecules, doubtful cases included, showed questionable single stranded parts. Incorporation and electronmicroscopical studies indicate that replication did not occur. The above results are in accordance with those reported (9,10), with the exception that specific activities were not zero. This could result from repair taking place or artificial ring closure of nuclear DNA.

TABLE 1

Stage	μg cmtDNA	Nuclear DNA	dpm	$\frac{dpm}{\mu g}$
morula	0.96		2859	2978
mesenchyme blastula	0.43		2100	4883
		13.4	2036284	151961
early gastrula	0.19		572	3010
prism stage	0.27		295	1092

Specific activities of closed circular mtDNA. Incubation in ^{3}H-thymidine, 4 hrs, 1μc ^{3}H-thymidine/ml, other procedures as described (3).

mtproteins: Synthetic rates of mtmembrane proteins were measured in two experiments. In the first experiment membrane fractions were lysed and lysates analysed by electrophoresis. In Fig.1 electropherograms are shown and in Table 2, their evaluation. As seen from the table, labelling occurred but specific activities did not indicate significantly increased synthetic rates in later stages. - The majority of mtpolypeptides in the bands are synthesized in the cytoplasm. Inhibition of their synthesis by cycloheximide in order to detect synthesis of polypeptides within the mitochondria was not possible since the antibiotic did not penetrate into the embryos. In one preliminary experiment polypeptides were extracted with chloroform-methanol from swimming blastula and early pluteus stages. The ratio of the specific activities blastula/pluteus stage of 0.46 did not favour increased synthetic rates in the pluteus stage. - To measure the dependence of the synthetic rates from the leucine concentration, a higher leucine concentration was added in the second experiment. Since specific acti-

* I thank Dr.A.Arnberg for spreading the DNA and its summary evaluation in the electronmicroscope.

vities were expected to be significantly lower, no gels were run. The results in Table 3 demonstrate that under these conditions also the rates were nearly constant.

FIG.1

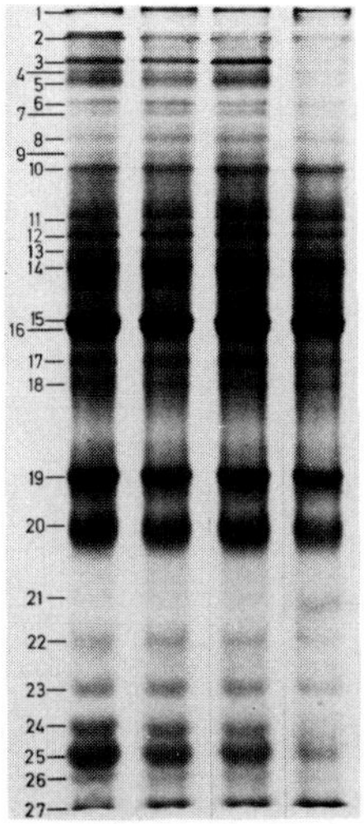

TABLE 2

Stage	Band No.	MW	dpm/area	Stage	Band No.	MW	dpm/area
1	8	120000	7317	1	19	34000	448
2			8442	2			503
3			6890	3			519
4			5496	4			492
1	10	108000	4268	1	20	29000	319
2			3629	2			363
3			2814	3			340
4			2150	4			400
1	14	74000	1758	1	22	18500	1474
2			2042	2			810
3			1681	3			1030
4			1599	4			1690
1	15	60000	359	1	25	12500	928
2	16	58000	400	2			441
3			475	3			310
4			564	4			256

Fig.1. Electropherograms of mtmembrane fractions. From left to right: Swimming blastula (15 hrs), mesenchyme blastula (20 hrs), gastrula (25 hrs), prism stage (40 hrs after fertilization).

Table 2. Specific activities of bands shown in Fig.1. Area: Band area in densitometric tracings. 5000 embryos/ml, 1l suspension, 0.5μc/ml 4.5 ^{3}H-leucine 50c/mM. Incubation 2 hrs.

TABLE 3

Stage	$\frac{dpm}{\mu g}$(membrane)	Ratio	$\frac{dpm}{\mu g}$(matrix)	Ratio
swimming blastula	184	1.00		
mesenchyme blastula	187	1.02	82	1.00
pluteus stage	239	1.30	148	1.80

Specific activities of mtmembrane and matrix fractions. Incubation as in Table 2 but in addition 2μg/ml unlabelled L-leucine were added.

Phospholipids: The synthetic rates of different phospholipids in different stages are demonstrated in Fig.2. Although no change in synthetic rates of phosphatidylcholine and phosphatidylethanolamine was observed, an increase in cardiolipin synthesis and, less pronounced, also in phosphatidylinositol syntheses could be measured.

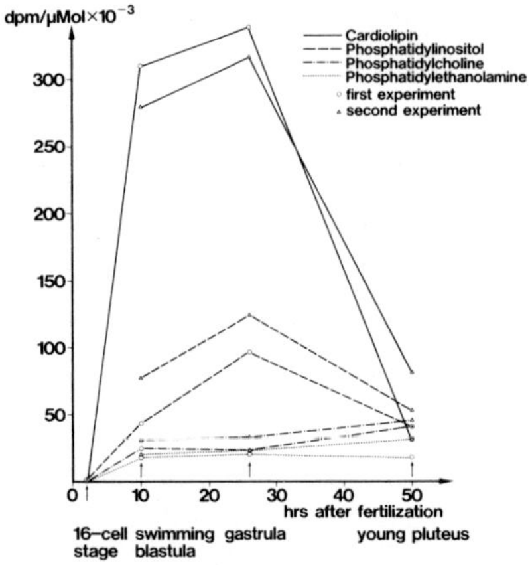

Fig.2. Synthesis of mtphospholipids in different stages. Incubation 2 hrs in glycerol-1(3)-^{3}H 2.8c/mM. 5000 embryos/ml, 1l suspension, 1μc/ml, 10μg/ml unlabelled glycerol. Antibiotics as in (3).

DISCUSSION

In these studies no indication of altered synthetic rates in relation to an increase in number of mitochondria could be demonstrated for mtDNA. No different synthetic rates in total membrane proteins and in resolved bands were measured. Increased synthetic rates of cardiolipin might indicate differentiation of the inner membrane. In this connection the study of the dynamics of proteins specifically located in the inner membrane will be of great interest. In combination with the reported increase in the number of the particles (1,2), however, increased synthetic rates could also indicate an increased incorporation of other phospholipids, basic structural elements of membranes. If phosphatidylcholine and phosphatidylethanolamine are obtained from yolk, no higher incorporation of the precursor would be indicated. - It might be argued that also mtproteins are taken directly from yolk, thus also no higher synthetic rates would be detected. However, this is not the case since yolk does not contain these components.

REFERENCES

1. Gustafson,T. and Lenicque,P. (1955) Exp.Cell Res. 8, 114-117
2. Shaver,J.R. (1956) Exp.Cell Res. 11, 548-559
3. Bresch,H. (1973) FEBS Letters 31, 233-236
4. Arnberg,A., Van Bruggen,E.F.J., Ter Schegget,J. and Borst,P. (1971) Biochim.Biophys.Acta 246, 353-357
5. Turner,G. (1973) Eur.J.Biochem. 40, 201-206
6. Neville,D.M.,Jr. (1971) J.Biol.Chem. 246, 6328-6334
7. Folch,J., Lees,M. and Sloane-Stanley,G.H. (1957) J.Biol.Chem. 226, 497-509
8. Hallermayer,G. and Neupert,W. (1974) Hoppe-Seyler's Z.Physiol.Chem. 355, 279-288
9. Piko,L. (1969) Amer.Zool. 9, 1118.
10. Matsumoto,L., Kasamatsu,H., Piko,L and Vinograd,J. (1974) J.Cell Biol. 63, 146-159

Genetics and Biogenesis of Chloroplasts and Mitochondria, Th. Bücher et al. eds.

INHIBITION OF CYTOPLASMIC PROTEIN SYNTHESIS BY MITOCHONDRIAL SOLUBLE FACTORS IN RAT LIVER AND WALKER CARCINOSARCOMA

Néstor González-Cadavid, Blas Dorta and Andrés Carmona

Departamento de Biología Celular, Facultad de Ciencias,

Universidad Central de Venezuela, Apartado 10098

Caracas, Venezuela

The coordination of the formation of mitochondrial proteins inside and outside the organelle is a complex process which requires several control mechanisms for its correct operation. One of the possible regulatory signals would consist of the export to the cytosol of products of mitochondrial origin which would switch on and off the translation of mRNAs coding for the synthesis of mitochondrial proteins on 80S ribosomes. Besides this "fine" control, specific for the needs of mitochondrial biogensis, the existence can be envisaged of mitochondrial exportable factors that would operate as a sort of "coarse" control affecting the overall rate of cytoplasmic translation. The second class of signals would imply that the organelle intervenes in the regulation of total protein synthesis in the eukaryotic cell in a process not based on mere variations of the ATP supply, a possibility as yet not investigated.

The hypothetical mitochondrial control factors would not necessarily involve mRNAs of mitochondrial origin, since they could simply act as positive or negative modulators of cytoplasmic protein synthesis. Any of the cell constituents already shown to regulate translation, such as low molecular weight RNAs[1-3], cyclic nucleotides[4], haemin[3], etc., could conceivably play that role provided they could be easily transported across the mitochondrial membrane. If these mitochondrial factors really operate in the normal cell, they could well differ, either in level or mechanism of action, from those of neoplastic mitochondria, since the structure and function of the organelle is considerably altered in the cancer cell (see ref. 5, 6).

The present report describes the isolation of high and low molecular weight soluble fractions from liver and Walker carcinoma mitochondria, which are able to inhibit protein synthesis by polyribosome systems from both sources. These factors are not tissue specific and can act reversibly on intact cells, but only the smaller size fraction has inhibitory activity devoid of RNAse contamination.

FRACTIONATION OF THE CRUDE MITOCHONDRIAL EXTRACT BY GEL FILTRATION

Our working hypothesis required the factors to be soluble and easily extractable from the organelle since they are assumed to be transferred in vivo to the cytoplasm according to the physiological needs of the cell. The mitochondria were isolated from rat liver by a technique[7] which practically eliminates contamination by cytoribosomes, and from the Walker carcinoma by a procedure[5] which although not achieving the same degree of purity is satisfactory for our purposes. The tumour was excised 10 days after implantation. Both preparation were then submitted to hypotonic shock in 10 mM tris buffer, pH 7.8 at 2°C, for 30 min, at a concentration of about 20 mg protein/ml, and ultrasonicated for 1 min. The membrane fragments were removed by centrifugation at 40 000 g for 30 min. The yield of soluble extract per 100 g of liver and tumour tissue was usually around 15 and 3 ml, respectively. Protein concentrations were 8-10 mg/ml (liver) and 6-8 mg/ml (carcinoma).

The effect on cytoplasmic protein synthesis was tested using both detergent-treated total polyribosomes from liver and free polyribosomes fron carcinoma. The latter represent most of the ribosomal population in the tumour and no detergent is required for their isolation, thus avoiding partial inactivation due to the high RNAse activity of this tissue. Variations due to the source of pH 5 fraction were eliminated by using only liver in both cases. Conditions were as previously described[8], except that final volumes were 0.1 ml and $[^{14}C]$leucine was present at 0.5 μC/ml.

Preliminary experiments showed that the "crude" extract from liver, added in the proportion of 200 μg protein (20 μl samples) per 25 μg of rRNA incubated, inhibited protein synthesis on liver total polyribosomes by 70% (control, with 20 μl 0,1 M tris: 39.4 d.p.m./μg rRNA). By gel filtration on Sephadex G-25 two main fractions were separated, one containing all the protein and the other the low molecular weight components up to a K_{av} of 0.86. The fractions were freeze-dried, dissolved in the same volume of crude extract submitted to gel filtration, and added to the test system in identical proportion (20 μl). Both fractions inhibited protein synthesis by 50 to 60%, irrespective of the source of ribosomes (liver or carcinoma) or the type (total "detergent-treated", or free ribosomes).

In order to achieve a better resolution of the soluble factors, the "crude" liver mitochondrial extract was passed through a column of Sephadex G-50 and the fractions collected were assayed with liver detergent-treated ribosomes after suitable concentration. Fig. 1-a shows that the inhibitory effect on protein synthesis was present in two zones. One, named MSF-I, corresponded to the elution profile of 95% of the protein and four fractions in the peak of inhibition decreased protein synthesis by 70 to 80%. The second zone, named MSF-II, extended from a K_{av} of 0.8 to 1.1 and showed three peaks of activity, from 60 to 75% inhibition.

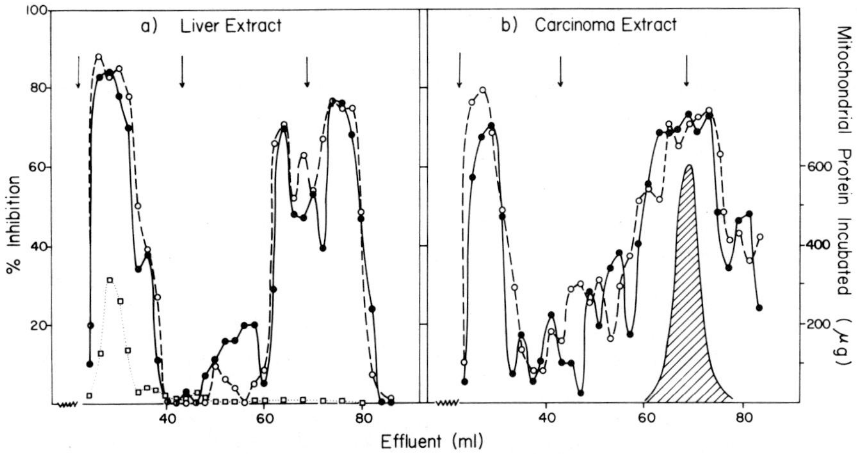

Fig. 1. Fractionation of protein synthesis inhibitory factors from liver and carcinoma mitochondria by gel filtration. Samples from 'crude' extracts, containing 42 and 9 mg of soluble protein respectively, were passed through 1.7 x 30 cm Sephadex G-50 columns equilibrated with 10 mM tris buffer, pH 7.8 at 2°C, and 2 ml fractions were collected. After freeze-drying, residues were dissolved in 0.45 ml (liver) or 0.1 ml (carcinoma) of water. Inhibition of cytoplasmic protein synthesis was tested by adding 20 μl samples to incubations of either detergent-treated polyribosomes from liver (o---o) or free polyribosomes from Walker carcinoma (●——●). Mean control values were 46.8 and 26.5 d.p.m./μg of rRNA, respectively. Arrows indicate the Vo (void volume) and Ve (elution volumes) for cytochrome c and [^{3}H]leucine. Protein contents are represented by □ ····· □ . Hatched area: elution of [^{3}H]leucine added as internal standard.

The central section coincided with the elution profile of [^{3}H]leucine added as internal standard (K_{av}: 0.98), the left section corresponded to higher molecular weights, and the right section included substances with delayed elution probably due to partial absorption. No difference was noticed when carcinoma free ribosomes substituted for liver total ribosomes in the test system.

The lack of tissue specificity is shown in Fig. 1-b, since the soluble extract from carcinoma mitochondria displayed a similar pattern to that of the liver extract inhibiting also to the same extent the activity of both kinds of ribosomes.

The pool of fractions corresponding to MSF-I from the liver extract inhibited protein synthesis by 50% with only 45 μg protein added to the system (5 μl), and the effect reached a plateau of 80% inhibition from 15 μl (Fig. 2). Supplementation with saturating amounts of initiation factors from rabbit reticulocytes[9] increased the inhibition to 90%, and even at 2.5 μl (23 ug protein) the decrease of protein synthesis was 65%. No significant difference was observed when tumour ribosomes substituted for liver ribosomes, either in the presence or the absence of initiation factors.

The plateau of inhibition for MSF-II in the unsupplemented liver system was

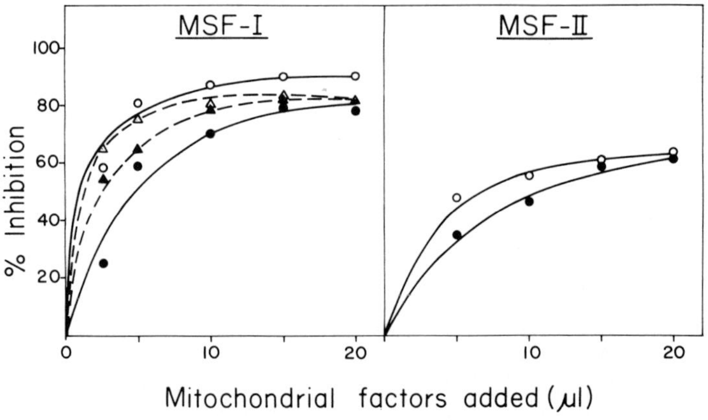

Fig. 2. Effect of increasing concentrations of high and low molecular weight factors from liver mitochondria, on protein synthesis by cytoplasmic ribosomes under initiating and non-initiating conditions. The peak fractions of Fig. 1 were pooled and the two resulting fractions were named MSF-I and MSF-II, respectively. Protein concentration in MSF-I was 9 mg/ml. The test systems (in triplicate) were as in Fig. 1, except that half of the tubes contained 50 μg of rabbit reticulocyte initiation factors (RIF) and that mitochondrial factors were only from liver. Control values in d.p.m./μg rRNA appear in brackets. Liver ribosomes with RIF (o —— o, 74.6) and without RIF (● —— ●, 29.5); carcinoma ribosomes with RIF (△----△, 33.8) and without RIF (▲----▲, 22.6).

achieved with a similar volume (15 μl) but reached a lower level: 62%. Since this fraction is devoid of protein, it is difficult to express concentrations in meaningful terms, but it can be calculated that the volume causing a 50% inhibition (10 μl) contains the low molecular weight extractives from 200 ug of total mitochondrial protein. Supplementation with initiation factors increased the inhibitory effect slightly. No comparison was made using tumour ribosomes.

ASSESMENT OF RIBONUCLEASE CONTRIBUTION TO THE INHIBITORY ACTIVITY

Mitochondria contain several ribonucleases which can be assumed to be extracted, at least partially, under the conditions used in the present work. It is important to rule out this contamination as a first step in demonstrating physiological significance to any inhibitory factor of protein synthesis. We tested the RNAse activity of liver MSF-I and II by measuring the increase of absorbance at 260 nm soluble in 6% perchloric acid-0.18% uranyl acetate[10], in incubations of tRNA at 37°C for 30 min. Although no difference was found as compared to the controls, the possibility remained that traces of RNAse enough to break mRNA would not be detected by this procedure. For that reason we incubated liver polyribosomes at 37°C for 10 min, with concentrations of MSF-I and II causing a 60% inhibi-

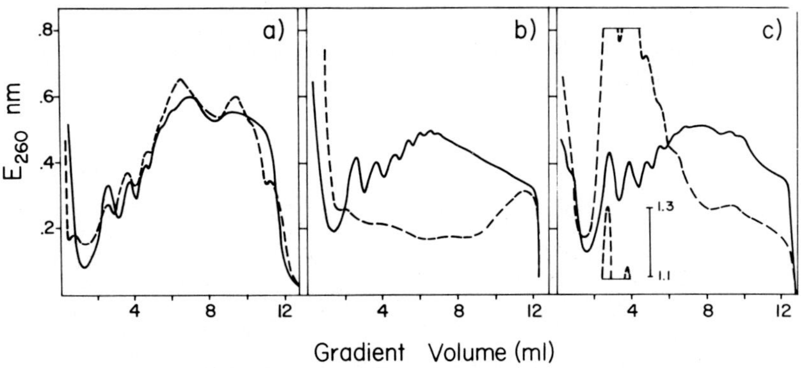

Fig. 3. Dissagregation of the polyribosomal profile caused by liver MSF-I. Detergent-treated ribosomes from liver (0.25 mg rRNA) were incubated for 10 min at 37°C (except one control at 0°C) in 1 ml of the usual medium, under the following conditions: a) control, 37°C: ——; control, 0°C: ----; b) cycloheximide, 100 µg/ml: ——; pancreatic ribonuclease, 20 µg/ml: ----; and c) MSF-I, 50 µl, plus cycloheximide, 100 ug/ml: ----; MSF-II, 200 ul, plus cycloheximide, 100 µg/ml: ——. The polyribosomal profile was analysed by sucrose gradient centrifugation for 75 min at 200 000 g as previously described[8]. Top of gradient on the left side.

tion of protein synthesis. The incubations were carried out in the presence of cycloheximide and without energy source and pH 5 fraction, to avoid running off under non-initiating conditions.

Fig. 3 shows that the ribosome profile dissapeared completely with 20 µg/ml of ribonuclease but was not affected by cycloheximide. MSF-I dissasembled the polyribosomes into mono and disomes, suggesting a low level of RNAse activity. In contrast MSF-II did not affect at all the ribosome profile, thus showing that its protein synthesis inhibitory action is not due to RNAse.

EFFECT ON PROTEIN AND NUCLEIC ACID SYNTHESIS IN INTACT CELLS

To see whether these factors can pass through the cell membrane and act _in vivo_, we investigated the incorporation of $[^{14}C]$leucine, $[^{3}H]$thymidine, and $[^{3}H]$ uridine, into trichloroacetic acid precipitable material by incubations of trypsinized carcinoma cells[11] under increasing concentrations of liver MSF-I and II. The heterologous cell system was used instead of liver cells simply because tumour cells were easier to prepare, but the experiments will be repeated with liver cells obtained by the collagenase-perfussion procedure. MSF-I (75 µl) inhibited protein synthesis by 50% (Fig. 4) at concentrations equivalent in the ribosomal system to the addition of 7.5 µl causing 60% inhibition, implying a

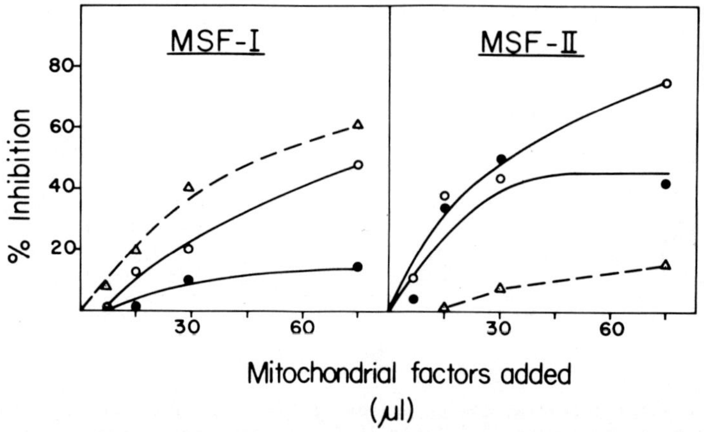

Fig. 4. Effect of liver MSF-I and MSF-II on protein and nucleic acid synthesis by Walker carcinoma cells. Pieces of tumour were trypsinized in the presence of 56 μg/ml of deoxiribonuclease[11], the cells harvested washed with Earle's medium containing DNAse and finally suspended in the same medium at 10^6 cells/ml. Half of the suspension received [^{3}H]thymidine (0.5 μC/ml) plus [^{14}C]leucine (0.05 μC/ml), and the remainder, [^{3}H]uridine (0.5 μC/ml). Aliquots of 1 ml were incubated in triplicate at 37°C for 2 h under aseptic conditions, either in the presence or absence of the mitochondrial soluble factors prepared in the experiment of Fig. 2. Protein and nucleic acids were precipitated and washed twice with 10% trichloroacetic acid (TCA), residues dissolved in 0.1 ml formic acid, and the radioactivity determined by the double channel procedure. Control values in d.p.m./10^6 cells appear in brackets. Curves describe TCA-insoluble incorporation of [^{14}C]leucine (o —— o, 10 030), [^{3}H]thymidine (● —— ●, 23 160), and [^{3}H]uridine (△----△, 30 900).

similar sensitivity for both systems. The RNA radioactivity was also considerably decreased, but this effect probably reflects a partial hydrolysis due to the RNAse content rather than a direct interference with RNA synthesis. The incorporation of [^{3}H]thymidine was not significantly affected.

MSF-II was even more active on cell protein synthesis than it was with the subcellular system, since concentrations causing a 40% inhibition in the ribosomal incubations, induced a 75% decrease of cell protein radioactivity. As expected, no considerable effect on RNA synthesis occured at the concentrations tested, but DNA synthesis was impaired by 40%. It has to be decided whether the latter effect was direct or secondary to the inhibition of protein synthesis.

An important point for investigating the possible physiological role of the inhibitory activity was to decide whether normal levels of synthesis could be obtained by removal of the factors from the medium. The cell suspensions were incubated for 30 min at the meximum concentrations used in the previous experiment and then the medium was replaced by one without factors. Radioactivity was added and the incubations proceeded for 90 min more. Protein and DNA synthesis were restored to the control values both with MSF-I and II, thus showing a complete reversibility and suggesting that MSF-I does not impede polyribosomal re-assembly. In contrast, the incorporation of $[^3H]$uridine into RNA remained low when preincubation was done in the presence of MSF-I.

PRELIMINARY CHARACTERIZATION OF MSF-II

The low molecular weight fraction deserves special attention for the following reasons: a) easy diffusibility across membranes; b) higher activity on intact cells; c) no contamination by RNAse or protein; and d) parallel inhibitory effect on DNA synthesis. As we have been recently interested in characterizing the transcriptional products of carcinoma mitochondria[5,6] in an initial approach to their possible intervention as regulators of cytoplasmic translation, we tried to isolate any low molecular weight RNA that could be present in MSF-II. However, all the RNA extracted by the phenol procedure was confined to MSF-I (<4 S_e in acrylamide electrophoresis), and no absorbance at 260 nm was precipitated by ethanol from the upper phase extract of MSF-II.

The substantial decrease in protein radioactivity caused by MSF-II does not seem to arise simply from the presence of a large leucine pool diluting the $[^{14}C]$ leucine added to the incubation, as suggested by three independent experiments. The degree of inhibition in the ribosomal system reached a plateau at 55-60% in spite of a two-fold increase in the amount of MSF-II added (Fig. 2). The intact cells contain more leucine than the pH 5-fraction ribosomal system but the peak of inhibition was higher and required less MSF-II (Fig. 4). Finally, the profile of inhibitory activity in the MSF-II zone of the Sephadex filtration (Fig. 1 b) exceeded on both sides the elution curve of added $[^3H]$leucine.

The concentrations of ATP and ADP contributed by 10 ul of MSF-II, determined in a nucleotide analyzer, were 45 and 70 pmoles, respectively, in the 0.1 ml ribosomal incubation. These traces do not affect significantly the concentration of ATP optimum for protein synthesis (2mM). Moreover, the constituents of MSF-II passing through the resin column during sample loading, retained most of the inhibitory activity, as well as the absorbance at 256 nm. Neither the mitochondrial ions, particularly K^+ and Mg^{2+}, calculated from published data[12] to be present in the extract, would modify substantially the composition of the incubation medium. The inhibition of protein synthesis cannot be due therefore to a mere quantitative change of the medium by adding MSF-II and should be ascribed to factors of more specific effect.

We are at present trying to identify and purify those substances present in MSF-II responsible for the effect on protein synthesis and to determine whether they affect preferentially the formation of either mitochondrial or non-mitochondrial proteins. The physiological implications of our findings will depend obviously on the demonstration of the transport of these factors across the mitochondrial membrane and their intervention in controlling the rate of cytoplasmic translation according to functional states of the organelle.

ACKNOWLEDGEMENTS

Nucleotide analysis was kindly performed by Dr. Itala Becemberg. This work was supported by a grant from the Fund for Overseas Research Grants and Education.

REFERENCES

1. Heywood, S.M., Kennedy, D.S. and Bester, A.J. (1975) Eur. J. Biochem. 58, 587-593.
2. Hunter, T., Hunt, T., Jackson, R.J. and Robertson, H.D. (1975) J. Biol. Chem. 250, 409-417.
3. Beuzard, Y. and London, I.M. (1974) Proc. Nat. Acad. Sci. USA 71, 2863-2866.
4. Klaipongpan, A., Bloxham, D.P. and Akhtar, M. (1975) FEBS Lett. 58, 81-83.
5. González-Cadavid, N.F. and Pérez, J.L. (1975) Biochem. J. 146, 361-373.
6. González-Cadavid, N.F. and Pérez, J.L. (1976) Cancer Res. 36, 222-228.
7. O'Brien, T.W. and Kalf, G.F. (1967) J. biol. Chem. 242, 2172-2179.
8. González-Cadavid, N.F. and Herrera Quijada, F. (1974) Biochem. J. 138, 129-141.
9. Crystal, R.G., Elson, N.A. and Anderson, W.F. (1974). In Methods in Enzymology (Moldave, K., eds. and Grossman, L.) vol 30, part F, pp 101-127, Academic Press, New York and London.
10. Uchida, T. and Egami, F. (1967). In Methods in Enzymology (Grossman, L. and Moldave, K., eds.) vol 12, part A, pp 228-239, Academic Press, New York and London.
11. Norrby, K., Knutson, F. and Ludin, P.M. (1966) Exptl. Cell Res. 44, 421-428.
12. Munn, E.A. (1974) The Structure of Mitochondria, Academic Press, New York and London, 465 pp.

Genetics and Biogenesis of Chloroplasts and Mitochondria, Th. Bücher et al. eds.

MAMMALIAN EMBRYOS: A MODEL FOR STUDYING THE DEPENDENCE OF GROWTH AND DIFFERENTIATION PROCESSES ON MITOCHONDRIAL BIOGENESIS AND FUNCTION

Rolf Bass
Pharmakologisches Institut der Freien Universität
- Abt. Embryonalpharmakologie - 1000 Berlin 33, F.R.G.

During recent years a number of studies on conditions which lead to impairment of mitochondrial biogenesis and function have been reported. Initially it was observed that mitochondria from yeast can be manipulated both in biochemical composition and morphological appearance by cutting down the oxygen supply (1,2). The use of antibiotics which specifically inhibit mitochondrial or extramitochondrial protein synthesis led to the discovery of some mitochondrial enzymes,which,for their role fulfilment in oxidative phosphorylation, are dependent on mitochondrial synthesis reactions (cf.3). Although the inhibited functions are necessary for mitochondrial electron transport and oxidative phosphorylation, drastic effects on energy requiring extramitochondrial processes have not been reported. This can be due to several reasons: the obligation of energy production is taken over extramitochondrially, the remaining mitochondria are sufficient in quantity and function, or, the energy requiring processes are discontinued to await the reactivation of the mitochondria. The use of mammalian embryos at the organogenesis stage of development has shown that a block in mitochondrial biogenesis and function - induced by the *in vivo* inhibition of mitochondrial protein synthesis - not only leads to mitochondrial malfunction, but also expresses itself as inhibition of growth processes up to embryolethality (4,5).

Mammalian embryos during organogenesis are composed of very rapidly proliferating and differentiating tissues. Their DNA content is doubled between approximately 6-8 hrs. (rat embryos, day 10 of gestation) and about 22 hrs. (day 14). During this period of rapid development a shift in energy-conserving pathways from extramitochondrial towards mitochondrial ones occurs at the placentational stage (in the rat about day 12 of gestation). At this stage the embryo exceeds the limiting thickness guaranteeing nourishment via diffusion (6) and demands a functioning cardiovascular system. The contractile activity of isolated rat embryonic hearts becomes sensitive to inhibitors of oxidative phosphorylation and less sensitive to inhibitors of glycolysis (7). Glucose utilization by whole rat embryos, as measured under aerobic conditions *in vitro*, decreases by a factor of five (8). At the same time, the high capacity for mitochondrial protein synthesis *in vitro* also steeply declines (9). These data suggest that mature mitochondria are synthesized before placentation. When the chorioallantoic placenta starts functioning, mitochondria are then able to fulfil the task of highly efficient energy production to sustain the fast proliferation rate.

To investigate the necessity of mitochondrial function for embryonic development during late organogenesis, the specific inhibitor of mitochondrial protein synthesis, chloramphenicol (CAP), and its analogue thiamphenicol (TAP) were given to pregnant rats. *In vitro*, mitochondria from adult rat liver and from rat embryos were shown to be equally susceptible to both substances (9,10). Mitochondrial function was measured as cytochrome oxidase activity, the enzyme partially being synthesized by the mitochondrion (cf.3). The activity of this enzyme can be estimated employing whole homogenates. In this way the tedious work of isolating mitochondrial fractions from embryonic tissue is avoided (4). Overall embryonic development was followed by measuring the DNA content of the embryos. Some of the data obtained with TAP are presented in Fig.1. Interestingly, a given degree (75% used in this case) of inhibition of mitochondrial protein synthesis with prolongation of treatment leads to more pronounced impairment of growth processes. The effects

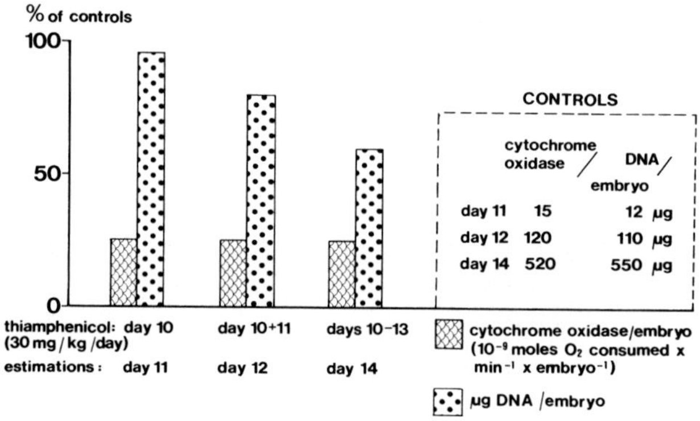

Fig.1. Effect of TAP on mitochondrial and overall embryonic development. Estimations were performed on the day following the treatment period indicated. TAP was given on: day 10 of gestation (24 hrs. ≙ 4 replications), day 10+11 (48 hrs. ≙ 7 replications), days 10-13 (96 hrs. ≙ 9.5 replications).

on both parameters are dose-dependent (5). To further examine the link between growth retardation and inhibition of mitochondrial biogenesis and function as the rate-limiting step in energy supply for replication, the following parameters were tested after the application of TAP on day 10 and 11 of gestation: respiration of whole, naked rat embryos *in vitro*, and measurement of ATP levels. The respiratory capacity of the embryos was measured in the absence and presence of dinitrophenol using the Waeburg method. The dose-dependent increase in inhibition of respiration well agrees with the declining content in ATP (Table 1). Treatment with 125 mg per kg TAP or more, given on day 10 and 11 of gestation so drastically inhibits the development of mitochondrial function that embryolethality results.

TABLE 1

Effect of thiamphenicol treatment on respiration and ATP content in rat embryos

amount of TAP given (day 10+11)	Q_{O_2} remaining (% of controls)	pmoles ATP per µg DNA remaining (% of controls)	µg DNA remaining (% of controls)
30 mg/kg/day	81%	74%	79%
60	60%	69%	71%
100	38%	67%	65%
200	35%	64%	50%

Q_{O_2} (+dinitrophenol) of controls: 22; ATP (measured according to (14)) in controls: 230 pmoles/µg DNA), DNA content of controls: 110 µg per embryo. All estimations were performed at the end of the treatment period, day 12 of gestation.

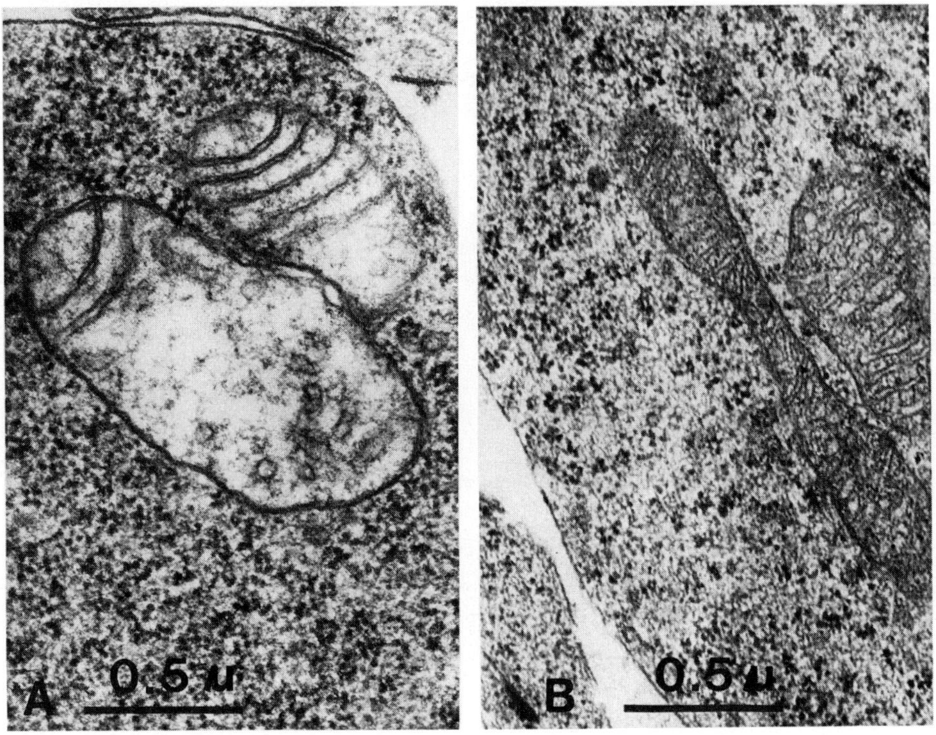

Plate 1. A:Swollen mitochondria of a proliferating neuroepithelium. The cristae are decreased in number, the matrix is translucent. B:Unaltered mitochondria of a neuroblast (stage of differentiation). TAP (200mg/kg/day) was given on day 12+13 of gestation, samples were taken on day 14. Pictures were taken by Prof. H.J. Merker, Anatomical Institute, Abt. Embryonalpharmakologie, Freie Universität, Berlin.

To find out whether mitochondria from tissues at different stages of differentiation during organogenesis are equally susceptible to an inhibition of mitochondrial protein synthesis, electronmicroscopic investigations were performed. The effect of TAP treatment on two types of cells are shown in Plate 1. Mitochondria contained in rapidly proliferating neuroepithelial cells are swollen with few visible cristae and a translucent matrix. However, in neighboring differentiating neuroblasts the mitochondria appear unaltered.

It can be said that our model system, using rat embryos during organogenesis, is suitable to follow the effects of inhibition of mitochondrial protein synthesis on extramitochondrial processes. Based on DNA content such embryos, or other rapidly proliferating tissues (e.g. hemapoetic cells), contain about 20 fold less cytochrome oxidase activity than do slowly growing tissues (e.g. adult liver) (4,11). It is suggested that this lower content of functioning mitochondrial components in embryonic cells is the reason for the sensitivity of undifferentiated tissues to CAP/TAP. But adult rat liver, even after stimulation of growth by partial hepatectomy (one replication in about 24 hrs.), has enough mitochondrial components remaining to permit regeneration under CAP (12).

These results demonstrate that the rate at which an energy deficit develops is dependent upon the type of tissue treated, the availability of energy from extramitochondrial sources, and the necessity of growth for further development. Depending on these three variables, cells should either hibernate until energy production is resumed or die - as in the case of embryonic tissue under conditions which inhibit mitochondrial biogenesis and function.

REFERENCES

1. Criddle, R.S. and Schatz, G. (1969) Biochem. 8, 322.
2. Plattner, H. and Schatz, G. (1969) Biochem. 8, 339.
3. Schatz, G. and Mason, T.L. (1974) Ann. Rev. Biochem. 43, 51.
4. Örter, D. and Bass, R. (1975) Naunyn Schmiedebergs Arch. Pharmacol. 290, 175.
5. Bass, R. (1975) In: New approaches to the evaluation of abnormal embryonic development, Thieme, Stuttgart pp. 524.
6. Spielmann, H. and Lücke, I. (1971) Naunyn Schmiedebergs Arch. Pharmacol.270,10.
7. Cox, S.J. and Gunberg, J. (1972) J. Embryol. exp. Morph. 28, 591.
8. Neubert, D. et al., (1971) Naunyn Schmiedebergs Arch. Pharmacol. 268, 235.
9. Jäger, E. and Bass, R. (1975) Naunyn Schmiedebergs Arch. Pharmacol. 290, 161.
10. Nijhof, W. and Kroon, A.M. (1974) Postgrad. Med. 50 (Suppl.5), 53.
11. Örter, D. et al. (1974) Postgrad. Med. 50 (Suppl.5), 65.
12. Kroon, A.M. and de Vries, H. (1969) FEBS letters 3, 208.
13. Strehler, B.L. (1970) In: Methoden der enzymatischen Analyse, Vol II, Verlag Chemie, Weinheim pp. 2036.

This project was supported by grants awarded to Sonderforschungsbereich 29 by the DFG. I would like to express my appreciation to Ms. K. Bordasch and I. Schütte for their expert technical assistance.

Genetics and Biogenesis of Chloroplasts and Mitochondria, Th. Bücher et al. eds.

UNMASKING OF MITOCHONDRIAL PRECURSORS STORED IN THE YOLK PLATELETS OF *Artemia salina* DORMANT GASTRULAE

CARMEN G.VALLEJO and ROBERTO MARCO
Instituto de Enzimología del C.S.I.C.
Facultad de Medicina de la Universidad Autónoma
Madrid-34, Spain

INTRODUCTION

The crustacean *Artemia salina* under unfavourable conditions lays cysts stopped in their development at an early gastrula stage. Swimming larvae (nauplii) hatch in 15-20 hours when the dried cysts are rehydrated in a suitable saline medium. The gastrulae are abundant in yolk platelets which have been so far considered as passive food stores. These organelles are metabolized during the posterior development. In good correlation, important changes in the amount and properties of mitochondria along this time period have been reported previously (Marco et al. *10th FEBS Meeting* (1975) 308). The occurrence of mitochondrial precursors stored in the yolk platelets as well as their *in vitro* unmasking is now presented.

RESULTS

The behaviour of mitochondria enzymes activities during *Artemia* development: After 24 hours of the resumption of development by rehydration of the dried cysts, a 30-fold increase in mitochondrial enzimatic activities can be observed (expressed in units/10^6 animals)

TABLE 1

Enzymes	Cysts (time 0)	Nauplii (time 24)
Cytochrome oxidase	1.7	57.1
Succinate cytochrome c reductase	0.7	13
Malate dehydrogenase	28.6	836.5
Isocitrate dehydrogenase (NADP)	2.8	86
Fumarase	3.6	96.7

The unraveling of mitochondria stored in yolk platelets (YP): We have been led to think that a considerable part of this rapid increase is due to unraveling of mitochondrial precursors stored in YP, based on the following data.

a) Passage through sucrose gradients: The YP fraction was obtained by centrifuging homogenates of 0 time cysts at 500 g x 10 min. This subcellular fraction comprises exclusively yolk platelets and nuclei as can be seen in Plate I. The YP fraction was layered on

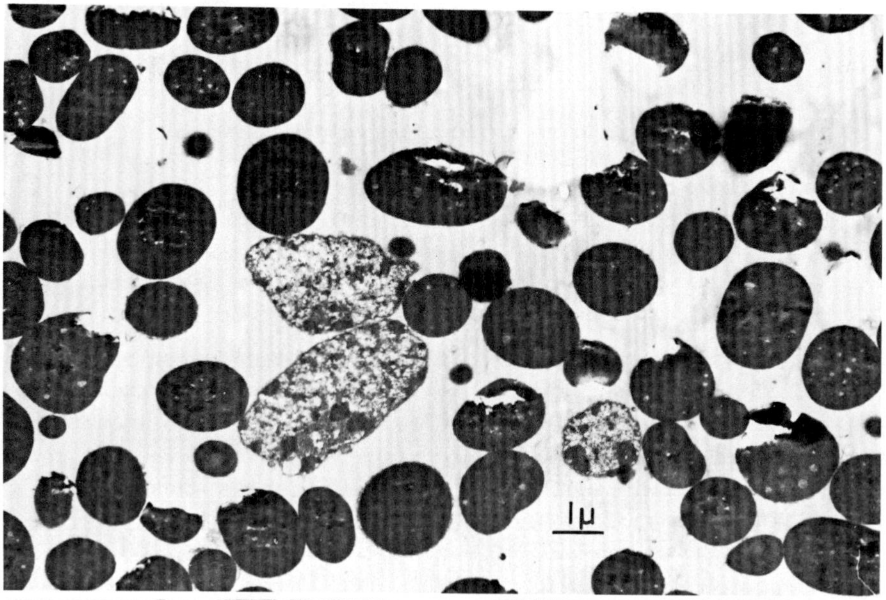

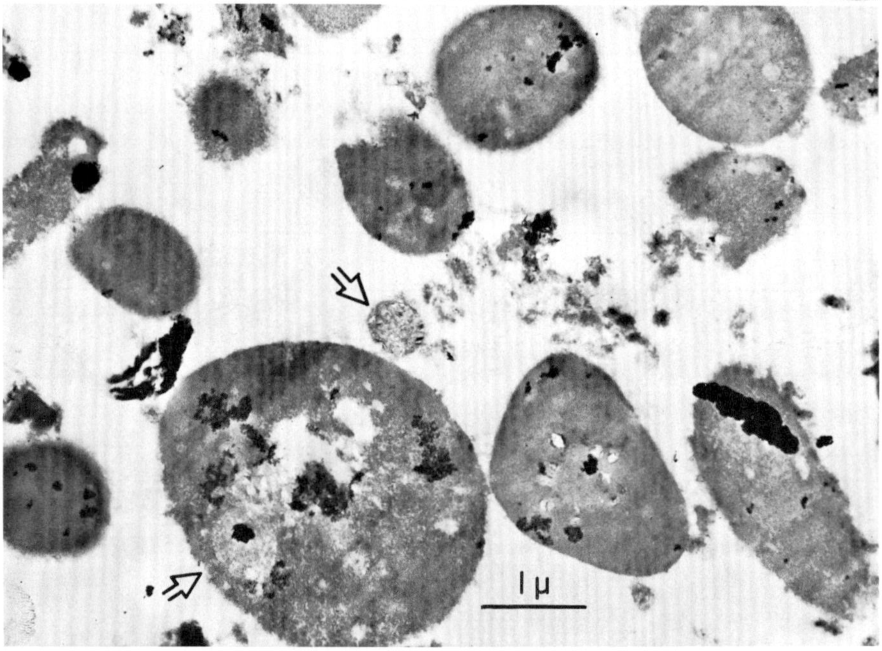

Plate I. Electronmicrograph of a yolk platelets fraction from cysts just hydrated (0 time cysts)

Plate II. Electronmicrograph of the pellet obtained by passage of the fraction shown in Plate I through a sucrose gradient.

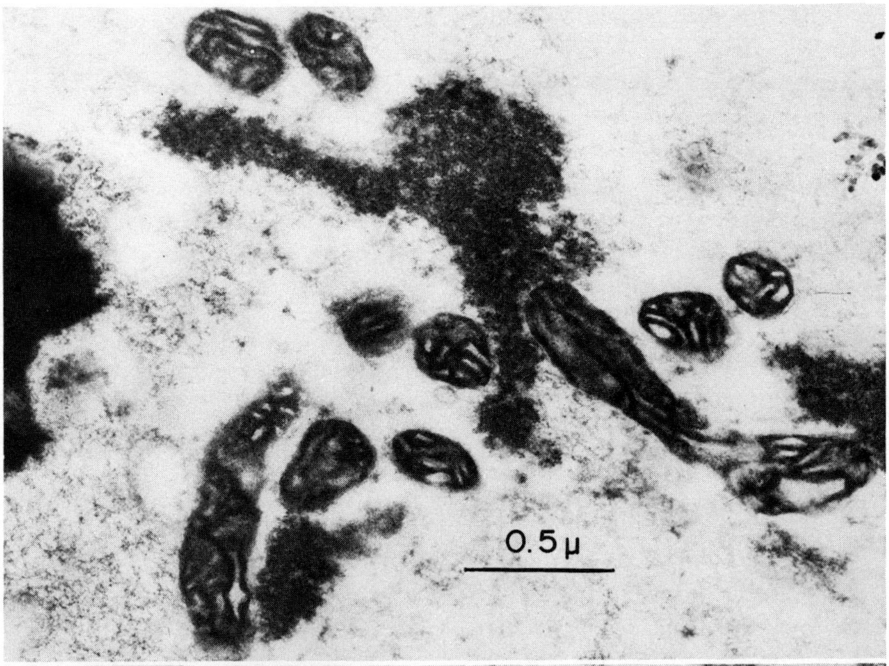

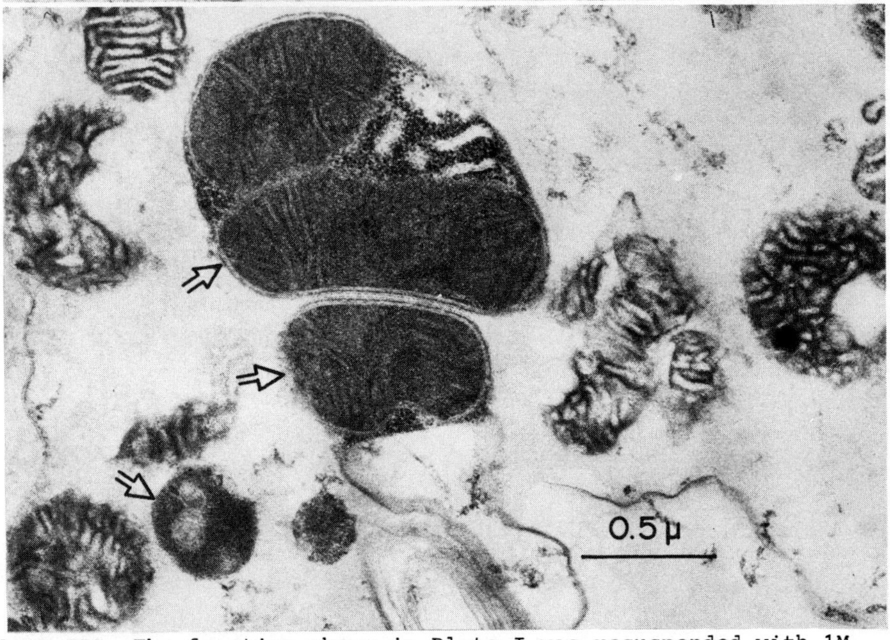

Plate III. The fraction shown in Plate I was resuspended with 1M NaCl, spinned at 134,000 g x 75 min and the pellet observed at the electron microscope.

Plate IV. Some yolk platelets partly metabolized, in the way of releasing mitochondria, are shown.

the top of a continuous sucrose gradient (2.0 M to 0.96 M), Figure 1A shows a drawing of the centrifuge tube after an overnight centrifugation at 90,000 g and Figure 1B, the distribution of cytochrome oxidase after the fractionation of the gradient. As can be observed, cytochrome oxidase is located exclusively in the dark pellet observed at the botton of the tube. This sucrose pellet includes only nuclei (not seen in Plate II) and yolk platelets somewhat damaged by the treatment. The activity recovered is 4 times the one detected in the original fraction and amounts to 60% of the total activity determined at this stage.

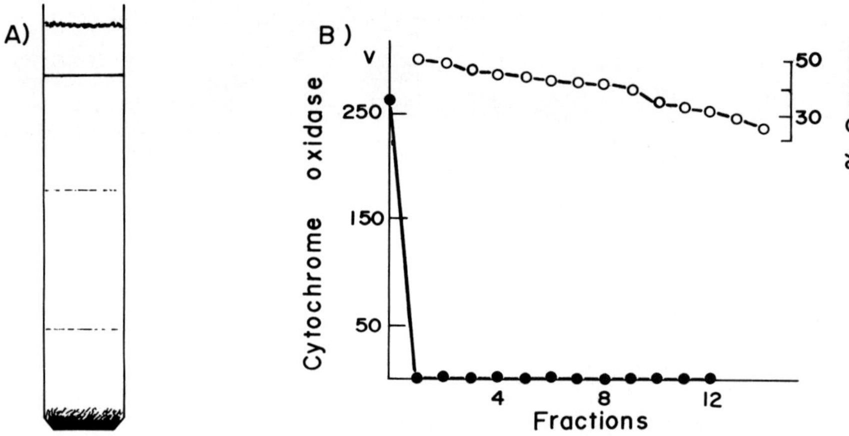

FIGURE 1

b) <u>The unmasking of mitochondria by NaCl treatment</u>: YP proteins have been reported to become soluble by 1M NaCl. As shown in Plate III, after treatment of YP with salt, mitochondria can be seen coming out from the dark masses of disrupted YP.

c) <u>The *in vivo* observation of intermediate stages of yolk platelets degradation</u>: At about hatching time and afterwards, structures resembling partly mitochondria and partly YP have been found (Plate IV)

Genetics and Biogenesis of Chloroplasts and Mitochondria, Th. Bücher et al. eds.
© 1976 Elsevier/North-Holland Biomedical Press, Amsterdam, The Netherlands.

SCREENING TESTS FOR SUPPRESSORS OF RESPIRATORY DEFICIENT MUTANTS IN *Schizosaccharomyces pombe* AND MODEL FOR A MITOCHONDRIAL PARTIAL SUPPRESSION OF A NUCLEAR PLEIOTROPIC STRAIN

André Goffeau, Françoise Labaille, Oscar Mohar and Anne-Marie Colson
Laboratoire d'Enzymologie, Université de Louvain
1348 Louvain-la-Neuve, Belgium

Schizosaccharomyces pombe is a "petite-negative" yeast unable to produce stable rho$^-$ mitochondrial mutants. Particularly stable nuclear pleiotropic respiratory-deficient mutants are however produced at high frequency either spontaneously or after mutagenesis (see reference 1 for a review). We wish to demonstrate that *S. pombe* is particularly suitable for the study of mitochondrial suppressors of nuclear respiratory-deficient mutations.

Sensitivity to bongkrekic acid of growth on glucose

When grown on 10% glucose or 10% raffinose, a non-repressive carbon source, the growth of *S. pombe* is not inhibited by the addition to the culture medium of 0.85 μM bongkrekic acid, an inhibitor of mitochondrial translocation of nucleotides[2] (Table 1). However strains in which the respiration is restricted either physiologically or genetically by the nuclear respiratory-deficient mutation pet1, do carry out a smaller number of cell generations in the presence of bongkrekic acid.

TABLE 1
EFFECTS ON CELLULAR MULTIPLICATION OF BONGKREKIC ACID AND ANTIMYCIN OF *Schizosaccharomyces pombe* STRAINS GROWN ON RAFFINOSE

Additions	Wild type (10^6 cells/ml)	pet1 (10^6 cells/ml)	pet1 [SUP2] (10^6 cells/ml)
None	258	135	189
Bongkrekic	238	7	164
Antimycin	168	25	140
Bongkrekic + antimycin	40	1	9

Cells per ml after 48h in 10% raffinose, 2% yeast extract pH 4.5. Inoculum 10^6 cells/ml. When indicated 0.18 μM antimycin A and/or 0.85 μM bongkrekic acid were added.

Similar sensitivity to bongkrekic acid of the growth in glucose of respiratory-deficient *Saccharomyces cerevisiae* strains has been reported[3,4] and was interpretated as indicative of a requirement for cellular multiplication of an unidentified process or component which is diluted out in a few generations in the absence of intramitochondrial ATP.

Reduced-growth by bongkrekic acid of respiratory-deficient strains cannot be used *per se* as a selection procedure in *S. cerevisiae* because, under these conditions, up to 100% rho$^-$ mutants are induced by bongkrekic acid[3].

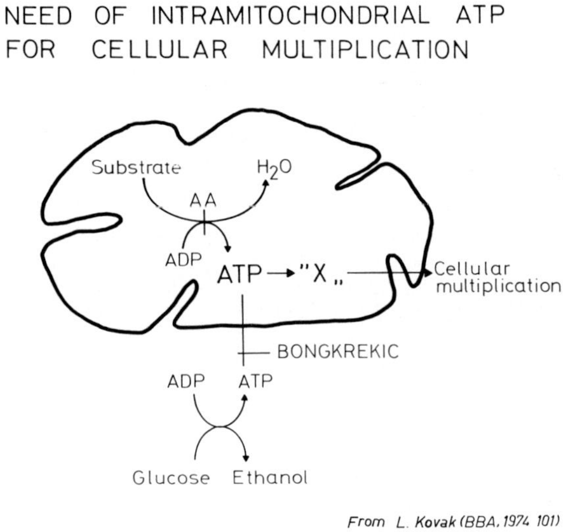

Fig. 1. Need of intramitochondrial ATP for cellular multiplication

However as suggested by Kovac, the formation of rho$^-$ by bongkrekic acid can be used as a screening test for *S. cerevisiae* deficient in oxidative phosphorylation[5]. Since *S. pombe* is a "petite-negative" yeast, the complication to study rho$^-$ formation does not exist and the sensitivity to bongkrekic acid for growth on glucose can be used in this species as a much simpler test for the detection of mutants deficient in respiration and/or oxidative phosphorylation.

Sensitivity to antimycin A of growth on glucose

Cellular multiplication on glucose or raffinose of the respiratory-deficient mutant pet1 is also reduced by the addition to the growth medium of 0.18 μM antimycin A, or appropriate concentration of other respiratory inhibitors. Since pet1 exhibits very low but significant "antimycin-sensitive" respiration, which is also observed in the wild type grown in the presence of antimycin[6], it is concluded that a minimum level of "normal" respiration (antimycin-sensitive) is required for continuous cell multiplication in *S. pombe*. This conclusion is supported by the fact that *S. pombe* does carry out only a limited number of generations under anaerobiosis[7].

In species like *Neurospora*[8] or *Paramecium*[9], the need for intramitochondrial oxidation could possibly be fullfilled by the mitochondrial antimycin and cyanide-insensitive "alternative" respiration which probably do not exist in *S. pombe*. Indeed the antimycin-insensitive respiration of *S. pombe* is 20 to 50 times lower than in *Neurospora* and *Paramecium*. In contrast to the latter species, it is insensitive to hydroxamic acids (SHAM) and probably extra-mitochondrial in *S. pombe*[6].

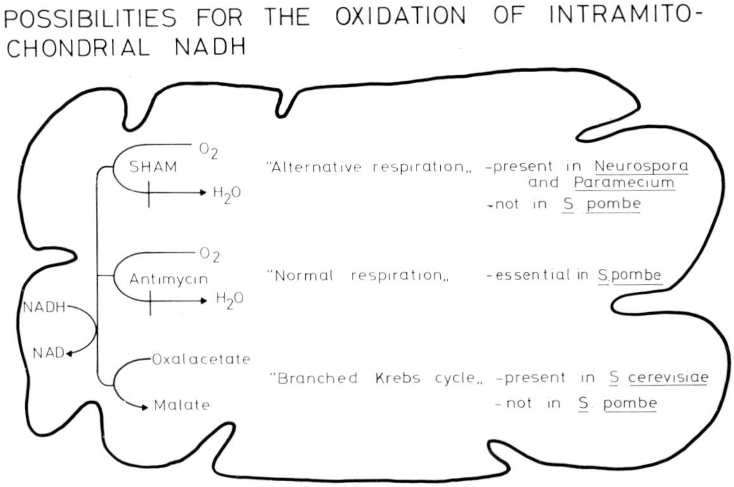

Fig. 2. Possibilities for the oxidation of intramitochondrial NADH

In respiratory-deficient *S. cerevisiae*, oxidation of the intramitochondrial pyridine nucleotides could be carried out by the reductive part of an anaerobic "branched Krebs cycle"[10]. The latter pathway does probably not develop in *S. pombe* since anaerobic cellular multiplication is limited in this species[7].

Screening for suppressors of respiratory-deficient mutants

In cultures of *S. pombe*, respiratory-deficient mutants, incubated in glucose (or raffinose) containing growth medium supplemented with antimycin or bongkrekic acid, there is a strong selective pressure in favor of the inhibitor-resistant strains that could be induced. The latter strains are either revertants or suppressors. They can be distinguished by backcrosses with the wild type and screening in the segregants for growth on glycerol as well as for resistance to respiratory inhibitors of growth on glucose. The two characters dissociate frequently in the case of extra-genic suppression; they largely cosegregate in the case of reversion and intra-genic suppression. Extra-genic suppressors strains modified in the permeability for the inhibitor or in the inhibitors-target binding properties can be eliminated by checking that in strains resistant to each of the two single inhibitors, the combination of antimycin and bongkrekic acid still reduces growth on glucose (or raffinose) as in the wild type (Table 1). Partial suppressors are inhibitor-resistant for growth on glucose but do not grow on glycerol. In total suppressions, growth on glycerol is also restored. Functional are distinguished from informational suppressions by the fact that the latter also suppress known ochre or amber mutations. Mitochondrial and extra-nuclear suppressors are distinguished by nuclear ones by the fact that they do not exhibit Mendelean meiotic segregation and the transmission of the resistant trait into diploids is reduced by ethidium bromide treatment of the parents[11,12].

Partial restoration of oxidative phosphorylation in a nuclear pleiotropic mutant by a mitochondrial mutation

Following the above criteria, several suppressors of the nuclear respiratory-deficient pleiotropic mutation pet1 were obtained[6,12]. Among them the SUP2 gene was demonstrated to be a partial, extra-genic suppressor of extra-nuclear (probably mitochondrial) heredity[12]. It is though to be a functional rather than an informational suppression. pet1 (previously referred to as M126) is a stable single-gene nuclear mutant that do not grow on glycerol[13]. When grown on glucose (or raffinose), its antimycin-sensitive and antimycin-insensitive respirations are very low (QO_2 of 1.5 and 2.1 $\mu lO_2 \times h^{-1} \times mg^{-1}$ DW respectively). The isolated pet1 mitochondria do not oxidize significantly NADH, α-ketoglutarate, succinate, malate, isocitrate or α-glycerophosphate. No oxidative phosphorylations can be detected. Cytochrome aa_3, cytochrome b_{560} (at -196°), cytochrome c oxidase, NADH: cytochrome c reductase and oligomycin-sensitive ATPase are either not detected or present in very low amounts[13]. The mitochondrial protein synthesis is functioning[13] but its products are abnormal; the high molecular weight peptides being particularly reduced[14]. In many respects *S. pombe* pet1 ressembles the pleiotropic mutants pet936[15] and N9-168[16] of *S. cerevisiae*.

As expected, the cellular multiplication of pet1 in glucose or raffinose is strongly reduced by bongkrekic acid or antimycin (Table 1). Each resistance is restored in the suppressed strain pet1 [SUP2] eventhough combination of the two inhibitors reduces markedly the growth-yield (Table 1). The cellular antimycin-sensitive respiration of glucose-grown pet1 [SUP2] is only 13% that of the wild type but is markedly derepressed in raffinose, reaching QO_2 of 34 $\mu lO_2 \times h^{-1} \times mg^{-1}$ DW (compared to 50 for the wild type and 1.5 for pet1) Oxidations rates of about one third that of the wild type are observed for all respiratory substrates in mitochondria isolated from glucose-grown pet1. ATP/O of 1.0 (compared to 1.95 in the wild type) is observed during oxidation of succinate. All mitochondrial cytochromes are restored in raffinose-grown pet1 [SUP2] and NADH: cytochrome c reductase, cytochrome c oxidase and oligomycin-sensitive ATPase specific activities in mitochondrial fraction isolated from raffinose-grown pet1 is 40 to 80% restored compared to the wild type[6]. Also, the leaky mitochondrial protein synthesis of pet1 is largely restored in pet1 [SUP2] and mitochondrial products of 43 000, 38 000, 27 000 belonging to ATPase and cytochrome c oxidase are much more prominent than in pet1[14]. None of the suppressions is however total and pet1[SUP2] is unable to grow on glycerol[12].

Model of mitochondrial-nucleo interaction

All the respiratory-deficiencies observed in pet1, can be explained as secondary consequences of the leaky mitochondrial protein synthesis.

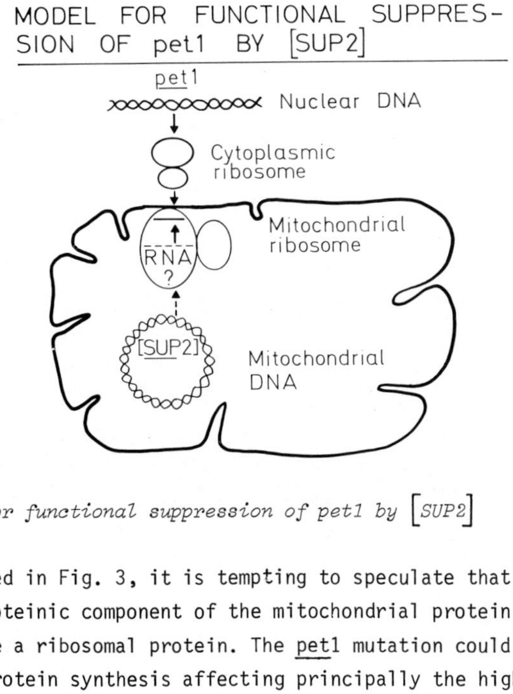

Fig. 3. Model for functional suppression of pet1 by [SUP2]

As illustrated in Fig. 3, it is tempting to speculate that the PET1 gene product is a proteinic component of the mitochondrial protein synthesis machinery, for instance a ribosomal protein. The pet1 mutation could produce a leaky mitochondrial protein synthesis affecting principally the high molecular weight mitochondrial products. The primary effect of the mitochondrial suppressor gene [SUP2] would then be to restore partly the leaky mitochondrial protein synthesis. A possible candidate for the [SUP2] locus could be a ribosomal RNA. In this model, we have to imagine that a genetic modification of a mitochondrial rRNA compensates a ribosomal deficiency due to a modified nuclear-coded ribosomal protein. Several variations of the above hypothesis and many alternatives models can be proposed. An important argument in favor of the above hypothesis of a primary deficiency in the mitochondrial protein synthesis is the high frequency in *S. pombe* of non-allelic nuclear pleiotropic mutants similar to pet1[13]. The large number of nuclear-coded peptides in the complex mitochondrial protein-synthesis apparatus implyes a large number of possibilities for mutations modifying the mitochondrial transcription or translation. In this respect, it would be interesting to test whether the nuclear pleiotropic respiratory-deficient mutant cl1 of *Paramecium* which is also suppressed by a mitochondrial mutation[17] does like *S. pombe* pet1 exhibit leaky mitochondrial protein synthesis.

REFERENCES

1. *Goffeau, A., Briquet, M., Colson, A.M., Delhez, J., Foury, F., Labaille, F., Landry, Y., Mohar, O. and Mrena, E. (1975) In Membrane Biogenesis (Tzagoloff, A., ed.) pp 63-97, Plenum Press.*
2. *Henderson, P.J.F. and Lardy, H.A. (1970) J. Biol. Chem. 245, 1319-1326.*
3. *Subik, J., Kolarov, J. and Kovac, L. (1972) Biochem. Biophys. Res. Commun. 42, 192-198.*
4. *Subik, J. (1974) FEBS Letters 42, 309-313.*
5. *Kovac, L. (1974) Biochim. Biophys. Acta 346, 101-135.*
6. *Labaille, F., Petit, L., Colson, A.M. and Goffeau, A. (in preparation).*
7. *Heslot, H., Goffeau, A. and Louis, C. (1970) J. Bact. 104, 473-481.*
8. *Lambowitz, A. and Slayman, C.W. (1971) J. Bact. 108. 1087-1096.*
9. *Sainsard, A., Claisse, M. and Balmefrezol, M. (1974) Mol. Gen. Genet. 130, 113-125.*
10. *Machado, A., Nunez de Castro, I. and Mayor, F. (1975) Mol. Cell. Biochem. 6, 93-99.*
11. *Wolf, K., Burger, G., Lang, B. and Kaudewitz, F. (1976) Mol. Gen. Genet. 144, 67-74.*
12. *Colson, A.M., Labaille, F. and Goffeau, A. (1976) Mol. Gen. Genet. (in press).*
13. *Goffeau, A., Landry, Y., Foury, F. and Briquet, M. (1973) J. Biol. Chem. 248, 7097-7105.*
14. *Mohar, O. and Goffeau, A. (in preparation)*
15. *Ebner, E. and Schatz, G. (1973) J. Biol. Chem. 248, 5379-5384.*
16. *Tzagoloff, A., Akai, A. and Needleman, R. (1975) J. Biol. Chem. 250, 8228-35.*
17. *Sainsard, A. (1975) Nature 257, 312-314.*

Genetics and Biogenesis of Chloroplasts and Mitochondria, Th. Bücher et al. eds.
© 1976 Elsevier/North-Holland Biomedical Press, Amsterdam, The Netherlands.

RESPIRATION DEFICIENT MUTANTS WITH INTACT MITOCHONDRIAL GENOMES: CASTING A WIDER NET

Henry R. Mahler, Tomasz Bilinski, Donald Miller and Deborah Hanson
Chemistry Department, Indiana University, Bloomington, Indiana 47401
and
Philip S. Perlman and Catherine A. Demko
Department of Genetics, Ohio State University, Columbus, Ohio 43210

INTRODUCTION

The study of mutants with a well defined phenotype provides the most conclusive and discriminating evidence for the nature of the information encoded in mitochondrial DNA. Several classes of such mutants in bakers' yeast (Saccharomyces cerevisiae) have been reported and are still under active investigation. These include various antibiotic resistance mutants[1,2], electrophoretic variants [3,4] and several novel types of respiration-deficient mutants[5-7]. Recently, very significant advances in our knowledge of the mitochondrial genome have been made possible by the discovery of mit- mutants[8,9] and their genetic characterization by Slonimski and Tzagoloff[10].

Mit- mutants meet three criteria: i) they are respiration deficient, ii) their inheritance is mitochondrial and iii) they retain the ability to synthesize proteins on mitoribosomes. It is this third criterion that sets them apart from classical petites (rho- or rho-o). However, although sufficient for this purpose, it may be unnecessarily restrictive since it excludes mutants in genes required for mitochondrial protein synthesis (RNAs and perhaps protein factors) or, possibly, transcription. In order to include such mutants, and perhaps others, we have devised a more general set of criteria and a screening procedure based on them, that, we hope, will detect respiration deficient mutants retaining a virtually complete mitochondrial genome. These mutants are designated as "genomic integrity" (gin) mutants; we expect that mit- mutants will constitute a subset of the class of gin mutants.

CRITERIA

The criteria that we have selected in order to define these mutants are listed in Table 1. In our initial studies, A1 constituted the basis of our screening procedure, as outlined below. It is based on our experience with literally thousands of petite strains, none of which retains, in a stable manner, even the three linkage groups defined by the chl, oli1 and par loci[1,2]. To provide additional discrimination we also required proof of the retention of the oli2 locus (i.e. the fourth linkage group). Criteria A2 and B1 are necessary and sufficient to define an extranuclear respiratory mutation. If positive, criteria A3a or b and B2 are sufficient to assign a mutant into the gin class. These criteria could miss some types of potentially useful mutants (e.g. double mutants

TABLE 1

CRITERIA TO BE MET BY *gin* MUTANTS

A. Genetic

1. Retention of the 4 linkage groups *chl*, *oli1*, *oli2*, *par*
2. Non nuclear (no *rho+* progeny when crossed with *rho*^{o})
3. Retrievable loss of mitochondrial information
   a. Revert spontaneously; if negative
   b. W.t. progeny when crossed to collection of *rho-*

B. Phenotype

1. Respiration deficient (no growth specifically on glycerol)
2. Mitochondrial protein synthesis, present or absent

C. Molecular

1. Retention of ≥90% w.t. base sequence (from restriction pattern)

or fairly large or complex deletions); mutants meeting criterion A1 but failing A3a, b and B2 would be retained only if they meet criterion C1. After some experience with this screening protocol we have changed the order of application of the criteria; we now use the petite restoration test[10] (criterion A3b) as the initial test. This change was made because reliance on transmission of *chl-r*, *oli-r* and *par-r* would fail to or only weakly detect mutations in or closely-linked to those loci; also the petite restoration test has proven to be rapid and reliable.

MUTANT ISOLATION

The procedure and the genetic criteria used for the isolation of mutants is outlined in Figure 1. Strain ID41/161 (*a ade lys rho+ ome- chl-r ery-s oli1-r oli2-s par-r*) was used as the parent in all experiments. A culture was treated with 10 mM $MnCl_2$ for 5 hours in YEPD[11], cells plated on solid YEPD, and the respiration deficient clones subjected to the tests outlined. Strains meeting criterion A1 or A3b were subjected to genetic purification by subcloning, monitored by scoring for transmission of the three resistance markers indicated or else by scoring for ability to be restored to respiration sufficiency when crossed to a mixture of petite mutants. At least some of the mutants can be purified quite readily since the petite restorable members of those clones yield large tan colonies on YEPD plates; to date all such mutants have mitochondrial protein synthesis. The petite restoration test used in these studies is a modification of that used by Slonimski and Tzagoloff[10]; instead of performing hundreds of crosses for each presumptive *gin* isolate, we prepared a mixture of 300 clones of petite mutants derived from cells treated with ethidium bromide. Each of those clones is itself quite heterogeneous[12] and so the mixture should contain

FIGURE 1

MUTANT ISOLATION PROTOCOL

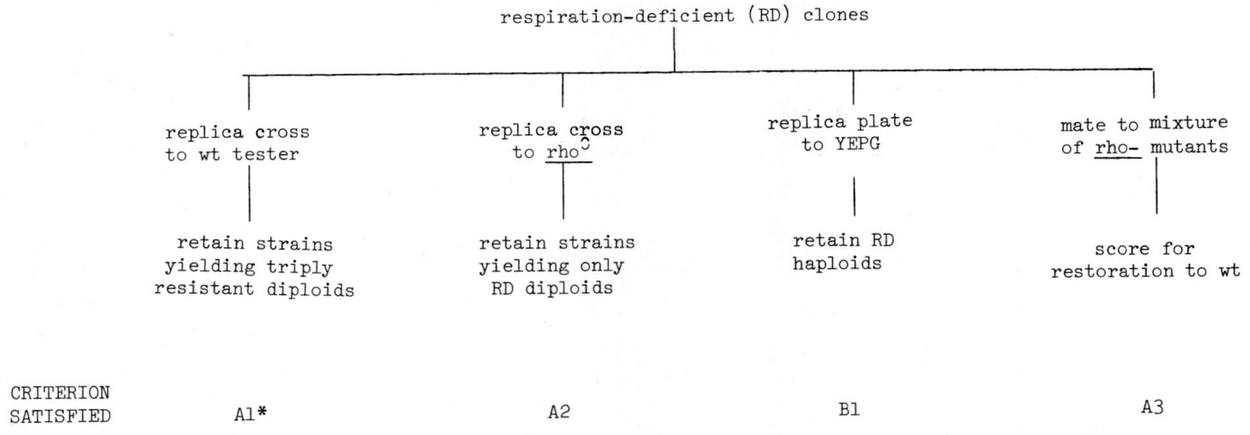

* Test for retention of oli2 is performed later after subcloning

at least some petites representing virtually all segments of the mitochondrial genome.

GENETICS

As is summarized in Table 2 we have isolated a number of mutants that retain all markers and are petite restorable; some revert to rho+ spontaneously while others do not. We have devised a preliminary mapping protocol that is a simplified version of the methods used by Slonimski and Tzagoloff[10]. Wild-type segregants from the petite restoration test are analyzed for alleles of the drug markers by replica-plating onto drug-containing solid media. For example, if a gin mutation is located near the par locus it will be restored by at least some petites retaining that marker and so the progeny tested will be par-r (from the gin mutant) and par-s (from the petite mixture). More distant drug markers (in this illustration, olil, chl and ery) are rarely retained together with par and so the cells tested will have mostly or exclusively alleles from the gin mutant. Analysis of the data in Table 2 suggests that we have detected at least 4 different loci. So far all of the petite restorable isolates have mitochondrial protein synthesis (see next section) and so they are mit- mutants.

TABLE 2

SUMMARY OF GENETIC RESULTS

	Transmission of alleles from petite parent to wild-type progeny of restoration test				Mutants Isolated
	chl	ery	olil	par	
I. Revert to rho+					
	-	-	-	++	M38, M375, M37
	-	-	++	-	M113
	-	-	++	++	M117, M44
	++	-	-	++	M372
II. Do not revert to rho+					
	-	-	-	++	M70, M80, C245
	-	-	++	-	M357, M45

BIOCHEMISTRY

All strains have been examined for their ability to carry out mitochondrial protein synthesis, both by simple incorporation and by electrophoretic pattern, for cytochrome spectra and for the enzymes indicated in Table 3 which summarizes the results. It is clear that the screening procedure employed detects mit- mutants apparently indistinguishable from those previously reported[8-10]. Based on the genetic evidence it appears that mutants in class A represent at least 3 distinct loci but are otherwise indistinguishable by these phenotypic tests. Mutant M45 is strongly petite restorable and has an apparently normal ATPase but grows much more slowly than do the other mutants on minimal or YEP media containing galactose. Mutants in class C are especially interesting because their lesion does not affect the major components of the oxidation and phosphorylation apparatus.

The mitochondrial proteins synthesized by the different mit- mutants have been analyzed on acrylamide slabs by means of the high resolution acrylamide system described by Laemmli[13] as modified by Douglas and Butow[3] (Figure 2). Although these mutants synthesize mitochondrial proteins to roughly the same extent as does their wild-type parent several alterations in band pattern were found among the twelve mutants studied. The band pattern of mutant C245 is typical of most members of class A (Table 3) and closely resembles the pattern of the wild-type. M44 is an exception in that it shows increased intensity of one band (marked by 1 in the figure). The class B mutant (M45) resembles the majority of class A ones. The two class C mutants both show an altered profile: M375 has increased intensity of one band (2); M113 has at least two additional bands (3 and 4).

TABLE 3

SUMMARY OF BIOCHEMICAL RESULTS

Mutants	Protein Synthesis	Cytochromes			Enzymes	
		aa_3	b	$C:O_2$	NADH:C	OS, VS-ATPASe*
A. C245, M36, M37, M38, M44, M70, M80, M117, M357	+	−	+	−	+	+
B. M45	+	−	−	−	−	+
C. M113, M375	+	+	+	+	+	+

* Sensitive to 20 μg/ml of oligomycin and 5 μg/ml of venturicidin. The parent strain is oli-r and ven-s in vivo but the ATPase is sensitive to both drugs in vitro.

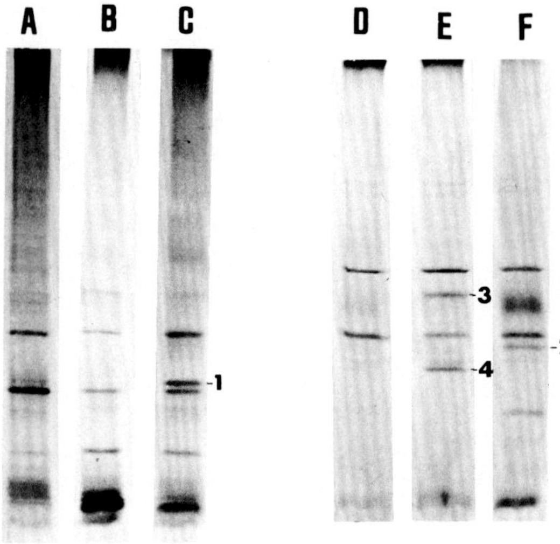

FIGURE 2

AUTORADIOGRAPHS OF $^{35}SO_4^{=}$ LABELLED PRODUCTS OF MITOCHONDRIAL PROTEIN SYNTHESIS FROM mit- MUTANTS

A, ID41/161; B, C245; C, M44; D, ID41/161; E, M113; F, M375. Gels A-C are from one experiment and D-F from another. Numerals indicate bands different from the wild-type pattern.

DISCUSSION AND SUMMARY

In the INTRODUCTION to this report we have postulated the existence of a novel and quite general class of mitochondrial mutants rendered deficient in respiration only by virtue of a minor lesion in an otherwise intact mitochondrial genome. This class of mutants has been defined by a set of criteria which we expect to result in the identification of strains exhibiting an extremely wide diversity of phenotypic alterations and which includes the mit- mutants as a subset. The data reported here suffice to show that our screening protocol is both appropriate and sufficiently sensitive to meet our aims. So far the small number of mutants investigated includes only mit- mutants; we are currently examining a much larger set of mutants and hope to obtain some gin mutants that lack mitochondrial protein synthesis. Our preliminary biochemical analysis of twelve mit- mutants has already revealed several interesting alterations in the products of mitochondrial protein synthesis; we are especially encouraged by those mutants that appear to over-produce one of the products since that is one possible consequence of certain types of regulatory mutations.

ACKNOWLEDGEMENTS

The highly competent technical assistance of Mary Townsend, Mary Ensign and Margaret Lloyd is acknowledged with pleasure and gratitude. Supported by research grant GM 12228 from NIH-NIGMS to HRM and GM 19607 from NIH-NIGMS to PSP. HRM is a recipient of Research Career Award K06 05060 from the same institute.

REFERENCES

1. Birky, C. W., Jr. (1975) in Genetics and Biogenesis of Mitochondria and Chloroplasts (Birky, C. W., Perlman, P. S. and Byers, T. J., eds.) pp. 182-244, Ohio State University Press, Columbus.
2. Gillham, N. W. (1974) Ann. Rev. Genetics 8, 347-392.
3. Douglas, M. G. and Butow, R. A. (1976) Proc. Nat. Acad. Sci. USA 73, 1083-1086.
4. Douglas, M. G., Strausberg, R. L., Perlman, P. S. and Butow, R. A. (1976) this volume.
5. Flury, U., Mahler, H. R. and Feldman, F. (1974) J. Biol. Chem. 249, 6130-6137.
6. Storm, E. M., and Marmur, J. (1975) Biochim. Biophys. Res. Comm. 64, 752-759.
7. Handwerker, A., Schweyer, R. J., Wolf, K. and Kaudewitz, F. (1973) J. Bact. 113, 1307-1310.
8. Tzagoloff, A., Akai, A. and Needleman, R. B. (1975) J. Bact. 122, 826-831.
9. Tzagoloff, A., Akai, A., Needleman, R. B. and Zulch, G. (1975) J. Biol. Chem. 250, 8236-8244.
10. Slonimski, P. P. and Tzagoloff, A. (1976) Eur. J. Biochem. 61, 27-41.
11. Putrament, A., Baranowska, H. and Prazmo, W. (1973) Mol. Gen. Genet. 126, 357-366.
12. Perlman, P. S. (1975) in Genetics and Biogenesis of Mitochondria and Chloroplasts (Birky, C. W., Jr., Perlman, P. S. and Byers, T. J., Eds.) pp. 136-181, Ohio State University Press, Columbus.
13. Laemmli, U. K. (1970) Nature 227, 680-685.

Genetics and Biogenesis of Chloroplasts and Mitochondria, Th. Bücher et al. eds.

ASSEMBLY OF THE CYANIDE-INSENSITIVE RESPIRATORY PATHWAY IN *Neurospora crassa*

David L. Edwards,[+] John H. Chalmers Jr.,[≠*] Howard J. Guzik,[+] and Joseph T. Warden[≠]
[+]Department of Biological Sciences, State University of New York, Albany, New York 12222. [≠]Department of Genetics, University of California, Berkeley, California 94720. [≠]Biochemistry Program, Department of Chemistry, Rensselaer Polytechnic Institute, Troy, New York 12181.

SUMMARY

A method is described that selects for mutants that are deficient in the mitochondrial cyanide-insensitive respiratory pathway of *Neurospora crassa*. Analysis of some of these mutants by complementation demonstrates at least two nuclear genes coding for components of the pathway. These have been designated *alx*-1 and *alx*-2. An analysis of the *cni*-1 mutant by electron spin resonance gives evidence for an iron-sulfur protein that is a component of the pathway. This protein gives a new esr signal with an approximate g-factor component of $g \sim 2.020$.

INTRODUCTION

The cyanide-insensitive pathway has been shown to be a component of the mitochondrial respiratory system of several mutants of the fungus *Neurospora crassa*.[1-5] Mutants exhibiting this pathway have a branched mitochondrial respiratory chain with one portion being the classical cytochrome chain that is inhibited by cyanide or antimycin A and the other portion being a novel respiratory system that is not inhibited by these inhibitors.[6] This alternate pathway can be inhibited by iron chelating agents such as 8-hydroxyquinolin[7] or substituted hydroxamic acids.[1,8]

There is presently little information available about the nature of the components of the cyanide-insensitive pathway or the mechanism of oxygen reduction. A cytochrome does not appear to be involved[6] and it has been speculated that flavoproteins[9] or iron-sulfur proteins[8,10,11] may be components of the pathway.

---

*Present address: Department of Biochemistry, Baylor College of Medicine, Houston, Texas 77030

A most useful approach in the study of cyanide-insensitive respiration has been the induction of the pathway in wild-type cells.[12,13] Inhibition of mitochondrial transcription or translation in wild-type cells results in the rapid appearance of the cyanide-insensitive pathway. This pathway also disappears rapidly when the inhibitory block is removed.[13] Studies of this type have shown that the structural genes for the cyanide-insensitive pathway are encoded in the nucleus and that a regulatory gene for the pathway resides in the mitochondrial genome.[13]

Our approach to the study of the cyanide-insensitive pathway has been to select mutants that are defective in either the structural or regulatory genes for the pathway and to examine these by biochemical and physical methods. In this report we describe the selection of mutants that cannot produce the pathway, a complementation analysis of some of these, and an examination of a mutant that has the pathway by electron spin resonance.

MATERIALS AND METHODS

Strains: Strains inl-89601 (497), wild-type strain 74OR23-1A (987), and leu-1 (162) were obtained from the Fungal Genetics Stock Center, Arcata, California. F.G.S.C. stock numbers are in parenthesis. The isolation of mutant cni-1 has been described previously.[3]

Selection of mutants: Mutants were isolated in a colonial strain of Neurospora, SCIP-14 ALB-3, which carries the markers sn (C136); cr-1 (B123); inos (JH319); and pyr-3 (KS43); and an unmapped albino allele. Although mixed in origin it is of predominately Oak Ridge background and has the same heterokaryon compatibility as other St. Lawrence derivatives. SCIP-14 ALB-3 has colonial morphology which allows replica plating of colonies and is female fertile. Mutants isolated in this strain were then backcrossed to the appropriate wild-type or auxotrophic strains to improve growth, and obtain mycelial morphology.

The procedures used for the selection of mutants are a modification of the methods of Lester and Gross.[14] Conidia from 10-14 day old cultures of SCIP-14 ALB-3 were filtered through cotton, washed by centrifugation in sterile distilled water, and irradiated with ultraviolet light to 20% survival. The conidia, at a final concentration of 10^6/ml, were then inoculated into 20 ml of Vogel's Medium

N^{15} supplemented with 2% sucrose, 100 µg/ml uridine, and 0.25 µg/ml antimycin A. Cultures were incubated at 30° for 5 days in a reciprocating water bath. Cultures were filtered through 4 layers of sterile cheesecloth when they became cloudy, usually on the second day of incubation. At the end of 5 days the ungerminated conidia were collected on a millipore filter, resuspended in 5 ml of sterile distilled water and spread on 50 plates of Vogel's Medium N^{15} supplemented with inositol and uridine but without antimycin A.

The plates were incubated at 34° for 2-3 days and replicated with dry velvet onto plates containing antimycin A according to the method of Maling.[16] Colonies growing in the absence but not in the presence of antimycin A were considered putative mutants. These colonies were transferred to slants and vegetatively purified by streaking and replica plating.

The putative mutants isolated in the manner described above were then tested to see whether cyanide-insensitive respiration could be induced by treatment with chloramphenicol. Cultures were grown at 30° for 24 hrs, and then chloramphenicol was added at a final concentration of 2 mg/ml. Putative mutants that failed to exhibit cyanide-insensitive respiration after 24 hrs in the presence of chloramphenicol were retained and crossed to mycelial strains. All of the mutants tested segregated in a Mendelian manner in such crosses. Mycelial progeny from the crosses were recovered and again tested for cyanide-insensitive respiration using the method described by Edwards et al.[12] Mutants that failed to produce the cyanide-insensitive pathway under these conditions were called ANT and designated by an isolation number. This nomenclature does not imply a genetic locus for these mutants. Thus, mutants ANT-1 and ANT-3 simply represent two independent isolates of antimycin A sensitive strains which may or may not be allelic.

Electron spin resonance studies: Studies of submitochondrial particles from some of these strains by electron spin resonance were carried out on a Varian Associates E-9 spectrometer equipped with an Air Products LTD-3-110 liquid helium cryostat for making measurements at low temperatures. Measurements were made in 3 mm quartz tubes on 0.3 ml samples containing 30-50 milligrams of protein/ml. Temperature was measured by a carbon resistance thermometer.

RESULTS AND DISCUSSION

Genetic studies: Studies of the effect of chloramphenicol on the production of cyanide-insensitive respiration in inl-89601 and in two of the ANT mutants, ANT-1 and ANT-3, are shown in Fig. 1. Treatment of inl-89601 with chloramphenicol results in the rapid appearance of cyanide-insensitive respiration.[12,13] Chloramphenicol does not induce the cyanide-insensitive pathway in the ANT mutants, however, even if the cells are grown in the presence of the drug for 24 hours. We have carried out detailed studies of the respiratory properties, phosphorylation rates and cytochrome content of mutants ANT-1 and ANT-6. These parameters do not differ significantly from those of wild-type mitochondria and it seems likely that the mutations causing ANT-1 and ANT-6 reside in genes that are unique to the cyanide-insensitive pathway and are not common to the cyanide-sensitive pathway.

We have carried out complementation studies between some of the ANT mutants using forced heterokaryons between various ANT mutants in either the inl-89601 or leu-1 nuclear backgrounds. Our results indicate that mutants ANT-1 and ANT-6, fail to complement each other as do mutants ANT-1 and ANT-4. When a heterokaryon is formed between ANT-1 and ANT-3, however, complementation occurs and the

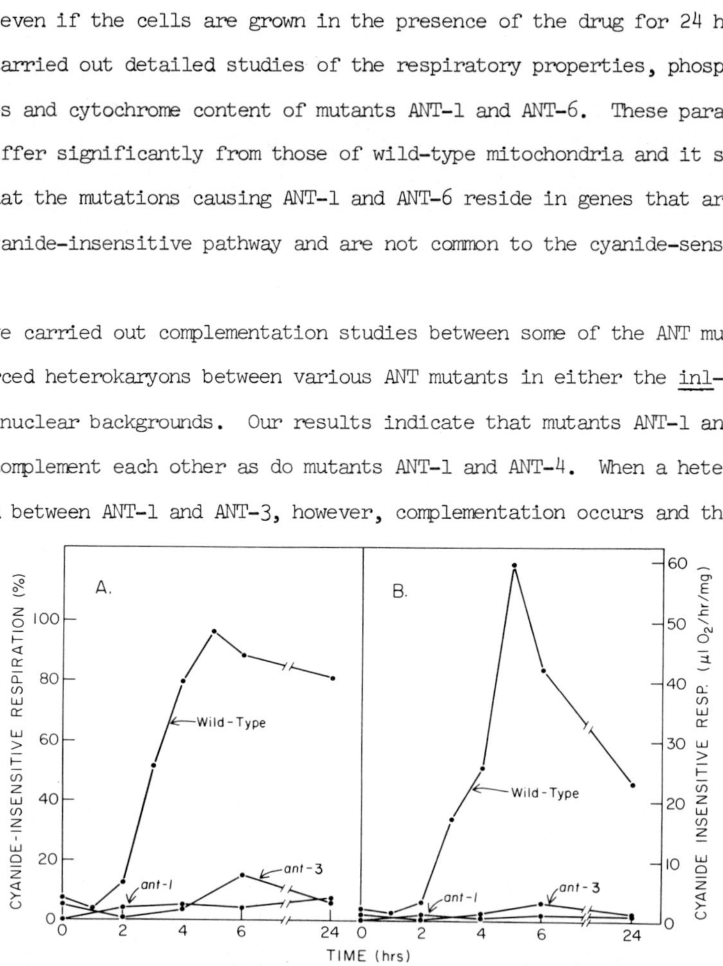

Fig. 1. Induction of cyanide-insensitive respiration in wild-type (74OR23-1A) and in two ANT mutants of Neurospora. Cells were grown for 12 h and the pathway was then induced by the addition of chloramphenicol according to the method of Edwards et al.[12] A; Cyanide-insensitive respiration plotted as a % of the total respiration rate. B; Cyanide-insensitive respiration plotted as µl O_2/hr/mg. Temperature is 30°.

cyanide-insensitive pathway can be induced in these cells. The induction of cyanide-insensitive respiration by chloramphenicol treatment of a heterokaryon between ANT-1 and ANT-3 is shown in Fig. 2. While the induction process in the heterokaryon takes a longer time than in inl-89601, the pathway becomes fully induced. This lag could be due to the number of gene copies present in the heterokaryon or to the physiological age of the culture. We have not yet mapped the ANT mutant but it seems unlikely to us that this complementation is due to an intragenic effect since the rate of cyanide-insensitive respiration is as high as that produced by wild-type cells. A much more likely conclusion is that at least two genetic loci code for components of the cyanide-insensitive pathway. Based on these complementation studies we have designated these loci alx-1 and alx-2. Locus alx-1 is represented by isolates ANT-1, ANT-4 and ANT-6. Locus alx-2 is represented by isolate ANT-3.

Physical studies: Since the branch point for the cyanide-insensitive pathway is believed to be in a region of the respiratory chain that contains iron-sulfur proteins and flavoproteins,[8-11] and since iron-chelating agents are inhibitors of the pathway[1,7,8] we have initiated a study of the components of Neurospora

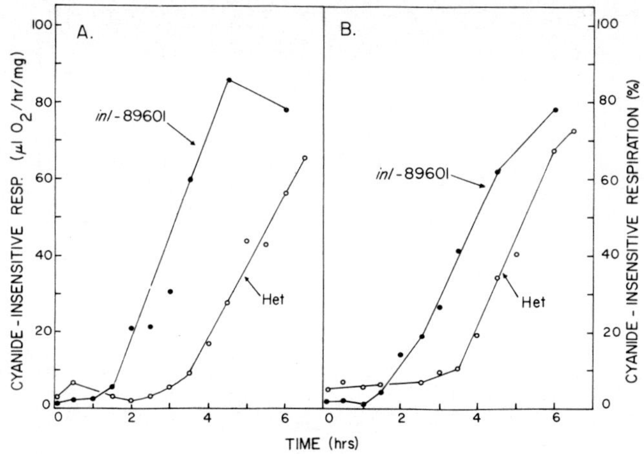

Fig. 2. Induction of cyanide-insensitive respiration in strain inl-89601 and in a heterokaryon between mutants ANT-1 and ANT-3. Details of the experimental procedure are the same as in Fig. 1. Mutant ANT-1 was crossed into an inl-89601 nuclear background while mutant ANT-3 was in a leu-1 nuclear background. The heterokaryon was formed by spotting conidia from both strains on a slant of minimal medium. Abbreviation used: Het = heterokaryon formed between inl-89601 - ANT-1 and leu-1 - ANT-3.

mitochondria that are detectable by electron spin resonance (esr). Fig. 3 shows an esr spectrum of the NADH-reducible iron-sulfur proteins in submitochondrial particles from inl-89601. On the basis of studies in other organisms[17,18] and of temperature-dependence studies of our own (J.T. Warden and D.L. Edwards in preparation) we have made tentative assignments of the signals seen in Neurospora submitochondrial particles to the iron-sulfur centers previously described. Fig. 4 shows a comparison of esr spectra of submitochondrial particles from inl-89601 and from early log phase cni-1. The cni-1 mutant is deficient in signals at g ∿ 1.92 and g ∿ 2.05 (arrows) which are due to iron-sulfur center 2. We have noted, however, that there is a rapid flux of reducing equivalents through the cyanide-sensitive respiratory chain of cni-1 mitochondria even in the absence of this iron-sulfur center (J. Klein and D.L. Edwards in preparation). We have also noted that chloramphenicol-treatment of inl-89601 results in a rapid decrease in the center 2 signal suggesting that a mitochondrial translation product might be a component of this iron-sulfur complex.

In order to determine whether the cyanide-insensitive pathway contains any new

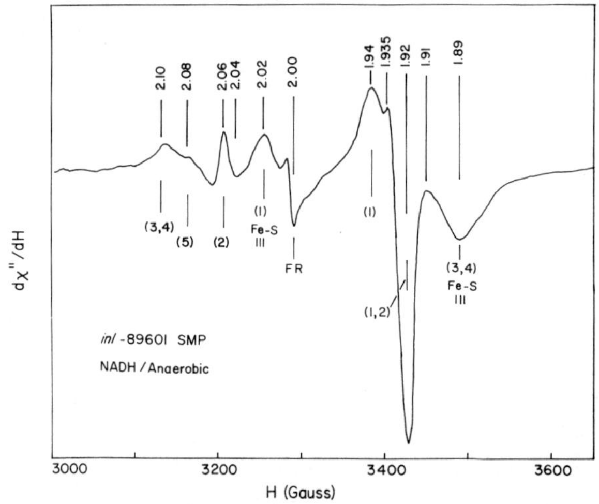

Fig. 3. Esr spectrum of submitochondrial particles from inl-89601. Particles were reduced by the addition of NADH (10 mM) and allowed to become anaerobic under an atmosphere of nitrogen. Numbers at the top of the trance are g - factors calculated for the various signals. Numbers in parenthesis and letters below the trance are the tentative assignments of the signals to iron-sulfur centers and other paramagnetic species. FR = free radical; Fe-S III = iron-sulfur component of respiratory complex III. Protein concentration is 47 mg/ml. Temperature is 13.5K. Power, 40 mW; modulation amplitude, 10 G, scan rate, 250 G/min.

paramagnetic components, we have measured the esr spectrum of submitochondrial particles from cni-1 reduced with NADH in the presence of salicyl hydroxamic acid. These results are shown in Fig. 5. Reduction of cni-1 in the presence of salicyl hydroxamic acid results in the appearance of a new signal at g ∿ 2.020 (arrow) and a change in the signal profile in the region g ∿ 2.11 (3125 Gauss). These results suggest the presence of a new iron-sulfur species that is reducible by NADH in the presence of salicyl hydroxamate.

In order to test the possibility that a novel iron-sulfur center may be a component of the cyanide-insensitive pathway, we have attempted to induce the pathway with chloramphenicol in inl-89601 cells using medium that is deficient in either iron or sulfur. Under these experimental conditions the appearance of cyanide-insensitive respiration is greatly retarded compared with control cultures. These data lend support to the contention that an iron-sulfur protein is a component of the cyanide-insensitive pathway.

Using the data that we have presented here and elsewhere[13] we suggest that the cyanide-insensitive respiratory pathway of Neurospora mitochondria consists of at

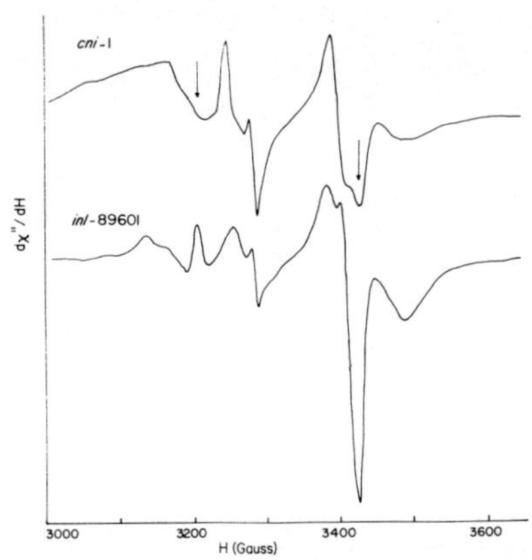

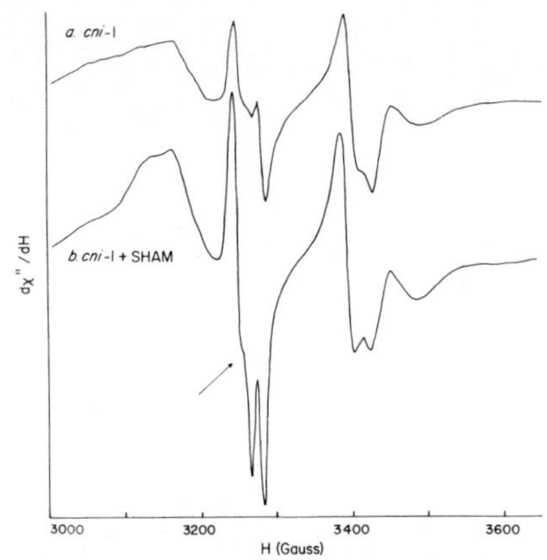

Fig. 4. Esr spectra of submitochondrial particles from inl-89601 and cni-1. Reduction was with NADH as described in Fig. 3. The arrows indicate the field position of the signals for iron-sulfur center 2. Protein concentrations inl-89601, 47 mg/ml; cni-1, 27 mg/ml. Temperature is 13.5K. Esr parameters are given in Fig. 3.

Fig. 5. Esr spectra of cni-1 reduced by NADH in the presence or absence of salicyl hydroxamic acid (SHAM). Details of reduction and esr parameters are in Fig. 3. The arrow indicates a new signal at g ∿ 2.020. Final SHAM concentration = 120 μg/ml. Protein concentration = 27 mg/ml. Temperature is 13.5K.

least two components coded for by nuclear genes <u>alx</u>-1 and <u>alx</u>-2. We believe that one of these components is an iron-sulfur protein giving an esr signal at g ∿ 2.020. The synthesis of these components is then regulated by a mitochondrial gene product.

REFERENCES

1. Lambowitz, A. and Slayman, C.L. (1971) J. Bacteriol. 108, 1087-1096.
2. Weiss, H., von Jagow, G., Klingenberg, M. and Bucher, T. (1970) Eur. J. Biochem. 14, 75-82.
3. Edwards, D.L.; Kwiecinski, F. and Horstmann, J. (1973) J. Bacteriol. 114, 164-168.
4. Rosenberg, E., Mora, C. and Edwards, D.L. (1976) Genetics 83, 11-24.
5. Klein, J., Edwards, D.L. and Werner, S. (1975) J. Biol. Chem. 250, 5852-5858.
6. Lambowitz, A., Slayman, C.W., Slayman, C.L. and Bonner, W.D., Jr. (1972) J. Biol. Chem. 247, 1536-1544.
7. Bahr, J.W. and Bonner, W.D., Jr. (1973) J. Biol. Chem. 248, 3441-3445.
8. Schonbaum, G.S., Bonner, W.D., Jr., Storey, B.T. and Bahr, J.T. (1971) Plant Physiol. 47, 124-128.
9. Erecinska, M. and Storey, B.T. (1970) Plant Physiol. 46, 618-624.
10. Cammack, R. and Palmer, J.M. (1974) Ann. N.Y. Acad. Sci. 222, 816-823.
11. Palmer, J.M. (1976) Ann. Rev. Plant Physiol. 27, 133-157.
12. Edwards, D.L., Rosenberg, E. and Maroney, P. (1974) J. Biol. Chem. 249, 3551-3556.
13. Edwards, D.L. and Rosenberg, E. (1976) Eur. J. Biochem. 62, 217-221.
14. Lester, H.E. and Gross, S.R. (1959) Science 129, 572.
15. Vogel, H.J. (1964) Amer. Natur. 98, 435-446.
16. Malling, B. (1960) J. Gen. Microbiol. 23, 257-260.
17. Orme-Johnson, N.R., Orme-Johnson, W.H., Hansen, R.E., Beinert, H. and Hatefi, Y. (1971) Biochem. Biophys. Res. Comm. 44, 446-452.
18. Ohnishi, T., Asakura, T., Yonetani, T. and Chance, B. (1971) J. Biol. Chem. 246, 5960-5964.

ACKNOWLEDGMENTS

This work was supported by research grant GM-194110 to D.L.E. from the National Institute of General Medical Sciences. D.L.E. is also supported by a Research Career Development Award (K04-GM-00018). J.H.C. was supported by N.I.H. training grant TOI GM 00367. J.T.W. acknowledges support by a Cottrell Grant from the Research Foundation. The authors would like to thank Dr. Patricia St. Lawrence, for advice and encouragement and Susan Jenesel for preliminary studies on some of these mutants.

Genetics and Biogenesis of Chloroplasts and Mitochondria, Th. Bücher et al. eds.

PHYSIOLOGICAL AND GENETICAL ANALYSIS OF THE RESPIRATORY CHAIN OF PARAMECIUM

Jacques Doussière[x], André Adoutte[△], Annie Sainsard[△], Françoise Ruiz[△], Janine Beisson[△] and Pierre Vignais[x]
[x] Departement de Recherche Fondamentale Laboratoire de Biochimie
Centre d'Etude Nucléaires 38000 GRENOBLE (France)
[△] Centre de Génétique Moléculaire
C.N.R.S. 91190 Gif-sur-Yvette (France)

Paramecium tetraurelia is a strictly aerobic organism. Because of this strict requirement for oxygen, the recent isolation of a mutant (1) displaying more than a ten-fold decrease in cytochrome oxidase while retaining respiratory rates equivalent to those of wild-type cells raised the question of the structure of the respiratory chain in this very "ancient" eucaryote. It suggested the occurence, in Paramecium, of an alternate respiratory pathway, analogous to the cyanide-insensitive pathway described mostly in plants (2). It was found indeed that wild-type cells display a low but constant level of cyanide-insensitive respiration (10 to 20%) and that this level was much increased in the cl_1 strain (60 to 80%). On this basis we have started a detailed characterization of the respiratory chain of Paramecium and have devised a method to isolate more mutants affecting the respiratory chain. Mutations affecting the respiratory chain in Paramecium can serve two main purposes: firstly, allow the "dissection" of the respiratory chain in this organism, and secondly provide opportunities to study nucleo-mitochondrial interactions. Gene cl_1 is an example of such a possibility since on the one hand it leads to a 10-fold decrease in cytochrome oxidase and on the other hand it displays a strong incompatibility when confronted to wild-type mitochondria leading to a severe disorganization of these organelles (1, 3).

This paper summarizes the data on the characterization of the respiratory chain in wild-type Paramecium tetraurelia, describes the isolation of new mutants affecting the respiratory chain and discusses the possible physiological basis of these mutations.

I. CHARACTERIZATION OF WILD-TYPE RESPIRATORY CHAIN:

Well coupled mitochondria can be readily isolated from Paramecium in 0,5 M Mannitol, 0,2% B S A, 1mM EDTA, and 2mM MOPS pH7.3, after disruption of the cell membrane in a cell homogeneizer (A. Thomas & Co). Krebs cycle intermediates are oxidized by isolated Paramecium mitochondria yielding P/O ratios similar to those of higher eucaryotes (Figure 1) and indicating the occurence of three, two and one phosphorylation sites when pyruvate, succinate and ascorbate respectively are used as subtrates.

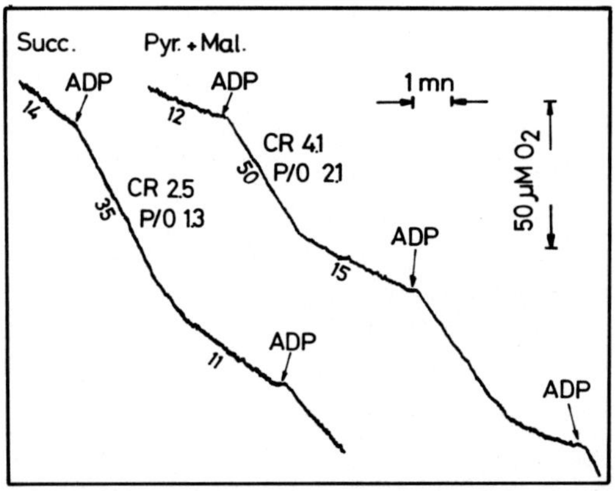

Fig. 1 - Respiratory activity of Paramecium mitochondria. To 2ml of the incubation medium described in the text 10mM Pi, 5mM $MgCl_2$ and 10mM substrate (succinate or pyruvate) were added. The ADP additions were of 268nM. Mitochondria were at a concentration of 0.5 mg/ml for pyruvate and 0.8 mg/ml for succinate. The figures along the curves indicate the respiratory activity expressed as n-moles O_2/mn/mg prot. Recording was carried out in a Gilson oxygraph with a Clark oxygen electrode.

A substantial fraction (10 to 20%) of the respiration in whole cells, isolated mitochondria and gradient purified mitochondria, remains insensitive to cyanide at concentrations higher than 10^{-3}M (Fig. 2).

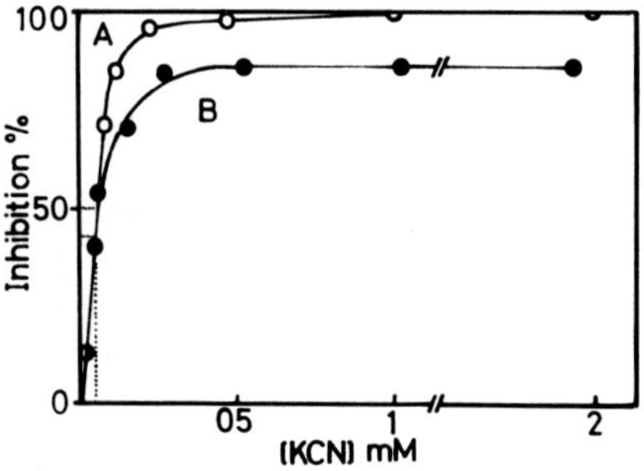

Fig. 2 - The effect of cyanide on the respiration of isolated mitochondria from Paramecium and rat liver.
A. Rat liver mitochondria (2mg/ml) were in 2ml. of medium containing 152mM KCl, 0.2% BSA, 2mM MOPS pH 7.4; 5mM $MgCl_2$ and 10mM succinate.
B. Paramecium mitochondria were in the medium described in Fig. 1; KCN was prepared as a fresh solution and brought to pH7 in MOPS buffer.

The sensitivity to cyanide of the cyanide-sensitive respiration is similar to that observed in mammalian mitochondria (half-inhibition reached at 40µM KCN). After cyanide has been added to respiring mitochondria, the remaining cyanide-insensitive respiration is completely inhibited by SHAM, an inhibitor of the alternate respiratory pathway in plants (4). SHAM added alone, however, has no effect on the respiration rate.

Low temperature spectra of purified mitochondrial preparations indicate the presence of a, b and c types cytochromes with bands situated at 607, 553-557 and 549 nm respectively (Fig. 3A). After extraction of cytochrome c by washing mitochondria in a hypotonic KCl solution, the band at 549 nm disappears while the bands at 553 nm and 557 nm are retained and a band at 551 nm emerges, which may correspond to cytochrome c_1.

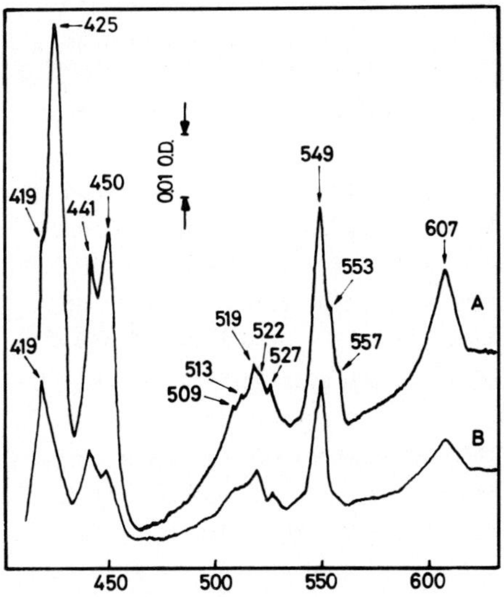

Fig. 3 - Low temperature spectra of mitochondrial preparations.
3A. Differential (succinate reduced minus H_2O_2 oxidized) spectra of purified mitochondria (4mg prot/ml).
3B. Same as above +5µg antimycin/mg prot. in the reduced compartment.

Antimycin has absolutely no inhibitory effect on respiration. However, in its presence, succinate fails to reduce both b type cytochromes while cytochromes c and a are partially reduced (Fig. 3B) suggesting that in Paramecium the b cytochromes can be reoxidized through an alternate pathway leading to oxygen.

Low temperature spectra of reduced + CO against reduced mitochondrial preparations indicate the presence of a least two different carbon monoxide reacting pigments (Fig. 4A).

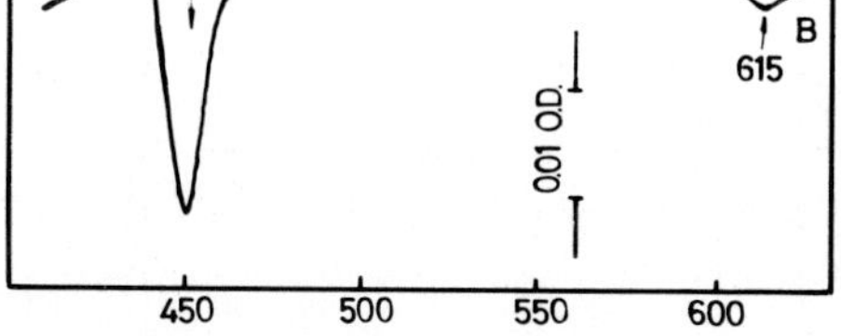

Fig. 4 - CO-reacting pigments of purified mitochondria.
4A. Difference spectra of gradient purified mitochondria (reduced + CO minus reduced).
4B. Difference spectra of reduced + CO preparation minus reduced + CO + photoirradiation at 600 - 650 nm. Reduction was always achieved by dithionite.

Photoirradiation at 600 - 650 nm results in the specific dissociation of the CO - cytochrome a_3 complex (fig. 4B). The second CO-binding pigment has properties quite similar to those defined by Castor and Chance for the b type cytochrome called cytochrome o (5). It may therefore correspond to an autoxidizable cytochrome b. More experiments are necessary, however, before considering this pigment as the alternative oxidase.

In the presence of KCN, P/O ratios slightly higher than 1 with pyruvate as substrate as well as persistence of phosphorylation with succinate have been repeatedly observed. Furthermore, mitochondria extracted from the cl_1 strain in the exponential phase of growth, although completely devoid of cytochrome oxidase, oxidize pyruvate and succinate with P/O ratios of 1.1 - 1.2 and 0.5 - 0.6 respectively; 1mM cyanide has no effect on the P/O ratios and on the rate of respiration while 1mM SHAM blocked respiration completely. These results suggest that the second phosphorylation site is present in the cyanide-insensitive pathway in contrast to the situation described in plants (reviewed in ref. 2).

Finally it was observed that AMP and ADP had a marked effect on the chanelling of electrons between the two respiratory pathways: in the presence of ADP respiration becomes nearly completely sensitive to KCN; the residual respiration can, however, be stimulated 5 to 10 fold by AMP. Addition of SHAM prevents this AMP induced stimulation indicating that it is indeed due to an increased activity of the alternate pathway.

These results, and others to be detailed elsewhere, lead to the following tentative model of the respiratory chain in Paramecium tetraurelia.

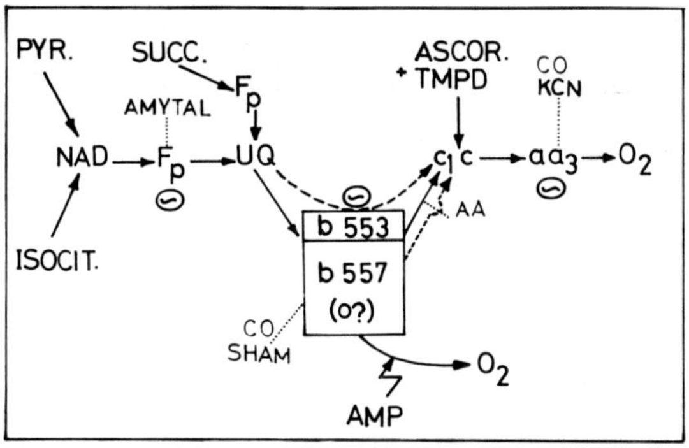

Fig. 5 - Tentative model of the respiratory chain of Paramecium tetraurelia.

This model suggests the possibility of obtaining mutations altering any of the various terminal steps of the "normal" pathway since the electrons can still be transported by the alternate pathway. A search for such mutants was carried out.

II. ISOLATION OF MUTATIONS AFFECTING THE RESPIRATORY CHAIN.

It is well known that respiring vs non-respiring colonies in yeast can be discriminated by staining the colonies with a tetrazolium salt (6) and Edwards et al. (7) have recently extended this method to the isolation of respiratory deficient strains of Neurospora. By analogy, a screening method was devised in Paramecium based on the use of 2, 3, 5 - triphenyltetrazolium chloride (TTC). It was found that under a set a precise conditions wild-type cells of Paramecium not only stain in red but die in TTC containing medium while the cl_1 strain as well as erythromycin-treated wild-type cells remain unstained and survive. Tetrazolium resistant cells were therefore selected for. After a nitrosoguanidine mutagenesis, out of about 50.000 cells, 27 stable tetrazolium resistant mutants were isolated. They can be grouped in three levels of resistance: low, medium or strong. Two of them may correspond to permeability mutations. The low temperature spectra carried out on whole cells have revealed that one mutant (TET 64 - 1) due to a single nuclear mutation, is strongly deficient in cytochrome c (Fig. 6).

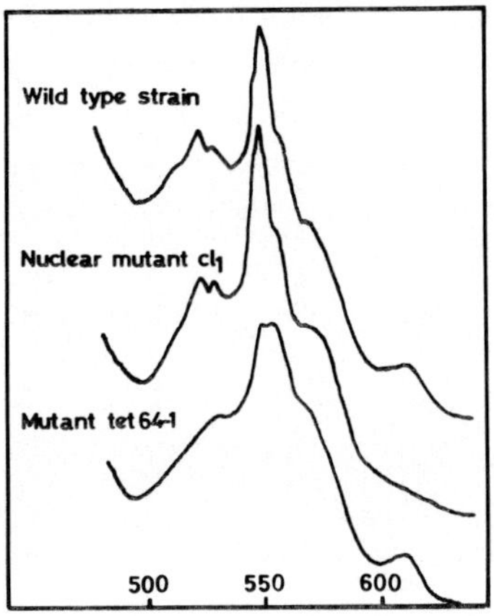

Fig. 6 - Whole cells absolute low temperature spectra (dithionite reduced) of wild-type (WT) and two nuclear mutants: cl_1, cytochrome oxidase deficient and TET 64-1, cytochrome c deficient.

Among the other strains three levels of cytochrome aa_3 are found: it can be normal, slightly or strongly decreased (Fig. 7).

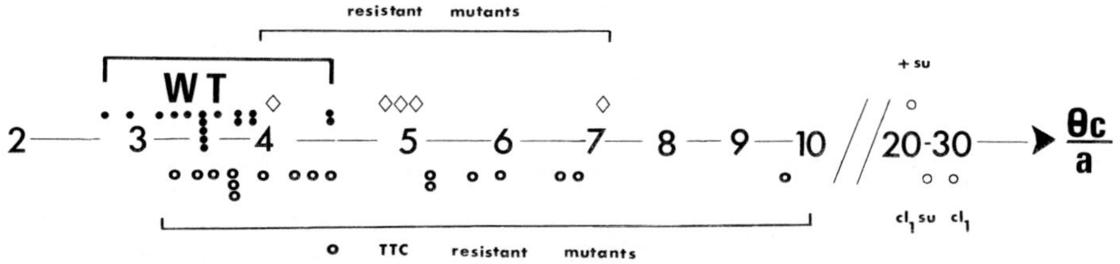

Fig. 7 - The distribution of mutants affecting mitochondria according to their content in cytochrome aa_3.

θc/a is the ratio of the amount of cytochrome c over cytochrome aa_3, determined spectroscopically. Several values have been given for wild-type (black dots). Each white dot represents a different TTC-resistant mutant and each lozenge a different antibiotic-resistant mutant.

One strain strongly deficient in cytochrome aa_3 was also found to result from a single nuclear mutation. An important proportion of the strains tested display highly increased levels of cyanide-insensitive respiration, specially among those having decreased amounts of cytochrome aa_3.

III. POSSIBLE BASIS OF THE MUTATIONS LEADING TO CYANIDE-INSENSITIVE RESPIRATION

There are at present two types of mutant strains displaying cyanide-insensitive respiration along with cellular resistance to tetrazolium:

1 - Nuclear mutant displaying either profound deficiencies in one cytochrome (cl_1, TET 64-1) or less profound deficiencies, essentially in cytochrome oxidase (see fig. 7). The molecular basis of these mutations is unknown.

2 - Mitochondrial mutants, previously isolated (8-9) on the basis of their resistance to erythromycin (E^R) or chloramphenicol (C^R). These mutants display more or less decreased levels of cytochrome oxidase. This deficiency most probably results from an impairment of mitochondrial protein synthesis since:
a) the mutations are known to affect the mitochondrial ribosomes (10) and
b) phenocopies of the mutations are obtained by growing wild-type cells in low concentrations of erythromycin (30-60 μg/ml).

The respiratory chain of both types of mutants is presently under analysis. Some data concerning one of the mitochondrial mutants, a doubly-resistant one, $(CE)_2^R$ - b, are given on table 1.

Strain	Generation time	Amount of cyt aa_3	Inhibition of respiration by		
			SHAM	KCN	SHAM + KCN
CE_2^R - b expon.	9h	30%	0%	0%	100%
CE_2^R - b stat.		90%	0%	80%	100%
Wild-type	6h	100%	0%	80-90%	100%

Table I - Properties of strain CE_2^R - b.
Expon and stat designate mitochondria extracted respectively from cells in exponential or stationary phase of growth. The wild-type yields similar results in both conditions. The amount of cytochrome aa_3 was determined spectroscopically.

These data show striking similarities to those obtained with the *Neurospora* mutants "poky" and *cni* (11, 12, 13, 14, 15, 16): deficiency in cytochrome oxidase and very high levels of cyanide-insensitive respiration in the exponential phase of growth, recovery of a quasi wild-type phenotype in the stationary phase of growth, and possibility of chanelling the electrons in either respiratory pathways with equal efficiency. Furthermore decrease in P/O ratios with several substrates was observed after addition of cyanide suggesting that in the absence of KCN the electrons tend to be chanelled into the normal pathway in agreement with the results and model of Bahr and Bonner (18, 19) and the results on Neurospora (20).

SUMMARY AND CONCLUSIONS.

This paper summarizes the present knowledge of the respiratory chain of Paramecium. A number of features are common to the few protozoan species studied so far (21, 22, 23, 24, 25). Furthermore striking similarities with plants (2) and, to some extent, with some bacteria (26) are disclosed. The major similarity is the occurence of a branched respiratory chain consisting of a rather "classical" pathway and an alternate cyanide-insensitive pathway. The region of the b cytochromes however has a number of original properties in Paramecium. First, respiration is completely insensitive to antimycin even in cyanide-sensitive mitochondria although it can be demonstrated spectroscopically that antimycin affects b type cytochromes. Second, it is most likely that the second phosphorylation site is retained in the alternate pathway suggesting that the branching of the cyanide-insensitive pathway is posterior to cytochrome b. These homologies and differences raise a number of interesting evolutionary problems.

The occurence of this alternate pathway allows the isolation of mutations causing profound alterations of the terminal portion of the normal respiratory chain, a possibility previously restricted to facultative aerobes.

Using a screening method based on resistance to a tetrazolium salt, an array of mutants have been isolated. Several strains display abnormal cytochrome spectra along with highly increased levels of cyanide-insensitive respiration. The spectral alterations may be due either to a relatively direct effect of the mutation on one cytochrome or to an indirect effect resulting from perturbation of mitochondrial protein synthesis. These mutants should yield further information on the structure and functioning of the respiratory chain in Paramecium. On the other hand these nuclear mutations are presently being combined, either by conjugation or by microinjection (27) to mitochondrial mutations affecting the mitochondrial ribosomes in a search for specific nucleo-mitochondrial interactions. Finally the nuclear mutations may be used to obtain new mitochondrial markers by isolating mitochondrial mutations suppressing their effects as already found in the case of gene cl_1 (28).

AKNOWLEDGMENTS:

This work was supported by grants form the Délégation Générale à la Recherche Scientifique et Technique and from the Centre National de la Recherche Scientifique (RCP 284).

REFERENCES:

1. Sainsard, A., Claisse, M. and Balmefrézol, M. (1974) Mol. Gen. Genet. 130, 133
2. Henry, M.F. and Nyns E.J. (1975) Sub-Cell. Biochem. 4, 1.
3. Sainsard, A. (1976). Mol. Gen. Genet. 145, 28.
4. Schonbaum, G.R., Bonner, W.D., Storey, B.T. and Bahr, J.T. (1971) Plant Physiol. 47, 124.
5. Castor, L.N. and Chance, B. (1959) J. Biol. Chem. 234, 1587.
6. Ogur, M., St John, R. and Nagai, S. (1957) Science, 125, 928.
7. Edwards, D.L., Kwiecinski, F. and Hortsmann, J. (1973) J. Bact. 114, 164.
8. Adoutte, A. and Beisson, J. (1972) Nature, 235, 393.
9. Adoutte, A. (1974) in: The Biogenesis of Mitochondria, Ed. Kroon, A. and Saccone, C. Acad. Press, p. 263.
10. Tait, A. (1971) FEBS Lett 24, 159.
11. Lambowitz, A. and Slayman, C. (1971) J. Bact. 108, 1087.
12. Von Jagow, G. and Klingenberg, M. (1972) FEBS Lett 24, 278.
13. Lambowitz, A., Smith, E. and Slayman, C. (1972) J. Biol. Chem. 247, 4850.
14. Lambowitz, A., Smith, E. ans Slayman, C. (1972) J. Biol. Chem. 247, 4859.
15. Von Jagow, G., Weiss, H. and Klingenberg, M. (1973) Eur. J.Biochem. 33, 140.
16. Klein, J., Edwards, D. and Werner, S. (1975) J. Biol. Chem. 250, 5852.
17. Edwards, D. and Kwiecinski, F. (1973). J. Bact. 116, 610.
18. Bahr, J.T. and Bonner, W.D. (1973) J. Biol. Chem. 248, 3441.
19. Bahr, J.T. and Bonner, W.D. (1973) J. Biol. Chem. 248, 3446.
20. Edwards, D., Rosenberg, E. and Maroney, P. (1974) J. Biol. Chem. 249, 3551.
21. Turner, G., Lloyd, D. and Chance, B. (1971) J. Gen. Microbiol. 65, 339.
22. Kusel, J.P. and Storey, B.T. (1973) Biochem. Biophys. Acta, 305, 570.
23. Sharpless, T.K. and Butow, R. (1970). J. Biol. Chem. 245, 58.
24. Hill, G. and Cross, G. (1973). Biochem. Biophys. Acta 305, 590.
25. Edward, C. and Llyod, D. (1973) J. Gen. Microbiol. 79, 275.
26. Weston, J.A. and Knowles, C.J. (1974) Biochem. Biophys. Acta 333, 228.
27. Knowles, J.K.C. (1974) Exp. Cell Res. 88, 79.
28. Sainsard, A. (1975) Nature 257, 5524.

Genetics and Biogenesis of Chloroplasts and Mitochondria, Th. Bücher et al. eds.

GENETIC CONTROL OF GLYCEROL-3-PHOSPHATE DEHYDROGENASE SYNTHESIS IN NEUROSPORA

James B. Courtright
Department of Biology
Marquette University
Milwaukee, Wisconsin, USA

INTRODUCTION

Recent experiments have shown that the synthesis and assembly of the flavin linked glycerol-3-phosphate dehydrogenase (GPDH) into the mitochondrial inner membrane requires the active translation of genetic messages on both cytoplasmic and mitochondrial ribosomes[1,2]. These results, as well as those with other mitochondrial enzyme systems[3], indicate that the synthesis and assembly of membrane bound enzymes involves interactions of both nuclear and mitochondrial gene products, and accordingly, genetic studies have been undertaken to identify the number and mode of action of the various genetic loci which are specifically necessary for the synthesis and assembly of this mitochondrial enzyme. In this communication, the effect of mutations in both the nuclear and mitochondrial genome on the synthesis of GPDH is described.

MATERIALS AND METHODS

Genetic strains: The glycerol non-utilizing strains of Neurospora were derived from a glp-3 inl^t strain[4] and are characterized by normal growth on acetate, glucose, or sucrose, but not glycerol. Cytoplasmically inherited strains were obtained from the Fungal Genetics Stock Center, Arcata, CA, and the strain employed in these studies were as follows: mi-1 (mi-1-1-.4), mi-3 (3754), SG-1 (RL3202-23), and SG-3 (RL3120-10). Strain 245 was obtained from Dr. H.G. Kølmark[5].

Enzyme assays: GPDH was assayed by means of either the phenazine methosulfate-MTT tetrazolium assay previously described[1] or by employing DCPIP as the electron acceptor. GPDH was extracted from the mitochondrial by incubation in 0.01 M G3P and 0.5% Na-cholate, and electrophoresed in 20 mM Tris-4 mM G3P, pH 8.7 on agarose gels (0.5%) containing 2.5 mg/ml Triton-X100. GPDH was visualized by immersing the gels in a staining solution containing the components of the assay solution[1] but substituting Nitro Blue tetrazolium (0.5 mM).

RESULTS

Mutations at the glp-2 locus, located near aro-1 on the right arm of chromosome II result in a complete inability of the mutant to grow on minimal glycerol media and a nearly complete loss of mitochondrial GPDH activity under all conditions of growth (Table 1). Revertants of these original mutants have been obtained and found to have identical electrophoretic mobilities to those found in induced wild type cells. However, the fact that the K_m for G3P in both mutants and revertants was different from that measured in wild type suggests that this locus represents a structural gene for this enzyme. In agreement with this interpretation of the role of this locus, it was found that glp-2$^-$ mutants are recessive in both heterokaryons and in partial diploids constructed by use of the insertional translocation, T (II-V) ALS 176.

TABLE 1

PROPERTIES OF GPDH IN NUCLEAR MUTANTS

Strain	Growth medium	Specific Activity	Electrophoretic mobility[d]	K_m (mM)
74-4[a]	Acetate	0.051	1.0	-
74-4	Glycerol	0.726	1.0	18
glp(14)	Glycerol	0.008	n.d.[c]	7200
glp(17)	Glycerol	0.012	n.d.	7200
14-1[b]	Glycerol	0.856	1.0	33
17-1[b]	Glycerol	0.919	1.0	29

[a] Wild type

[b] Revertants of glp(14) and glp(17), mutant alleles of the glp-2 gene.

[c] n.d. = not detected.

[d] Electrophoretic mobility expressed relative to wild type (R_f = 1).

Similar analysis of the GPDH in the cytoplasmically inherited mutants (Table 2) revealed that for most mutants, there is a reduction in the activities of the GPDH detected as well as an alteration in the electrophoretic mobility of components with GPDH activity. These alterations have been noted in all the maternally derived ascospore progeny when mi-3 is used as the protoperithecial parent.

TABLE 2

PROPERTIES OF GPDH IN CYTOPLASMIC MUTANTS

Strain	Growth medium	Specific Activity	Electrophoretic mobility	K_m (mM)
245	Glycerol	0.273	1.0	-
mi-1	Glycerol	0.459	1.0	-
mi-3	Glycerol	0.163	0.9*	-
SG-1	Glycerol	0.151	1.0	-
SG-3	Glycerol	0.173	1.0	-

* This mutant also possesses minor GPDH components (R_f = 0.4 and 2.2)

DISCUSSION

While mutations at the glp-2 locus result in alterations in the kinetic properties of the GPDH, the possibility that these changes are due to incomplete or defective assembly of additional subunits of the enzyme has not been ruled out. On the other hand, the synthesis and assembly of this enzyme may be under a type of genetic control similar to that described for the cni-1 locus in *Neurospora*[6]. Additional glp^- mutants have been isolated in our laboratory which should permit an unambiguous assignment for the structural gene(s).

In previous studies[2], it was found that the incorporation of this enzyme into mitochondrial membranes was inhibited by incubation of cells with chloramphenicol. Since the induction of GPDH is incomplete in all cytoplasmic mutants analyzed and is accompanied by physical alterations of components of the enzyme in the mi-3 strain (Table 2), it therefore seems likely that one or more mitochondrially synthesized polypeptides are required for the normal assembly of this enzyme into mitochondrial membranes. Although not all of these maternally inherited mutations have been examined with regard to alterations in other mitochondrial enzymes, the slow growth phenotype of these mutants on all carbon suggests pleiotropic effects similar to those for mutants of Saccharomyces.[7]

SUMMARY

a) a nuclear gene, glp-2, has been identified which affects the flavin linked glycerol-3-phosphate dehydrogenase;

b) all cytoplasmically inherited mutants of Neurospora examined to date lack fully induced levels of GPDH;

c) a cytoplasmically inherited mutant, mi-3, possesses a GPDH with altered electrophoretic mobility.

REFERENCES

1. Courtright, J. (1975) Arch. Biochem. Biophys. 167: 21.
2. Courtright, J. (1975) Arch. Biochem. Biophys. 167: 34.
3. Tzagoloff, A. et al. (1975) Proc. Nat. Acad. Sci. USA 72: 2054.
4. Courtright, J. (1975) J. Bacteriol. 124: 497.
5. Nilheden, E. et al. (1975) Hereditas 79: 239.
6. Klein, J. et al. (1975) J. Biol. Chem 250: 5852.
7. Tzagoloff, A. et al. (1975) J. Biol. Chem. 250: 8236.

Addendum: In recent experiments, additional glycerol non utilizing mutants have been isolated. The complementation analysis of these mutants indicates that there may be as many as three genes at the glp-2 locus.

Genetics and Biogenesis of Chloroplasts and Mitochondria, Th. Bücher et al. eds.

PRIMARY ANTIMITOCHONDRIAL ACTIVITY OF CARCINOGENS IN SACCHAROMYCES CEREVISIAE

V. Egilsson, I. H. Evans and D. Wilkie

Department of Botany and Microbiology, University College London

INTRODUCTION

The hypothesis that mitochondrial change is a first step in oncogenesis has been proposed from time to time ever since Warburgs early work in which he discovered an abnormally low respiratory rate in cancer cells[1]. In promoting this theory there are two main problems: 1) how can primary changes in mitochondria be detected and 2) how can mitochondrial alteration (presumably heritable) bring about the series of changes that characterize cancer cells. In considering the action of chemical carcinogens as we do in this paper, an answer to the first problem is readily available if the test organism is the eukaryotic yeast cell.

MATERIALS AND METHODS

Test for selective inhibition of mitochondria. Full details are available in ref 2. Briefly, the test exploits the capacity of yeast cells to dispense with mitochondrial respiration (the organism is a facultative anaerobe) and grow on the products of glycolysis. Thus, a drug which selectively blocks mitochondria either at the level of function or of biogenesis, will have little or no effect on cell growth provided glucose is available. However, if the carbon source is non-fermentable, that is, can only be metabolized *via* the krebs cycle (e.g. glycerol), the mitochondrial inhibition would preclude growth. Inhibition acting on biogenesis could be blocking 1) protein synthesis in the mitochondria directly at the level of translation or at transcription. Such inhibition would be detectable as failure of cytochromes $\underline{aa}_3$ and $\underline{b}$ of the respiratory chain to be synthesized and appear during growth in glucose medium, 2) normal replication of mito chondrial DNA which usually leads to genetic lesions particularly that of the mutation known as petite. Inhibitors of function would have a direct effect on oxygen consumption either of intact cells or

or of isolated mitochondria. Inhibitors of biogenesis would of course lead to inability of cells to take up oxygen.
Growth medium contained 1% yeast extract with carbon source either 2% glucose or 4% glycerol. Difco agar (2%) was added when required. Haploid yeast strains of this laboratory were used and a device for multiple inoculation was employed[2].
$CaCrO_4$ was obtained from BDH, thioacetamide (TAA) from Sigma, $CdCl_2$ from Koch-Light and 4-nitroquinoline-1-oxide (NQO) from K & K (California). Carcinogens were added from stock solutions to autoclaved medium prior to pouring if agar and prior to use if liquid.

Absorption spectra. Cells were harvested at stationary phase in liquid glucose medium. Heavy suspensions (approx 10^9/ml) were used to obtain absorption spectra in the SP 1800 recording spectrophotometer.

Oxygen uptake. Oxygen uptake by both whole yeast cells and of isolated rat liver mitochondria were measured polarographically in a Rank-type oxygen electrode. Carcinogens were added as indicated and their effects recorded. Yeast cells were suspended in the electrode chamber in Hepes buffer, pH 6.8. Isolated mitochondria were treated as described in ref 3. Volume of suspending fluid in the electrode chamber was 2 ml.

RESULTS

Effects of carcinogens on growth

Minimum inhibitory concentrations (MICs) to arrest growth of yeast strains are recorded in Table 1 for each of the carcinogens. It is clear that cells are much more sensitive to the inhibition when the carbon source is glycerol compared to glucose. This was most marked in the case of TAA. These results indicated that the mitochondrial system was the primary target of the carcinogens, the organelle being on average 4, 3, 3 and 24 times more sensitive than any other essential component of the cell to $CaCrO_4$, $CdCl_2$, NQO and TAA respectively. It is also evident that strains whose mitochondria are most senitive to one carcinogen are not necessarily the most sensitive to any other or, in other words, there is no cross-correlation. This indicates that the mechanism of inhibition is different for each carcinogen. K_2CrO_4 had the same activity as $CaCrO_4$. MICs which blocked visible growth on glycerol-agar permitted a limited amount of budding. This was established by micromanipulation of single cells onto glycerol-agar blocks containing carcinogen and observing cell division under the microscope. This showed that arrest of growth was not immediate, suggesting that mitochondrial biogenesis rather than function was being directly affected.

Table I. Minimum inhibitory concentrations (MIC) of carcinogens to arrest growth of yeast cells utilising dextrose (D) or glycerol (G) as carbon sources.

Strain	$CaCrO_4$ (mM)		$CdCl_2$ (mM)		NQO (μM)		TAA (mM)	
	G	D	G	D	G	D	G	D
D6	0.3	1.0	1.0	1.0	2	6	5	180
D4	0.3	1.5	0.6	1.0	2	6	10	180
45B	0.1	1.0	0.025	0.1	6	20	5	150
D18	0.1	1.0	0.5	1.0	0.6	2	2	130
A15	0.3	1.5	0.05	0.6	2	6	10	180
B41	0.3	1.0	0.3	0.6	0.6	2	10	180
D22	0.3	0.7	0.025	0.05	0.6	2	5	100
D26	0.3	1.5	0.05	0.2	2	6	15	200
B21	0.1	0.7	0.6	1.0	0.6	2	2	100
A285	0.2	1.0	0.025	0.1	2	6	10	150
188	0.3	0.3	0.05	0.1	6	20	10	150
D15	0.3	0.7	0.3	1.0			10	180
A30	0.3	1.5	0.04	0.4	2	6	10	100
D11	0.4	1.0	0.025	0.05	6	20	10	100
B27	0.3	1.0	0.6	1.0	2	6	5	130
Average $\frac{MIC(D)}{MIC(G)}$	4		3		3		24	

Absorption spectra

Selective inhibition of mitochondria was further apparent by the finding that cells growing in glucose medium in the presence of carcinogen that inhibited growth in glycerol,failed to develop cytochrome $\underline{aa}_3$ and to a lessor extent, cytochrome $\underline{b}$ (Fig.1). This gave a clear indication that the carcinogens were acting to arrest protein synthesis in mitochondria while apparently having little or no effect on other cellular biosynthetic processes.

Oxygen uptake.

Oxygen uptake of whole yeast cells was unaffected by the addition of even high concentrations of chromate or of TAA to the electrode system. Isolated rat liver mitochondria were likewise unaffected by TAA up to a concentration of 150 mM (maximum used), but chromate when added to give a concentration of 0.8 mM, depressed oxygen uptake by about 50% with pyruvate as substrate. At a concentration of 2.4 mM chromate there was total inhibition of oxygen uptake by the mitochondria. Although intact mammalian cells were not tested, it was tentatively concluded that sites were exposed in isolated mitochondria to chromate activity which were not so readily available in intact cells.

Genetic effects .

It has been shown that total blockage or prolonged partial blockage of mitochondrial protein synthesis leads to aberration of mitochondrial DNA replication and gives rise to the petite mutation in yeast cells[4]. The effeiciency with which the carcinogens induced these genetic lesions in mitochondrial DNA was tested and results are shown in Table 2. There are reports elsewhere in the literature[5,6] that Cd and NQO induce the petite mutation and this was also the case in our experiments. Only chromate gave results that were incon clusive and further work is being done with cells exposed to the carcinogen for longer periods in the expectation that petite induction will occur.

Cells were isolated from petite and normal colonies of the TAA-treated strain A285 by micromanipulation onto glucose agar medium containing 60 mM TAA. The rate of cell division was slightly faster for the normal cells indicating that the higher frequency of petite cells in cultures of A285 in the presence of TAA was due to induction and not selection. Exposure to the carcinogen had no detectable effect on the viability of these cells.

Effects on cell metabolism.

All parental strains were able to utilize the fermentable sugars galactose, methylglucoside and maltose for growth. When TAA was present in concentrations ranging from 2-20 mM, most strains were unable to use one or other of these sugars. Results similar to

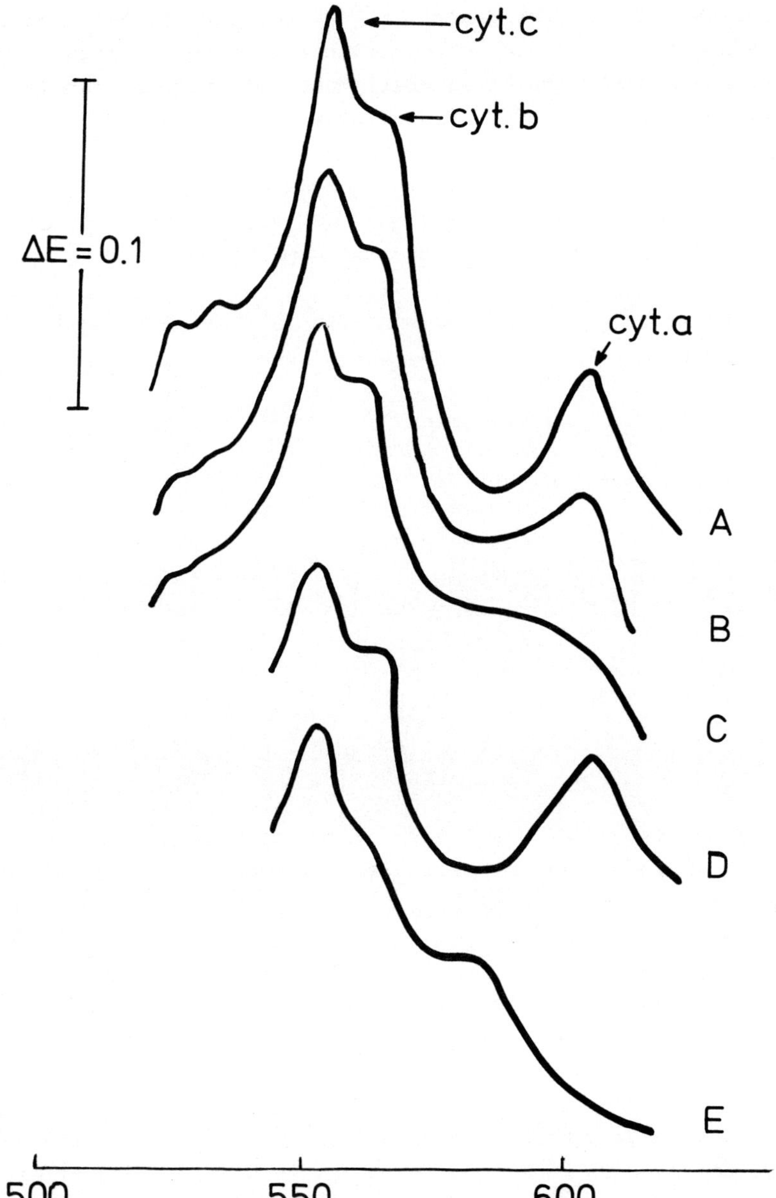

Fig.1. Absorption spectra of strain D4 grown in 0, 0.4 mM and 0.6 mM $CaCrO_4$ (curves A, B, C). D,E, strain D18 in 0 and 5 mM TAA, respectively.

these were obtained by treatment with low concentrations of acriflavine [7], a specific inhibitor of mitochondrial transcription. Results with chromate and NQO indicated that these carcinogens also led to the arrest of growth of most strains using these sugars as carbon sources. Since these inhibitors at the concentrations used, have no significant effect on the growth of cells utilizing glucose, the conclusion is drawn that, in analogy with acriflavine, the carcinogens interfere in a selective manner with mitochondrial transcription.

Table 2. Induction of the petite mutation in yeast strains grown in glucose medium containing carcinogens

Strain	Treatment	% petite*
D4	Control	1
	0.3 mM $CaCrO_4$	2
	0.6 mM $CdCl_2$	15
	2 uM NQO	25
D6	Control	4
	60 mM TAA	10
D18	Control	1
	60 mM TAA	10
A285	Control	5
	60 mM TAA	30

*Cells were harvested at stationary phase and plated. Petite colony was scored on the basis of small size and inability to grow when transferred to glycerol medium.

Other carcinogens.

The number of carcinogens used in these studies was extended to include benzidine, ethionine, lead acetate and nickel chloride. Benzidine and ethionine showed selective activity against mitochondria and inhibited growth of all strains in glycerol medium at concentrations permitting growth in glucose medium. This was particularly marked in the case of benzidine which was on average, ten times more potent as an inhibitor of mitochondria than of the cell. Lead and nickel on the other hand, were selectively antimitochondrial only in a few strains and then only marginally so.

DISCUSSION

These results indicate that mitochondria are the primary targets of some, possibly many carcinogens in yeast cells. Their significance in cancer depends on how readily the results can be extrapolated to mammalian cells. Comparatively few studies have been made of genetic effects of carcinogens on mammalian mitochondria but Graffi and co-workers [8] found that nitrosodimethylamine preferentially alkylates mitochondrial DNA of liver and kidney after administration to rats and hamsters. Nitrosamine has also been studied by Zimmerman and associates [9] with respect to yeast cells and they report a higher induction rate of petite mutation compared with nuclear mutation following treatment with this compound. Similarly, Graffi *et al.* conclude that mutation would occur more readily in mitochondria than in the nucleus of mammalian cells treated with the alkylating agent. We can reiterate the question as to how mitochondrial mutation can affect, in a genetic sense, extra-mitochondrial characteristics. We have found in yeast cells that genetic lesions in mitochondria can result in loss of function of nuclear genes specifying cellular characteristics. These include apparent alteration of plasma membrane [10, 11] and non-inducibility of enyzmes in the metabolism of sugars [12, 13, 7]. In this connection, Nass and coworkers [14] have shown that ethidium bromide, a highly selective inhibitor of mitochondrial DNA replication and transcription in yeast and mammalian cells, affects permeation of hamster cells in culture.

We conclude that mitochondria may be directly involved in oncogenesis at least in some cases, a point which we have discussed elsewhere [15].

Acknowledgement. This work is supported by a grant from the Cancer Research Campaign to D.W. and is gratefully acknowledged.

REFERENCES

1. Warburg, O. (1956) Science, 123, 309.
2. Wilkie, D. (1972) Med. Biol. Illustr., 22, 119.
3. Chappell, B.A. (1972) In Subcellular Components (ed. G.B. Birnie) Butterworth, London, p. 77.
4. Williamson, D.H., Maroudas, N. and Wilkie, D. (1971) Mol. Gen. Genet., 111, 209.
5. Lindegren, C.C. and Lindegren, G. (1973) Mutation Res., 21, 315.

6. Morita, T. and Mifuchi, I. (1974) GANN, 65, 27.
7. Evans, I.H., Egilsson, V. and Wilkie, D. (1976) In: The Function of Mitrochondrial DNA (Eds. C. Saccone and A.B. Kroon) North-Holland Press, Amsterdam, p. 209.
8. Wunderlich, V., Tetylaff, I. and Graffi, A. (1971) Chem. Biol. Interactions, 4, 81.
9. Schwaier, R., Nashed, N. and Zimmermann, F. (1968) Mol. Gen. Genet., 102, 290.
10. Linstead, D., Evans, I.H. and Wilkie, D. (1974) In: The Biogenesis of Mitochondria (Eds. A.M. Kroon and C. Saccone) Academic Press, New York, p. 179.
11. Evans, I.H. and Wilkie, D. (1975) In: Molecular Biology of Nucleocytoplasmic Relations (Ed. A. Puisseux-Dao) Elsevier, Amsterdam, p. 179.
12. ——— (1976) Genet. Res. Camb., 27, 89.
13. Puglisi, P. and Algeri, A. (1974) In: The Biogenesis of Mitochondria (Eds. A.M. Kroon and C. Saccone) Academic Press, New York, p. 169.
14. Soslau, G., Fuhrer, J.P., Nass, M.K. and Warren, L. (1974) J. Biol. Chem., 249, 3014.
15. Wilkie, D., Egilsson, V. and Evans, I.H. (1975) Lancet, 1 (7908) 697.

Author Index